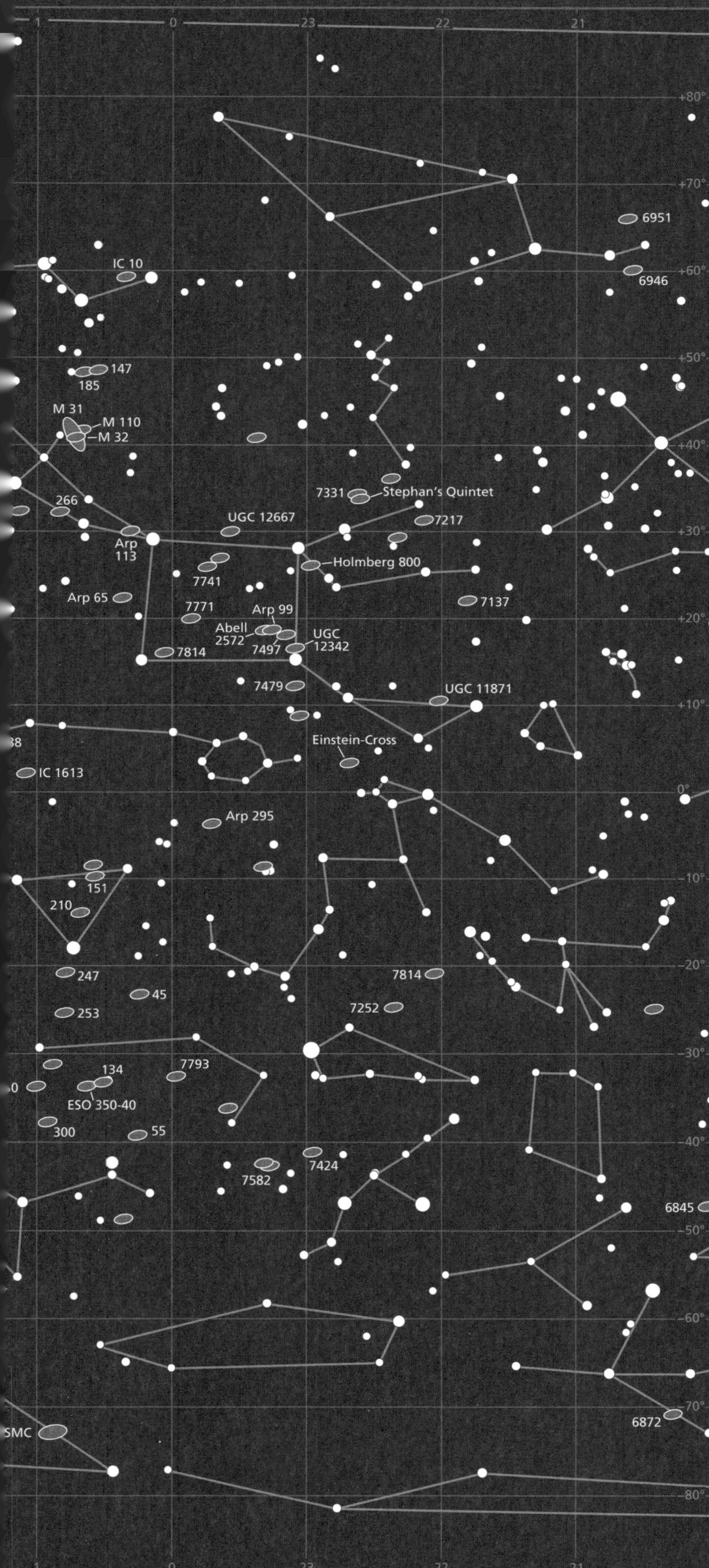

Galaxy	Right Ascension h m	Declination
Abell 262	01 53	[illegible]
Abell 426	03 20	[illegible]
Abell 1060	10 37	[illegible]
Abell 1367	11 45	[illegible]
Abell 1656	13 00	[illegible]
Abell 2065	15 23	[illegible]
Abell 2151	16 05	[illegible]
Abell 2218	16 36	[illegible]
Abell 2572	23 18	+18 44
Arp 65	00 22	+22 24
Arp 84	13 59	+37 26
Arp 99	23 15	+19 00
Arp 113	00 18	+30 05
Arp 148	11 04	+40 51
Arp 254	15 22	–07 23
Arp 273	02 22	+39 22
Arp 286	14 20	+03 58
Arp 295	23 42	–03 38
Copeland's Septet	11 38	+21 58
Draco Dwarf	17 20	+57 55
Einstein Cross	22 41	+03 22
ESO 138-IG29	17 29	–62 27
ESO 350-040	00 38	–33 43
ESO 351-030	01 00	–33 43
Holmberg II	08 19	+70 43
Holmberg 124	09 21	+64 13
Holmberg 224	11 03	+18 04
Holmberg 800	22 58	+26 08
HCG 61	12 12	+29 12
HCG 68	13 54	+40 19
HS 1216+5032	12 19	+50 16
IC 10	00 20	+59 18
IC 239	02 36	+38 58
IC 342	03 47	+68 06
IC 1296	18 53	+33 04
IC 1613	01 05	+02 07
IC 2574	10 28	+68 25
Leo 1	10 08	+12 18
LMC	05 24	–69 45
M 31	00 43	+41 16
M 32	00 43	+40 52
M 33	01 34	+30 40
M 49	12 30	+08 00
M 51	13 30	+47 14
M 58	12 38	+11 49
M 60	12 44	+11 33
M 61	12 22	+04 28
M 63	13 16	+42 02
M 64	12 57	+21 41
M 65	11 19	+13 06
M 66	11 20	+12 59
M 74	01 37	+15 47
M 77	02 43	–00 01
M 81	09 56	+69 04
M 82	09 56	+69 41
M 83	13 37	–29 52
M 85	12 25	+18 11
M 87	12 31	+12 23
M 88	12 32	+14 25
M 89	12 36	+12 33
M 90	12 37	+13 10
M 91	12 35	+14 30
M 94	12 51	+41 07
M 95	10 44	+11 42
M 96	10 47	+11 49
M 98	12 14	+14 54
M 99	12 19	+14 25
M 100	12 23	+15 49
M 101	14 03	+54 21
M 102	15 06	+55 46
M 104	12 40	–11 37
M 105	10 48	+12 35
M 106	12 19	+47 18
M 109	11 58	+53 22
M 110	00 40	+41 41
Maffei 1/2	02 37	+59 39
NGC 45	00 14	–23 11

Galaxy	Right Ascension h m	Declination
[illegible]	00 48	–25 17
NGC 266	00 50	+32 17
NGC 300	00 55	–37 41
NGC 488	01 22	+05 15
NGC 520	01 25	+03 47
NGC 536	01 26	+34 42
NGC 613	01 34	–29 25
NGC 660	01 43	+13 39
NGC 678	01 49	+22 00
NGC 772	01 59	+19 00
NGC 891	02 23	+42 21
NGC 925	02 27	+33 35
NGC 1073	02 44	+01 23
NGC 1097	02 46	–30 16
NGC 1187	03 03	–22 52
NGC 1232	03 10	–20 35
NGC 1291	03 17	–41 06
NGC 1300	03 20	–19 25
NGC 1313	03 18	–66 30
NGC 1316	03 23	–37 12
NGC 1357	03 33	–13 40
NGC 1365	03 34	–36 08
NGC 1399	03 38	–35 27
NGC 1530	04 23	+75 18
NGC 1532	04 12	–32 52
NGC 1961	05 42	+69 23
NGC 2146	06 19	+78 21
NGC 2207	06 16	–21 22
NGC 2276	07 27	+85 45
NGC 2403	07 37	+65 36
NGC 2442	07 36	–69 32
NGC 2444	07 47	+39 02
NGC 2460	07 57	+60 21
NGC 2537	08 13	+45 59
NGC 2595	08 28	+21 29
NGC 2683	08 53	+33 25
NGC 2685	08 56	+58 44
NGC 2775	09 10	+07 02
NGC 2841	09 22	+50 59
NGC 2859	09 24	+34 31
NGC 2903	09 32	+21 30
NGC 2964	09 43	+31 51
NGC 2997	09 46	–31 11
NGC 3079	10 02	+55 41
NGC 3169	10 14	+03 28
NGC 3184	10 18	+41 25
NGC 3190	10 18	+21 50
NGC 3227	10 24	+19 52
NGC 3239	10 25	+17 10
NGC 3256	10 28	–43 54
NGC 3310	10 39	+53 30
NGC 3338	10 42	+13 45
NGC 3344	10 44	+24 55
NGC 3396	10 50	+32 59
NGC 3521	11 06	–00 02
NGC 3621	11 18	–32 49
NGC 3656	11 24	+53 51
NGC 3718	11 33	+53 04
NGC 3938	11 53	+44 07
NGC 3953	11 54	+52 20
NGC 4038	12 02	–18 52
NGC 4088	12 06	+50 32
NGC 4216	12 16	+13 09
NGC 4236	12 17	+69 28
NGC 4278	12 20	+29 17
NGC 4350	12 24	+16 42
NGC 4395	12 26	+33 33
NGC 4438	12 28	+13 01
NGC 4449	12 28	+44 06
NGC 4490	12 31	+41 39
NGC 4535	12 34	+08 12

— The Cambridge Photographic Atlas of

Galaxies

Galaxies – the Milky Way's siblings – offer a surprising variety of forms and colours. Displaying symmetrical spiral arms, glowing red nebulae or diffuse halos, even the image of a galaxy can reveal much about its construction. All galaxies consist of gas, dust and stars, but the effects of gravity, dark matter and the interaction of star formation and stellar explosions all influence their appearances. This volume showcases more than 250 of the most beautiful galaxies within an amateur's reach and uses them to explain current astrophysical research. It features fantastic photographs, unique insights into our knowledge, tips on astrophotography and essential facts and figures based on the latest science. From the Andromeda Galaxy to galaxy clusters and gravitational lenses, the nature of galaxies is revealed through these stunning amateur photographs. This well illustrated reference atlas deserves a place on the bookshelves of astronomical imagers, observers and armchair enthusiasts.

MICHAEL KÖNIG has a doctorate in astrophysics and studied galaxies professionally for many years. As an amateur astronomer he photographs galaxies from his own observatory, see: www.astro-images.de.

STEFAN BINNEWIES has more than 30 years of experience as an astrophotographer; his images are published in magazines and books all over the world. His website is: www.capella-observatory.com.

PHILLIP HELBIG worked in cosmology and gravitational lensing at Hamburg and Jodrell Bank Observatories and the Kapteyn Astronomical Institute.

ACKNOWLEDGEMENTS TO THE GERMAN EDITION

This book would not have been possible without the great effort of all those involved. We extend our special thanks to the many astrophotographers from many countries who have allowed us to use their pictures. Their "amateur images" are the result of years of experience in astrophotography and diligent work in the details of image processing. These add something special to this Atlas.

We would like to especially thank our editors: Dr. Wolfgang Steinicke critically read the introductory chapter and Peter Riepe was the scientific editor of the entire text and made valuable contributions. Many thanks are also due to Agathe Schmid-König who was responsible for the layout and for all illustrations which were modified or newly produced for the book.

Finally, we thank the publisher Kosmos for taking on this large project and investing much work in its realization. In particular we mention Sven Melchert and his co-workers who made sure that all parts of the book form a coherent whole.

Rimbach and Much, March 2016
Michael König and Stefan Binnewies

— The Cambridge Photographic Atlas of

Galaxies

MICHAEL KÖNIG
STEFAN BINNEWIES

Translated by PHILLIP HELBIG

CAMBRIDGE
UNIVERSITY PRESS

CAMBRIDGE
UNIVERSITY PRESS

University Printing House, Cambridge CB2 8BS, United Kingdom

One Liberty Plaza, 20th Floor, New York, NY 10006, USA

477 Williamstown Road, Port Melbourne, VIC 3207, Australia

4843/24, 2nd Floor, Ansari Road, Daryaganj, Delhi – 110002, India

79 Anson Road, #06–04/06, Singapore 079906

Cambridge University Press is part of the University of Cambridge.

It furthers the University's mission by disseminating knowledge in the pursuit of education, learning and research at the highest international levels of excellence.

www.cambridge.org
Information on this title: www.cambridge.org/9781107189485

Originally published in German as Binnewies/König, *Bildatlas der Galaxien*, 2016

In cooperation with Oculum-Verlag GmbH, Erlangen

First English edition published 2017

Printed in the United Kingdom by Clays, St Ives plc

A catalogue record for this publication is available from the British Library

ISBN 978-1-107-18948-5 Hardback

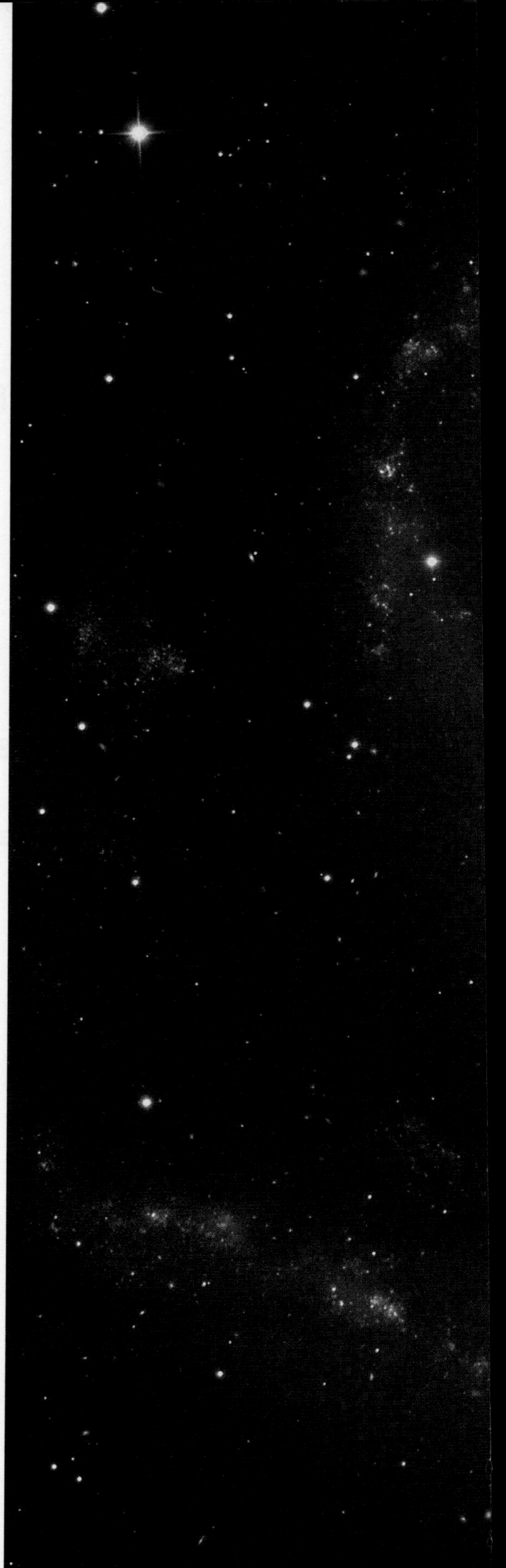

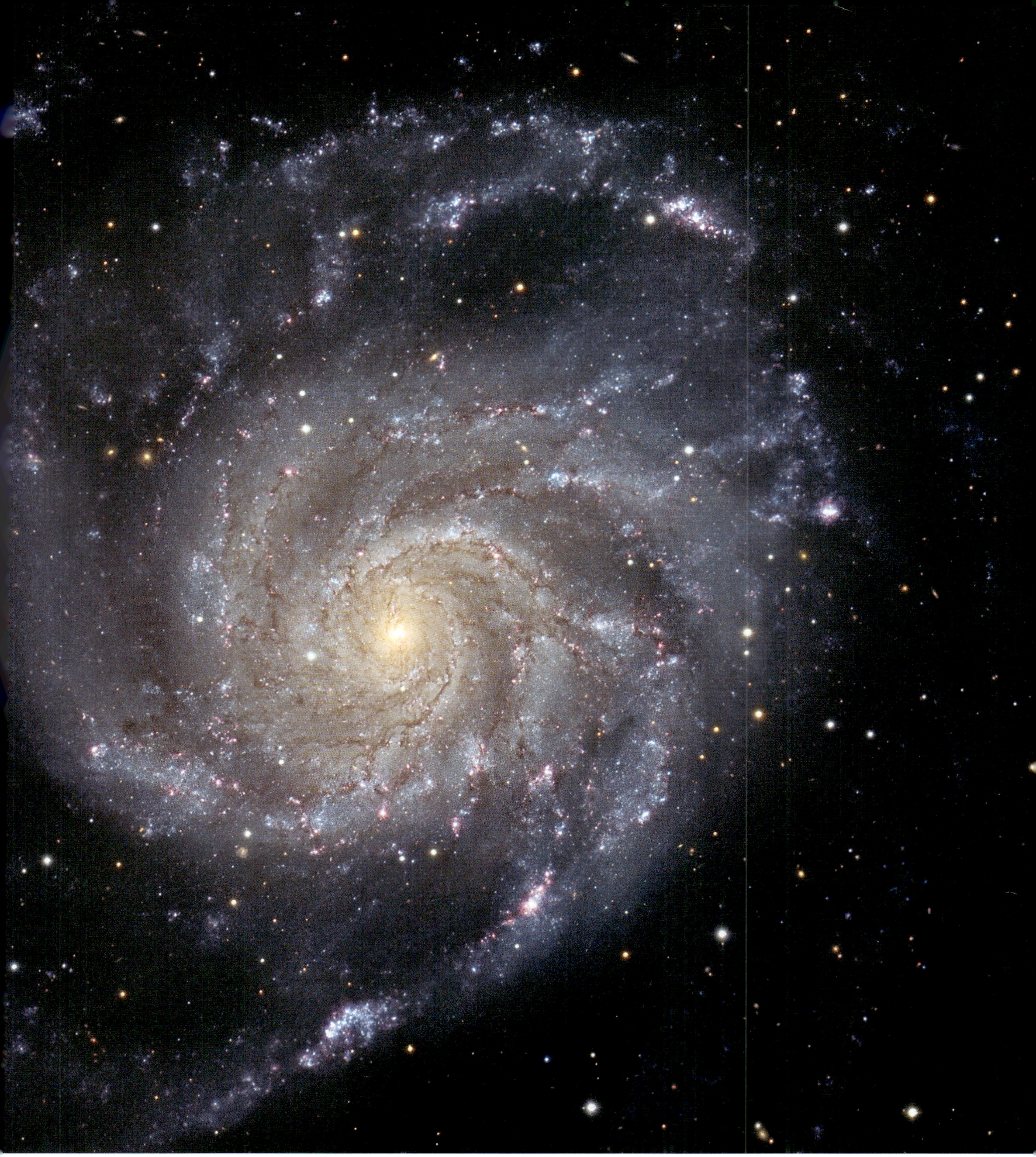

☞ *Contents*

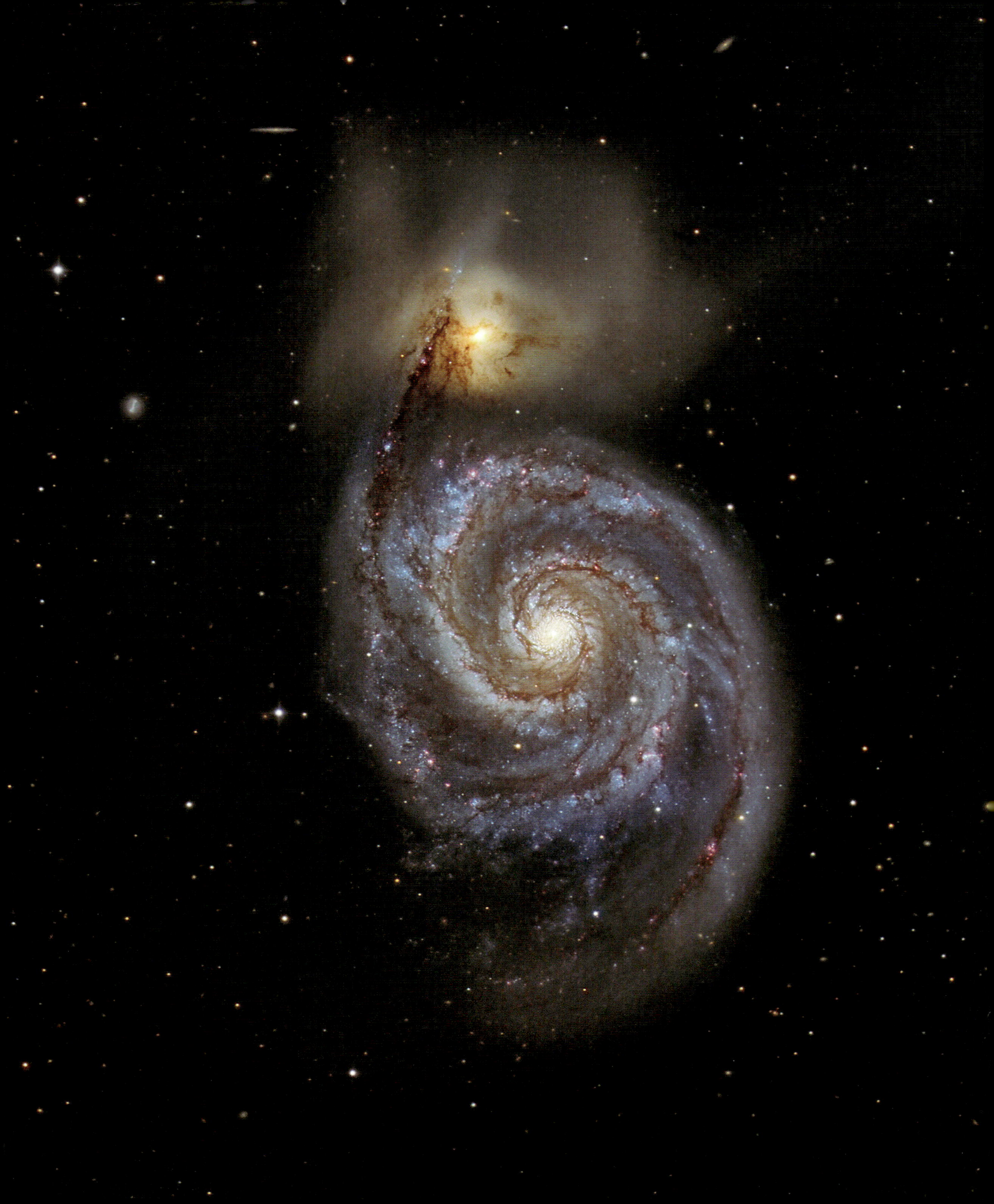

INTRODUCTION

Galaxies are one of the most fascinating phenomena in the universe. The diversity of their forms is an indication of various interaction processes which take place within or between galaxies.

Galaxies can be described abstractly as collections of gas, dust, and stars. However, a look at the astrophysics of galaxies allows one to better understand their morphology. Only shortly after their discovery, the study of the forms and evolution of galaxies led to fundamental questions about the nature of the universe.

THE FIRST GALAXY CATALOGUES

One of the early telescopic observers was Charles Messier, who catalogued the first galaxies in the second half of the eighteenth century. His motivation was to avoid confusing these specks of light with new comets. The first edition of the Messier catalogue was published in 1774 and extended in 1780 and 1781. In the Messier catalogue, there was no classification of nebulae according to their appearance. The first galaxy in the Messier catalogue is M 31, the Andromeda Galaxy. In the original catalogue, Messier describes M 31 as "the beautiful nebula in the belt of Andromeda, in the form of a spindle, similar to two opposite cones or pyramids whose bases touch". The galaxies M 32 and M 33 follow in the catalogue. M 32 is described as a "small, round nebula" and one reads about M 33 that "the nebula is of white light which is almost uniform, but somewhat brighter at two-thirds of its diameter". These prosaic texts describe the objects, but do not allow a comparison or classification since such descriptions depend on the individual impressions at the telescope and the size of the instrument.

A further pioneer in the observations of galaxies was Friedrich Wilhelm Herschel (known as William Herschel after his emigration to England) who, aided by his sister Caroline, actively looked for further such nebulae and star clusters. Their observations led to several catalogues of "new Nebulæ, nebulous Stars, planetary Nebulæ, and Clusters of Stars" which were published in 1786, 1789, and 1802. William Herschel writes in the preface to his third catalogue that "our stock of materials is now so increased, that we should begin to arrange them more scientifically". He also writes: "The classification adopted in my catalogues, is little more than an arrangement of the objects for the convenience of the observer, and may be compared to the disposition of the books in a library, where the different sizes of the volumes is often more considered than their contents". About his method of classification, Herschel says that he had "to examine the nature of the various celestial objects that have been hitherto discovered, in order to arrange them in a manner most conformable to their construction". Nebulae were described by William Herschel as "curious objects" which must be especially far away. He estimated the distance of the Andromeda Nebula completely wrongly at 2000 times the distance of Sirius, although William Herschel was possibly aware of the fact that this nebula could be composed of many individual stars.

A further milestone in the cataloguing of galaxies was the publication in 1888 of John Louis Emil Dreyer's *New General Catalogue of Nebulae and Clusters of Stars* (NGC) containing 7840 objects. He also made use of older catalogues by William Herschel which had been extended by his son John Frederick William Herschel. In 1895 and 1907 followed the *Index Catalogues, IC I* and *IC II*, with a further 5386 nebulae and star clusters. In spite of the size of these initial catalogues, the nature of the nebular specks was unknown. William Parsons, 3rd Earl of Rosse, speculated that nebulae are large, distant whirls of gas. (John Louis Emil Dreyer had worked as an assistant for Parsons's son Laurence, 4th Earl of Rosse.) This was motivated by the spiral patterns visible to him at the telescope. Some astronomers at the time considered the galaxies to be distant stellar systems, similar to the Milky Way. Other astronomers assumed that the Milky Way was a special object in the universe and that the nebulae were a part of it.

The Whirlpool Galaxy M 51 displays a broad spiral pattern with two spiral arms. At the northern end of M 51 one can see the companion NGC 5195, an SB0/a galaxy (see page 196). Image credit: Josef Pöpsel, Stefan Binnewies (600-mm reflector).

THE NATURE OF GALAXIES

These two points of view came to a head in 1920 in the so-called Great Debate between the two astronomers Harlow Shapley and Heber Doust Curtis. In earlier papers about the irregular distribution of globular clusters, Shapley had estimated the diameter of the Milky Way at 300 000 light years. This value was ten times larger than that derived by Jacobus Kapteyn at the turn of the century. In Harlow Shapley's one-galaxy model, the Solar System was located 60 000 light years from the centre of this huge Galaxy. The debate was decided by Cepheid variable stars. In 1908, Henrietta Swan Leavitt had discovered the relation between their period and their luminosity. Henrietta Swan Leavitt observed Cepheids in the Magellanic Clouds, which are all at about the same distance. This means that the apparent luminosity is proportional to the absolute luminosity and the relation discovered is known today as the period–luminosity relation for Cepheids. The first generalization of the process was due to Ejnar Hertzsprung, who estimated the absolute luminosity from the proper motion of nearby Cepheids and measured the associated periods. Using this strategy, in 1923 Edwin Powell Hubble used the then most powerful reflecting telescope at the Mount Wilson Observatory to measure the distance to the Andromeda Nebula at 980 000 light years. This ended the debate – galaxies are very distant, independent stellar systems.

Looking at the result of the debate in more detail, one must note that Harlow Shapley was not completely wrong. The relative offset between the Solar System and the Galactic centre turned out to be right. With regard to the dimension of the Milky Way, both parties were wrong by the same factor. Harlow Shapley overestimated the size of the Milky Way by a factor of three while Heber's value was a third of the value established today.

Hubble used the Cepheid calibration of Ejnar Hertzsprung, which had also been used by Harlow Shapley. In the 1950s, it was shown that the Cepheids in the spiral arms of the Andromeda Nebula discovered by Hubble belong to the younger population I. They are about four times as bright as the older Cepheids of population II which one had found in globular clusters and at the centre of the Milky Way. Using the latter period–luminosity relation, the calculated distance of the Cepheids in M 31 had been underestimated. Work by Walter Baade cleared up the confusion and use of the proper Cepheid calibration resulted in a doubling of the distance of the Andromeda Nebula and all other extragalactic objects.

After the nature of the nebulae had been resolved, Edwin Hubble and his assistant Milton Humason turned to the search for Cepheids in further galaxies. They made use of spectra which had been taken on photographic plates with an objective-prism spectrograph. This had already been done by Vesto Slipher in 1912. The optical spectrum of the Andromeda Nebula showed a blueshift in Slipher's measurements. Interpreted as a Doppler effect, M 31 would be approaching the Milky Way. Already in 1915, Vesto Slipher presented the spectra of 15 galaxies, some of which were receding from the Milky Way at more than 1000 km/s.

The more galaxies Edwin Hubble and Milton Humason studied, the more they noticed a preponderance of redshifts.

THE HUBBLE LAW

The mathematical relation discovered by Edwin Hubble between the measured redshift z of spectral lines in galaxy spectra and the distance D of a galaxy, $c \cdot z = H_0 \cdot D$, is an approximation which delivers good results up to about $z = 0.5$. Hubble used apparent magnitude as a measure of distance. One must keep in mind that local velocities can change the theoretical recession-velocity values and thus distort the result.

It seems plausible to interpret this law as the result of the Doppler effect, but the actual cause of the redshift is the expansion of the universe. The Hubble constant H_0 can be interpreted as the current value of a time-dependent Hubble parameter H(t). (The name is not a misnomer; it refers to the mathematical constant in the equation above, which also relates velocity and distance and is valid for all velocities and distances at a given cosmic time. The term "constant" does not imply that it is constant in time; in general, the value of this constant changes with cosmic time.) One makes use of the Friedmann–Robertson–Walker model of the cosmos which defines H(t) as $\dot{R}(t)/R(t)$. The quantity R(t) is the scale factor of this cosmological model, the Hubble parameter H(t) the relative rate of change of the scale factor. Mathematically, H_0 is the slope of the curve of R(t) at a given cosmic time. The linearity follows from the homogeneity and isotropy of the expansion; at a given cosmic time, proper distance and the corresponding velocity are related by the value of the Hubble constant at that cosmic time. (However, the linearity between observable quantities such as redshift and luminosity distance generally holds only for small redshifts.) At each point in the universe, an observer has the impression that the galaxies are moving away from him; there is no special point and thus no centre of this expansion.

In his original paper, Edwin Hubble estimated the value of the constant H_0 at 500 to 550 km/s/Mpc. The value of H_0 according to current measurements, however, is between 68 and 72 km/s/Mpc; the dimension of H_0 is an inverse time. H_0 corresponds to the recession velocity at a distance of a megaparsec, which is equal to 3.26 million light years. The reciprocal of H_0 is known as the Hubble time. Assuming a uniform expansion of the cosmos, this can be interpreted as the age of the universe. With a value of $H_0 = 72$ km/s/Mpc, one thus obtains a Hubble time of 13.6 billion years. In reality, the expansion is influenced by dark energy and (dark) matter, which in general leads to a departure of the Hubble time from the true age of the universe.

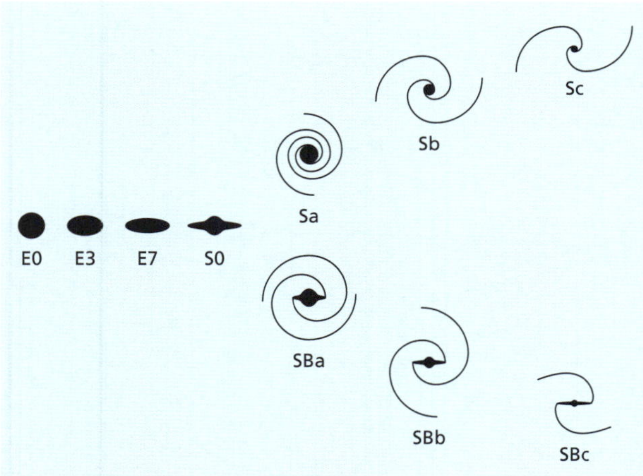

Figure I.1: Hubble's tuning-fork diagram (after a figure in The Sequence of Nebular Type, *E. Hubble, 1936).*

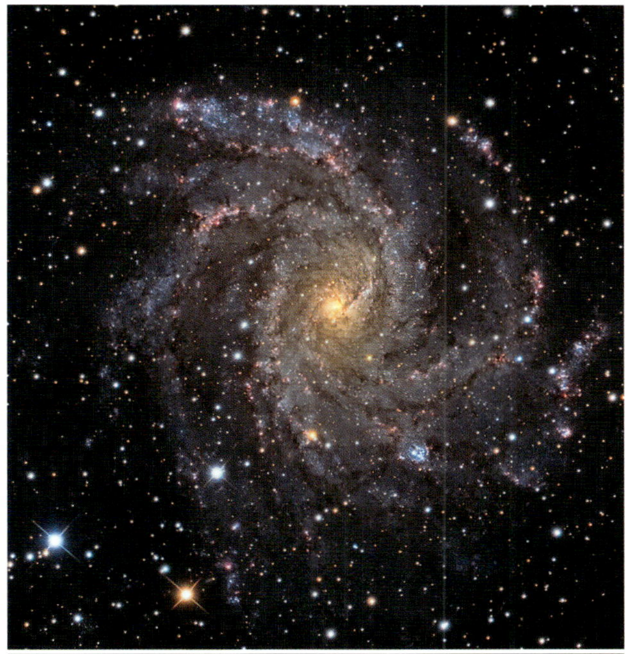

Figure I.2: NGC 6946 is observed almost completely face-on (see page 65).

In 1929, Hubble published the relation between the redshift and the distance of galaxies: the calculated recession velocity increases proportionally with distance. The constant of proportionality H_0 is known as the Hubble constant. This fundamental relation confirmed a cosmological theory initially proposed in 1922 by Alexander Friedmann: the universe is expanding.

THE PATH TO THE CLASSIFICATION OF GALAXIES

The use of photographic plates in astrophotography provided the basis for comparative studies of the morphology of galaxies. In the first attempts at galaxy classification, Edwin Hubble used a scheme which Max Wolf had published in 1908. Since 1903, the latter had published several *Köngistuhl-Nebel-Listen* (lists of nebulae observed at the observatory at Königstuhl, Heidelberg) which contained hundreds of "prominent nebular specks". There were descriptions of the details of the objects with regard to form, size, and luminosity, but planetary nebulae and galaxies were treated the same in the list, making a generalization for galaxies difficult. Edwin Hubble distinguished four types of galaxies based on form: elliptical galaxies (E), spiral galaxies (S), barred spiral galaxies (SB) and irregular galaxies (Irr). He arranged the first three in a tuning-fork scheme in which the two arms contain the S and SB types. Within the arms, there is a linear a/b/c subclassification. This depends on how prominent the central region is and how narrow or wide the spiral pattern appears. The subclassification of elliptical galaxies uses the measured ellipticity and ranges from round E0 to flattened E7 types. They are at the base of the tuning fork in the diagram. Edwin Hubble postulated that this series is also an evolutionary sequence of E and S/SB galaxies, leading from "early" through "intermediate" to "late" types. This sequence is left to right in the diagram. Hubble also noted that "a more exact classification, perhaps as a decimal number, is not justified, because current knowledge does not allow this". Nevertheless, historically these terms became established, but they cannot be used to indicate an age sequence. Just to the left of the junction of the two arms of the tuning fork, Hubble placed S0 galaxies, which display no spiral structures and which he referred to as "more or less hypothetical". In Figure I.1 there is a lenticular form sketched at this position; in the original there is a circle. Galaxies without organized structures were termed "irregular" by Hubble. Edwin Hubble had also used the symbol Q for galaxies which were too weak to be classified.

The famous tuning-fork diagram was published by Edwin Hubble in 1936 in his book *The Realm of the Nebulae*. Although there were some other classification schemes proposed by other astronomers, over the years Hubble's classification became the standard due to its simplicity. Hubble's work on the classification of galaxies was continued in the 1950s by Allan Rex Sandage. A milestone of his contribution is the *Hubble Atlas of Galaxies*, which appeared in 1961.

In 1959, Gérard de Vaucouleurs introduced a revised version of the Hubble classification intended to take the characteristic aspects of the dynamics of galaxies into account. For example, he extended the subtypes by the category d. He also introduced the additional specification (m) if the galaxy had a form similar to the Magellanic Clouds and in his opinion was to be positioned at the transition between the regular and irregular spiral galaxies. If there were hints of a bar, this was assigned to the intermediate classification SAB, lying between Hubble's S and SB galaxies. S0 galaxies were subdivided by de Vaucouleurs into $S0^-$, $S0^0$, $S0^+$. His extensions also included a finer distinction of the spiral patterns, which were assigned the additional specifications (s) for spiral-shaped and (r) for ring-shaped. If the pattern of the spiral arms suggested a ring which, however, was incomplete, this was described by the intermediate form (rs). Thus the types S, SAB, and SB as well as (r), (rs), and (s) resulted in nine possible combinations, which de Vaucouleurs divided into eight segments. At the centre of this pattern is the intermediate type SAB(rs).

For a sequential representation, Gérard de Vaucouleurs used the arrangement E0–E6, $S0^-$, $S0^0$, $S0^+$, Sa, Sb, Sc, Sd, Sm, and Im. At each point of this axis there is a plane containing the eight segments (with the ninth, centre, segment along the axis). The variation within the corresponding plane appeared to him to be different for different types of galaxies.

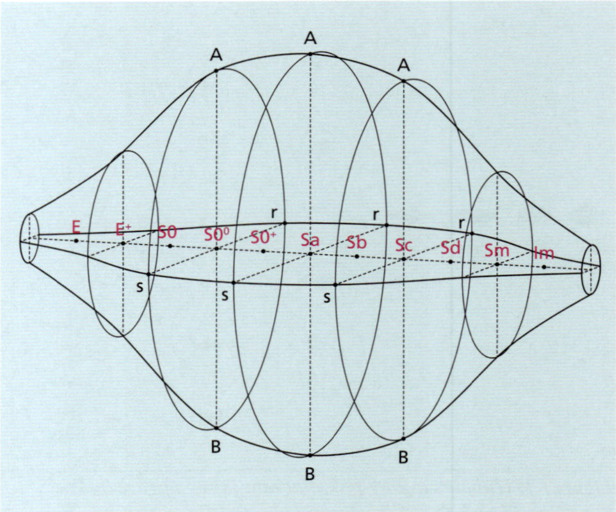

Figure I.3: 30D classification model by Gérard de Vaucouleurs (after a figure in Classification and Morphology of External Galaxies, *G. de Vaucouleurs, 1959).*

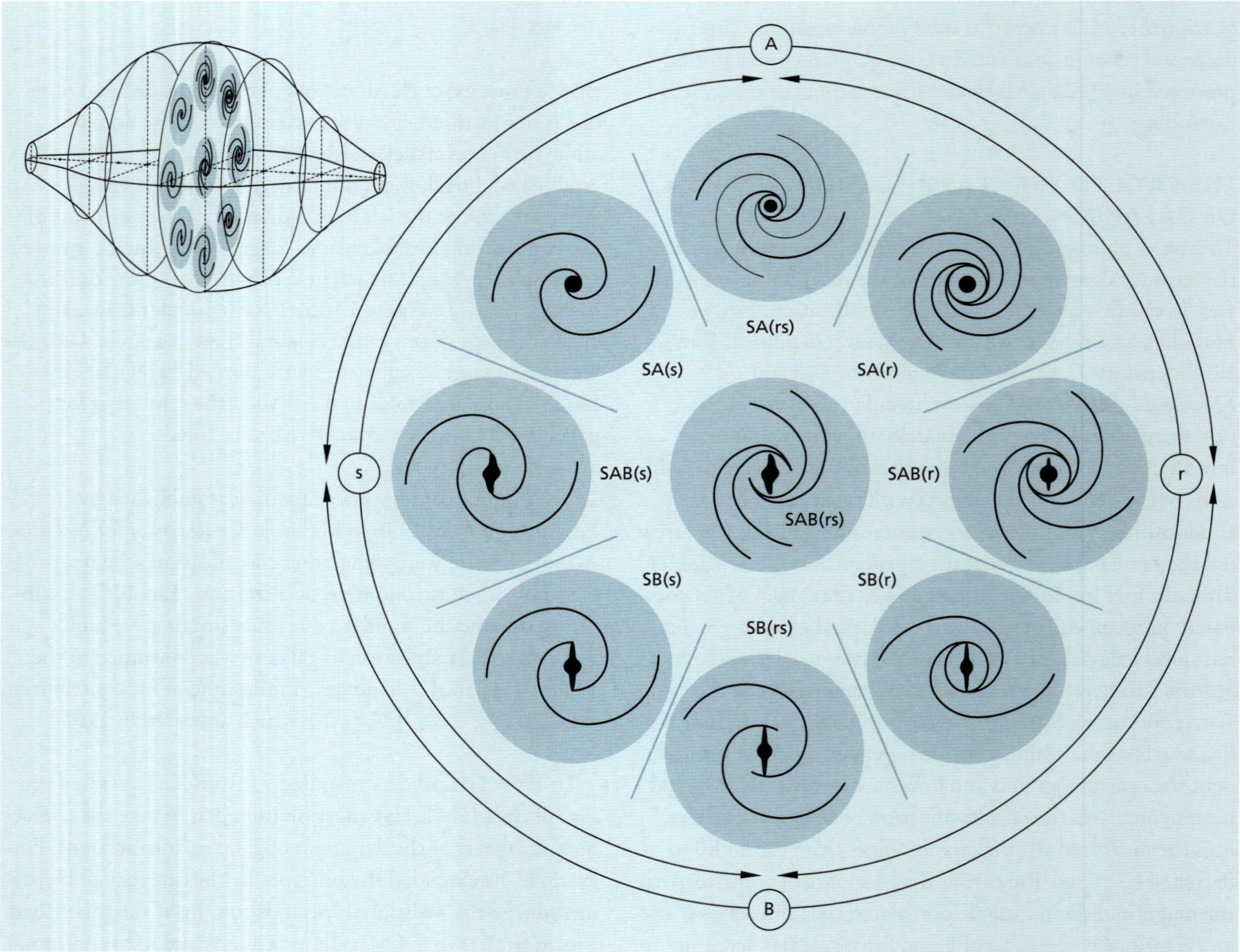

Figure I.4: Cross-section of the 3-D classification model by Gérard de Vaucouleurs (after a figure in Classification and Morphology of External Galaxies, *G. de Vaucouleurs, 1959).*

This is represented by the three-dimensional spindle-like form with the largest volume corresponding to S0 and Sa galaxies while the differences vanish for the elliptical and irregular galaxies at the ends.

The first, simplified form of this system was used by de Vaucouleurs in the *Reference Catalogue of Bright Galaxies* which was published in 1964 and contains 2593 galaxies. This nomenclature, abbreviated "RC", is found in many astronomical databases. A revised version, RC2, was published in 1976 with 4362 galaxies and the RC3 catalogue followed in the mid-1990s with more than 23 000 galaxies. In order to make the classification of Gérard de Vaucouleurs accessible not only to astronomers but also to students and amateurs, in 2007 his colleagues Ronald J. Buta, Harold G. Corwin, and Stephen C. Odewahn published *The de Vaucouleurs Atlas of Galaxies* containing extensive illustrations of the various morphological types, illustrated with detailed CCD images. The atlas suffers from the selection effect that it contains galaxies with relatively high surface brightness. Also lacking is a representative picture of dwarf galaxies, today known to be the most common type of galaxy.

THE CAMBRIDGE PHOTOGRAPHIC ATLAS OF GALAXIES

With this book, we would like to document impressive images taken by amateur astronomers and comment on the astrophysics revealed by them. The chapters correspond to the main galaxy types with the addition of dwarf, irregular, interacting, and ring galaxies. The two chapters on clusters and groups and on quasars give an idea of the photographic possibilities available to amateurs today. The descriptions of the images highlight the astrophysical aspects of the galaxies or their surroundings shown and also offer additional information to the general descriptions in the introductory chapters. The appendix contains references to the literature and web links for each of the image descriptions, allowing the interested reader the possibility to learn more about the galaxies shown. The web links point to published articles or websites. We would like to emphasize the relevant terms and conditions when using the sources. The order of objects in the individual chapters follows their right ascension; depending on the size of the image, there can be slight variations from this scheme.

The data on distance and size of the galaxies were obtained from the NASA/IPAC Extragalactic Database (NED). The distance measures have uncertainties which in some cases can be as large as 20 per cent. It should be emphasized that the angular-diameter values of a galaxy are not exact: often these data are derived from photographs, and therefore depend on the wavelengths used for the photo. A galaxy's measured size can differ depending on the wavelengths used, i.e. the diameter of a galaxy will be different in blue and in red light. In order to standardize this, we have used NED as a primary reference. Note also that there are types of galaxies where the NED diameters are misleading, such as when the bright inner part of a galaxy is surrounded by a ring or by filaments with lower brightness. Here, we have tried to use the image of the galaxy to measure the angular diameter, and then calculate the absolute size using the NED distance value. We know that these values are also not exact, but simply give a reasonable estimate of the galaxy's true dimensions.

BIBLIOGRAPHY

Baade, W.: *The Period-Luminosity Relation of the Cepheids*, Publications of the Astronomical Society of the Pacific, **68**, 1956

Bartusiak, M.: *The Cosmologist left behind*, 2009, www.marciabartusiak.com/uploads/8/5/8/9/8589314/cosmologist_left_behind.pdf

Buta, R. et al.: *The de Vaucouleurs Atlas of Galaxies*, Cambridge University Press, 2007

de Vaucouleurs, G.: *Classification and Morphology of External Galaxies*, Handbuch der Physik, **53**, 1959

Frommert, H.: *Original Messier Catalog of 1781*, 2014, http://messier.seds.org/xtra/Mcat/mcat1781.html

Herschel, W.: *Catalogue of 500 New Nebulae, Nebulous Stars, Planetary Nebulae, and Clusters of Stars; With Remarks on the Construction of the Heavens*, Philosophical Transactions of the Royal Society of London, **92**, 1802

Hubble, E.: *The Classification of Spiral Nebulae*, The Observatory, **50**, 1927

Hubble, E.: *The Realm of the Nebulae*, 1936, https://archive.org/details/TheRealmOfTheNebulae

Jet Propulsion Laboratory / California Institute of Technology: *NASA/IPAC Extragalactic Database*, 2014, http://ned.ipac.caltech.edu

NASA website: *The Shapley-Curtis Debate in 1920*, 2014, http://apod.nasa.gov/diamond_jubilee/debate_1920.html

Sandage, A. R.: *The Hubble Atlas of Galaxies*, Carnegie Institution, Washington, 1961

Steinicke, W.: *Nebel und Sternhaufen: Geschichte ihrer Entdeckung, Beobachtung und Katalogisierung*, Books on Demand, Norderstedt, 2009

Wolf, M.: *Königstuhl-Nebel-Liste 8. Mittlere Örter, Beschreibung und Helligkeitsvergleichung von 770 Nebelflecken bei +33,2174° Ursae*, Publikationen des Astrophysikalischen Instituts Königstuhl-Heidelberg, **3**, 1908

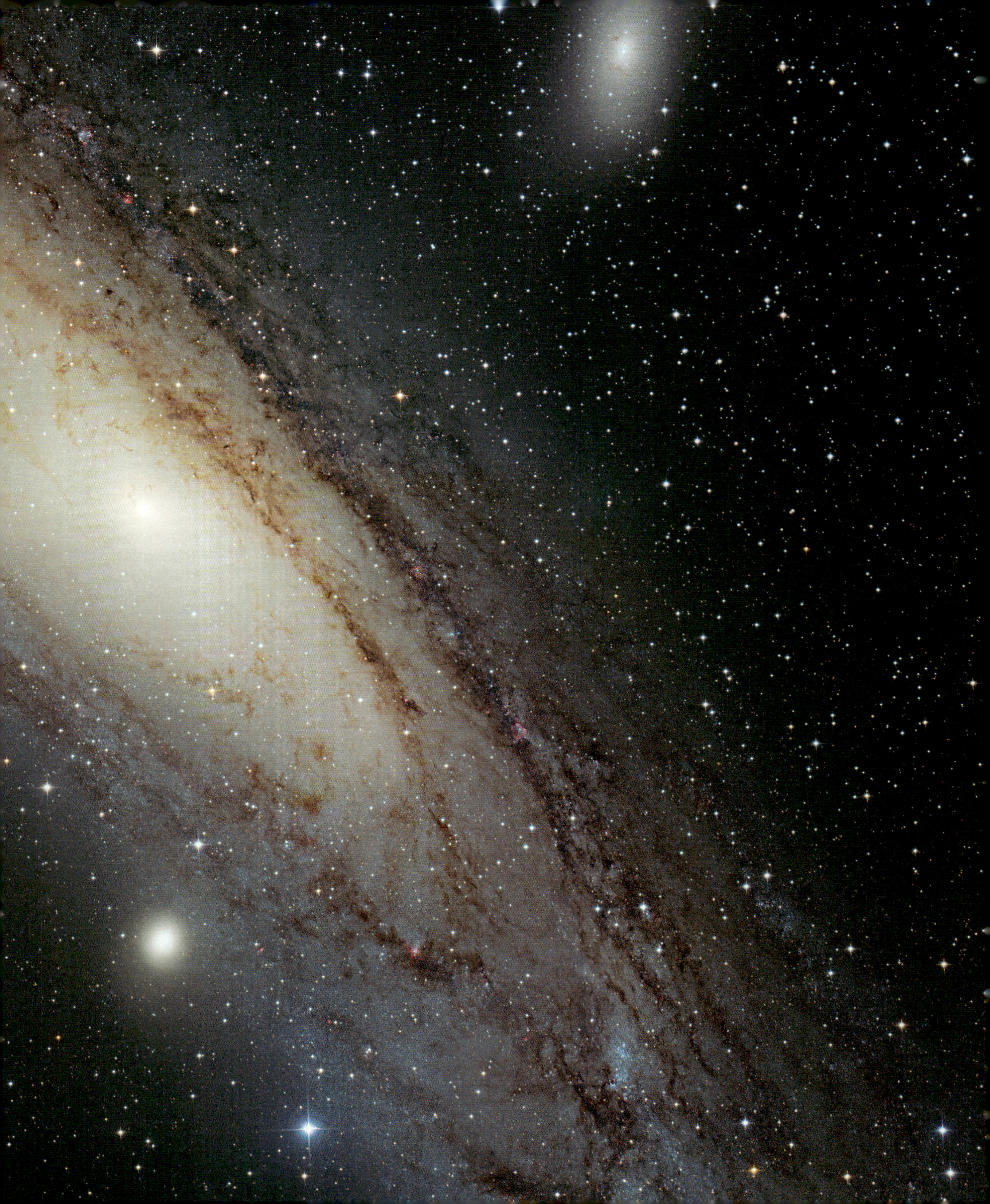

SPIRAL GALAXIES

For many astrophotographers, the spiral galaxy M 31 was their first extragalactic subject – the first image remains in memory. This is a good point of introduction to the astrophysics of these spiral systems.

THE CLASSIFICATION OF SPIRAL GALAXIES

Spiral galaxies with their spiral arms are prototypical galaxies. These impressive forms allow even those with little knowledge of astronomy to instantly identify spirals as "galaxies" in their images. Spiral arms have thus played a deciding role in the various classification schemes for this type of galaxy. Note that all classification schemes are based on the view onto the main plane of the galaxy. In other words, the "face-on" view is what matters for the classification. Galaxies seen from an oblique inclination are rotated in order to simulate a face-on view. The classification of edge-on galaxies is difficult.

An edge-on spiral galaxy allows a view into its galactic plane with large amounts of gas and dust. However, this makes it more difficult to say anything about the spiral arms. This dilemma also applies to the classification of our Milky Way, since Earth is located within the main plane which thus doesn't afford a view onto this plane. Further conclusions require complex measurements in wavelengths which are able to penetrate the thick clouds of gas and dust. Today, the consensus among astronomers is an SB classification with four spiral arms. However, the Milky Way did not always have this four-arm structure; it is suspected that tidal interactions induced disturbances in the plane of the disc which led to the splitting of two dominant spiral arms.

The Hubble classification, which was mentioned in the Introduction, has proved useful over many years and distinguishes spiral types based on the compactness of the pattern of the arms. The nomenclature uses, in addition to the letter S, the suffixes a, b, and c, for example describing as Sa a galaxy the arms of which appear very tight and as a rule can be followed over one or two full revolutions. Sb denotes types the arms of which are not as tightly wound and more coarsely structured compared to Sa. In the case of Sc galaxies, the arms are the least tightly wound and have the smallest radius of curvature. The spiral arms of Sc types manage only about one revolution from the centre to the outskirts of the disc.

In the literature one often finds expressions such as "less developed" or "more developed" Hubble types. Caution is required here since, despite the impression created by the series of types, this is not an evolutionary sequence. This was not known in the mid-1930s, when Edwin Hubble published his classification. One should thus not overlook this historical aspect when considering the physical aspects of the classification. This was the time of the "Great Debate" which was about the distances to photographed "nebulae". Were these part of the Milky Way or more distant, independent galaxies? In order to decide this, Edwin Hubble made use of Cepheids which he could identify in the Andromeda Nebula (M 31) and M 33. For these giant pulsating variables, there is a relation between the length of the period and their absolute luminosity. This was discovered in 1912 by Henrietta Leavitt, an astronomer at Harvard College Observatory. She made use of Cepheids in the Magellanic Clouds, two satellite galaxies of the Milky Way; since these Cepheids are all at approximately the same distance, their apparent and absolute magnitudes are proportional. Edwin Hubble, in 1924, was the first to determine the distance to the more distant galaxies M 31 and M 33, using Cepheids as distance indicators. It was noticed later that Cepheids are not ideal distance indicators, since the luminosity of these giant pulsating variables depends on the amount of heavy elements they contain, i.e. on their metallicity. The Cepheids in the Andromeda Nebula, like those in the Milky Way, are more metal-rich than those in the Magellanic Clouds and are about 1.5 magnitudes brighter for the same pulsation period. The German-American astronomer Walter Baade studied these different

The Andromeda galaxy M 31 is the largest galaxy in the Local Group and has several satellite galaxies (see page 16). Image credit: Stefan Hentz, Stefan Binnewies (600-mm reflector).

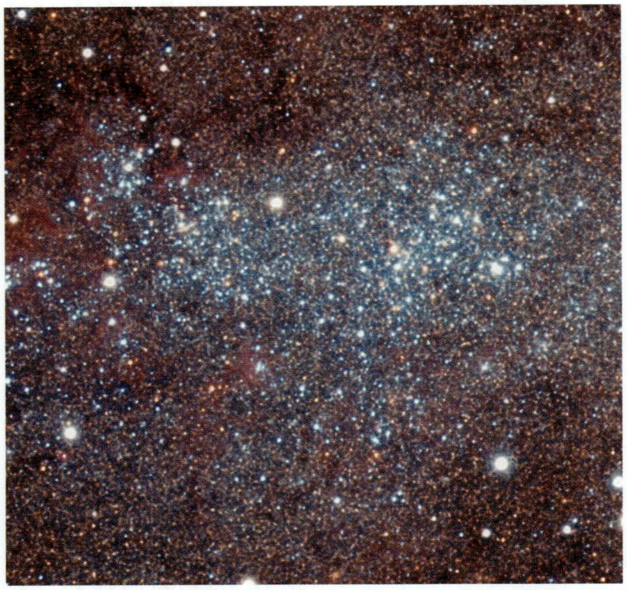

Figure 1.1: Star-formation regions are located in spiral arms, as here in NGC 206 (see page 18).

types of Cepheids in the 1950s. He was able to show that the calibration of the period–luminosity relation which Hubble used, based on the Cepheids in the Magellanic Clouds, was incorrect and delivered distances which were too small, not only for the Andromeda Nebula and M 33. The required correction increased all extragalactic distances by a factor of two.

The lintel-shaped S0 systems are a special type of spiral galaxy. They consist of a core and a disc component, but no spiral arms are visible in the disc of an S0. There are also at most hints of a bar in the centre. In addition, the core is relatively brighter compared to the disc than is the case for other types of spirals. In some edge-on S0 systems, one can see conspicuous dust lanes in the plane which are also used to distinguish subtypes of S0s.

Various factors play a role in the form and structure of spiral arms; most important is the presence of interaction partners. This doesn't refer to direct interaction; rather, an approach within ten galactic radii is enough to cause visible disturbances in the spiral arms. The spiral arms are very sensitive indicators even of very small disturbances of this kind. Since classification is always a simplification and a reduction of the variety of forms to a few characteristics, it is difficult to take into account all possible deviations from the ideal form of unperturbed spiral arms. In the 1960s, Gérard-Henri de Vaucouleurs introduced a refinement of the Hubble classification by adding a third dimension, distinguishing between ring- and s-shaped types. This formal extension of the description allowed for mixed types, and the terms Sa(s) and Sa(r) differ in that in the Sa(s) case one can follow conspicuous spiral arms from the centre to the outer reaches while an Sa(r) galaxy has a ring in the inner area from which the spiral arms emanate.

One also has to take the spectral bandwidth of the film or detector into account during classification. Since the spiral arms are dominated by young, hot stars, these structures are particularly prominent in the UV band, which could even influence the Hubble classification. Over the years, the original classification scheme has lost its importance for current science, since today more sophisticated analyses are necessary in order to take into account and explain the multitude of information available for a galaxy. Today, integral and spectral properties are usually used for the scientific classification of galaxies.

THE MORPHOLOGY OF SPIRAL GALAXIES

The spiral arms which one can see in images of galaxies indicate the fact that the density is higher here than on average. This is evidence of regions of compressed gas. The matter density in spiral arms is 10–15 per cent higher than the average in the disc and leads to a local increase in the star-formation rate. Our home galaxy, the Milky Way, is a nearby example of what one finds in distant spiral galaxies as well: many young, hot stars with a dominant blue-white colour are responsible for making the spiral arms visible and emphasize these active regions in the galactic disc both visually and photographically. A nice example is NGC 206, a star-forming region in a spiral arm of the Andromeda galaxy. About 4000 light years in size, this collection of young, hot stars shines in a conspicuous blue light. This is due to the hot O and B stars within which have their maximum emission in the blue spectral range. Such collections are also known as OB associations and are a sure sign that star formation is taking place. In addition to the blue colour, one also recognizes in spiral arms prominent fine structures with violet or red colours. In these so-called H II regions, hot stars ionize the surrounding hydrogen gas, causing emission via recombination. Since the interstellar matter in spiral arms is 90 per cent hydrogen, the emission spectra are primarily characterized by the Balmer spectrum of hydrogen with the lines Hα, Hβ, Hγ .... The strongest of these lines is the red Hα line with a wavelength of 656.3 nm. Spiral galaxies can contain thousands of these H II regions, which occur primarily about halfway along the spiral arms.

In contrast to the disc, the core region of a galaxy appears yellow-red. This is an indication of the presence of many old population II stars, which radiate mainly in the yellow-red spectral range. The light of many such stars thus emphasizes the longer wavelengths and shifts the colour of the core

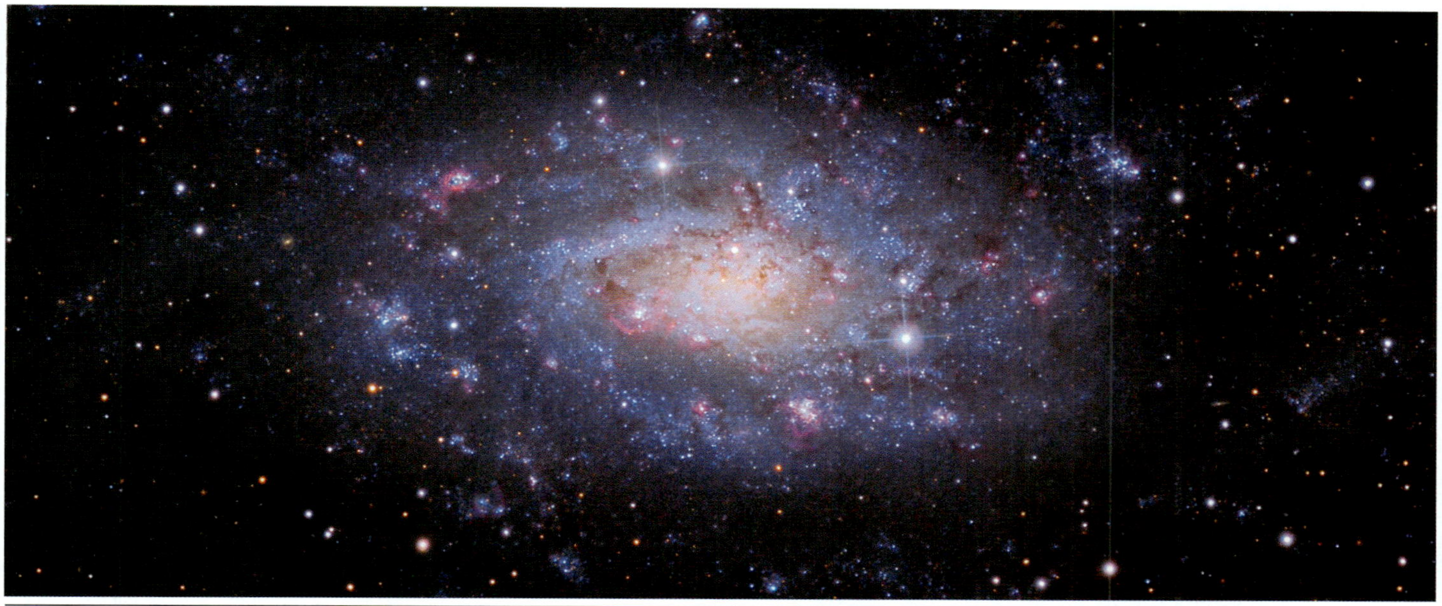

Figure 1.2: The spiral galaxy NGC 2403 has many active regions in its main plane which are visible as blue and red structures (see page 33).

correspondingly from blue to yellow-red. The transition from the spiral arms to the core of a spiral galaxy corresponds to a change in the stellar population and indicates a similarity between the core and an elliptical galaxy.

Spiral galaxies are not only morphologically similar but also similar in size (taking distance into account to convert apparent to physical size). Most spiral galaxies have a diameter of 65 000–80 000 light years. With a diameter of 100 000 light years, the Milky Way is above average in size and thus also in mass. One can see this as an indication of the fact that the current Milky Way has evolved from a normal spiral galaxy. During this evolution, smaller galaxies in the environment of the Milky Way merged with it and led to an increase in the disc and the total mass. This process continues today and can be seen in the Magellanic Clouds, which are visible from the southern hemisphere. These satellite galaxies of the Milky Way are dwarf galaxies which are interacting with the Galactic gravitational field and will be completely assimilated in the far future. A more dramatic example is provided by the Sagittarius dwarf galaxy, at 70 000 light years only half as distant as the Magellanic Clouds. This dwarf galaxy has already orbited the centre of the Milky Way several times and released several million solar masses of material in the process.

One can also study spiral arms purely geometrically. Such an analysis shows a strong similarity to logarithmic spirals. This construction principle occurs often in nature and can be described by noting that each revolution increases the distance from the centre by the same factor. This is found in high- and low-pressure zones in the atmosphere, in the arrangement of sunflower seeds, and in snail shells. Mathematically, logarithmic spirals have the property that a ray from the centre always cuts the spiral at the same angle. Thus, such a spiral is also known as an equiangular spiral. One defines the pitch angle as the difference between the angle of intersection and 90°. Studies of spiral galaxies show that typical pitch angles are 10–40° with a strong maximum around 20°. Larger pitch angles correspond to more open spirals and thus a more relaxed structure of the spiral arms.

One can also classify the spiral arms themselves based on their appearance. About ten per cent of spiral galaxies have two well defined spiral arms; these are often called "grand design" spirals, unperturbed textbook cases. In 60 per cent of cases, more than two spiral arms occur. This "multiple-arm structure" is often accompanied by splitting and modification. It is often difficult to determine the number of spiral arms due to the fact that they become less distinct as one moves away from the centre. Finally, about 30 per cent of spiral galaxies, so-called flocculent spirals, have arms which are almost completely dissolved, consisting only of small-scale arrangements of arm fragments.

Looking at the form of the disc plane in detail, one finds that in more than half of all spiral galaxies this plane is warped. The form reminds one of a sombrero, the broad brim of which is not flat but can be thought of as constructed from concentric circles which are slightly tilted with respect to one another. This asymmetry occurs independently of the

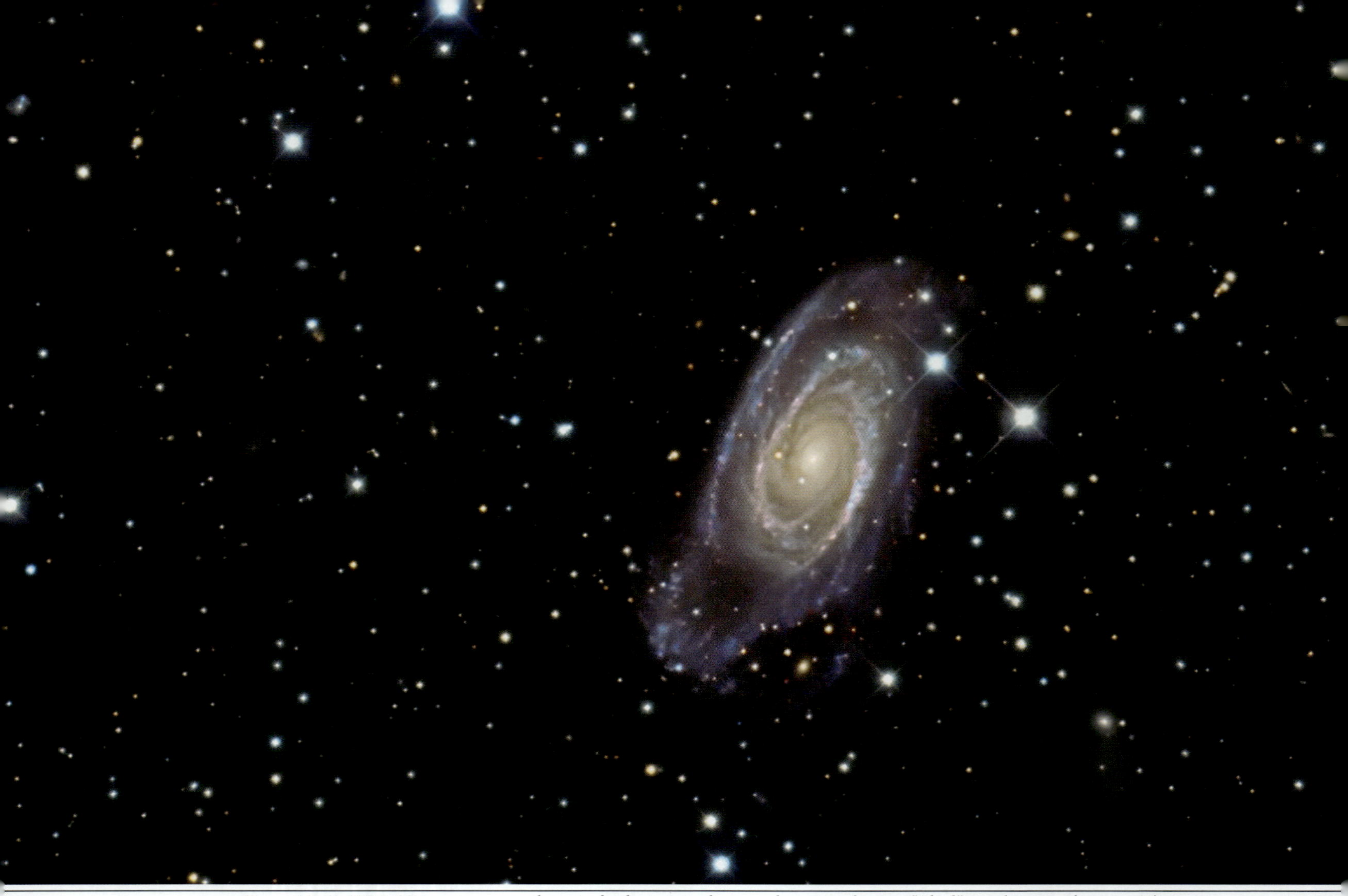

Figure 1.3: In NGC 6632, as in many other spiral galaxies, one discovers deviations from an "ideal" spiral pattern (see page 63).

classification of the spiral galaxy, and there is a relation between the radius of the warped disc and the amount of warp. In the case of strong warps, one suspects the cause to be a large-scale perturbation due to gravitational interaction with a companion galaxy. A slightly warped disc, on the other hand, can have several causes, such as an asymmetric flow of gas into the centre of the galaxy. These mechanisms lead one to the interesting conclusion that a flat disc in a spiral galaxy can form only if many large gas clouds combine and the successive overlap of small gas clouds takes place such that the main angular momentum dominates. This growth process differs from "hierarchical galaxy growth" in which large galaxies are formed by the merger of smaller galaxies, leading to a characteristic bulge in the central region. If the assembly of the galaxy is not disturbed by interactions, the result is a galaxy which is as flat as a pancake, known as a "pure-disc galaxy" in the literature.

THE ASTROPHYSICS OF SPIRAL GALAXIES

Together with knowledge about the various spectral classes and the blue-red range of the colour of stars, detailed studies of images can reveal physical details about the entire galaxy. Young, hot stars have larger masses compared to dimmer stars; the mass–luminosity relation for main-sequence stars is $L \sim M^{3.5}$. Star-forming regions are always bluer, and thus also the spiral arms where they are located, which is easily seen in the typical blue colour of the disc population of S/SB types. In contrast, an S0 type is much redder due to the lack of spiral arms with young, hot stars. The S0 type is thus similar to an elliptical galaxy which is dominated by dimmer, red, but also longer-lived types of stars.

Spiral galaxies are rich in dust and gas which can be seen in the form of dust lanes and quite distinctly in the case of edge-on galaxies. In this perspective, the dust lanes are in front of the spiral arms, leading to maximum contrast. Due to the strong extinction due to gas and dust, light is dimmed and in addition reddened. The reddening is due to wavelength-dependent scattering effects in which blue light is scattered more strongly, reducing the amount of blue in

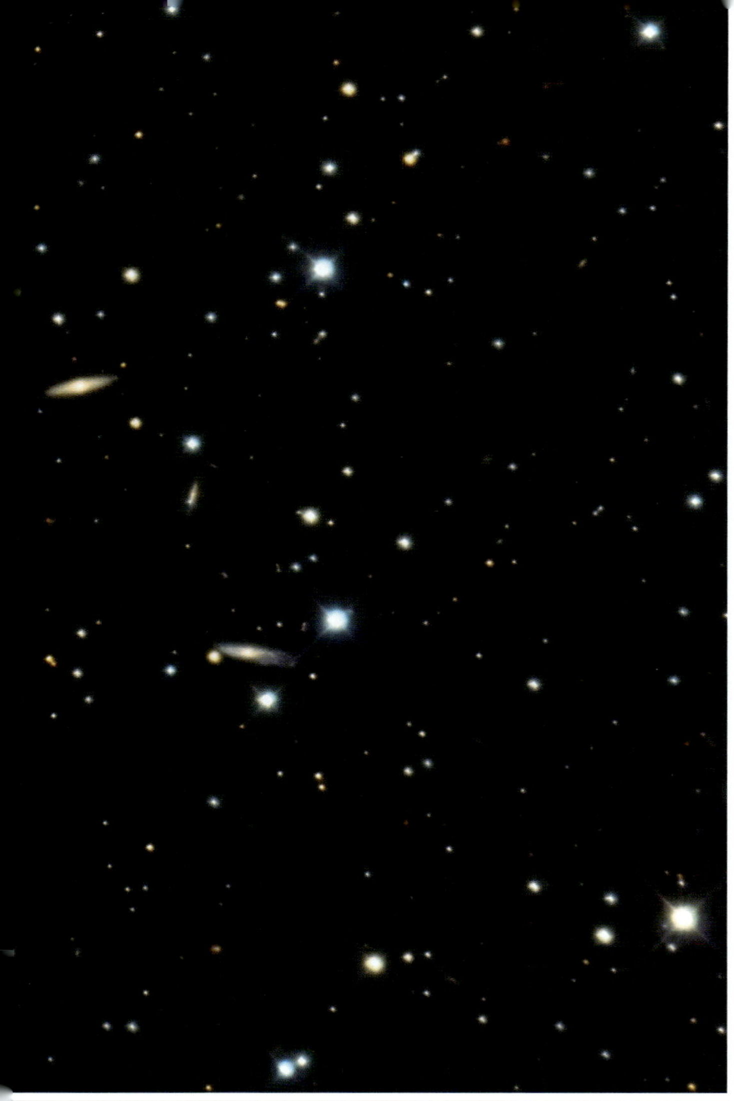

the observed object. In good images, one can study this in detail via the colours of stars whose light travels through the edges of dust lanes.

In the case of edge-on spiral galaxies, one can use the Doppler shift of characteristic lines to exactly determine the speed of rotation. In general, the angle of inclination of the galactic plane of the spiral galaxy is taken into account in the measurement. In the case of a face-on galaxy, it is not possible to determine the radial component of the speed of rotation. At large distances, rotation curves of spiral galaxies show no decline in the speed of rotation with distance from the centre such as one would expect from a Kepler orbit about a centre of mass; rather, the curve is flat. This is an indication that spiral galaxies do not have a simple mass distribution dominated by the core, such as is the case in our Solar System. In the last several decades, many spiral galaxies have been studied, and it was found that they are embedded in halos of dark matter, i.e. a smoothly distributed mass component which interacts only gravitationally. The model with a central mass in the middle of the spiral is not realistic. More appropriate is a thick disc or torus of dark matter in which the spiral plane is embedded.

One can use traditional physical methods to study the mechanisms in the plane of the disc, with differential rotation

THE PHYSICS OF VORONTSOV-VELYAMINOV ROWS

In the spiral arms of many galaxies, one finds straight segments which are known as Vorontsov-Velyaminov rows. The origin of these rows is not completely understood. It can't be explained via the gravitational potential of a spiral galaxy nor by tidal torques. This is due to the fact that in these gravitational fields there are no straight structures of constant force. In the 1990s, astronomers learned with the help of simulations that co-rotating perturbations in the plane of the spiral are a possible cause. In order to understand the physical basis of Vorontsov-Velyaminov rows, one has to model the gas dynamics in the spiral plane. An important hint is given by the dust lanes which one can observe along the spiral arms. These lanes are compressed layers of gas and dust which indicate where the shock fronts of the density waves are located. Since one observes linear dust formations in Vorontsov-Velyaminov rows, one can conclude that the corresponding local shock fronts are straight and differ from the curved form of the density waves propagating in the plane of the spiral.

How can local, straight areas arise in the global spiral-shaped density waves? The physical cause responsible for the origin of the rows also causes the rows to be stable, even if this leads to perturbations in the gas flow. If a perturbation leads to a concave form of the shock front, the gas flows behind the perturbation are compressed and thus accelerated. The dent in the shock front is thus compensated and the straight form is restored. A convex perturbation, on the other hand, leads to a deceleration of the gas flows and similarly the straight form is restored.

Astronomers have simulated the gas dynamics of density waves and shock fronts based on a differentially rotating gas disc and a spiral-shaped density wave. The density wave has a well defined amplitude in the plane of the spiral and a pitch angle such that it rotates like a solid body with constant angular velocity. In the simulations, one can observe a three-step process. This begins with shock fronts which are still spiral shaped. Then shear flows occur which interact with the density waves in the spiral arms and cause instabilities in the shock fronts. These shear flows are followed by so-called Kelvin–Helmholtz instabilities, which first cause wavy, small-scale segments to be formed, which then lead to straight secondary structures in the spiral arms (see Figure 1.4). Kelvin–Helmholtz instabilities are small eddies which occur along the boundary between two flows with different velocities. Kelvin–Helmholtz instabilities grow with time within this boundary and are visible as whirl-shaped patterns in the boundary layer.

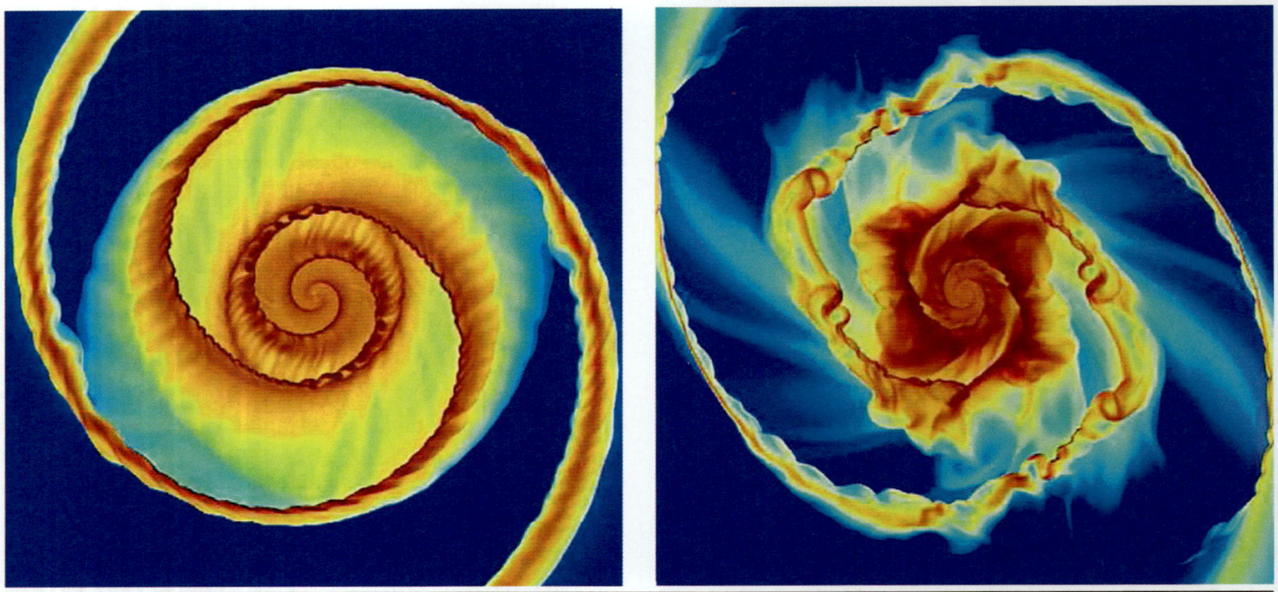

Figure 1.4: Colour illustration of hydrodynamical simulations. Left: View onto the spiral plane with two spiral-shaped density waves and initial perturbations due to shear flows. Right: Polygonal structures arise from perturbations of the gas dynamics. Colours indicate the density of gas in the plane of the spiral (red: high gas density, blue: low gas density). These simulations make use of a curvature radius of 15° and a density-wave amplitude of 0.1 (courtesy of S. A. Khosperskov).

as the starting point. Material in the disc at a distance r rotates with an angular velocity $\omega(r)$. Material closer to the centre rotates somewhat faster, that further away somewhat slower. How can the typical spiral patterns form via this differential rotation?

The first answer was given in 1925 by Bertil Lindblad, and in the mid-1960s the density-wave theory was further developed by Chia Chiao Lin and Frank Shu. This theory is based on a model for the formation of density waves which propagate in the plane and increase the surface density by 10–20 per cent compared to the average density.

The location of this increase in density defines the location of a spiral arm. The appearance of spiral arms, however, cannot be directly related to the orbital motion of stars about the galactic centre, since then differential rotation would have led to tighter winding of the spiral arms with time and hence to the destruction of the spiral structure.

How, then, do density waves originate? Every closed orbital curve has an angle of 180° between the minimum and maximum distance from the centre. For orbits within the disc, there are variations from this simple model, since the environment influences the gravitational potential and hence the motion of the body. The orbit of a star in the galactic plane is thus not closed and the angle between closest and farthest distance is less than 180°. The resulting form of the orbit corresponds to successive epicycles. The driving force of the density waves is the weak, distance-dependent drift of the orbital maxima. This drift transforms a leading wave, i.e. a wave whose outer edge travels faster than its inner edge, into a trailing wave. In such a trailing wave, the outskirts propagate more slowly than the more inner regions. Self-gravitation amplifies the surface density and hence the density wave itself. This is followed by feedback with the wave which travels through the galactic centre and thus becomes a leading wave. An almost stable spiral structure is formed in the disc, whereby, due to gravitational amplification, trailing waves are observed in most spiral systems. In a co-rotating coordinate system, stellar orbits appear almost closed and the minimum and maximum distances are separated by 90°. The existence of two spiral arms is a consequence of the symmetry of this simple model. From the point of view of an individual star, the action of the density wave results in an increase in the speed of the star when a wave is approaching and a decrease when it is moving away. The density wave leads to a local increase in the stellar density, which in turn leads to the forerunner of a later-forming spiral arm. This is because only such compression of gas leads gas clouds to collapse, which in turn allows star formation to begin in the spiral arms.

In the so-called "grand design" spirals of type Sbc/Sc, a phenomenon occurs in the spiral arms which was first discovered and studied in the 1950s by the Russian astronomer Boris Vorontsov-Velyaminov. He noticed straight portions in otherwise curved spiral arms, later known as Vorontsov-Velyaminov rows. Occasionally, a spiral arm has more than one of these rows, connected by a conspicuously constant angle, forming a polygon-like structure in the spiral arm. These spiral galaxies are otherwise regular and show no other abnormalities. The value of the angles between Vorontsov-Velyaminov rows lie between 110° and 140°, with 120° being most common. In the case of 120° angles, eight to ten such rows are observed.

If the angle is less or more than 120°, fewer rows are found in the spiral arms. One also finds that the length of a row increases linearly with distance from the centre. This is a hint that the cause of the rows is related to the large-scale density waves and not to small-scale perturbations. Examples of these rows can be found in M 51, M 101 and in NGC 2223. If one observes several Vorontsov-Velyaminov rows, one can broadly distinguish two types: on the one hand rows which are brighter than a normal spiral arm, but similar in form and width, and on the other hand conspicuously narrow rows in which the star-formation regions are more strongly segmented. There are also spiral galaxies such as NGC 1232 which contain both types.

The number of Vorontsov-Velyaminov rows depends on the pitch angle; a small angle of 5–20° leads to a large number of rows (5–7); for large angles of more than 40°, only one or two rows arise. The most important, form-determining parameter is the angular velocity of the spiral-shaped density waves, i.e. of the spiral pattern. An angular velocity of $\omega \leq$ 25 km/s/kpc has been found to be favourable for the formation of rows. Our Milky Way lies exactly within this range, so that we can assume that our home Galaxy has Vorontsov-Velyaminov rows.

BIBLIOGRAPHY

Chernin, A. D.: *Vorontsov-Velyaminov's rows in giant spiral galaxies: geometrical properties and physical interpretation*, Astrophysics, **41**, 1998

Chernin, A. D. et al.: *Vorontsov-Velyaminov rows: Straight segments in the spiral arms of galaxies*, Astronomy Letters, **26**, 2000

Filistov, E. A.: *Polygonal structure of spiral galaxies*, Astronomy Reports, **56**, 2012

Hubble, E.: *Cepheids in Spiral Nebulae*, Popular Astronomy, **33**, 1925

Khoperskov, S. A. et al.: *Polygonal Structures in the Gaseous Disk: Numerical Simulations*, Astronomy Letters, **37**, 2011

Kupka, F.: *Galaxien: Entwicklung*, 2005, www.mpa-garching.mpg.de/~fk/TUM

Lin, C. C. and Shu, F.: *On the Spiral Structure of Disk Galaxies*, Astrophysical Journal, **140**, 1964

Macri, L. M. et al.: *A Cepheid is No More: Hubble's Variable 19 in M33*, Astrophysical Journal, **550**, 2001

Pérez-Villegas, A. et al.: *Stellar orbital studies in normal spiral galaxies*, Astrophysical Journal, **772**, 2013

Pickering, E. C and Levitt, H.: *Periods of 25 Variable Stars in the Small Magellanic Cloud*, Harvard College Observatory Circular, **173**, 1912

Shu, F.: *Density Waves and Chaos in Spiral Galaxies*, 2005, www.phys.nthu.edu.tw/~colloquium/file/Chaos%20in%20Spiral%20Galaxies.pdf

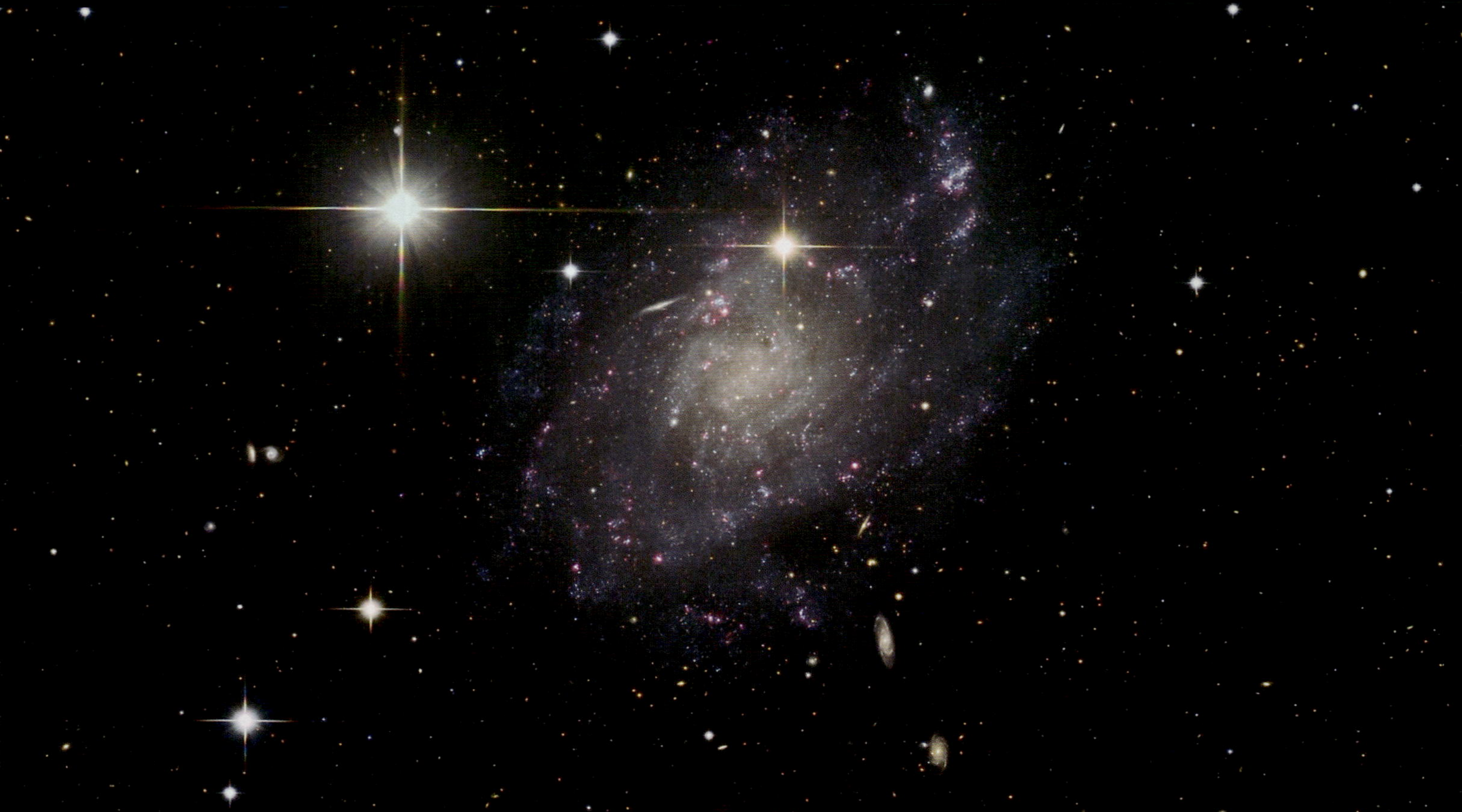

NGC 45

The view of NGC 45 is hampered by two bright stars of magnitude 10.6 and 6.5 which are located in front of and next to the galactic disc of this SA(s)dm galaxy. This disc of size 8.5′ × 5.9′ has a low surface brightness, below that of M 31. NGC 45 is 30 million light years away and the maximum transverse disc diameter is 74 000 light years. The inner isophotes indicate an inclination of 55° and a position angle of the major axis of 145°. NGC 45 has no prominent core, almost no bulge, and the spiral pattern consists of arm fragments. In the lower northern half of the disc, one can see a Vorontsov-Velyaminov row in the spiral arm located there. NGC 45 has almost no dust lanes or concentrations, so that one can conclude that this galaxy is not only faint but also contains little matter. Easily recognized background galaxies are also an indication of this transparency.

NGC 45 is a so-called "low surface brightness" galaxy (LSBG). About 10–20 per cent of all disc galaxies belong to this group. In addition to the low surface brightness, LSBGs are bluer than comparable SC types and have a much lower star-formation rate. Their low number of newly formed stars is nevertheless sufficient for them to appear blue. LSBGs are not, however, fading spiral galaxies whose star formation has ceased, but rather it is suspected that LSBGs are very slowly evolving spiral galaxies. A combination of optical and radio observations has delivered a rotation curve for NGC 45 which, after a steep increase, has a stable value of 100 km/s for distances greater than 4′. Models indicate a mass-to-light ratio of 5.2 and demonstrate the dominance of the dark-matter component in NGC 45. This is typical of LSBGs and one assumes that their delicate stellar discs are stable only because of the large amount of dark matter, which also protects them from external perturbations.

OBJECT	NGC 45
CONSTELLATION	Cetus
RIGHT ASCENSION	$00^h\ 14^m\ 04^s$
DECLINATION	−23° 10′ 56″
BRIGHTNESS	11.6 mag
TYPE	SA(s)dm
PHOTOGRAPHERS	Philipp Keller, Konstantin Buchhold, Bernd Flach-Wilken, Johannes Schedler, Volker Wendel (Chart 32-Team)
TELESCOPE	800-mm reflector
CAMERA	FLI Proline 16803
EXPOSURE TIME	1120 min
LOCATION	CTIO, Chile

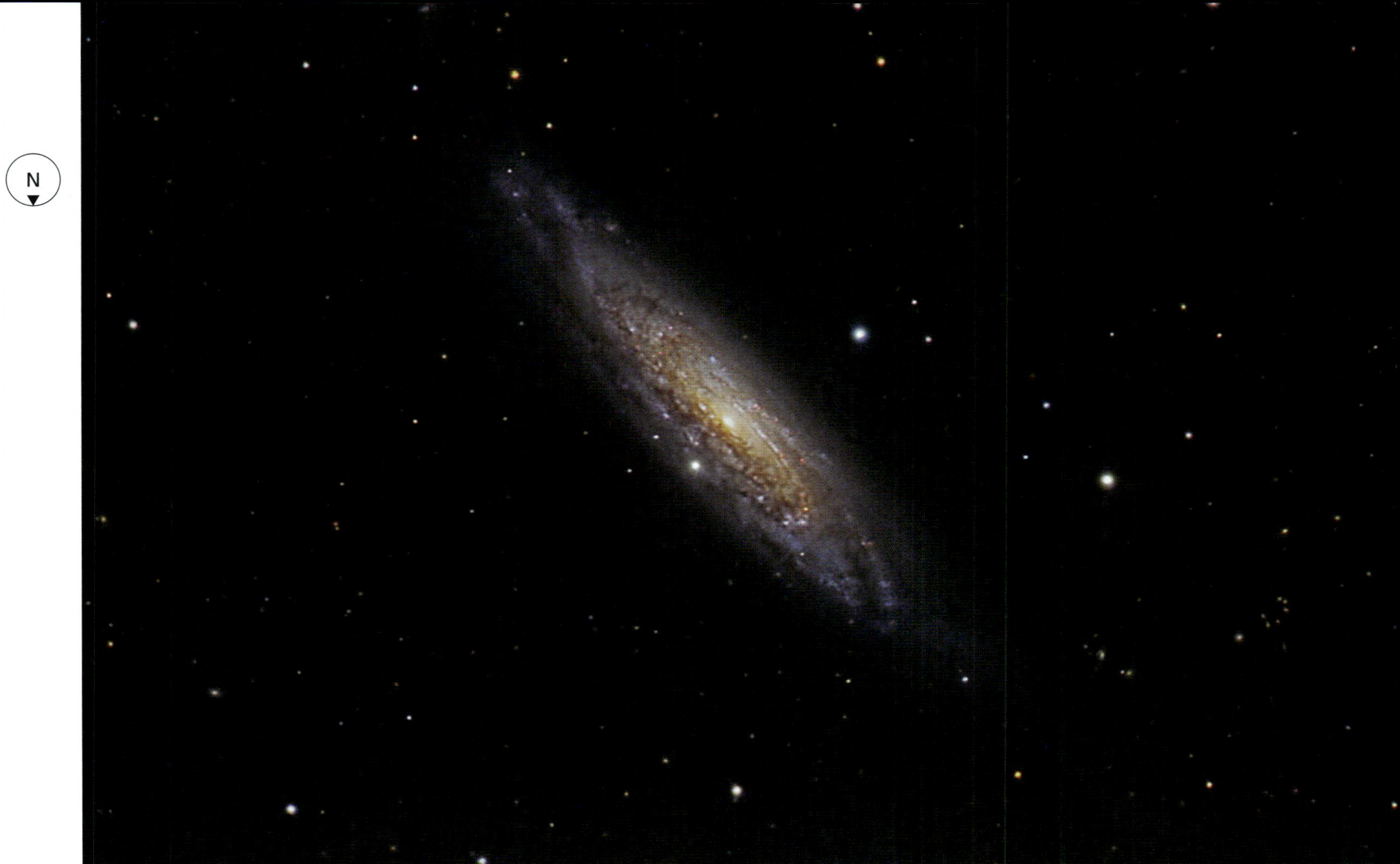

NGC 134

This SAB(s)bc galaxy lies in the southern part of the constellation Sculptor. Its distance from the Milky Way is 62 million light years and its angular size of 8.5′ × 2.0′ corresponds to projected 15 000 and 36 000 light years. The inclination of the spiral plane of NGC 134 is 78°, thus the inner part of the disc is partially obscured. This led to the SAB classification under the assumption that the spiral arms of NGC 134 begin at a bar rather than extending to the centre. In the southeast spiral arm, located to the right above the centre, one can recognize several bright H II regions. If one follows this spiral arm to the extreme right edge of the disc, one finds a few active regions at the end. This edge activity is also seen at the opposite side of the disc.

The analysis of the disc isophotes indicates that NGC 134 has a warped disc plane. The visible arc of the disc follows almost exactly the axis defined by the line of sight from the Milky Way. This perturbation is probably caused by the passage of a compact galaxy which like NGC 134 belongs to the Telescopium/Grus Group. The faint tidal tail at the lower right of the disc points to such a tidal perturbation and also delivers a plausible explanation for the large disc diameter of NGC 134.

OBJECT	NGC 134
CONSTELLATION	Sculptor
RIGHT ASCENSION	$00^h\ 30^m\ 22^s$
DECLINATION	−33° 14′ 39″
BRIGHTNESS	11.4 mag
TYPE	SAB(s)bc
PHOTOGRAPHERS	Bernd Flach-Wilken, Volker Wendel
TELESCOPE	400-mm reflector
CAMERA	SBIG ST-10XME
EXPOSURE TIME	150 min
LOCATION	Tivoli Farm, Namibia

M 31, The Andromeda Galaxy

The SA(s)b galaxy M 31 (NGC 224) is, along with the Milky Way, the second large spiral galaxy in the Local Group. The distance from the Milky Way is 2.57±0.06 million light years, and the angular size of 190′ × 60′ leads to a diameter of 142 000 million light years. M 31 is thus larger than the Milky Way; the number of stars is around 10^{12} while in the Milky Way it is around 3×10^{11}. The relative velocity of the Milky Way and M 31 is about −300 km/s, and one suspects that in about 3.9 billion years a first close encounter will occur, and after a further two billion years the two large spiral galaxies will merge, together with M 33.

If one looks directly onto the plane of the disc of a spiral galaxy, it is easy to determine whether or not a bar exists. This is more difficult in the case of an edge-on perspective, in which astronomers can nevertheless use kinematic data or bulge deformations viewed from the side to detect bars. The inclination of M 31 is 77° and the isophotes indicate a boxy bulge. Simulations of the stellar densities of various forms of bulge can be used to determine the basic parameters. These models correspond well to the isophotes if one assumes a 22′ long bar in M 31. The angle between the semi-major axis of the disc and that of the bar is 20°. The bar itself lies within the innermost dust lane, visible above the centre on the image.

The image shows also the two dwarf satellite galaxies, M 32 and M 110, which are described in detail in the chapter on dwarf galaxies. M 32 lies directly below the centre of M 31; M 110 is larger and is located above and to the right, to the northwest of the centre, at the diffuse edge of the disc. In 2006, NASA's Spitzer Space Telescope studied the distribution of dust in M 31 and discovered a bright, complete dust ring in the infrared, which in turn contains further spiral structures which extend to the centre. Astronomers discovered small, ring-shaped holes in the dust disc, which are seen as proof that M 32 has passed through the disc. M 32 travels along an almost polar orbit around M 31 and, according to the models, the last collision occurred 210 million years ago. Since M 31 is much more massive than M 32, and since the orbit is almost perpendicular to the disc, the spiral structure of M 31 is to a large extent preserved. Interaction with the halo, however, leads to a torque on the disc, and this is seen as the cause of the warp in the disc plane, which is clearly visible in M 31. In the image, the upper, northern end appears to be bent up and the southern end of the disc bent down.

Due to its close distance, M 31 presents the possibility to study the centre of a spiral galaxy in detail. X-ray satellites detect about a dozen bright X-ray sources within 1′ of the centre. These are binary stars which in X-rays are brighter than the central supermassive black hole. In addition to the supermassive black hole, with a mass of 1.4×10^9 solar masses, in 1993 the Hubble Space Telescope discovered a second emission region about five light years away, which was first classified as a component of a double nucleus. It was later found to be an eccentric disc of old stars which orbits around the black hole. Further images from the Hubble Space Telescope brought proof of a second, much more compact disc of about 200 A stars with a radius of about only a light year. These blue stars are younger than 200 million years and thus formed in this central region. Measurements of the speed of rotation of the blue stars, which indicate values of 1000 km/s, definitively exclude astrophysical alternatives to a black hole in M 31.

OBJECT	M 31, NGC 224
CONSTELLATION	Andromeda
RIGHT ASCENSION	$00^h\ 42^m\ 44^s$
DECLINATION	+41° 16′ 09″
BRIGHTNESS	4.4 mag
TYPE	SA(s)b
PHOTOGRAPHERS	Stefan Binnewies, Lucas Binnewies
TELESCOPE	300-mm objective
CAMERA	Canon EOS 5D
EXPOSURE TIME	100 min
LOCATION	Eifel, Germany

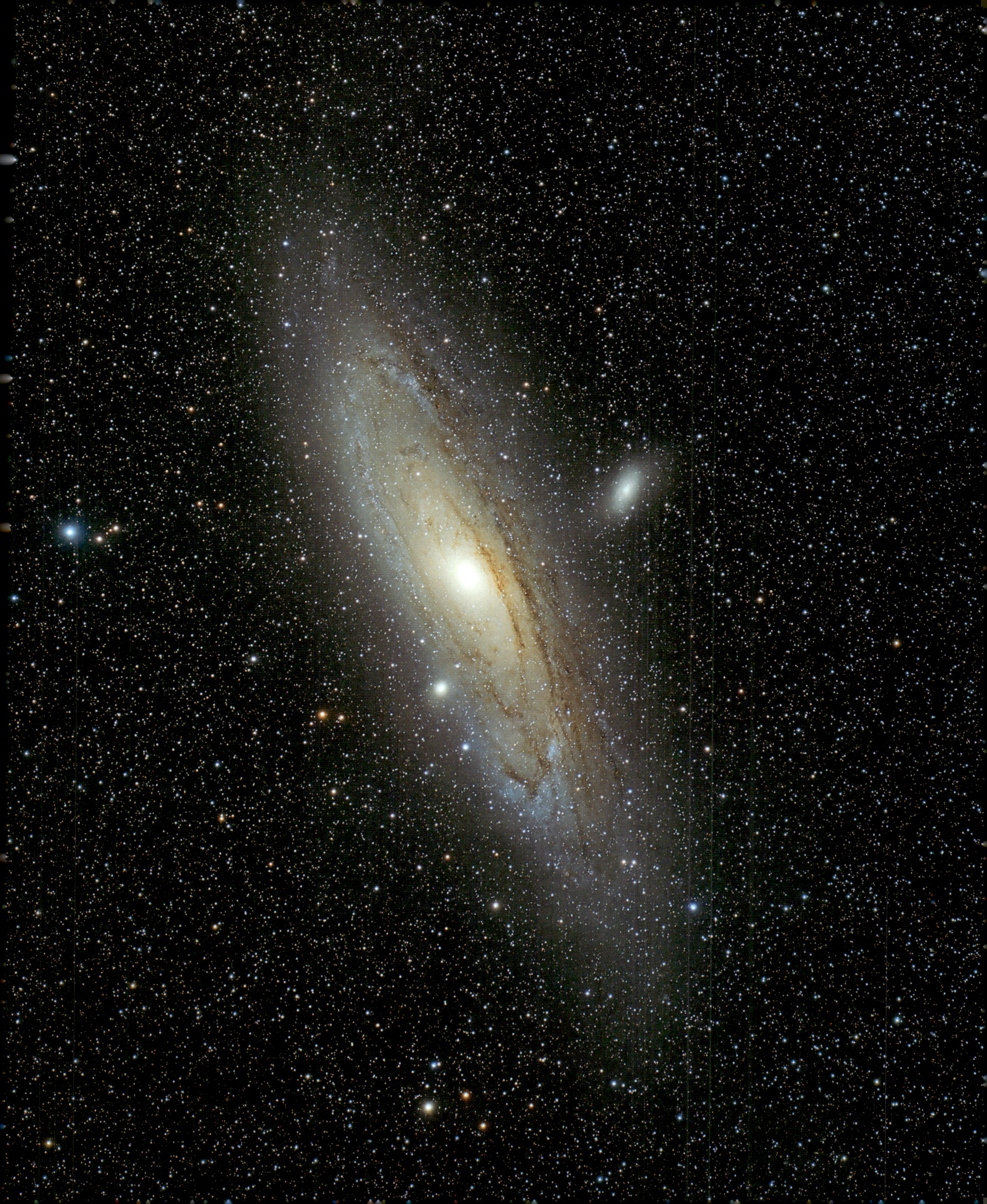

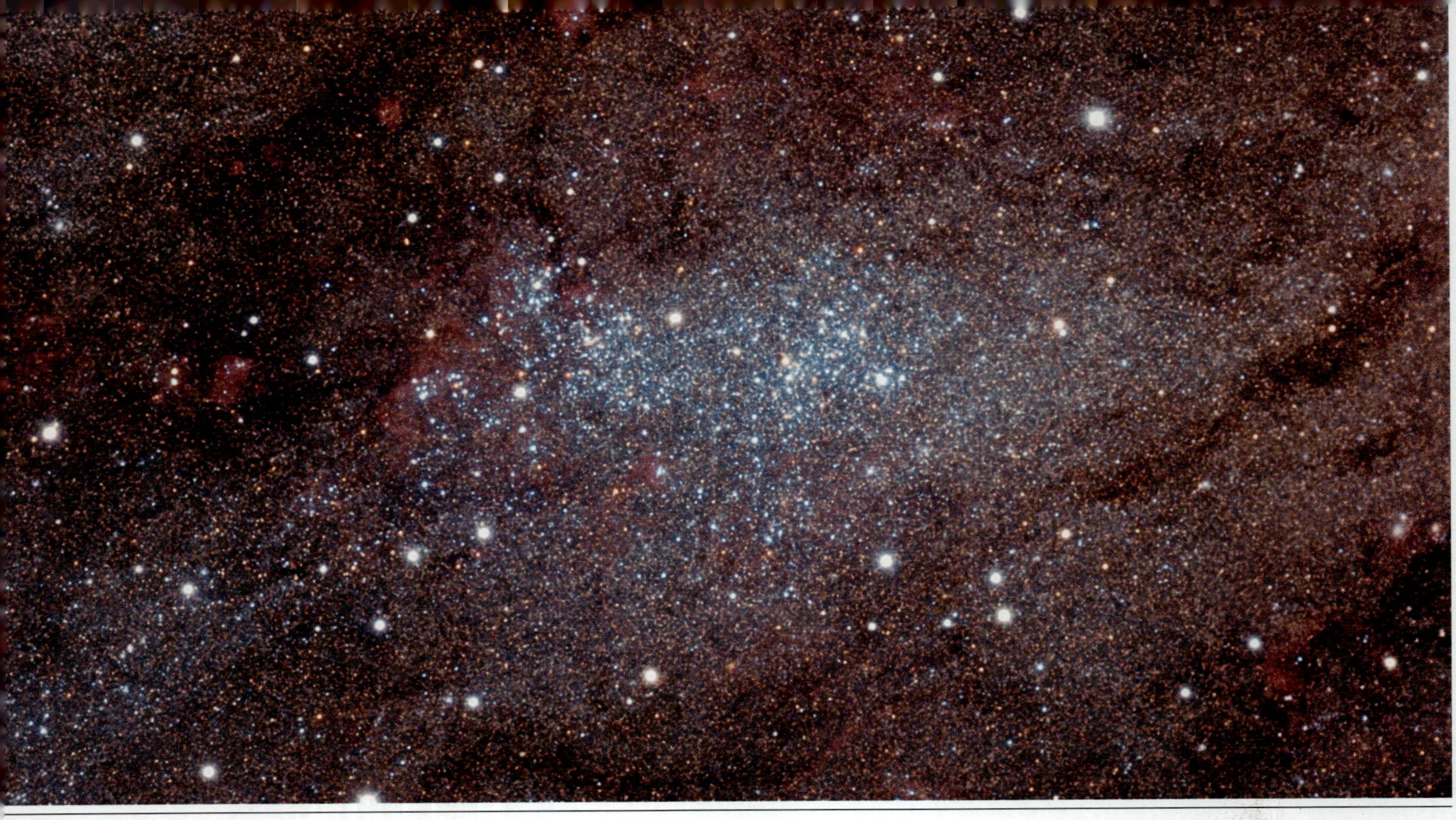

NGC 206

NGC 206 is the largest OB association (OB 78) in the southern half of the M 31 disc. Its angular size of 4.5′ × 3.0′ corresponds to a projected physical size of almost 3500 light years. The distance from the centre of M 31 to NGC 206 is about 35 000 light years. It is conspicuous that these bright stars in NGC 206 are not surrounded by a bright nebula, as is often the case in OB associations. A possible explanation are the stellar winds of the first, very massive stars which formed in NGC 206, which blew the gas out of the association. This is confirmed by the observed deficiency of H I, which differentiates NGC 206 from other OB associations in M 31 which are surrounded by emission nebulae. Mathematical models indicate that about ten per cent of the original 2×10^6 solar masses of neutral hydrogen gas was used in the formation of OB stars. Further 25 per cent of the gas was ionized and the rest was removed by stellar winds. Part of this gas makes up an H I disc around this OB association.

NGC 206 contains also luminous blue variables (LBV). These very massive stars (50–100 solar masses) attain a luminosity 10^6 times that of the Sun and live for only a few million years before they explode as supernovae. Their active LBV phase lasts only a few 100 000 years and is characterized by eruptions and stellar winds via which these stars lose a significant amount of mass. The young blue stars in NGC 206 contrast with the older yellow stars in the surrounding disc population. Astronomers suspect that the brightest blue stars in M 31 are to be found in NGC 206. They are brighter than the OB stars by more than 2 mag. With an age of about 30 million years, NGC 206 is a young object located at the intersection of two spiral arms in M 31 and could have formed from a local overdensity of matter located there.

OBJECT	NGC 206
CONSTELLATION	Andromeda
RIGHT ASCENSION	$00^h\ 40^m\ 31^s$
DECLINATION	+40° 44′ 21″
BRIGHTNESS	16.5 mag
TYPE	–
PHOTOGRAPHERS	Makis Palaiologou, Stefan Binnewies
TELESCOPE	1.3-m reflector
CAMERA	SBIG STL-6303
EXPOSURE TIME	220 min
LOCATION	Skinakas Observatory, Crete, Greece

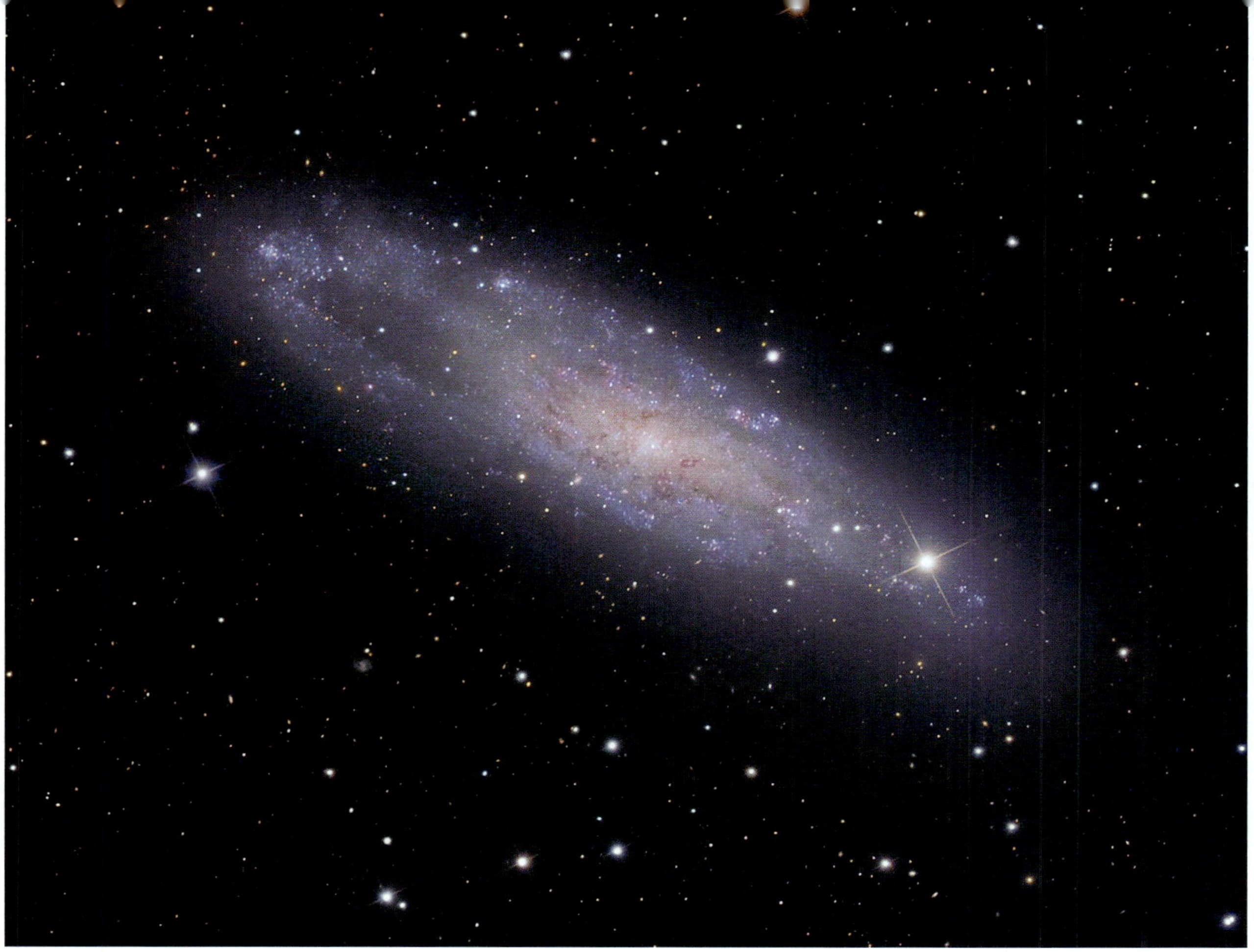

NGC 247

The SAB(s)d galaxy NGC 247 is part of a filament in the Sculptor Group which is located near the south Galactic pole of the Milky Way. At a distance of only 12 million light years, NGC 247 has an angular size of 21.4′ × 6.9′, which corresponds to a maximum transverse disc diameter of 75 000 light years. NGC 247 shows typical properties of late-type spirals: the disc plane has a low surface brightness, there is no bulge, and in the centre is a compact, bright core. In the inner part of the disc, dust filaments extending to the core are connected with bright H II regions. At the outer edge of the disc are many star clusters; particularly prominent are the ones at the left, northern edge of NGC 247.

A study of stellar populations in NGC 247 plotted the disc stars in a colour–magnitude diagram. The main-sequence turnoff (MSTO) of the stars in the inner part of the disc led to an age of the population of six million years. The inner region is dominated by young, hot stars; at greater distances, main-sequence stars have different ages. The MSTO value of the youngest stars at the edge of the disc at a radius of 55 000 light years is more than 40 million years. In these regions, the mean density of hydrogen is very low. One thus assumes that star formation is possible here only because compact hydrogen clouds exist. These local concentrations could be due to an interaction between NGC 247 and the large neighbouring galaxy NGC 253, whose transverse distance from NGC 247 is about 900 000 light years.

OBJECT	NGC 247
CONSTELLATION	Cetus
RIGHT ASCENSION	$00^h\ 47^m\ 09^s$
DECLINATION	−20° 45′ 37″
BRIGHTNESS	9.9 mag
TYPE	SAB(s)d
PHOTOGRAPHERS	Robert Gendler
TELESCOPE	368-mm reflector
CAMERA	SBIG STL-11000
EXPOSURE TIME	720 min
LOCATION	Fair Dinkum Skies, Australia

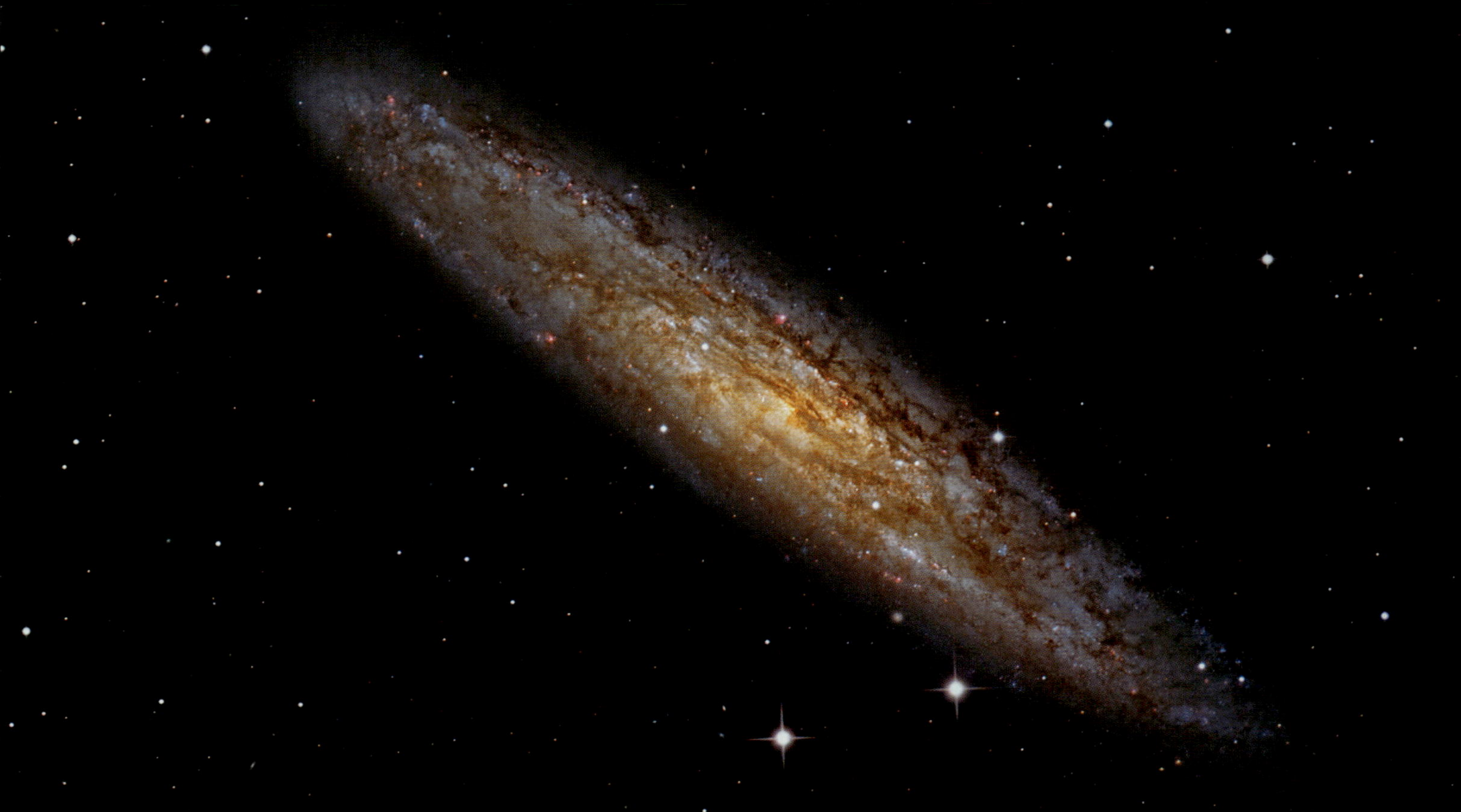

NGC 253

The starburst galaxy NGC 253 is the most massive galaxy in the Sculptor Group. Its distance from the Milky Way is 11.4 million light years and its angular size of 27.4′ × 6.8′ implies a projected physical diameter of 91 000 light years. Despite its southern location, NGC 253 can be found in the *Observer's Handbook* of the Royal Astronomical Society of Canada, and its description there as the "silver-coin galaxy" has, along with the "Sculptor galaxy", become established among amateur astronomers. NGC 253 is an SAB(s)c galaxy whose compact nucleus is covered by dust. The inner part of the disc contains many star-forming regions. This large number of newly forming stars is related to the warm dust nearby, which also explains the strong infrared radiation which one can observe in NGC 253. The large amount of dust causes both a dimming and a reddening of the disc stars and explains the yellowish inner part.

The angle of inclination of NGC 253 is 80° and its position angle 230°. The upper, northwestern side of the disc is closer to the Milky Way. Studies of the gas dynamics reveal that the northwestern side (see figure) is approaching the Milky Way at about 200 km/s; the opposite side is correspondingly receding at 200 km/s. More detailed studies show that one must distinguish between the rotation of atomic hydrogen and Hα emission. While the rotation curve of atomic hydrogen shows a slight increase from 190 km/s to 230 km/s at radii larger than 20 000 light years, the Hα rotation curve is significantly flatter and drops to 170 km/s between 10 000 and 20 000 light years. Both rotation curves can be modelled by a combination of a stellar disc and a gas disc. No extended halo of dark matter is needed in order to understand the rotation curve. As is also the case in NGC 7793, a further member of the Sculptor Group, the dark-matter halo appears to extend to only about 30 000 light years.

OBJECT	NGC 253
CONSTELLATION	Sculptor
RIGHT ASCENSION	$00^h\ 47^m\ 33^s$
DECLINATION	−25° 17′ 18″
BRIGHTNESS	7.1 mag
TYPE	SAB(s)c
PHOTOGRAPHERS	Josef Pöpsel
TELESCOPE	600-mm reflector
CAMERA	SBIG ST-10XME
EXPOSURE TIME	85 min
LOCATION	Amani Lodge, Namibia

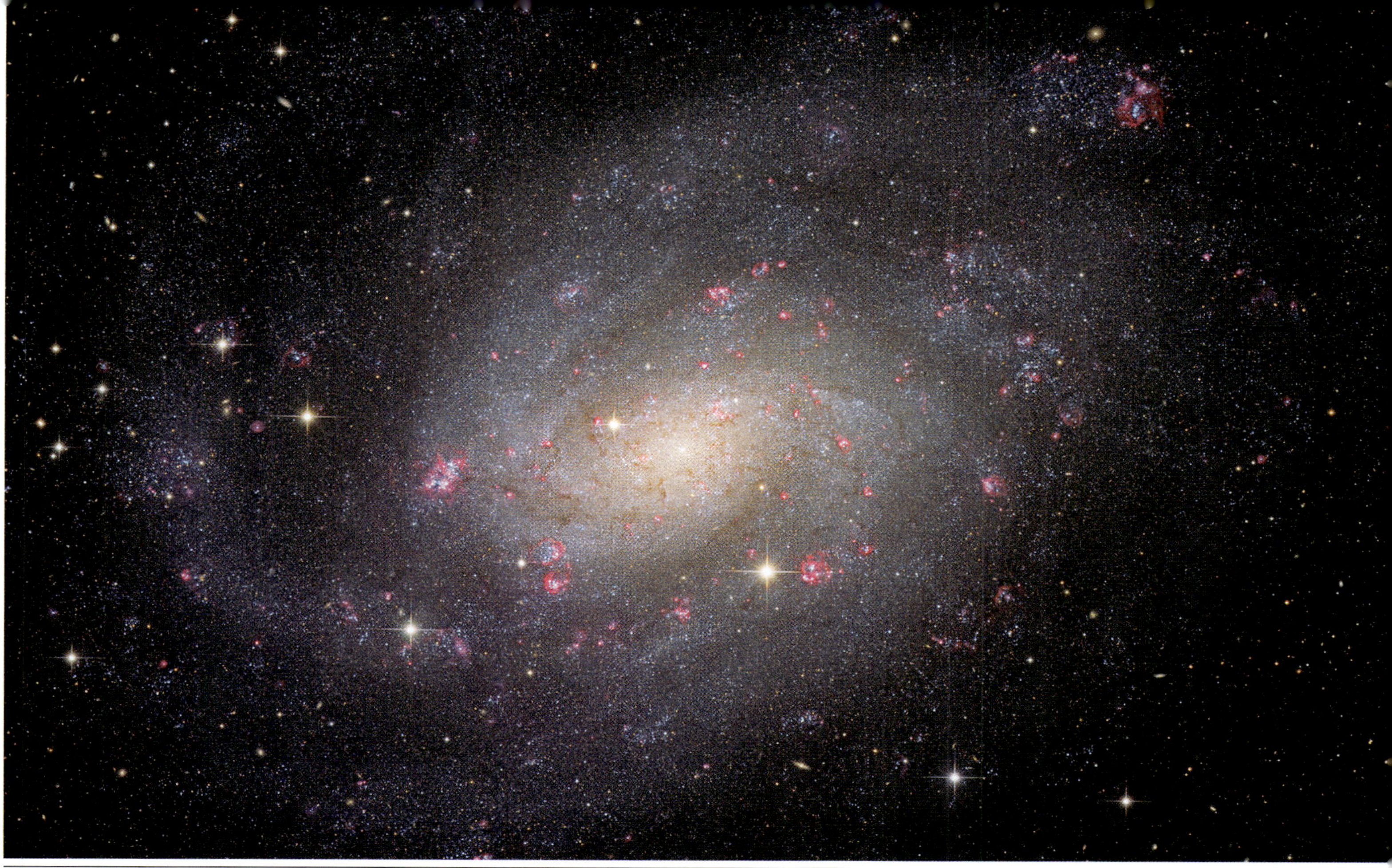

NGC 300

NGC 300 is one of the brightest galaxies in the Sculptor Group. Classified as SA(s)d, it is a late-type spiral with a large extent of 21.9′ × 15.5′ on the sky. The spiral pattern in NGC 300 is composed of two spiral arms extending from a stellar, compact nucleus. The arms are less well defined in the outer parts of the disc and in the left, eastern half of the disc there is a third spiral arm which extends to the northern edge of the disc in a large arc. The images from the Palomar Observatory Sky Survey (POSS), however, show almost nothing of the outer half of the disc. It is only via image processing that the dim arm structures become visible.

The Hubble Space Telescope has been used to study Cepheids in NGC 300. As described in the introduction to this chapter, this allows one to use the period–luminosity relation to determine the absolute luminosity and thus the distance to NGC 300. The problem here is the calibration of the relation, which involves the determination of the metallicity Z of the variables, which can also vary within a galaxy. This problem can be solved with the help of another distance indicator, the tip of the red-giant branch (TRGB) in the colour–magnitude diagram. This statistical technique makes use of the fact that the tip has a maximum absolute luminosity (M), depending on the age and metallicity of the red giants. For NGC 300, scientists derived a distance modulus $(m - M) = 26.37$ mag and thus a distance of 6.13 million light years.

OBJECT	NGC 300
CONSTELLATION	Sculptor
RIGHT ASCENSION	$00^h\ 54^m\ 53^s$
DECLINATION	−37° 41′ 04″
BRIGHTNESS	9.0 mag
TYPE	SA(s)d
PHOTOGRAPHERS	Philipp Keller, Konstantin Buchhold, Bernd Flach-Wilken, Johannes Schedler, Volker Wendel (Chart 32-Team)
TELESCOPE	800-mm reflector
CAMERA	FLI Proline 16803
EXPOSURE TIME	1680 min
LOCATION	CTIO, Chile

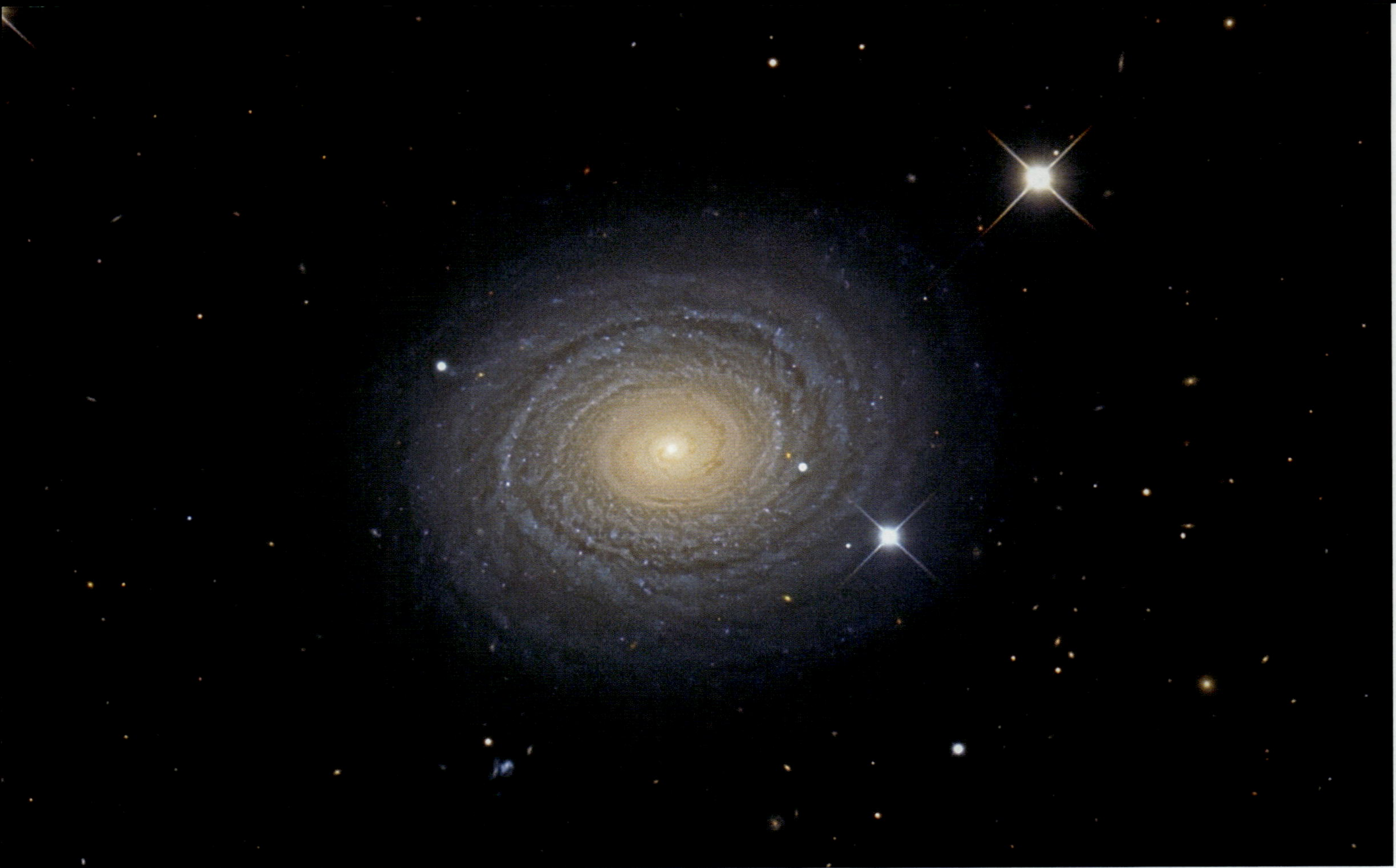

NGC 488

The SA(r)b galaxy NGC 488 is located at a distance of 100 million light years from the Milky Way and its angular size of 5.2′ × 3.9′ corresponds to projected physical distances of 151 000 and 113 000 light years. The bright nucleus of NGC 488 is surrounded by a large, slightly elliptical bulge. One can distinguish three areas in the disc of NGC 488. In the yellowish centre of the disc one can recognize two spiral-shaped dust lanes which extend to the nucleus. In the outer regions of the disc are less prominent spiral structures. They nevertheless have a regular form and constant pitch angle. In this respect, NGC 488 departs from the behaviour of late-type spirals in which one often observes a broadening of the spiral arms.

Measurements of the relative velocities of Hα and N II emission lines at the Cerro Tololo Inter-American Observatory in 1978 were used to derive a rotation curve for NGC 488. The double emission lines visible in the spectrum are a sign of a high velocity of rotation. Correcting for an inclination of 40°, one obtains a rotation velocity of 360 km/s at a distance of 60 000 light years. The form of the rotation curve is rising and thus typical for spiral galaxies with a massive halo. Such high rotation velocities are rare and indicate an unusually large amount of dark matter. This fits well with the facts that NGC 488 is surrounded by dwarf galaxies and that there were probably mergers in the past which caused the halo mass to increase.

OBJECT	NGC 488
CONSTELLATION	Pisces
RIGHT ASCENSION	$01^h\ 21^m\ 47^s$
DECLINATION	+05° 15′ 24″
BRIGHTNESS	11.2 mag
TYPE	SA(r)b
PHOTOGRAPHERS	Makis Palaiologou, Stefan Binnewies
TELESCOPE	1.3-m reflector
CAMERA	SBIG STL-6303
EXPOSURE TIME	230 min
LOCATION	Skinakas Observatory, Crete, Greece

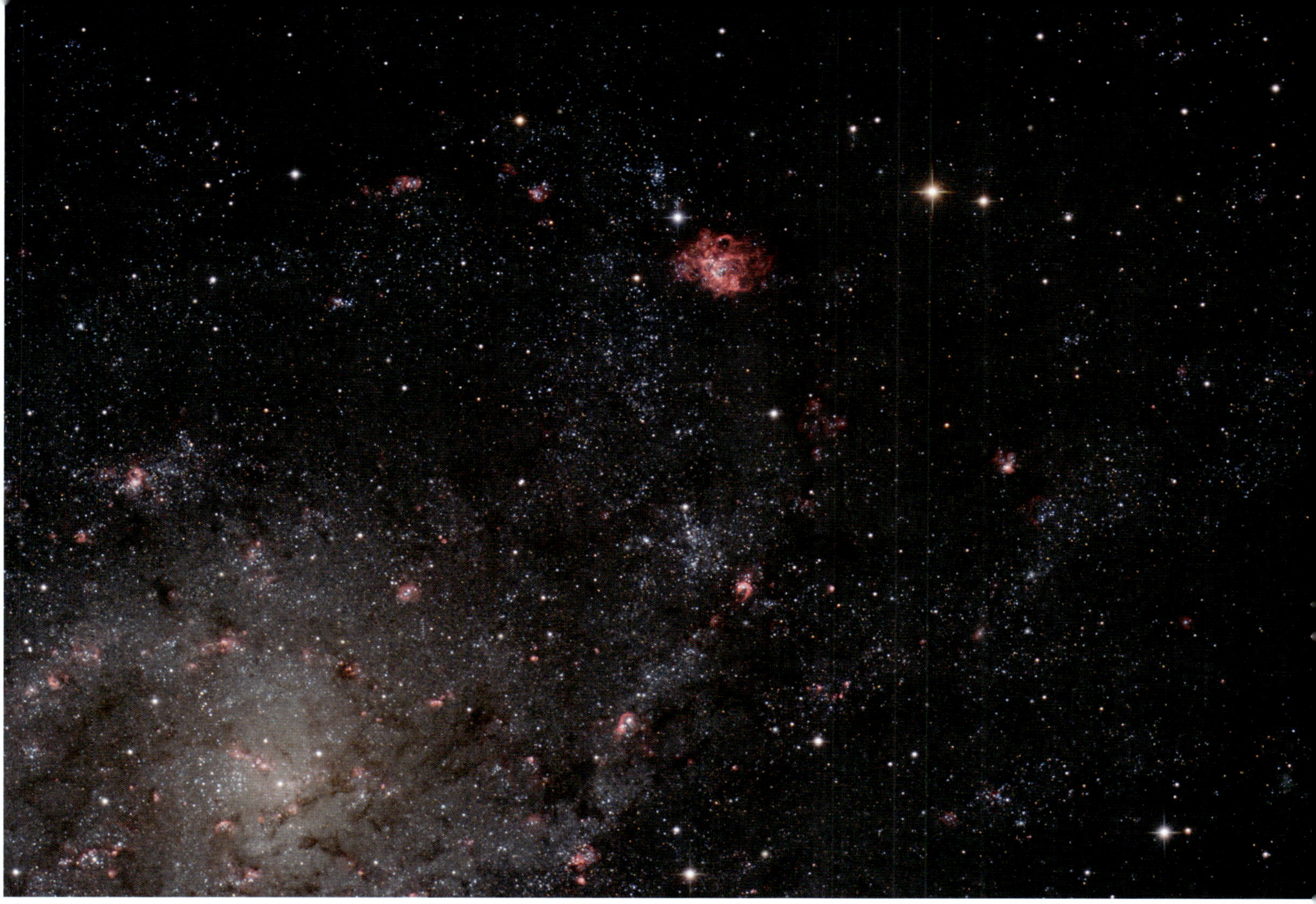

NGC 604

NGC 604 is a large H II region in M 33 whose luminosity is due to a cluster of young, massive stars. These are about 200 O- and Wolf-Rayet stars, only a few million years old. Astronomers distinguish two types of such large H II regions: on the one hand are super star clusters (SSC) which contain a compact star cluster, and on the other hand scaled OB associations (SOBA) which are larger and contain distributed substructures. This distinction is important when considering the longevity of a star cluster since the extended SOBAs are more easily destroyed via tidal forces. SSCs are more robust and can further evolve into globular clusters. NCG 604 is a SOBA, often cited as the prototype.

Studies of the brightest stars in NGC 504 with the Hubble Space Telescope show three O-stars whose spatially resolved spectra indicate masses around 120 solar masses. The spectral characteristics could also be explained, however, via aggregated Wolf-Rayet stars. The resolution of the Hubble Space Telescope is not sufficient to definitively decide whether these are individual, unusually massive stars or compact star clusters.

OBJECT	M 33
CONSTELLATION	Triangulum
RIGHT ASCENSION	$01^h\ 34^m\ 33^s$
DECLINATION	+30° 47′ 06″
BRIGHTNESS	14.0 mag
TYPE	–
PHOTOGRAPHERS	Stefan Heutz, Josef Pöpsel
TELESCOPE	600-mm reflector
CAMERA	SBIG STL-11000
EXPOSURE TIME	420 min
LOCATION	Skinakas Observatory, Crete, Greece

N

M 33

The galaxy M 33 (NGC 598) is the third-largest galaxy in the Local Group and the image shows the disc of this SA(s)cd galaxy with an angular size of 71′ × 42′. The name “Pinwheel Galaxy”, which it shares with M 101, describes well the spiral patterns. The spiral arms extend to the centre and one recognizes a two-arm structure here as well. M 33 is located at a distance of 2.9 million light years and its disc diameter of 60 000 light years is only half as large as that of the Milky Way. The number of stars in M 33 is about 5×10^{10} and is thus only about 1/10 of the number of stars in the Milky Way and 1/25 of that in M 31. The spiral arms are surrounded by bright star-forming regions and a large number of H II regions, some of which are visible as a red ring. The largest of these H II regions is NGC 604 with a diameter of 1400 light years, which can be seen on the image at the upper left edge of the disc, to the northeast of the centre. Such active regions are distributed throughout the entire disc and one always finds dust filaments nearby. Dust forms via coagulation of material from stellar winds and supernova explosions and is an important component of interstellar matter (ISM). The Spitzer Space Telescope was used to study this dust in detail in the infrared at a wavelength of 3.5–24 μm. These infrared images at a wavelength of 8 μm show the distribution of warm dust at a temperature of 5 °C; at a wavelength of 24 μm one sees dust at a temperature between −70 and −120 °C. The Spitzer images show how the warm dusk is closely associated with the spiral structure and that its concentration is higher in the inner part of the disc than in the outskirts.

The centre of M 33 has a compact, almost stellar, form and, in contrast to large spiral galaxies, the upper limit of the central black hole at the centre of M 33 is only 3000 solar masses. This black hole would thus be 1000 times less massive than that at the centre of the Milky Way. Interestingly, however, the most massive stellar black hole known is in M 33, the X-ray binary M 33 X-7. In this system, a blue main-sequence star of about 70 solar masses is orbited every 3.45 days by a black hole of 15.65 solar masses. This star of 70 solar masses is also unusual, since it is the most massive known companion in an X-ray binary and has accreted more than half of its current mass from the stellar wind and mass loss of the precursor to the black hole. M 33 X-7 is located in the middle part of the southern spiral arm, below the centre in the image.

As is the case in other spiral galaxies, the rotation curve of M 33 at first rises rapidly with increasing distance from the centre before levelling off. In M 33, this happens at a velocity of about 100 km/s at a distance of 10 000 light years. Interesting, however, is that between 10 000 and 50 000 light years the rotational velocity is not constant, but increases to 150 km/s. Interpreted in the context of a dark-matter model, this implies an extended dark-matter halo surrounding the visible part of M 33. Such models imply a halo mass of 5×10^{11} solar masses. The extent of the dark halo is about 600 000 light years and is thus comparable to the distance between M 33 and M 31.

OBJECT	M 33, NGC 598
CONSTELLATION	Triangulum
RIGHT ASCENSION	$01^h\ 33^m\ 51^s$
DECLINATION	+30° 39′ 37″
BRIGHTNESS	6.3 mag
TYPE	SA(s)cd
PHOTOGRAPHERS	Johannes Schedler
TELESCOPE	400-mm reflector
CAMERA	SBIG STL-11000
EXPOSURE TIME	1800 min
LOCATION	Wildon, Austria

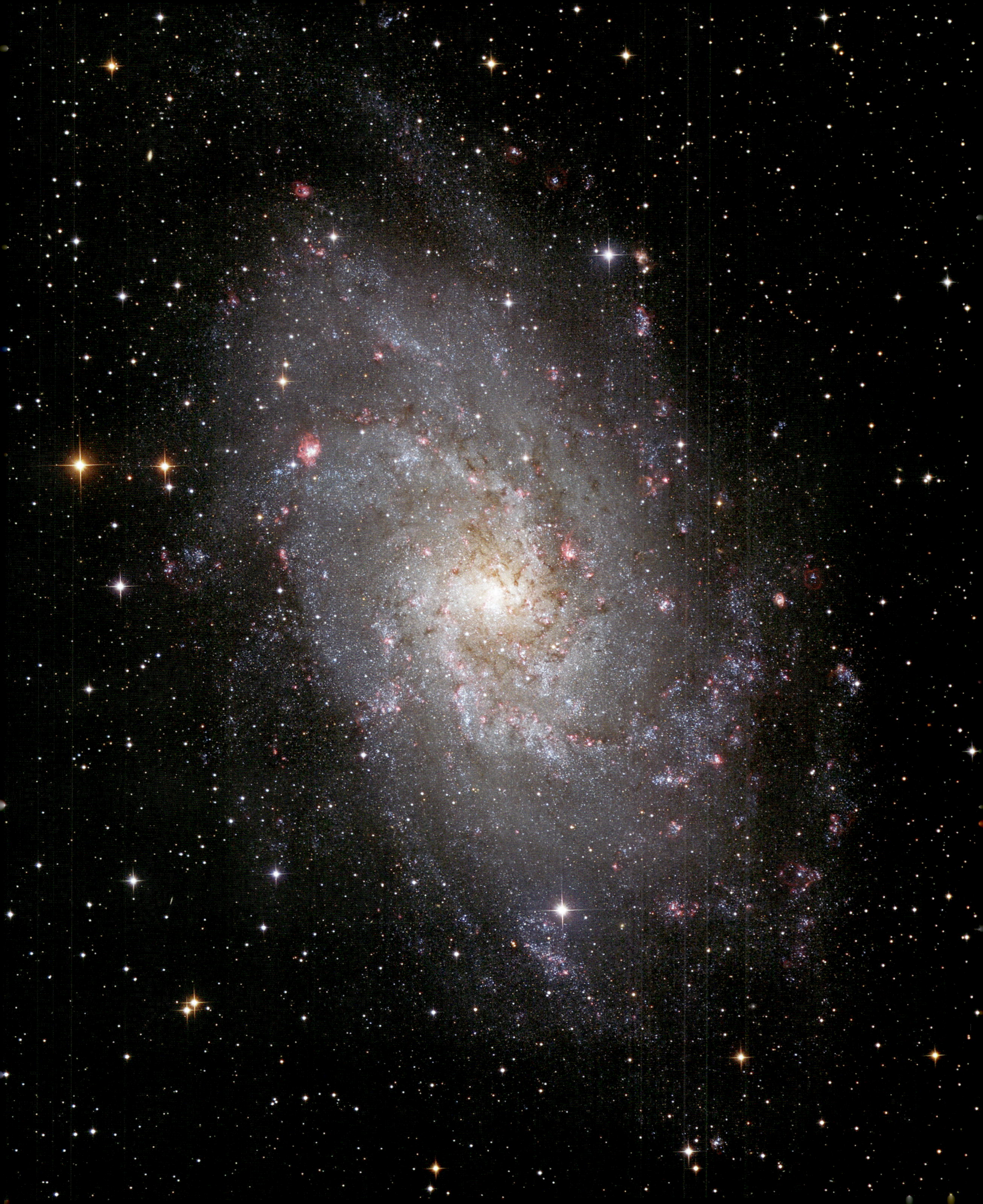

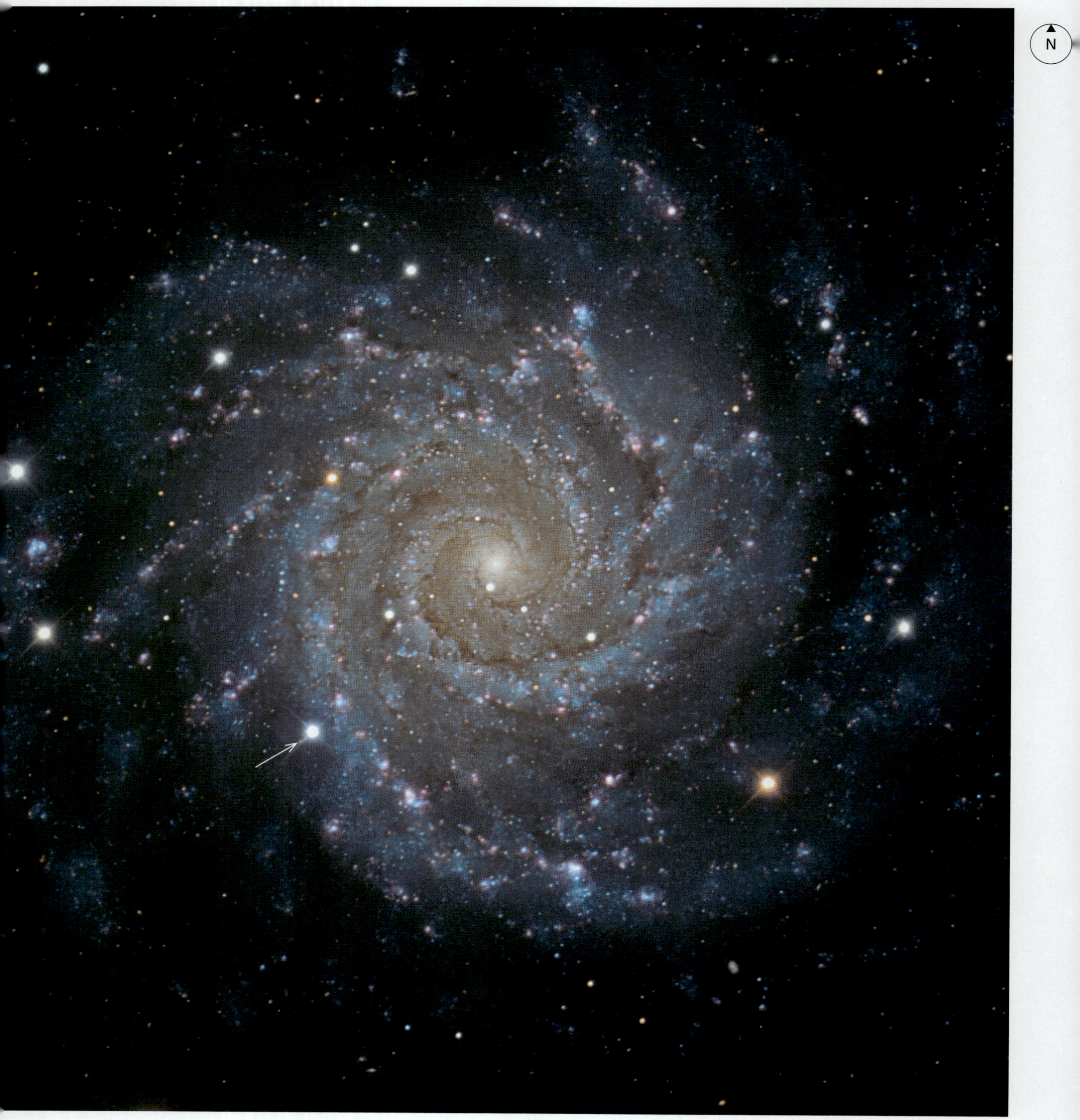
N

M 74

The SA(s)c galaxy M 74 is a grand-design spiral since in the face-on view one can clearly follow the spiral arms from the centre to the dim outer regions. In only about ten per cent of spiral galaxies does one observe such widely extended spiral-arm structures. M 74 is 30 million light years from the Milky Way and due to its large size is an ideal field of study. The angular size of 10.5′ × 9.5′ corresponds to a maximum transverse diameter of 92 000 light years. A large number of H II regions are visible in the image, in the outer regions of the spiral plane along the arms and also on their fragmented ends. The largest H II regions reach a diameter of 600 light years. When studying the morphology of M 74 in more detail, one notices that of the two spiral arms in the lower part of the disc, the southern one is stronger. This perturbation could have been caused by the action of two satellite galaxies (UGC 1176 and UGC 1171). At a distance of 1°, they are not visible in the image. Despite their distance of more than 400 000 light years from M 74, more angular momentum is present in this satellite system than in M 74 itself and astronomers see this as an indication of tidal forces acting in the spiral plane of M 74.

On this exposure of M 74 an arrow indicates supernova SN2013ej which exploded on 25 July 2013. The time of observation of this type II supernova was shortly after maximum brightness was reached and shows its almost white colour. Comparing the colour of the supernova with spectral classes of stars, the colour of the supernova typically evolves from a blue-white A to a white F star in the first two weeks. During the subsequent drop in brightness the colour further changes through yellow to red, and after two to three months the supernova is similar to an M star. The second, smaller image of M 74 is composed of exposures up to October 2013; here the expected reddening of the supernova can be seen.

M 74 was one of the first galaxies for which complete high-resolution spectral coverage was obtained (2006–2008, 3.5-m telescope at the Calar Alto Observatory, Spain). A two-dimensional map of the luminosity and metallicity was derived from more than 11 000 individual spectra in the wavelength range 370–700 nm. Here, one discovered a negative gradient in the ages of the population – regions lying farther in consist of old stars while those farther out consist of a larger fraction of young stars. This trend reverses only near the centre, within a radius of 3000 light years. Here one finds a ring of young stars which is associated with a circumnuclear star-forming region as is typical for spiral galaxies. The star-formation rate is an order of magnitude greater in the outer than in the inner H II regions. The mean star-formation rate is 2.4 solar masses per year and corresponds to a typical value for this type of spiral.

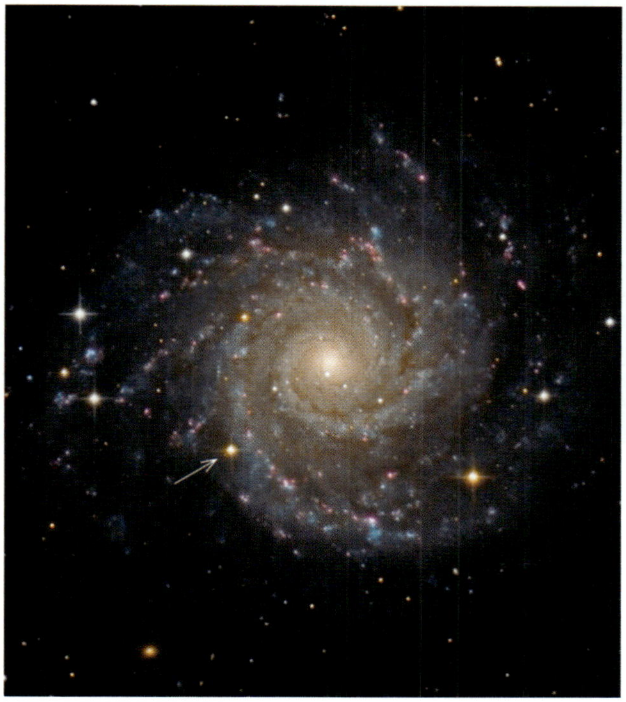

M 74 in an image by Wolfgang Ries and Stefan Heutz. The arrow indicates supernova SN2013ej which showed a significant reddening in October 2013.

OBJECT	M 74
CONSTELLATION	Pisces
RIGHT ASCENSION	$01^h\ 36^m\ 42^s$
DECLINATION	+15° 47′ 01″
BRIGHTNESS	9.9 mag
TYPE	SA(s)c
PHOTOGRAPHERS	Makis Palaiologou, Stefan Binnewies
TELESCOPE	1.3-m reflector
CAMERA	Andor DZ 436
EXPOSURE TIME	112.5 min
LOCATION	Skinakas Observatory, Crete, Greece

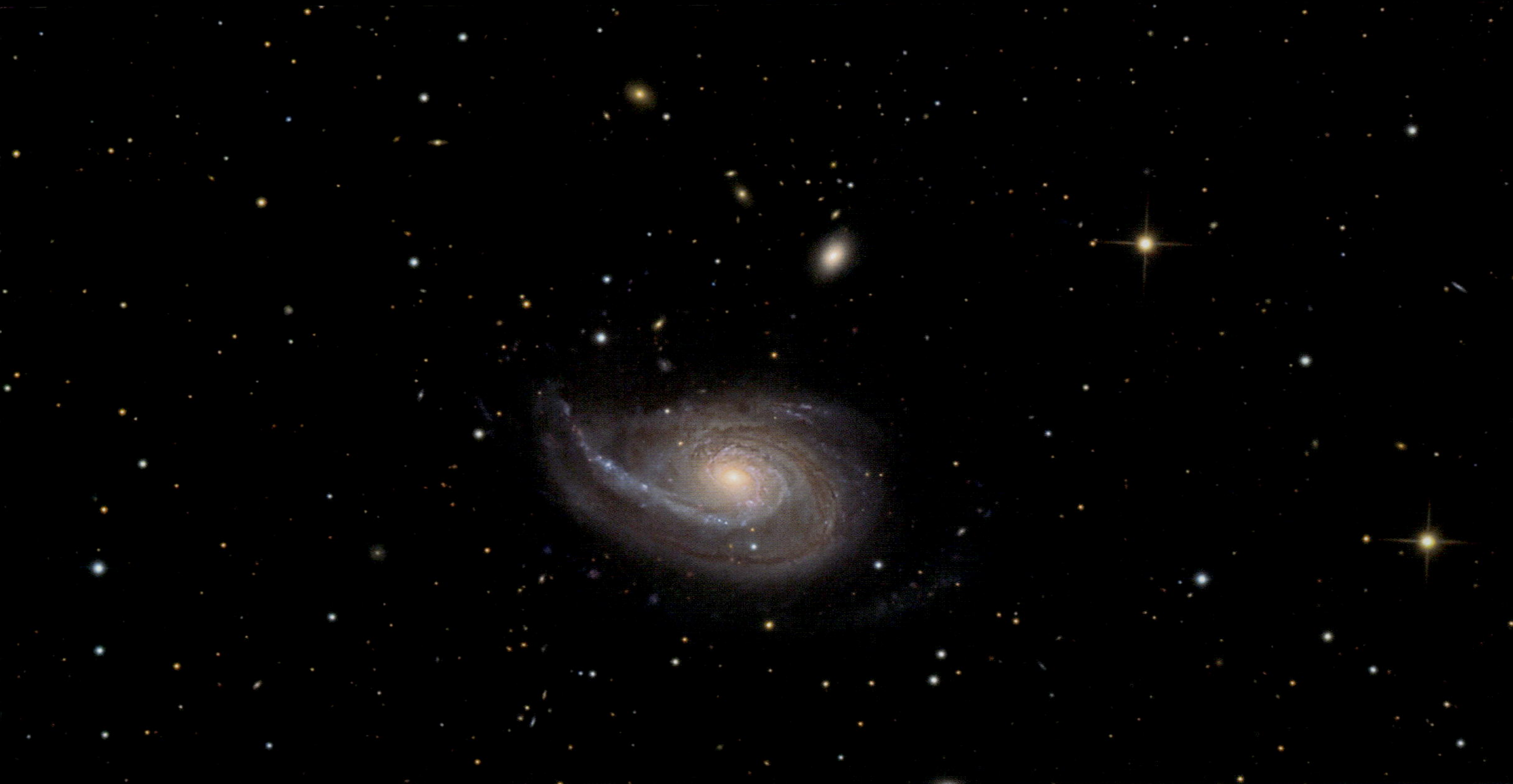

NGC 772

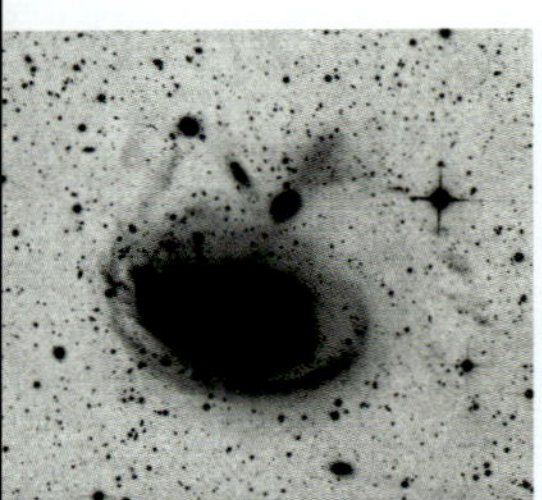

The SA(s)b galaxy NGC 772 (Arp 78) forms a pair with the smaller E3 galaxy 3.3′ to the southwest. Both galaxies lie at a distance of 115 million light years, and since the difference in their radial velocities is only 14 km/s, we are looking onto the plane of the orbit. The galactic plane of NGC 772 is inclined at 54° relative to the orbital plane. The galaxy has an asymmetric spiral structure which is dominated by a bright spiral arm which extends about 3′ to the northwest and contains almost all young, active regions in this galaxy. This spiral arm displays Vorontsov-Velyaminov rows. The opposite, southeastern side shows a spiral arm defined by a brown dust lane, which after half a revolution runs parallel to the bright spiral arm. In a quasar catalogue (1989, A. Hewitt and G. Burbidge), the source [HB89]0156+187 is listed, which is only about 30″ from the nucleus of NGC 772. Since its redshift implies a distance of 10.8 billion light years, one can assume that this quasar is behind NGC 772. However, since there is no corresponding optical identification, there is speculation that there could be a position error in the catalogue data of [HB89]0156+187.

The angular size of 7.2′ × 4.3′ for NGC 772 refers to the bright disc with the spiral pattern and corresponds to the transverse distances of 240 000 and 144 000 light years. The enormous extent of this disc is exceeded by two dim structures in the surroundings of NGC 772 which are visible in the contrast-enhanced negative image. The projected distance of the companion NGC 770 is 100 000 light years and the small image shows a tidal tail in NGC 770 which is directed away from NGC 772. Also, below the two bright stars to the right of NGC 772 one recognizes a faint arc which could be due to the interaction of a compact companion with the disc of NGC 772. The distance of this arc from the centre is more than 220 000 light years and extends to the edge of the large disc of atomic hydrogen of NGC 772, which has a radius of 240 000 light years.

OBJECT	NGC 772, Arp 78
CONSTELLATION	Aries
RIGHT ASCENSION	$01^h\ 59^m\ 20^s$
DECLINATION	+19° 00′ 27″
BRIGHTNESS	10.0 mag
TYPE	SA(s)b
PHOTOGRAPHERS	Adam Block
TELESCOPE	600-mm reflector
CAMERA	SBIG STL-11000
EXPOSURE TIME	405 min
LOCATION	Mount Lemmon SkyCenter/ University of Arizona, USA

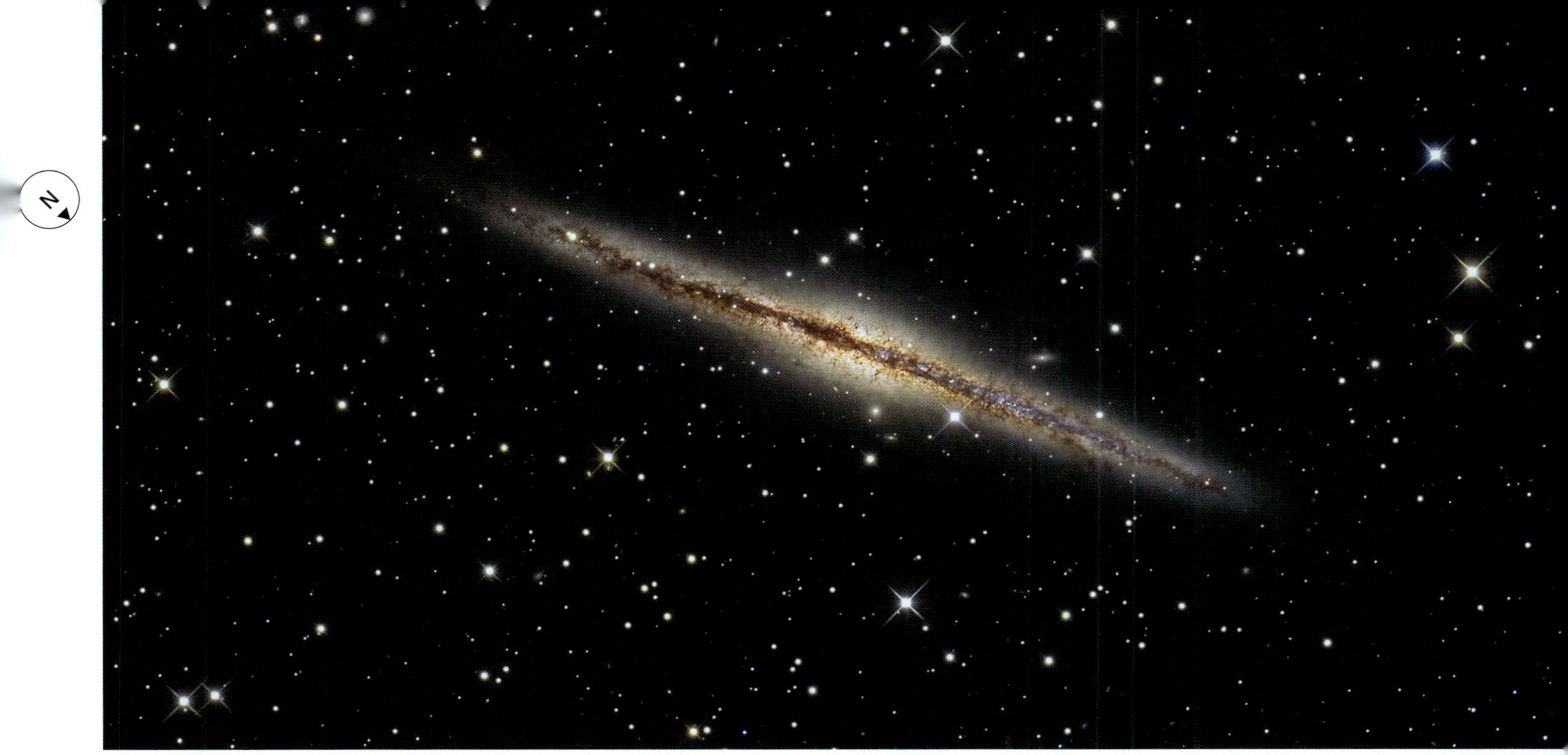

NGC 891

About 3° east of the star Alamak in the constellation Andromeda is the SA(s)bc galaxy NGC 891, which due to its angular size of 13.5′ × 2.5′ and its prominent dust lane is a popular subject for astrophotographers. NGC 891 is at a distance of 32 million light years and its edge-on disc plane has a diameter of 126 000 light years. NGC 891 is thus larger than the Milky Way. In direct comparison, it contains substantially more molecular gas, has a higher star-formation rate and its radio luminosity exceeds that of the Milky Way by 50 per cent. The increased activity in NGC 891 is presumably due to tidal perturbations, which have also led to a warp of the disc plane as one can see by following the dust lanes in the image. The dust lane is bent downward in the left (southern) half of the image while on the right side one can see an upward bend. Visible active regions there indicate that one is looking at the end of a spiral arm; on the left side the view into the plane is blocked by the dust lane.

The H I rotation curve in NGC 891 attains a velocity of 210 km/s within 1′ on each side. Slightly farther out, this falls to 180 km/s. This is an indication of a small fraction of dark matter in the halo of NGC 891. The relative lack of dark matter allows tidal forces to work less perturbed, which also explains the warped disc. Observed in the radio at the wavelength of atomic hydrogen, a lopsided disc is revealed. The H I emission of the southern side of the disc reaches almost twice as far out as the optical radius while on the north side the extra extent of the radio disc is only 15 per cent. According to recent articles, astronomers see the interaction with the small, gas-rich galaxy UGC 1807 as a possible cause of these perturbations in NGC 891. UGC 1807 is not visible in the image since it is at a distance of 25′, corresponding to 250 000 light years, from NGC 891. UGC 1807 has about ten per cent of the mass of NGC 891, and the relative position and velocity of only 100 km/s suggest a flyby of UGC 1807 along the plane of NGC 891 as the possible cause of the perturbation.

OBJECT	NGC 891
CONSTELLATION	Andromeda
RIGHT ASCENSION	$02^h\ 22^m\ 33^s$
DECLINATION	+42° 20′ 57″
BRIGHTNESS	10.8 mag
TYPE	SA(s)b
PHOTOGRAPHERS	Stefan Binnewies, Volker Wendel
TELESCOPE	600-mm reflector
CAMERA	SBIG STL-11000
EXPOSURE TIME	480 min
LOCATION	Skinakas Observatory, Crete, Greece

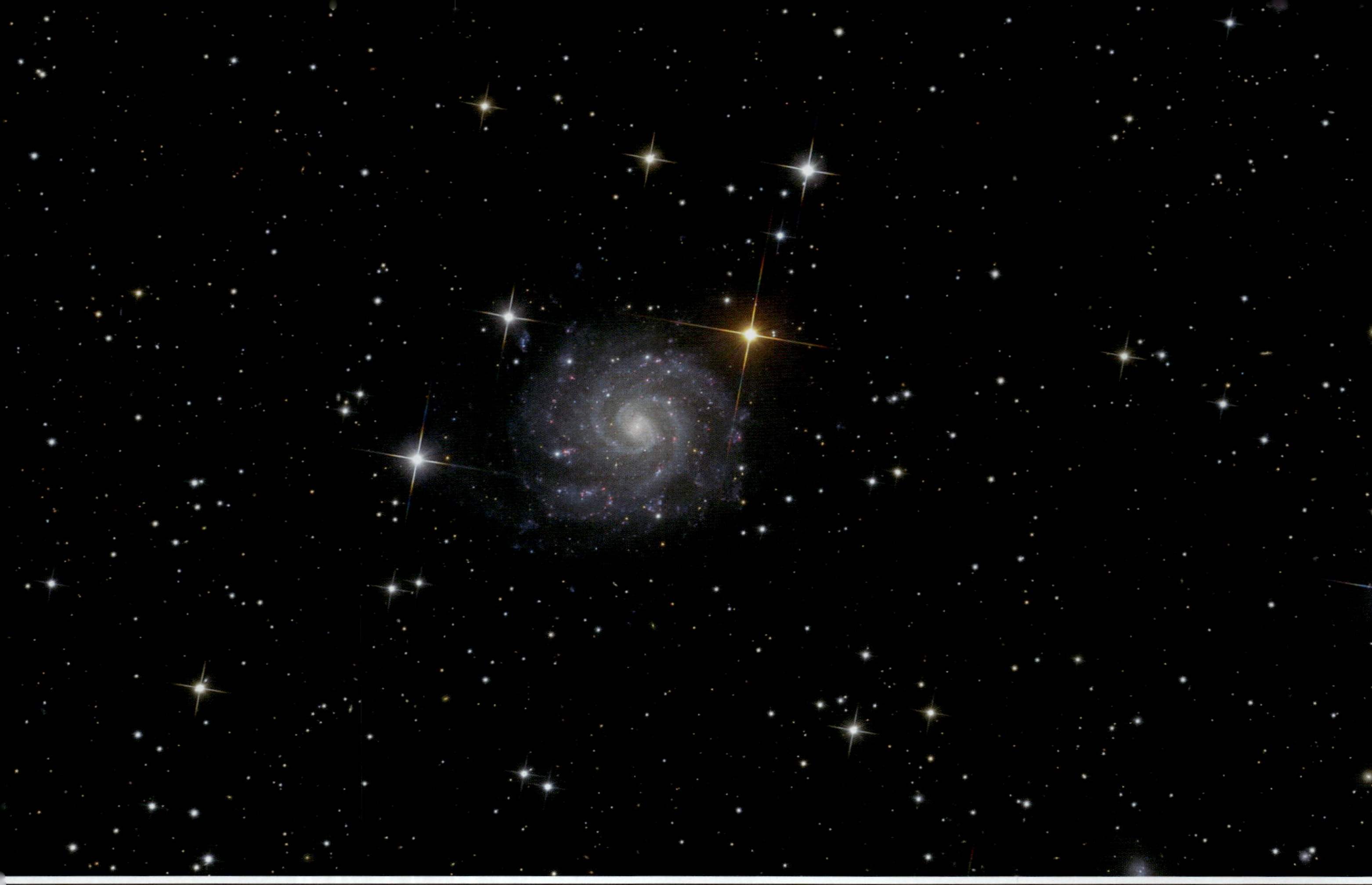

IC 239

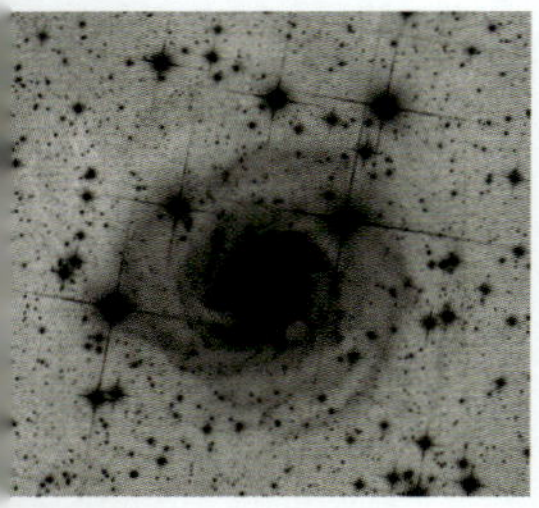

In front of the SAB(rs)cd galaxy IC 239 is a star cluster whose brightest member is an 8.5-mag red M star (SAO 55698). Compared to this star, particularly apparent is the blue colour of the spiral arms of IC 239, whose pattern can be followed out to the edge of the disc at a distance of 7′. Dim spiral arms are connected to a bright central region and even contain a few active regions (see figure at left). At the right edge of the disc, not one but two Vorontsov-Velyaminov rows are visible in the southern spiral arm, thus clearly distinguishing this from a tidal tail. The neutrally coloured centre of IC 239 consists of an oval, bar-like structure from which two spiral arms emerge. IC 239 is a low-surface-brightness galaxy.

IC 239 belongs to the NGC 1023 Group which is located at a mean distance of 32 million light years from the Milky Way. This group consists of ten larger galaxies and eight times as many dwarf galaxies. Comparison of the recession velocities of the brightest members indicates that IC 239 is a prominent outlier, since the scatter in the velocities, i.e. the width of their distribution, in the NGC 1023 Group is ±136 km/s. When viewed in three dimensions, the NGC 1023 Group is seen to be a filament extending from the Local Group to the Perseus–Pisces Cluster. This position between two large mass concentrations explains the relatively high velocities and suggests that IC 239 could be a galaxy passing through the NGC 1023 Group.

OBJECT	IC 239
CONSTELLATION	Andromeda
RIGHT ASCENSION	$02^h\ 36^m\ 28^s$
DECLINATION	+38° 58′ 12″
BRIGHTNESS	11.8 mag
TYPE	SAB(rs)cd
PHOTOGRAPHERS	Adam Block
TELESCOPE	800-mm reflector
CAMERA	SBIG STX-16803
EXPOSURE TIME	720 min
LOCATION	Mount Lemmon SkyCenter/ University of Arizona, USA

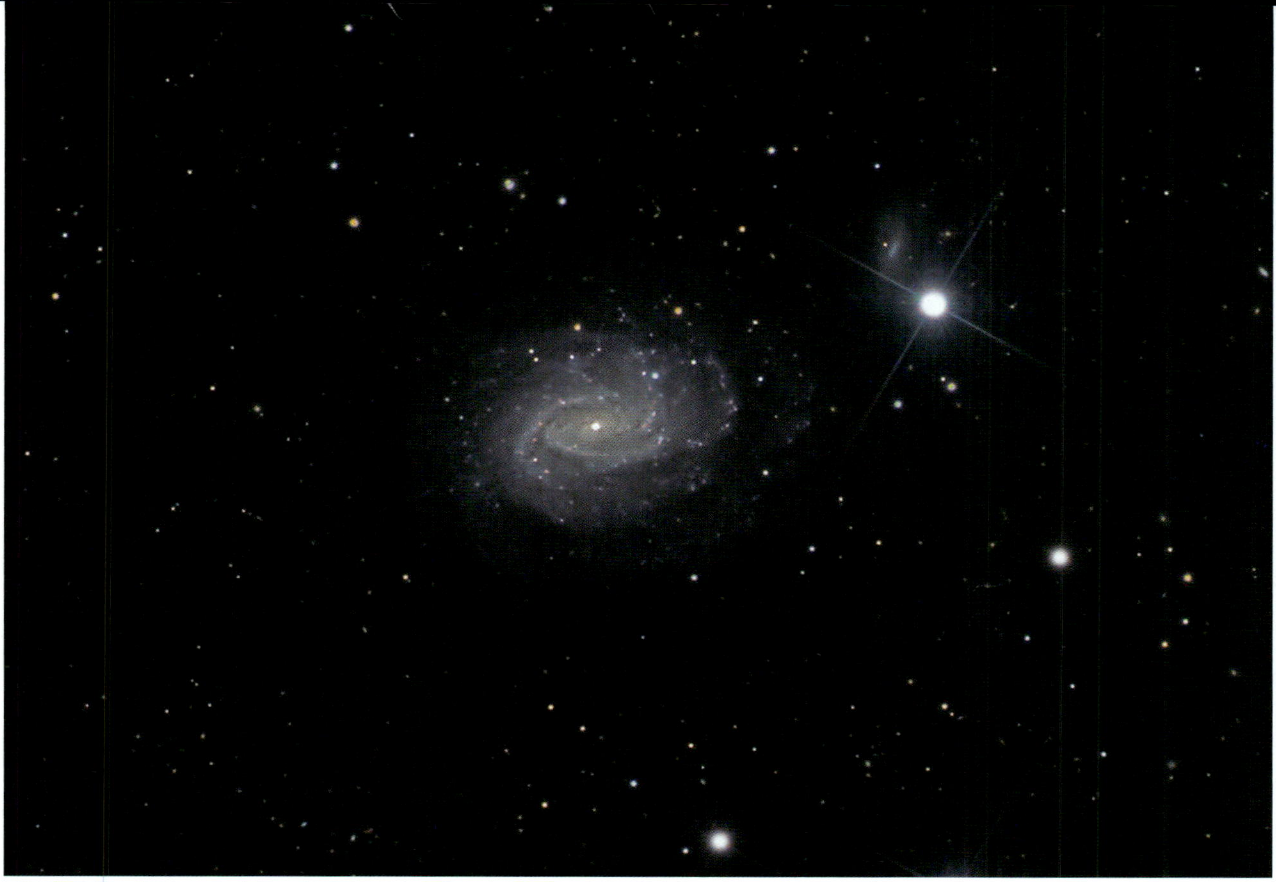

NGC 1187

The galaxy NGC 1187 belongs to the Eridanus Cluster. This contains about 60 brighter galaxies whose recession velocity lies in the range of 1400–1500 km/s. At 1390 km/s, NGC 1187 lies close to the mean value and thus at a distance of 60 million light years. The angular size of 5.5′ × 4.1′ corresponds to a transversal disc diameter of almost 100 000 light years. The inner spiral pattern is determined by three spiral arms which begin near the compact core. Two of these arms can be followed through 270° and make NGC 1187 a grand-design spiral of type Sab. An alternative SBc classification is based on a bar onto which two arms are attached. Many H II regions can be seen in the spiral arms; the largest of these regions with an angular size of 2″ reach a size of 500 light years. At a distance of 4.5′ to the northwest is the small satellite galaxy PGC 11469, only about 1′ away from a bright 8.7-mag foreground star (SAO 168248). The recession velocity of this satellite galaxy is 1445 km/s and it is thus also a member of the Eridanus Cluster. Its absolute diameter is only a quarter of that of NGC 1187.

The strength of the curvature of the spiral arms is also used in the classification of spiral galaxies. The measurement of the pitch angle allows one to learn something about currents of matter within the plane of the spiral galaxy. Astronomers have shown that there is a correlation between the size of the pitch angle and the mass concentration in the galaxy. Galaxies with a larger central mass concentration have smaller pitch angles, their spiral arms are tighter and more compact. The pitch angle in NGC 1187 of 16° is at the lower end of the distribution. Apart from a few exceptions, most galaxies have values of 15–35°. The mass concentration in NGC 1187 is thus comparatively high and allows one to conclude that shear currents exist in the disc which are twice as large as those in galaxies with a larger pitch angle.

OBJECT	NGC 1187
CONSTELLATION	Eridanus
RIGHT ASCENSION	$03^h\ 02^m\ 38^s$
DECLINATION	−22° 52′ 02″
BRIGHTNESS	11.7 mag
TYPE	SB(r)c
PHOTOGRAPHERS	Bernd Flach-Wilken
TELESCOPE	400-mm reflector
CAMERA	SBIG STL-6303
EXPOSURE TIME	165 min
LOCATION	Tivoli Farm, Namibia

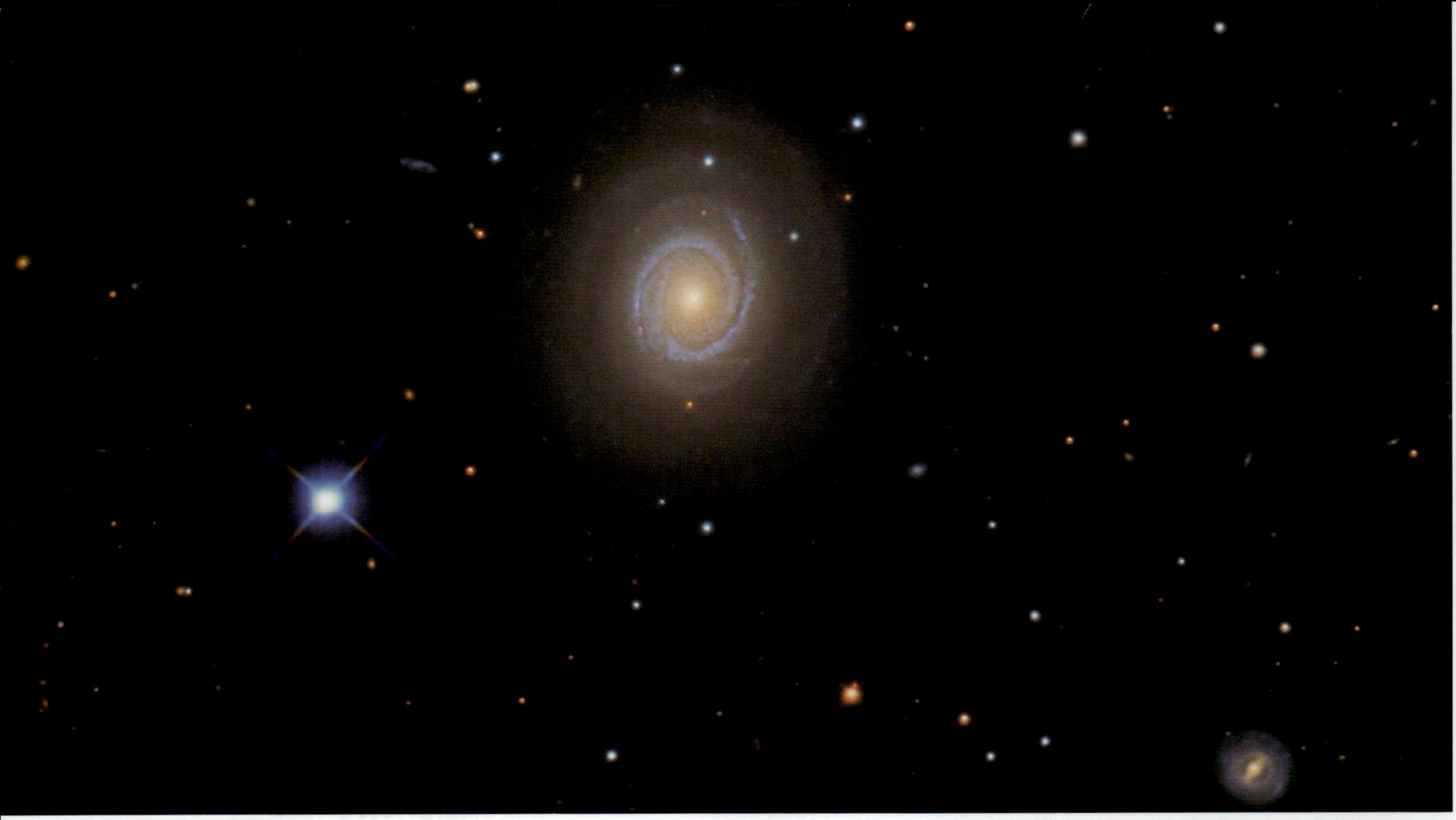

NGC 1357

At the eastern, sparse edge of the Eridanus Cluster is the SA(s)AB galaxy NGC 1357. In the middle, one can see two spiral arms with many young stars which give them a prominent blue colour. The tight, almost ring-like form of the spiral arms is responsible for the SA(s)AB classification, even though the arms are visible only in part of the disc. The angular diameter of this ring is 1′ and at a distance of 85 million light years this corresponds to a projected diameter of 25 000 light years. The outer disc reveals a diffuse form in which one can follow the spiral path of concentrations. One can recognize in the image that the diffuse disc at the upper, western and lower, eastern edges turns into two tidal tails. The diffuse disc has a diameter of 160 000 light years at this point. The SB galaxy (PGC 933903) to the lower right of NGC 1357 is 6.5 times as far from the Milky Way and has no interaction with NGC 1357.

If one can measure the rotation curve of the galaxy, i.e. the radial distribution of rotation velocities, well, then one can deduce the amount of shear currents in the plane of the disc. These shear currents are formed via differential rotation and correspond to radial changes in the flow of matter in the disc. There are analyses which show that strong shear currents occur in early-type spirals. Related to this is a higher star-formation rate and a smaller pitch angle of the spiral arms than in late types. The shear currents can be quantified via the differential shearing rate, which in NGC 1357 has the value of 0.56. Sd galaxies typically have rates of 0.45 and lower, while in Sa types the values extend to more than 0.60. Shearing rates of more than 0.70 are not observed and this critical limit corresponds to a minimum pitch angle of about 11°, corresponding to the tightest observed spiral arms.

OBJECT	NGC 1357
CONSTELLATION	Eridanus
RIGHT ASCENSION	$03^h\ 33^m\ 17^s$
DECLINATION	–13° 39′ 51″
BRIGHTNESS	12.8 mag
TYPE	SA(s)ab
PHOTOGRAPHERS	Adam Block
TELESCOPE	800-mm reflector
CAMERA	SBIG STL-11000
EXPOSURE TIME	300 min
LOCATION	Mount Lemmon SkyCenter/ University of Arizona, USA

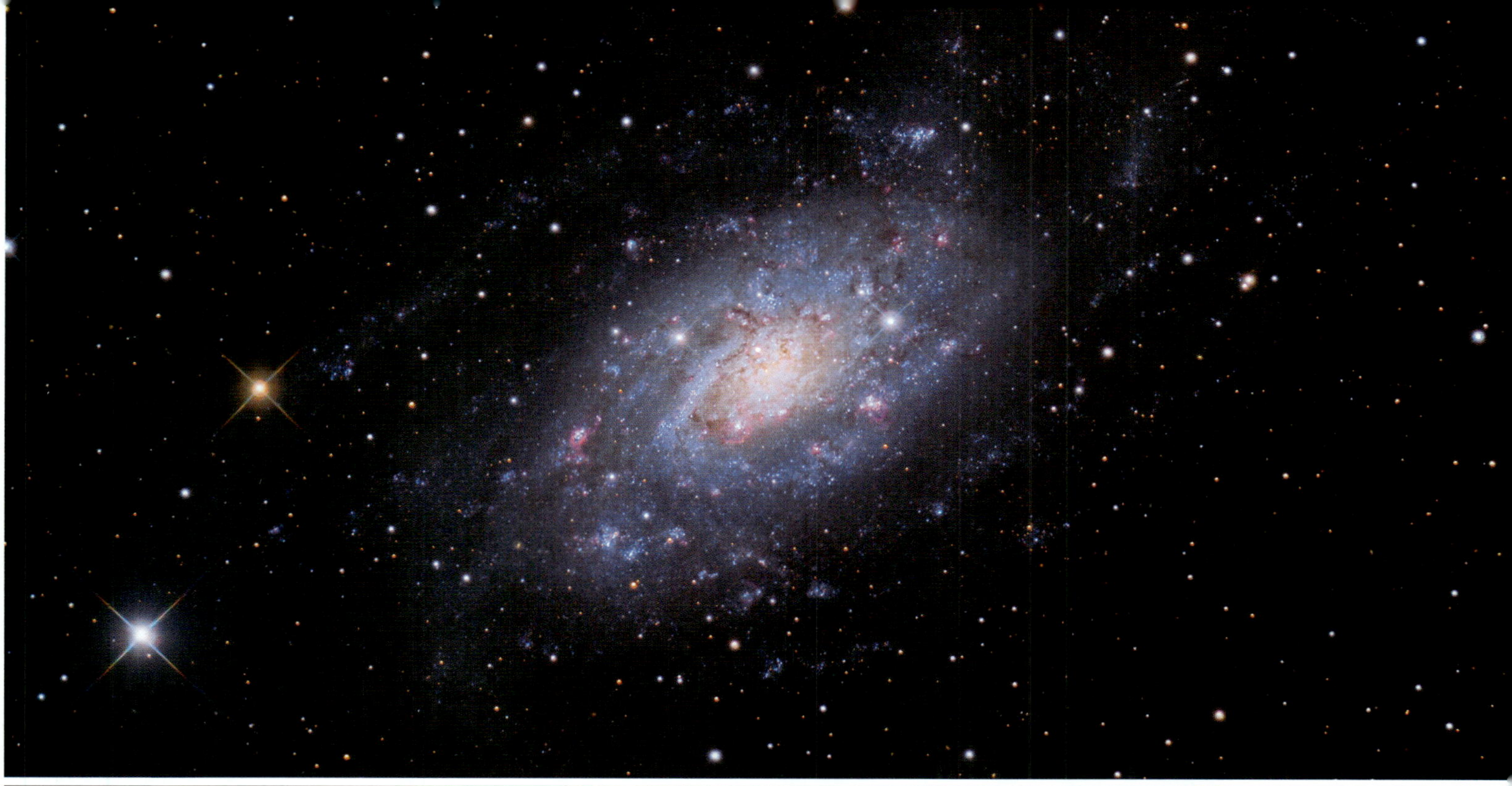

NGC 2403

The SAB(s)cd galaxy NGC 2403 is a member of the outer M 81 Group and at a distance of only twelve million light years is one of the closest neighbours of the Milky Way. The angular diameter of 22′ × 13′ corresponds to a projected disc diameter of 77 000 × 45 000 light years. NGC 2403 has no stellar bulge. The very small, compact centre is surrounded by light brown dust filaments, which appear denser at the upper (southwest) side of the disc. Areas of activity and a few H II regions can be found along the diffuse spiral arms. These large, red gas bubbles, up to 2000 light years in size, contain OB associations which power emission nebulae. Due to these prominent H II regions, NGC 2403 is often compared with M 33. NGC 2403 has a mostly unperturbed spiral-arm structure; only in the outskirts of the disc do the spiral arms fan out, visible as dim filaments in the image.

Since it is nearby, radio data deliver a detailed picture of neutral hydrogen in the disc and halo of NGC 2403. The radio disc of NGC 2403 is twice as large as the optical disc and the H I rotation curve is symmetric with the typical plateaus at radii larger than 5′ with a velocity of ±120 km/s. The lower left part of the galaxy in the image is approaching the Milky Way. The H I line in NGC 2403 is broadened and reveals the existence of a second gaseous component. This component has a lower velocity and in some regions has a different direction of rotation. The mass of this anomalous component which surrounds the disc is only one tenth of the total hydrogen mass of NGC 2403. Astronomers surmise that the hydrogen gas in this component was not accreted via interaction within the group, but is rather the result of active star-formation regions. The winds of the hot stars transport the gas out of the plane of the disc, forming a so-called galactic fountain, the gas then forming the observed second gaseous component. It appears that this mechanism occurs often in spiral galaxies; its direct detection in NGC 2403 is due to its small distance from the Milky Way and sufficiently sensitive radio telescopes.

OBJECT	NGC 2403
CONSTELLATION	Camelopardalis
RIGHT ASCENSION	$07^h\ 36^m\ 51^s$
DECLINATION	+65° 36′ 09″
BRIGHTNESS	8.1 mag
TYPE	SAB(s)cd
PHOTOGRAPHERS	Adam Block
TELESCOPE	600-mm reflector
CAMERA	SBIG STL-11000
EXPOSURE TIME	360 min
LOCATION	Mount Lemmon SkyCenter/ University of Arizona, USA

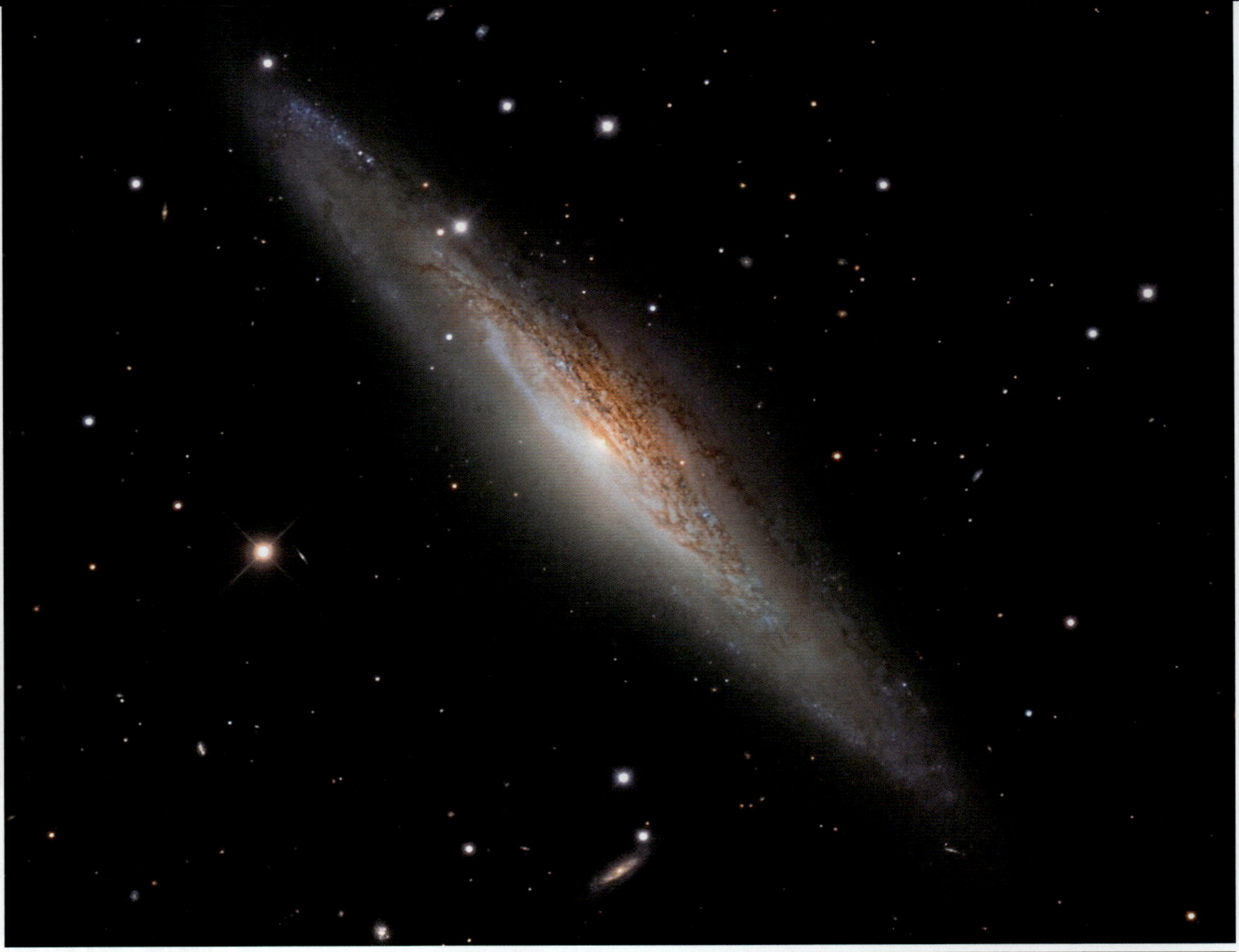

NGC 2683

The SA(rs)b galaxy NGC 2683 is about 6° east of the star Alpha Lyncis. It is part of the sparse environment of the Lynx–Cancer Void, a nearby filament of about 100 galaxies, many of them dwarfs. The distance of NGC 2683 from the Milky Way is about 32 million light years. The angular size of 9.3′ × 2.2′ corresponds to a projected disc diameter of 87 000 light-years. The inclination of this Sb galaxy is 78° and the edge-on view shows a granulated dust lane in the northwestern part of the disc. The centre appears compact and is surrounded by a bright blue spiral arm in the inner part of the disc opposite the dust lane. The outer parts of the disc are tilted compared to the inner parts. At the upper left edge one can notice a star-forming region with a few clusters of young stars; at the opposite side of the disc one can follow a spiral arm to the edge of the disc.

Analyses of the rotation curve via Hα and N II spectra of NGC 2683 show typical velocities of ±200 km/s in the outer regions. However, at radii less than 50″, there is a complex double structure. Astronomers associate these two curves with two independently rotating gaseous components. One of the components could be influenced by a bar. Numerical simulations show that this bar must be tilted by about 6° compared to the plane of NGC 2683. This suspicion goes well with the slightly boxy form of the bulge in NGC 2683, which could also be due to a bar, albeit a weak one.

OBJECT	NGC 2683
CONSTELLATION	Lynx
RIGHT ASCENSION	$08^h\ 52^m\ 41^s$
DECLINATION	+33° 25′ 18″
BRIGHTNESS	10.6 mag
TYPE	SA(rs)b
PHOTOGRAPHERS	Makis Palaiologou, Stefan Binnewies
TELESCOPE	1.3-m reflector
CAMERA	Andor DZ 436
EXPOSURE TIME	52.5 min
LOCATION	Skinakas Observatory, Crete, Greece

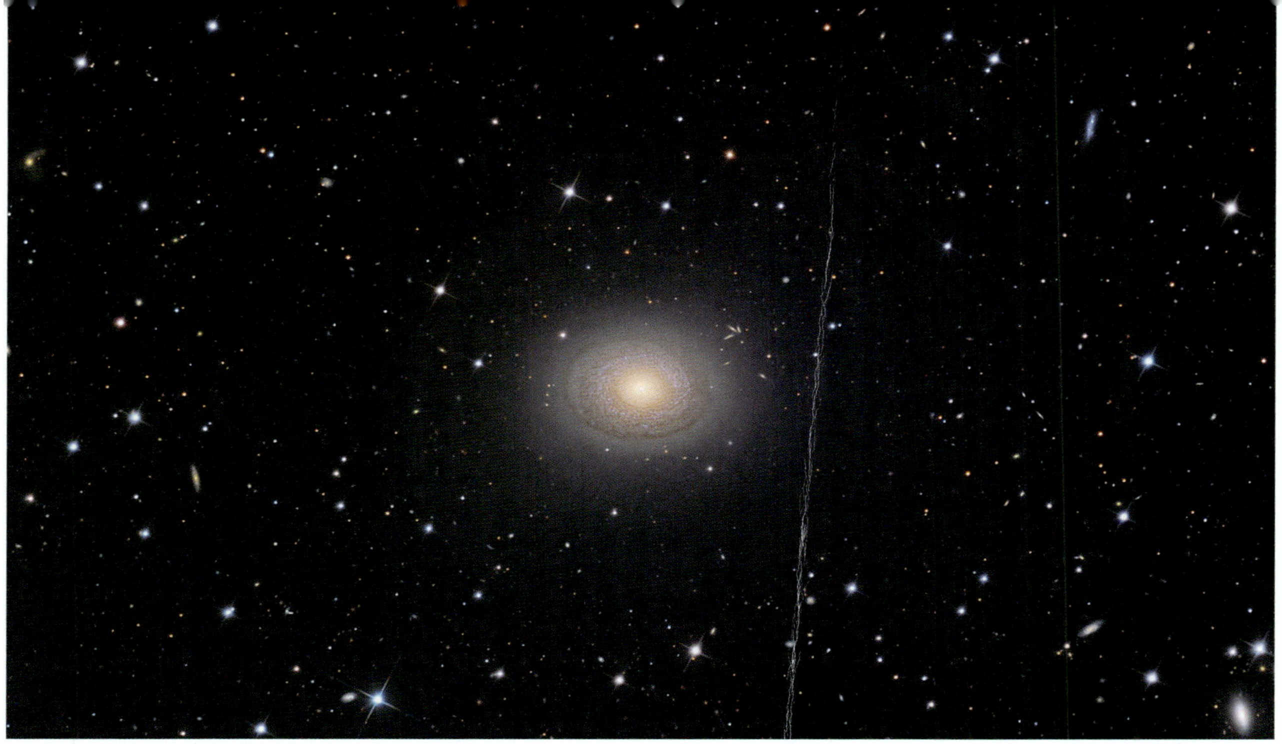

NGC 2775

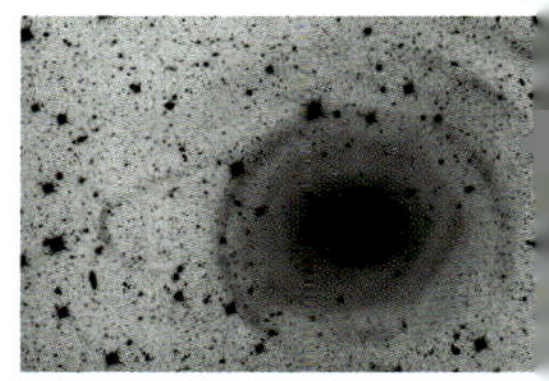

The galaxy NGC 2775 is at a distance of 55 million years from the Milky Way. Catalogues give an angular size of 4.3′ × 3.3′, corresponding to a transverse extent of 67 000 × 53 000 light years. The image has the north–south direction along the diagonal, from upper right (north) to lower left. The bright, compact nucleus is surrounded by a yellowish bulge. The bulge is uniformly bright and shows no spiral structure. Not until the surrounding disc zone, beginning at a radius of 0.4′, and extending to a maximum of 1.4′ at the southeast, is a flocculent arm pattern visible, bordered by a dust ring at the outer edge. The ring symmetry of this zone deviates from that of the nucleus: the left side of the dust ring appears more extended and the dust structures in the spiral-arm fragments cover a larger area than on the other side of the ring. There are also no active regions along the left edge which could be interpreted as continuations of spiral arms. Despite the star-forming regions in the flocculent spiral arms, yellowish colours dominate in NGC 2775, the colour index corresponding to that of E/S0 types. Thus, NGC 2775 is classified as an SA(r)ab galaxy.

The high-contrast image reveals complex structures in the halo of NGC 2775. The major axis of these elliptical structures is rotated by about 30° relative to the major axis of the disc plane. At the edges, structures which are presumably due to tidal tails contain condensations whose angular distances from the centre reach 4.5′ and thus transversal distances of 72 000 light years. Particularly prominent is a structure located 7.8′ further to the left. Its arc-like shape could be due to a track caused by a satellite galaxy which has orbited the larger galaxy several times, losing mass while doing so. According to this hypothesis, this mass resides in the shell-like structures visible today.

The elliptical form of this track and the disc around NGC 2775 lead to the conclusion that the orbital plane of the satellite galaxy was tilted by about 20° with respect to that of NGC 2775. This accretion model is also supported by the fact that close approaches of the satellite galaxy could lead to perturbations in the disc plane of NGC 2775, which would explain the flocculent spiral-arm pattern.

OBJECT	NGC 2775
CONSTELLATION	Cancer
RIGHT ASCENSION	$09^h\ 10^m\ 20^s$
DECLINATION	+07° 02′ 17″
BRIGHTNESS	11.0 mag
TYPE	SA(r)ab
PHOTOGRAPHERS	Adam Block
TELESCOPE	800-mm reflector
CAMERA	SBIG STL-11000 and STX-16803
EXPOSURE TIME	350 min
LOCATION	Mount Lemmon SkyCenter/ University of Arizona, USA

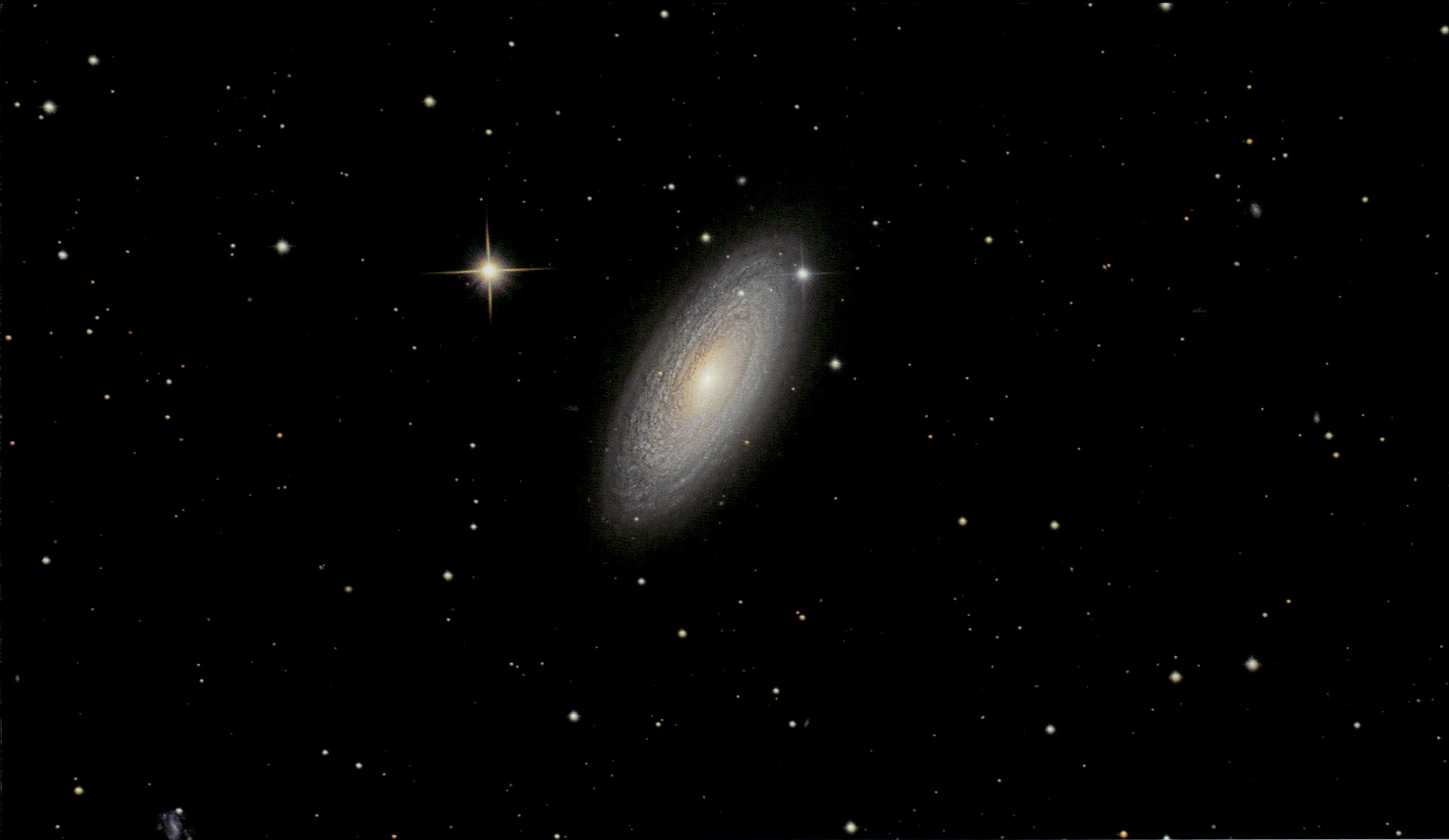

NGC 2841

The SA(r)b galaxy NGC 2841 is at a distance of 45 million light years from the Milky Way. With an angular size of 8.1′ × 3.5′, it is one of the top-five brightest and largest galaxies in the constellation Ursa Major. The transverse dimensions of the disc of NGC 2841 are 106 000 × 46 000 light years. Its spiral-arm structure is marked by a number of shorter spiral-arm fragments which can be followed through only 20–30°. At the left, northeast edge of the disc the dust components of this flocculent structure are easily recognizable as is the weak pattern of two longer spiral arms. The bulge of NGC 2841 has almost the same colour as the 8th-magnitude K star SAO 27227 2.4′ away. This proves that an old stellar population is dominant in the bulge, contrasting clearly with the bright blue stars in the flocculent spiral-arm zone.

Four supernovae have been observed in NGC 2841 since 1912. This relatively large number is an indication of currently active evolutionary processes. This is confirmed by an outer H I gaseous disc which extends from the optical disc of NGC 2841; all observed supernovae lie in the transition region. The NASA UV satellite GALEX (Galaxy Evolution Explorer) found many complete spiral arms in these outer regions which are not visible in the optical image. Astronomers thus classify NGC 2841 as an extended ultraviolet galaxy (XUV-disc galaxy). This property is found in about 30 per cent of spiral galaxies and one suspects that these discs are formed from accreted gas, and episodes of star formation are then stimulated by weak interactive processes. The occurrence of such XUV-disc galaxies could thus be a typical phase in the evolution of spiral galaxies.

OBJECT	NGC 2841
CONSTELLATION	Ursa Major
RIGHT ASCENSION	$09^h\ 22^m\ 03^s$
DECLINATION	+50° 58′ 36″
BRIGHTNESS	9.5 mag
TYPE	SA(r)b
PHOTOGRAPHERS	Johannes Schedler
TELESCOPE	400-mm reflector
CAMERA	SBIG STL-11000
EXPOSURE TIME	940 min
LOCATION	Wildon, Austria

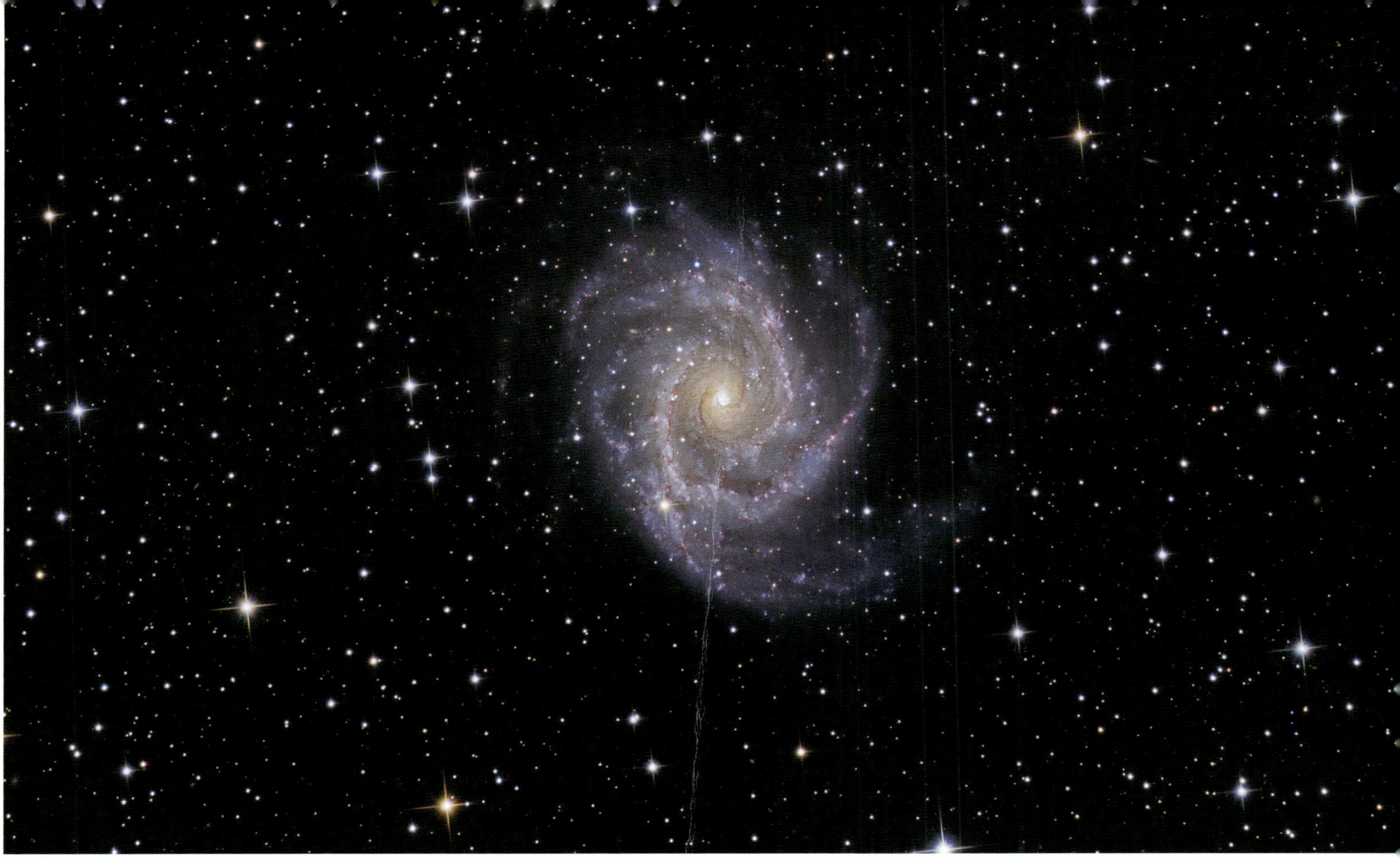

NGC 2997

The SAB(rs)c galaxy NGC 2997 is at a distance of 39 million light years and its projected disc diameter of 100 000 light years classifies it as a large spiral galaxy, 30 per cent larger than average. The grand-design spiral pattern of NGC 2997 is marked by two spiral arms which have clearly visible dust lanes in the inner part of the disc and which can be followed through 1.5 revolutions. Further out, the spiral arms broaden and have many active portions with a large number of H II regions. The brightest of these regions can be resolved into gaseous complexes with several centres.

NGC 2997 belongs to the loose galaxy group LGG 180. Its mass, luminosity, and star-formation rate are similar to both the Milky Way and M 31. Due to its relatively low distance and moderate inclination of 33°, NGC 2997 is well suited for comparative analyses, for example of neutral hydrogen gas which does not surround the disc. Thick H I clouds are located in this non-planar gas and their velocities can differ from that of their local environment by up to 200 km/s. Formally, astronomers refer to high-velocity clouds (HVCs) in the case of differential velocities in excess of 50 km/s. Since some of the H I clouds have hints of tidal tails, they are formed not only via processes such as supernova explosions which transport gas out of the disc, but also presumably via accretion of gas-rich dwarf galaxies. The image confirms this, since dim tidal tails, due to such interaction, are visible in the outer spiral arms. The masses of the H I clouds in NGC 2997 are around 10^7 solar masses and thus exceed those of the clouds in the Milky Way.

OBJECT	NGC 2997
CONSTELLATION	Antlia
RIGHT ASCENSION	$09^h\ 45^m\ 39^s$
DECLINATION	−31° 11′ 28″
BRIGHTNESS	10.1 mag
TYPE	SAB(rs)c
PHOTOGRAPHERS	Johannes Schedler
TELESCOPE	500-mm reflector
CAMERA	SBIG STL-11000
EXPOSURE TIME	140 min
LOCATION	Hakos Farm, Namibia

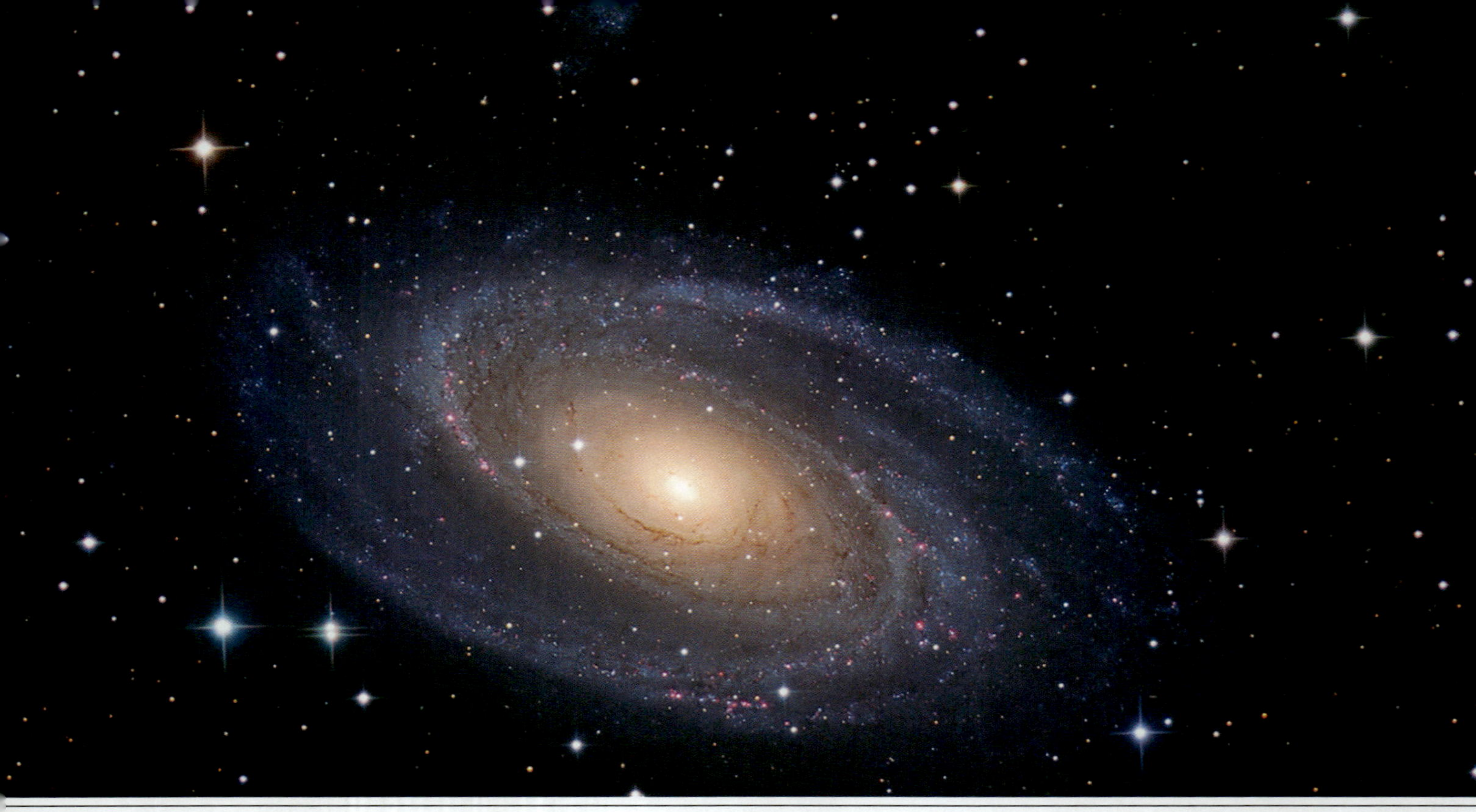

M 81

The SA(s)ab galaxy M 81 (NGC 3031) is part of an interacting triplet also containing M 82 and NGC 3077 and is the largest galaxy of a group with 34 members. The distance of M 81 from the Milky Way is 11.7±0.4 million light years. M 81 is approaching the Milky Way at 34 km/s. Its enormous angular size of 26.9′ × 14.1′ corresponds to an absolute disc diameter of 92 000 light years and a scale of about 3400 light years per minute of arc. The disc of M 81 has a mean inclination of 59° and the major axis of the disc is rotated by 30° relative to the line of sight from the Milky Way. Near the upper part of the image, one can recognize the dwarf irregular galaxy Holmberg IX, which is a few minutes away from the eastern spiral arm of M 81 and has an angular size of 3.5′ × 2.8′. Despite large uncertainties in various distance measurements, it is probably located behind M 81, and astronomers assume that it was formed via the condensation of a tidal tail. The high fraction of young, massive stars explains its blue colour and it is suspected that Holmberg IX was formed only a few 100 million years ago.

The plateau in the rotation velocity which is typical for spiral galaxies lies at 240 km/s. After a maximum of 255 km/s at a radius of 25 000 light years, it declines steadily and ends at 190 km/s at a distance of 48 000 light years from the centre. This decline is an indication of a relatively low fraction of dark matter in the halo of M 81. The maximum of the rotation curve is caused by the prominent grand-design spiral arm. Information about the distribution of gas and stars in the spiral plane allows one to make a mass model for M 81. This results in 6.9×10^{19} solar masses for the stellar disc. For the bulge, one obtains a mass 19 per cent of that of the disc. Models also indicate that the inclination of M 81 varies with radius, which indicates that the outer disc, relative to the inner part of the disc, is warped. Since opposing edges of the disc are also not parallel, this is evidence of appreciable tidal forces from the larger group members.

OBJECT	M 81, NGC 3031
CONSTELLATION	Ursa Major
RIGHT ASCENSION	$09^h\ 55^m\ 33^s$
DECLINATION	+69° 03′ 55″
BRIGHTNESS	7.9 mag
TYPE	SA(s)ab
PHOTOGRAPHERS	Johannes Schedler
TELESCOPE	400-mm reflector
CAMERA	SBIG STL-11000
EXPOSURE TIME	300 min
LOCATION	Wildon, Austria

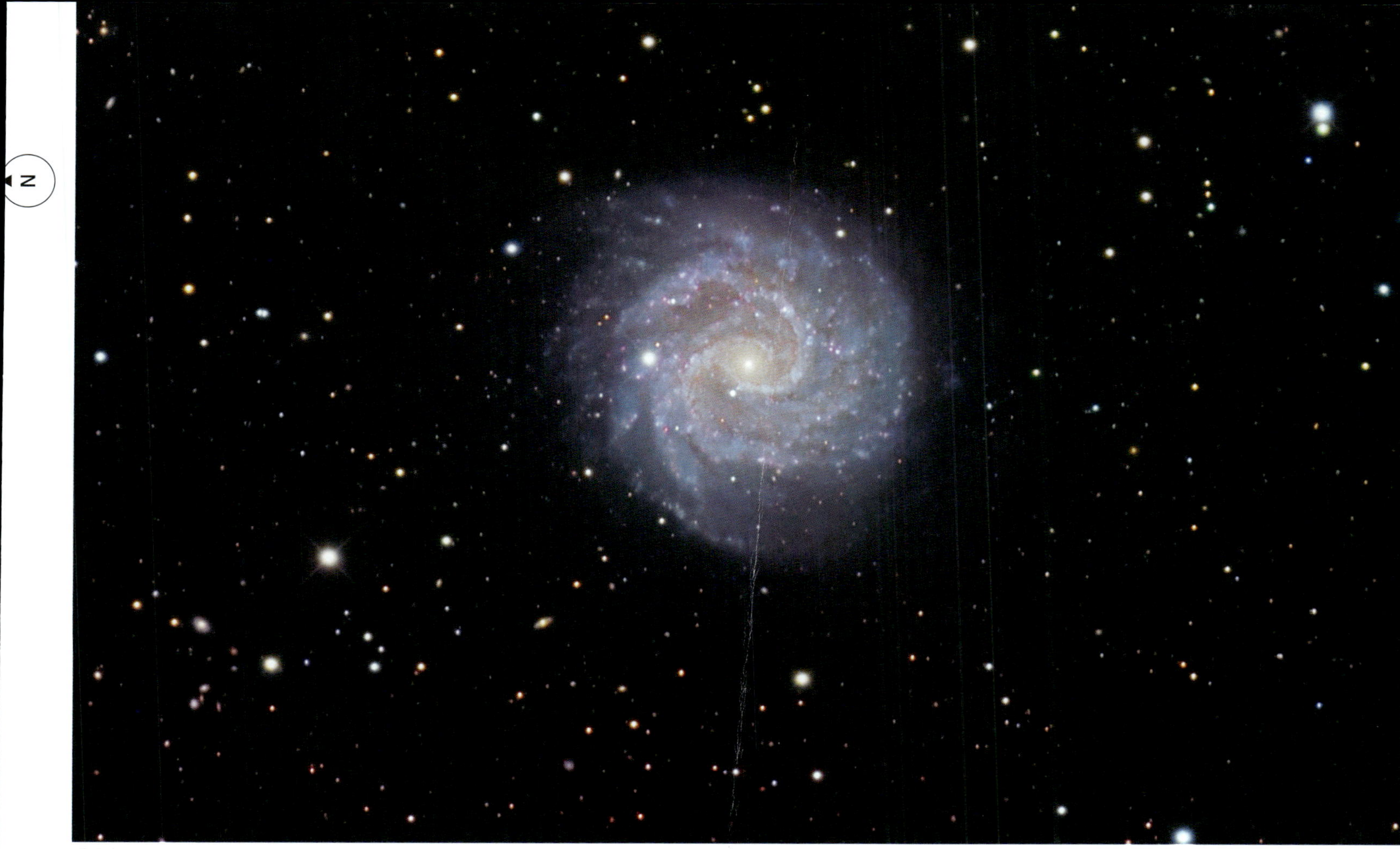

NGC 3184

The SAB(rs)d galaxy NGC 3184 is located in the southern part of the constellation Ursa Major, at the border with Leo Minor. The distance of 30 million light years implies that the angular size of 7.4′ × 6.9′ corresponds to projected dimensions of 65 000 × 60 000 light years. The spiral structure of NGC 3184 is marked by two spiral arms which emerge from an almost neutrally white, small disc. Vorontsov-Velyaminov rows are visible in both arms, forming a hexagonal border around the disc. There are two H II regions in the western spiral arm of NGC 3184, in the upper half of the disc in the image. These have their own catalogue numbers, NGC 3180 and NGC 3181.

In NGC 3184 a dim gaseous disc was discovered whose size exceeds that of the optical disc by a factor of 1.7. More than 1000 compact objects were found in this outer disc. They were identified as star clusters of 100–10 000 solar masses and ages of up to a billion years. Despite their low surface brightness of 30–32 mag/arcsec^2, astronomers were able to detect colour differences between these star clusters. The red clusters are located at larger radii than the blue clusters. Since the positions of these star clusters are correlated with the spiral structure of neutral hydrogen gas in the disc, one suspects that most of them were formed in this outer region. These observations show that in the outskirts of the discs of spiral galaxies continuous star formation can take place which is part of the intrinsic galactic evolutionary history and doesn't need to rely on any external interaction.

OBJECT	NGC 3184
CONSTELLATION	Ursa Major
RIGHT ASCENSION	$10^h\ 18^m\ 17^s$
DECLINATION	+41° 25′ 27″
BRIGHTNESS	10.3 mag
TYPE	SAB(rs)cd
PHOTOGRAPHERS	Robert Gendler
TELESCOPE	317- and 600-mm reflector
CAMERA	SBIG STL-11000
EXPOSURE TIME	1200 min
LOCATION	Nighthawk Observatory, New Mexico, USA

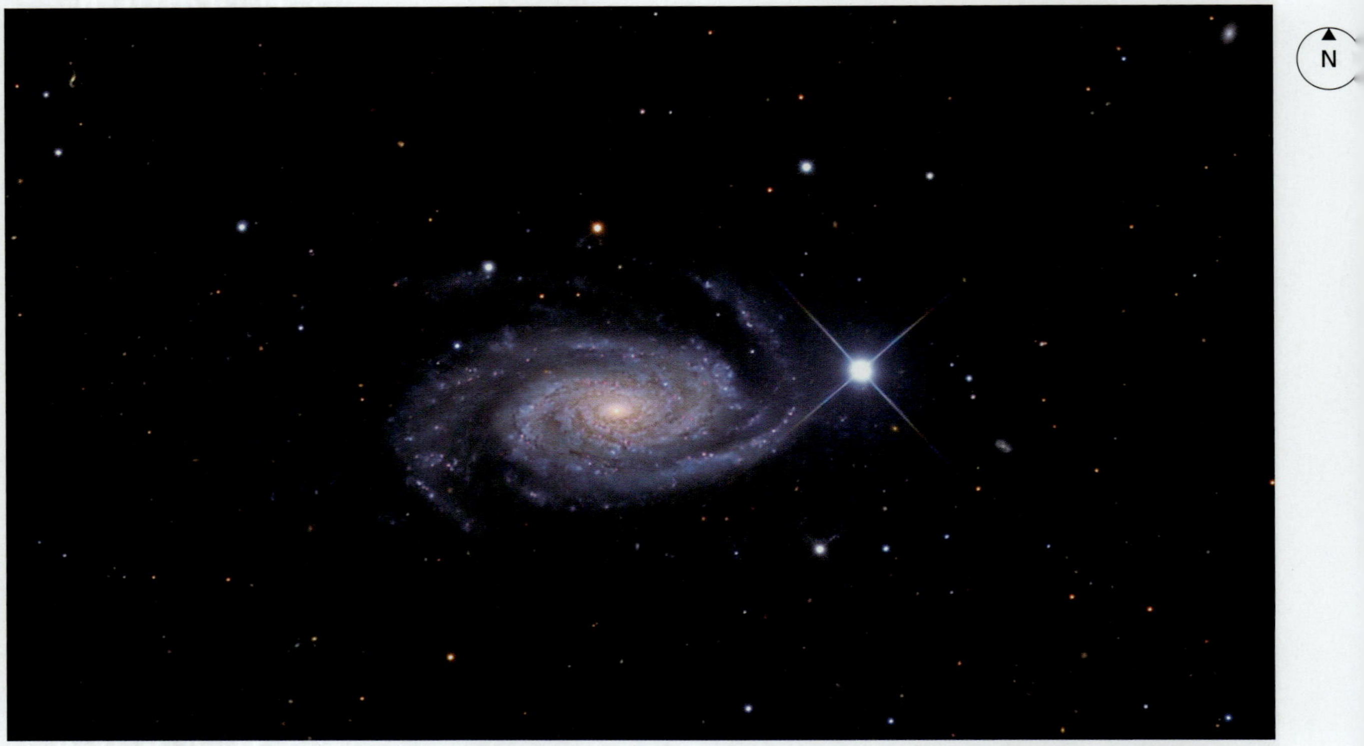

NGC 3338

The SA(s)c galaxy NGC 3338 is located 2.5′ east of the 8.9-magnitude bright A star SAO 99253. The distance of NGC 3338 from the Milky Way is 54 million light years; the angular size of the bright optical disc is 3.4′ × 1.3′ and corresponds to a projected size of 53 000 × 20 000 light years. At the centre of NGC 3338 is a point-like source surrounded by a small elliptical bulge. Connected to a ring surrounding the bulge are two spiral arms which are fragmented in their inner parts. After half of a revolution the spiral arms become brighter and their structure more diffuse. There are many active regions even in the outer parts of the arms. The brightness of NGC 3338 is 12.8 mag. This applies only to the central region; the ends of the spiral arms and the tidal tails which are connected to these reach all the way to the bottom of the image.

According to the NASA/IPAC Extragalactic Database (NED), within 15′ of NGC 3338 there are almost 3000 extragalactic objects, most of which were identified by the Sloan Digital Sky Survey. Only three of these are located in the cosmic neighbourhood of NGC 3338 (recession velocity 1302 km/s). These are at 8′ distance SDSS J104140.96+134929.5 with a velocity of 1271 km/s (17 mag, 0.3′ diameter), at 10.9′ SDSS J104252.43+134427.8 with 1145 km/s (17.4 mag, 0.3′), and SDSS J104226.50+135725.7, farthest away at 13.4′ (17 mag, 0.6′). The sizes of these galaxies are 10 000, 9 000, and 15 000 light years and thus all are classified as dwarf galaxies with a maximum transverse distance to NGC 3338 of less than 250 000 light years. One can thus assume that these galaxies are gravitationally bound to NGC 3338 and interact with it. Close approaches between these galaxies and NGC 3338 could be responsible for its tidal tails.

OBJECT	NGC 3338
CONSTELLATION	Leo
RIGHT ASCENSION	$10^h\ 42^m\ 08^s$
DECLINATION	+13° 44′ 49″
BRIGHTNESS	12.8 mag
TYPE	SA(s)c
PHOTOGRAPHERS	Adam Block
TELESCOPE	800-mm reflector
CAMERA	SBIG STL-11000
EXPOSURE TIME	620 min
LOCATION	Mount Lemmon SkyCenter/ University of Arizona, USA

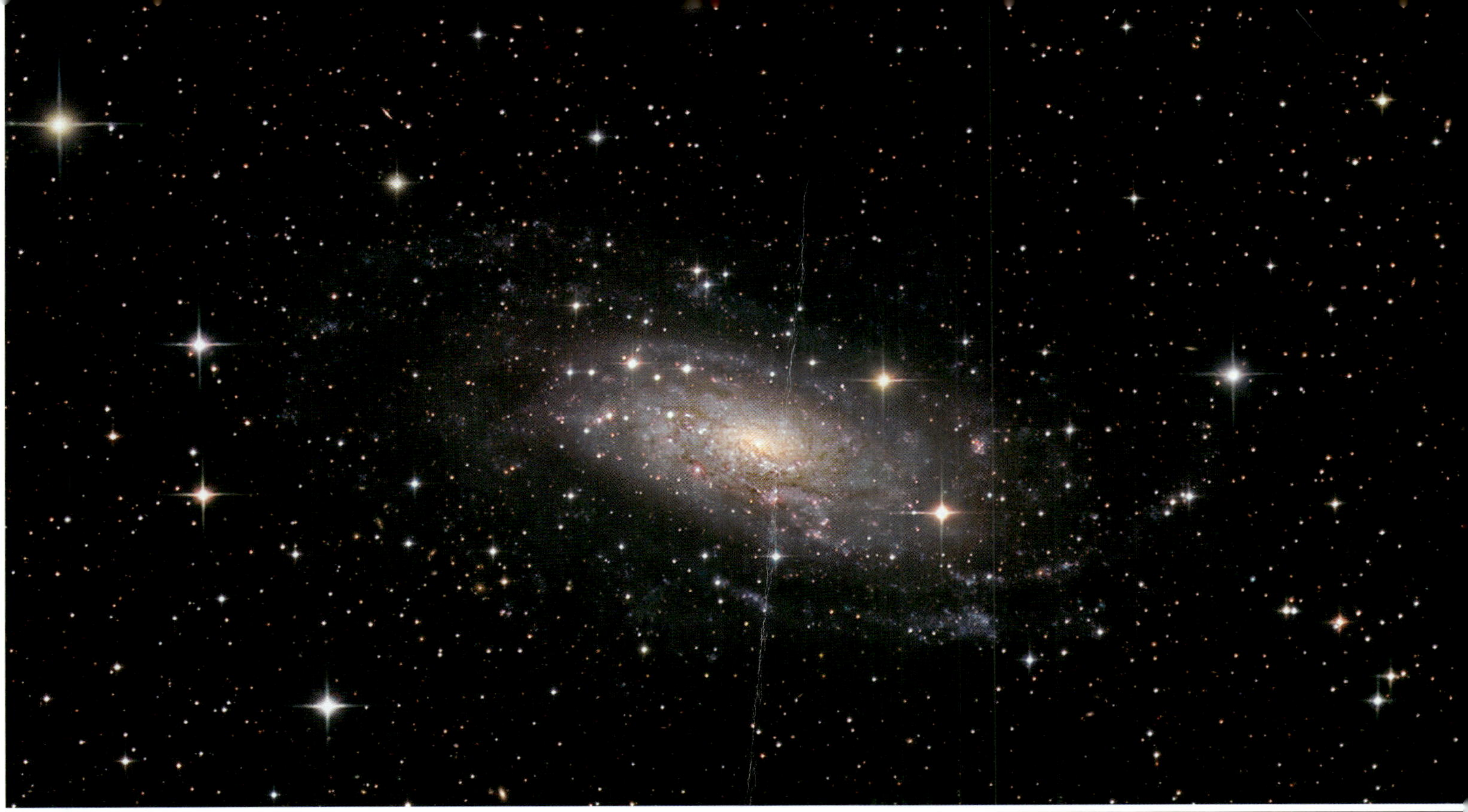

NGC 3621

NGC 3621 is an isolated SA(s)d galaxy at a distance of 24 million light years from the Milky Way. The nearest neighbour with a similar recession velocity is more than 3° away which corresponds to a distance of 1.3 million light years. The large angular size of 12.3′ × 7.1′ is a result of the relatively small distance. Based on the absolute size of the optical disc of 86 000 light years, NGC 3621 is considered to be a normal spiral galaxy. The moderate increase in brightness towards the centre indicates that NGC 3621 does not have a bulge and that its flat disc is dominated by long spiral arms. NGC 3621 is a pure-disc galaxy. Its unperturbed, flat shape is an indication that it has not experienced any interactions.

In galaxy-evolution models there is a correlation, due to the increasing stellar velocity dispersion, between the increase in mass of the central supermassive black hole due to accretion and the size of the bulge. It thus seems plausible that most galaxies with active nuclei have large and prominent bulges.

NGC 3621 is an exception to this rule. It has a Seyfert nucleus although it has no appreciable bulge. The central black hole has a mass of 3.5×10^5 solar masses and thus far below the masses which occur in active galaxies with prominent bulges. Observations with the Chandra X-ray satellite detected, in addition to the central source very close to the supermassive black hole, two further point-like X-ray sources. Their X-ray luminosities indicate masses of a few thousand solar masses. They are thus intermediate mass black holes (IMBHs) between stellar black holes and supermassive black holes in galaxy cores. One can thus assume that NGC 3621 is moving to the next stage in the evolution to a Seyfert galaxy and could develop a bulge in the future.

OBJECT	NGC 3621
CONSTELLATION	Hydra
RIGHT ASCENSION	11^h 18^m 17^s
DECLINATION	−32° 48′ 51″
BRIGHTNESS	10.2 mag
TYPE	SA(s)d
PHOTOGRAPHERS	Josef Pöpsel, Beate Behle
TELESCOPE	600-mm reflector
CAMERA	SBIG ST-10XME
EXPOSURE TIME	210 min
LOCATION	Amani Lodge, Namibia

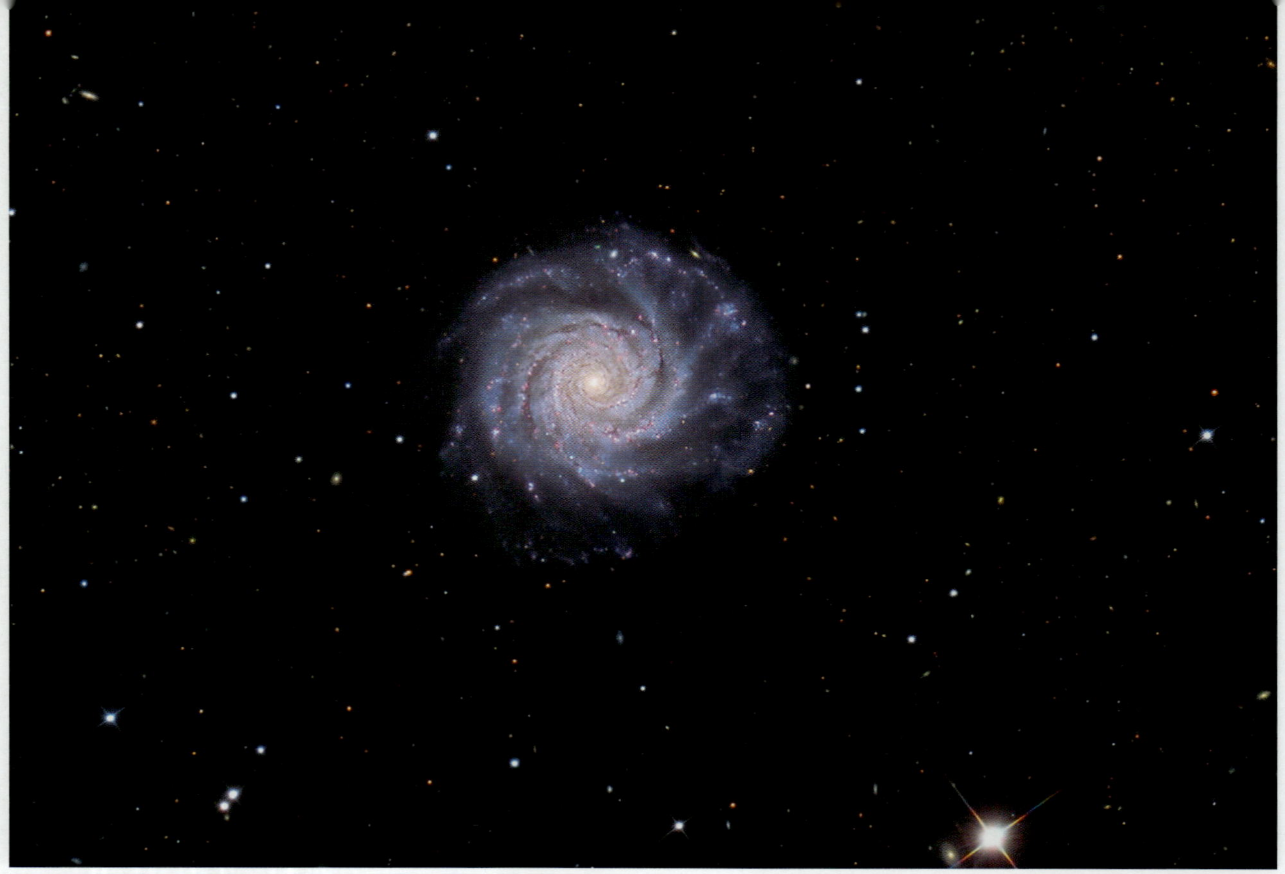

NGC 3938

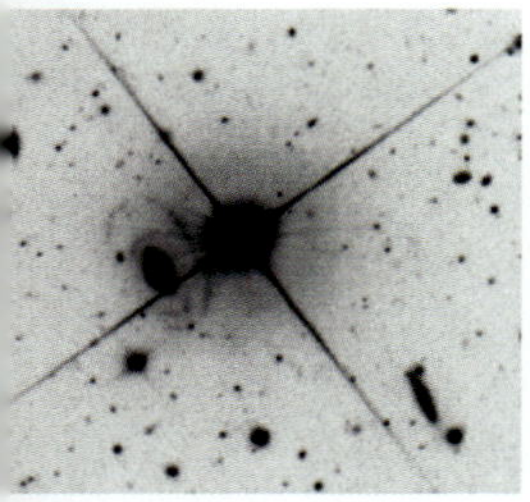

The SA(s)c galaxy NGC 3938 belongs to the southern part of the Ursa Major Cluster (UMa I) and is at a distance of 60 million light years from the Milky Way. It is one of the largest members of this galaxy cluster and its angular size of 5.4′ × 4.9′ corresponds to a transverse distance of 94 000 × 86 000 light years. The inner part of the disc is marked by two spiral arms which are easy to make out due to the large number of H II regions. The spiral arms broaden in the outer part of the disc, leading to a more complex arm structure. The ends of these arms contain active regions which are concentrated on the northern edge of the disc (on the right in the figure). These lead to the impression of a complete, thick band surrounding the disc. This band is not circular but polygonal and the enclosed area is not symmetric with respect to the centre of NGC 3938.

In the image, near the ninth-magnitude bright G star to the lower right of NGC 3938 one can just see the galaxy SDSS J115216.24+441308.8. This galaxy at a distance of more than 900 million light years is characterized by two dim rings which surround it. The outer ring has a diameter of 320 000 light years (see the high-contrast negative image).

The northern part of the Ursa Major Cluster is receding at 200–300 km/s relative to the southern part and NGC 3938. This suggests that the cluster could have been formed by the merger of two smaller clusters. This is supported by the large number of spirals and could explain why NGC 3938 shows signs of interaction but at the same time still has a mostly intact spiral pattern as compared to other cluster galaxies. The rotation curve of NGC 3938 shows a steep rise at radii 0.5–2.5′ from 120 km/s to 160 km/s. This indicates a massive dark-matter halo in NGC 3938 which also helps to keep the spiral pattern intact despite the dynamics of the cluster.

OBJECT	NGC 3938
CONSTELLATION	Ursa Major
RIGHT ASCENSION	$11^h\ 52^m\ 49^s$
DECLINATION	+44° 07′ 15″
BRIGHTNESS	10.8 mag
TYPE	SA(s)c
PHOTOGRAPHERS	Adam Block
TELESCOPE	800-mm reflector
CAMERA	SBIG STX-16803
EXPOSURE TIME	510 min
LOCATION	Mount Lemmon SkyCenter/ University of Arizona, USA

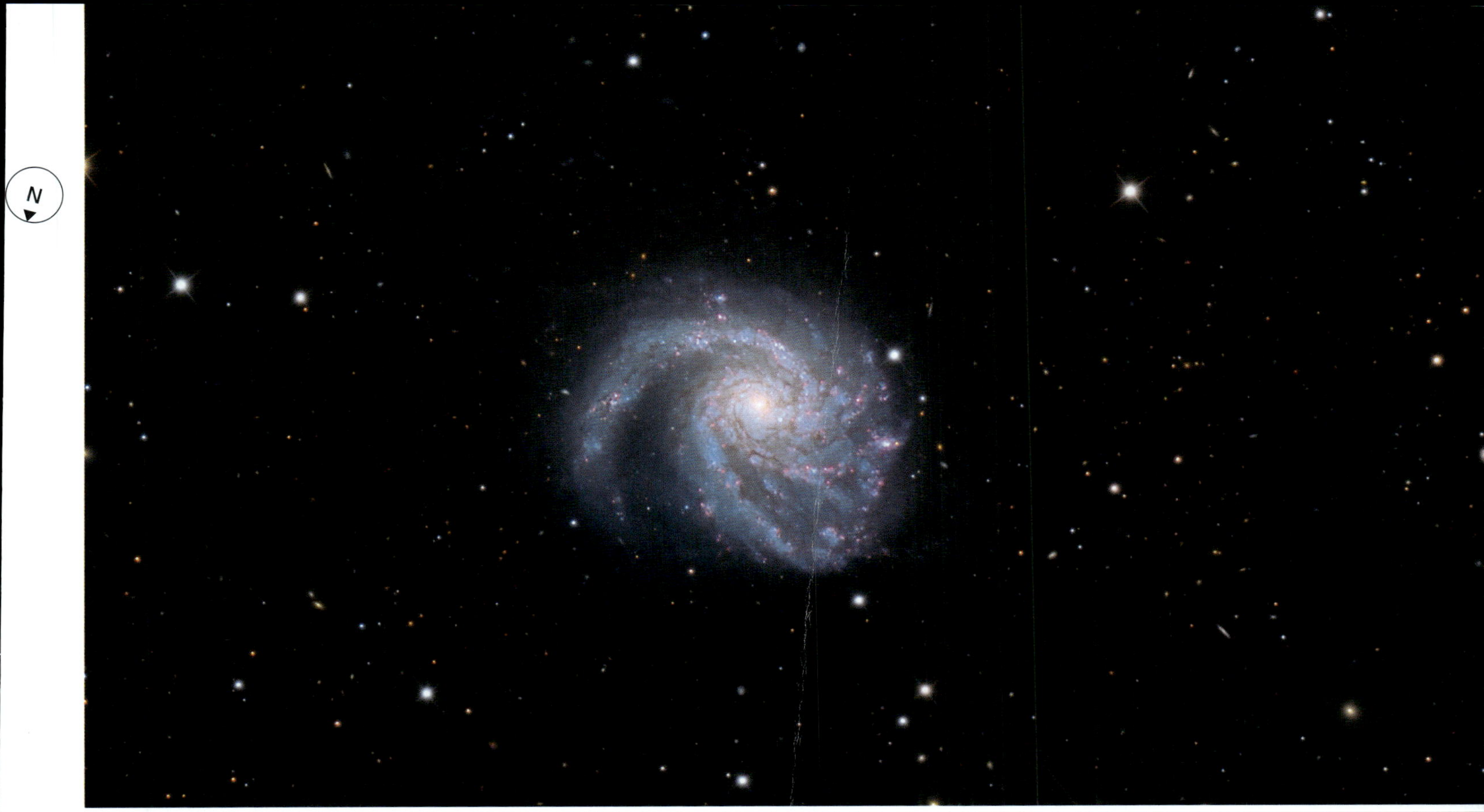

M 99

M 99 (NGC 4254) is a bright SA(s)c galaxy found at the northwestern edge of the Virgo Cluster. Its angular distance from the centre of the cluster is 3.7° and it lies at a distance of 50 million light years from the Milky Way. The disc diameter of 5.4′ corresponds to a projected distance of 79 000 light years. M 99 has a small bulge surrounded by a two-arm spiral pattern. The prominent bright blue arms indicate large-scale star formation and contain many H II regions. The left (western) spiral arm is asymmetric with decreasing pitch angle. Beneath it is a sparse zone in the plane of the disc, which almost makes M 99 a Magellanic-type galaxy.

The image also shows many small, dim filaments above the bright disc. These could be made of gas, dust and stars which were pulled out of the outer spiral arms via interactions with other galaxies.

M 99 was one of the galaxies studied in the SAURON project (Spectroscopic Areal Unit for Research on Optical Nebulae, Roque de los Muchachos Observatory, La Palma, Spain). Spatially resolved spectroscopy was used to study the distribution of characteristic emission lines in the plane of the disc and this in turn was used to determine the distribution of metallicity, which astronomers describe with the parameter Z. In the 18 Sb and Sc spirals studied, it was found that these have lower Z values than ellipticals or earlier-type spirals. Stellar-evolution models can be used to determine the age of the stellar population in spiral galaxies. Metal-rich galaxies have short timescales while young, metal-poor galaxies have longer values, which is also the case for M 99 with 2.8 billion years.

OBJECT	M 99, NGC 4254
CONSTELLATION	Coma Berenices
RIGHT ASCENSION	$12^h\ 18^m\ 50^s$
DECLINATION	+14° 24′ 59″
BRIGHTNESS	10.4 mag
TYPE	SA(s)c
PHOTOGRAPHERS	Adam Block
TELESCOPE	800-mm reflector
CAMERA	SBIG STX-16803
EXPOSURE TIME	460 min
LOCATION	Mount Lemmon SkyCenter/ University of Arizona, USA

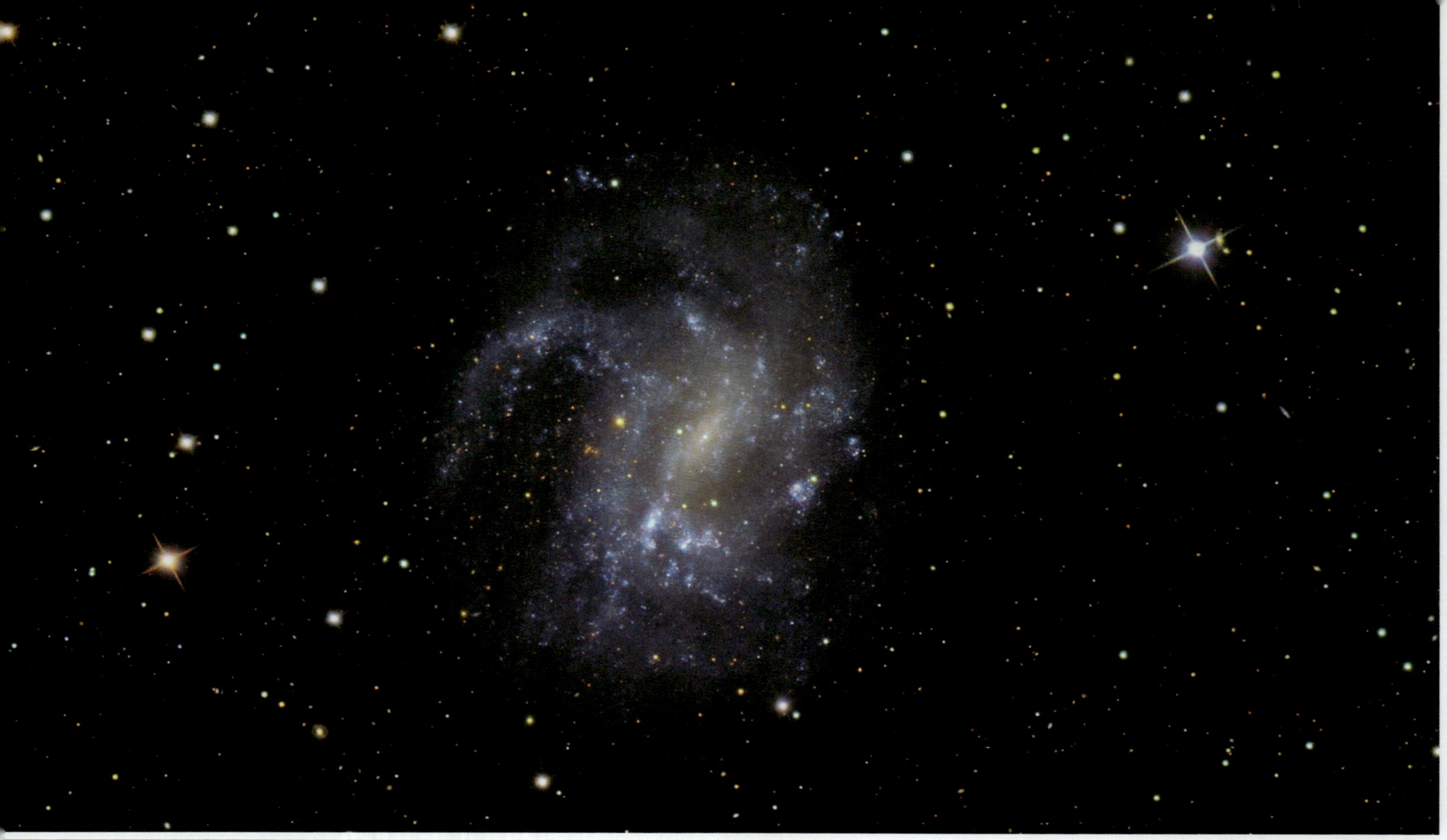

NGC 4395

NGC 4395 is a low-surface-brightness galaxy (LSBG) at a distance of only 14 million light years from the Milky Way. Within the dim spiral structure are a few bright, active regions, which due to suspected separation were classified as NGC 4399, NGC 4400, and NGC 4401.

Consensus is that LSBGs are dominated by dark matter, which is also responsible for the formation of small-scale structures. The large angular size of 13.2′ × 11.0′ results in a projected disc diameter of 54 000 × 45 000 light years. In the optically unremarkable centre of NGC 4395 is a Seyfert-1 nucleus – the closest one to the Milky Way. Its luminosity is low since the black hole at the centre with a mass of 10^5 solar masses is relatively lightweight. Interestingly, the strongest radio source lies about 6′ away from the core. It is thus assumed that this source does not belong to NGC 4395.

Despite its low luminosity, the core of NGC 4395 is significantly variable in X-rays. Astronomers have measured an increase in X-ray luminosity by a factor of two within 300 seconds. Measurement of such stochastic variation indicates that long-scale variability is more common than short-scale. In time-series analysis, in order to quantify the variability, astronomers use a maximum timescale over which variations are uniformly distributed. Its value for NGC 4395 is about 540 seconds and is thus appreciably smaller than the corresponding values of other Seyfert-1 galaxies, which have timescales of 10^4–10^7 seconds. There is a general correlation in Seyfert-1 galaxies between this timescale and the mass of the central black hole and NGC 4395 defines the dim end of this distribution. One thus obtains a value of 10^4–10^5 solar masses, which agrees well with other determinations of the mass of the central black hole in this lightweight Seyfert-1 galaxy.

OBJECT	NGC 4395
CONSTELLATION	Canes Venatici
RIGHT ASCENSION	$12^h\ 25^m\ 49^s$
DECLINATION	+33° 32′ 49″
BRIGHTNESS	10.6 mag
TYPE	SA(s)m
PHOTOGRAPHERS	Johannes Schedler
TELESCOPE	400-mm reflector
CAMERA	SBIG STX-16803
EXPOSURE TIME	920 min
LOCATION	Wildon, Austria

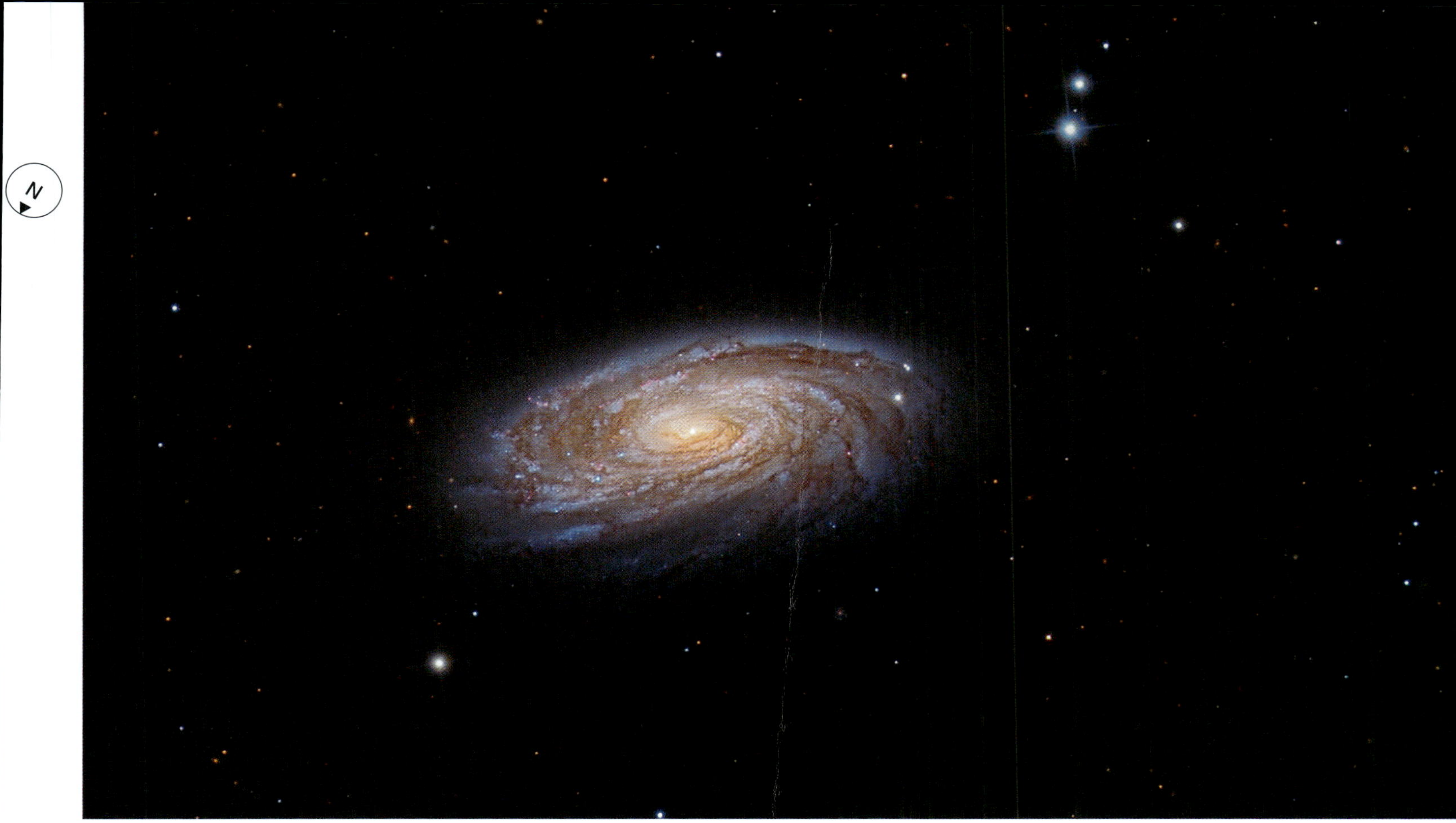

M 88

M 88 (NGC 4501) is an SA(rs)b galaxy which is part of a subgroup of the Virgo Cluster. In a region with a diameter of 12 million light years, about a dozen galaxies surround NGC 4486. The measured recession velocity of M 88 of 2281 km/s is influenced by the Virgo infall and thus the naively calculated cosmological distance of 120 million light years is vastly exaggerated. Using alternative distance measures, astronomers derive a distance of 61±16 million light years. The angular size of M 88 is 6.9′ × 3.7′ resulting in a transversal diameter of the disc plane of 122 000 light years. No significant halo is visible surrounding the stellar core of M 88. This allows an unhindered view into the core, and images from the Hubble Space Telescope show that the spiral structure continues all the way to the core of M 88.

In a study of optical spectra in Hα, velocity maps were obtained for the bright members of the Virgo Cluster. Using these, astronomers could assign a relative velocity to each part of a galaxy due to the Doppler effect which shifts the Hα lines of the moving hydrogen gas. For M 88, this Hα map indicates a regularly formed rotating disc which shows small shear currents near the minor axis. These perturbations are visible in the image in the broadening of the lower (northeastern) spiral-arm structure. The rotation curve of M 88 with ±230 km/s has the largest velocity range of all the studied galaxies, which is interpreted as an indication of a large halo mass in M 88.

OBJECT	M 88, NGC 4501
CONSTELLATION	Coma Berenices
RIGHT ASCENSION	$12^h\ 31^m\ 59^s$
DECLINATION	+14° 25′ 13″
BRIGHTNESS	9.6 mag
TYPE	SA(rs)b
PHOTOGRAPHERS	Adam Block
TELESCOPE	600-mm reflector
CAMERA	SBIG STL-11000
EXPOSURE TIME	300 min
LOCATION	Mount Lemmon SkyCenter/ University of Arizona, USA

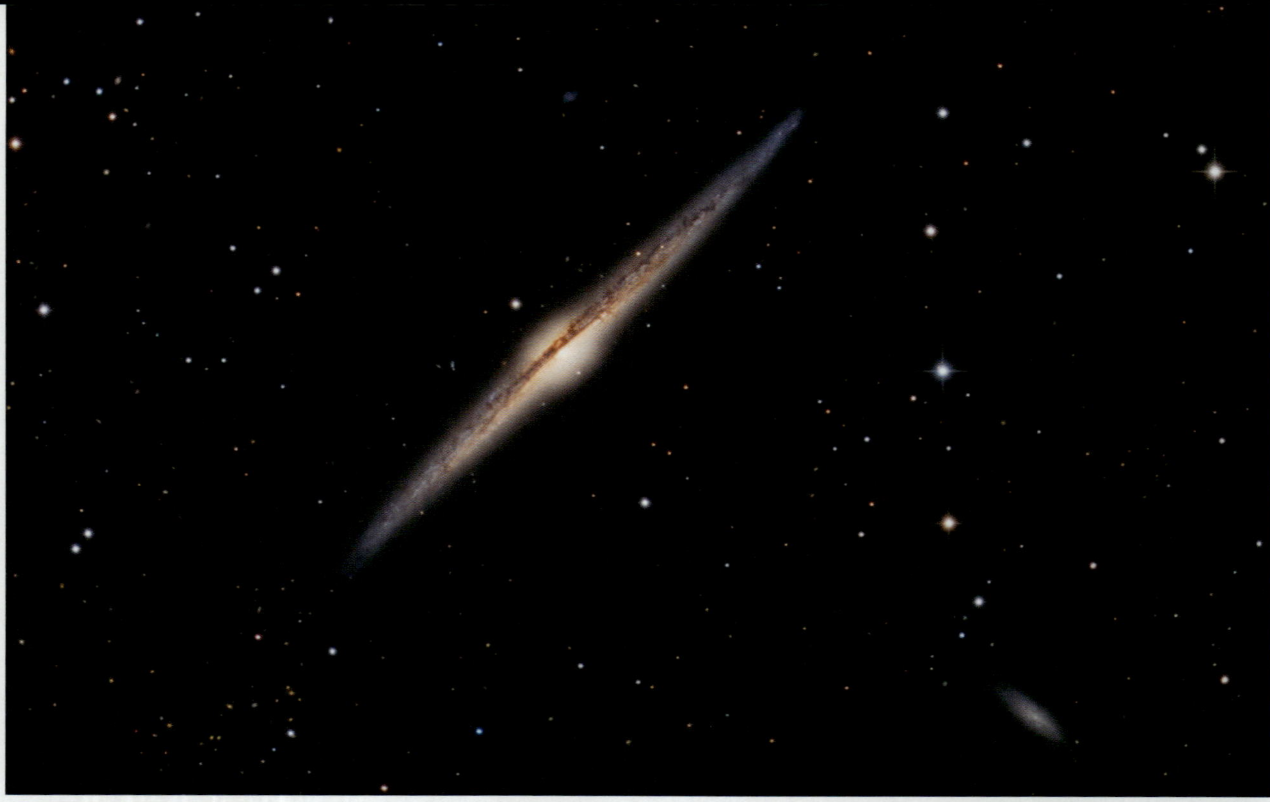

NGC 4565, The Needle Galaxy

The SA(s)b galaxy NGC 4565 belongs to the Coma Cluster where it is part of a subgroup with a dozen other galaxies. The galaxies in the Coma Cluster have random relative velocities of several hundred km/s, so the distance to NGC 4565 can be given only approximately at 40–60 million light years. Alternative distance measures which do not use the recession velocity indicate that NGC 4565 is at a distance of 35 million light years. This implies that the angular size of 15.9′ × 1.9′ corresponds to projected sizes of 162 000 light years for the diameter and 20 000 light years for the thickness of the box-like bulge. The dust lane in NGC 4565, at an inclination of 88°, runs almost through The middle of the core region and is the source of the name The Needle Galaxy. To the lower right of NGC 4565, 13.2′ to the southwest, is the small SB galaxy NGC 4562. It is moving away from NGC 4565 at a radial velocity of +123 km/s.

The edge-on view allows one to see a warp in the disc. The lower left edge of the disc is bent down and the opposite edge is bent up. This warp is also visible in the H I radio data. The radial extent of neutral hydrogen exceeds that of the optical disc by about 15 per cent. No extraplanar radio structures are visible; only at the upper edge of NGC 4565 are the radio isophotes shifted in the direction of the small satellite IC 3571. This dwarf galaxy with a size of 7000 light years is located 5.8′ above the centre of NGC 4565 and has almost the same recession velocity. Warped disc are observed in about 50 per cent of spiral galaxies. This phenomenon is more common in cluster galaxies, and thus in all spiral galaxies of the Local Group. A possible cause is the interaction with a small satellite such as IC 3571, causing oscillations in the disc. Mathematical models indicate that there are two basic modes of oscillation in the disc which in superposition explain the warp well.

OBJECT	NGC 4565
CONSTELLATION	Coma Berenices
RIGHT ASCENSION	$12^h\ 36^m\ 21^s$
DECLINATION	+25° 59′ 16″
BRIGHTNESS	10.4 mag
TYPE	SA(s)b
PHOTOGRAPHERS	Richard Müller
TELESCOPE	317-mm reflector
CAMERA	Artemis 4021
EXPOSURE TIME	300 min
LOCATION	Lohmar, Germany

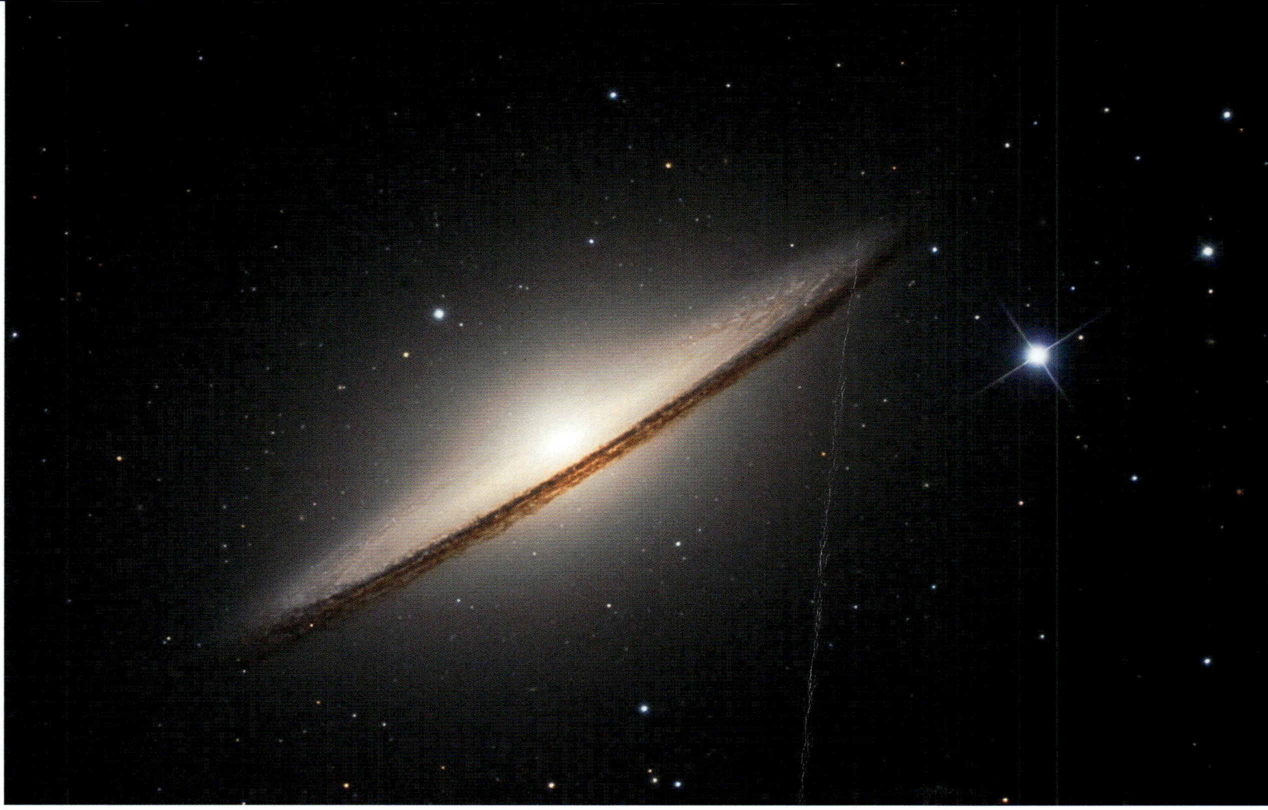

M 104, The Sombrero Galaxy

The SA(s)a galaxy M 104, known as the Sombrero Galaxy, is 34 million light years from the Milky Way. Its angular diameter of 8.4′ corresponds to a distance of 83 000 light years. M 104 is a transitionary type between early- and late-type spirals with a few special features. One of these is the large, prominent bulge, which contains an active, supermassive black hole with a mass of 109 solar masses. The bright light of the bulge illuminates the northern side of the dust ring, tilted by about 6° to the line of sight. M 104 has a large bulge-to-disc ratio of about six, even larger than the typical values of three to four for S0 galaxies. For comparison, this ratio between the stellar masses of the bulge and disc is only 1 to 4 for Sc galaxies.

Detailed infrared observations of the dust ring in M 104 with the Spitzer Space Telescope revealed both a significant warp and compact regions at the edges which indicate regions of star formation. Interpreting the optical image in light of the infrared observations, the left side of the dust ring is convex and the right side concave. This also explains why the illumination by the bulge extends farther on the right. Further observations show that the rotation curve of M 104 indicates an increasing mass-to-light ratio with increasing radius and thus a massive dark-matter halo. Kinematic studies of the 2000 globular clusters in M 104 indicate that their radial velocities in the halo lie between 1000 and 1200 km/s. However, these are not orbital motions of the globular clusters, but rather uncoordinated motions in the halo which are unrelated to the rotation of the disc. This could be a hint that the globular clusters did not form with the disc, but rather were accreted by the halo in several mergers. This explanation agrees with the fact that the Sombrero Galaxy is one of the most massive galaxies in the southern part of the Virgo Cluster. Its total mass of 8×10^{11} solar masses is 4.4 times as large as the mass of the Milky Way, although its diameter is only 75 per cent of that of the Milky Way. The high density of M 104 is also indicated by the supermassive black hole with a mass of up to 10^9 solar masses which is suspected at the centre.

OBJECT	M 104
CONSTELLATION	Virgo
RIGHT ASCENSION	$12^h\ 39^m\ 59^s$
DECLINATION	−11° 37′ 23″
BRIGHTNESS	9.0 mag
TYPE	SA(s)a
PHOTOGRAPHERS	Rainer Sparenberg, Stefan Binnewies
TELESCOPE	600-mm reflector
CAMERA	SBIG ST-10XME
EXPOSURE TIME	185 min
LOCATION	Amani Lodge, Namibia

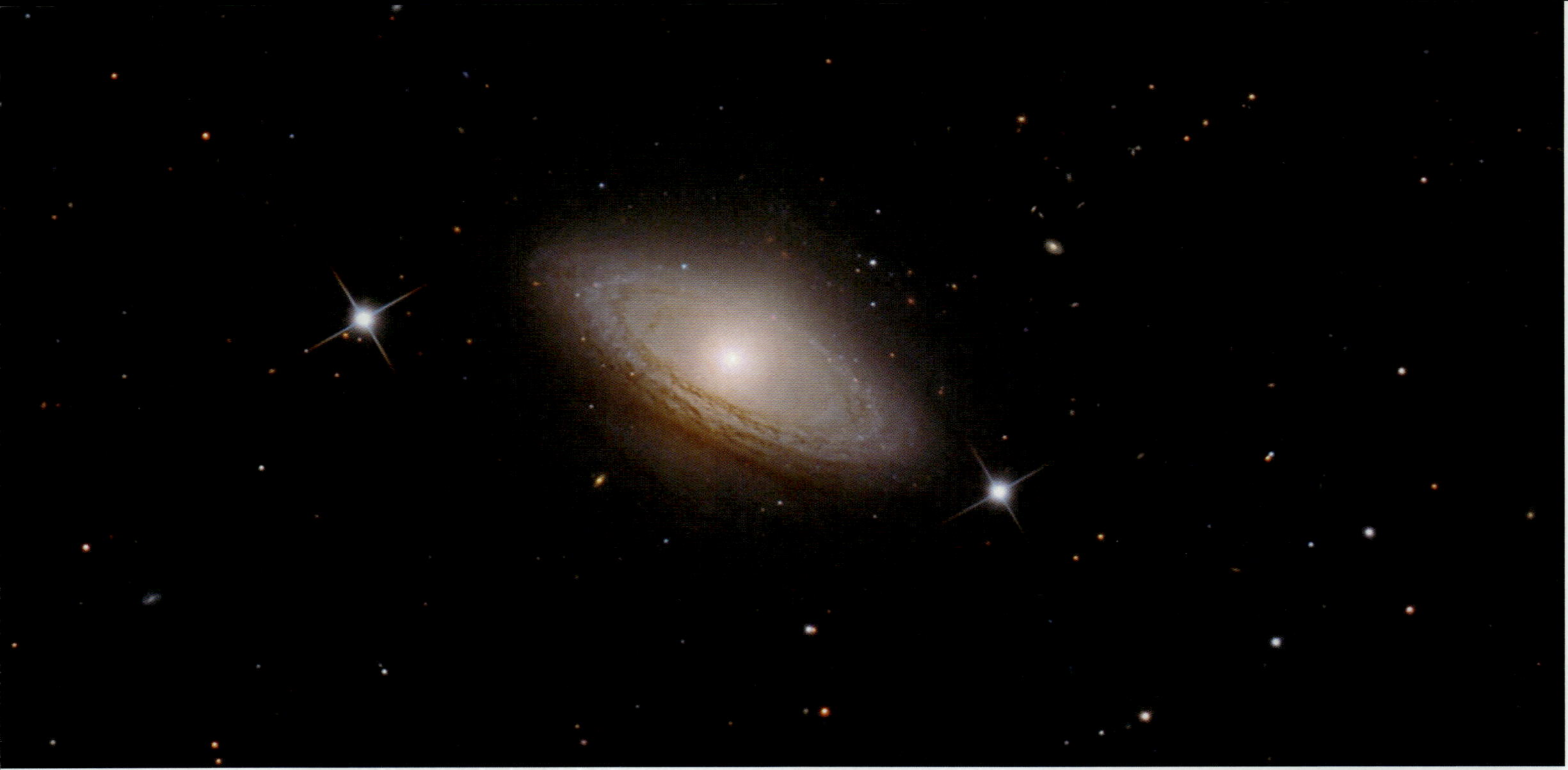

NGC 4698

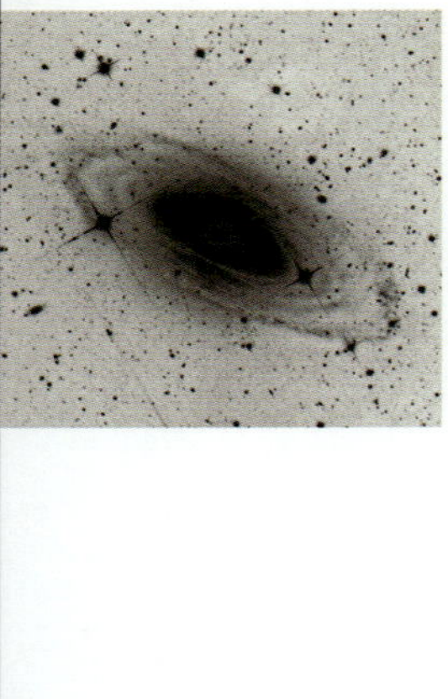

The SA(s)ab galaxy NGC 4698 is located at the southeastern edge of the Virgo Cluster. It is difficult to confirm the membership of NGC 4698 in this cluster. On the one hand, the recession velocity corresponds to a distance of 43 million light years, which would mean that NGC 4698 lies in front of the Virgo Cluster. Alternative distance measures place the galaxy at 77 million light years and hence behind the Virgo Cluster. NGC 4698 has a bright core which is embedded in a diffuse spherical bulge. It is surrounded by a fragmented spiral-arm pattern whose dust lanes contrast with the background light of the bulge. The angular size of this dust formation is 3.2′. A dim outer disc extends from its edges, reaching an angular size of 8.5′. In this outer disc, one can recognize prominent light blue rings with active regions. The brightest concentration is visible in the image at the right edge, the northern side of the disc. The high-contrast image shows even a fine granulation in this bright area. In the Palomar Observatory Sky Survey (POSS) image of NGC 4698, the outer disc is completely invisible.

An analysis of the isophotes of NGC 4698 indicates a disc inclination of 52° and a phase angle of 170°. In a detailed photometric study of the core region of NGC 4698, astronomers were able to use distortions in the isophotes to detect a second, nuclear disc which at 77° is more tilted and whose phase angle of 71° departs from that of the outer disc. This multipart construction in the central regions is confirmed by further kinematic observations. Additionally, the rotation axis of the bulge lies within the plane of the disc. Thus, the angular momenta of bulge and disc are perpendicular. These lead to the assumption that NGC 4698 must have formed in several steps. At the beginning, two spiral galaxies could have merged to form an S0/E galaxy. The accretion of matter lost by the precursor galaxies could have led to the formation of the outer disc. This scenario would also explain the very large disc diameter, which extends as much as 105 000–190 000 light years, depending on its true distance.

OBJECT	NGC 4698
CONSTELLATION	Virgo
RIGHT ASCENSION	$12^h\ 48^m\ 23^s$
DECLINATION	+08° 29′ 15″
BRIGHTNESS	11.6 mag
TYPE	SA(s)ab
PHOTOGRAPHERS	Adam Block
TELESCOPE	800-mm reflector
CAMERA	SBIG STX-16803
EXPOSURE TIME	950 min
LOCATION	Mount Lemmon SkyCenter/ University of Arizona, USA

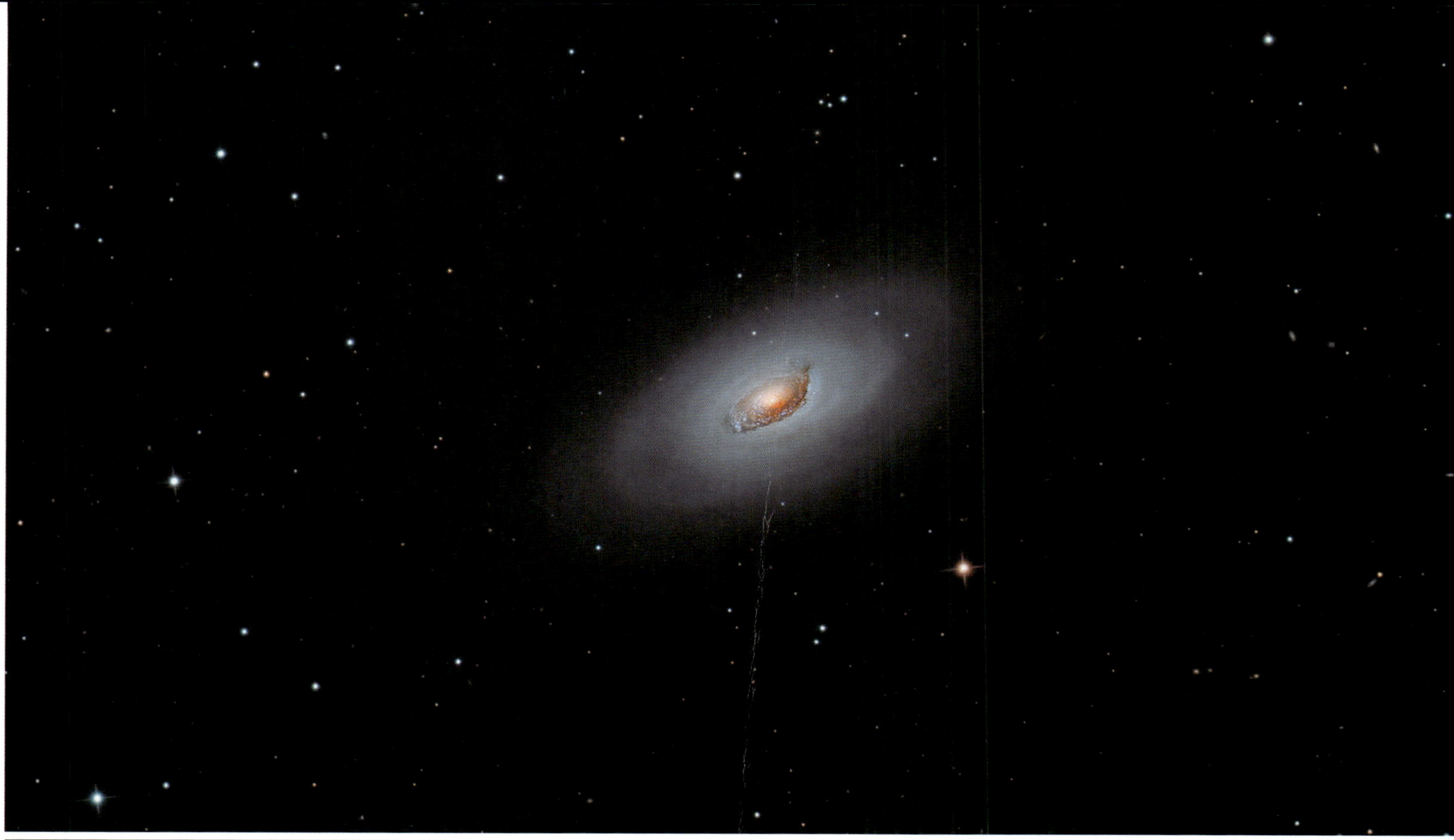

M 64, The Black Eye Galaxy

M 64 (NGC 4826) is known as the Black Eye Galaxy. This describes well the division of the disc into an inner region with spiral-arm fragments with active regions and, separated by a wide dust lane, an outer region which shows no spiral pattern. This outer disc is marked by a diffuse blue glow, with various details visible in the image. For example, the brightness doesn't decrease smoothly from inside to out, but rather in steps in at least three ring zones. On the right below the dust lane in the inner, brightest zone one can see a spiral-shaped concentration which appears to have weak dust structures at the right edge. The outer disc of M 64 has an angular size of 10.0′ × 5.4′; the angular diameter of the inner part is 1.8′. M 64 lies at a distance of 17 million light years, which results in a projected diameter for the disc of 49 000 light years and for the inner part of 9000 light years.

Radio observations show an additional special feature of this spiral galaxy. The measured velocity field of rotating atomic hydrogen covers the range 225–595 km/s and shows an appreciable difference in the direction of rotation between the innermost edge and the rest of the gas disc. While in the inner parts the gas rotates together with the stars and interstellar medium, the gas farther out moves in the opposite direction. The inner edge of this gas disc corresponds to the inner part of the optical disc which is bordered by the dust lane. The outer gas disc, however, is larger than the optical disc and the neutral-hydrogen disc reaches a diameter of 25′. Astronomers interpret this as the result of a billion-year-old merger or accretion of a dusty spiral galaxy which was rotating in the opposite direction to M 64.

OBJECT	M 64, NGC 4826
CONSTELLATION	Coma Berenices
RIGHT ASCENSION	$12^h\ 56^m\ 44^s$
DECLINATION	+21° 40′ 59″
BRIGHTNESS	9.4 mag
TYPE	(R)SA(rs)ab
PHOTOGRAPHERS	Rainer Sparenberg, Josef Pöpsel
TELESCOPE	600-mm reflector
CAMERA	SBIG STL-11000
EXPOSURE TIME	180 min
LOCATION	Skinakas Observatory, Crete, Greece

NGC 4945

At a declination of almost 50° south between two stars of magnitude 4 and 5 (SAO 223909, SAO 223870) separated by 40′ in the constellation Centaurus one finds the galaxy NGC 4945. This SB(s)cd galaxy belongs to the Centaurus Group and has an enormous angular size of 20′ × 9.3′, which at a distance of 13 million years corresponds to a disc diameter of 76 000 light years. At a distance of 1° are the galaxies NGC 4945A, NGC 4976 and PGC 45373 which are not visible in the image. The mean distance of these is four times farther from the Milky Way than NGC 4945. At the left edge of the disc in the image is the galaxy PGC 45371 at a distance of 380 million light years. About 4′ above the core of NGC 4595 is the galaxy LEDA 3097829 (CFC97 Cen 05) which is still often described as a dwarf galaxy in the Centaurus Group. However, the image clearly identifies LEDA 3078929 as a barred spiral galaxy, and assuming a typical size leads to a distance estimate of 350–400 million light years.

At the upper (northwest) edge of the disc is a prominent red star which is the supernova SN2005af, which exploded in NGC 4945 about 4 months before the time the image was taken. When discovered on 8 February 2005 its visual magnitude was 8.4 and its spectrum indicated that SN2005af was a type II supernova. Supernovae of type II are due to the collapse of a massive star of at least eight solar masses in which nuclear fusion in the core has ceased. The resulting lack of radiation pressure leads to gravitational collapse in which the outer shells fall onto the iron core, leading to thermonuclear reactions. The core is further compressed, which can lead to a neutron star or black hole. The hydrogen-rich envelopes of the precursor impact onto the compact core and rebound at speeds of several 1000 km/s. During this process, heavier elements are formed via neutron- and proton capture. The spectrum of SN2005af shows the hydrogen emission lines characteristic of type II supernovae. The luminosity of SN2005af did not decay linearly, but rather passed through a plateau of approximately constant brightness lasting about three weeks. This leads to SN2005af being classified as subtype II-P. In this case, the high kinetic energy of the envelope is compensated by expansion-induced cooling, which delays the subsequent drop in brightness. Interestingly, type II supernovae occur almost only in spiral galaxies. Astronomers believe that this is because the precursor stars are of population I, which are found in spiral galaxies.

OBJECT	NGC 4945
CONSTELLATION	Centaurus
RIGHT ASCENSION	$13^h\ 05^m\ 27^s$
DECLINATION	−49° 28′ 06″
BRIGHTNESS	9.3 mag
TYPE	SB(s)cd
PHOTOGRAPHERS	Josef Pöpsel, Beate Behle
TELESCOPE	600-mm reflector
CAMERA	SBIG ST-10XME
EXPOSURE TIME	120 min
LOCATION	Amani Lodge, Namibia

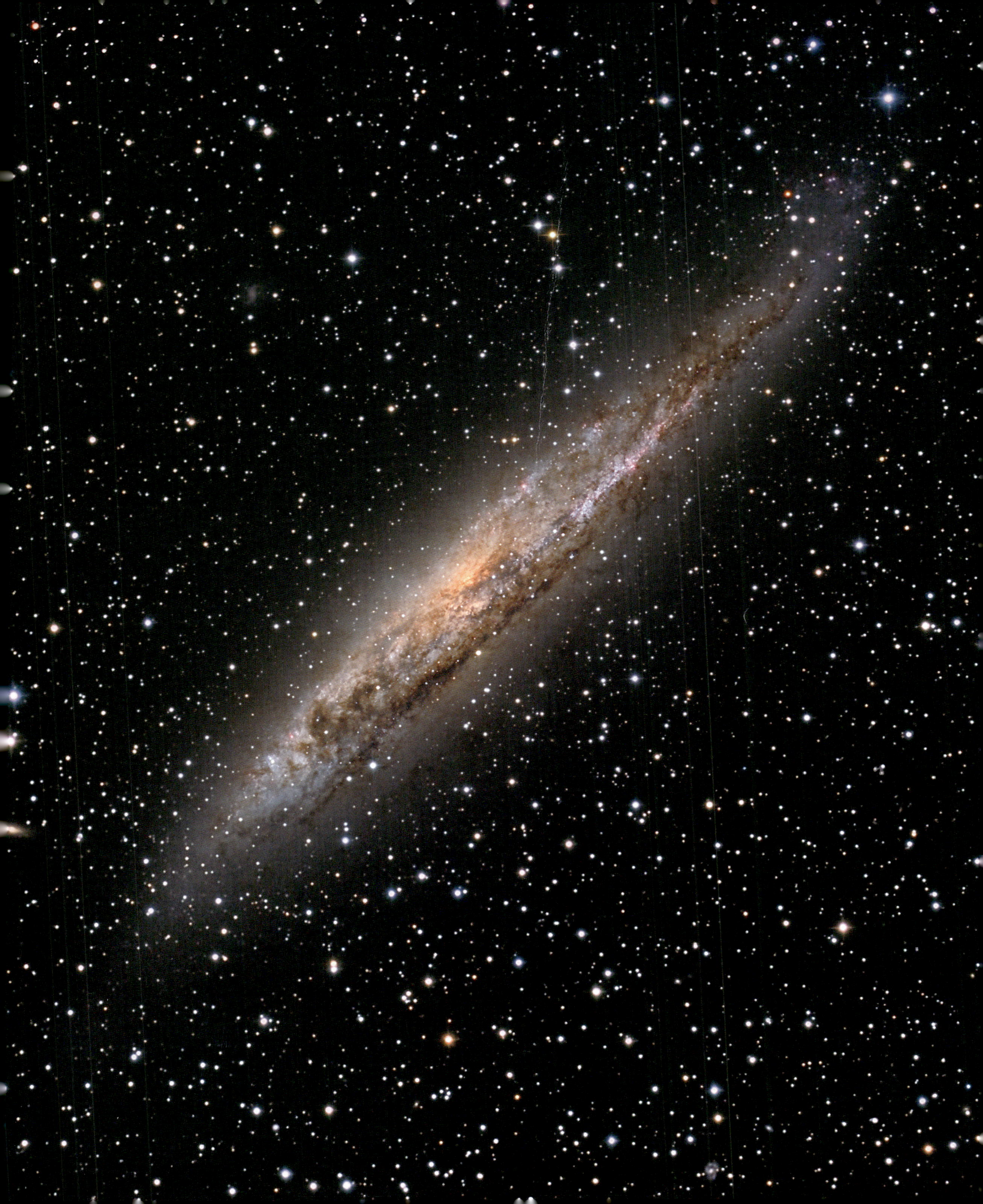

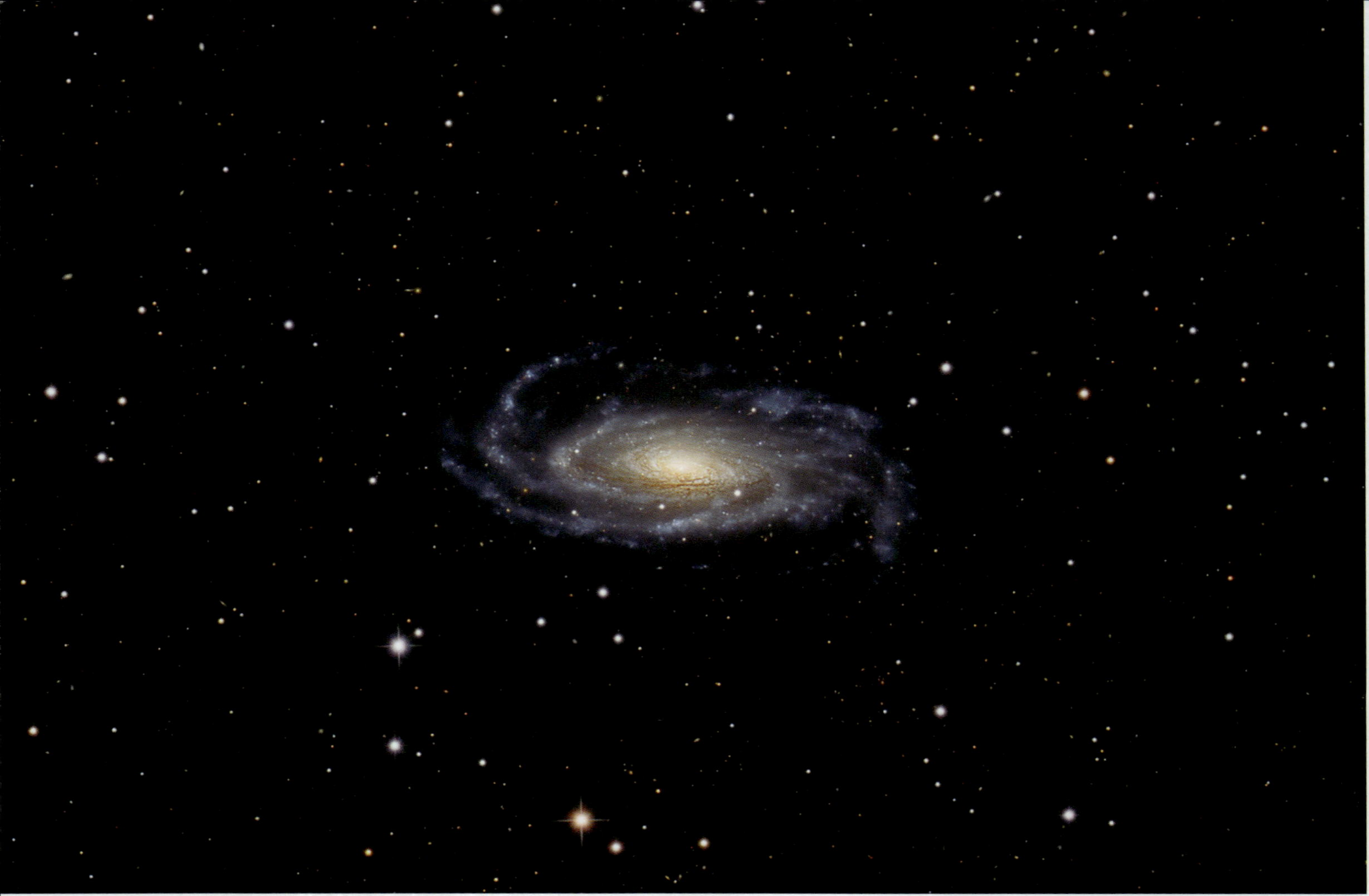

NGC 5033

The spiral galaxy NGC 5033 is classified as SA(s)c and its spiral arms have a very pronounced structure, especially in the outer parts. The prominent blue spiral arms with many active regions define an angular size of 10.7′ × 5.0′, which, at NGC 5033's distance of 40 million light years, corresponds to a projected disc diameter of 125 000 light years. The compact, luminous core is surrounded by thick dust lanes which connect to the prominent spiral arms. Some studies suggested that the visible asymmetry of the dust lanes was due to several bars in the core, but this suspicion turned out to be wrong. At the lower edge of the disc one can see the begin of a tidal tail which can be followed further to the left as a dim trail.

The centre of NGC 5033 contains a Seyfert-1 galaxy core whose spectrum displays the typical broad emission lines. The relative displacement of these lines, due to the Doppler effect, demonstrates the dynamic nature of the emission regions. The ionized gas forms a double cone which astronomers explain by the central ionization source being partially covered by dense matter. One has also found that in NGC 5033 the centre of this emission differs from the centre of rotation of the disc. This could be due to a past merger with a lower-mass, compact satellite, which could also explain the warped disc and the spiral structure in NGC 5033.

OBJECT	NGC 5033
CONSTELLATION	Canes Venatici
RIGHT ASCENSION	$13^h\ 13^m\ 27^s$
DECLINATION	+36° 35′ 38″
BRIGHTNESS	10.8 mag
TYPE	SA(s)c
PHOTOGRAPHERS	Johannes Schedler
TELESCOPE	400-mm reflector
CAMERA	SBIG STX-16803
EXPOSURE TIME	980 min
LOCATION	Wildon, Austria

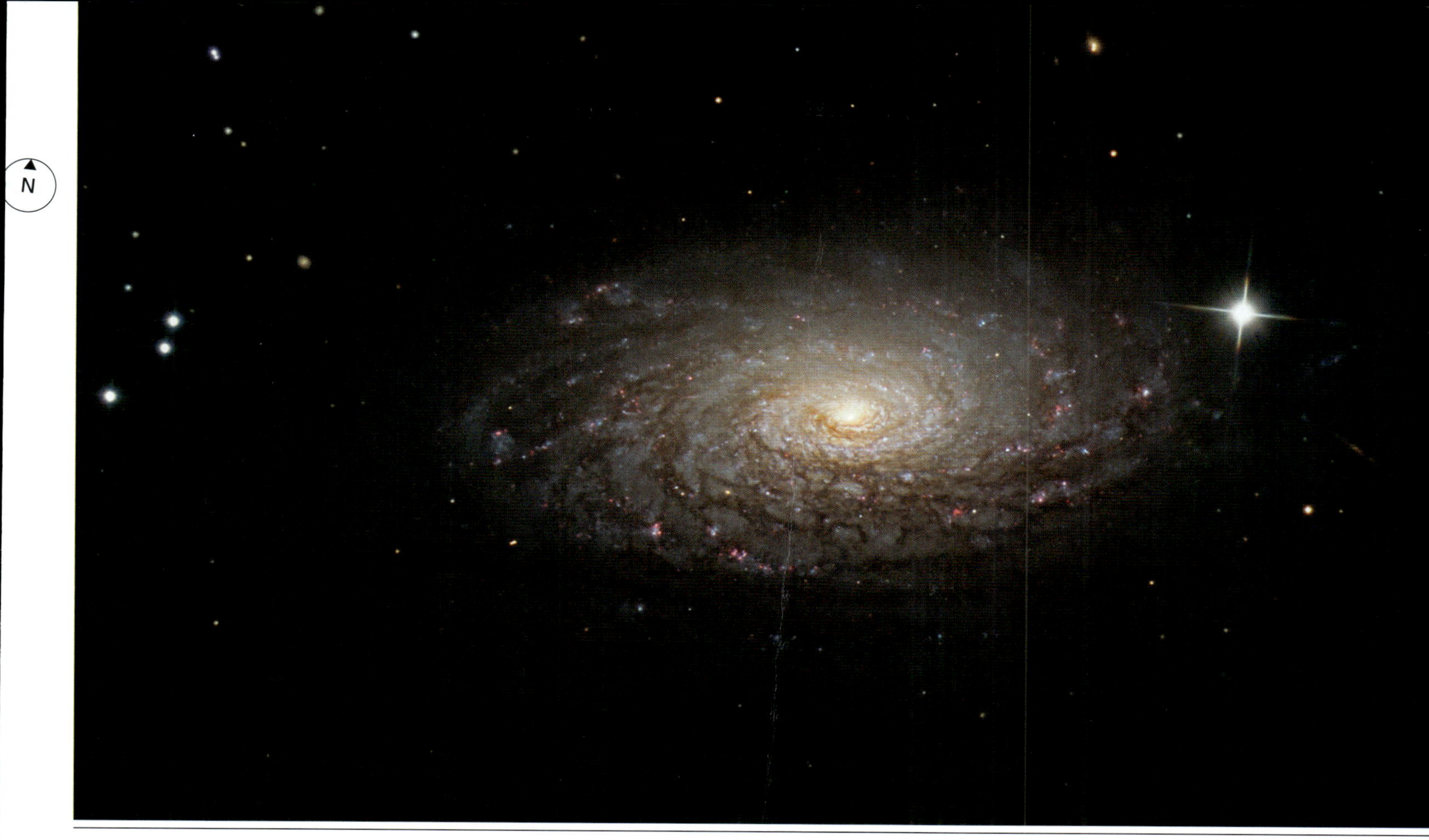

M 63

The SA(rs)bc galaxy M 63 lies at a distance of 27 million light years from the Milky Way. Its angular size of 12.6′ × 7.2′ corresponds to projected lengths of 99 000 × 57 000 light years. Near the northwestern edge of the disc is the 9.3-magnitude star SAO 44530 and close by, as well as along the entire edge of the disc, one can see spiral-arm filaments which indicate that the disc has a dim extension. At the southeastern edge of the disc are almost horizontal dust lanes which are presumably not part of the disc but rather in front of it, made visible by backlighting. Radio observations of M 63 confirm the existence of a large disc of atomic hydrogen surrounding the galaxy. This gas disc has a radius of 150 000 light years and is thus more than twice as large as the optical radius. From measurements of gas velocities, astronomers were led to a model of a strongly warped disc whose outer regions are tilted by 20° compared to the inner, optical part.

The spiral pattern of M 63 is mostly fragmented and doesn't have a uniform luminosity. Such flocculent patterns are seen in about a quarter of all spiral galaxies. In contrast to the density-wave theory described at the beginning of this chapter, these small-scale structures are due to local mechanisms. Models assume that supernovae explode in the star-formation regions and their shock fronts induce star formation in the neighbouring regions. This continues and the differential disc rotation finally leads to these fragmented, active regions spreading out over the disc. There is, however, no global mode as in the density-wave model.

OBJECT	M 63
CONSTELLATION	Canes Venatici
RIGHT ASCENSION	$13^h\ 15^m\ 49^s$
DECLINATION	+42° 01′ 45″
BRIGHTNESS	9.3 mag
TYPE	SA(rs)bc
PHOTOGRAPHERS	Stefan Binnewies, Josef Pöpsel
TELESCOPE	600-mm reflector
CAMERA	SBIG STL-11000
EXPOSURE TIME	345 min
LOCATION	Skinakas Observatory, Crete, Greece

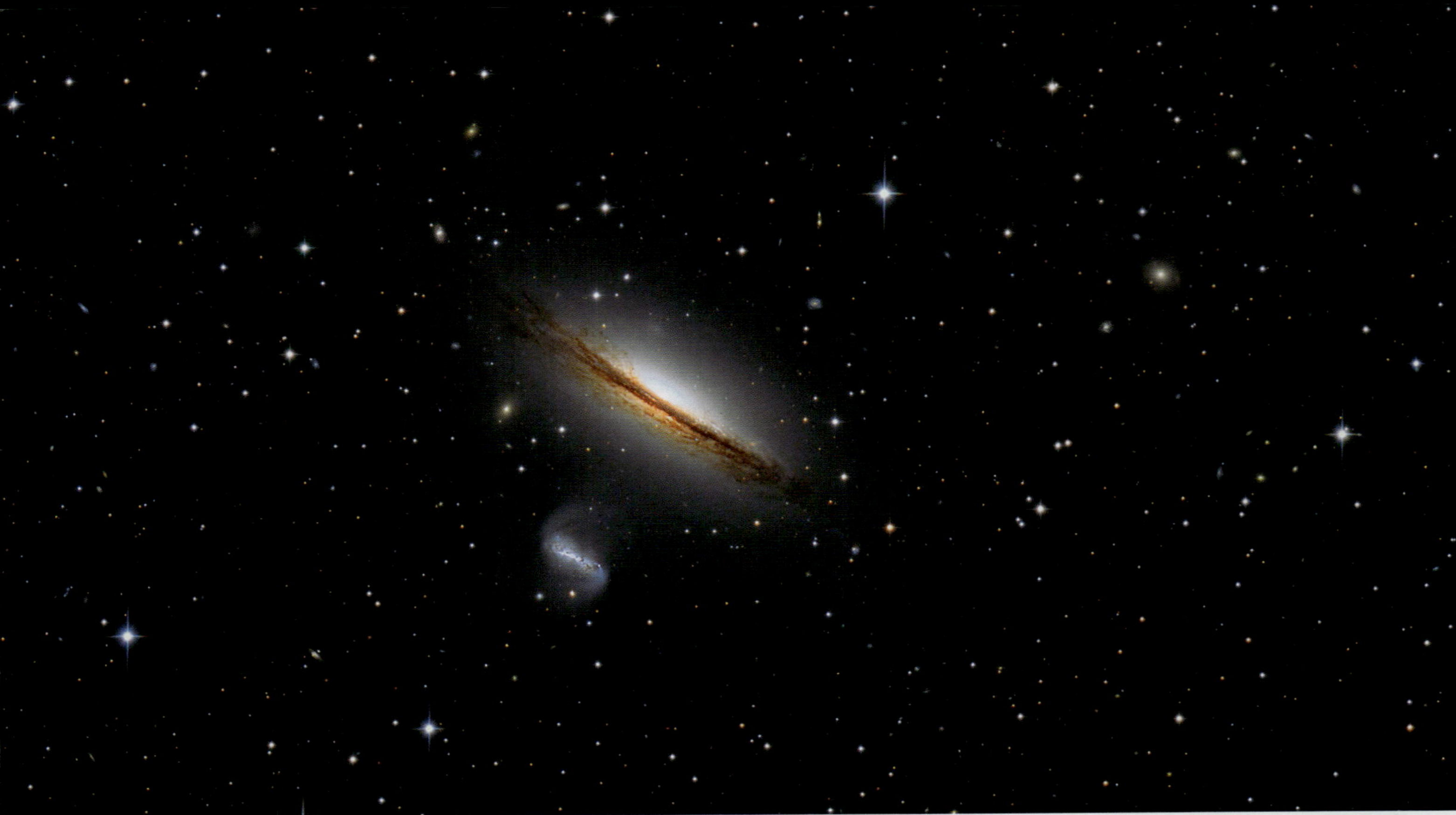

NGC 5078

The SA(s)a galaxy NGC 5078 is located in the southern hemisphere just a few degrees north of the Centaurus Cluster. At a distance of 90 million light years, it is spatially separated from this elongated galaxy cluster whose members are at distances of 140–200 million light years from the Milky Way. NGC 5078 is dominated by a large, unusual bulge. The angular size of the bright bulge ellipsoid is 4.0′ × 1.9′ which corresponds to a transverse diameter of 105 000 light years and a thickness of the bulge of 50 000 light years. The dust lanes, whose detailed silhouette one can follow in front of the diffuse luminous background, extend beyond the bright bulge. They have a span of 6.4′ which means that the diameter of the optically invisible outer disc is 170 000 light years.

The SB galaxy IC 879 located at a distance of 2.3′ to the west (here below left) has a recession velocity 220 km/s higher than that of NGC 5078. Although this velocity difference would also be compatible with spatial proximity with opposite relative motions, several arguments are in favour of IC 879 being about ten million light years farther away than NGC 5078. In the former case, the size of IC 879 would be 35 000 light years and thus IC 879 would be only about 40 per cent the size of an average SB galaxy. The envelope of NGC 5078 shows neither a perturbed morphology nor a variation in brightness in the direction of IC 897, which would be expected as the first sign of an interaction. No bridge of material is visible between the two galaxies in the radio. These data show that NGC 5078 has, in addition to the disc component, a second component perpendicular to the disc visible in the radio out to about 0.6′ on both sides. This could be explained either by weak radio jets or strong stellar winds which transport matter out of the disc and stimulate radio emission.

OBJECT	NGC 5078
CONSTELLATION	Hydra
RIGHT ASCENSION	$13^h\ 19^m\ 50^s$
DECLINATION	−27° 24′ 37″
BRIGHTNESS	11.8 mag
TYPE	SA(s)a
PHOTOGRAPHERS	Philipp Keller, Konstantin Buchhold, Bernd Flach-Wilken, Johannes Scheder, Volker Wendel (Chart 32-Team)
TELESCOPE	800-mm reflector
CAMERA	FLI Proline 16803
EXPOSURE TIME	980 min
LOCATION	CTIO, Chile

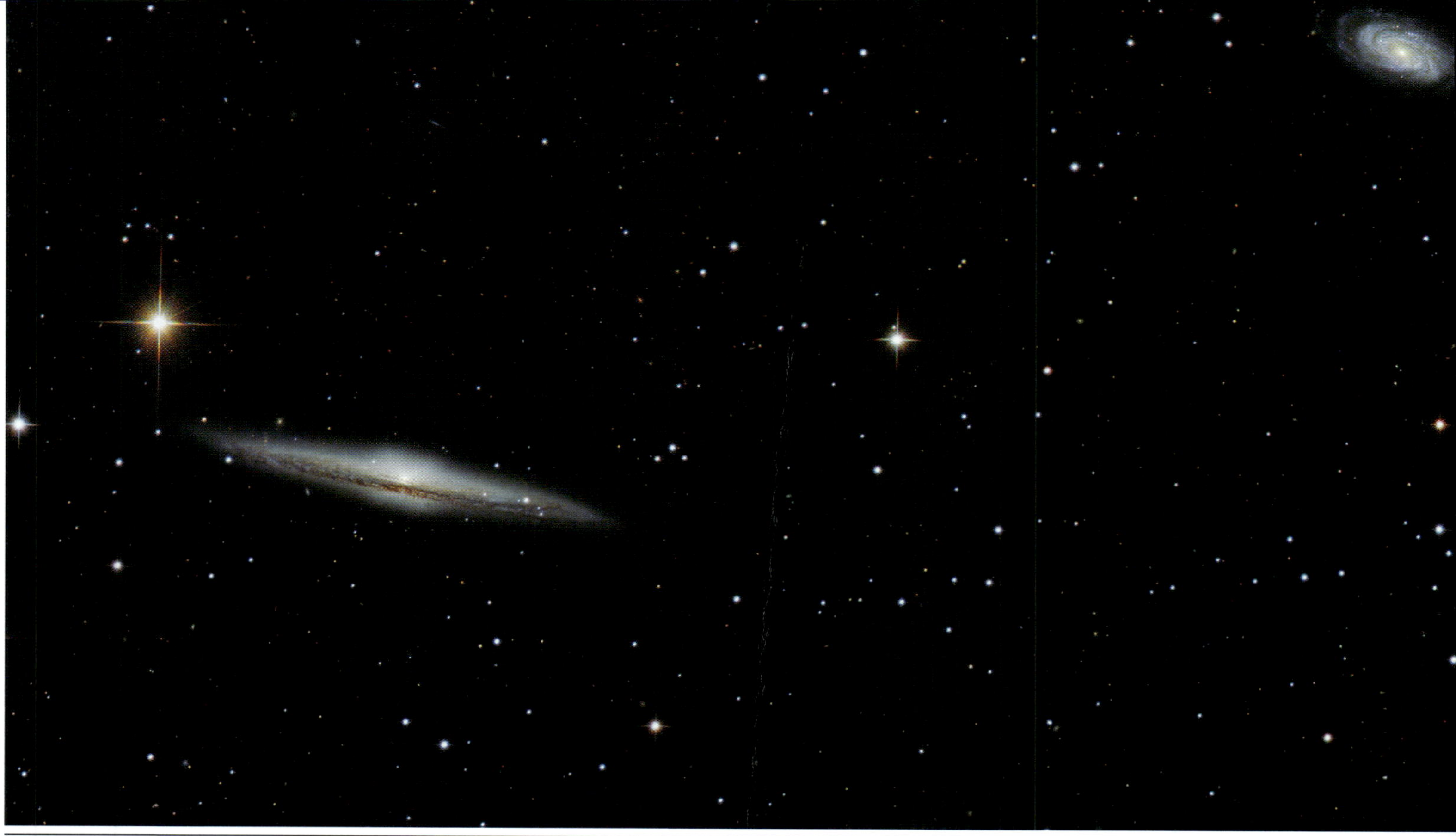

NGC 5746

At the left side of the image, near an 8.5-magnitude star (SAO 120633), lies the SAB(rs)b galaxy NGC 5746. Its recession velocity is 1724 km/s, which corresponds to a distance of 77 million light years. At a distance of 18′ to the southwest, at the upper right of the image, is the galaxy NGC 5470, which in catalogues is assigned the same classification. The smaller galaxy NGC 5470 is moving away at 1572 km/s. Relative to NGC 5746, NGC 5740 is moving at +152 km/s towards the Milky Way. If one assumes that both galaxies are at the same distance, then NGC 5740 is only one-third the size of NGC 5746. This large galaxy has neither starburst regions nor an active nucleus. The transverse disc diameter of NGC 5746 is 165 000 light years and its most prominent characteristic is the boxy bulge.

Studies of the structure of spiral galaxies have shown that there is a relation between the bulge and the disc. This shows up in the bulge-to-disc brightness ratio, which is greatest for early-type spirals. In late-type galaxies (Sc,d) with smaller, compact bulges, the surface brightness of the bulge is about 2–3 mag smaller than in S0/Sa types. The brightness distribution in Sb/Sc types can be best modelled if both bulge and disc have exponential luminosity profiles. One assumes that dominant bulges with a $1/(r^4)$ luminosity profiles are due to mergers. Smaller bulges, in comparison, evolve along with the disc. The example NGC 5746 shows that instabilities due to bar structures play an important role in this process. The boxy bulge of NGC 5746 is thus interpreted by astronomers as a bar observed from the end. If one could view NGC 5746 face-on, it would presumably display a form similar to that of NGC 5740.

OBJECT	NGC 5746
CONSTELLATION	Virgo
RIGHT ASCENSION	$14^h\ 44^m\ 56^s$
DECLINATION	+01° 57′ 18″
BRIGHTNESS	11.3 mag
TYPE	SAB(rs)b
PHOTOGRAPHERS	Josef Pöpsel, Stefan Binnewies
TELESCOPE	600-mm reflector
CAMERA	SBIG STL-11000
EXPOSURE TIME	450 min
LOCATION	Skinakas Observatory, Crete, Greece

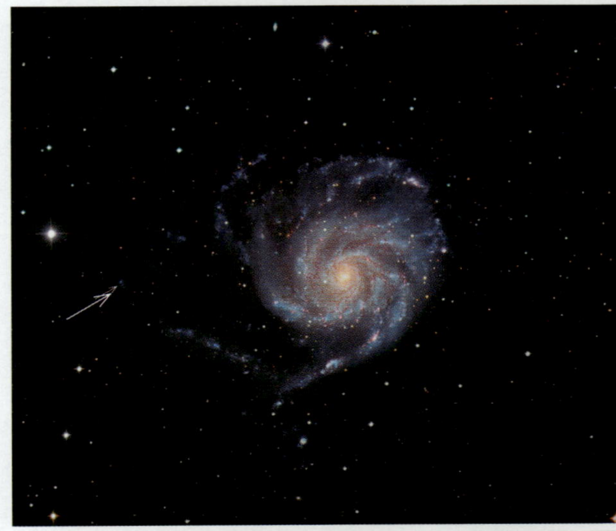

The H II region NGC 5471

M 101, The Pinwheel Galaxy

M 101 (NGC 5457) is an SAB(rs)cd galaxy which is part of a group of nine relatively large galaxies at a mean distance of 24 million light years from the Milky Way. This galaxy group is similar to the Local Group, except that it is dominated by only one large spiral galaxy, M 101. M 101 has a grand-design spiral pattern which explains the name "Pinwheel Galaxy". The enormous angular size of 28′ × 26′ also includes the dim outer parts of the spiral arms and corresponds to a maximum disc diameter of 200 000 light years. In comparison, the image covers 160 000 light years. The longest spiral arm in M 101 can be followed through a full revolution and has two prominent Vorontsov-Velyaminov rows in the lower, eastern part of the disc. In the penultimate row one can recognize the bright H II regions NGC 5461 and NGC 5462. The second, northern spiral arm of M 101 is to the left of the centre and isn't as long as the first one. Vorontsov-Velyaminov rows occur in this arm as well, but are shorter due to the smaller distance from the centre. The penultimate row in this spiral arm also has more active regions than the last one. In the upper right, southwest quadrant of the disc are spiral-arm fragments which are surrounded by the bright H II regions NGC 5447, NGC 5450 and NGC 5455.

The bright H II regions are powered by star clusters with young, massive stars. These giant extragalactic H II regions (GEHR) in M 101 are important indicators of star formation and thus for the evolutionary state of the galaxy. One of the brightest regions, NCG 5471, is marked in the small image. NGC 5471 has a diameter of 17″ and a substructure of 5 bright knots. With a diameter of 2000 light years, NGC 5471 is comparable to NGC 604 in M 33 and one of the largest GEHRs. Using a combination of UV, optical, and infrared data, astronomers have arrived at two solutions, depending on the assumed amount of dust, for the stellar age in NGC 5471: For an average amount of extinction one obtains an age of eight million years, for an extremely low amount 60 million years. This simple model with a single population is, however, too simple for the complex stellar-evolution history in NGC 5471. The colour–magnitude diagram displays OB main-sequence stars with an age of only four million years and redder stars with an age of 15–70 million years. The activity in NGC 5471 is concentrated in the last 100 million years, with a maximum 20 million years ago. The star-formation rate in NGC 5471 exceeds that of NGC 604 by more than a factor of two, which indicates a larger reservoir of gas which fuels star formation and thus justifies the classification as a GEHR.

OBJECT	M 101, NGC 5457
CONSTELLATION	Ursa Major
RIGHT ASCENSION	$14^{h}\ 03^{m}\ 13^{s}$
DECLINATION	+54° 20′ 56″
BRIGHTNESS	8.3 mag
TYPE	SAB(rs)cd
PHOTOGRAPHERS	Makis Palaiologou, Stefan Binnewies
TELESCOPE	600-mm reflector
CAMERA	SBIG STX-16803
EXPOSURE TIME	230 min
LOCATION	Skinakas Observatory, Crete, Greece

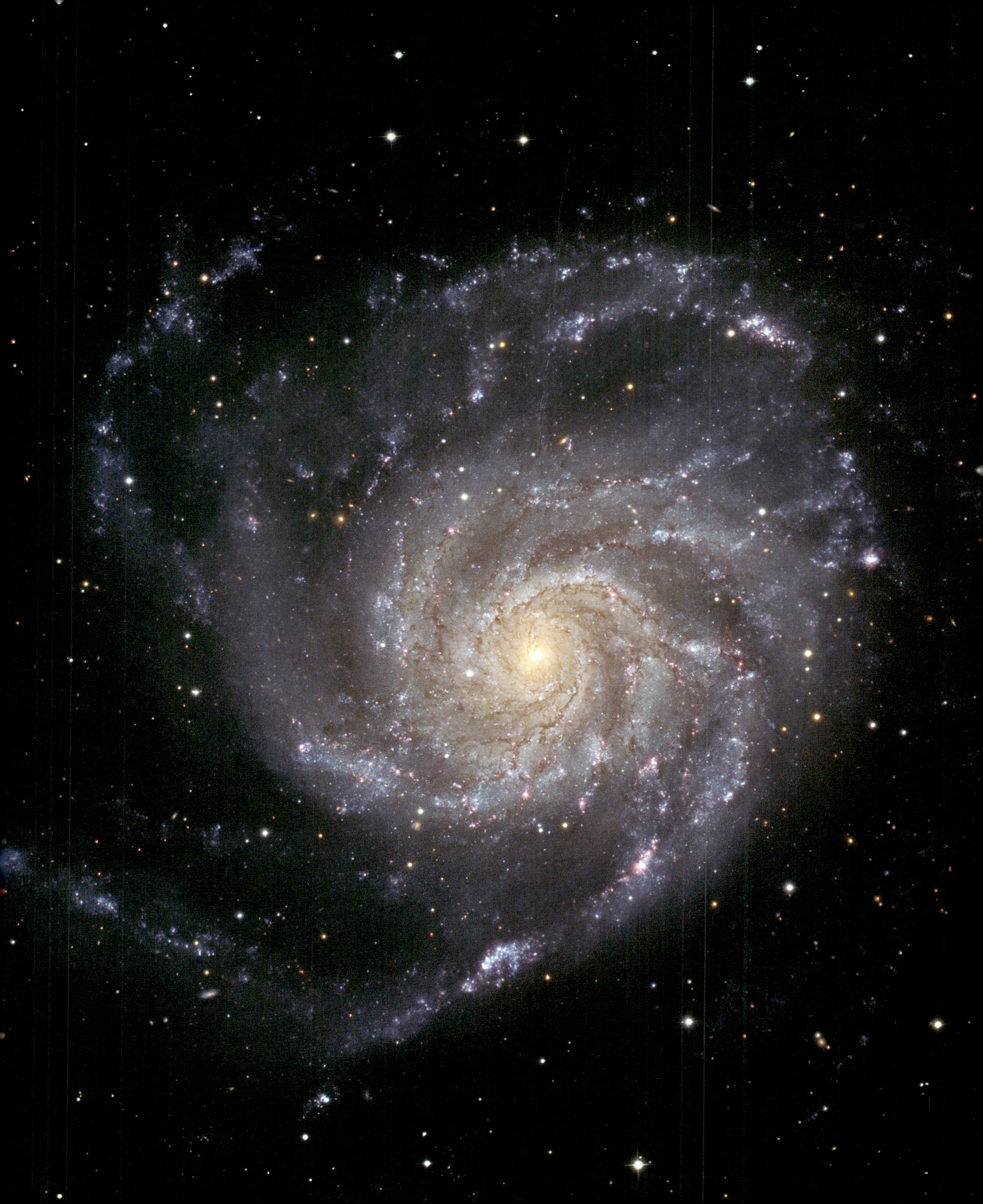

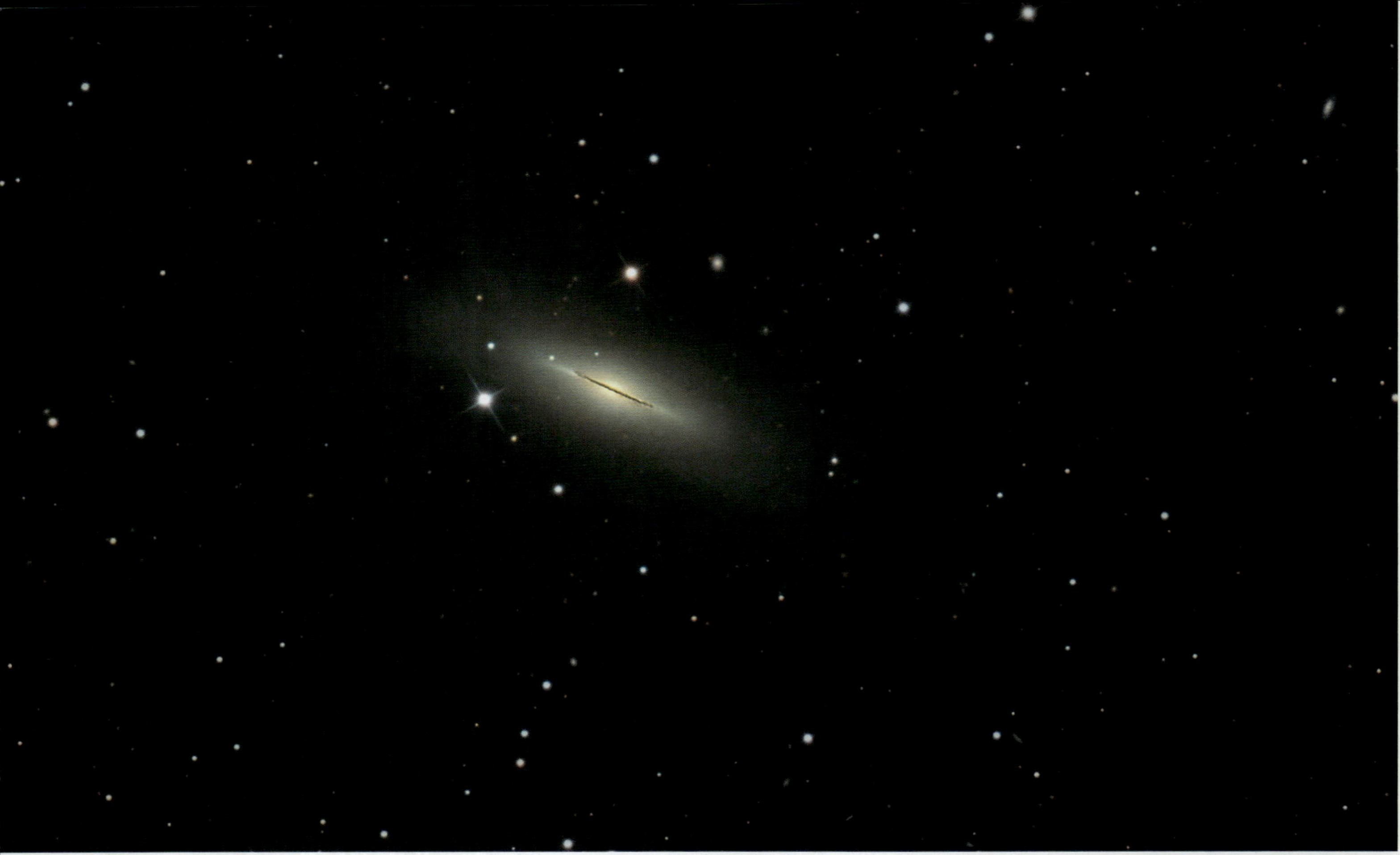

M 102

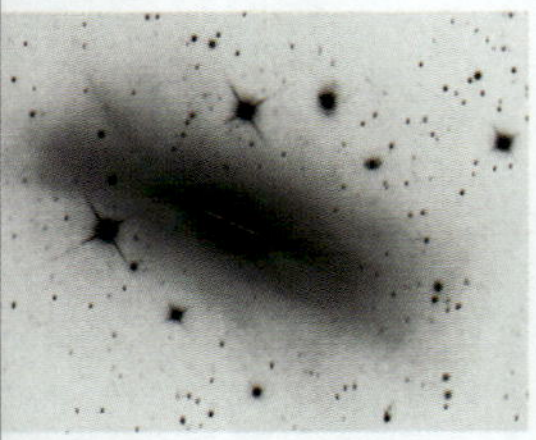

The S0 galaxy M 102 (NGC 5866) is 42 million light years from the Milky Way. The lens-shaped galaxy is seen edge-on and a narrow dust lane lies in front of the yellowish central bulge. This dust lane extends across 1′ and thus corresponds to 12 000 light years. At each end of the dust lane, long, slightly blue extensions are visible. These are a hint of a further, dim disc plane. Relative to this, there is a slight warp in the dust lane.

Images from the Hubble Space Telescope have shown that the dust lane in M 102, which is only 300 light years thick, contains dust fingers – long dust filaments perpendicular to the plane of the dust lane. These filaments are created by shock waves from supernovae and extend about 1000 light years, but are astronomically short-lived and collapse, leaving the extremely thin dust lane intact. More detailed studies of M 102 show that the dust lane is also stable with respect to external perturbations. The halo doesn't exhibit any spherical structure, but rather appears boxy, as if composed of several rectangles. These structures could be explained by tidal tails seen in perspective, formed by the merger of dwarf galaxies with NGC 5866. This is particularly prominent in an elongated luminous trail which emerges from the core at the above left and below and is easily seen in the negative, contrast-enhanced image of M 102.

OBJECT	M 102, NGC 5866
CONSTELLATION	Draco
RIGHT ASCENSION	$15^h\ 06^m\ 30^s$
DECLINATION	+55° 45′ 48″
BRIGHTNESS	10.7 mag
TYPE	S0
PHOTOGRAPHERS	Stefan Binnewies, Volker Wendel
TELESCOPE	600-mm reflector
CAMERA	SBIG STL-11000
EXPOSURE TIME	315 min
LOCATION	Skinakas Observatory, Crete, Greece

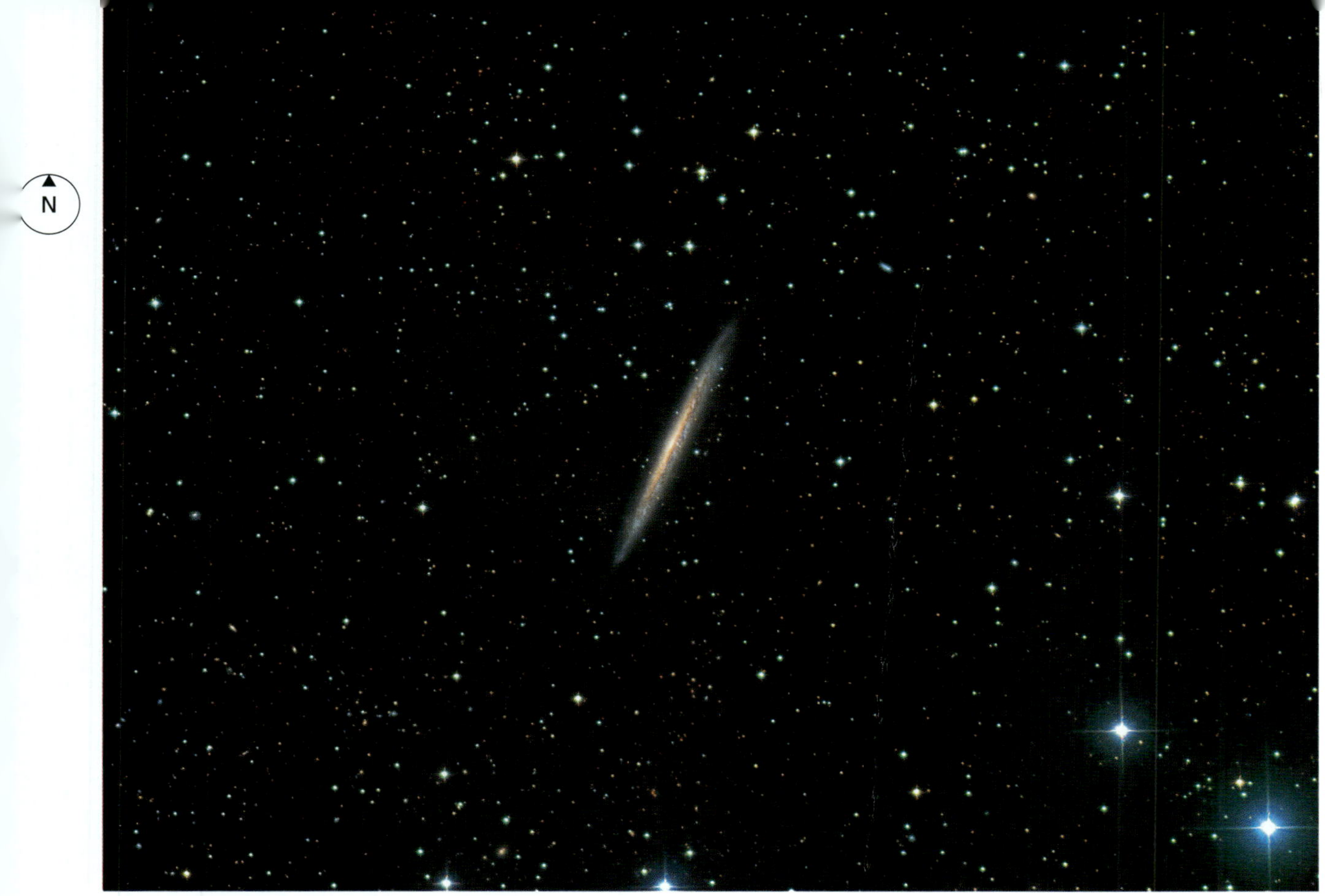

NGC 5907

The Sc galaxy NGC 5907 belongs to the NGC 5866 Group at a distance of about 40 million light years. Its almost exact edge-on orientation has given it the names "Splinter Galaxy" and "Knife-Edge Galaxy". The angular size of 12.8′ corresponds to a visible disc with a diameter of 149 000 light years. NGC 5907 has no bulge and the core is almost completely hidden by dust structures in the disc. The disc is warped and this deformation continues into the diffuse outer edges of the disc, giving the edge-on appearance of NGC 5907 the form of an integral sign. The rotation curve of neutral hydrogen in NGC 5907 is characterized by a constant plateau at 220 km/s and suggests a halo dominated by dark matter surrounding the galaxy.

In the year 2000, astronomers observed a dim, arc-like signature of a stellar stream in the halo of NGC 5907 and in 2008, with the help of deep images by amateur astronomers, the complete picture was revealed and studied astrophysically (see the negative, contrast-enhanced image). Simulations showed that these stellar streams can be explained by four orbits of a dwarf galaxy. During these orbits the dwarf galaxy lost more and more matter along its orbit and after 3.6 billion years it was completely dissolved. The orbit leads to a mass of about 2×10^8 solar masses for the dwarf galaxy, which corresponds to less than one per cent of the mass of NGC 5907. Further detailed studies of edge-on galaxies revealed that the dim tidal streams in NGC 5907 are not a special case. Irregularities were found in almost 20 per cent of edge-on galaxies, and six per cent had these well defined tidal streams in their halos.

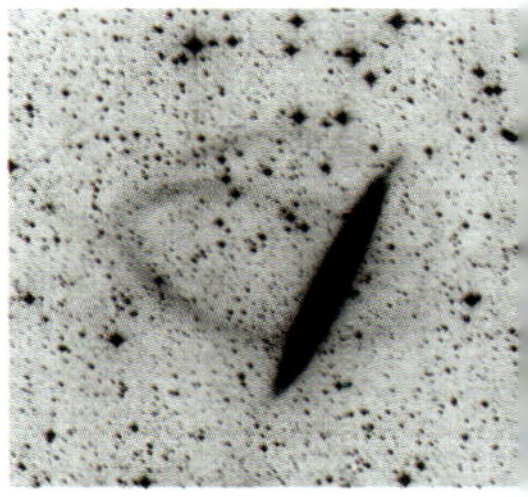

OBJECT	NGC 5907
CONSTELLATION	Draco
RIGHT ASCENSION	$15^h\ 15^m\ 54^s$
DECLINATION	+56° 19′ 44″
BRIGHTNESS	11.1 mag
TYPE	SA(s)c
PHOTOGRAPHERS	Stefan Binnewies, Josef Pöpsel
TELESCOPE	600-mm reflector
CAMERA	SBIG STL-11000
EXPOSURE TIME	360 min
LOCATION	Skinakas Observatory, Crete, Greece

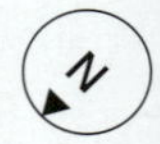

NGC 5963

The image shows at the left edge the two spiral galaxies NGC 5963 and NGC 5965, separated by only 9′ on the celestial sphere. The galaxy NGC 5963 is located to the southwest above NGC 5965, an edge-on Sb type. The small angular separation, however, doesn't correspond to physical proximity, since NGC 5963 is located at a distance of 85 million years from the Milky Way but NGC 5965 at 160 million. This results in an enormous disc diameter of 240 000 light years for NGC 5965 and a diameter of 100 000 light years for NGC 5963. NGC 5963 is a low-surface-brightness galaxy (LSBG). Its outer, blue spiral arms are very dim due to the weak stellar component there. The gas densities in LSBGs are lower than the values in normal spiral galaxies, making star formation in NGC 5963 very difficult. LSBGs also lack active regions with young, hot and thus massive stars. There are thus no places where heavy elements can be produced, which leads to a low metallicity. This is an indicator of the slow evolution of an LSBG.

At the right edge is the Sa galaxy NGC 5971, even farther away at a distance of 200 million light years, but of the same absolute size as NGC 5963. The contrast-enhanced negative image shows a tidal plume in NGC 5971, a mushroom-shaped tidal tail, presumably caused by interaction with a dwarf galaxy.

The rotation curve of the disc gas in NGC 5963 shows a constant value of ±120 km/s out to the edge of the disc and is

OBJECT	NGC 5963
CONSTELLATION	Draco
RIGHT ASCENSION	$15^h\ 33^m\ 28^s$
DECLINATION	+56° 33′ 35″
BRIGHTNESS	13.1 mag
TYPE	S pec
PHOTOGRAPHERS	Josef Pöpsel, Stefan Binnewies
TELESCOPE	600-mm reflector
CAMERA	SBIG STL-11000
EXPOSURE TIME	810 min
LOCATION	Skinakas Observatory, Crete, Greece

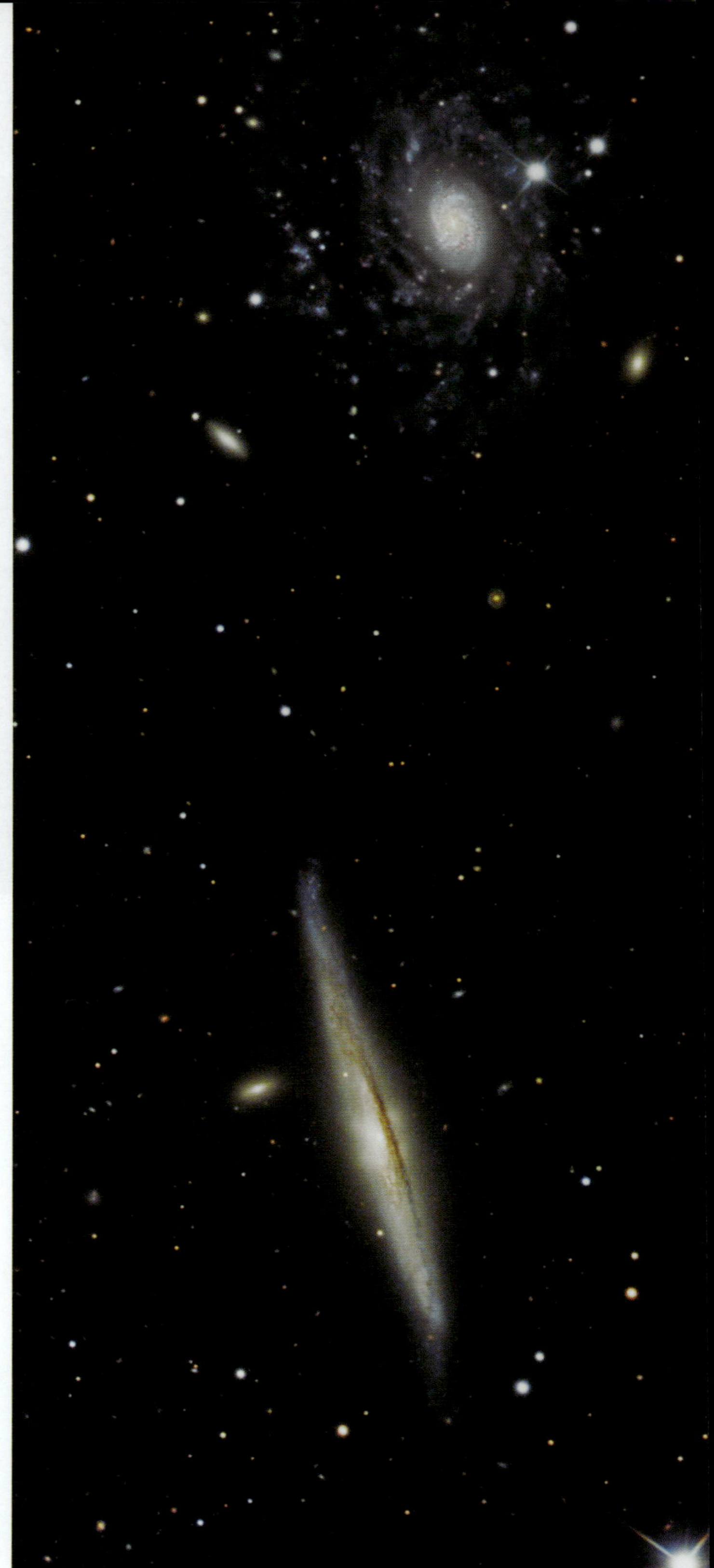

thus an indication of a large amount of dark matter. This typical property of LSBGs is often used by astronomers to test hypotheses related to dark matter. For example, an interesting question is in what form dark matter is self-interacting. Self-interaction would lead to a constant mass

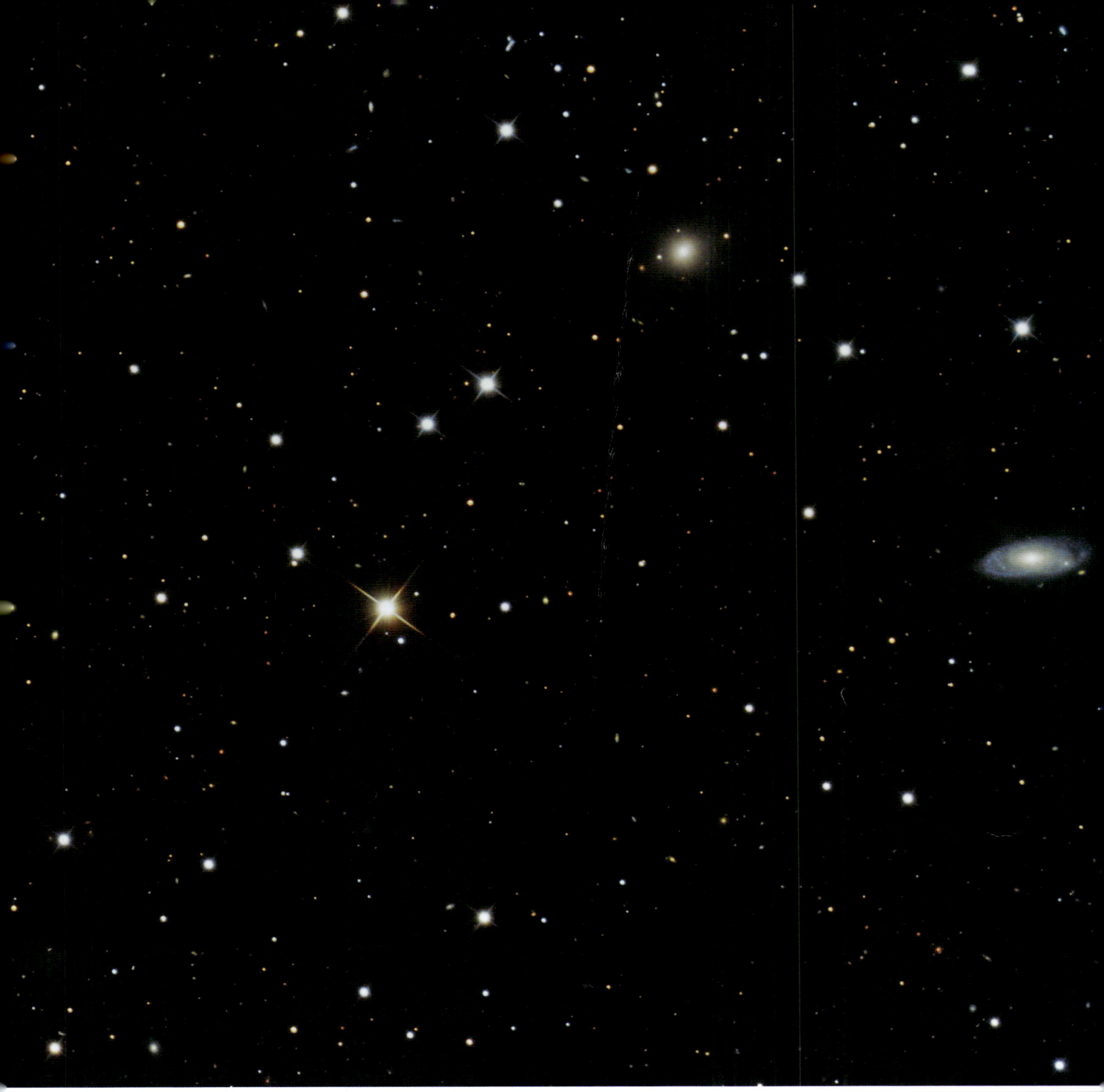

density in the inner halos of LSBGs. The results of the analysis for NGC 5963, however, in contrast to this, show a high mass concentration in the inner halo. A possible explanation is that the dark matter, whose true nature is still unknown, interacts not only gravitationally with the baryons in the halo but that particles of dark matter can influence the concentration of luminous matter via the strong nuclear force. Such a postulated particle interaction could explain structures on small cosmic scales, such as the mass concentration in the halo of NGC 5963.

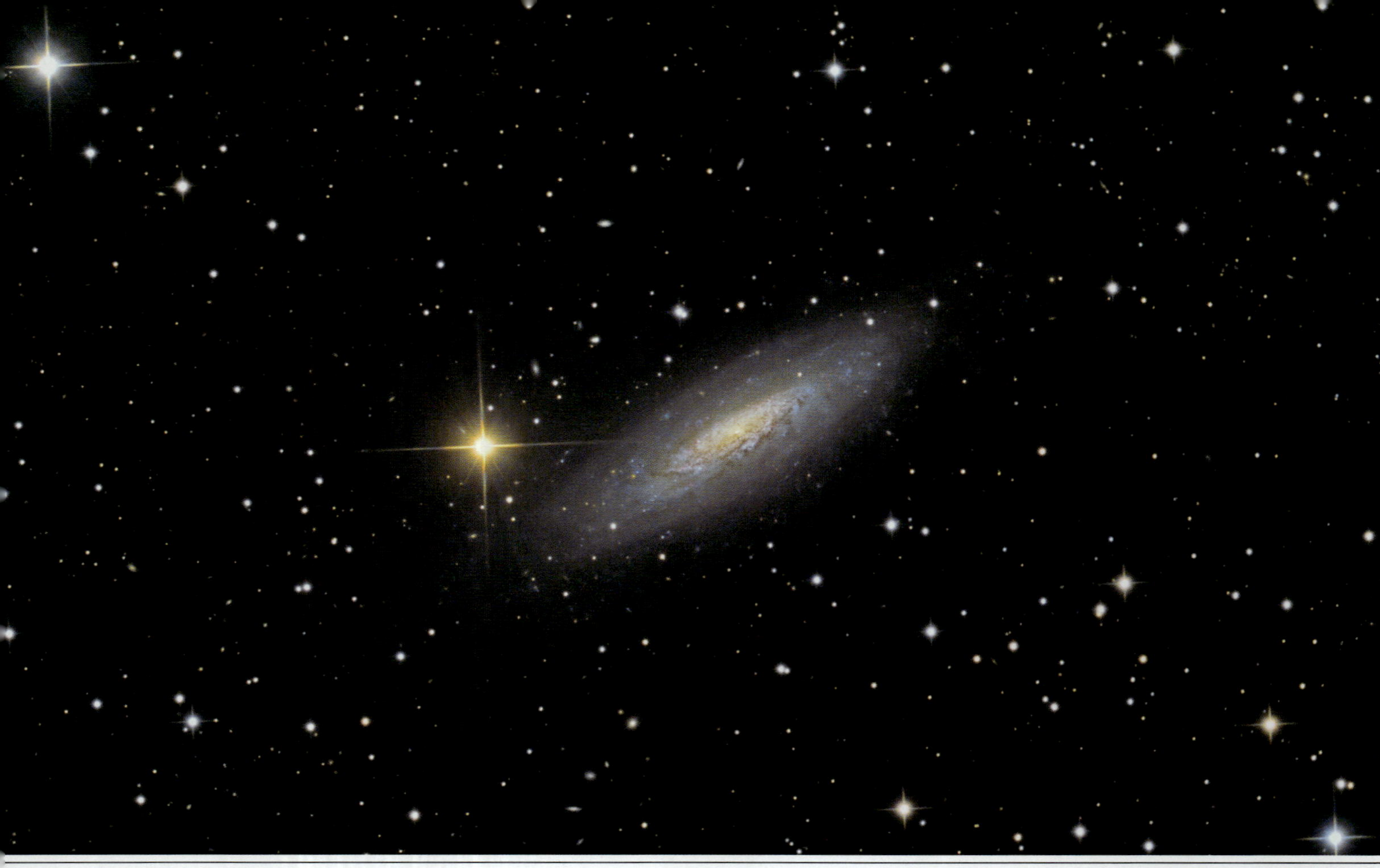

NGC 6503

The SA(s)cd galaxy NGC 6503 is at the edge of the Local Void, an almost empty region surrounded by the Local Group and the Hercules and Coma Clusters. The distance from the Milky Way to the centre of this zone is 75 million light years; the distance to NGC 6503 is only 18 million light years. The angular size of NGC 6503 is given as $7.1' \times 2.4'$, thus corresponding to a transverse disc diameter of 37 000 light years. This means that NGC 6503 is only about 40 per cent of the standard size of a spiral galaxy and is classified as a dwarf galaxy in some catalogues. The core of NGC 6503 is compact, the surrounding spiral-arm structure only weakly defined and appears partially fragmented in the middle region of the disc.

Radio observations show no H I traces at all around NGC 6503 and thus confirm the isolated status of this small spiral galaxy. The measured H I intensity and distribution lead to a mass of neutral hydrogen of 1.3×10^9 solar masses, which is about one per cent of the typical value for an Sc galaxy. The analysis shows also that NGC 6503 can be described by a two-part disc model, consisting of a large, thin disc and a smaller, thicker disc with lower gas density. One possibility astronomers see for the origin of this extraplanar thick gas disc are the active regions whose stellar winds and supernovae transport gas out of the disc. This galactic fountain produces thin, hot gas in the transition to the halo and is seen in NGC 6503 as a dim glow in X-rays.

OBJECT	NGC 6503
CONSTELLATION	Draco
RIGHT ASCENSION	$17^h\ 49^m\ 26^s$
DECLINATION	+70° 08′ 40″
BRIGHTNESS	10.9 mag
TYPE	SA(s)cd
PHOTOGRAPHERS	Makis Palaiologou, Volker Wendel
TELESCOPE	600-mm reflector
CAMERA	SBIG STX-16803
EXPOSURE TIME	450 min
LOCATION	Skinakas Observatory, Crete, Greece

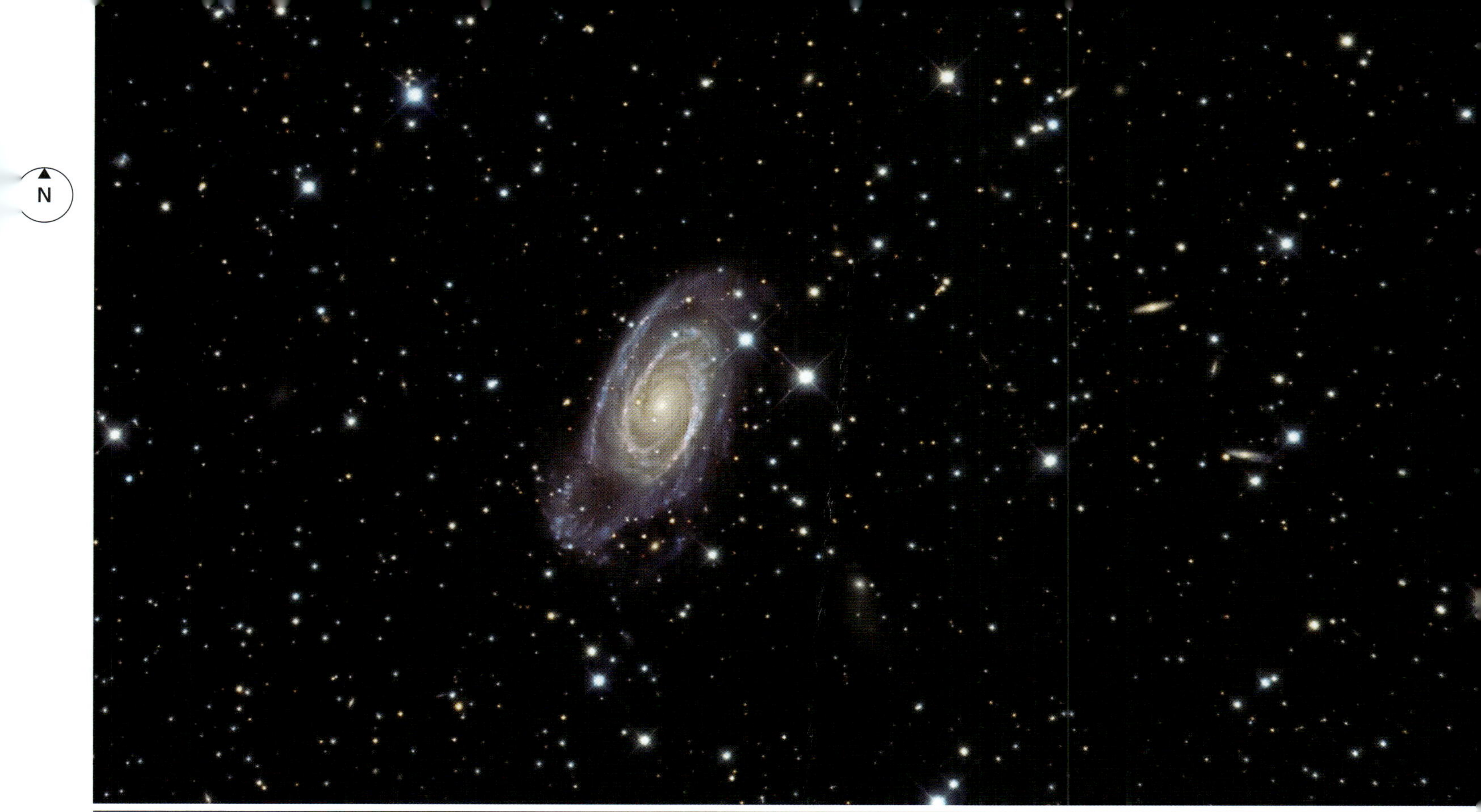

NGC 6632

The galaxy NGC 6632 is located 6′ south of the ninth-magnitude K star SAO 86018 and lies at a distance of 220 million light years from the Milky Way. The angular size of 3.0′ × 1.4′ corresponds to a projected size of the galactic disc of 192 000 × 90 000 light years. The classification of NGC 6632 as SA(rs)bc is often simplified to Sb. It is conspicuous that the spiral pattern in the plane of the disc is not uniform. The inner region is marked by two prominent, bright spiral arms which differ in colour from the bluish arm fragments which are visible at the edge of the disc. At upper and right edges of the disc, the fragments join to form an outer spiral arm while in the lower part of the disc the form is disturbed. The plane of the disc appears to be broken here, its surface brightness is significantly less, and in comparison with the other side more active regions are found in this disturbed zone.

A structural analysis of the spiral pattern in NGC 6632 demonstrates the existence of a double spiral pattern. When this occurs, it is usually the result of an interaction with a compact companion which has passed through and disturbed the disc. However, the mass of the companion was too small to lead to a merger and completely destroy the spiral structure. In the contrast-enhanced image, the companion is still visible 2.5′ to the lower right (southwest) of NGC 6632 as a diffuse object and is catalogued as 2MASXJ18245534+2730302. The contrast enhancement also reveals dim but visible tracks at the upper edge of the disc which indicate that the companion probably made several revolutions around NGC 6632 before passing through the disc. In this process, it lost much matter, leaving a small, compact core with a comet-like tail.

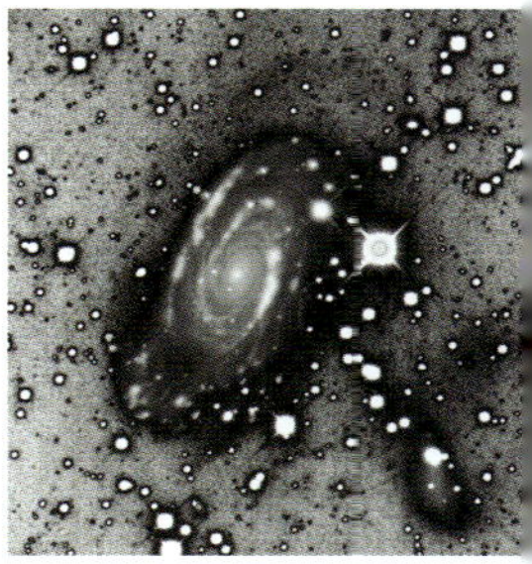

OBJECT	NGC 6632
CONSTELLATION	Hercules
RIGHT ASCENSION	18h 25m 03s
DECLINATION	+27° 32′ 07″
BRIGHTNESS	12.9 mag
TYPE	SA(rs)bc
PHOTOGRAPHERS	Makis Palaiologou, Stefan Binnewies
TELESCOPE	1.3-m reflector
CAMERA	SBIG STX-16803
EXPOSURE TIME	330 min
LOCATION	Skinakas Observatory, Crete, Greece

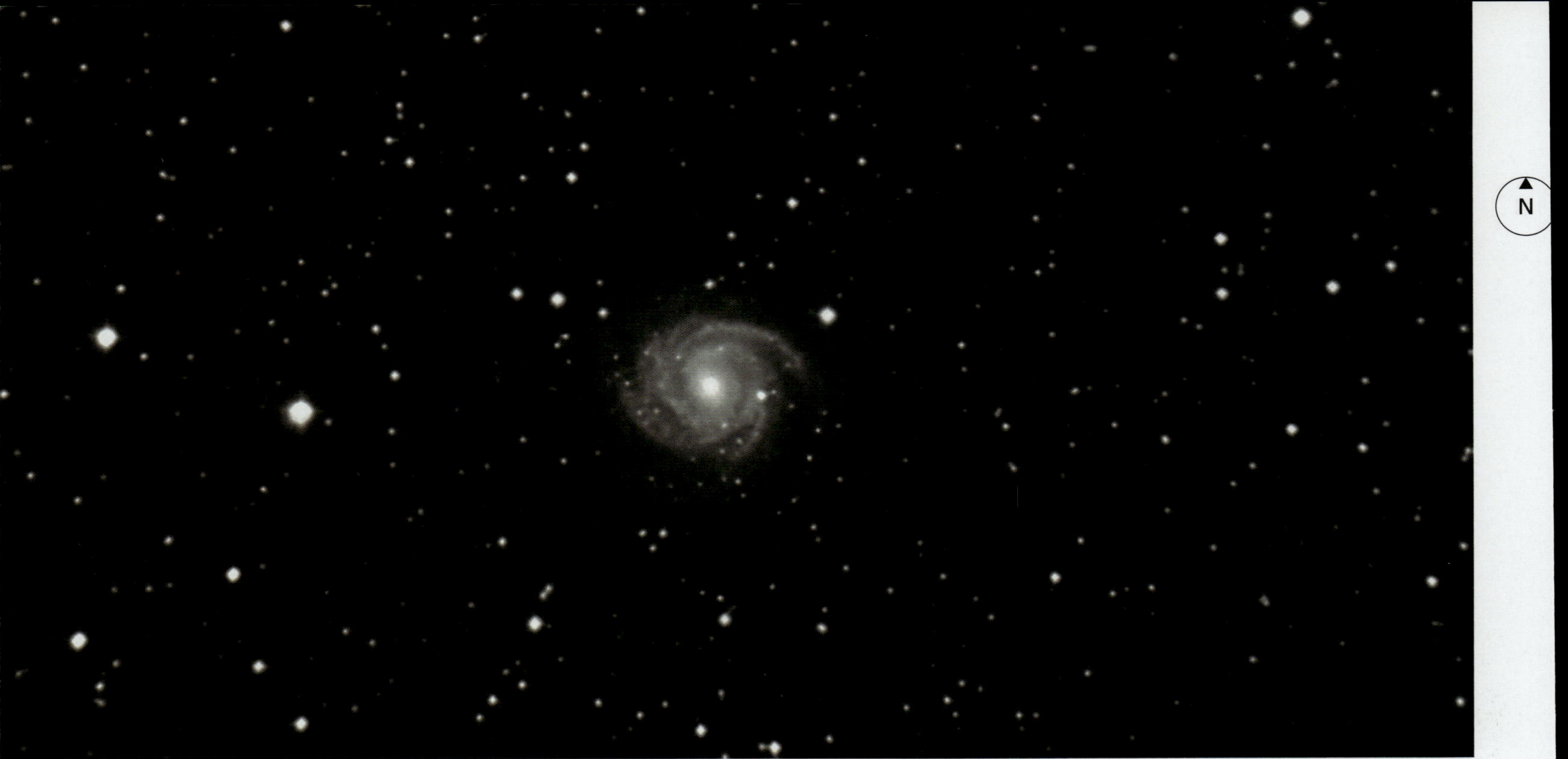

NGC 6814

The SAB(rs)bc galaxy NGC 6814 in the constellation Aquila lies near the star-studded Milky Way. At a distance of 75 million light years, its angular diameter of 3‘ corresponds to a transverse disc diameter of 65 000 light years. The form of the galaxy is marked by a compact core and two spiral arms, the lower (southern) one splitting in two. NGC 6814 is a known Seyfert galaxy and was observed already in the 1980s by several X-ray satellites. Although it was known that active galactic nuclei can be highly variable in X-rays, something out of the ordinary was found in NGC 6814: the variation had a periodic component with a period of 3.4 hours. The problem was then to develop a mechanism to explain periodic behaviour in the core of an active galaxy. For example, the hypothesis was investigated that a star or star cluster could periodically pass through the accretion disc and thus cause periodic X-ray outbursts. It was found, however, that relativistic effects near the supermassive black hole would have quickly forced such an orbit into the main plane. The answer to the riddle came from ROSAT observations of NGC 6814. A team of scientists from the University of Tübingen could show that the regular X-ray pulses originated not in the galaxy but in a nearby X-ray binary. Early satellites did not have sufficient resolution to distinguish between the two sources separated by 37′. There were also problems at the beginning of the ROSAT observations since the new source was blocked by one of the detector struts. This previously unknown X-ray binary RXJ 1940.1-10.25 is part of the Milky Way and consists of a white dwarf as the compact component and a main-sequence star. Matter is transferred from this star to the compact companion and, guided by the strong magnetic field, impacts onto the pole of the white dwarf. Before the impact, this matter is strongly heated and emits X-rays. The orbital motion modulates this emission and the measured 3.4-hour period is the period of the X-ray binary. Using a 40-cm telescope to measure variability, researchers at the University of Tübingen were able to observe the orbital motion in the optical as well.

OBJECT	NGC 6814
CONSTELLATION	Aquila
RIGHT ASCENSION	$19^h\ 42^m\ 41^s$
DECLINATION	−10° 19′ 25″
BRIGHTNESS	12.1 mag
TYPE	SAB(rs)bc
PHOTOGRAPHERS	Michael König
TELESCOPE	280-mm reflector
CAMERA	SBIG ST-10XME
EXPOSURE TIME	120 min
LOCATION	Rimbach, Germany

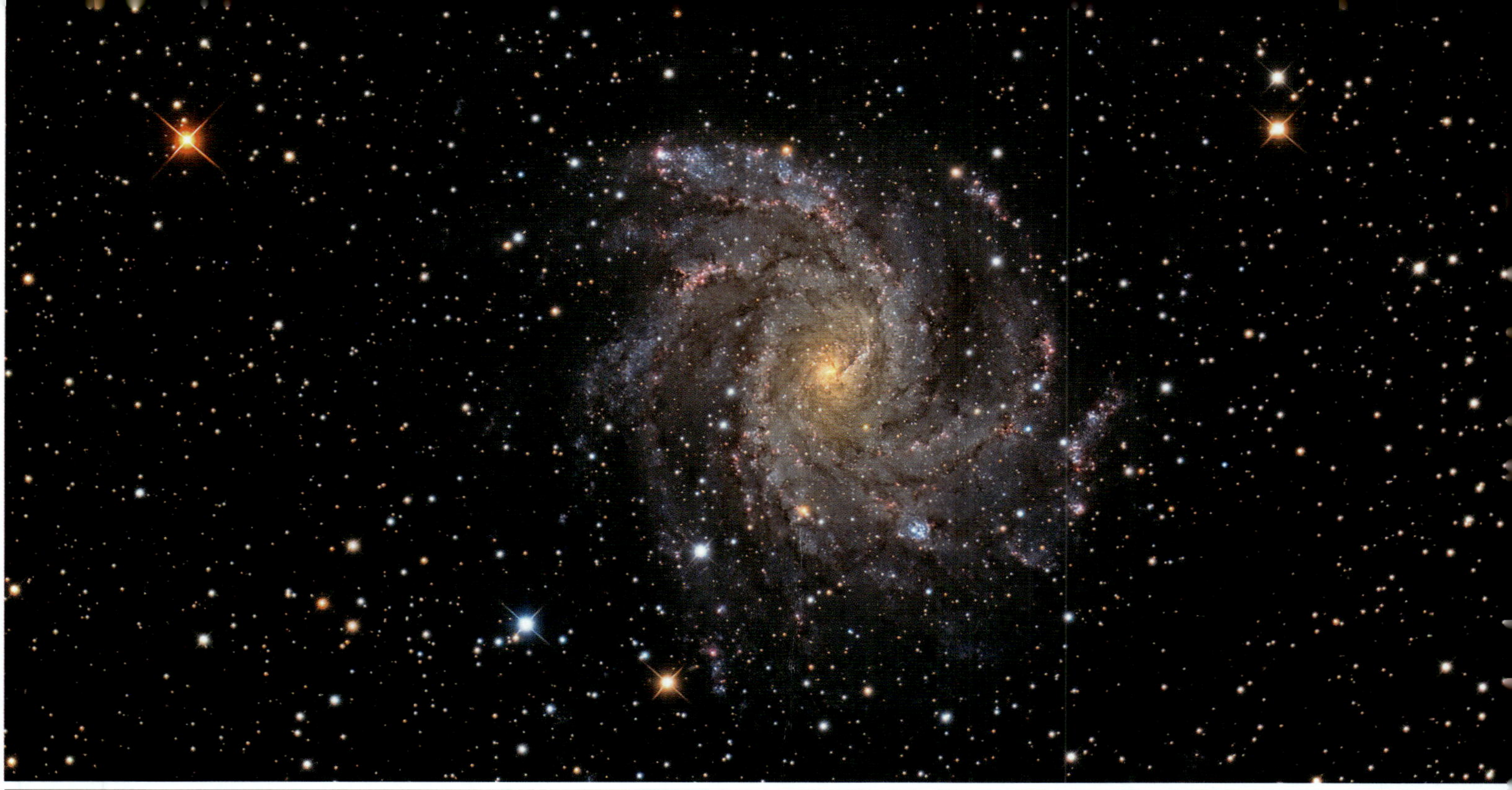

NGC 6946

The SAB(rs)cd galaxy NGC 6946 is one of the closest neighbours of the Local Group, receding from its centre at 344 km/s. Its distance from the Milky Way is only 18 million light years and due to its large angular diameters of 11.5′ and 9.8′ it is a favourite subject for amateur astrophotographers. The projected disc diameter is 60 000 light years. Due to the low inclination of 31°, NGC 6946 offers a detailed view of the many active regions and of the H II regions in the spiral arms with sizes of up to 10″. This explains the name Fireworks Galaxy which one can find in the literature on NGC 6946. At the eastern edge (upper left), in which H II regions can be seen along the spiral arm like pearls on a string. The spiral arms are well defined only in the outer regions of the disc and especially in the southern (left) half of the disc long Vorontsov-Velyaminov rows are visible.

Since the galactic latitude of NGC 6946 is about 12°, the line of sight contains much Galactic dust. In addition to a dimming of 1.6 magnitudes, this leads to a reddening of the light. In this reddened galaxy the bright blue complex of several clusters with young, hot stars with a size of about 2000 light years is thus particularly prominent. This is about 3′ to the right, below and to the right of the centre. The luminosity of this super star cluster implies a mass of 10^6 solar masses. Most of the stars in the almost two dozen star clusters have masses in the range 10–20 solar masses. The interesting things about this complex are the high stellar density and its circular shape. Astronomers suspect that it could have formed from the collapse of a large amount of gas in a spiral arm. The colour–magnitude analysis of the stars dates this possible collapse at 30 million years ago and also demonstrates that at least two evolutionary steps have taken place. These perturbations in the spiral arms could be due to tidal forces of a companion. Proof of the external causes of the perturbations are the dim tidal tails of NGC 6946 which extend beyond the edge of the image and can be seen in the contrast-enhanced image.

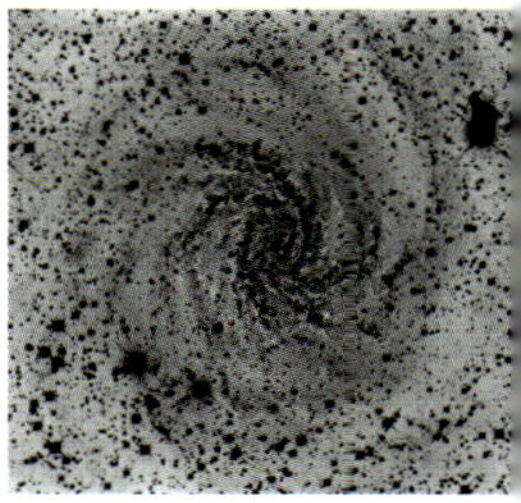

OBJECT	NGC 6946
CONSTELLATION	Cygnus
RIGHT ASCENSION	$20^h\ 34^m\ 52^s$
DECLINATION	+60° 09′ 14″
BRIGHTNESS	8.2 mag
TYPE	SAB(rs)cd
PHOTOGRAPHERS	Josef Pöpsel, Stefan Binnewies
TELESCOPE	600-mm reflector
CAMERA	SBIG STL-11000
EXPOSURE TIME	540 min
LOCATION	Skinakas Observatory, Crete, Greece

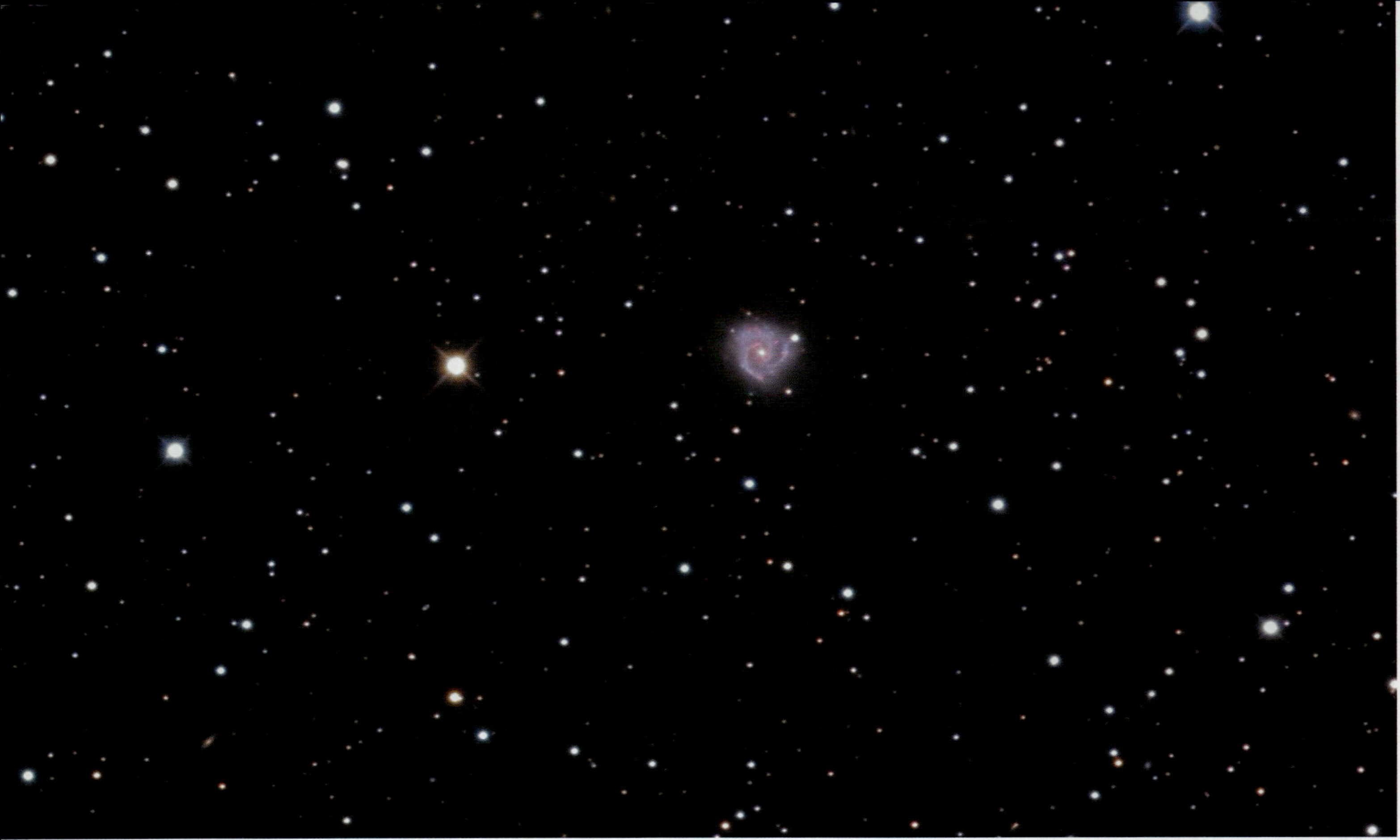

NGC 7137

The galaxy NGC 7137 lies at a distance of 82 million light years and its angular diameters of 1.6′ × 1.4′ correspond to a disc diameter of 38 000 light years. This SAB(rs)c galaxy is thus a small example of an Sc galaxy. In spite of the small size, one can recognize in NGC 7137 well defined Vorontsov-Velyaminov rows which, together with a broad dust lane, led to NGC 7137 being catalogued in the 1960s as the three-armed galaxy. The spiral arms contain many H II regions which colour the arm fragments reddish blue. Unfortunately, one can also see the disturbing influence of scattered light in the image. This two-colour effect is explained by two different positions of a variable glass plate in the beam path used in exposing the different colour channels. This glass plate is part of a system in front of the camera which tries to correct the seeing via rapid tilting motions in order to improve the image quality. The chip size of the camera corresponded to a sky area of about 30′ × 20′ at the time of exposure; the star responsible for the reflection (SAO 89978, 7.1 mag) was 15′ from the edge of the CCD.

From simulations of spiral patterns, astronomers have determined that three-armed spirals can arise. These structures can form under certain resonance conditions in a model with a rotating, asymmetric bar. Unfortunately, these models don't correspond very closely with the actual appearance of NGC 7137. For example, the Vorontsov-Velyaminov rows were not visible and the simulation exhibited only the typical spiral arms which get weaker with increasing distance from the core. This, however, is not what NGC 7137 looks like – its morphology is determined mainly by its compactness and can thus not be explained by the influence of a bar in the centre.

OBJECT	NGC 7137
CONSTELLATION	Pegasus
RIGHT ASCENSION	$21^h\ 48^m\ 13^s$
DECLINATION	+22° 09′ 35″
BRIGHTNESS	13.1 mag
TYPE	SAB(rs)c
PHOTOGRAPHERS	Michael König
TELESCOPE	350-mm reflector
CAMERA	SBIG STL-6303
EXPOSURE TIME	470 min
LOCATION	Rimbach, Germany

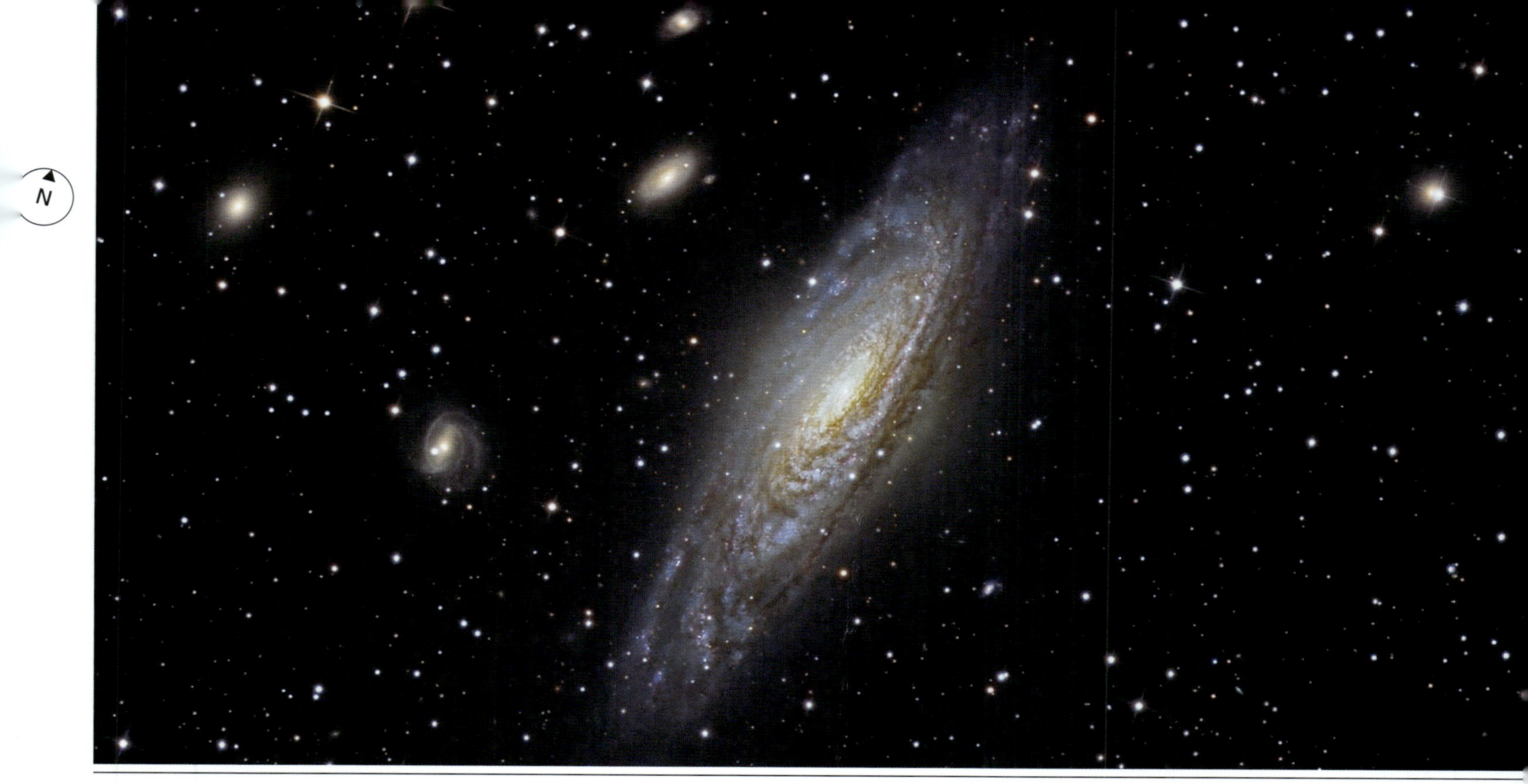

NGC 7331

The SA(s)b galaxy NGC 7331 is at a distance of 46 million light years from the Milky Way and due to the high inclination of 77°, part of the disc is blocked by broad dust lanes. The angular size of the disc of 10.5′ × 3.7′ corresponds to a transverse diameter of 140 000 light years; thus NGC 7331 is substantially larger than the Milky Way. Infrared images of NGC 7331 show a ring with a radius of 20 000 light years. This ring contains a large amount of dust and many star-formation regions lie in the ring. In the visible, the ring is barely seen due to the surrounding dust. Modelling the temperature, one can conclude that the mass of the ring is about 10^9–10^{10} solar masses. Depending on the details of the model, the ring makes up about 20–30 per cent of the total amount of disc gas in NGC 7331.

Above the eastern (left) edge of the disc is a group of a few background galaxies which are about 6.5 times as far away as NGC 7331. The lower SB galaxy NGC 7337 can be used as a starting point to find to faint dwarf galaxies which are part of the NGC 7331 system. One of the dwarf galaxies lies about 1′ below NGC 7337. The second, somewhat brighter dwarf lies in the direction of the lower, southern end of the disc of NGC 7331. Astronomers have found in deep images of NGC 7331 a further two dwarf galaxies which are outside the field of this image. All four dwarf galaxies are classified as dSph due to their low luminosity. In contrast to M 31, NGC 7331 has no dE dwarf galaxies, so that one can estimate that the stellar mass of the dwarf galaxies of NGC 7331 is at most two per cent of that observed in M 31. At the lower end of the disc one can see in contrast-enhanced images a faint, approximately 1.5′ long tidal tail. Since its colour is similar to the light blue of the dwarf galaxies, a physical connection seems probable. The prominent warp in the disc of NGC 7331 could thus have formed due to an interaction between the dwarf galaxies and the outer disc region.

OBJECT	NGC 7331
CONSTELLATION	Pegasus
RIGHT ASCENSION	$22^h\ 37^m\ 04^s$
DECLINATION	+34° 24′ 56″
BRIGHTNESS	10.4 mag
TYPE	SA(s)b
PHOTOGRAPHERS	Volker Wendel, Stefan Binnewies
TELESCOPE	600-mm reflector
CAMERA	SBIG STL-11000
EXPOSURE TIME	465 min
LOCATION	Skinakas Observatory, Crete, Greece

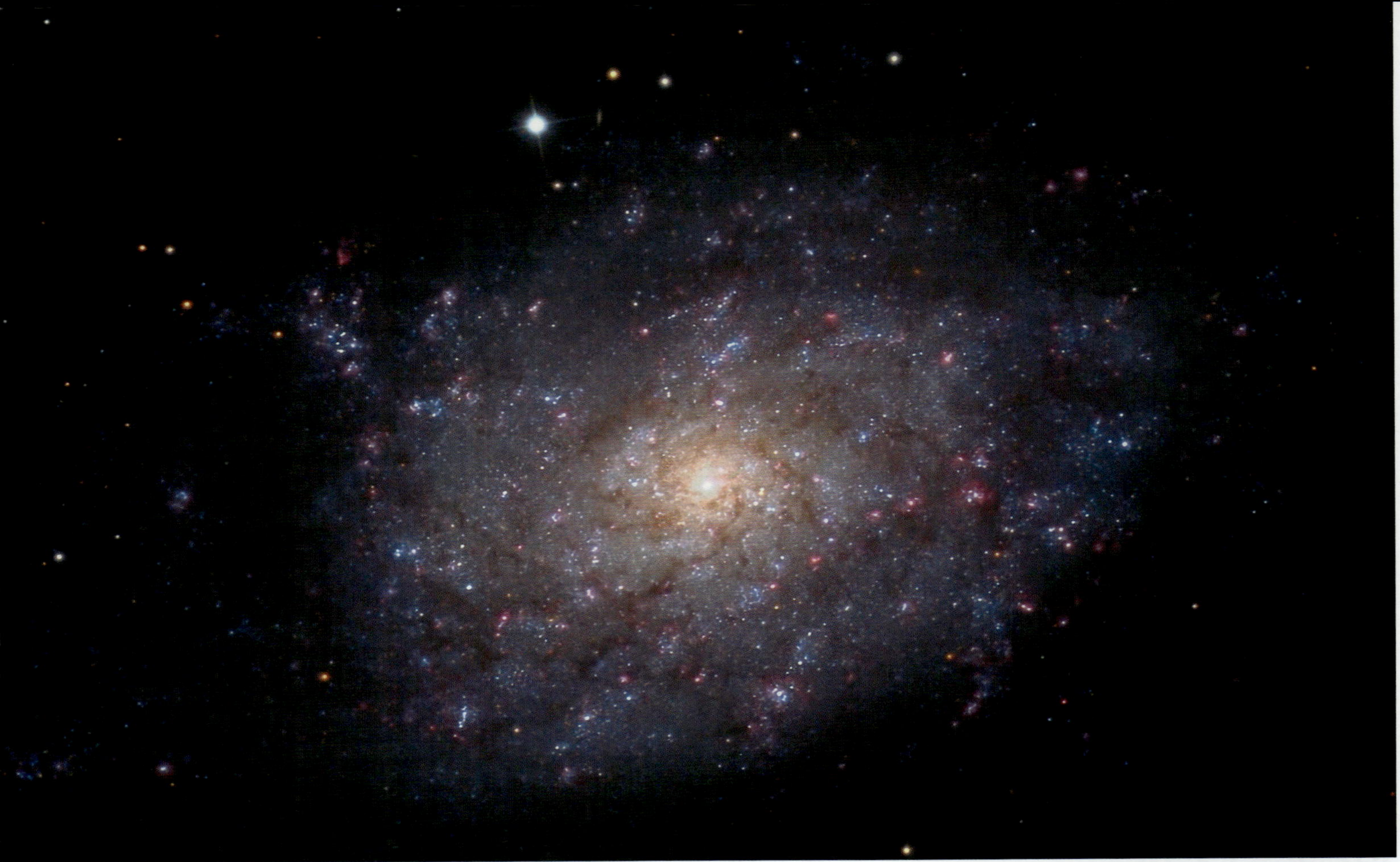

NGC 7793

The SA(s)d galaxy NGC 7793 is twelve million light years from the Milky Way and its angular size of 9.3′ × 6.3′ corresponds to a projected disc diameter of 32 000 light years. NGC 7793 is thus one of the smaller examples of an Sd galaxy. Its morphology is dominated by a compact nucleus and a strongly fragmented spiral-arm pattern which is visible in the ultraviolet, optical, and infrared parts of the spectrum. NGC 7793 is one of the five brightest members of the Sculptor Group, a loosely packed group stretching over 20° on the sky. This group could, however, be a filament located close to the Local Group. Observed along the long axis, the projection causes it to appear as a group.

In several observations of NGC 7793, the velocity of rotation of ionized hydrogen was measured. The curve shows prominent differences from the typical shape for spiral galaxies. After a moderate rise, the values reach a maximum of 130 km/s at a distance of 4′. After that, the rotation curve falls so that the rotating hydrogen gas has a velocity of 90 km/s at a distance of 6′ from the centre. Due to the weakly defined spiral structure, it is difficult to measure the inclination of NGC 7793. A value of 47° is given for the inclination and the variation of ±5° is explained by a warped disc. Models indicate a total mass of 1.4×10^{10} solar masses for NGC 7793 and a ratio of dark to luminous matter of 0.75–1.0. This indicates a relatively small amount of dark matter in NGC 7793, which nevertheless is needed in order to correctly describe the measured rotation curve.

OBJECT	NGC 7793
CONSTELLATION	Sculptor
RIGHT ASCENSION	$23^h\ 57^m\ 50^s$
DECLINATION	−32° 35′ 28″
BRIGHTNESS	10.0 mag
TYPE	SA(s)d
PHOTOGRAPHERS	Josef Pöpsel
TELESCOPE	600-mm reflector
CAMERA	SBIG ST-10XME
EXPOSURE TIME	120 min
LOCATION	Amani Lodge, Namibia

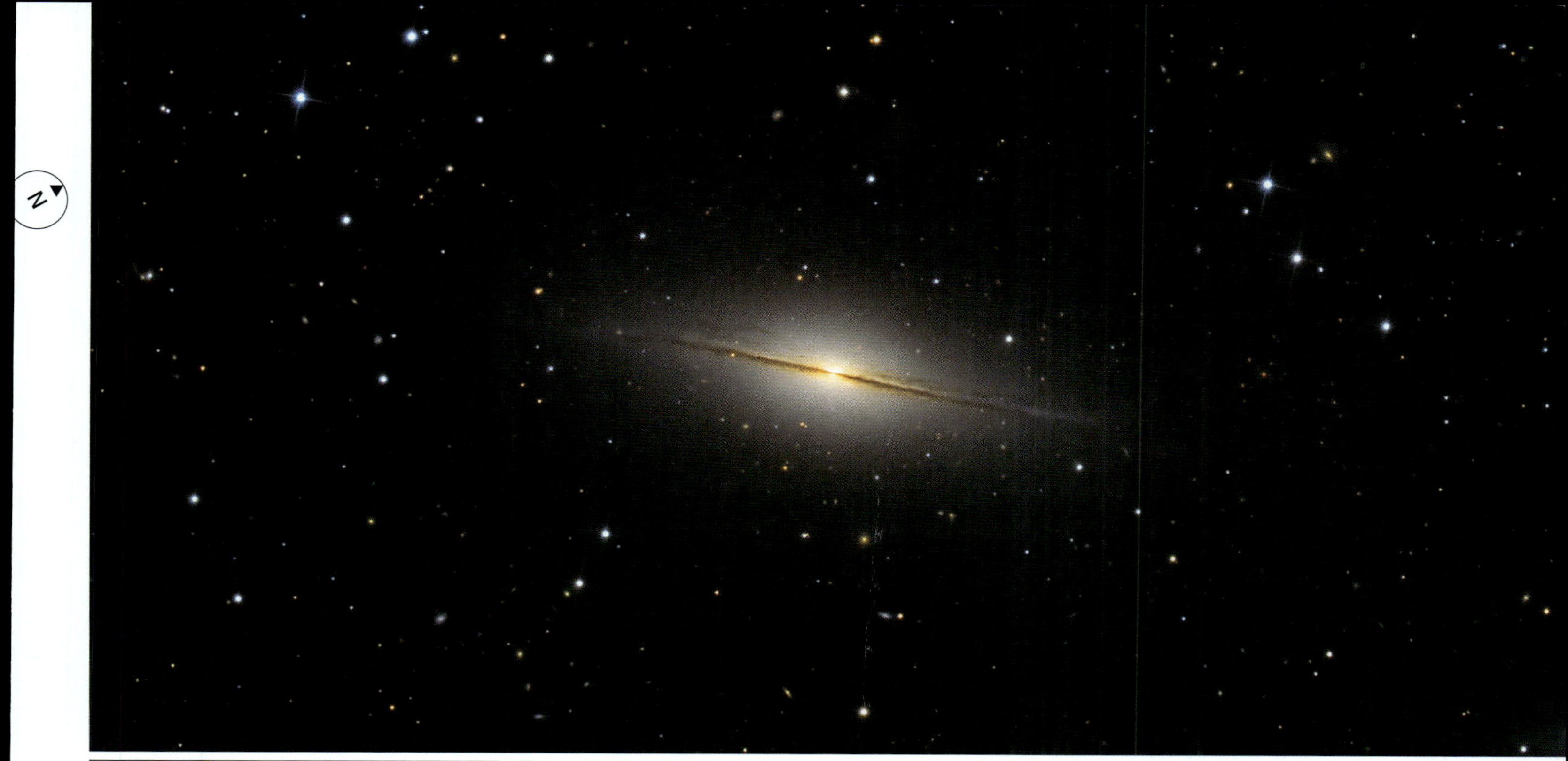

NGC 7814

The SA(s)ab galaxy NGC 7814 is located near the 2.8-magnitude star Algenib which marks the southeast corner of the Great Square of Pegasus. Its distance from the Milky Way is 53 million light years and the angular size of 5.5′ × 2.3′ corresponds to a projected diameter of 85 000 light years. The galaxy is seen almost edge-on and there is an apparently delicate dust lane in the plane of the disc. The high angle of inclination prevents one from saying anything about the spiral pattern of NGC 7814. At the upper northeast part of the disc, a few arc-shaped dust filaments are visible in the diffuse light of the bulge. The nucleus is a bright point source surrounded by an ellipsoidal bulge. The dominance of the bulge is responsible for NGC 7814 being designated a bulge-dominated galaxy. No active regions are visible, so only the visible dust lane prevents NGC 7814 from being classified as S0. The diffuse yellow light from the bulge lends weight to the hypothesis that the current star-formation rate NGC 7814 is low.

The stars in the bulge have a three-dimensional velocity distribution. The fact that the bulge has not yet collapsed to a disc is interpreted by astronomers to indicate that it is dynamically young and must have formed within the last 10^9 years. The rotation curve of neutral hydrogen in NGC 7814 consists of two components, a first plateau at 230 km/s in the range 1–2.5′ from the centre which then within 0.5′ drops to 210 km/s and retains this value to the edge of the disc. The radio data extend 4′ beyond the radius of the optical disc. The first plateau of the rotation curve lies within the bulge and indicates its high fraction of dark matter. This is confirmed by a very faint brightening 3.4′ above the core of NGC 7814, which could be a satellite galaxy. This dim dwarf galaxy is not described in any catalogues. Based on the distance to NGC 7814, one obtains a diameter of 8000 light years. The dark matter in the bulge protects the disc against strong tidal forces and is responsible for the fact that no warp is visible in the bulge zone of NGC 7814.

OBJECT	NGC 7814
CONSTELLATION	Pegasus
RIGHT ASCENSION	$00^h\ 03^m\ 15^s$
DECLINATION	+16° 08′ 44″
BRIGHTNESS	11.6 mag
TYPE	SA(s)ab
PHOTOGRAPHERS	Adam Block
TELESCOPE	800-mm reflector
CAMERA	SBIG STL-11000
EXPOSURE TIME	350 min
LOCATION	Mount Lemmon SkyCenter/ University of Arizona, USA

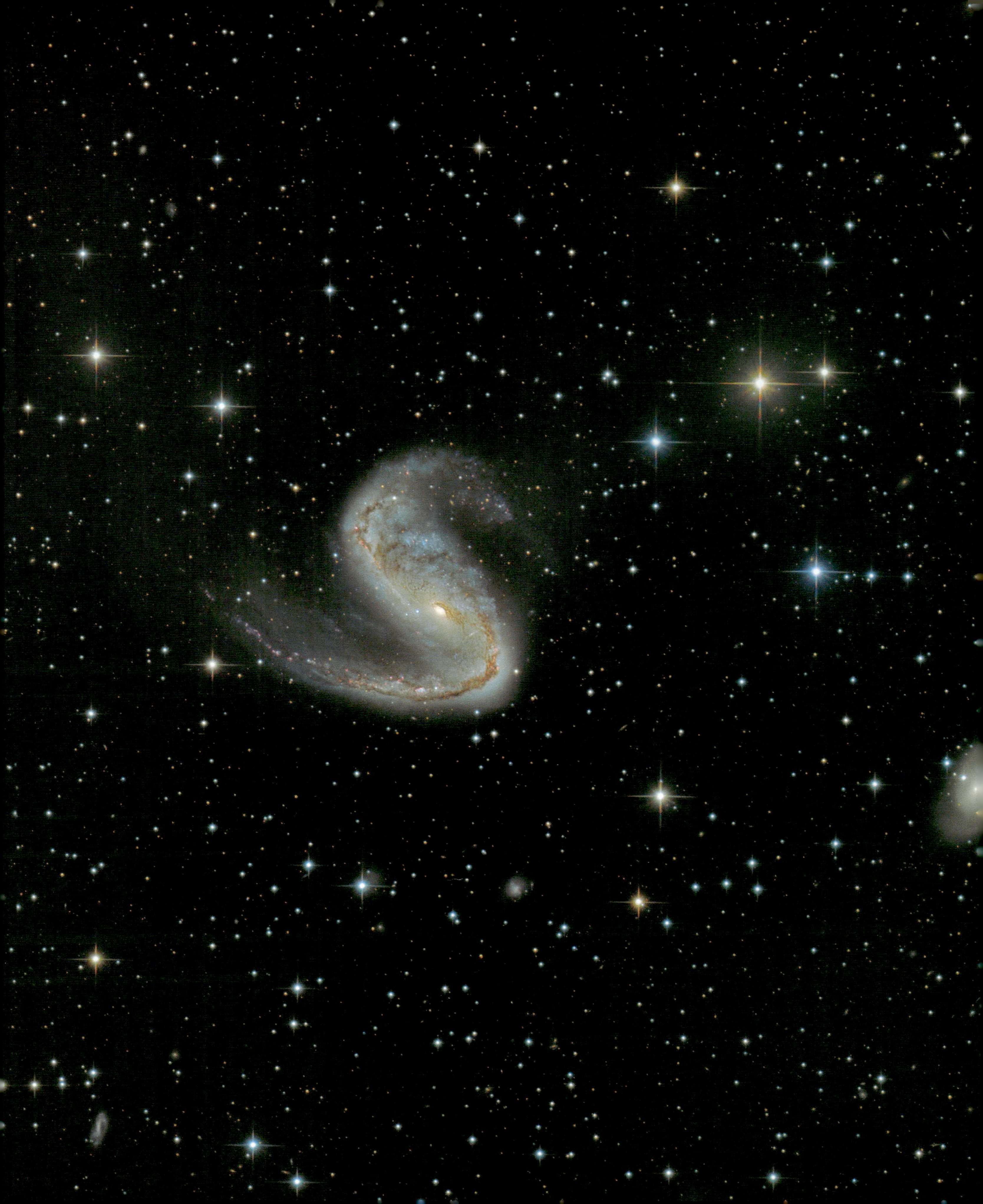

BARRED SPIRAL GALAXIES

Barred spiral galaxies raise the question of what differentiates them from normal spiral galaxies. In this chapter, special features of barred spirals are explained, which indicate dependence of their forms on time.

THE CLASSIFICATION OF BARRED SPIRALS

Barred spiral galaxies contain, in contrast to normal spiral galaxies, a straight stellar bar which is symmetrical about the core and whose ends connect to the spiral arms. The term "bar" goes back to Edwin Hubble, who in 1936 introduced the classification "SB" for "spiral barred", in order to distinguish between S and SB types. The classification of barred spirals is the same as that of normal spirals. For example, an SB type classified as SBa is a galaxy with a tight spiral pattern and bright nucleus. Moving to "later" SB types, the pitch angle increases and the core at the middle of the bar becomes more compact and less prominent.

In the galaxy classification introduced by Gérard de Vaucouleurs in 1959, the Hubble classification was extended by the SAB type. In this system, galaxies with weak bars are classified as SAB. In addition, de Vaucouleurs introduced the supplementary classifications (s), (r), and (rs) to describe the transition region between bar and spiral arms. This made it possible to distinguish between pure spiral patterns and galaxies with an inner ring connected to the bar region.

Bars often have a diffuse appearance and show less structure than spiral arms. Thus, there is no finer classification based on the bar. Only with the advent of modern methods of astronomical research was it determined that there are measurable differences in the bar structures. For example, the form of the bar can be more boxy or more disc-like. An important quantity for the assessment of the dynamics of a barred spiral galaxy is the axial ratio of the bar, i.e. the ratio of width to length. In the earlier Hubble types SBa to SBb, the relative length of the bar is greater than in the later types (SBc to SBd). One can visualize this as an SBa galaxy having, compared to the spiral arms, a longer bar with a smaller ratio, surrounded by tightly wound SBa spiral arms. This can give the impression of a complete ring.

Comparative studies often quote a fraction of barred spirals of 30–40 per cent of all spiral galaxies. If one considers as well the colour of the galaxy, which is correlated with the star-formation rate, then one can further quantify the probability for the appearance of a bar. Among "blue" galaxies with relatively many young stars, only about 20 per cent are barred. Among "red" galaxies, which contain an older stellar population, one finds a bar in about 50 per cent. If, in addition to the colour, the absolute luminosity is used as a criterion, then dim, red galaxies have the highest rate at 70 per cent.

THE MORPHOLOGY OF BARRED SPIRALS

The early Hubble types SBa and SBb have bars of uniform brightness. The spiral arms become dimmer as one follows them away from the bar. In the literature, this is attributed to a comparatively low fraction of luminous matter, making the early-type spiral arms appear dim. In the later types SBc and SBd, the brightness profile is no longer constant, but rather there is an exponential drop as one moves from the centre to the ends of the bar. This drop in the brightness profile along the bar expressed in magnitudes per surface area corresponds to a linear decrease in surface brightness. In contrast, the spiral arms show a constant or even increasing brightness. As far as the length of the bar goes, longer bars are found more often in early SB types. This can be explained by more stars having strongly eccentric orbits in these types of galaxies. The bar consists of stars, thus it displays an average over their orbits. In early SB types, the length of the bar in relation to the disc diameter is in the range 35–43 per cent. The ratio drops in late types and reaches values of 25 per cent and less. A rule of thumb based on many studies is that the bars in early-type spirals are twice as long as in late types.

The galaxy NGC 2442 in the constellation Volans displays a prominent bar and wide spiral arms (see also page 90). Image: CHART 32-Team (800-mm reflector).

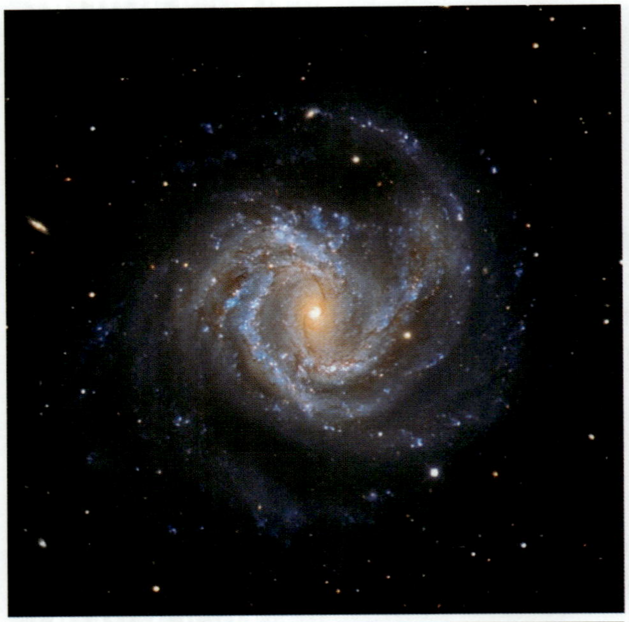

Figure 2.1: M 61 is an SAB(rs)bc type with an inner ring (see page 102).

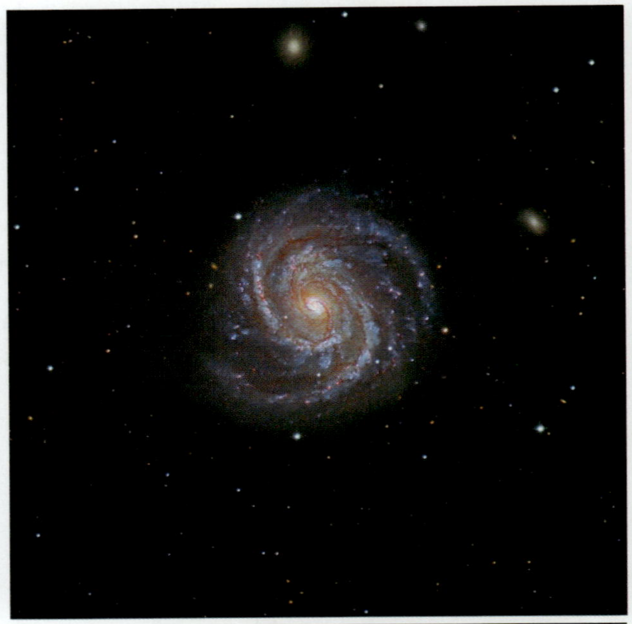

Figure 2.2: M 100 is an SAB(s)bc type, the spiral arms more prominent than in M 61 (see page 104).

Surveys for barred spiral galaxies indicate that bars are rarely seen in edge-on cases, though based on the relative probability one would expect to find some. This means that the bars must be as flat as the discs. They don't have a round cross-section, which means that barred spirals are under-represented among edge-on galaxies.

THE ASTROPHYSICS OF BARRED SPIRALS

Density waves, described in the chapter Spiral Galaxies and responsible for the spiral structure, play an important role in barred spirals as well. The question is why bars arise in these but not in normal spirals.

Estimates vary as to the fraction of inner bars in spiral galaxies. The assumption is that about one-third of spiral galaxies are pure barred spirals of type SB and another third of the transitionary type SAB. Under the assumption that bars are a time-dependent phenomenon in the evolutionary history of spiral galaxies, then the frequency of bars indicates that these are long-lived and/or recurring phenomena.

Observations of stars in bars have shown that these follow highly eccentric orbits about the galactic centre. Their highly elongated orbits form the bar and give a first hint at the astrophysical cause of bar formation. Simulations show that bars arise spontaneously. An explanation is that due to small variations in the density (e.g. an increase because of chance gatherings of stars), further stars from nearby move toward a local sink in the gravitational potential. This leads to more matter and an even deeper potential.

As is the case with the normal spiral structure, the barred-spiral structure also owes its structure on larger scales to density waves. The elongated shape of the density wave forces almost all stars into the bar, leaving almost empty regions perpendicular to it. An individual star in the bar no longer moves in a circular orbit about the centre, as is the case for stars near the centre of a barless spiral galaxy, but rather has a highly eccentric orbit within the bar. The sum of the orbital tracks of these stars thus defines a self-preserving bar structure.

In the main plane of a galaxy, there is a constant exchange of gas and dust. This follows the gravitational potential and in the central regions is an inward flow. Within a barred spiral galaxy, the spiral arms are attached to the ends of the bar. Thus, in this region there must be an exchange process between the matter in the bar and the inflowing material from the disc. Matter can flow through this join into the core, however, only if its angular momentum is lost. The orbital angular momentum of a star, which is conserved during its orbit, is reduced when the orbital motion is braked. This takes place in regions of increased matter density – thus exactly at the end of the bar. In this process, known as dynamical friction, stars move onto highly elongated orbits and hence enhance the bar structure.

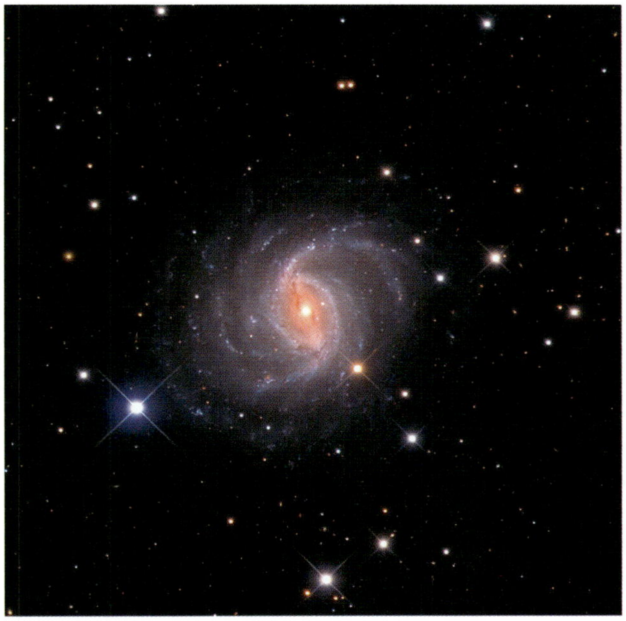

Figure 2.3: NGC 5921 is an SB(r)bc type with prominent bar (see page 118).

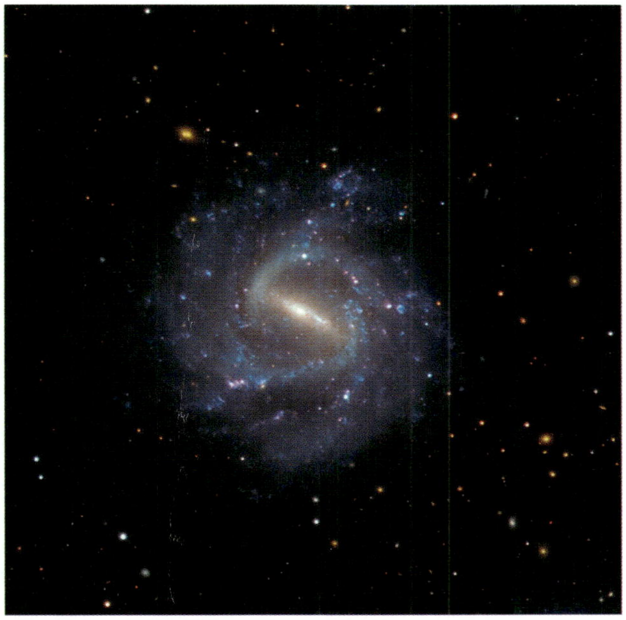

Figure 2.4: NGC 1073 is an SB(rs)c type, similar to M 61, but with a more prominent bar (see page 81).

Computer simulations can accurately model these processes and the evolution of barred spirals. For example, one can observe bars forming and growing. Simulations show as well, though, that an increase in the gravitational field of the bar with time stops the flow of matter towards the centre. The transport of material can even be reversed, moving matter to the edge of the disc. This area is known among astrophysicists as the "region of the outer Lindblad resonance". Thus there is a lack of matter flowing to the centre, which leads to a reduction in the bar. Other models show complete dissolution of the bar. This in turn means the loss of the bar-shaped gravitational field, which means that matter can flow towards the centre again. This leads to an increase in the surface density and hence gravitation of the disc.

The game begins again with this new flow of material and again a bar is formed. However, since there is now more mass in the inner regions, this new bar has a different form. Its length is less and the circular velocity of the co-rotating bar is thus higher. This rotation corresponds to that of a solid body.

Viewing this from outside, one notices that the spiral arms start farther in. This second bar in the life of the galaxy appears less prominent in the disc plane than the first one. The process of creation and destruction of a bar structure can take place many times. Simulations show that three or four bar phases are possible within a Hubble time. A bar is stable for about two billion years, which corresponds well to the observed ratio of barred to normal spiral galaxies.

In addition to the description of the astrophysical mechanism, a second question is of interest: on the one hand, what is the ultimate cause of bars, and on the other hand, what prevents bars from forming in normal spiral galaxies? When considering this, one should not forget that the disc is embedded in the halo of the galaxy and this halo is filled with dark matter. The halo mass is more than that of the luminous matter, i.e. stars, gases, dust, of the rest of the galaxy, by about a factor of ten. The mass of the halo of dark matter in the Milky Way is estimated to be 6×10^{11} to 1×10^{12} solar masses and that of the luminous matter 9×10^{10}. The ratio of the two masses is thus in our galaxy between 7 and 11.

It has been shown that this ratio of halo to disc mass is an indicator for the occurrence of a bar: the disc mass is closely correlated with the surface density in the spiral plane. A low surface density leads to a large ratio and in the case of a low-mass disc no or only small amplification effects take place. Thus, the radial symmetry of the gravitational field is preserved; the result is a spiral galaxy without a bar. The lower the ratio, the more the amplification effects in the disc become visible, and more often a bar is formed. The same is true for spiral galaxies with a small amount of dark matter. Here as well there is a small ratio and one does indeed find that bars often occur is these galaxies.

THE SECULAR EVOLUTION OF BARRED SPIRALS

As described in this chapter, bars are not permanent properties of a disc galaxy. Thus, one can ask what other properties of galaxies also change with time.

In barred spiral galaxies, resonance effects in the vertical component of the motion of stars can lead to thickening of the bar. Viewed edge-on, this can be seen as club-like widening at the ends of the bar. In the literature, this is referred to as "peanut-shaped" or, if viewed perpendicularly, as "box-shaped". Due to the rotating bar, torques are applied to the gas in the disc. As described in this chapter in the discussion of the origin or bars, material can be transported to the region of the outer Lindblad resonance where it builds up. The result can be an outer ring, such as those observed in some SB galaxies. Resonance effects play a role here as well and in the plane of the disc two larger, overlapping rings can arise and form a lentil shape enclosing the bar region.

A further notable feature of SB types are small, compact bars which one can detect in the cores of some barred spiral galaxies. This substructure is termed a "nuclear bar" and is usually surrounded by a nuclear star-formation ring, easily visible due to its high infrared luminosity.

Simulations hint that the dissolution of a bar takes place in two phases. First, there is a weakening of the bar structure and the less prominent bar rotates more quickly than the original one. The reason for the dissolution are the increasingly chaotic orbits of the stars in the bar regions. Simulations also confirm further real phenomena: as the bar disappears, the ends brighten, forming so-called ansae (Latin for "handles") at its ends, and at the centre of the bar an axially symmetric potential remains. A second, smaller bar can form here. It is, however, difficult to make precise statements about the evolution of the bar in a galaxy, since interactions with the halo and the dark matter it contains can influence changes in the bar. One idea discussed in connection with these processes assumes that S0 galaxies could form from barred spirals which have lost their bars. Their lentil-shaped cores would then be the relics of dissolved bars.

The Milky Way probably contains a central bar structure with two large and two to four smaller spiral arms; our Solar System is located in one of these small spiral arms (Local Arm, Orion Arm). Although classification of the Milky Way is difficult, many radial-velocity measurements in the radio and other observations lead to the conclusion that type SBc is probable. The two large spiral arms, known as the Norma Arm and Sagittarius Arm, are connected to the bar. The Milky Way's bar is about 10 000 light years long and the bar region spans the Galactic latitude ±5°.

Barred spirals are also found among distant spiral galaxies. One has to keep in mind, though, that these galaxies are smaller and more compact than nearby galaxies with redshifts of $z < 0.5$. The difficulty in classifying galaxies at $z = 1$ is that these are comparatively "clumpy". Such galaxies are also known as "protospiral galaxies". The clumpy structures are large star-forming regions with masses in the range 10^8–10^9 solar masses. Distant galaxies usually have five to ten such regions. Observations indicate that the star-formation rate in these regions of 20 solar masses per year was much larger in the past than in present galaxies. Larger and thinner discs as seen in today's galaxies are formed only after the interaction of these clumps. The prominent bars shown in this chapter are probably a modern phenomenon, mainly seen among nearby ($z < 0.5$) galaxies. Their bars have a longer dynamic time scale and thus one can assume that the bar structures in nearby galaxies will dissolve only slowly with time.

In some studies, the relation between the occurrence of barred spirals and redshift was investigated. Astronomers used data from several surveys and thus covered a range ($0 < z < 0.8$) corresponding to seven billion years. The trend is that the fraction of SB types decreases with increasing redshift. For all identified barred spirals, the fraction dropped from 0.65 to 0.20. For those with strong bars, the value drops from 0.3 to 0.1. For closer galaxies ($0 < z < 0.3$) the drop is weaker, which indicates a longer time scale in nearby barred spirals. These data, which allow us to observe a universe which was half as old as it is today, indicate that the familiar models of classification can't be used here. Only about 20 per cent of the galaxies had become dynamically stabilized and thus reached a form which is still found today.

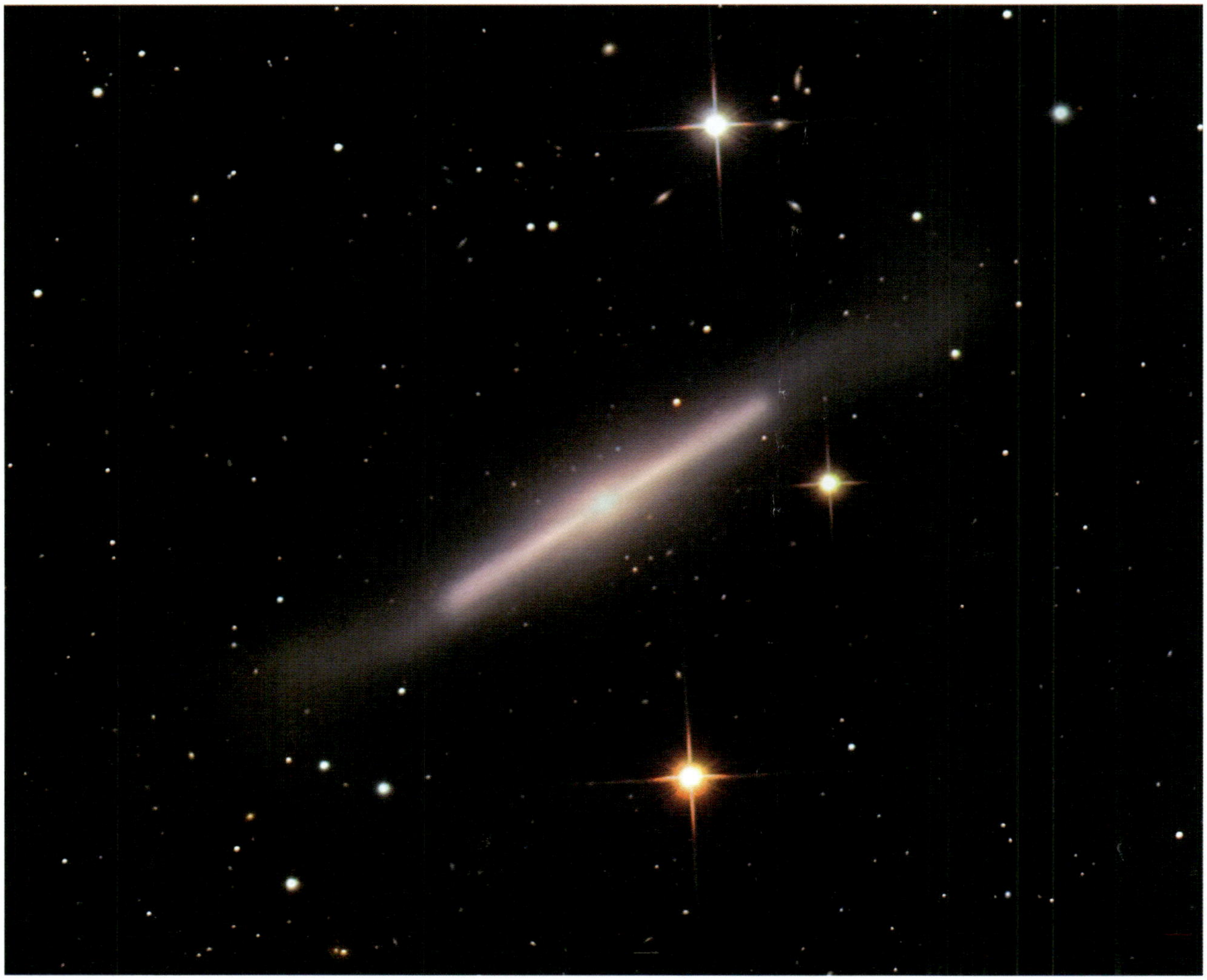

Figure 2.5: 4762 is difficult to classify due to the edge-on perspective (see page 111).

BIBLIOGRAPHY

Combes, F.: *Dynamics of Galaxies – Bars and AGN fueling*, 2012, http://aramis.obspm.fr/~combes/post.../PDEA2-en.ppt

Elmegreen, B.: *Bars, Spiral Structure, and Secular Evolution in Disk Galaxies*, 2005, www.astro.rug.nl/~islands/Elmegreen_Terschelling.pdf

Erwin, P.: *How Large Are the Bars in Barred Galaxies?* 2005, www.mpe.mpg.de/~erwin/research/papers/bar-sizes.pdf

Hoyle, B. et al.: *Galaxy Zoo: Bar Lengths in Local Disc Galaxies,* Monthly Notices of the Royal Astronomical Society, **415**, 2011

Masters, K. L. et al.: *Galaxy Zoo: Bars in Disc Galaxies,* Monthly Notices of the Royal Astronomical Society, **411**, 2011

Mihos, C.: *Galaxies and Cosmology – Spiral Bars*, 2011, http://burro.astr.cwru.edu/Academics/Astr222/Galaxies/Spiral/bars.html

Nair, P. B. and Abraham, R. G.: *On the Fraction of Barred Spiral Galaxies*, The Astrophysical Journal Letters, **714**, 2010

Sellwood, J. A. and Wilkinson, A.: *Dynamics of Barred Galaxies*, Reports on Progress in Physics, **56**, 1993

Sheth, K. et al.: *Evolution of the Bar Fraction in COSMOS: Quantifying the Assembly of the Hubble Sequence*, The Astrophysical Journal, **675**, 2008

Tully, R. et al.: *Our Peculiar Motion Away from the Local Void*, The Astrophysical Journal, **676**, 2008

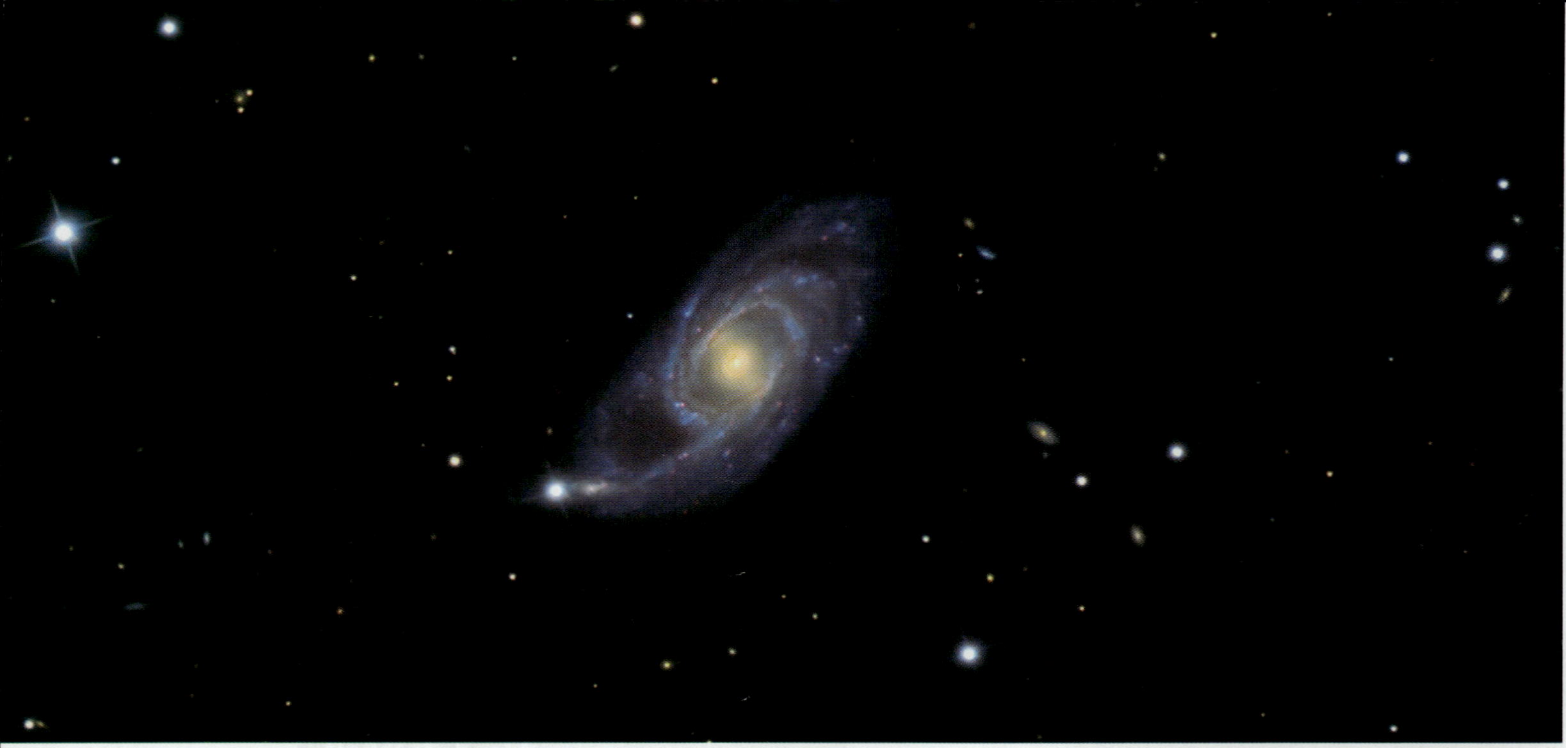

NGC 151

The SB(r)bc galaxy NGC 151 is located at a distance of 170 million light years from the Milky Way. Its angular size of 2.6′ × 1.2′ corresponds to a maximum transverse disc diameter of 130 000 light years. Due to an unrecognized duplicate observation in 1886, NGC 151 is also present in the New General Catalogue under the name NGC 153. NGC 151 has a bar of length 20″. In addition to its contrast in colour to the spiral arms, due to an older population of stars in the bar, it is notable for the rectangular form of the bulge, whose major axis is perpendicular to that of the bar. The spiral structure in NGC 151 is determined by two large spiral arms which begin not at the ends of the bar but 20° before. The resulting ring impression is typical for barred spiral galaxies. In the outer regions, the spiral arms fragment, and at the lower left, eastern, edge of the disc, part of a spiral arm extends towards the galaxy 2MASXJ00340814, which lies 1.4′ from the core of NGC 151. Close by is a 13th magnitude bright foreground star (GSC 5269 612).

So-called "M 51 types" make up a special group of binary galaxy systems. These consist of a large spiral galaxy and a small companion located at the end of a spiral arm. In the case of NGC 151 and 2MASXJ00340814, spectroscopic observations were used to determine whether it is an M 51 type. Astronomers placed the slit along a line passing through both galaxies. The analysis for NGC 151 showed the plateaus at ±200 km/s typical of spiral galaxies, whereby the values had stronger variation in the direction of the possible companion. The rotational velocity of the outer spiral-arm segment NGC 151 which lies close to 2MASXJ00340814 corresponds to the plateau value; the measured value for 2MASXJ00340814 however jumps to 280 km/s within a few arcseconds. These observations indicate that NGC 151 and 2MASXJ00340814 form an M 51 system. The close position, however, contradicts the recession velocity of 2MASXJ00340814, which is 1216 km/s higher than that of NGC 151. This contradiction is found in several M 51 systems. Astronomers explain this via a possible influence of background galaxies on the measurements or wrong catalogue data for the companions.

OBJECT	NGC 151, NGC 153
CONSTELLATION	Cetus
RIGHT ASCENSION	$00^h\ 34^m\ 03^s$
DECLINATION	−09° 42′ 19″
BRIGHTNESS	12.8 mag
TYPE	SB(r)bc
PHOTOGRAPHERS	Adam Block
TELESCOPE	800-mm reflector
CAMERA	SBIG STL-11000
EXPOSURE TIME	290 min
LOCATION	Mount Lemmon SkyCenter/ University of Arizona, USA

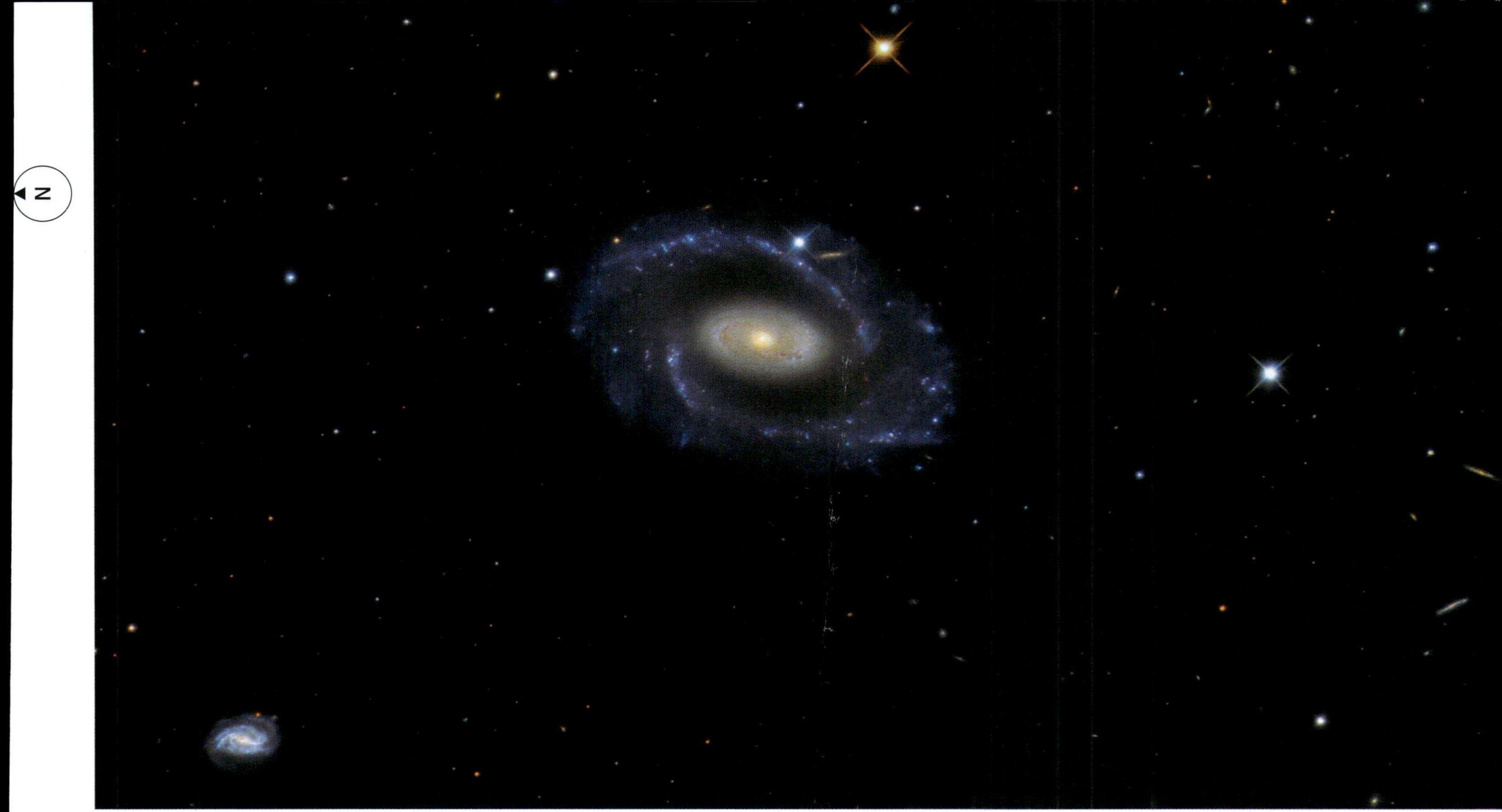

NGC 210

NGC 210 is the brightest galaxy in a small group located at a distance of 75 million light years from the Milky Way. The measured angular size of 5.0′× 3.3′ results in a projected distance of 110 000 light years for the diameter of the disc. At a distance of 7.6′ at the lower left is MCG-02-02-082, an SBab galaxy which is five times farther away than NGC 210. In the NED catalogue, NGC 210 is classified as SAB(s)b, but the morphology is not uniform. The spiral pattern in the outer regions is determined by two spiral arms which contain many active regions. The western spiral arm in the upper half of the disc can be divided into three Vorontsov-Velyaminov rows; this is not possible for the lower spiral arm, which appears more fragmented. Both spiral arms begin at the bright inner region of the disc and their opposite placement underlines the barred appearance. The inner region appears oval due to the inclination of the disc, since in this zone an additional, weakly defined spiral pattern has formed, whose two spiral arms begin at a bulge with a compact nucleus.

The distribution of neutral hydrogen in NGC 210 shows a rotating disc with a rotational velocity of ±140 km/s. The left side of the disc is receding, the right approaching the Milky Way. At the lower edge of the disc there is a further tail-like structure with a width of 10′ which runs toward the east below the right edge of the disc. Its cause is presumably a dwarf galaxy which approached NGC 210 too closely and was completely destroyed in the process. The H I structure of 6×10^8 solar masses contains about ten per cent of the total H I mass in NGC 210. The gas velocities of the structure follow the rotation of the disc and demonstrate an inward flow of hydrogen gas in the outer regions of the disc. This also explains the high star-formation rate in the spiral arms.

OBJECT	NGC 210
CONSTELLATION	Cetus
RIGHT ASCENSION	$00^h\ 40^m\ 35^s$
DECLINATION	−13° 52′ 22″
BRIGHTNESS	12.5 mag
TYPE	SAB(s)b
PHOTOGRAPHERS	Adam Block
TELESCOPE	800-mm reflector
CAMERA	SBIG STL-11000
EXPOSURE TIME	330 min
LOCATION	Mount Lemmon SkyCenter/ University of Arizona, USA

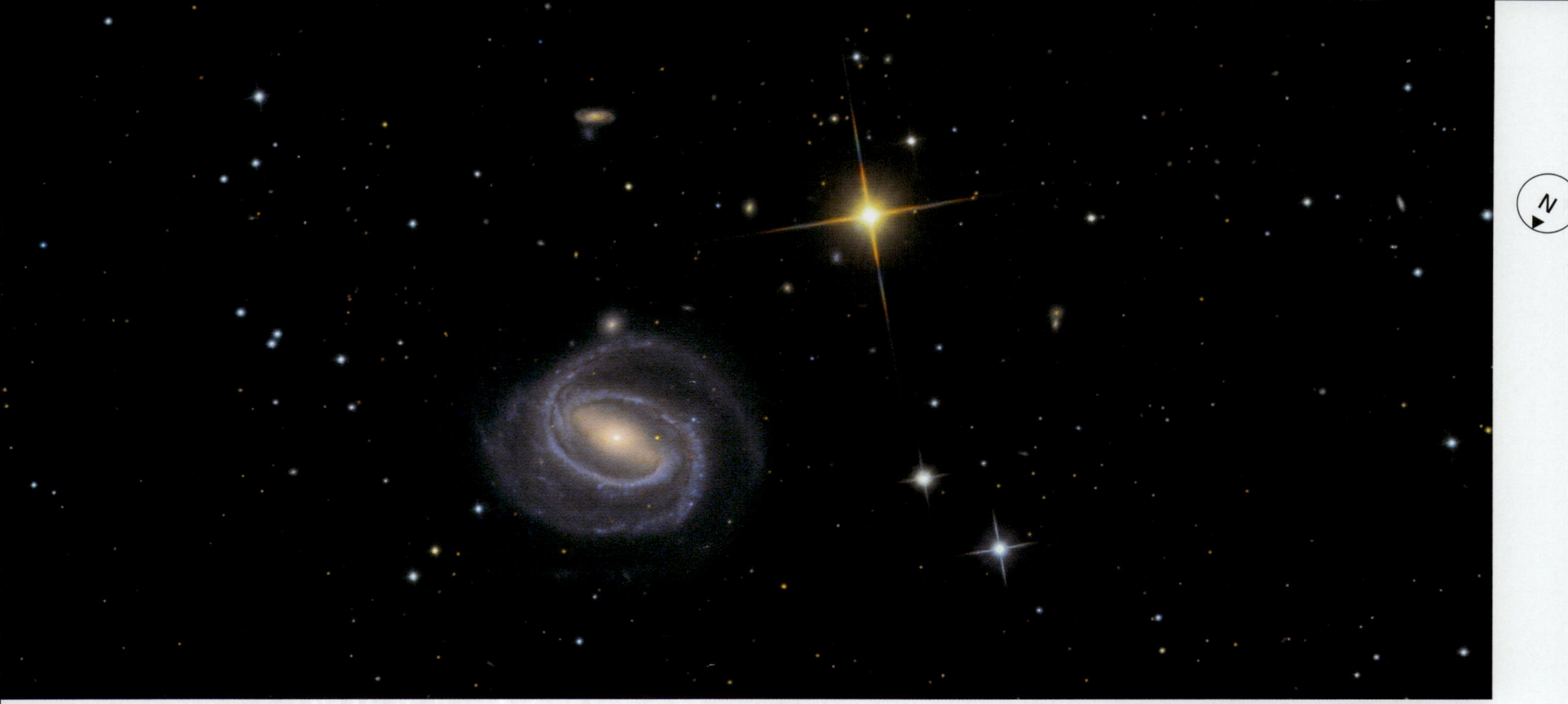

NGC 266

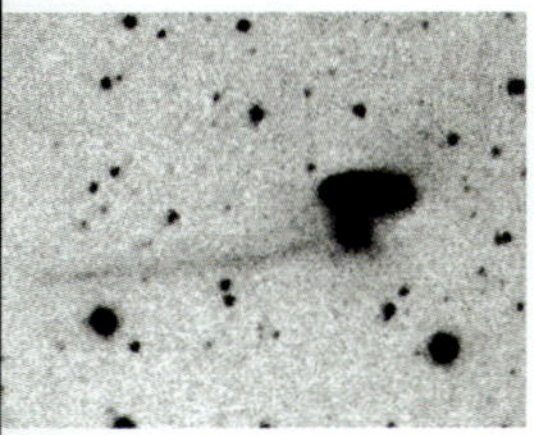

The SB(rs)ab galaxy NGC 266 is only 3.6′ away from the 8th-magnitude K star SAO 54174, at the border between the constellations Pisces and Andromeda. The recession velocity corresponds to a distance of 215 million light years and the angular size of the optical disc of 3′ to an extremely large diameter of 190 000 light years. Two spiral arms connect to the bar and each can be followed through a full rotation. In the outer reaches the spiral arms are more diffuse but still so well defined that Vorontsov-Velyaminov rows can be discerned. One can see that as a result the radial distance between the two arms doesn't continuously increase, but rather oscillates.

Above NGC 266 is the spiral galaxy 2MASX J00493722+3213548, which together with the K star forms an equilateral triangle. Since no distance data are present in the catalogues, one can use the angular size of 0.25′ and the assumption that 2MASX J00493722+3213548 is a typical spiral galaxy to deduce a distance of about 1.1 billion light years. The image shows a peculiarity of 2MASX J00493722+3213548: just below the spiral galaxy there appears to be a remarkably blue companion galaxy which has a long tidal tail. This can be followed as far as 1.6′ and, assuming that it is at the same distance as 2MASX J00493722+3213548, corresponds to a length of more than 500 000 light years (see contrast-enhanced image).

NGC 266 is not an isolated galaxy but rather a member of the far-reaching Perseus–Pisces Supercluster. Astronomers have determined that NGC 266 lies at the edge of a large cloud of neutral hydrogen which surrounds the active galaxy NGC 262 (Mrk 348). This intergalactic gas cloud has the enormous size of 600 000 × 375 000 light years. The angular distance between the two galaxies is 23′ and the relative velocity −113 km/s. Due to the similar recession velocities, interaction appears probable and mathematical models indicate a time of 10^9 years for a close encounter of the two galaxies. The lower amount of H I in NGC 266 compared to NGC 262 is an indication that the neutral hydrogen gas in the gas cloud comes from NGC 266. This missing hydrogen gas also explains the weak star formation in the outer spiral arm of NGC 266.

OBJECT	NGC 266
CONSTELLATION	Pisces
RIGHT ASCENSION	$00^h\ 49^m\ 48^s$
DECLINATION	+32° 16′ 40″
BRIGHTNESS	12.5 mag
TYPE	SB(rs)ab
PHOTOGRAPHERS	Adam Block
TELESCOPE	800-mm reflector
CAMERA	SBIG STL-11000
EXPOSURE TIME	450 min
LOCATION	Mount Lemmon SkyCenter/ University of Arizona, USA

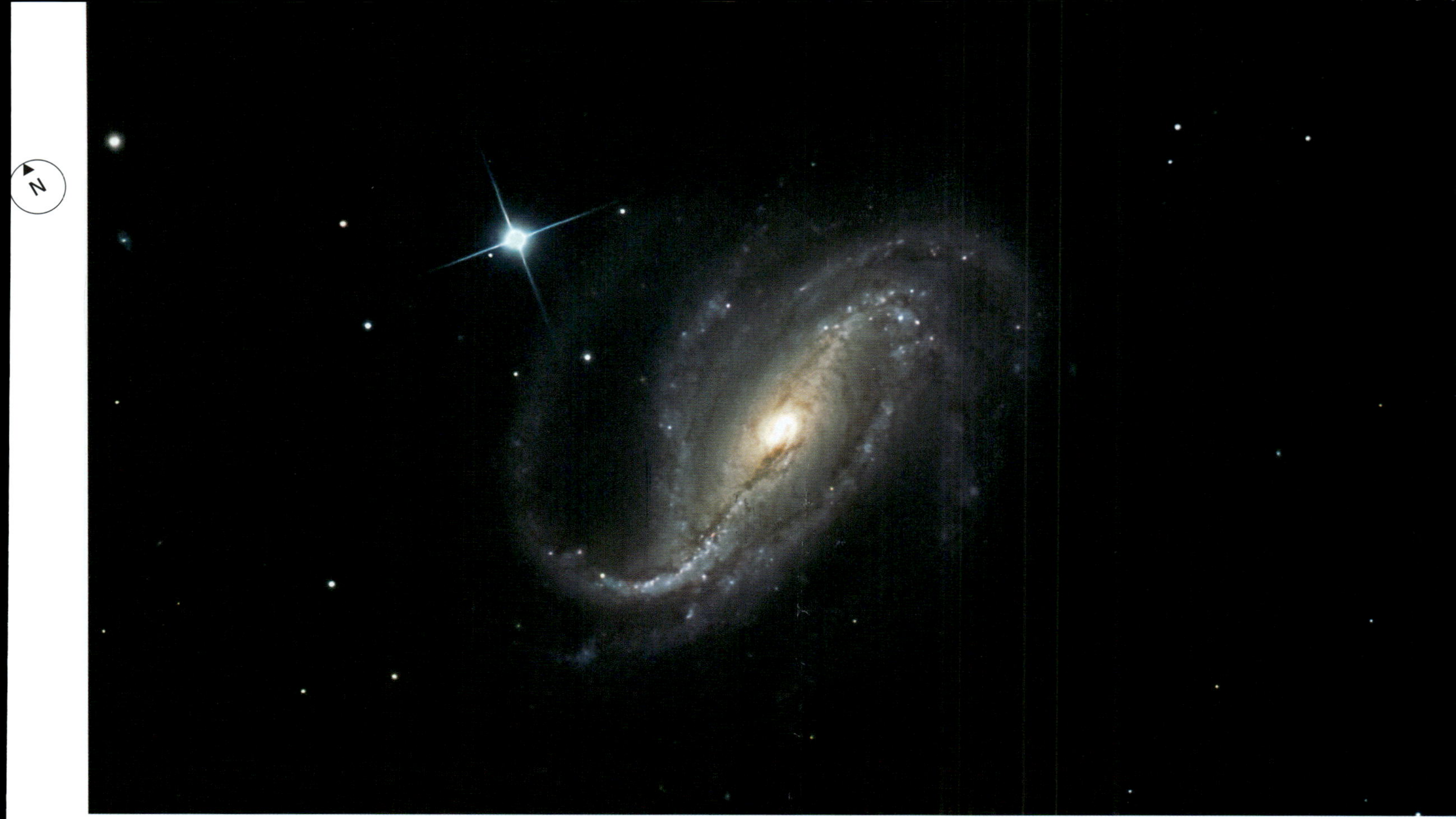

NGC 613

The SB(rs)bc galaxy NGC 613 is 65 million light years from the Milky Way and the angular size of 5.5′ × 4.2′ corresponds to a transverse angular size of 100 000 light years. The form of NGC 613 is determined by a broad, 3.1′ long bar which is crossed by dust lanes. In the lower, southeast half of the bar the dust is prominently visible. Moving along the bar, one notices a jump in the dust lanes near the centre, which contains an active galactic nucleus. Two spiral arms are connected to the ends of the bar, with many active regions in the transition zones. The spiral pattern of NGC 613 is irregular, and the tight structures of the inner arms form an oval ring which surrounds the bar.

Radio observations of NGC 613 in the NRAO VLA Sky Survey (NVSS) show, in addition to the bright radio source, two sources which lie within ±2° of the axis of the optical bar. These are located at a distance of 1.4′ from the centre and correspond to the inner ends of the spiral arms. Their radio luminosity is eight times lower than that of NGC 613. In addition to the radio data, observations by the Hubble Space Telescope provide further hints of the effects of the active nucleus. For example, a circumnuclear ring consisting of several active region was detected in the central region of NGC 613. With a diameter of about 1500 light years, this ring surrounds a region of highly concentrated molecular hydrogen gas. Astronomers could show that the age of star clusters in the circumnuclear ring is correlated with the hydrogen concentration. The lower the H_2 concentration, the older are the stars in the star clusters. The dust structures in the bar are tangential to this ring and this explains the observed jump in the bar of NGC 613.

OBJECT	NGC 613
CONSTELLATION	Sculptor
RIGHT ASCENSION	$01^h\ 34^m\ 18^s$
DECLINATION	−29° 25′ 06″
BRIGHTNESS	10.7 mag
TYPE	SB(rs)bc
PHOTOGRAPHERS	Josef Pöpsel
TELESCOPE	600-mm reflector
CAMERA	SBIG ST-10XME
EXPOSURE TIME	120 min
LOCATION	Amani Lodge, Namibia

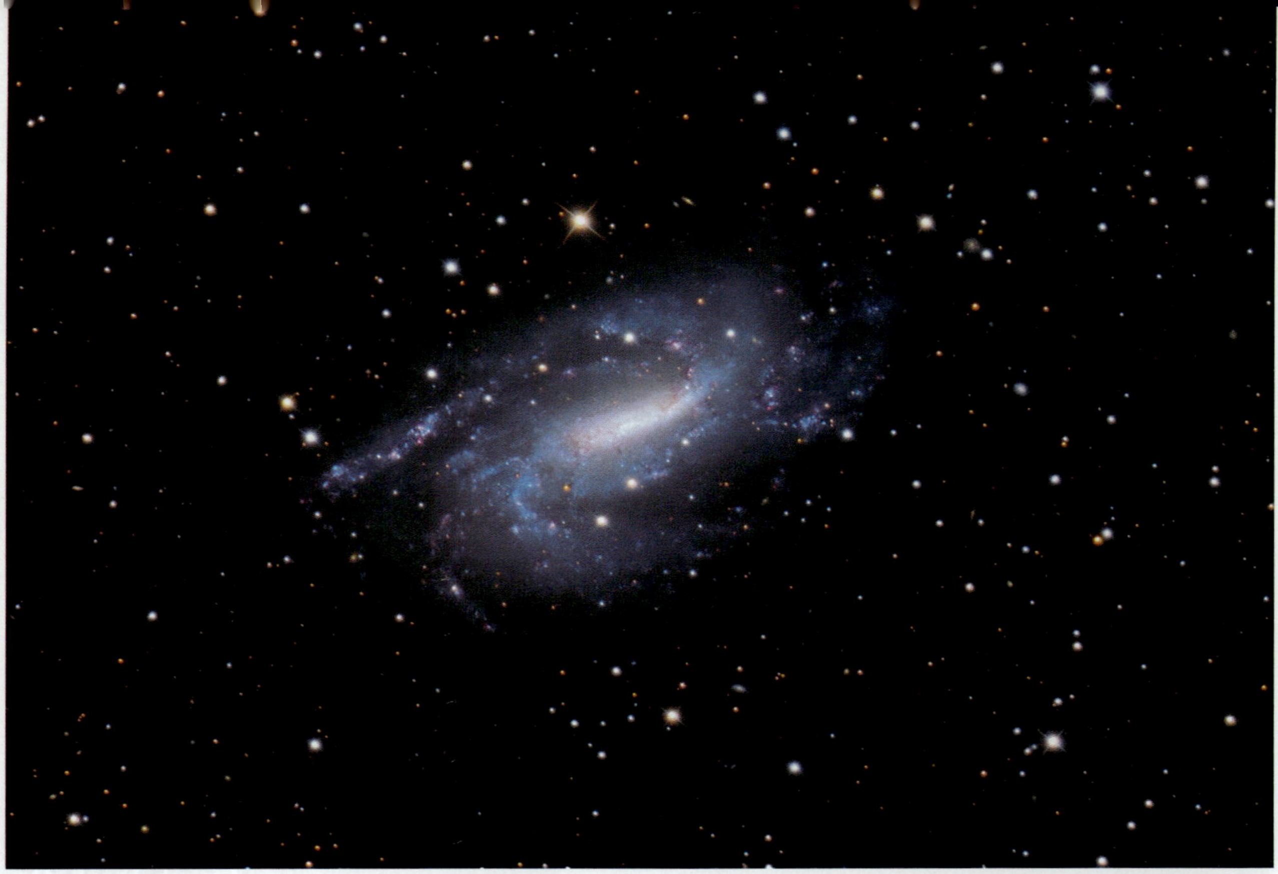

NGC 925

The SAB(s)d galaxy NGC 925 lies at a distance of 28 million light years and is part of a galaxy group in the constellations Triangulum and Andromeda which also contains NGC 891 and NGC 1023. The optical diameter of 10.5′ corresponds to a projected disc diameter of 86 000 light years. The bar in NGC 925 has a particularly blue-white colour and thus deviates from the standard bar colour of yellow. Additionally, there is no compact bright core at the centre of the bar. The two spiral arms are rich in active regions and one can recognize a number of H II regions. The largest of these complexes has an angular size of 5″ and hence 700 light years. The H II regions define the course of the spiral arms and extend to the edge of the disc. The southern spiral arm at the left edge of the disc has more of these regions and is better defined as the second, flocculent spiral arm. Due to its proximity to the Milky Way, single stars brighter than 20 mag in blue can be resolved in NGC 925.

The asymmetry in the spiral arms in NGC 925 is also seen in radio observations of neutral hydrogen and in Hα observations of ionized hydrogen gas. In the southern spiral arm, the strongest H I and Hα emission regions coincide, which is not the case in the portion of the northern spiral arm on the opposite side. According to the density-wave theory, a colour gradient occurs perpendicular to spiral arms, which is due to an age gradient of the stars in the arm. This gradient can usually be seen as a shift between the maxima in the red and blue luminosity in the arm cross-section. In the spiral-arm fragments in the northern part of the disc, in contrast, no colour gradient is seen. Star formation in these fragments is thus determined by local processes and there is no large-scale organization such as a spiral-arm pattern. The dominance of the southern spiral arm, as well as the form and colour of the bar, indicate that NGC 925 is a transitionary type which will further evolve into a Magellanic galaxy.

OBJECT	NGC 925
CONSTELLATION	Triangulum
RIGHT ASCENSION	$02^h\ 27^m\ 17^s$
DECLINATION	+33° 34′ 45″
BRIGHTNESS	10.7 mag
TYPE	SAB(s)d
PHOTOGRAPHERS	Adam Block
TELESCOPE	800-mm reflector
CAMERA	SBIG STL-11000
EXPOSURE TIME	250 min
LOCATION	Mount Lemmon SkyCenter/ University of Arizona, USA

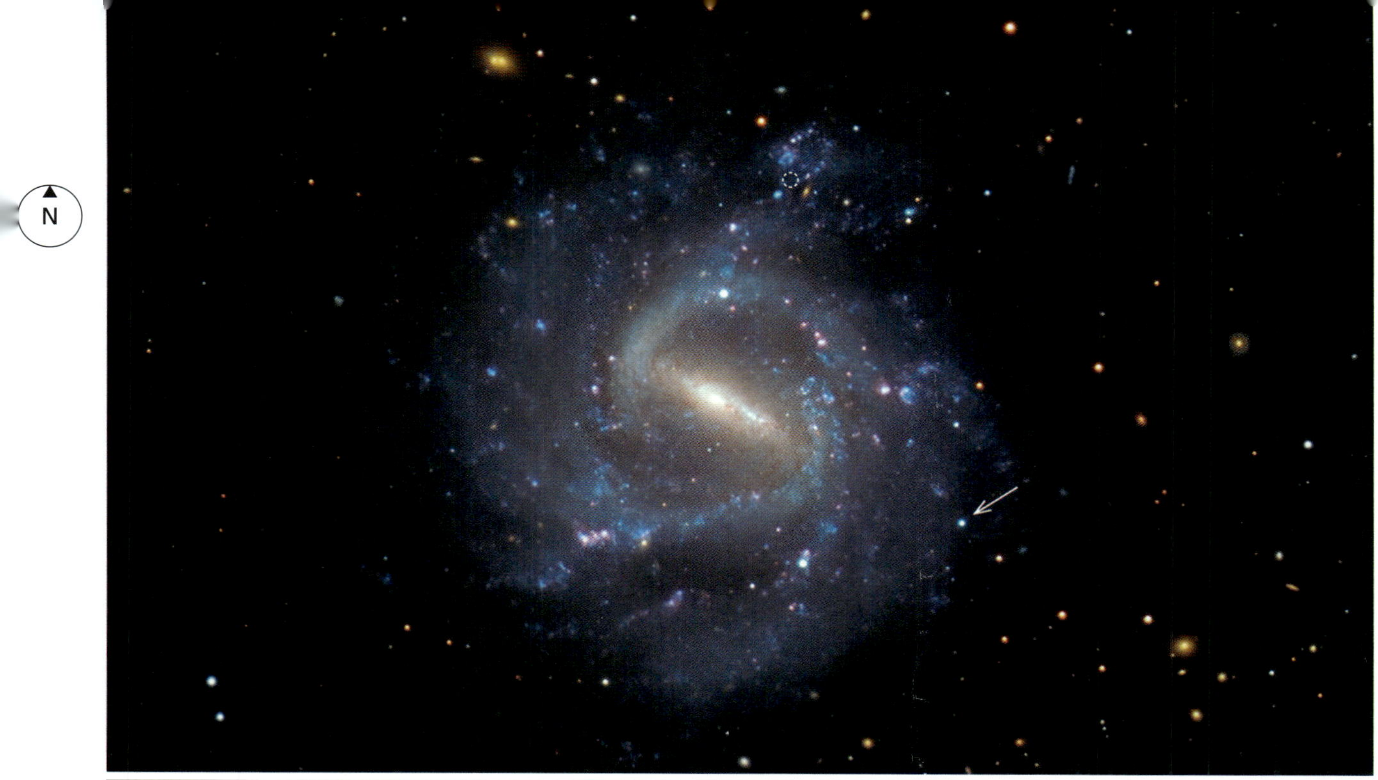

NGC 1073

The SB(rs)c galaxy NGC 1073 is 48 million light years from the Milky Way and belongs to the NGC 1068 galaxy group. The almost face-on disc has an angular size of 5.5′ × 4.8′ and its projected diameter of 77 000 light years lies in the standard range for spiral galaxies. The bar is narrow and one recognizes an asymmetry in the star-formation regions along the axis of the bar. The dust structures in this half are more prominent and partially cover the compact bright nucleus of NGC 1073. In the western spiral arm, which begins at this active side of the bar, more active regions can be discerned than in the opposite arm. Both spiral arms break up after half a revolution and are very dim in the outer parts of the disc.

Several background galaxies can be seen through the outer edge of the disc of NGC 1073. Prominent is the galaxy 2MASSX J02434668+012225077 near the upper left edge of the disc. In addition to the yellow colour of this galaxy of size 14″, notable are the two nuclei, catalogued as the Chandra source CXO J24346.62+012508.9. X-ray data show the quasar [HB89] 0240+011 (Hewitt+Burbidge QSO compilation) as the brightest source, visible in the image as a bright point source (see arrow). The redshift of [HB89] 0240+011 is z = 0.599 and its distance thus 8 billion light years. The nucleus of the galaxy has only one-sixth the X-ray luminosity of this quasar. An additional X-ray source, 20 per cent brighter than the nucleus, is CXO J024338+012411 or, for short, IXO5 (Intermediate Luminosity X-ray Object).

The position of IXO5 is marked by a dotted circle at the upper edge of the disc. During the search for the optical counterpart to IXO5, with the Hubble Space Telescope two stars were found within the error radius of 0.3″ which, due to their spectral classes of B and A, are candidates for companion stars in the IXO5 system. IXO5 is an ultra-luminous X-ray binary which contains a black hole of 10–20 solar masses. Its X-ray emission, however, is not isotropic, but is beamed and thus appears highly amplified. The X-ray luminosity of IXO5 exceeds all known X-ray binaries in the Milky Way and Magellanic Clouds by a factor of several.

OBJECT	NGC 1073
CONSTELLATION	Cetus
RIGHT ASCENSION	$02^h\ 43^m\ 41^s$
DECLINATION	+01° 22′ 34″
BRIGHTNESS	11.5 mag
TYPE	SB(rs)c
PHOTOGRAPHERS	Makis Palaiologou, Stefan Binnewies
TELESCOPE	1.3-m reflector
CAMERA	SBIG STX-16803 and Andor DZ 436
EXPOSURE TIME	480 min
LOCATION	Skinakas Observatory, Crete, Greece

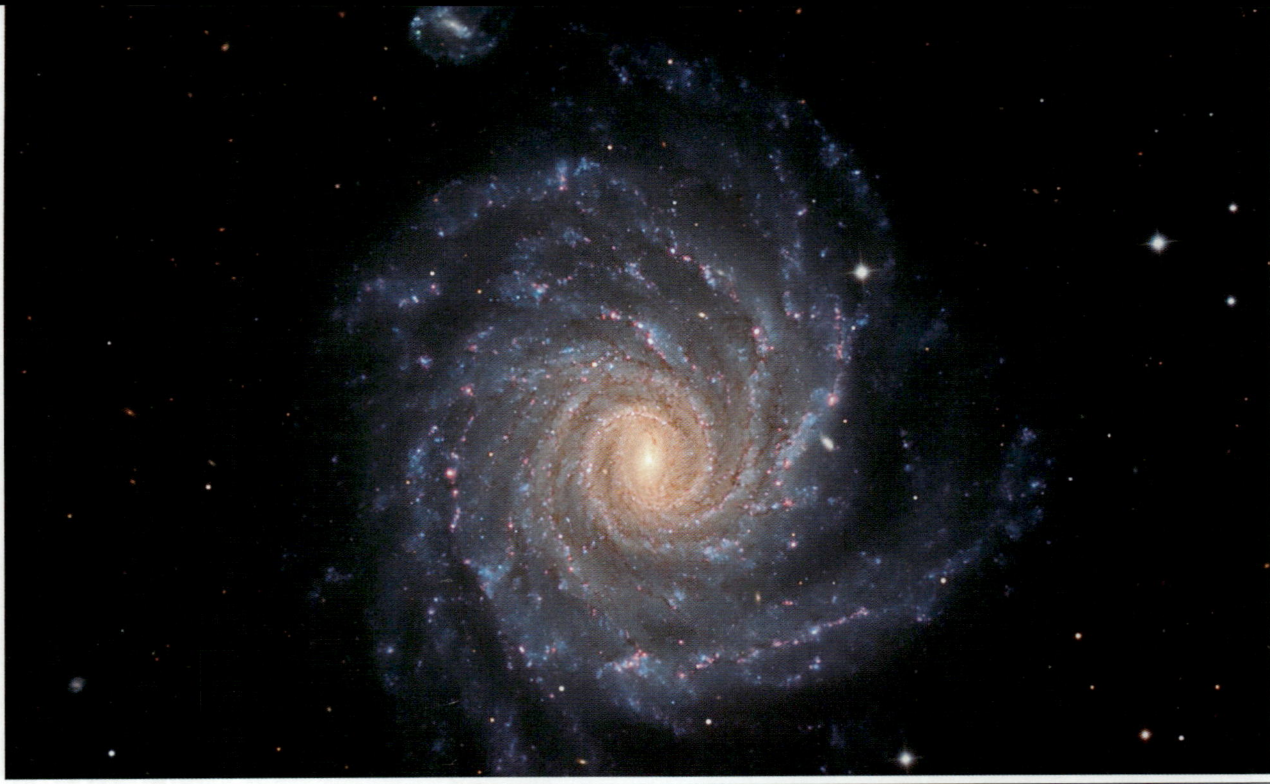

NGC 1232

The SB(rs)c galaxy NGC 1232 is 68 million light years from the Milky Way. The angular size of this grand-design spiral is 7.4′ × 6.5′ and corresponds to a projected disc diameter of 146 000 light years. The bright nucleus of NGC 1232 is compact and part of a short, weak bar which is oriented vertically in the image. Along the bar, however, the central region is elongated, and in the surrounding yellow bulge there are a few dust lanes which lead to the beginning of the spiral arms. The spiral pattern in NGC 1232 is determined by several spiral arms which, however, do not form a flocculent pattern, but rather can be followed through almost an entire revolution. The numerous H II regions in the spiral arms highlight Vorontsov-Velyaminov rows; due to the face-on view, their 120° angles and increasing length are particularly easy to recognize.

At the upper edge of the image one sees 4′ towards the east the galaxy NGC 1232A (PGC 011834). Since the recession velocity of this SB galaxy at 6599 km/s is greater than that of NGC 1232 at 1603 km/s, interaction is unlikely. The distance, based on the recession velocity of 290 million light years, means that the angular size of 0.9′ corresponds to a typical disc diameter of 75 000 light years for NGC 1232A.

Observations with the Chandra X-ray Observatory (NASA) reveal a peculiarity in X-rays. The data show a diffuse, comet-like emission cloud. Its luminosity centre doesn't correspond to that of NGC 1232 but rather lies 14 000 light years away in the southern half of the disc. In the image, this position is to the left of the centre of NGC 1232, but there is no optical counterpart. The X-rays are due to thermal emission of 6×10^6 K hot gas in the cloud, whose total mass is estimated at a few 10^6 solar masses. Astronomers suspect that the cloud is the remains of a dwarf galaxy which passed through NGC 1232 and was destroyed in the process. When passing through the halo of NGC 1232, the gas was heated by a shock wave and left a prominent trace. Since the spiral structure of the galaxy was not perturbed, it is probable that NGC 1232 contains a large amount of dark matter.

OBJECT	NGC 1232
CONSTELLATION	Eridanus
RIGHT ASCENSION	$03^h\ 09^m\ 46^s$
DECLINATION	−20° 34′ 46″
BRIGHTNESS	10.9 mag
TYPE	SAB(rs)c
PHOTOGRAPHERS	Josef Pöpsel
TELESCOPE	600-mm reflector
CAMERA	SBIG ST-10XME
EXPOSURE TIME	140 min
LOCATION	Amani Lodge, Namibia

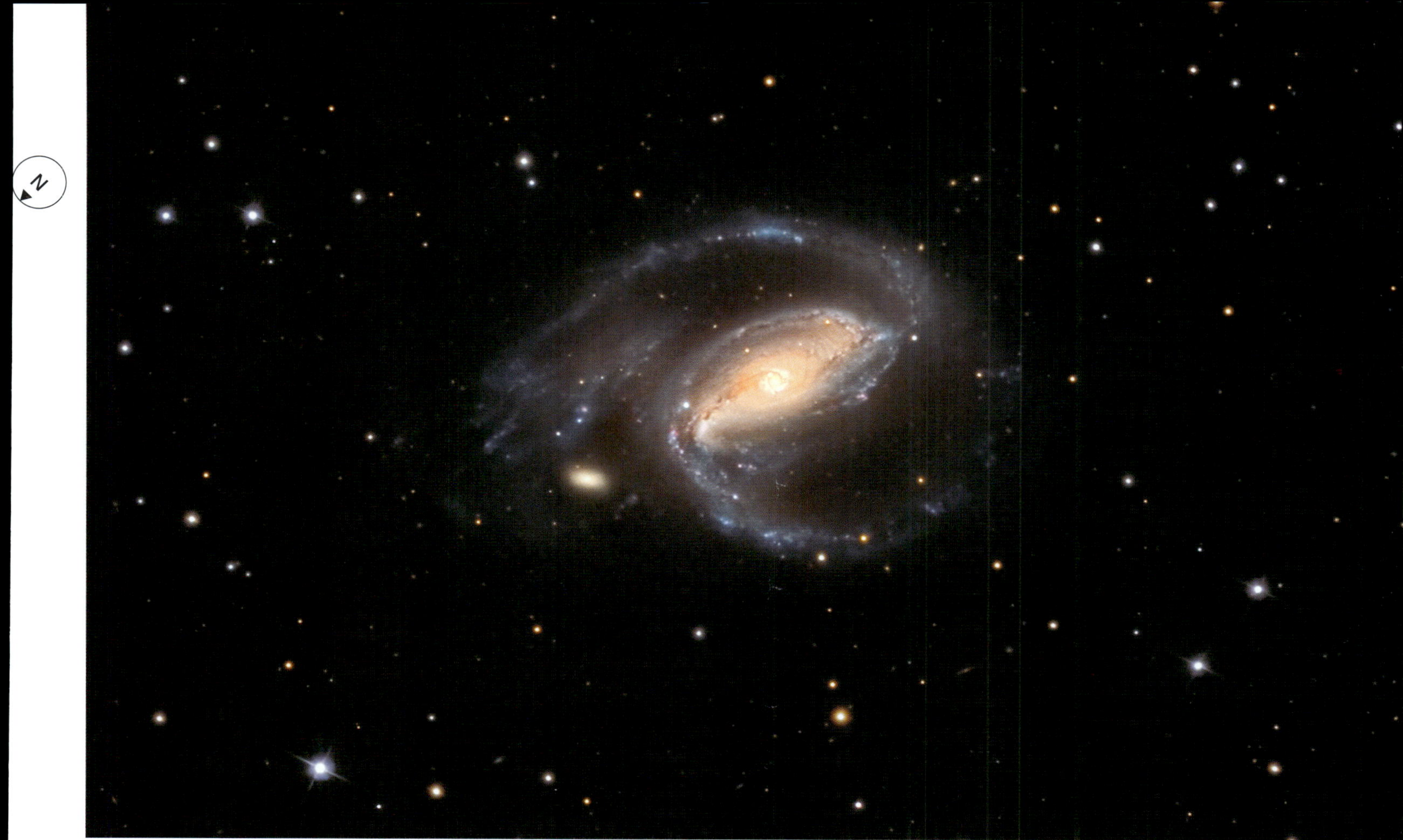

NGC 1097

The barred spiral galaxy NGC 1097 is located within the constellation Fornax and is the brightest galaxy of a galaxy group consisting of five members (LGG 075, Lyon Group of Galaxies Catalogue). Its distance from the Milky Way is 53 million light years and its angular size of 9.3′ × 6.3′ corresponds to a diameter of the optical disc of 148 000 light years. The bar of NGC 1097 contains dust lanes which end at a bright, ring-shaped central region. Two spiral arms are connected to the ends of the bar. They are characterized by dust lanes which can each be followed through half a revolution. The upper, western spiral arm fragments at the end and there contains a small elliptical companion (NGC 1097A). It velocity relative to NGC 1097 is only 97 km/s and the Spitzer Space Telescope shows that the dust structure in NGC 1097 has a hole at the position of the companion. NGC 1097A could have passed through the disc of NGC 1097 at this location. The contrast-enhanced image overleaf displays stellar streams in the environment of NGC 1097. The longest is the northeastern stream (lower right in the image), whose end, almost in the form of a right angle, is more than 200 000 light years from NGC 1097. The image shows only three streams; a short fourth stream opposite the northeastern stream can be discerned on deeper images. All of these streams display a prominent X-shaped form.

At the core of NGC 1097 is an active Seyfert core surrounded by a ring of several hundred star-forming regions with a diameter of 5000 light years. This region contains netlike substructures which indicate the transport of gas and dust to the core. Radio observations of NGC 1097 also confirm the active nucleus, but also show that the optical stellar streams have no H I emission. They are thus not hydrogen gas pulled by tidal forces out of a companion which came too close to the gravitational field of a large galaxy. Also, the stellar streams NGC 1097 are too blue for the visible companion NGC 1097A to be the source.

Thus, one has used computer simulations in an attempt to explain the origin of the stellar streams. The model has a massive disc potential and a small, spherical satellite galaxy

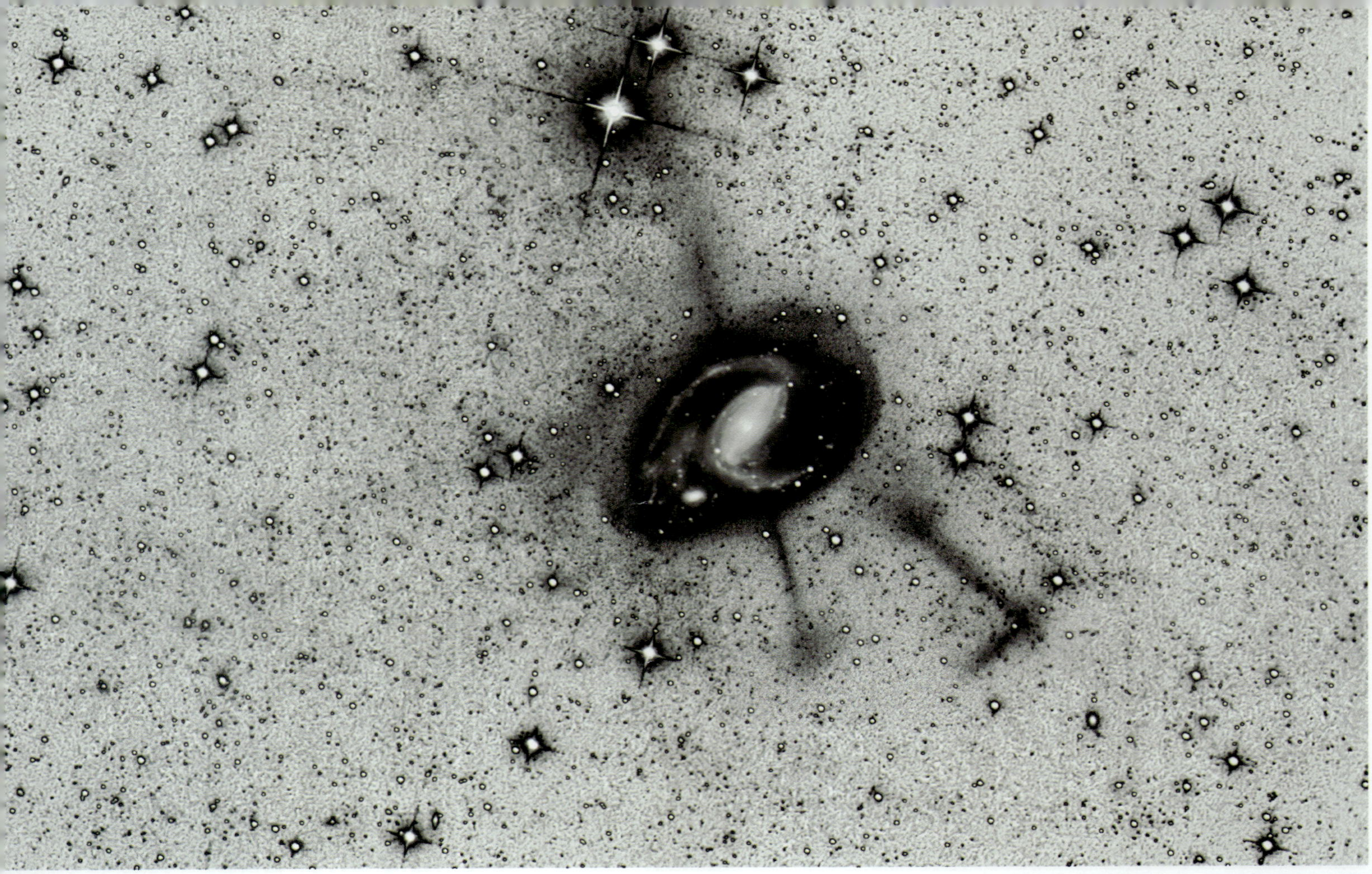

which collides several times with the disc and is cannibalized in the process. The model mass of this dwarf galaxy is 0.1 per cent of the mass of NGC 1097, its diameter 14 000 light years. When the companion galaxy crosses through the plane of the disc, this has almost no effect on NGC 1097, but the companion loses almost all of its interstellar medium to the gas disc in NGC 1097. This loss of gas causes star formation in the satellite galaxy to almost completely stop and in the next two to three billion years most of its stars will burn out. A billion years after passing through the disc, the companion galaxy reaches its maximum distance of 300 000 light years and falls back to the disc of NGC 1097, which is crossed for the second time after 1.5 billion years. Due to its angular momentum, the companion adopts a drop-like shape before the second collision. The various orbits of the stars at the outside of the drop are the basis for the stellar streams, which were formed after two further passages through the disc. After 3.7 billion years, a warped ring of stellar streams has formed which, seen in perspective, appears as an X-shaped structure. The simulations explain not only the X-shape but also the observed right-angle ends of the stellar streams observed in NGC 1097 and the differences in luminosity between the stellar streams. Only the interaction of several processes allows the streams to be visible in this special form and proves that such "NGC 1097 types" are very rare.

OBJECT	NGC 1097
CONSTELLATION	Fornax
RIGHT ASCENSION	$02^h\ 46^m\ 19^s$
DECLINATION	–30° 16′ 30″
BRIGHTNESS	10.2 mag
TYPE	SB(s)b
PHOTOGRAPHERS	Dietmar Böcker
TELESCOPE	600-mm reflector
CAMERA	SBIG ST-10XME
EXPOSURE TIME	90 min
LOCATION	Amani Lodge, Namibia
CONTRAST-ENHANCED IMAGE	Makis Palaiologou, Stefan Binnewies, 300-mm reflector, SBIG STL-6303, 330 min, Skinakas Observatory, Crete, Greece

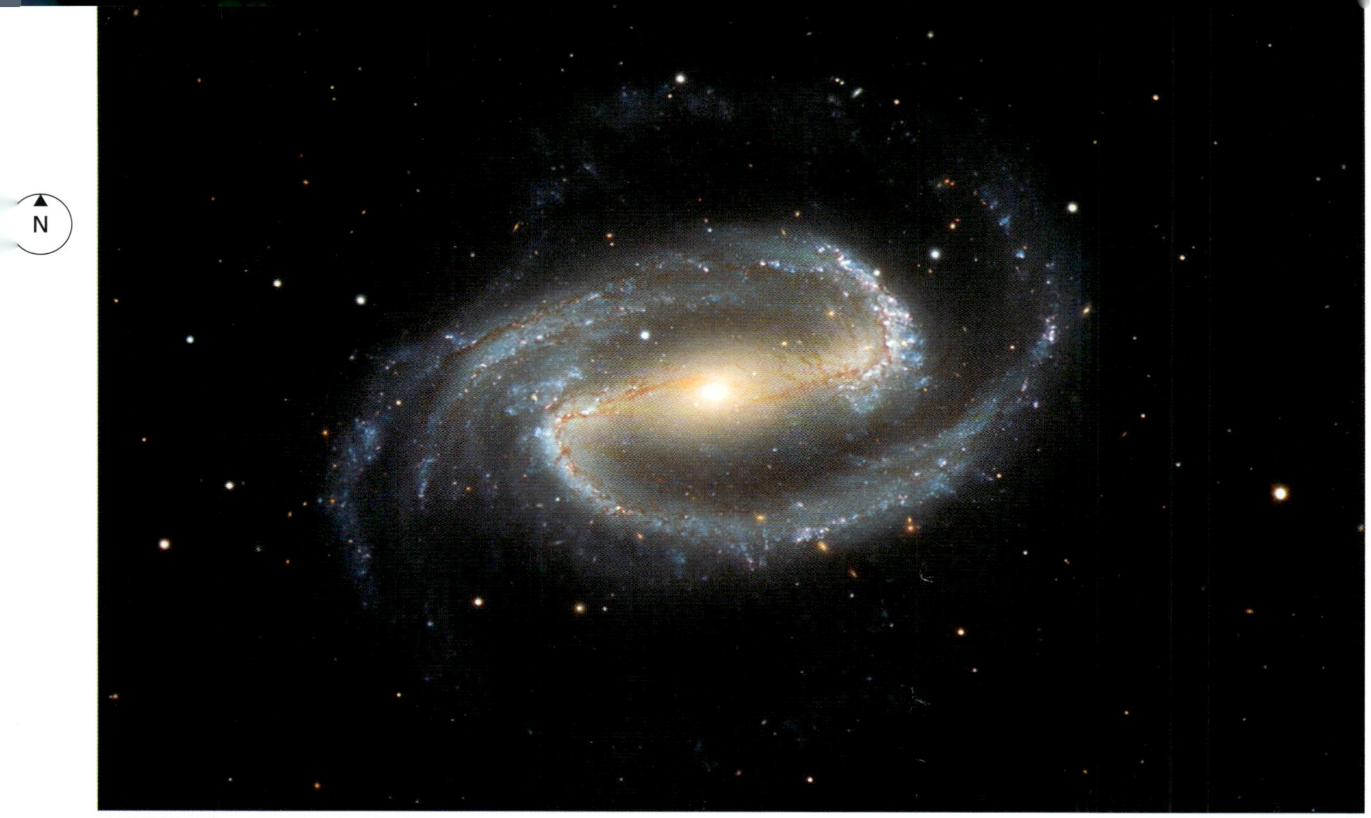

NGC 1300

The SB(rs)bc galaxy NGC 1300 – often used as the prototype of a barred spiral galaxy – is at a distance of 68 million light years from the Milky Way. Due to its recession velocity, it is considered to belong to the Fornax–Eridanus galaxy association. The angular size of 6.2′ × 4.1′ corresponds to a projected disc diameter of NGC 1300 of 123 000 light years. The prominent bar is 2.8′ long and contains brown dust lanes. These can be followed beyond the ends of the bar into the two spiral arms, where they surround active regions. In the spiral arms are several H II regions, the brightest of which can be seen at the beginning of the western, right spiral arm. The diffuse yellow bulge has the same orientation as the bar and has a bright, differentiated core region. It is notable that the dust lanes are not aligned with the axis of the bar, but rather depart from this toward the centre. The diameter of the bright central region is about 3500 light years.

It has been shown with the Hubble Space Telescope that a spiral structure has formed in the central region itself. The resolution of the image is not sufficient to display this pattern, but the miniature spiral arms can be seen as a ring-like brightening. The connected dust structure in NGC 1300 is interpreted as a flow of gas into the centre, which one can follow both on the large scale of the spiral arms and the large bar and on the smaller scale of the nuclear region. The gas dynamics indicates a mass of 7×10^7 solar masses for the supermassive black hole. This black hole, however, is not active; the activity in the nuclear region consists only of an elevated star-formation rate of 0.2 solar masses per year in a nuclear ring. This consists of 12 H II regions which are located at distances of 3–4″ (in the mean this corresponds to 1100 light years) from the centre. Another peculiar aspect of the image is the large number of background galaxies which can be seen through the disc of NGC 1300. These galaxies are recognized by their orange colour and, assuming the standard size of a spiral galaxy, one can estimate their distance at 1.5 billion light years.

OBJECT	NGC 1300
CONSTELLATION	Eridanus
RIGHT ASCENSION	$03^h\ 19^m\ 41^s$
DECLINATION	−19° 24′ 41″
BRIGHTNESS	11.4 mag
TYPE	SB(rs)bc
PHOTOGRAPHERS	Makis Palaiologou, Stefan Binnewies
TELESCOPE	1.3-m reflector
CAMERA	Andor DZ 436
EXPOSURE TIME	225 min
LOCATION	Skinakas Observatory, Crete, Greece

N

NGC 1365

NGC 1365 is an SB(s)b galaxy which belongs to the Fornax galaxy cluster and lies at a distance of 60 million light years from the Milky Way. Its angular size of 11.2′ × 6.2′ corresponds to a projected diameter of 200 000 light years. The prominent bar, about 3′ long, has a length of 53 000 light years and is characterized by the disruption of its otherwise straight shape on the left, eastern side. Here one can see an offset to the end of a spiral arm. There are also more H II regions than in the other spiral arm. Typical of SBb types is the opposite orientation of the dust lanes and the axis of the bar. At the left side of the bar the dust lane is below the axis of the bar, at the right side above.

NGC 1365 has a small, diffuse, elongated bulge at an angle of 45° to the axis of the bar. The active Seyfert nucleus of NGC 1365 can be seen as a compact source in the image and is surrounded by two dust lanes. Along the northern spiral arm, which extends towards the top of the image, one can see a dim, preceding spiral arm. This duplication occurs more weakly in the other spiral arm and explains why NGC 1365 is also described as a multi-armed galaxy.

The contours of H I radio observations of NGC 1365 follow the active regions in the spiral arms; in the active region almost no neutral hydrogen is detectable. The H I rotation curves extend to the optical edge of the disc and show a drop of ±200 km/s when moving outward in the region of the bar. Over 40 per cent of the radius, the shape of the curve corresponds to the Keplerian 1/r form. Compared to other Sb galaxies, NGC 1365 is not surrounded by an extensive H I envelope. From both observations, one can conclude that NGC 1365 has a small halo with a relatively small amount of dark matter. More recent observations with NASA's Chandra telescope have discovered a further peculiarity in the nucleus of NGC 1365. The absorption of X-rays is variable on timescales of a few minutes and astronomers interpret this as occultation of the central X-ray source by clouds of material near the accretion disc. Relativistic models of this scenario indicate that the central black hole has a mass of 2×10^6 solar masses and rotates very rapidly. This rotation in NGC 1365 is only slightly less than the theoretical maximum limit and is an indication that the flow of material onto the black hole is responsible not only for the increase in mass but also that the transfer of angular momentum can cause the rapid rotation of the black hole.

OBJECT	NGC 1365
CONSTELLATION	Fornax
RIGHT ASCENSION	$03^h\ 33^m\ 36^s$
DECLINATION	−36° 08′ 25″
BRIGHTNESS	10.3 mag
TYPE	SB(s)b
PHOTOGRAPHERS	Philipp Keller, Konstantin Buchhold, Bernd Flach-Wilken, Johannes Schedler, Volker Wendel (Chart 32-Team)
TELESCOPE	800-mm reflector
CAMERA	FLI Proline 16803
EXPOSURE TIME	1000 min
LOCATION	CTIO, Chile

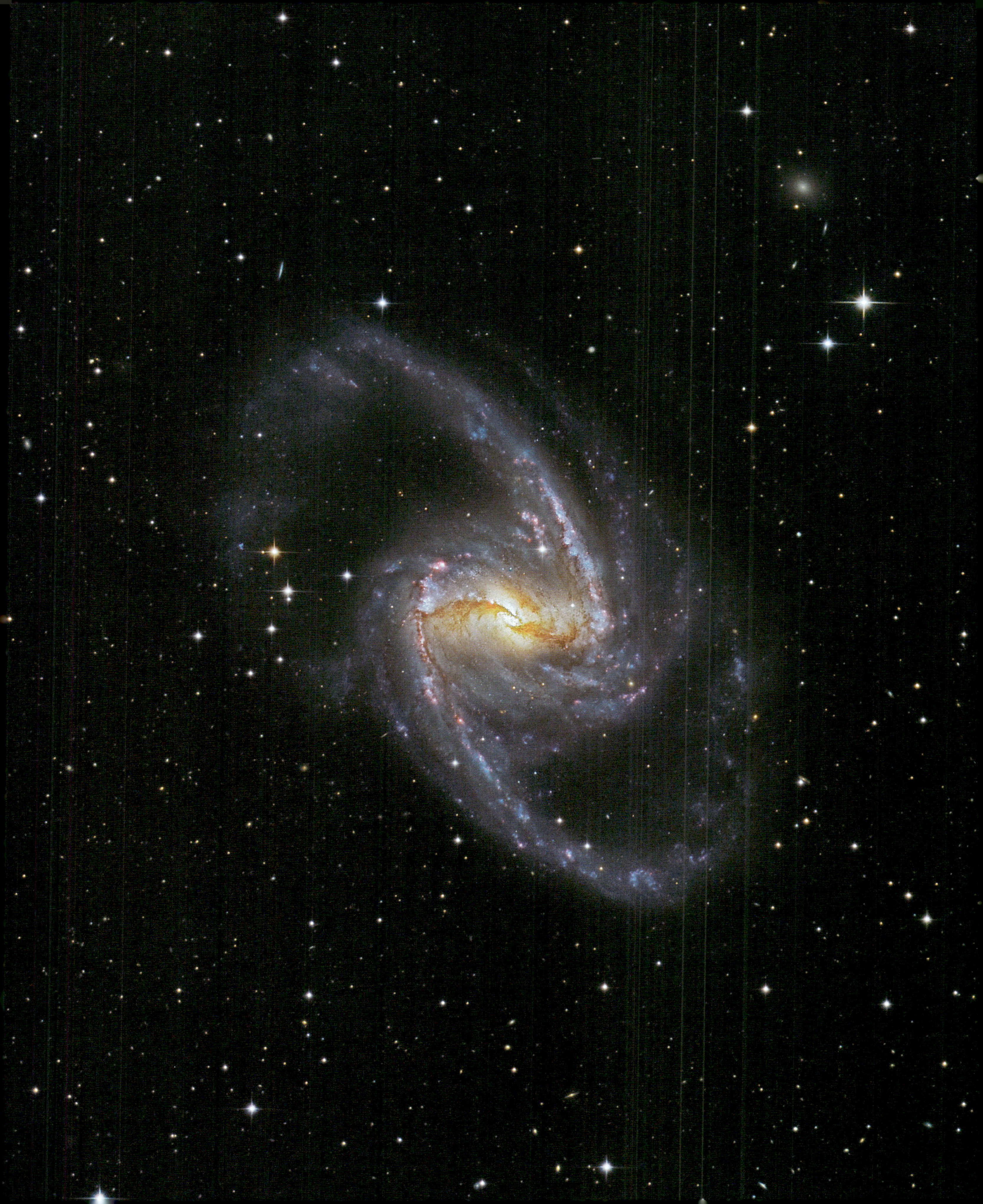

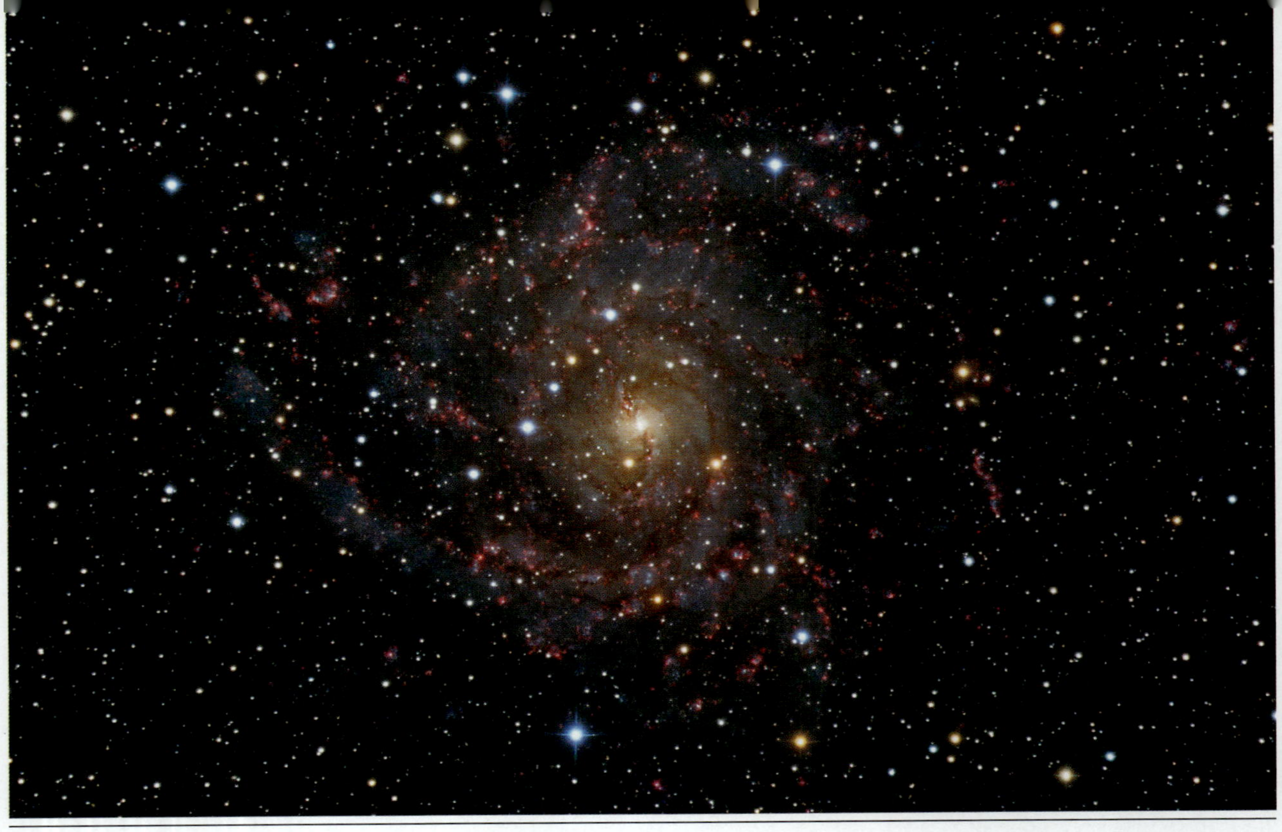

IC 342

The SAB(rs)cd galaxy IC 342 at a Galactic latitude of 10.6° is close to the plane of the Milky Way. It is the largest galaxy among the approximately dozen members of the Maffei/IC 342 galaxy group. The distance to IC 342 is 10 million light years and despite its large angular size of 21.4′ × 20.9′, its projected disc diameter of 62 000 light years is smaller than the standard size of a spiral galaxy (angular sizes are from the NED). Current studies indicate that the disc of neutral hydrogen with a diameter of 42′ is significantly larger. It is notable that spiral structure is also seen in this radio disc and deep amateur images even show star-forming regions in the dense areas. Based on the farthest H II regions, the diameter of IC 342 is 115 000 light years.

Due to obstruction by dust and gas, the spiral arms in IC 342 appear dim. In addition to the description "hidden galaxy", this also explains why the galaxy was not discovered until 1890, by the British astronomer William F. Denning. In infrared light, IC 342 appears much more prominent. Here one can see both a central bar and a two-arm spiral pattern which has many net-like connections between the spiral-arm fragments.

Barred spiral galaxies have, in addition to the bar, other special zones, such as the inner ring surrounding the bar or the ends of the bar. In the bar itself the star-formation rate of 0.1–0.4 solar masses per year is in the range typical for star-forming regions. There, gas and dust are transported and one finds that in bright bars usually straight dust structures are observed. In weaker bars, as in IC 342, one often sees warped shapes instead. Astronomers have been able to detect flows of molecular gas in the bar of IC 342 which occur primarily at the ends of the bar and are displaced by about 1000 light years compared to hydrogen gas. The reason for this is still unclear, but it is suspected that the gas flows in the bar are more complex than previously assumed and that both star formation and tidal forces can exert strong influences on the bar.

OBJECT	IC 342
CONSTELLATION	Camelopardalis
RIGHT ASCENSION	$03^h\ 46^m\ 49^s$
DECLINATION	+68° 05′ 47″
BRIGHTNESS	9.1 mag
TYPE	SAB(rs)cd
PHOTOGRAPHERS	Johannes Schedler
TELESCOPE	400-mm reflector
CAMERA	SBIG STL-11000
EXPOSURE TIME	1080 min
LOCATION	Wildon, Austria

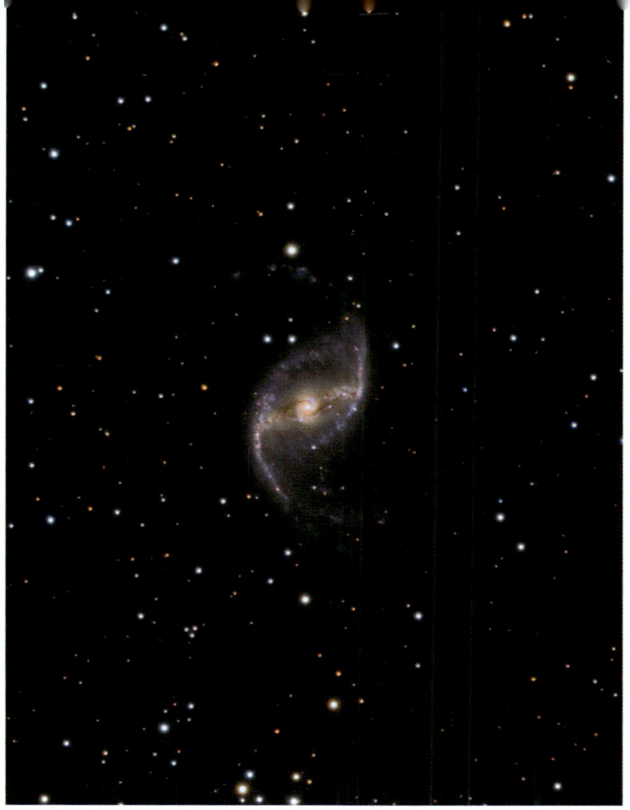

NGC 1530

NGC 1530 is an essentially isolated SB(rs)bc galaxy whose nearest neighbour NGC 1530A (IC 381) is about 19′ away. The angular size of NGC 1530 is 4.6′ × 2.4′. The recession velocity of 2461 km/s leads to a distance of 113 million light years and hence an enormous diameter of 154 000 light years. Alternative methods of distance determination lead to a smaller distance of only 80 million light years and thus a smaller projected disc diameter of almost 100 000 light years. In the case of the neighbour NGC 1530A, the various methods of distance determination are in better agreement. Its recession velocity of 2476 km/s results in a distance of 115 million light years and a disc size of 80 000 light years, which corresponds well to the standard for spiral galaxies. If one applies this standard to NGC 1530 as well, it would be only 80 million light years away and would be receding from the Milky Way faster than average.

NGC 1530 has a prominent bar with dust structures which run along the edges of the bar into the centre and there connect to a small spiral structure. This is only 12″ across and appears to be extended perpendicular to the axis of the bar. In the small spirals, as in the bar, a few active areas with large H II regions can be seen in the image. The spiral structure of NGC 1530 is determined by two spiral arms whose active regions in the transition zone from the bar to the arms are especially numerous. The spiral arms don't start at the transitions, but rather are shifted by about 120° and form a lentil-shaped structure which almost completely surrounds the bar zone. The contrast-enhanced image shows, in the outer regions of the disc, both dim tidal tails and a splitting of the lower, southern spiral arm with Vorontsov-Velyaminov rows.

The assumption among barred spiral galaxies is that star formation is weak in prominent bars. According to this theory, the flow of gas and dust in such bars is so strong that the collapse of molecular clouds, and thus the formation of new stars, is prevented by shock waves and shear flows. NGC 1530 is one of the few exceptions to this rule. In studies of the Hα emission in NGC 1530, astronomers found 17 bright H II emission regions in the bar near the inner, small spiral. The two brightest of these regions are in the right, northwestern half of the bar. One can show via star-formation simulations that H II regions which are farther from the dust lane are 1.5–2.5 billion years older than those close by. It could also be shown that the position of the H II regions are correlated with the dust filaments which are situated perpendicular to the axis of the bar. These long filaments are also visible in the image and are proof of the flow of matter from the lintel-shaped rand towards the axis of the bar. Compared to the flow of gas in the bar, in the dust filaments the gas density is higher and the transport velocity is lower. Thus, even in a barred galaxy with a prominent bar, as is the case in NGC 1530, star formation can take place.

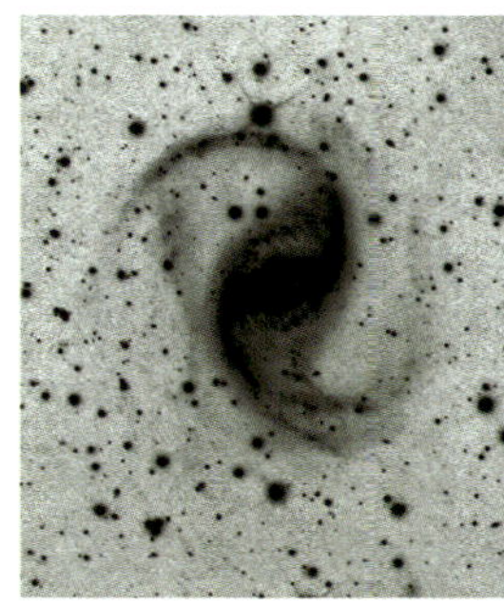

OBJECT	NGC 1530
CONSTELLATION	Camelopardalis
RIGHT ASCENSION	$04^h\ 23^m\ 27^s$
DECLINATION	+75° 17′ 44″
BRIGHTNESS	12.3 mag
TYPE	SB(rs)b
PHOTOGRAPHERS	Adam Block
TELESCOPE	600-mm reflector
CAMERA	SBIG STL-11000
EXPOSURE TIME	395 min
LOCATION	Mount Lemmon SkyCenter/ University of Arizona, USA

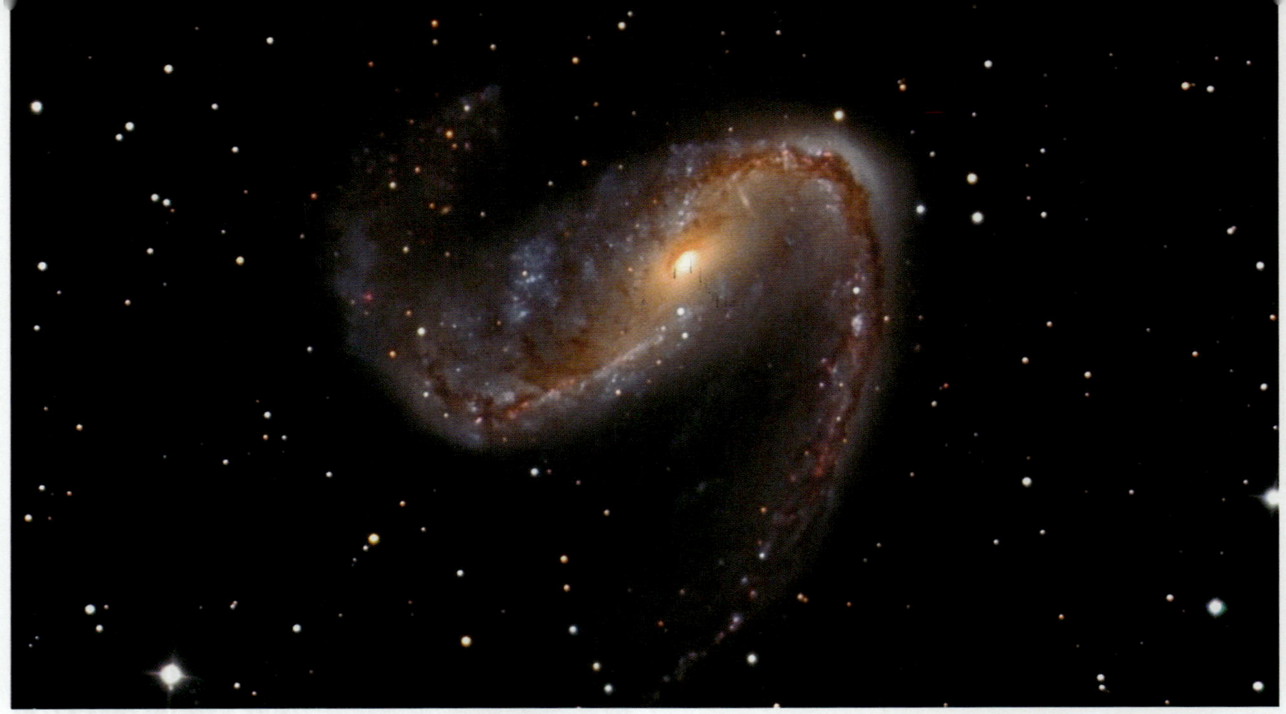

NGC 2442

NGC 2442 is a barred spiral galaxy which belongs to a small galaxy group in the southern constellation Volans. Its distance from the Milky Way is 56 million light years, its angular size 5.5′ × 4.9′, and its projected disc diameter is calculated to be 90 000 light years. The classification of NGC 2442 is given as SAB(s)bc, though often with the addition of "peculiar". NGC 2442 has a diffuse, spindle-like central region with a bright nucleus which is surrounded by a semicircle of dust. This inner region is surrounded by wide spiral arms. The bar morphology of NGC 2442 is due to the kinks in the spiral arms. This change in direction at the beginning of the arms is diametral, but farther along the arms are asymmetrical. The southern arm originates at the lower, northern end of the inner region, splits, bends through 120°, and changes into a broad, cloud-like structure at the end, rotated through 180° compared to the original direction. The northern spiral arm at the lower right shows less deformation and extends farther. The colour of NGC 2442 is dominated by red and brown hues. In stark contrast to these is a luminous, blue star-formation region, about 7000 light years in size, at the edge of the southern spiral arm.

The strength and direction of galactic magnetic fields can be studied via polarized radio emission. These magnetic fields are determined by the flow characteristics of interstellar gas and lead to polarization of synchrotron emission in the radio. Due to non-circular motion of the gas, in barred spirals overdensities and shock waves and thus strong magnetic fields are formed. In the case of NGC 2442, the map of polarized radio emission shows that the magnetic field in sections of the spiral arms is usually parallel to the spiral arm. In the northern spiral arm, near the kink in the arm one can detect a sudden change in the direction of the magnetic-field vector of 40° and a corresponding high magnetic-field strength of 26 μG (microgauss). In comparison, the magnetic field of the Earth is much stronger, at mid-northern latitudes the vertical component is about 0.44 gauss. The typical magnetic-field strengths in spiral galaxies are in the range 8–12 μG, in some starburst galaxies, local magnetic fields reach 50–100 μG. The intense gas flows in NGC 2442 are presumably caused by tidal interactions. In the near environment of NGC 2442 there are a few dwarf galaxies whose relative positions also point in the direction of the deformed spiral arms of NGC 2442 and thus confirm this assumption.

OBJECT	NGC 2442
CONSTELLATION	Volans
RIGHT ASCENSION	$07^h\ 36^m\ 24^s$
DECLINATION	−69° 31′ 51″
BRIGHTNESS	11.2 mag
TYPE	SAB(s)bc pec
PHOTOGRAPHERS	Dietmar Böcker
TELESCOPE	600-mm reflector
CAMERA	SBIG ST-10XME
EXPOSURE TIME	110 min
LOCATION	Amani Lodge, Namibia

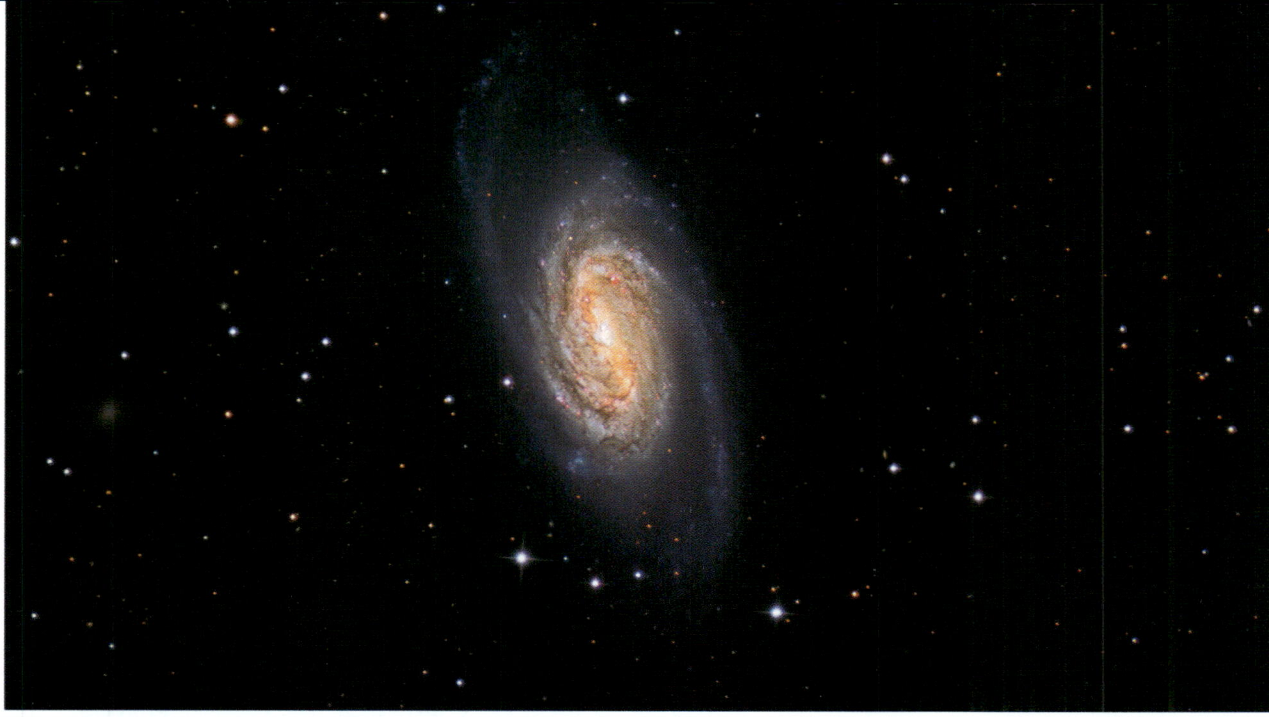

NGC 2903

The SAB(rs)bc galaxy NGC 2903 is one of the brightest galaxies in the northern hemisphere and at a distance of 24 million light years is part of the neighbourhood of the Local Group. Although there are five about equally bright Messier galaxies in the constellation Leo (M 65, M 66, M 95, M 96, M 105), NGC 2903 is not in the Messier catalogue. The angular size of NGC 2903 is 12.6′ × 5.5′ and the projected disc diameter is calculated to be 88 000 light years. In the image, the almost imperceptible bar is almost vertical and is defined by the central dust structure. Large amounts of dust also dominate the inner regions and the spiral arms which begin there. The number of arms is difficult to determine; in the outer regions the pattern is dominated by two spiral arms which extend to the edge. In the dusty inner regions, NGC 2903 has many active areas with large H II regions.

To the left of NGC 2903 in the image one can see the galaxy UGC 5086 located 9.3′ away. Due to the small relative velocity of only +46 km/s, UGC 5086 is seen as a companion of NGC 2903. In addition to UGC 5086, NGC 2903 has two further dwarf satellites which, however, are farther away and thus not in the image.

The star-formation rate of 2.2 solar masses per year in NGC 2903 is similar to that in the Milky Way (about 4 solar masses per year), and also the rotation curve of NGC 2903, with a maximum at 210 km/s followed by a drop to 180 km/s, corresponds over a wide range to that of the Milky Way. H I radio observations show a disc of neutral hydrogen which extends more than three times as far as the optical disc. Mathematical models indicate a halo mass of 1.8×10^{12} solar masses for NGC 2903. These models also give the distribution of halo matter on certain length scales. One can thus determine how many compact H I clouds should be observed – if the models corresponded to the distribution of dark matter. This number always depends, however, on the resolution of the radio telescope used. Based on the results of observations with the Arecibo radio telescope, one would expect 230 such clouds in NGC 2903. Only one further companion was found, however: NGC 2903-HI-1. Although the inclination of 65° reduces the expected number of satellites due to projection effects, the significant discrepancy indicates incorrect model assumptions. Astronomers thus believe that there are many companions but that these contain very little hydrogen and thus few stars.

OBJECT	NGC 2903
CONSTELLATION	Leo
RIGHT ASCENSION	$09^h\ 32^m\ 10^s$
DECLINATION	+21° 30′ 03″
BRIGHTNESS	9.7 mag
TYPE	SAB(rs)bc
PHOTOGRAPHERS	Johannes Schedler
TELESCOPE	400-mm reflector
CAMERA	SBIG STL-11000
EXPOSURE TIME	520 min
LOCATION	Wildon, Austria

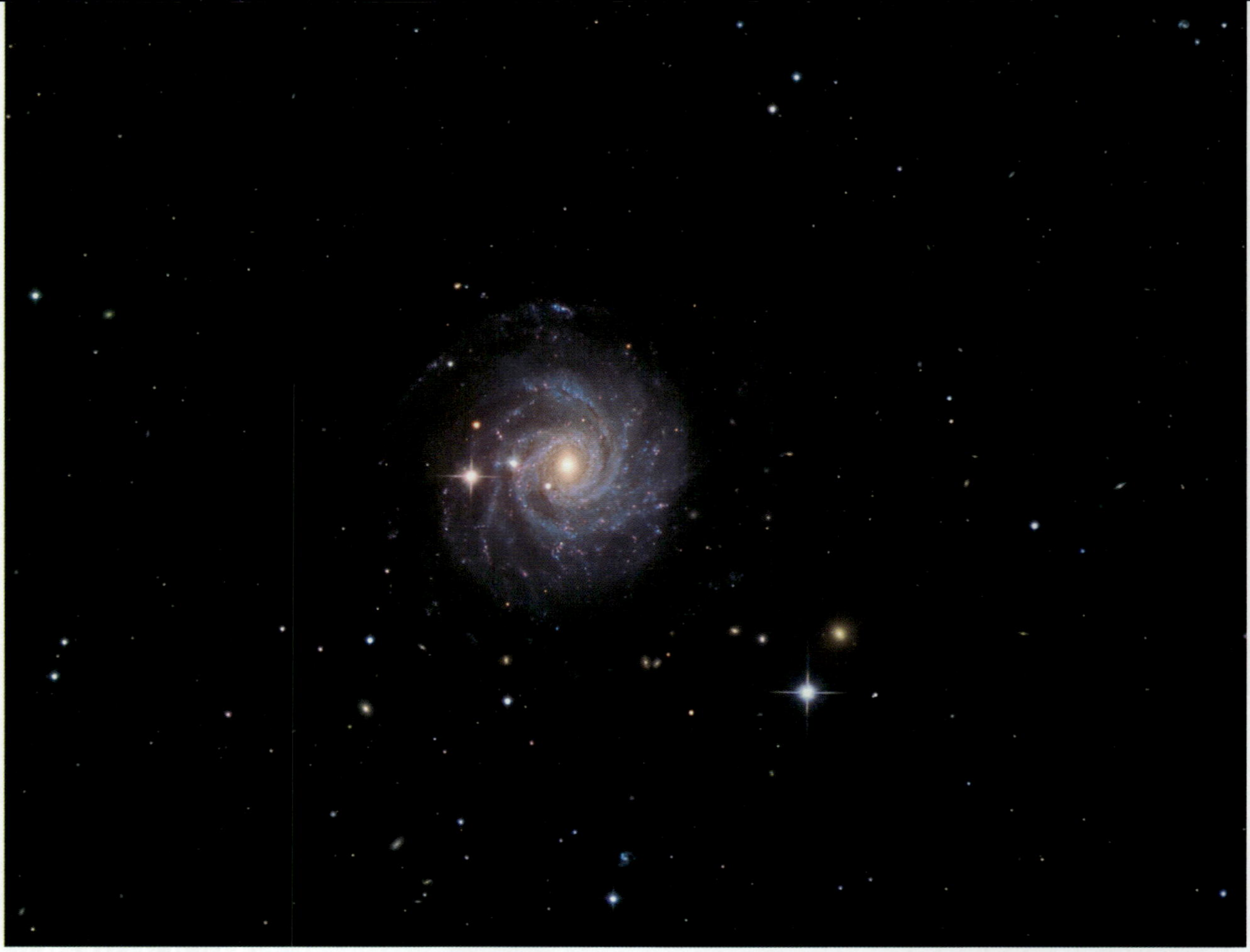

NGC 3344

The SAB(r)bc galaxy NGC 3344 is at a distance of 22 million light years from the Milky Way. Its angular size of 7.1′ × 6.5′ corresponds to a projected diameter of 45 000 light years. In relation to the disc, the bar appears strikingly small. Since the bulge is also relatively large, NGC 3344 was not classified as a barred spiral in earlier catalogues. The two larger spiral arms begin at the ends of the bar and, due to the tight spiral pattern, form an almost complete inner ring. The surrounding spiral-arm zone is filled with many active areas containing many H II regions and, in the outer third of the disc, fades into a dim outer strip which is surrounded by spiral-arm fragments. This gives the impression of an additional outer ring.

H I radio data on NGC 3344 indicate an asymmetry in the rotation of the disc. The inner disc rotates regularly while farther out the velocity field is inhomogeneous and can be described only by a model with a strongly warped disc. In the model, the inclination thus changes from 25° almost immediately to 45° in the outer third. An exception, however, is the bulge, whose direction of rotation differs from that of the disc. The rotation curve of NGC 3344 reaches a plateau at 150 km/s which is typical for smaller galaxies. Fitting for the mass distribution, one arrives at a large contribution from a halo of dark matter which dominates the outer half of the disc. Since NGC 3344 is an isolated galaxy, astronomers suspect that the ring structures and the small bar could have been formed via accretion events.

OBJECT	NGC 3344
CONSTELLATION	Leo Minor
RIGHT ASCENSION	$10^h\ 43^m\ 31^s$
DECLINATION	+24° 55′ 20″
BRIGHTNESS	10.5 mag
TYPE	SAB(r)bc
PHOTOGRAPHERS	Stefan Binnewies, Josef Pöpsel
TELESCOPE	600-mm reflector
CAMERA	SBIG-STL-16803
EXPOSURE TIME	375 min
LOCATION	Skinakas Observatory, Crete, Greece

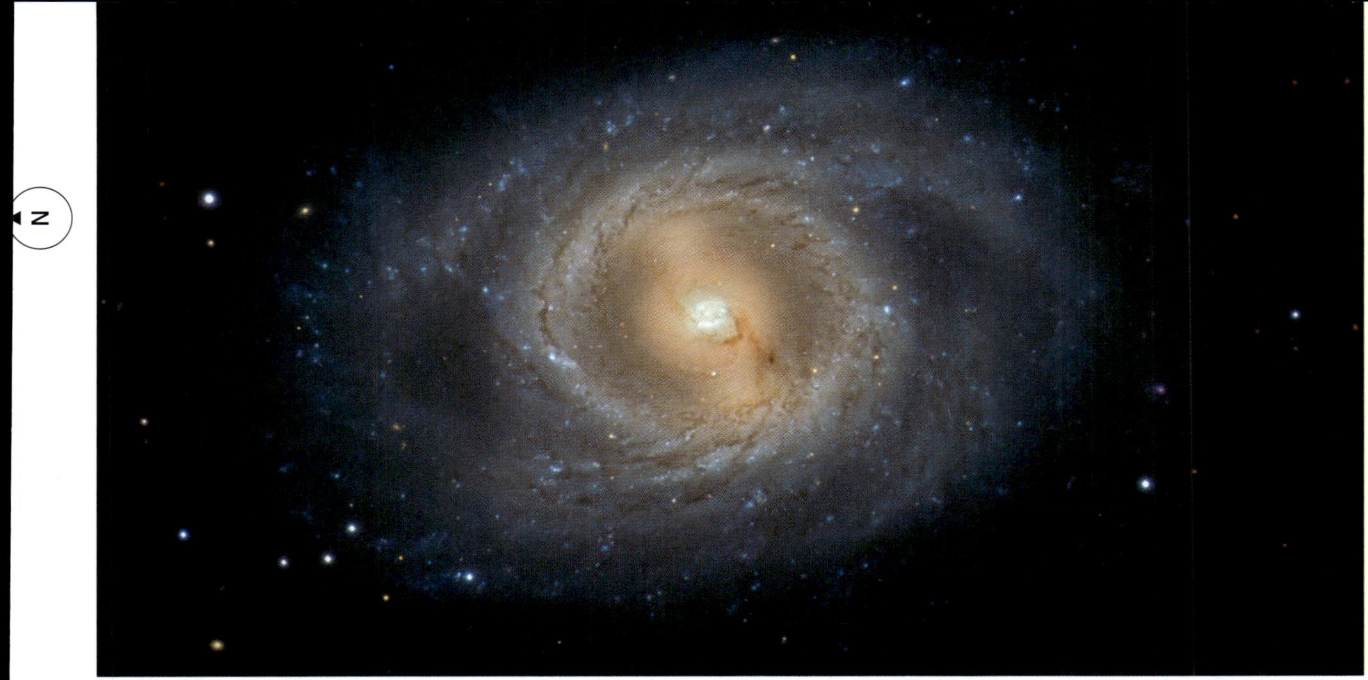

M 95

M 95 is an SB(r)b galaxy located at a distance of 32 million light years which is part of the Leo Group. The disc plane of M 95 is seen from the Milky Way almost face-on. The angular size of 7.4′ × 5.0′ corresponds to a maximum transverse diameter of almost 70 000 light years. M 95 has a stellar bar with prominent ansae at the ends. Dust structures which continue into the spiral arms are visible in the lower and thus eastern half of the bar. Both spiral arms form a complete ring in this transition zone. The outer arms appear dimmer and more fragmented; no arm segment can be followed through more than 90°. The galactic nucleus is surrounded by a ring of star-formation and H II regions, visible in the image as a ring-shaped series of bright white point sources. The diameter of this circumnuclear star-forming region (CNSFR) is 20″ and corresponds to a projected distance of 3100 light years. Compared to the CNSFR, the galactic nucleus is redder and is only 350 light years across. The brightest regions of the ring have a star-formation rate of 0.4–0.5 solar masses per year, significantly more than in the disc. Thus M 95 is known as a nuclear starburst galaxy.

More detailed studies of M 95 reveal three prominent H II complexes containing young O stars. Measured velocities indicate a radial flow into the core at 25 km/s. Outside of the central bar, the motion of the gas is more regular and is characterized by circular motion in the plane of the disc. The MERLIN radio telescope (Multi-Element Radio Linked Interferometer Network, Manchester, England) was used to search for supernova remnants (SNR) in M 95. Six typical SNR radio sources were found in the region of the disc, three of them at distances 0.9′–1.4′ from the nucleus. The analysis showed that probably only one of these is within M 95. The others are, based on the fact that they are extended sources, distant active galactic nuclei. Despite the high star-formation rate in M 95, astronomers have not been able to find any SNR there. This leads to the conclusion that signatures of SNR in the radio must be short-lived, due to the strong radial gas currents. The star-formation rate at the centre of this barred spiral takes place much faster than in the outer regions of spiral galaxies, where interaction processes can also induce star formation.

OBJECT	M 95
CONSTELLATION	Leo
RIGHT ASCENSION	$10^h\ 43^m\ 58^s$
DECLINATION	+11° 42′ 14″
BRIGHTNESS	11.4 mag
TYPE	SB(r)b
PHOTOGRAPHERS	Makis Palaiologou, Stefan Binnewies
TELESCOPE	1.3-m reflector
CAMERA	Andor DZ 436
EXPOSURE TIME	105 min
LOCATION	Skinakas Observatory, Crete, Greece

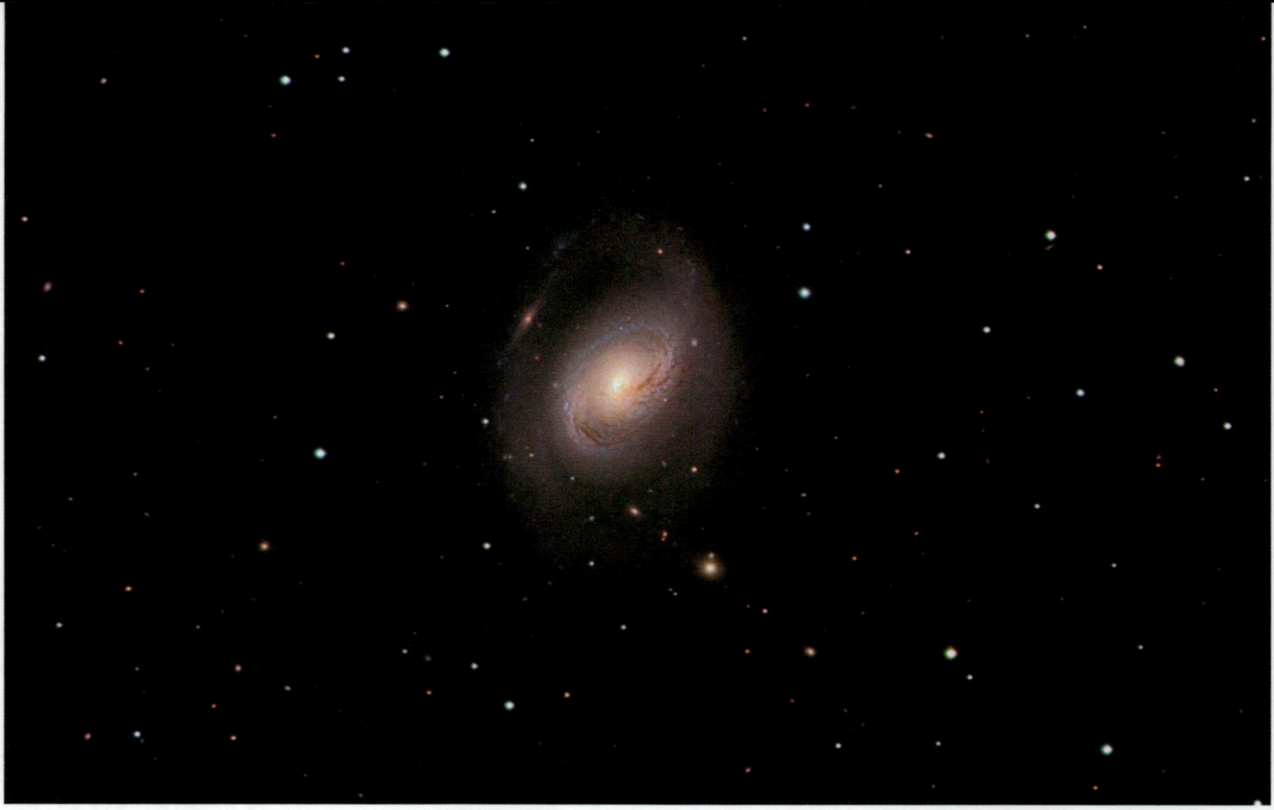

M 96

The SAB(rs)ab galaxy is also a member of the Leo I galaxy group, located at 35 million light years. Together with M 95 and M 105, M 96 forms a triplet of bright galaxies in the larger of the two Leo I subgroups. The distance to M 95 is 41.7′ and thus at least 435 000 light years; M 105 at 48.3′ is at least 500 000 light years away from M 96. The angular size of M 96 is 7.6′ × 5.2′ and corresponds to a projected size of 77 000 × 53 000 light years. The weak bar is almost completely surrounded by a ring-like spiral arm. Inside this ring are also several dust filaments which appear to be separating from the ring and moving to the centre. The edge of the disc is surrounded by a dust lane which contains some active regions at the upper, northern edge. Between the inner and outer ring structures, the disc has a low surface brightness. There is thus no continuous spiral pattern, and so some background galaxies can be seen through this gas-poor zone.

At the left, northeastern edge of the disc at a distance of 2′ from the centre is the spiral galaxy 2MFGC 08391 (2MASS Flat Galaxy Catalogue), which is seen exactly edge-on. At a distance of 700 million light years, 2MFGC 08391 lies far behind M 96 and its absolute size is three times that of M 96.

Analyses of the isophotes in the inner region reveal a double-bar structure in M 96. The form of this bar corresponds to an ellipsoidal configuration. The smaller bar has a diameter of 10″, the outer, larger one about 2′. The diameter of the inner ring, for comparison, is 2.8′. Data from the Hubble Space Telescope show a strong twisting of the isophotes in the centre of the small bar, only 2″ away from the nucleus. Astronomers conclude that it is due to a miniature disc of gas and dust. The inclination of M 96 is 45°, the phase angle of the symmetry axes of the inner ring and the edge of the disc differ by 35°. Spectroscopic studies of the disc rotation confirm the existence of two independent velocity fields. The complex form at the centre and the two disc components in M 96 are indications of past interactions within the Leo I Group.

OBJECT	M 96
CONSTELLATION	Leo
RIGHT ASCENSION	$10^h 46^m 46^s$
DECLINATION	+11° 49′ 12″
BRIGHTNESS	10.1 mag
TYPE	SAB(rs)ab
PHOTOGRAPHERS	Johannes Schedler
TELESCOPE	400-mm reflector
CAMERA	SBIG STL-11000
EXPOSURE TIME	330 min
LOCATION	Wildon, Austria

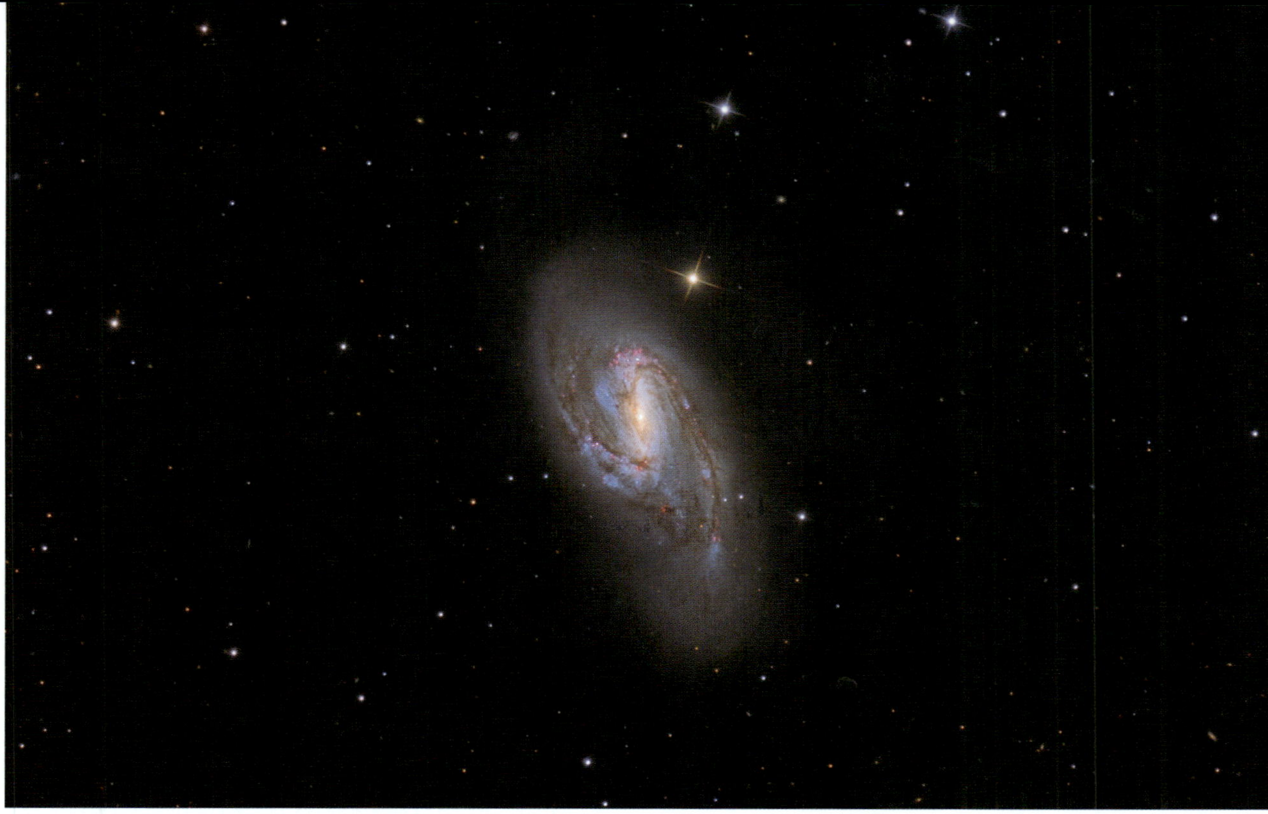

M 66

The SAB(s)b galaxy M 66 (NGC 3627) is the southern vertex of the Leo galaxy triplet which belongs to the Leo I Group (M 66/M 96 Group). The other triplet members are M 65 at a distance of 20.2′ and NGC 3628 at 35.9′. M 66 is at a distance of 35 million light years from the Milky Way. An angular size of 10′ corresponds to 100 000 light years. The angular size of M 66 is 9.1′ × 4.2′ and the transverse disc diameter is thus calculated to be 93 000 light years. At the centre of M 66 is a Seyfert-2 nucleus. The spiral structure is characterized by a large amount of dust whose filaments can also be seen between the two spiral arms. This is due to the diffuse light which illuminates the halo and makes the fine dust structures visible via contrast. The form of the spiral structure is asymmetric and the spiral arm beginning at the upper, northern end of the bar has more H II regions than the lower one. The upper right edge of the disc of the galaxy also has a dim tidal tail which could be a hint of past interactions.

Radio observations of the Leo galaxy triplets show no indication of existing bridges of hydrogen gas between M 66 and NGC 3628. The asymmetry of the spiral structure also indicates that there is no common interaction, but rather that M 66 is a product of its own evolution. In studies of the $H\alpha$ rotation curve of M 66, astronomers have discovered deviations from the typically constant form of the gas velocity. This can be modelled by a ring of hydrogen gas which is tilted by 15° with respect to the inner disc plane and has a diameter of 2′. Since the velocities in the spiral arms differ for radii larger than 1.5′, one can also conclude that the tilted ring of hydrogen gas is stronger in the lower, southern half of the disc. The cause of the asymmetry could be the irregular accretion of material from a dwarf galaxy which has orbited M 66 many times, being destroyed in the process. This suspicion is supported by the fact that several relatively large but also very dim dwarf galaxies have been found in the Leo I Group.

OBJECT	M 66, NGC 3627
CONSTELLATION	Leo
RIGHT ASCENSION	$11^h\ 20^m\ 15^s$
DECLINATION	+12° 59′ 30″
BRIGHTNESS	9.7 mag
TYPE	SAB(s)b
PHOTOGRAPHERS	Adam Block
TELESCOPE	800-mm reflector
CAMERA	SBIG STX-16803
EXPOSURE TIME	680 min
LOCATION	Mount Lemmon SkyCenter/ University of Arizona, USA

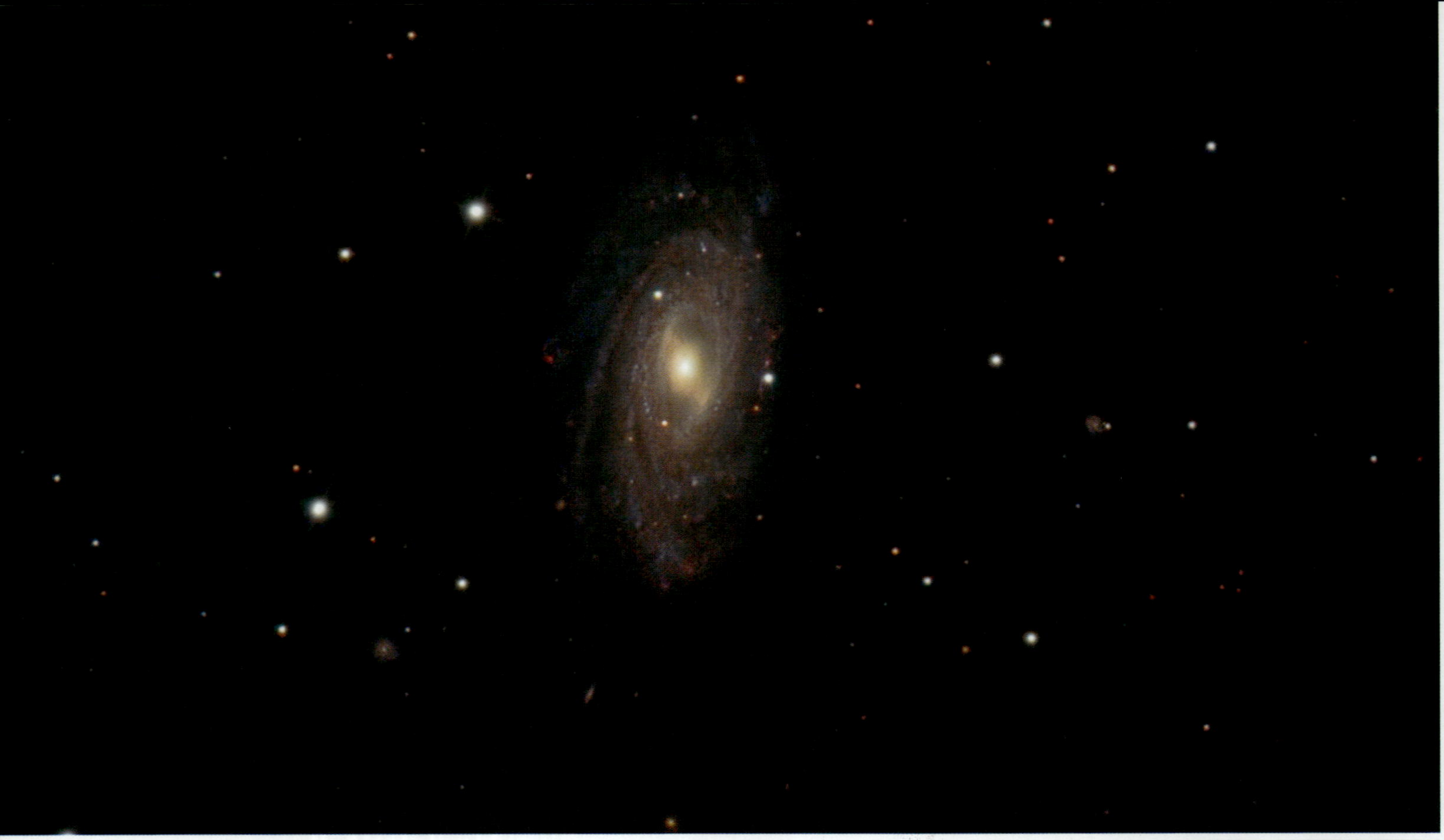

NGC 3953

The SB(r)bc galaxy NGC 3953 belongs to the southern part of the M 109 galaxy group and is at a distance of 55 million light years from the Earth. The angular size of 6.9′ × 3.5′ corresponds to a transverse diameter of 110 000 light years. In spite of the high inclination of the disc plane and the inconvenient position of the axis of the bar, the bar is nevertheless recognizable. Diffuse ansae glow at the ends of the bar. The eastern of the two ansae, above and to the left of the centre, is contrasted by a dust lane which runs from the spiral arm into the axis of the bar. Due to the projection, one observes an acute angle and a prominent kink in the dust lane. The inner spiral-arm pattern appears unordered and has a colour contrast to the outer disc, where the spiral arms appear bluer and better defined. The active edge in NGC 3953 can also be recognized via the luminous red H II regions which lie at the outside.

Based on optical morphology, one finds a strong bar in about 30 per cent of all spiral galaxies and a weak one in about 25 per cent. In the infrared, the fraction of barred spirals is, at 75 per cent, even higher. While the bar structure is due to similar rotationally asymmetric velocity fields, the form of the bars reveal, however, significant differences with regard to size, appearance and physical parameters. Two-dimensional Hα kinematics were investigated in a comparative study of 21 barred spiral galaxies, including NGC 3953. The data indicate a strong bar in NGC 3953 and a deficit of hydrogen gas at the centre, as found in some SB galaxies. With the addition of the Hα velocities, astronomers were able to obtain three-dimensional information and fit kinematic disc models. The result was that the values for the position angle of the major axis and the inclination of the disc correspond well to those derived from photometry.

OBJECT	NGC 3953
CONSTELLATION	Ursa Major
RIGHT ASCENSION	$11^h 53^m 49^s$
DECLINATION	+52° 19′ 36″
BRIGHTNESS	9.9 mag
TYPE	SB(r)bc
PHOTOGRAPHERS	Bernd Flach-Wilken, Volker Wendel
TELESCOPE	400- and 450-mm reflector
CAMERA	SBIG STL-6303 and SBIG ST-10XME
EXPOSURE TIME	250 min
LOCATION	Wirges and Weisenheim am Berg, Germany

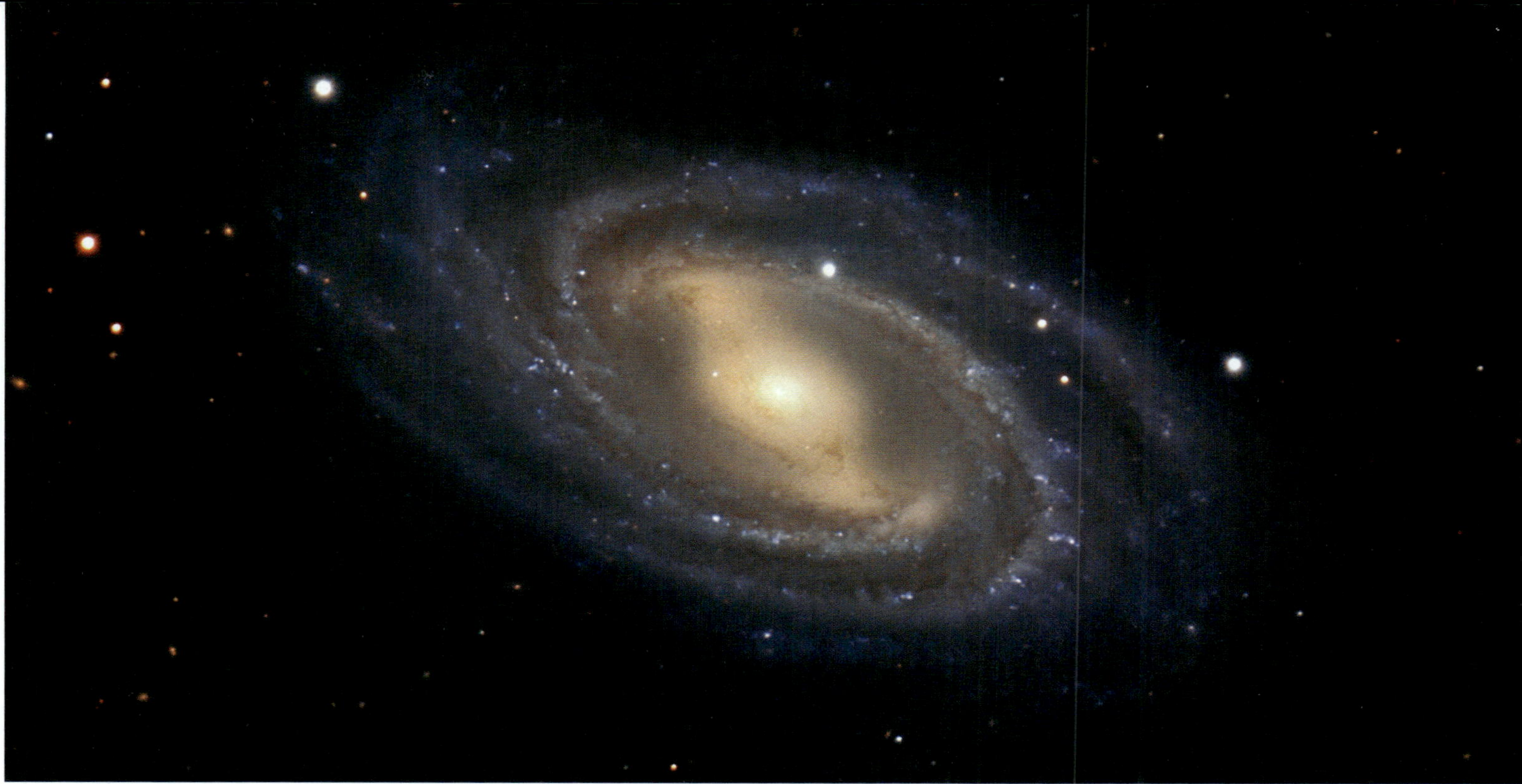

M 109

The SB(rs)bc galaxy M 109 (NGC 3992) is located about 1° southeast of the star Phekda (Gamma Ursae Majoris) and, at a distance of 55 million light years, is part of the Ursa Major galaxy cluster. The angular size of 7.6′ × 4.7′ corresponds to a projected disc diameter of 120 000 light years. In the area within 15′ of M 109 are three galaxies (UGC 6940, UGC 6969, and UGC 6923) whose recession velocities differ by at most 100 km/s from one another and which are classified as companions. These are dwarf galaxies with diameters less than 12 000 light years. The spiral pattern in M 109 is strikingly regular and determined by two spiral arms. The brightenings in the ansae at the ends of the bar merge into brightenings at the beginning of the spiral arms, which, however, are shifted by 15° relative to the axis of the bar. The spiral arms can be followed through an entire revolution. Only in the spiral arm which begins at the lower end of the bar and runs to the east of the centre is a perturbation, associated with a splitting of the arm, visible.

The dynamics of barred spiral galaxies are significantly determined by the angular velocity of the bar structure. Measurement of the rotation of the bar takes place via the Tremaine–Weinberg method which compares spectra taken at different zones of the bar. Astronomers have taken spectra of the middle and ends of the bar by changing the orientation of the spectrographic slit.

With an inclination of 57°, spectral data indicate a circular velocity of 22 km/s per 1000 light years distance from the centre. The comparatively rapid rotation of the bar is seen as an indication that the dynamic friction between the bar and the dark matter of the halo must be low in the bar zone. The distribution of dark matter would thus have a deficit at the centre and would be more like a ring than a ball. Radio observations of neutral hydrogen gas, which is almost undetectable in the bar zone, confirm this ring structure. Models of the H I rotation curve also work best when the dark matter is placed far outside.

OBJECT	M 109, NGC 3992
CONSTELLATION	Ursa Major
RIGHT ASCENSION	$11^h\ 57^m\ 36^s$
DECLINATION	+53° 22′ 28″
BRIGHTNESS	10.6 mag
TYPE	SB(rs)bc
PHOTOGRAPHERS	Makis Palaiologou, Stefan Binnewies
TELESCOPE	1.3-m reflector
CAMERA	Andor DZ 436
EXPOSURE TIME	150 min
LOCATION	Skinakas Observatory, Crete, Greece

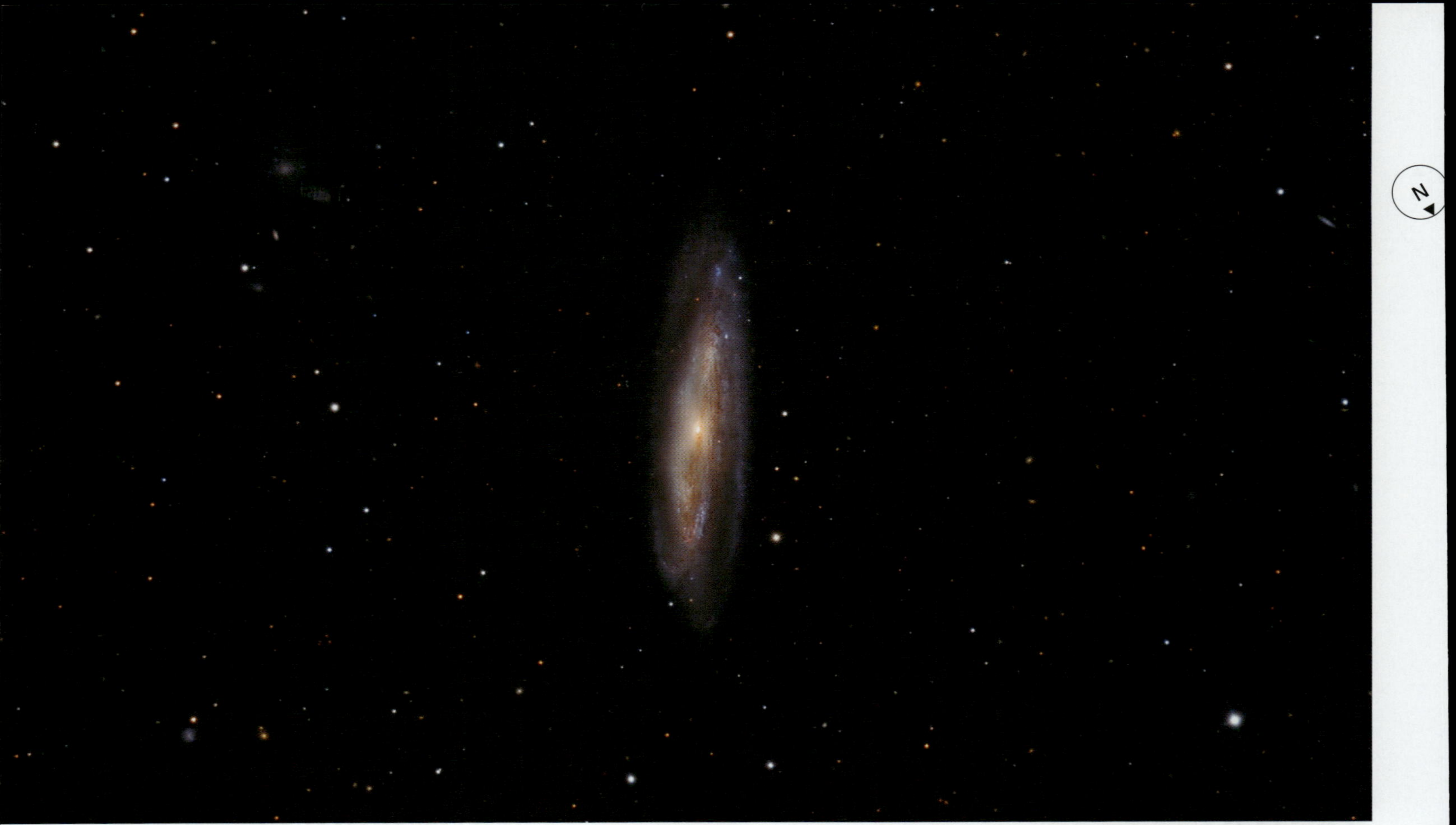

M 98

At the northwestern edge of the Virgo Cluster, 5′ away from Denebola (Beta Leonis), is the SAB(s)ab galaxy M 98 (NGC 4192). The same angular distance separates M 98 from M 87 at the centre of the cluster. Catalogues list its distance as 52 million light years and its angular size of 9.8′ × 2.8′ corresponds to a projected disc size of 148 000 × 42 000 light years. The disc plane is highly inclined at 83° and despite this angle the S-shaped form of the active regions can be recognized, indicating two dominating spiral arms. Deprojected onto a circle, it becomes clear that a dust filament passes close to the compact nucleus, which indicates a vertical bar. This is confirmed by an S-shaped luminosity distribution in the peanut-shaped bulge, visible to the left of the nucleus.

With a velocity of +142 km/s, M 98 is moving towards the Milky Way, which indicates a relative velocity of more than 1100 km/s. The H I radio map of M 98 shows, in addition to a warp at the edge of the disc, nearby H I clouds with velocities of 80–140 km/s. These H I high-velocity clouds could belong to the Milky Way. Astronomers interpret these, however, as a possible dynamic connection between M 98 and the Virgo Cluster. In this picture, neutral hydrogen gas flows out of M 98 and covers an area almost 1° across on the sky. Given the distance to M 98, this corresponds to a transverse distance of more than 900 000 light years. The long axis of the gas clouds points towards NGC 4254 and as simulations show, the clouds could indicate a high-speed encounter between M 98 and NGC 4254 in which both galaxies lost gas but experienced only moderate perturbations in the planes of their discs.

OBJECT	M 98, NGC 4192
CONSTELLATION	Coma Berenices
RIGHT ASCENSION	$12^h\ 13^m\ 48^s$
DECLINATION	+14° 54′ 01″
BRIGHTNESS	11.0 mag
TYPE	SAB(s)ab
PHOTOGRAPHERS	Johannes Schedler, Wolfgang Ries
TELESCOPE	400- and 450-mm reflector
CAMERA	SBIG STL-11000 and ST-10XME
EXPOSURE TIME	854 min
LOCATION	Wildon and Seng, Austria

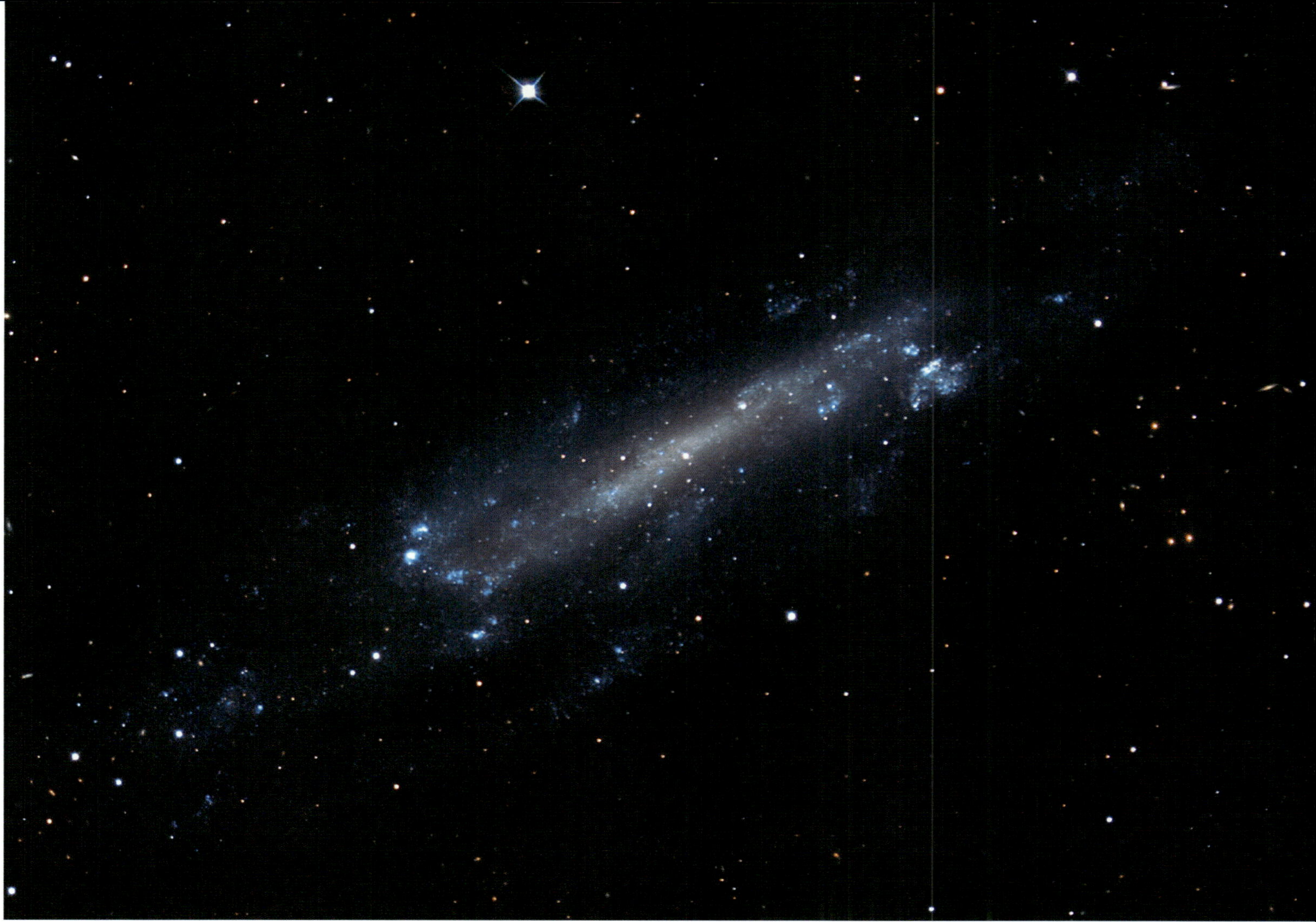

NGC 4236

NGC 4236 is 12 million light years away and is considered to be part of the M 81 Group. This SB(s)dm galaxy covers a large area of 21.9′ × 7.2′ but its projected disc diameter of 76 000 light years is in the typical range for spiral galaxies. NGC 4236 has a low surface brightness in spite of the high inclination of 75°. One recognizes a long, dim bar with no clearly defined centre.

The spiral pattern is characterized by particularly blue spiral arms. In the UV, the spiral structure in NGC 4236 is more pronounced than in the optical. This is especially visible in the active regions at the two ends of the bar. This morphological fine structure is also seen in the distribution of atomic hydrogen. The observed granulation shows the same distribution in H I radio observations as in the UV and is larger than the optical disc. NGC 4236 is considered a dim but extended UV-disc (XUV-disc) galaxy. XUV discs are notable for sustained star formation. These active regions are outside the optical disc and thus beyond the border of traditional star-forming regions in barred spiral galaxies.

OBJECT	NGC 4236
CONSTELLATION	Draco
RIGHT ASCENSION	$12^h\ 16^m\ 42^s$
DECLINATION	+69° 27′ 45″
BRIGHTNESS	10.1 mag
TYPE	SB(s)dm
PHOTOGRAPHERS	Rainer Sparenberg
TELESCOPE	1.12-m reflector
CAMERA	SBIG STL-11000
EXPOSURE TIME	220 min
LOCATION	Melle, Germany

M 106

The spiral galaxy M 106 (NGC 4258), at a distance of 24 million light years almost a neighbour of the Milky Way, is considered to be part of the Canes Venatici II galaxy group. M 106 is the brightest and largest galaxy in this group of 30. Its angular size of 18.6′ × 7.2′ corresponds to a large projected disc diameter of 130 000 light years. The prominent spiral arms in the inner region of the disc of this SAB(s)bc galaxy contain large H II regions up to a size of 10″, which corresponds to a diameter of 1000 light years. At the edge of the disc one can recognize spiral-arm fragments which also contain active regions. The right edge of the disc is also frayed. This perturbation is presumably due to tidal interaction with the neighbouring small, irregular galaxy NGC 4248 which is located only 5′ from the edge of the disc of M 106. It is not visible in the image. Its velocity relative to M 106 is −36 km/s and its transverse distance to the edge of the disc is only 35 000 light years.

M 106 has a Seyfert-2 nucleus which has been investigated in several studies. Spectral data from the radio to X-rays have allowed astronomers to estimate the mass of the supermassive black hole at 3.7×10^7 solar masses and the size of the surrounding gas disc at 500 light years. Material from the gas disc spirals into the black hole and emits X-rays. According to the standard model of active galaxies, in Seyfert-2 galaxies the line of sight to the black hole is blocked by a surrounding dust torus. Since the central disc in M 106 has a high gas density, scattered X-rays lead to the emission of microwave radiation. Thus M 106 was the first galaxy in which such H_2O maser radiation could be detected. A further peculiarity in M 106 are two opposing abnormal arm structures which can be seen as red filaments above and below the centre. The upper filament lies above the plane of the disc, the lower one below. These filaments are due to hot gas which is excited by the radio emission of two jets which emerge perpendicularly from the central disc. The angle between these jets and the plane of the galaxy is 30° and astronomers assume that these jets precess. According to this model, the period of precession of the jets in M 106 is 300 000 years.

OBJECT	M 106, NGC 4258
CONSTELLATION	Canes Venatici
RIGHT ASCENSION	$12^h\ 18^m\ 58^s$
DECLINATION	+47° 18′ 14″
BRIGHTNESS	8.4 mag
TYPE	SAB(s)bc
PHOTOGRAPHERS	Josef Pöpsel, Stefan Binnewies
TELESCOPE	600-mm reflector
CAMERA	SBIG STL-11000
EXPOSURE TIME	690 min
LOCATION	Skinakas Observatory, Crete, Greece

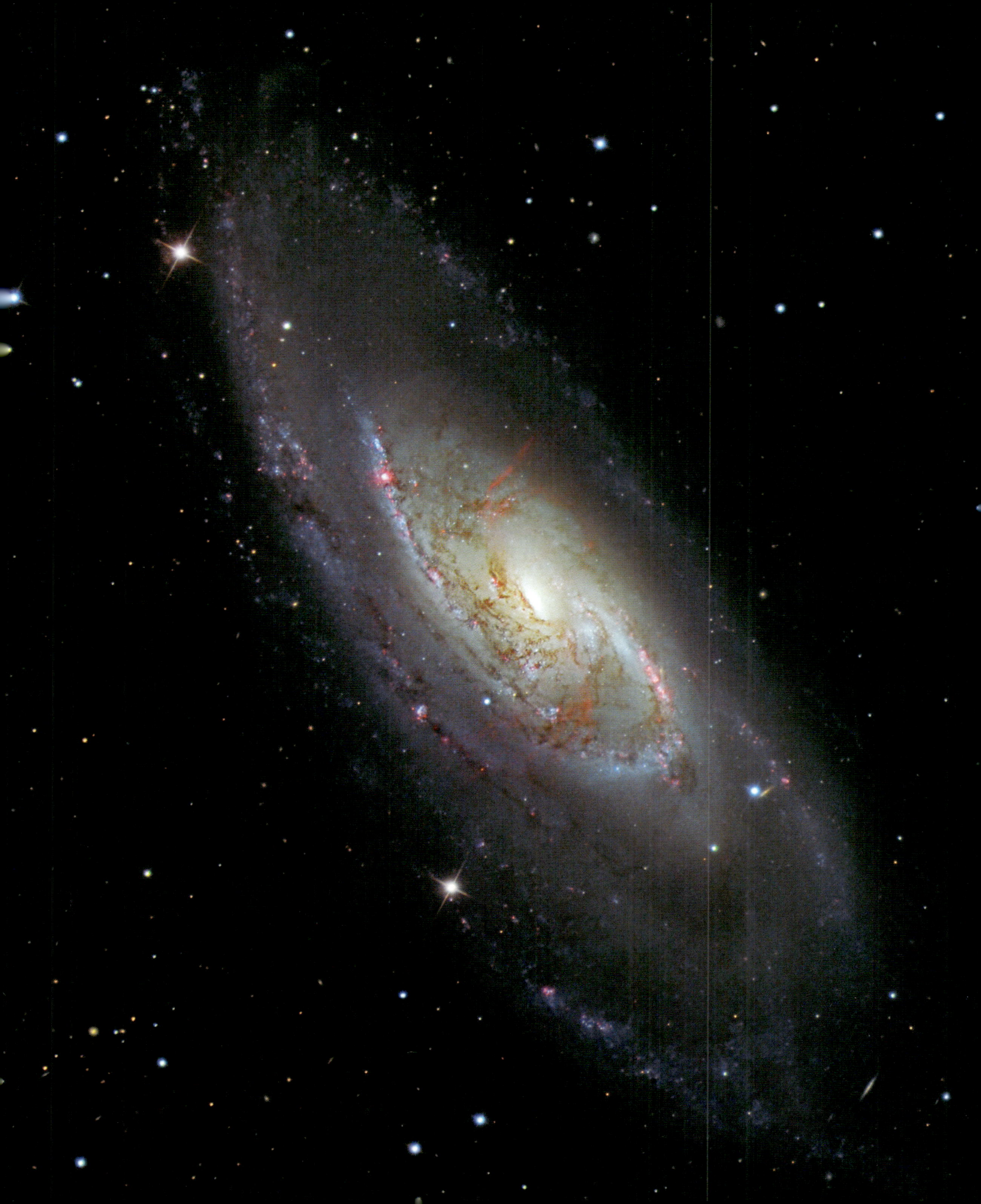

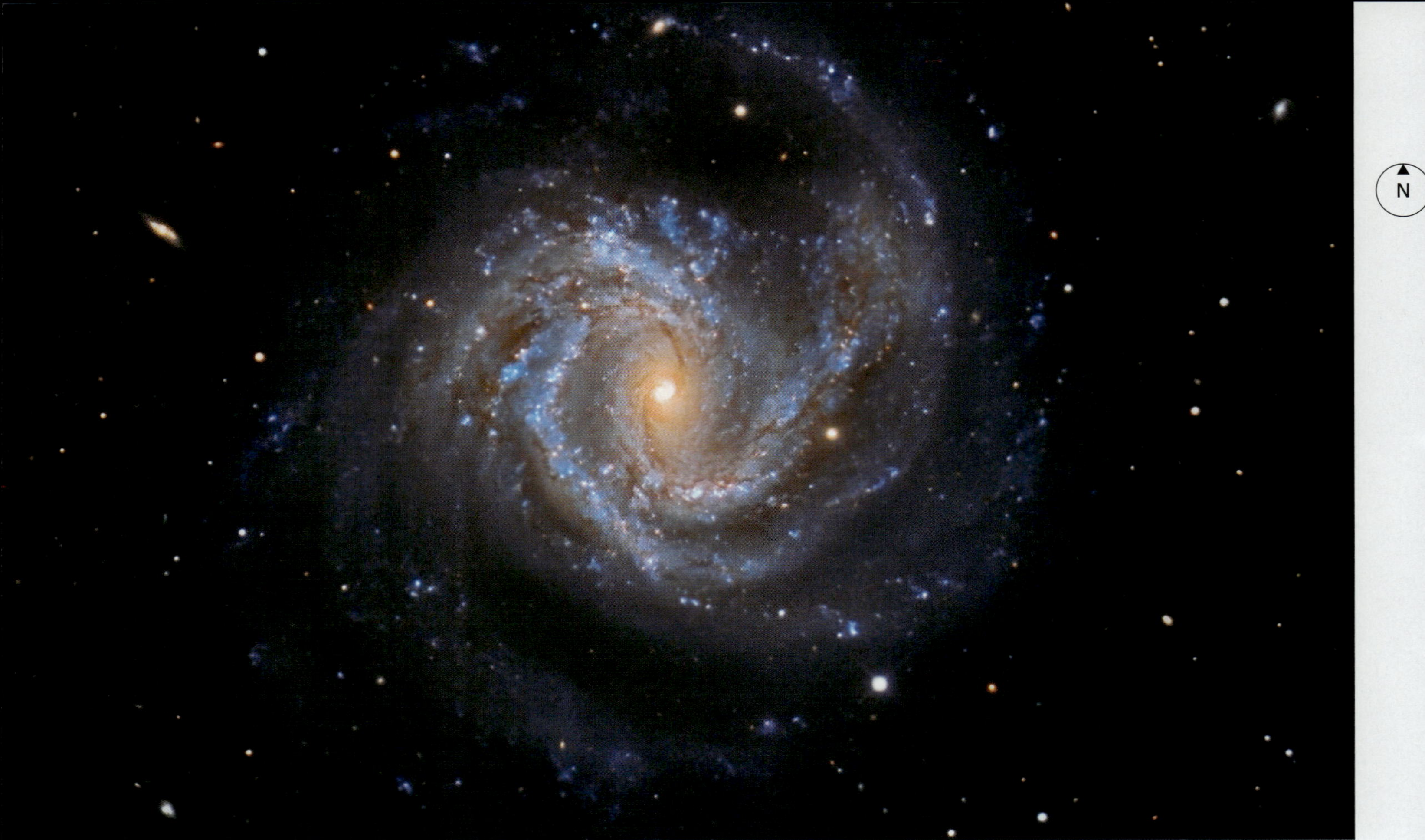

M 61

The SAB(rs)bc galaxy M 61 (NGC 4303) belongs to the Virgo Cluster and lies at a distance of 50 million light years. Its angular size of 6.5′ × 8.5′ corresponds to a maximum disc diameter of 95 000 light years. The centre of M 61 contains an active Seyfert nucleus which, however, has only a low luminosity. The inclination of the galaxy is 25° and thus allows one to view the grand-design spiral pattern. This is dominated by two arms which join onto a bar structure in the inner region. H II regions, of which there are many in the spiral arms, are not present in the bar. At the left side of the disc the spiral arms have pronounced Vorontsov-Velyaminov rows which are also recognizable in the dimmer outer ends of the arms.

The dust lanes in the bar do not pass through the centre, but rather end at a central disc whose diameter of 10″ corresponds to about 2400 light years. Here, astronomers have found a high concentration of molecular gas which rotates circularly except for perturbations at the outer edge. At the edge, the disc joins the bar, so strong velocity gradients occur in the gas flow here. Simulations show however, that molecular clouds can survive this passage and that later star clusters form. Although the star-formation rate is rather low, such star clusters can be detected in images taken by the Hubble Space Telescope. The clusters are not older than a million years, and have about 10^3–10^4 solar masses. The surrounding molecular clouds reach 10^4–10^6 solar masses.

OBJECT	M 61, NGC 4304
CONSTELLATION	Virgo
RIGHT ASCENSION	$12^h\ 21^m\ 55^s$
DECLINATION	+04° 28′ 25″
BRIGHTNESS	10.2 mag
TYPE	SAB(rs)bc
PHOTOGRAPHERS	Makis Palaiologou, Stefan Binnewies
TELESCOPE	1.3-m reflector
CAMERA	Andor DZ 436
EXPOSURE TIME	127 min
LOCATION	Skinakas Observatory, Crete, Greece

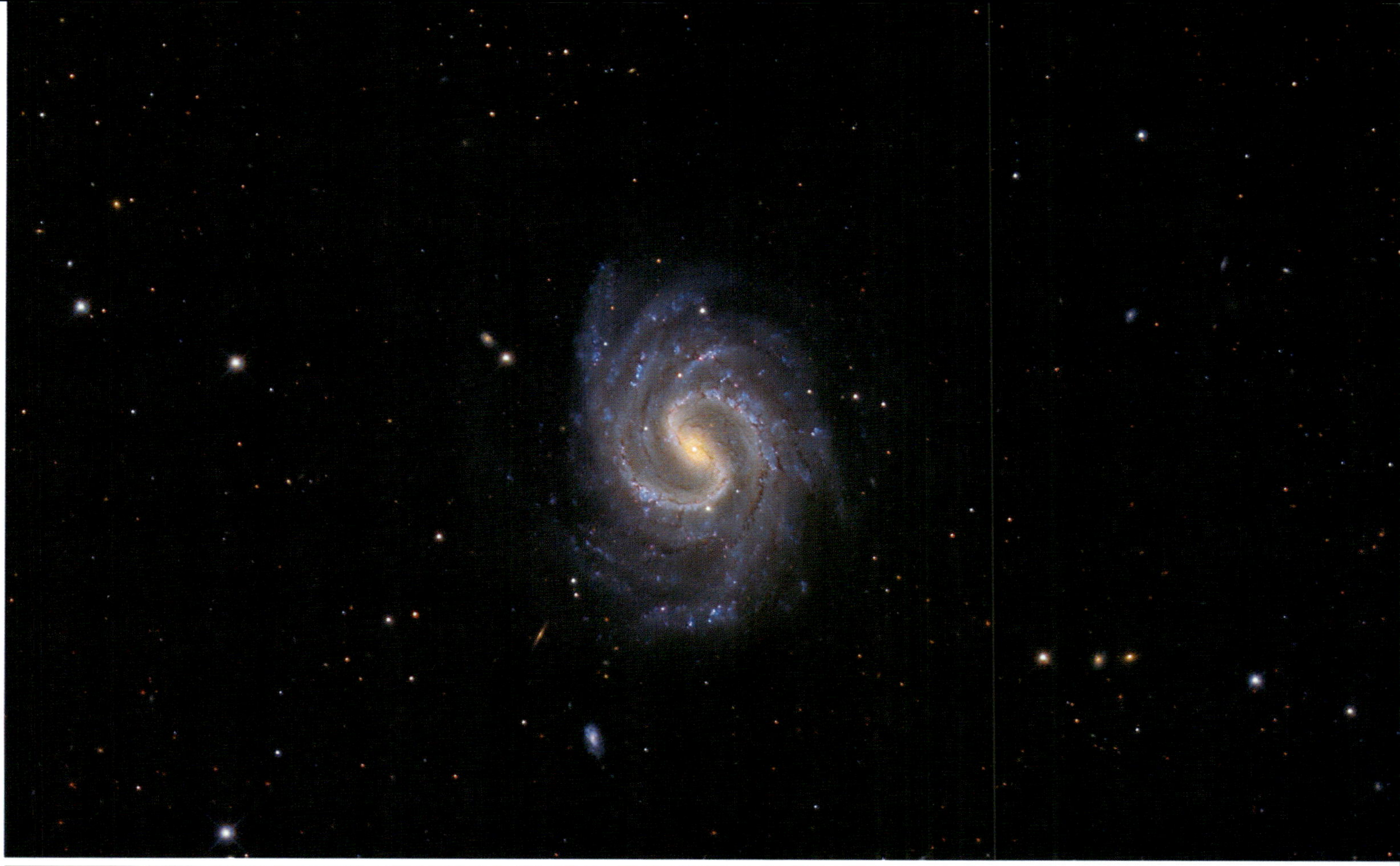

NGC 4535

The SAB(s)c galaxy NGC 4535 is one of the largest spiral galaxies in the Virgo Cluster and is located at the eastern edge of the southern part of the cluster, only 2° from M 49. The angular size of the spiral plane of 7.1′ × 5.0′ corresponds to a projected diameter of 112 000 light years at a distance of 54 million light years. NGC 4535 is thus one of the seven largest galaxies in the Virgo Cluster. Its bright, compact nucleus is embedded in an oval bulge. The bar is well defined, but the join with the S-shaped spiral arms appears softer than in other SAB galaxies due to the lack of a right angle at the transition. The regular beginning of the arms defines the inner spiral pattern, while farther out the morphology appears more irregular. The inner spiral arms are accompanied by displaced dust lanes. This is particularly prominent below and to the left of the centre, where the largest H II regions reach a size of 1″, corresponding to 250 light years. Connected to the edge of the disc are diffuse, dim, tidal-tail-like extensions containing elongated segments.

Studies of a sample of galaxies in the Virgo Cluster detected in some a high fraction of carbon monoxide (CO) in the nucleus. At millimetre wavelengths the central 1′ × 1′ region was mapped in high resolution, revealing a central CO intensity maximum in a third of the galaxies. NGC 4535 is one of these single-peak galaxies and the local extent of the CO maximum is 1500 light years. This CO concentration is related to the central star-forming regions which are crossed by dust lanes.

OBJECT	NGC 4535
CONSTELLATION	Virgo
RIGHT ASCENSION	$12^h\ 34^m\ 20^s$
DECLINATION	+08° 11′ 52″
BRIGHTNESS	10.6 mag
TYPE	SAB(s)c
PHOTOGRAPHERS	Adam Block
TELESCOPE	600-mm reflector
CAMERA	SBIG STL-11000
EXPOSURE TIME	285 min
LOCATION	Mount Lemmon SkyCenter/ University of Arizona, USA

M 100

The grand-design spiral galaxy M 100 (NGC 4321) is located in the Virgo Cluster at a distance of 55 million light years. It is of SAB(s)bc type with angular size of 7.4′ × 6.3′ and transverse disc diameter of 118 000 light years. The bar is in the bright core which in the image contains many bright blue star-forming regions arranged in a ring. The dust structures in the two spiral arms can be followed to the ends of the bar. The weakly contrasting bar measures 1.7′ and the axis of the bar lies along the diagonal of the image in southeast–northwest direction (lower left–upper right). At 15°, the pitch angle of M 100 is relatively small which is why the spiral arms are open only slightly. Two opposing tidal tails are connected to the stellar disc.

The image shows two smaller galaxies near M 100: at the left edge of the image is the SA0 galaxy NGC 4328 which lies 6.1′ away from the centre of M 100 and whose relative velocity is +1092 km/s. At the upper edge of the disc, the dwarf elliptical galaxy NGC 4322 is closer at 5.3′. Due to its relative velocity of 282 km/s it seems probable that it is interacting with M 100, which is also suggested by its warped shape. This assumption is confirmed by deep images which show that the dim part of the optical disc of M 100 extends almost to NGC 4328. The distance mentioned above leads to a diameter of 167 000 light years and thus a diameter 40 per cent larger than that mentioned in NED.

Radio observations of H I have allowed astronomers to measure the speed of rotation of the spiral pattern in this galaxy at 6 km/s per 1000 light years. Spiral galaxies have values between 3 km/s and 9 km/s per 1000 light years. This important quantity in density-wave theory implies a corotation radius of 25 000–35 000 light years for M 100. The image shows that outside this region the spiral arms fan out and widen. Density-wave theory also delivers a maximum disc radius of 59 000 light years which corresponds well to the bright optical disc of M 100.

OBJECT	M 100, NGC 4321
CONSTELLATION	Coma Berenices
RIGHT ASCENSION	$12^h\ 22^m\ 55^s$
DECLINATION	+15° 49′ 19″
BRIGHTNESS	9.8 mag
TYPE	SAB(s)bc
PHOTOGRAPHERS	Stefan Binnewies, Josef Pöpsel
TELESCOPE	600-mm reflector
CAMERA	SBIG STL-11000
EXPOSURE TIME	540 min
LOCATION	Skinakas Observatory, Crete, Greece

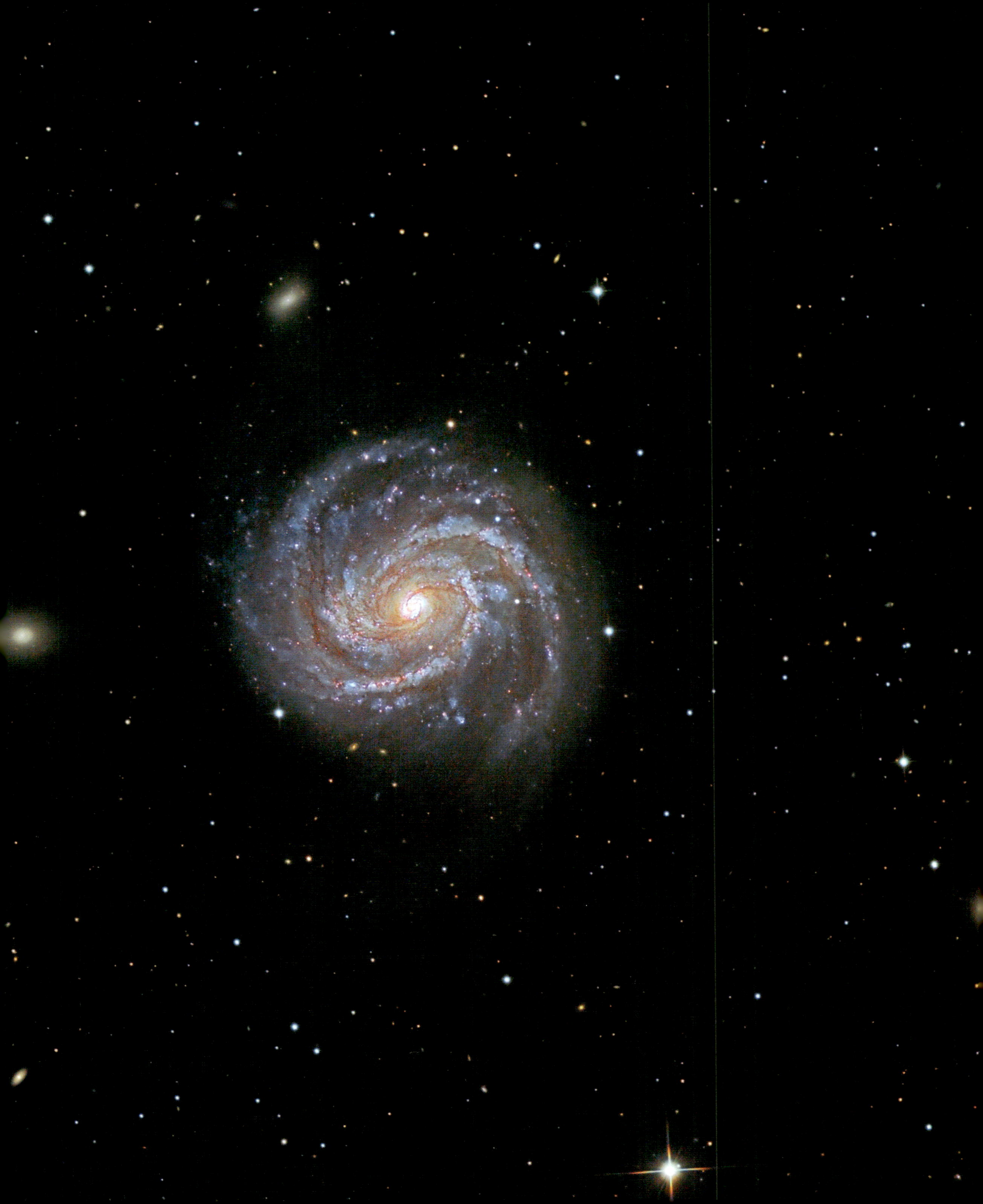

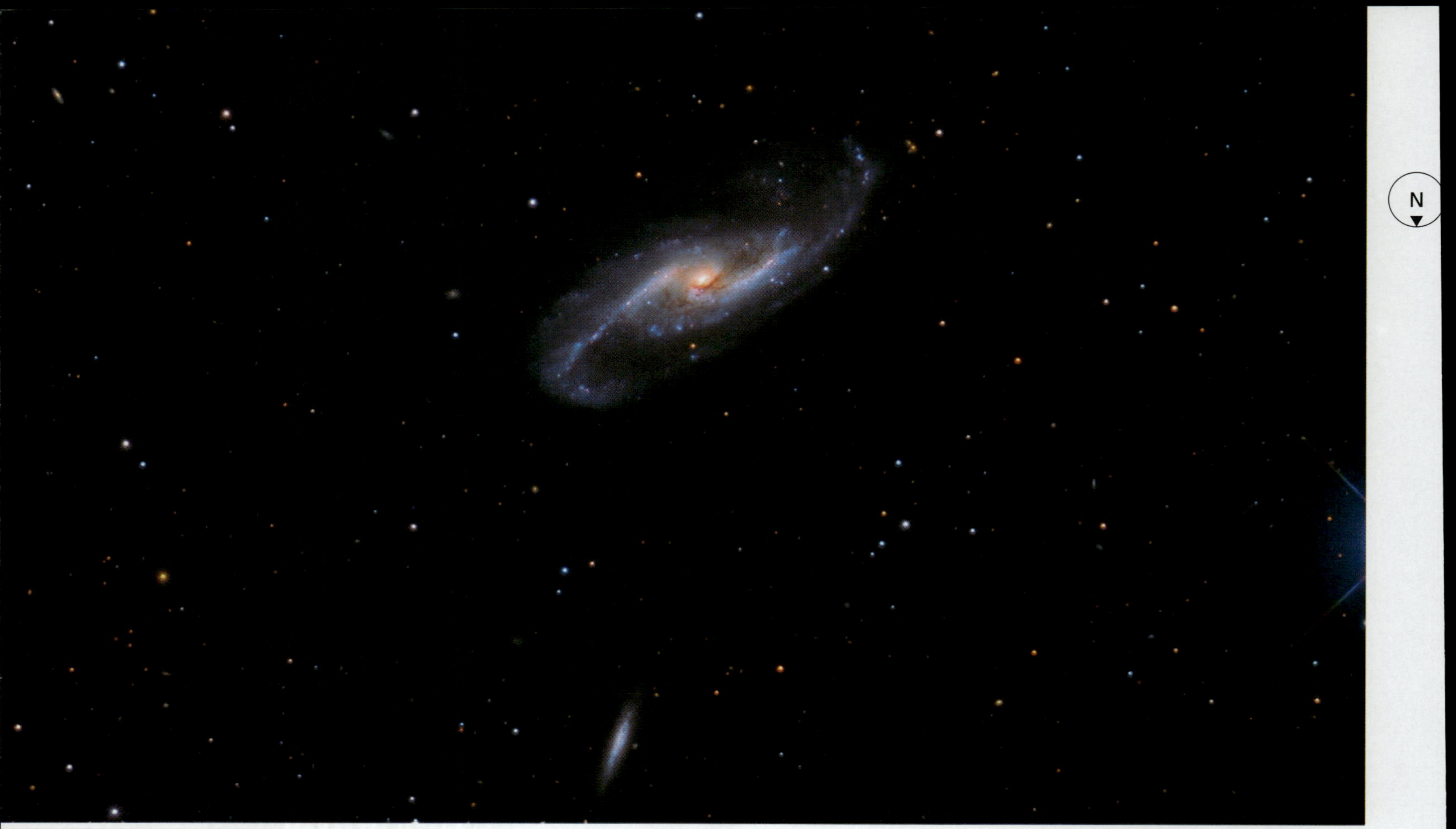

NGC 4536

The SAB(rs)bc galaxy NGC 4536 is 52 million light years distant and belongs to the southern periphery of the Virgo Cluster. The angular size of 7.6′ × 3.2′ corresponds to a transverse disc diameter of 115 000 light years for the optical disc plane. At the northern edge of the image is NGC 4533 which is 8.3′ away from NGC 4536 and relative to it is approaching the Milky Way at 53 km/s. A classical bulge cannot be observed in NGC 4536; the brightening in the centre is due to the perspective shortening the bar, whose axis is tilted by about 30° from the vertical. Below the centre is a prominent X-shaped dust structure. The transition from the bar to the spiral arms is only weak; a rhombus formed by the inner spiral-arm fragments can be clearly seen. In the outer part of the disc are two bright spiral arms whose active regions extend to the edge of the disc.

In most spiral galaxies which have a prominent budge there is also an active galactic nucleus. This is explained by a strong correlation between the mass increase of the central supermassive black hole and the growth of the bulge. In a study of 32 late-type spirals whose optical spectra had no definitive indication of an active galactic nucleus, data from the infrared spectrograph on the Spitzer Space Telescope found exceptions to this rule. NGC 4536 is one of seven galaxies in which emission lines of highly ionized neon and oxygen gas were detected. One can use luminosity models for these galaxies, which are transitioning to a phase of galactic activity, to estimate the central mass. This results in a mass of 2.5×10^6 solar masses of the black hole at the core of NGC 4536.

OBJECT	NGC 4536
CONSTELLATION	Virgo
RIGHT ASCENSION	$12^h\ 34^m\ 27^s$
DECLINATION	+02° 11′ 17″
BRIGHTNESS	11.2 mag
TYPE	SAB(rs)bc
PHOTOGRAPHERS	Adam Block
TELESCOPE	600-mm reflector
CAMERA	SBIG STL-11000
EXPOSURE TIME	320 min
LOCATION	Mount Lemmon SkyCenter/ University of Arizona, USA

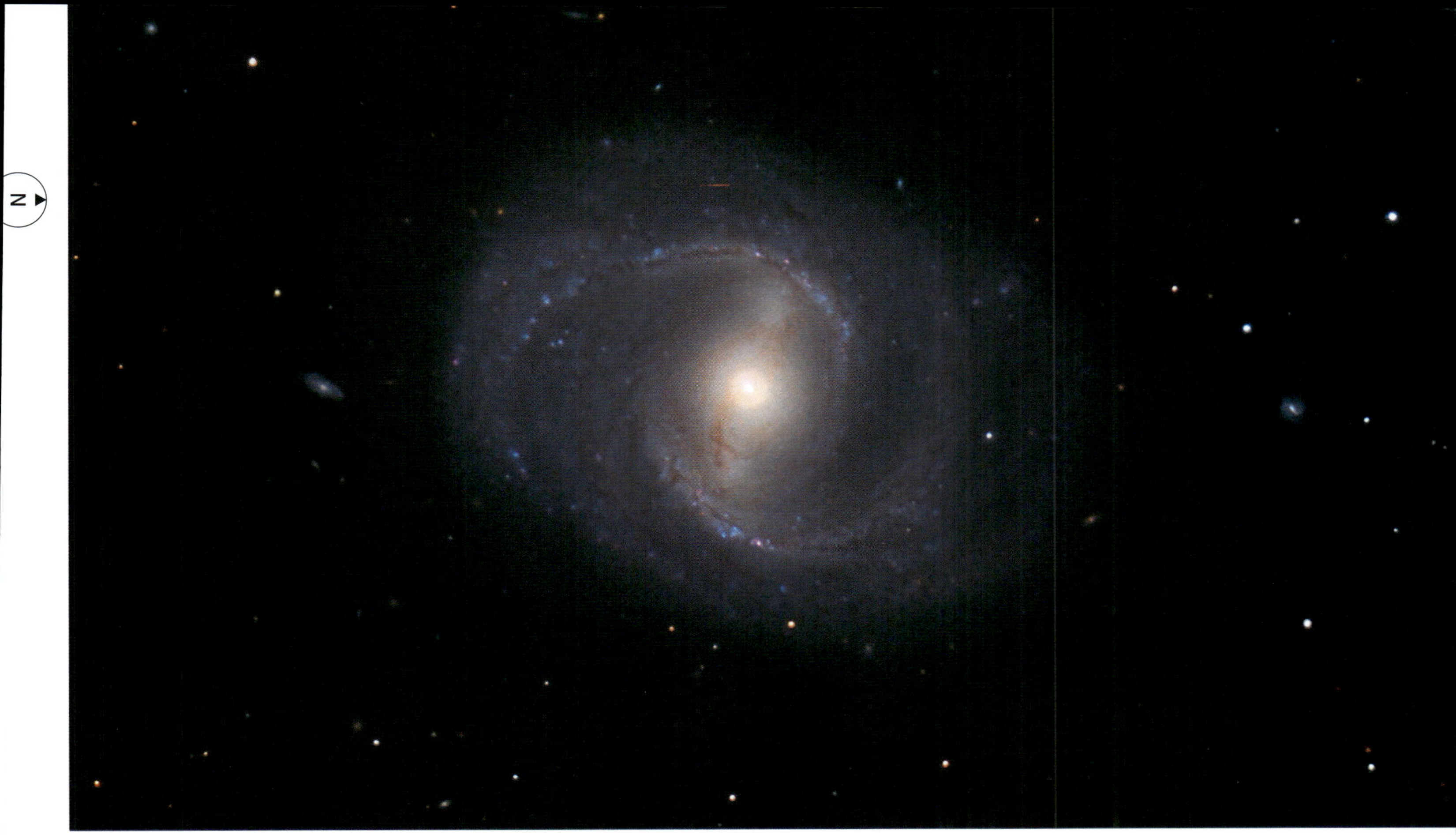

M 91

The SBb galaxy M 91 (NGC 4548) is part of the Virgo Cluster and lies at a distance of 53 million light years. The angular size of 5.4′ × 4.3′ corresponds to a projected disc diameter of 83 000 light years. M 91 is the dimmest object in the Messier catalogue. The galaxy was for a long time one of the lost Messier objects, since Charles Messier had recorded an incorrect position when it was discovered in 1781. Only in 1969 was it possible to associate M 91 with NGC 4548 (discovered in 1784 by W. Herschel). M 91 has a bright nucleus which is embedded in a compact bulge. The bar structure is almost vertical and in the lower, western part of the bar dust lanes are more easily recognized than in the upper. The two spiral arms are not connected to the ends of the bar, but rather their inner ends are rotated by 90° and located in the arc-like segments of an incomplete inner ellipse. The northern, in the image right-hand, spiral arm is shorter and appears compressed relative to the other arm.

The map of neutral hydrogen gas in M 91 shows a peculiarity: in the central part of the bar almost no hydrogen gas is detectable, since it is concentrated in a flat ring. The inner side of this ring is delimited by the ends of the bar, the outer side by the size of the stellar disc of M 91. The few active regions in the spiral arms can also be identified in the H I map of the ring distribution and most are located where the end of the bar joins the ring. Relative to the centre of the cluster, M 91 is moving at about 600 km/s through the Virgo Cluster towards the Milky Way. Ram pressure in the cluster medium has led to loss of hydrogen gas in M 91, leaving only the ring structure. The relative length of the disc plane to the direction of motion also explains the observed irregularities in the exposed northern side of the hydrogen ring.

OBJECT	M 91, NGC 4548
CONSTELLATION	Coma Berenices
RIGHT ASCENSION	$12^h\ 35^m\ 26^s$
DECLINATION	+14° 29′ 47″
BRIGHTNESS	11.0 mag
TYPE	SB(rs)b
PHOTOGRAPHERS	Rainer Sparenberg, Stefan Binnewies
TELESCOPE	600-mm reflector
CAMERA	SBIG ST-10XME
EXPOSURE TIME	120 min
LOCATION	Amani Lodge, Namibia

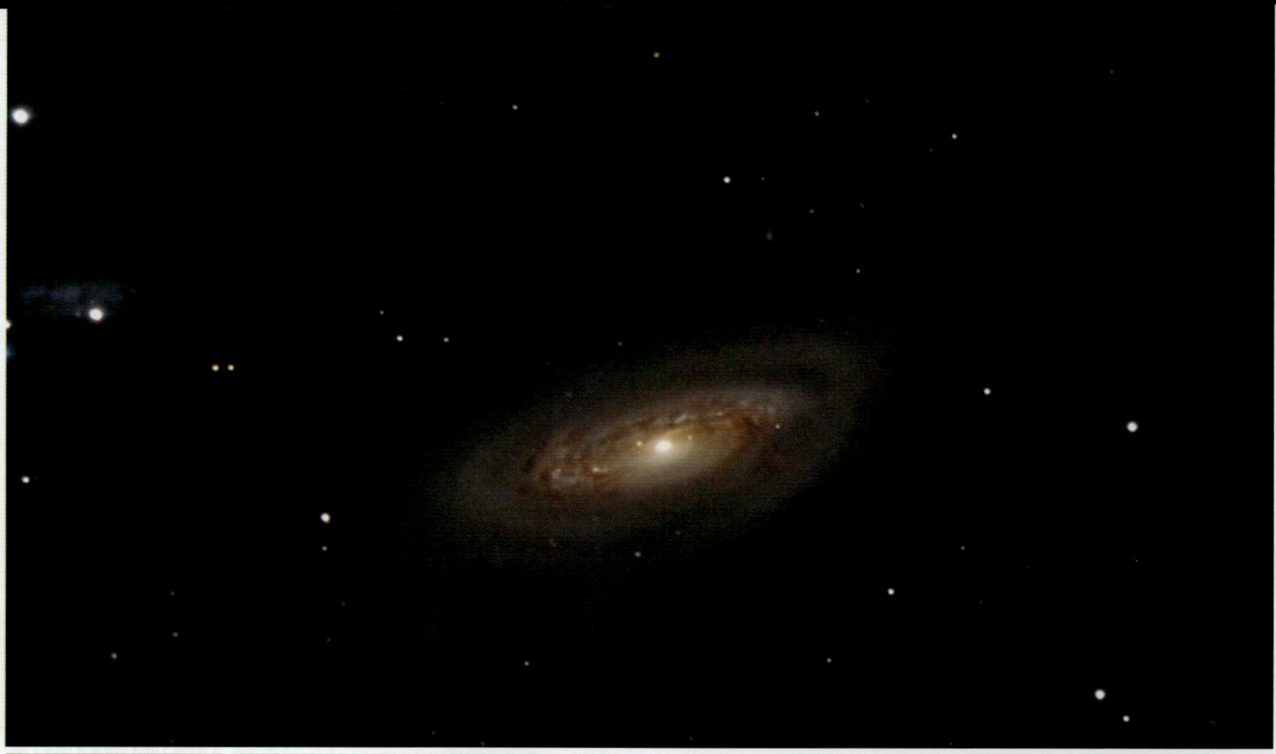

M 90

The SAB(rs)ab galaxy M 90 (NGC 4569) is a member of the Virgo Cluster, located at a distance of about 55 million light years. The angular size of 9.5′ × 4.4′ results in a transverse disc diameter of 150 000 light years. The velocity, deduced from shifted spectral lines, of −235 km/s means that M 90 is approaching the Milky Way and that its velocity differs by about 1200 km/s from the mean of the Virgo Cluster. The Doppler shift caused by motion within the Virgo Cluster is in the case of M 90 greater than the oppositely directed cosmological recession velocity. M 90 is 1.5 million light years from the centre of the Virgo Cluster and has a compact core with an active galactic nucleus to which an almost horizontal elongated structure is connected. The dust lanes and active regions visible here, however, do not continue in a spiral pattern, but rather lead to diffuse and partially ring-like structures which fill the inner disc and are similar to tidal tails.

At the left edge of the image to the north, 6′ from the centre, is the dwarf galaxy IC 3583. Due to the high relative velocity, one would at first not expect any interaction. However, since the dwarf galaxy appears stretched in the direction of M 90, interaction due to a close approach appears probable after all.

Within the Virgo Cluster, M 90 is an exception due to both its differing relative velocity and its hydrogen fraction. Radio observations indicate a large deficit of neutral hydrogen gas and comparison with other galaxies shows that M 90 must have lost almost 90 per cent of its neutral hydrogen gas. The H I distribution in the galactic plane is concentrated in the central region and corresponds to a disc rotation of at most 250 km/s assuming an inclination angle of 56°. Velocity gradients of this distribution enabled astronomers to test interaction models for M 90. The best results were given by a scenario in which M 90 passed through the centre of the Virgo Cluster 300 million years ago. Due to the associated ram-pressure stripping, M 90 lost its neutral hydrogen gas and thus the ability to form new stars in the outer disc. Simulations show also that these ring-shaped gas structures in the disc are short-lived and formed only after passing through the centre of the cluster.

OBJECT	M 90, NGC 4569
CONSTELLATION	Virgo
RIGHT ASCENSION	$12^h\ 36^m\ 50^s$
DECLINATION	+13° 09′ 47″
BRIGHTNESS	10.3 mag
TYPE	SAB(rs)ab
PHOTOGRAPHERS	Stefan Heutz
TELESCOPE	300-mm reflector
CAMERA	Starlight Xpress SXV-H9
EXPOSURE TIME	210 min
LOCATION	Velbert, Germany

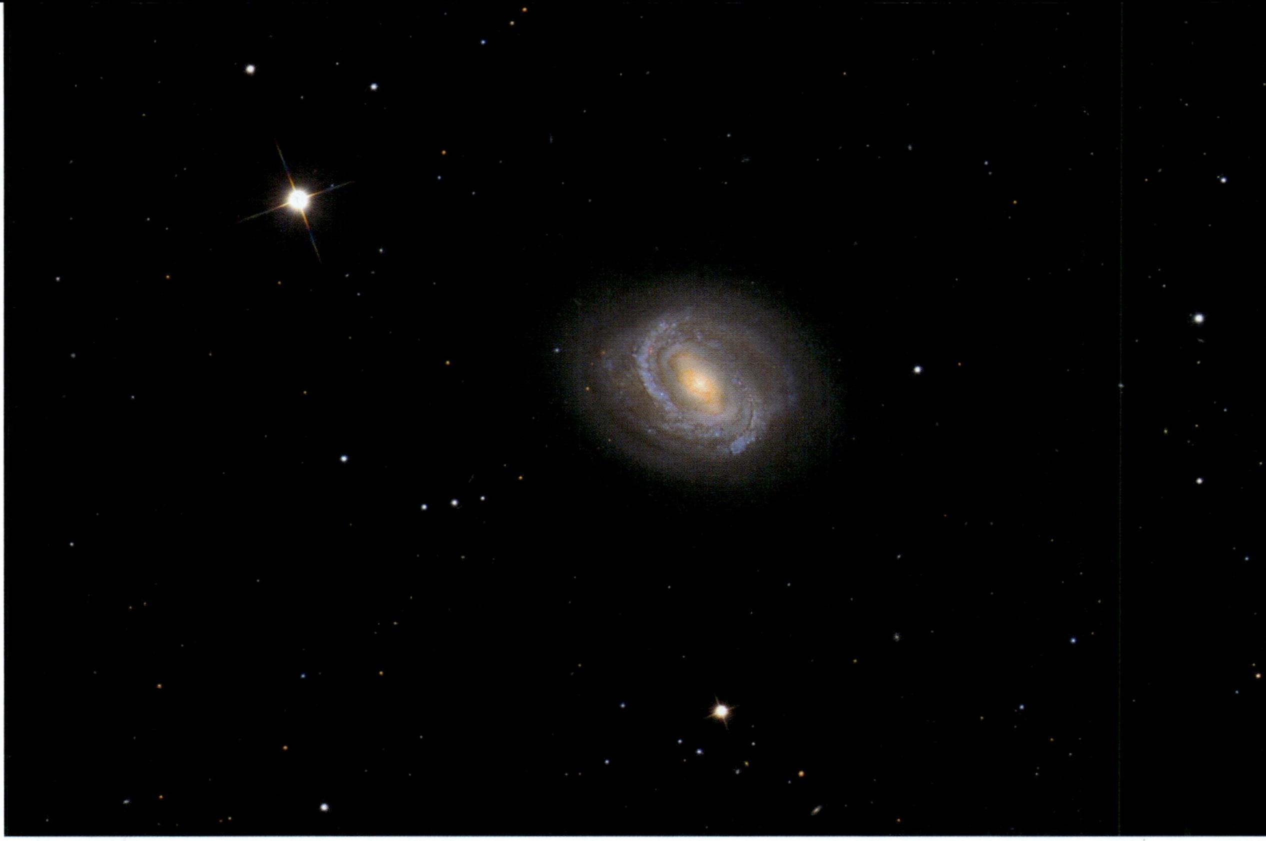

M 58

M 58 (NGC 4579) is an SAB(rs)b galaxy which belongs to the Virgo Cluster and whose recession velocity indicates a distance of 65 million light years. The angular size of 5.9′ × 4.7′ corresponds to a projected disc diameter of 110 000 light years. M 58 has an elliptical bulge which contains an active galactic nucleus, though this has only a low luminosity. The bar in M 58 is defined via a distribution of diffuse, yellow light. At the upper, southwestern end, the dust structures within the bar are shifted relative to one of the two bright ansae at the ends of the bar. The arm beginning at the southwest is thus appreciably longer and can be followed through 300°, the northeastern one only through 200°. The spiral pattern displays the typical, compact form of an early-type spiral and is surrounded by a disc periphery with a tidal tail at the right, eastern, side. Infrared data from the Spitzer Space Telescope show a continuation of the two-arm spiral pattern to the edge of the optical disc.

In comparison with other spiral galaxies, M 58 has little hydrogen gas. This gas deficit is presumably due to interaction with the cluster medium. In M 58, hydrogen is concentrated in the inner part of the disc and near the core star formation takes place in an ultra-compact nuclear ring (UCNR). Except for its size, it is similar to the nuclear star-forming regions which surround the cores of many spiral galaxies and whose diameters reach values of a few thousand light years. As shown in simulations, UCNR can be produced in barred spiral galaxies as a side effect of gas flowing into the centre. In M 58, the measured diameter of the UCNR is 1.5″ and corresponds to 250 light years. Since the ring is not completely closed and shifted relative to the centre, astronomers suspect a perturbation of the gas flow. It seems plausible that this perturbation also causes the measured variation with time of the luminosity of the core, since the UCNR is located directly next to the active galactic nucleus.

OBJECT	M 58, NGC 4579
CONSTELLATION	Virgo
RIGHT ASCENSION	$12^h\ 37^m\ 44^s$
DECLINATION	+11° 49′ 06″
BRIGHTNESS	10.5 mag
TYPE	SAB(rs)b
PHOTOGRAPHERS	Adam Block
TELESCOPE	800-mm reflector
CAMERA	SBIG STX-16803
EXPOSURE TIME	450 min
LOCATION	Mount Lemmon SkyCenter/ University of Arizona, USA

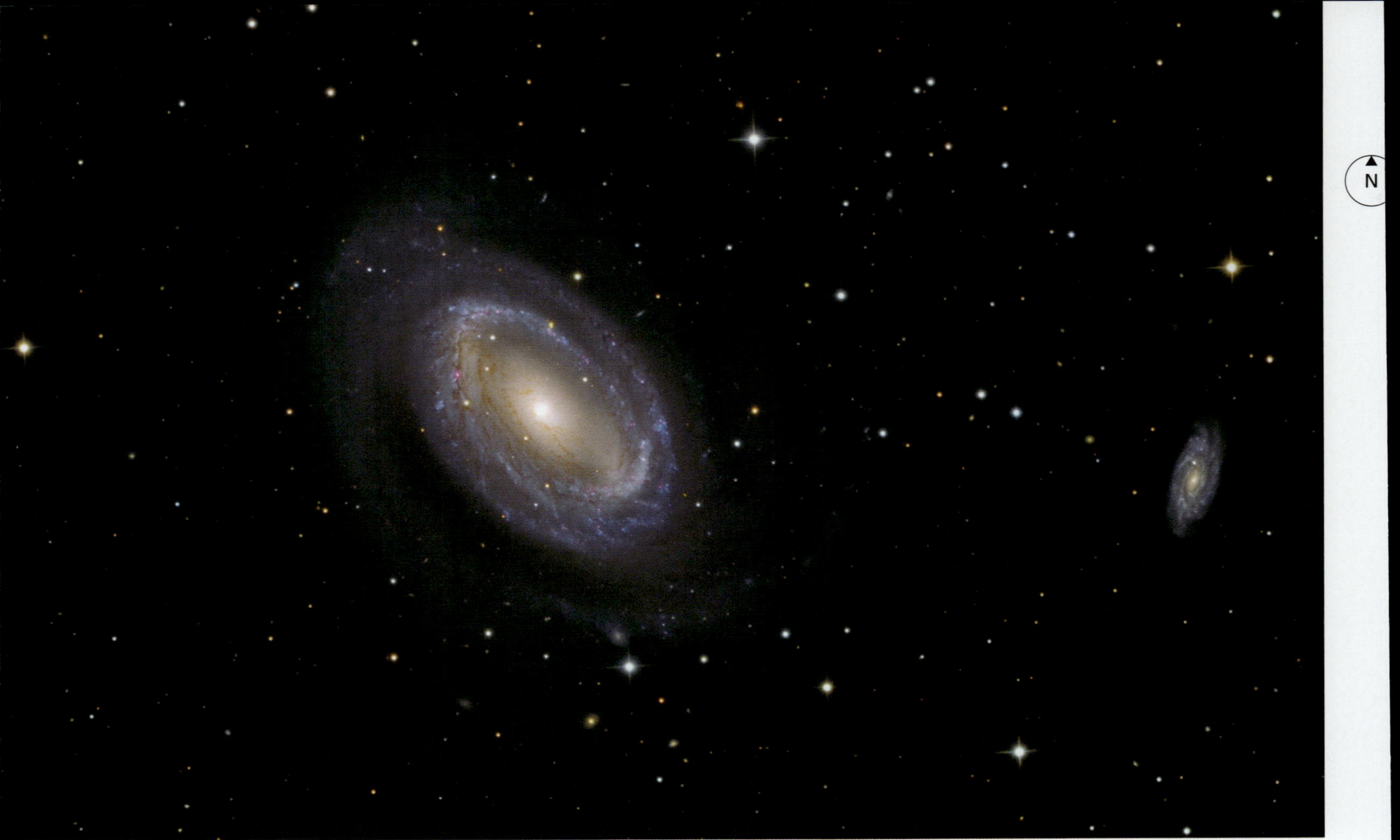

NGC 4725

Together with NGC 4747, which lies outside the image, the SAB(r)ab galaxy NGC 4725 forms an interacting pair which is part of the Coma–Sculptor Cluster. The mean distance of the pair is 42 million light years and for NGC 4725 the angular size of 10.7′ × 7.6′ corresponds to a projected disc diameter of 130 000 light years. The spiral pattern in NGC 4725 consists of not two but rather just one spiral arm whose brightest regions can be followed through 1.5 revolutions. Farther out, the arm fans out and changes into a broad, dim tidal tail. Even in the final pieces of the lower, southern part of the tail, active zones can still be found.

In about a third of all barred spiral galaxies one finds, in addition to the bar, a second, smaller bar structure in the centre. There is no particular angle between the two bars and neither is there a relation with the classification of the barred spiral. This independence of the two bars is confirmed by simulations which also indicate that the small bar plays an important role in the transport of gas into the centre. NGC 4725 also has a double-barred structure. The outer, larger bar is aligned with the diagonal of the image while the axis of the inner bar is rotated relative to it by 81° and its length of 12″ corresponds to 2500 light years. The local velocity distribution has maxima and minima which are aligned with the axis of the inner bar and also indicate the existence of a stellar disc near the centre of NGC 4725.

OBJECT	NGC 4725
CONSTELLATION	Coma Berenices
RIGHT ASCENSION	$12^h 50^m 27^s$
DECLINATION	+25° 30′ 03″
BRIGHTNESS	10.1 mag
TYPE	SAB(r)ab pec
PHOTOGRAPHERS	Stefan Binnewies, Josef Pöpsel
TELESCOPE	600-mm reflector
CAMERA	SBIG STL-16803
EXPOSURE TIME	300 min
LOCATION	Skinakas Observatory, Crete, Greece

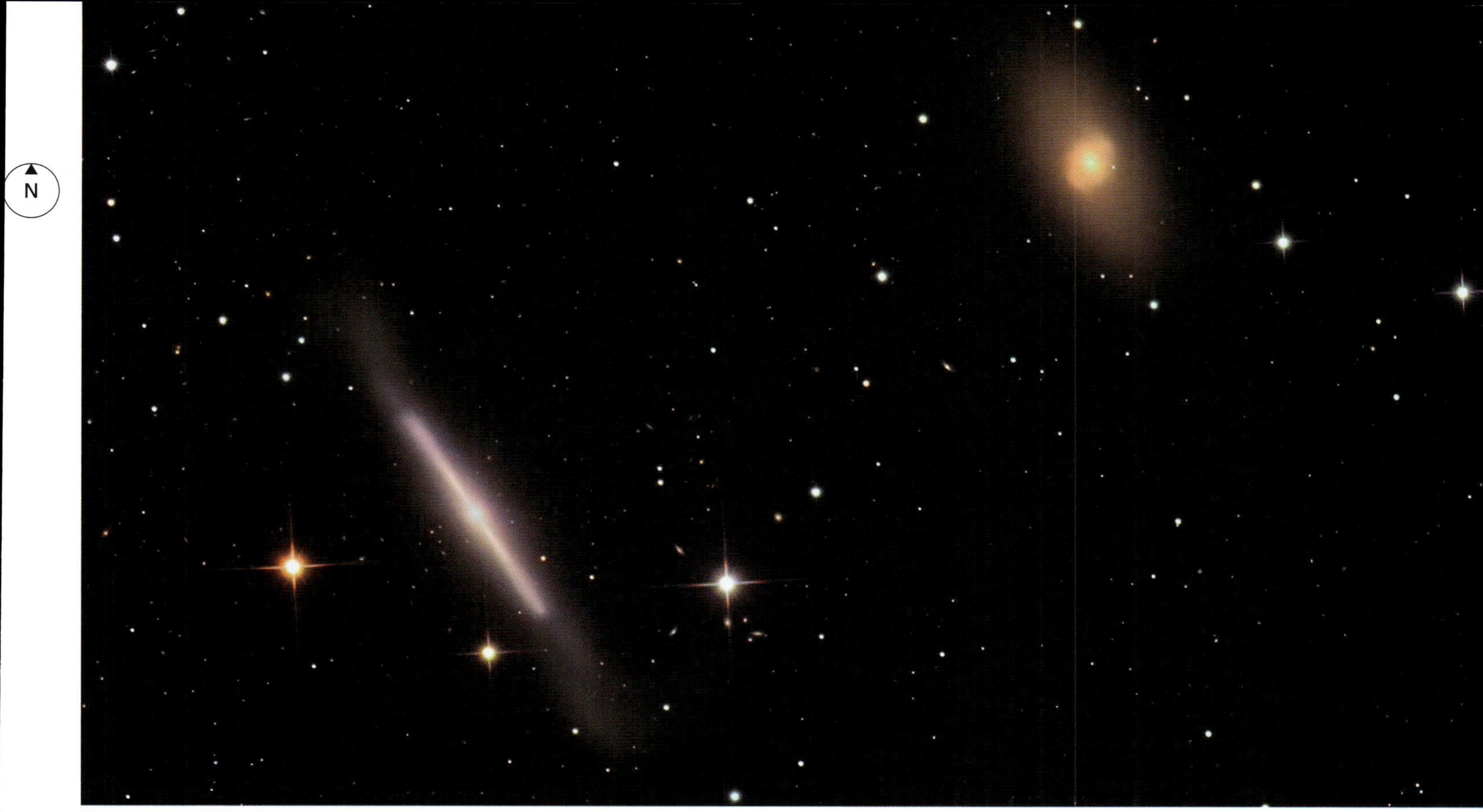

NGC 4762

The image shows the two galaxies NGC 4762 (left) and NGC 4754 (right) whose angular separation is 10.5′. The difference in recession velocity is 363 km/s and indicates that they are at different distances. Alternative methods of distance determination give a distance of 49 million light years for both NGC 4762 and NGC 4754. NGC 4762 is larger and its angular size of 8.7′ × 1.7′ corresponds to a projected transverse disc diameter of 134 000 light years and thus exceeds that of NGC 4754 by a factor of two. NGC 4762 is seen exactly edge-on so that the view is onto the bright inner edge of the disc. This inner disc of size 1.7′ × 0.15′ has no dust lanes and its thickness decreases uniformly towards the compact centre. The image shows bright extensions on both sides of the bright disc which make the diameter of the disc appear 25 per cent larger. The outer disc is dimmer. At both ends are tidal tails which lead out of the disc.

Analysis in several bands shows a red longitudinal structure with an extension of about 1′ which is embedded in the blue disc. This structure lies within the plane of the disc, has the same centre as the disc, and can be interpreted as a bar. For this reason, NGC 4762 is classified as SB0 in some catalogues. Until the 1980s, astronomers assumed that stellar bars had to be flat structures. Only with the advent of 3-D computer simulations did it become accepted that bars can be better described as boxy structures. Since NGC 4762 is the flattest known lenticular galaxy, there is doubt as to the existence of the bar. An explanation is given by the relatively small bulge which indicates that the bar could be at an early stage in its evolution. This theory is also supported by visible traces of interactions at the edge of the disc, which could have caused the formation of the central bar.

OBJECT	NGC 4762
CONSTELLATION	Virgo
RIGHT ASCENSION	$12^h\ 52^m\ 56^s$
DECLINATION	+11° 13′ 51″
BRIGHTNESS	11.1 mag
TYPE	$SB0^0$(r)? edge-on
PHOTOGRAPHERS	Josef Pöpsel, Stefan Binnewies
TELESCOPE	600-mm reflector
CAMERA	SBIG STL-11000
EXPOSURE TIME	435 min
LOCATION	Skinakas Observatory, Crete, Greece

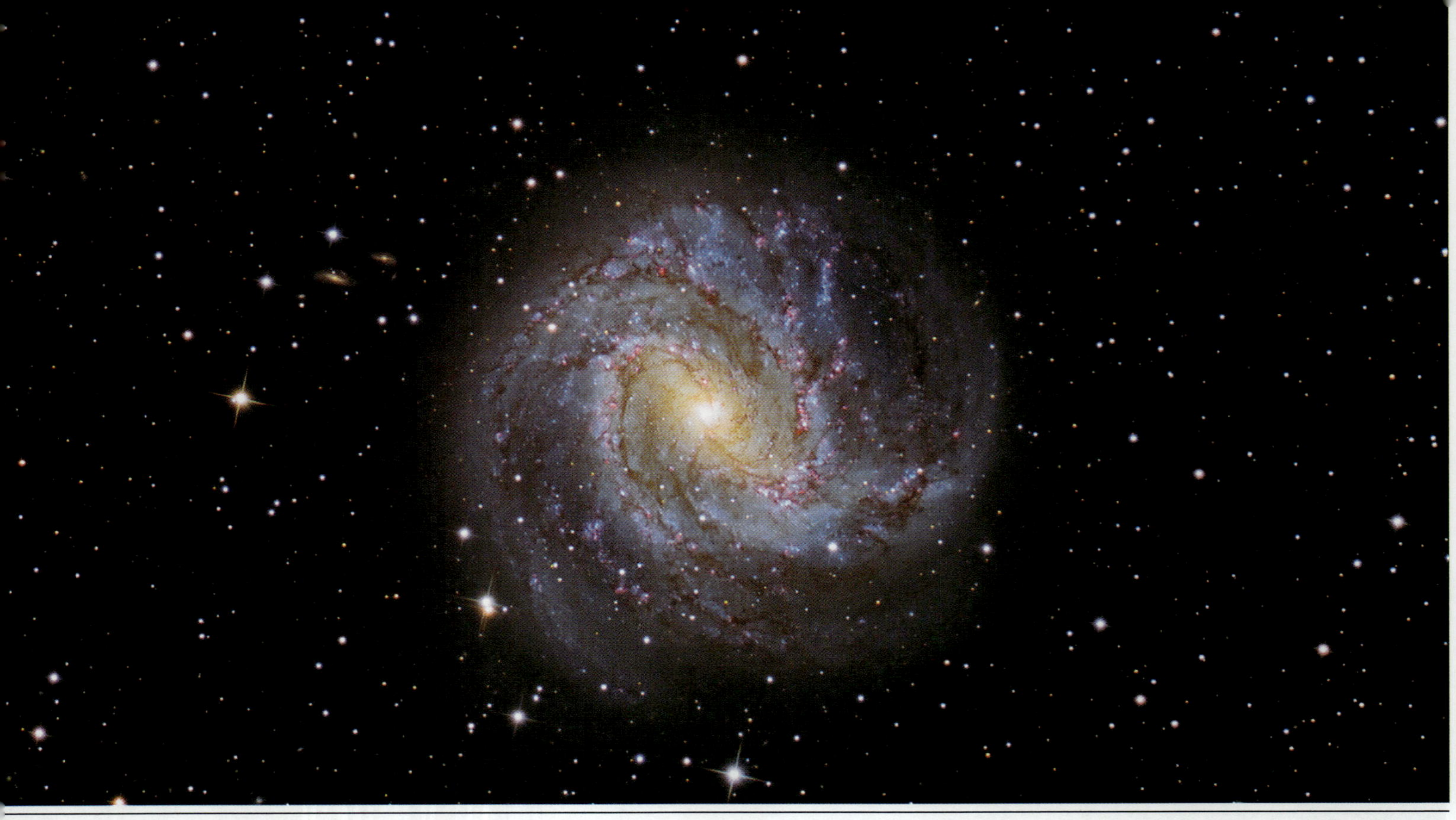

M 83, The Southern Pinwheel Galaxy

M 83 (NGC 5236) is a member of the M 83/NGC 5128 galaxy group, one of the closest neighbouring groups to the Local Group. The M 83 subgroup at 20 million light years is the smaller and farther part of the group, which consists of 45 members. The angular size of 12.9′ × 11.5′ corresponds to a projected disc diameter of 75 000 light years. Due to its southern position at almost −30° declination, M 83 is also known as the Southern Pinwheel Galaxy. The corresponding northern galaxy, M 101 (type SABcd), however, has a more open spiral structure than the SABc galaxy M 83. The bar structure in M 83 is defined by elongated dust lanes in the centre; the stellar bar is weak. M 83 is a prototype of the SAB classification which Gérard de Vaucouleurs introduced in 1959. Two spiral arms are connected to the bar, and arm fragments with many active regions appear to be separating from the one pointing upward in the right, western half of the disc.

The image shows two very faint asteroids as well: 146411 at the upper right (19.3 mag) and 88238 (19.1 mag) at the lower right edge of the disc. The weak, coloured track is due to the fact that the image is composed of a brightness and an RGB component.

Both UV and H I radio data indicate the existence of a large disc in M 83 which exceeds the diameter of the optical disc by a factor of five. In many studies, astronomers have used emission-line indicators to compare star-formation rates in H II regions in the inner and outer discs. Observations confirm the theory that the outer regions of spiral galaxies are less developed and the calculated star-formation rates in M 83 of 10^{-5} solar masses per year are more than three orders of magnitude smaller than in the inner regions. The slow evolution of the outer disc indicates that the gas is still increasing in density and thus suggests possible consequences of one or more interactions with the surroundings of M 83.

OBJECT	M 83, NGC 5236
CONSTELLATION	Hydra
RIGHT ASCENSION	$13^h\ 37^m\ 01^s$
DECLINATION	−29° 51′ 56″
BRIGHTNESS	8.2 mag
TYPE	SAB(s)c
PHOTOGRAPHERS	Johannes Schedler
TELESCOPE	500-mm reflector
CAMERA	SBIG STL-11000
EXPOSURE TIME	180 min
LOCATION	Hakos Farm, Namibia

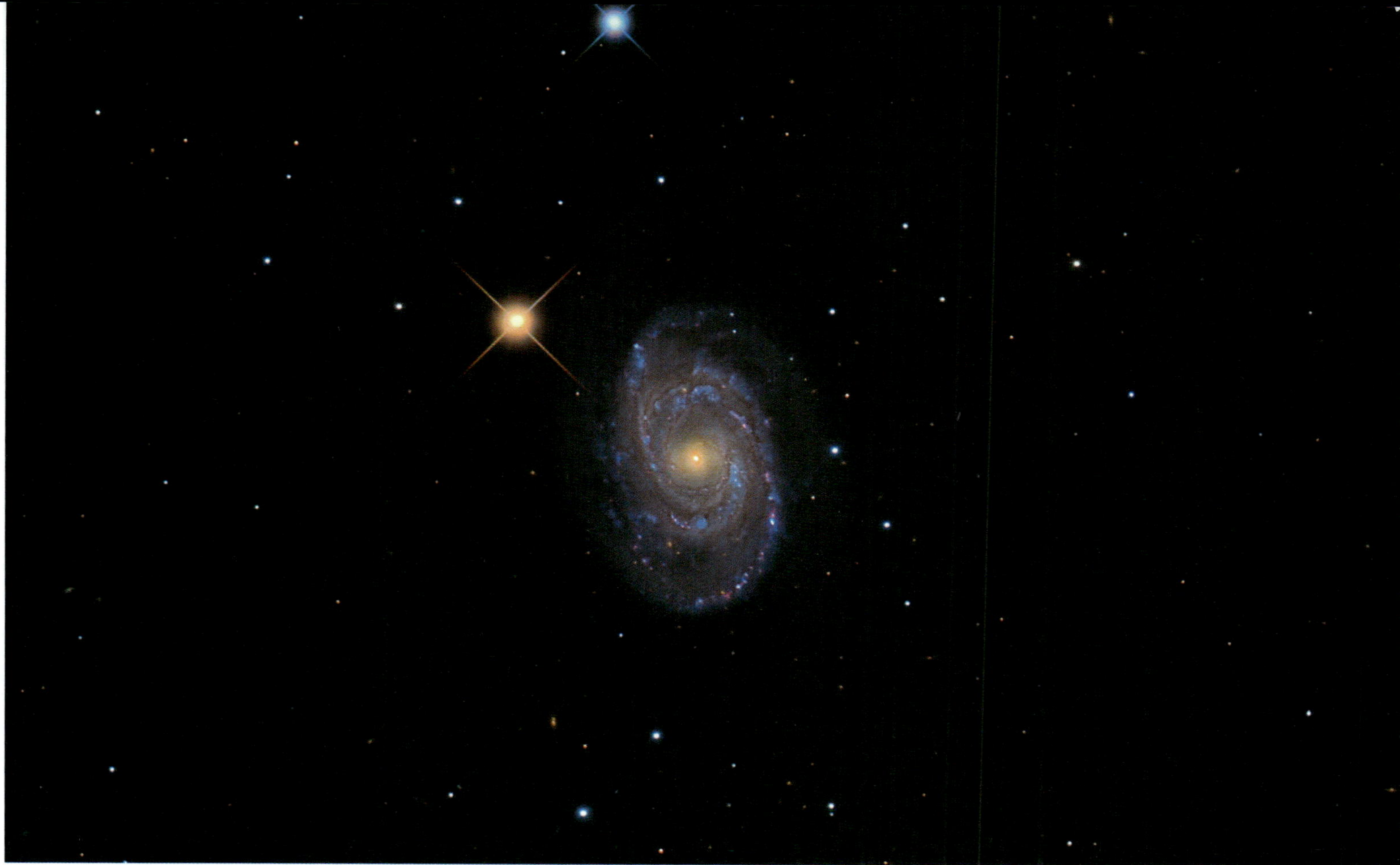

NGC 5371

The SAB(rs)bc galaxy NGC 5371 is at a distance of 100 million light years from the Milky Way. The angular diameters of 4.4′ and 3.5′ result in a projected disc size of 128 000 light years. Due to its classification as a LINER (low-ionization nuclear emission-line region), one can assume that this galaxy contains an active galactic nucleus, which appears in the image as a compact source in the orange-yellow, oval bulge. The spectra of LINER galaxy nuclei show prominent emission lines of weakly ionized atoms (O^+, N^+, S^+). At the right, western side of the bar a double spiral arm begins whose outer portion, after 270°, transforms into a tidal tail which can be followed through 180°. The spiral arm which begins at the eastern end of the bar overlaps with the inner part of the western spiral arm after 90°.

Most of the active regions are in the western spiral arm and point in the direction of the compact galaxy group Hickson 68, which is not visible in the image. The five members of this group are located at about the same distance as NGC 5371; the transverse distance to NGC 5371 is 900 000 light years. At the centre of the galaxy, emission lines of molecular carbon-monoxide gas (CO), which are used as an indicator of dust in star-forming regions, are not very intense. In these radio observations, only at the western end of the bar and in one star-formation region did one find a CO emission line which was above the noise level. The local suppression of star formation in NGC 5371 could be a consequence of tidal interaction with the Hickson 68 group.

OBJECT	NGC 5371
CONSTELLATION	Canes Venatici
RIGHT ASCENSION	13^h 55^m 40^s
DECLINATION	+40° 27′ 42″
BRIGHTNESS	11.3 mag
TYPE	SAB(rs)bc
PHOTOGRAPHERS	Adam Block
TELESCOPE	600-mm reflector
CAMERA	SBIG STL-11000
EXPOSURE TIME	190 min
LOCATION	Mount Lemmon SkyCenter/ University of Arizona, USA

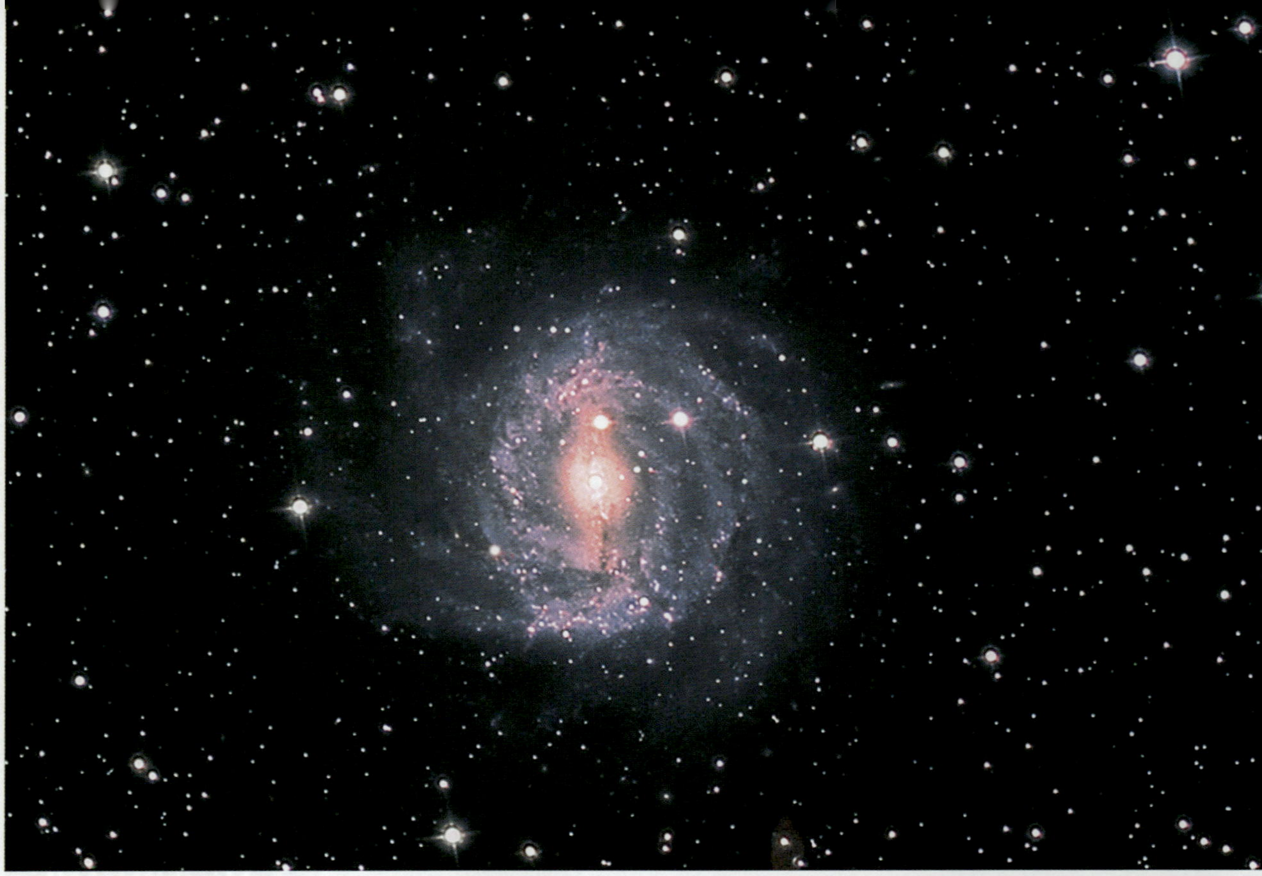

NGC 5643

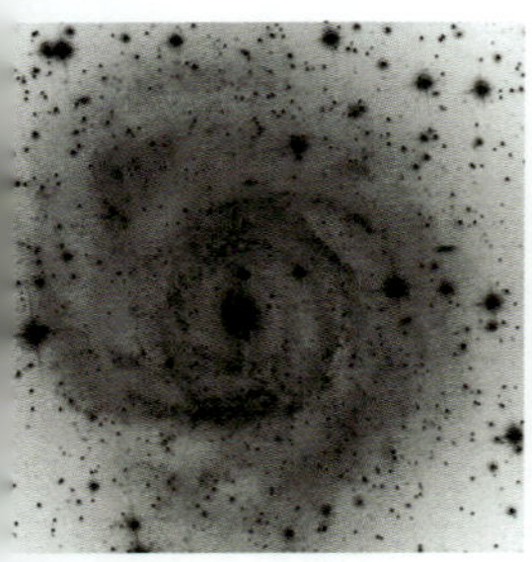

At a declination of −44°, the SAB(rs)c galaxy NGC 5643 is a southern-sky object. The distance from the Milky Way is 48 million light years and the angular size of 4.6′ × 4.0′ corresponds to a projected diameter of 64 000 light years for the optical disc. The vertical bar in NGC 5643 runs in east-west direction and contains both a compact nucleus and an elliptical bulge. The major axis of the bulge is rotated by about 30° compared to the axis of the bar. Connected to the upper, western end of the bar is a spiral arm whose brightness drops rapidly. At the other side of the bar, the beginning of the arm is less distinct; here there is a prominent active spiral-arm fragment farther out which one can associate with the upper spiral arm. The edge of the disc plane in NGC 5643 contains several dim spiral-arm ends which in the contrast-enhanced image display a few Vorontsov-Velyaminov rows.

NGC 5643 contains a Seyfert-2 nucleus. With narrow-band images made with the Hubble Space Telescope, astronomers have studied O III and Hα emission with the result that the line emission is not symmetric about the nucleus, but rather concentrated in a cone-shaped area to the east of the centre. The Hα emission extends about twice as far as O III, but the brightest emission zones are identical. This structure is interpreted as an ionization cone caused by energetic radiation from the central source. The fact that one sees the ionization cone only on the eastern side is explained by the fact that the western cone is behind the galactic disc of NGC 5643. The symmetry axis of the cone is tilted by about 60° to the line of sight. The direct view onto the central source is blocked by a dust torus along the line of sight. Further typical scattered-radiation effects in active galaxies, such as a blue continuum at the tip of the cone, confirm the Seyfert-2 model for NGC 5643.

OBJECT	NGC 5643
CONSTELLATION	Lupus
RIGHT ASCENSION	$14^h\ 32^m\ 41^s$
DECLINATION	−44° 10′ 28″
BRIGHTNESS	10.7 mag
TYPE	SAB(rs)c
PHOTOGRAPHERS	Josef Pöpsel, Beate Behle
TELESCOPE	600-mm reflector
CAMERA	SBIG ST-10XME
EXPOSURE TIME	180 min
LOCATION	Amani Lodge, Namibia

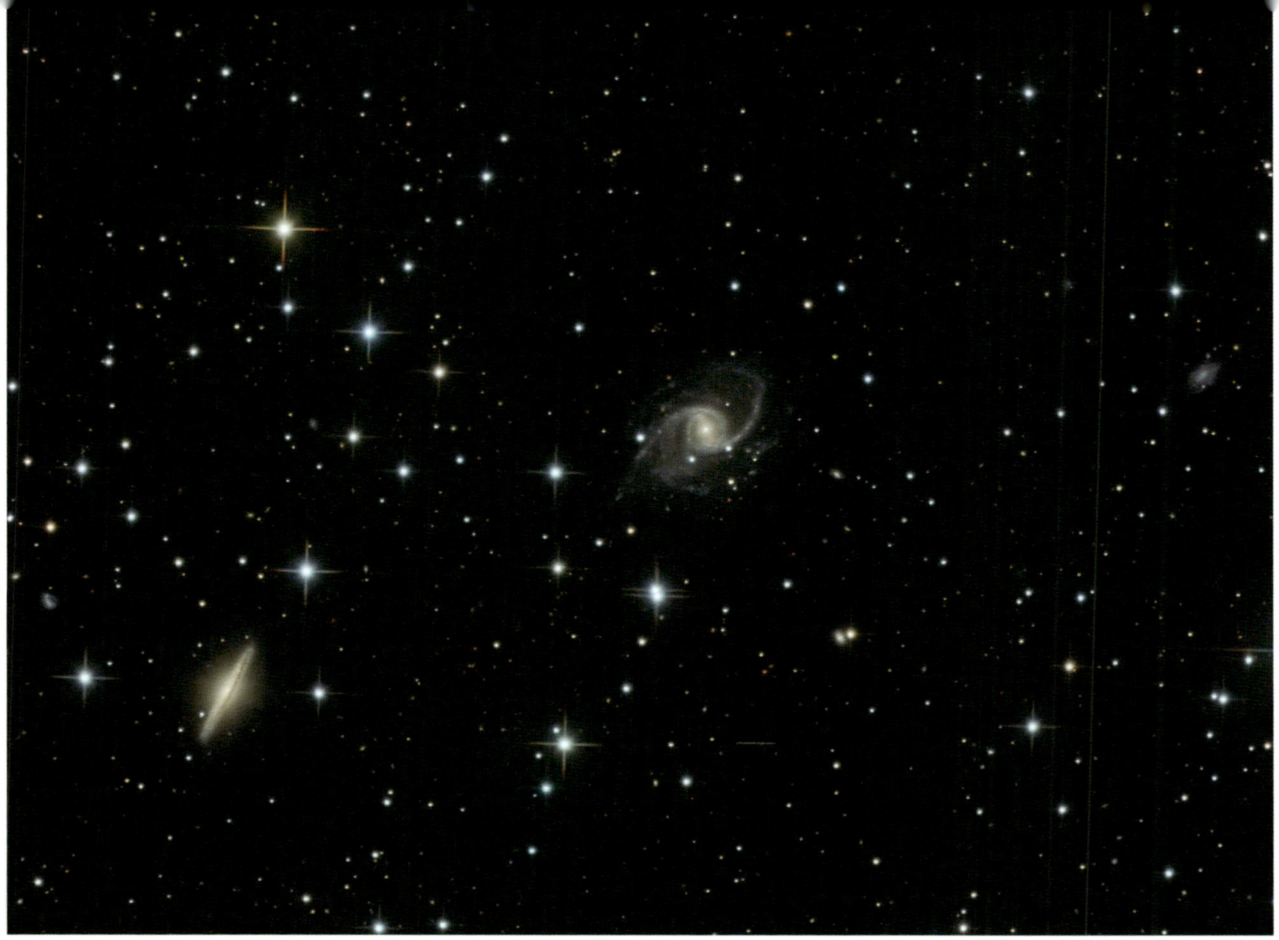

NGC 5905

At the centre of the image is NGC 5905, an SB(r)b galaxy whose distance from the Milky Way is 160 million light years. The angular size of 4.0′ × 2.6′ appears small but the corresponding projected disc diameter of almost 190 000 light years shows its enormous absolute size. The bar is tilted by about 30° from the vertical and the two spiral arms connected to it form an inner ring with a diameter of 0.9′. They have large pitch angles and reach the edge of the disc after 270°. The outer zone of the disc contains a few prominently blue arm fragments. At a distance of 13′ below and to the left of NGC 5905 is the SAb galaxy NGC 5908. Its recession velocity is only 84 km/s larger and thus suggests that the pair is gravitationally bound. There are, however, no visible signs of an interaction in the environment of the two galaxies.

NGC 5905 is catalogued as an H II galaxy and high-resolution spectra of the central source show narrow emission lines of hydrogen, nitrogen and oxygen. Due to these optical properties, NGC 5905 is often classified as a Seyfert-2 galaxy. In the centre itself, astronomers observe reduced Hα emission. This can be interpreted as an increase in absorbing hydrogen gas along the line of sight. If one varies the orientation of the spectrographic slit, one finds that the emission lines are broadened; this is a clear sign of a rotating gas disc in NGC 5905. The largest rotational velocities of ±100 km/s are reached at 2–3″ from the centre and correspond to a mean disc diameter of about 4000 light years. The nucleus of NGC 5905 is normally hidden by this disc and is visible only on rare occasions. This is possible when the disc gas is ionized by X-ray eruptions and thus becomes transparent in X-rays.

OBJECT	NGC 5905
CONSTELLATION	Draco
RIGHT ASCENSION	$15^h\ 15^m\ 23^s$
DECLINATION	+55° 31′ 03″
BRIGHTNESS	12.5 mag
TYPE	SB(r)b
PHOTOGRAPHERS	Richard Müller
TELESCOPE	318-mm reflector
CAMERA	Artemis 4021
EXPOSURE TIME	950 min
LOCATION	Lohmar, Germany

NGC 5945

The SB(rs)ab galaxy NGC 5945 can be seen in the upper part of the image. At a distance of 8.4′ to the south is the galaxy NGC 5943, which shows evidence of a merger. At the lower edge of the picture is the interacting pair NGC 5934 and NGC 5935. NGC 5945 belongs to a galaxy group consisting of about a dozen brighter galaxies stretched out over 40′ along a filament in the northern constellation of Boötes. The recession velocities of most of these galaxies lie within ±200 km/s of one another and indicate that they are gravitationally bound. The distance of NGC 5945 from the Milky Way is 230 million light years. With a diameter of 1′, the size of the prominent inner ring in NGC 5945 corresponds to 67 000 light years in projection. This is in the range of typical diameters of spiral galaxies, which is confirmed by the other galaxies in the field of the image. A few arcminutes to the east of the galaxy pair NGC 5934 and NGC 5935, one notices a cluster of small, orange-yellow galaxies. This is the cluster WHL J152840.8+425617, which contains 14 members and is located at an approximate distance of 3.7 billion light years.

The bar in NGC 5945 is almost horizontal and one recognizes two delicate, bright brown dust lanes. These run into the oval centre, whose longitudinal axis is perpendicular to that of the bar, and form a dust ring around this centre. The two spiral arms can be followed only through half a revolution and form two slightly offset halves of the inner ring. The spiral arms also seem to end in two dim tidal tails which form an outer, tilted ring. If this ring is considered to be part of the disc plane, then it attains the enormous diameter of 200 000 light years. It is suspected that NGC 5945 formed from the merger of a compact galaxy. Conspicuous is that the non-circular perturbation of the disc potential in NGC 5945 is very large. The bar in NGC 5945 measures about 65 000 light years and thus exceeds the length of the bar in the Milky Way by a factor of three.

OBJECT	NGC 5945
CONSTELLATION	Boötes
RIGHT ASCENSION	$15^h\ 29^m\ 45^s$
DECLINATION	+42° 55′ 07″
BRIGHTNESS	13.6 mag
TYPE	SB(rs)ab
PHOTOGRAPHERS	Stefan Binnewies, Josef Pöpsel
TELESCOPE	600-mm reflector
CAMERA	SBIG STL-16803
EXPOSURE TIME	425 min
LOCATION	Skinakas Observatory, Crete, Greece

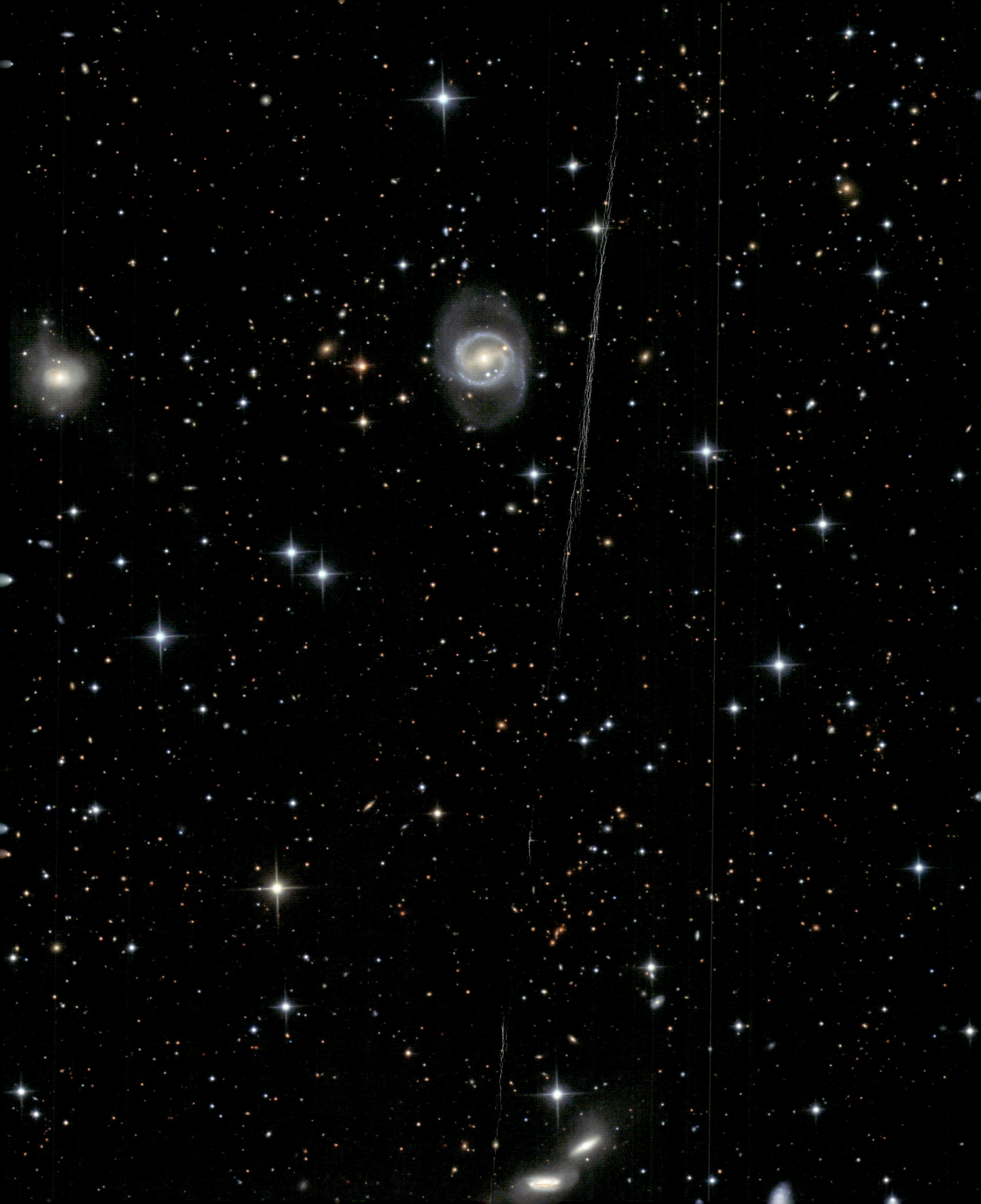

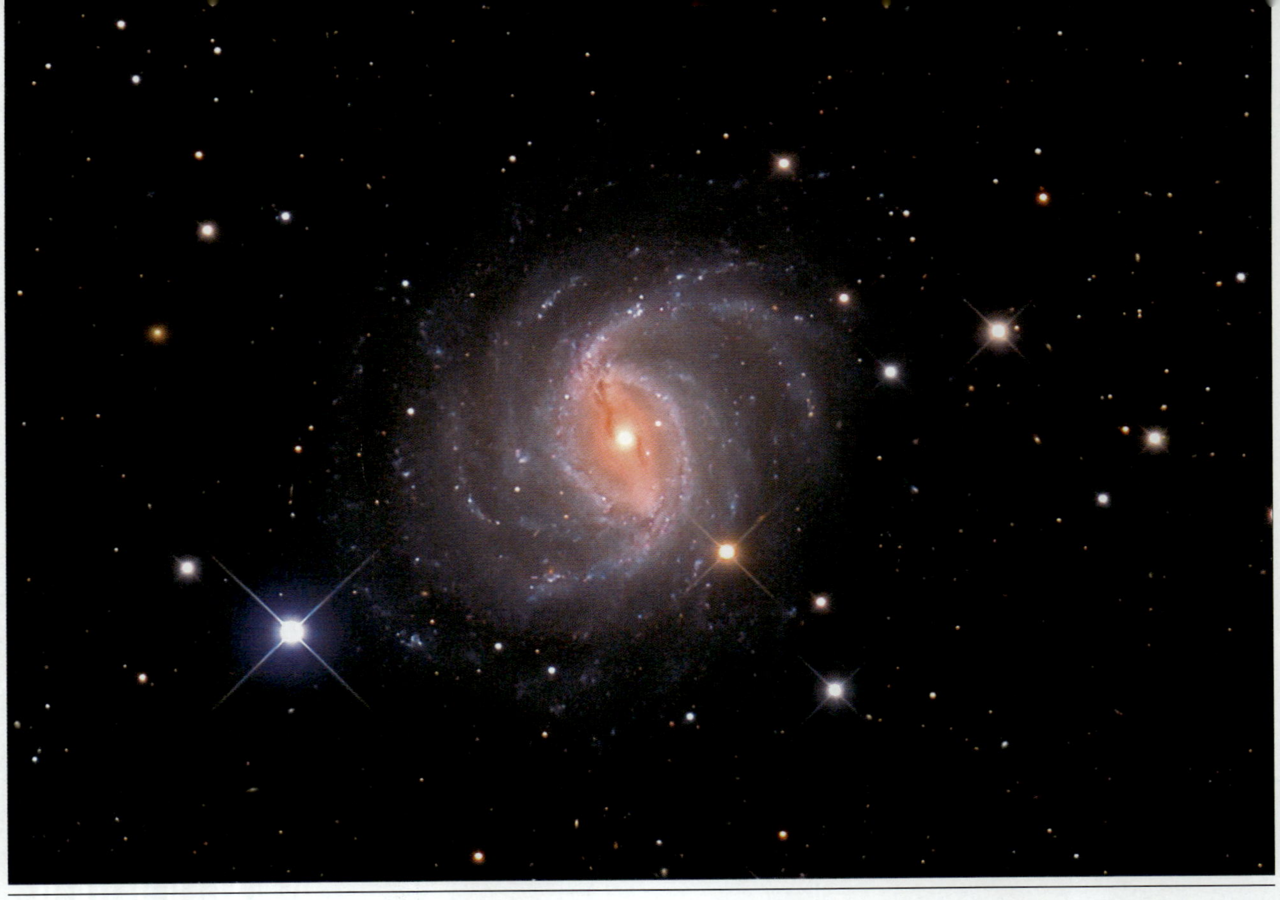

N

NGC 5921

NGC 5921 is an SB(r)bc galaxy located 3° north of the globular cluster M 5. This isolated field galaxy is at a distance of 68 million light years from the Milky Way. 1′ corresponds to a projected distance of about 20 000 light years. The angular size of 4.9′ × 4.0′ corresponds to a disc diameter of 97 000 light years. The bar in NGC 5921 is crossed by dust lanes. At the upper, northern end these are especially thick. They run diagonally compared to the longitudinal axis and the join at the bright nucleus is almost perpendicular to the axis. At the ends of the bars are two bright spiral arms with particularly small pitch angles. Due to the compact form of the two spiral arms, an oval frame of the bar zone is formed. Each of the spiral arms crosses the starting point of the other spiral arm after half a revolution. These regions of overlap are characterized by many H II regions. In the outer regions, the spiral arms open and fragment, causing the disc to be defined by many dim arm fragments.

Comparing red and blue images of NGC 5921, one can see that the spiral pattern is visible almost only in the blue. The bar dominates in the red. The exception are Hα images which document the location of the H II regions of the spiral arms. The bar itself is hardly visible in the Hα images, which indicates a possible deficit of hydrogen gas in NGC 5921. The measured rotation curve in ionized hydrogen gas, however, confirms that this deficit doesn't apply to the entire bar region. The wavy form of the rotation curve in the central region of ±0.5′ is an indication of large velocity differences in the flow of hydrogen gas in the bar. Studies of neutral hydrogen confirm the general hydrogen deficit and offer a reason why NGC 5921 is also classified as a galaxy with reduced star formation and thus low surface brightness.

OBJECT	NGC 5921
CONSTELLATION	Serpens
RIGHT ASCENSION	$15^h\ 21^m\ 57^s$
DECLINATION	+05° 04′ 14″
BRIGHTNESS	11.5 mag
TYPE	SB(r)bc
PHOTOGRAPHERS	Makis Palaiologou, Stefan Binnewies
TELESCOPE	1.3-m reflector
CAMERA	Andor DZ 436
EXPOSURE TIME	112 min
LOCATION	Skinakas Observatory, Crete, Greece

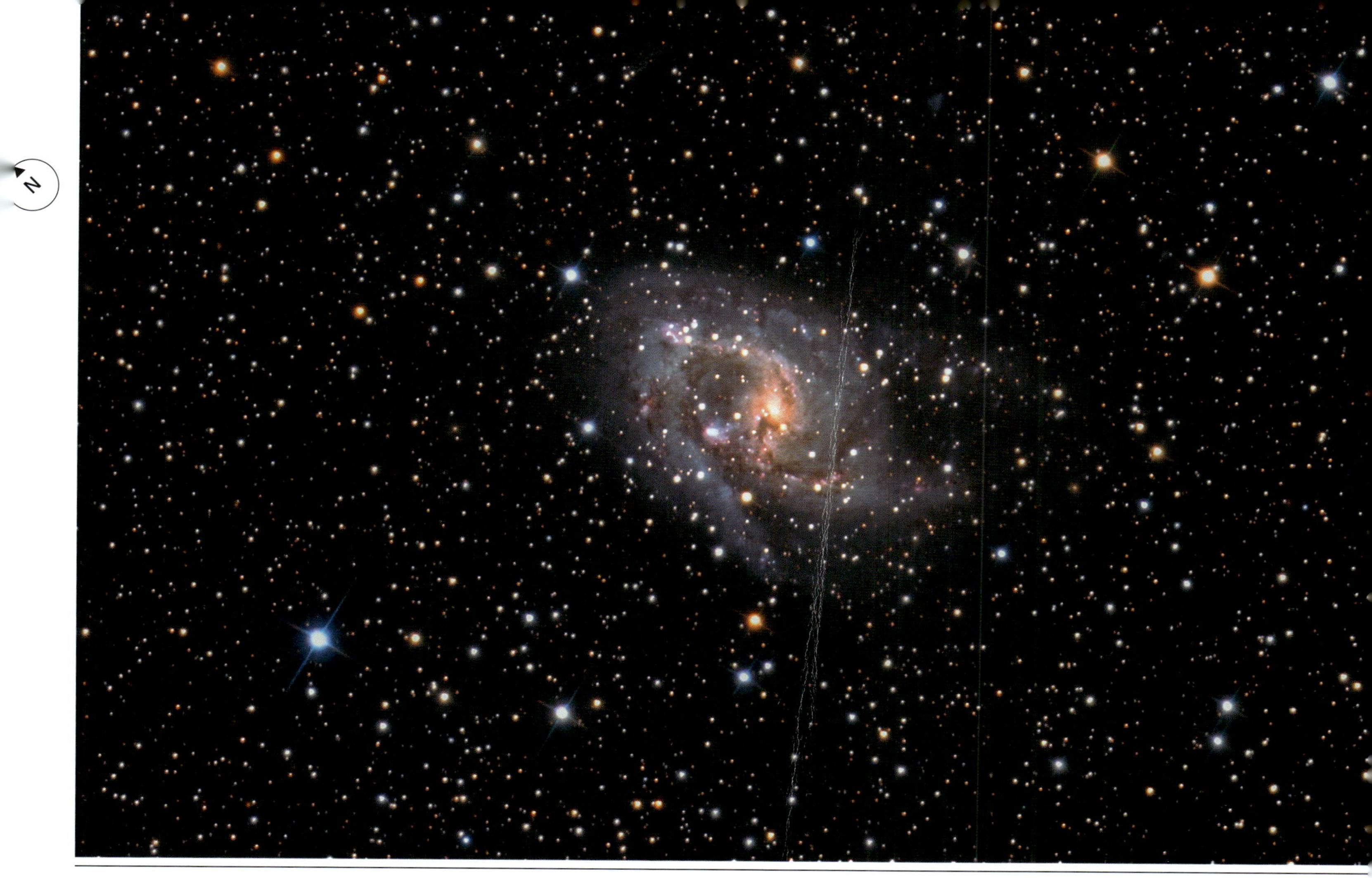

NGC 6221

The SB(s)c galaxy NGC 6221 is in the southern constellation of Ara and at a Galactic latitude of −9.6° lies near the plane of the Milky Way. It forms a pair with NGC 6215, which lies 19′ to the northwest and cannot be seen in the image. NGC 6221 is at a distance of 60 million light years; the maximum angular size of 3.5′ corresponds to a projected diameter of 61 000 light years. Its bright core lies within a broad bar which is crossed by dust lanes, showing an offset to the axis of the bar as is typical of SB galaxies. Two spiral arms connect to the ends of the bar; they have a few active regions, but don't form a large-scale spiral pattern. The disc is crossed by dust filaments and appears frayed at the right edge.

Detailed Hα studies reveal large, non-circular streams of ionized gas in the disc. The distribution of neutral hydrogen follows the optical emission, but appears more uniform. Radio data reveal an H I disc which is two to three times larger than the optical disc, two dwarf galaxies, and an indication of a bridge of material between NGC 6221 and NGC 6215. This structure stretches over 300 000 light years and contains 3×10^8 solar masses. This is only five per cent of the H I mass detected in NGC 6221. From the velocity difference one can estimate that there must have been a close encounter of the two galaxies 500 million years ago during which NGC 6221 was strongly disturbed.

OBJECT	NGC 6221
CONSTELLATION	Ara
RIGHT ASCENSION	$16^h\ 52^m\ 46^s$
DECLINATION	−59° 13′ 07″
BRIGHTNESS	11.2 mag
TYPE	SB(s)c
PHOTOGRAPHERS	Rainer Sparenberg, Stefan Binnewies
TELESCOPE	600-mm reflector
CAMERA	SBIG ST-10XME
EXPOSURE TIME	120 min
LOCATION	Amani Lodge, Namibia

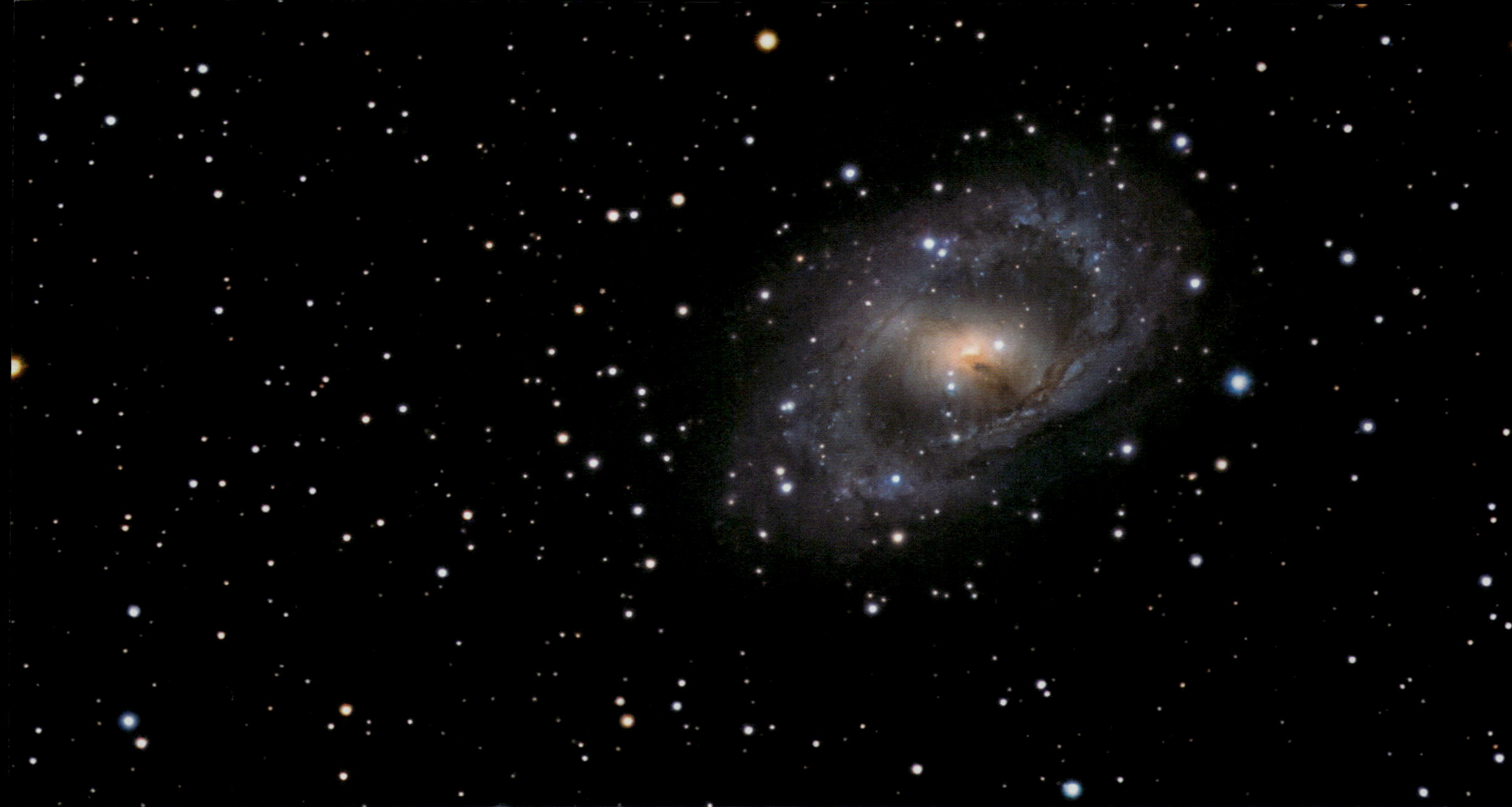

NGC 6300

NGC 6300, an SB(rs)b galaxy in the southern constellation of Ara, is a member of the galaxy group [TSK2008] 0349. Its distance from the Milky Way is 42 million light years, the angular size of 4.5′ × 3.0′ corresponds to a projected disc diameter of 55 000 light years. NGC 6300 is catalogued as an active Seyfert-2 galaxy. In the image, the major axis of the bulge is horizontal and tilted by 45° with respect to the axis of the bar. The bar is weak and covered by broad dust lanes. At the lower, southwestern end of the bar, the dust structure is easily recognizable and seems to surround the spiral arm that begins there. Both spiral arms in NGC 6300 have a broad, diffuse form and each form half of a ring which contains active regions that surround the zone of the bar. This bright, inner part of the disc is surrounded by a fainter, less blue outer region.

The variability of the active nucleus in NGC 6300 was studied in detail by the X-ray satellite XMM-Newton. The detectors registered a mean count rate of about one X-ray photon per second. The time series, approximately 12 hours long, shows a variable X-ray luminosity which contains a few peaks but is mainly determined by variability on timescales of 60–120 minutes. Periodograms were used to quantify the distribution of timescales. This typical X-ray variability can be described by a slope. For NGC 6300, one obtains a value of 1.75, which lies in the range 1–2, which is otherwise found only for Seyfert-1 nuclei. Seyfert-2 variability, on the other hand, is less characterized by short timescales and described by smaller slopes. Astronomers have concluded from these time-series data that NGC 6300 is a Seyfert-1 galaxy whose central regions, however, are covered by clouds of material, which led to the misclassification as Seyfert 2. The assumption of a cloud of material covering the nucleus is confirmed by the large amount of dust visible in this SB galaxy. The central regions of NGC 6300 thus differ from the standard model of an active galaxy.

OBJECT	NGC 6300
CONSTELLATION	Ara
RIGHT ASCENSION	$17^h\ 16^m\ 59^s$
DECLINATION	−62° 49′ 14″
BRIGHTNESS	11.0 mag
TYPE	SB(rs)b
PHOTOGRAPHERS	Josef Pöpsel, Beate Behle
TELESCOPE	600-mm reflector
CAMERA	SBIG ST-10XME
EXPOSURE TIME	200 min
LOCATION	Amani Lodge, Namibia

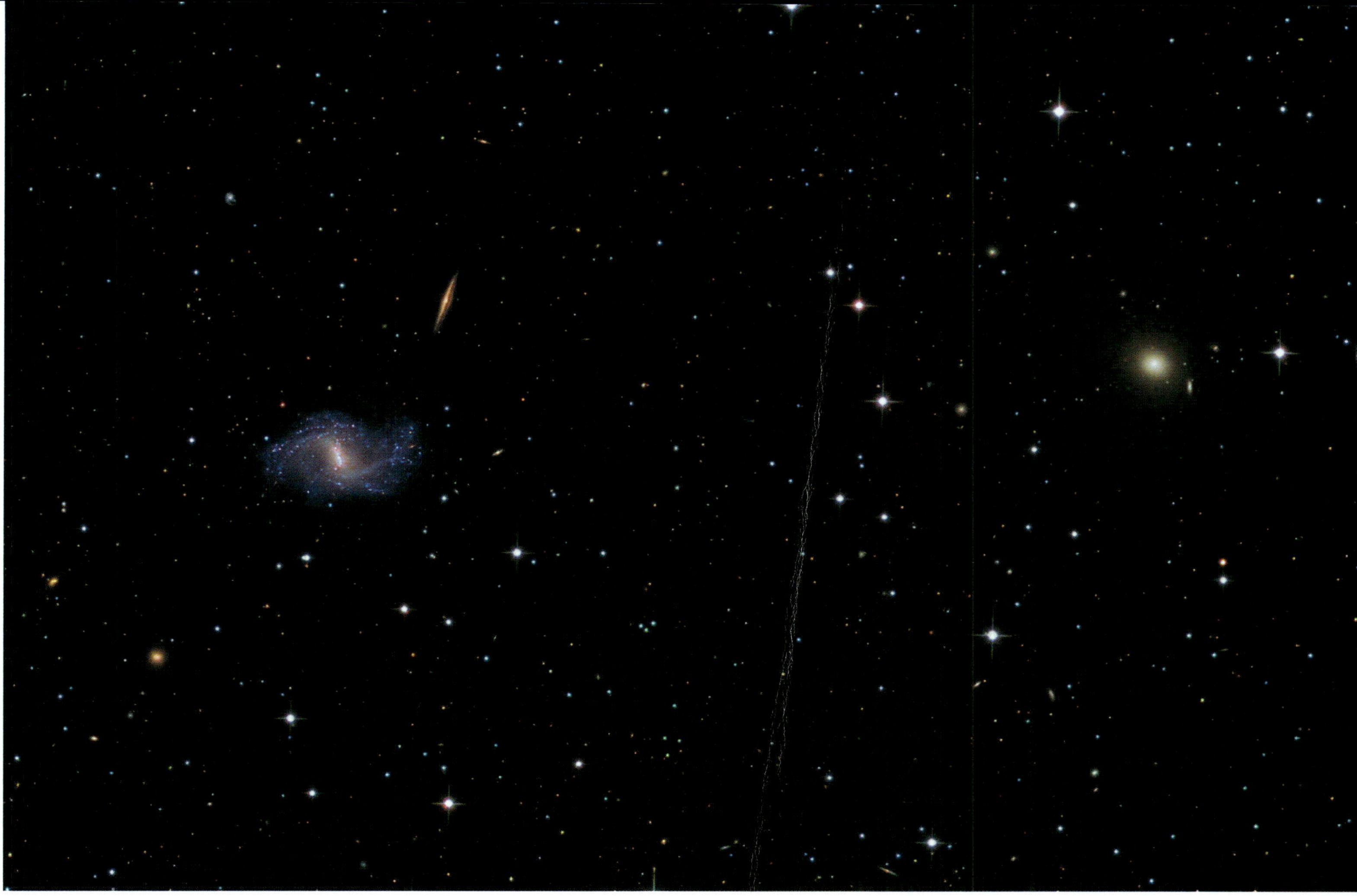

NGC 6339

In the left side of the image is the SBd galaxy NGC 6339. Above and close by is PGC 60007, which is, however, four times farther away and thus not interacting. NGC 6339 is at a distance of 100 million light years from the Milky Way. The angular size of 2.9′ × 1.7′ corresponds to a transverse diameter of 84 000 light years. At the right edge of the image, 12′ to the north of NGC 6339, is the galaxy NGC 6343, about as far away as PGC 60007. It is difficult to define a core in NGC 6339, since it has about the same brightness as the bar. H II regions are visible at both ends of the bar, and the two connected spiral arms have an open form. The spiral arms split several times on the way to the edge of the disc, with the fragments containing many star-formation regions like beads on a string.

H I radio data on NGC 6339 indicate a total of 3.4×10^9 solar masses of neutral hydrogen. This value corresponds to that typical for a gas-rich spiral galaxy. The radio map shows a uniform disc which at 5′ is about twice the size of the optical disc. Further H I sources are not visible in the radio data. The major axes of both discs are rotated by less than 10° relative to each other, so that one can assume that the neutral-hydrogen disc in NGC 6339 is almost flat. This suggests that this galaxy has not been perturbed and that its description as an isolated field galaxy is completely correct.

OBJECT	NGC 6339
CONSTELLATION	Hercules
RIGHT ASCENSION	$17^h\ 17^m\ 07^s$
DECLINATION	+40° 50′ 42″
BRIGHTNESS	13.3 mag
TYPE	SBd
PHOTOGRAPHERS	Josef Pöpsel, Stefan Binnewies
TELESCOPE	600-mm reflector
CAMERA	SBIG STL-11000
EXPOSURE TIME	525 min
LOCATION	Skinakas Observatory, Crete, Greece

IC 1296

The SBbc galaxy IC 1296 is probably photographed often by amateur astronomers only because of its proximity to the planetary nebula M 57. Compared to M 57, which is 2300 light years from the Sun, the distance to IC 1296 is 240 million light years. The angular size of 1.1′ corresponds to a projected disc diameter of 80 000 light years. The two spiral arms connected to the approximately 7″ long bar in IC 1296 end before the edge of the disc is reached. In addition, within a radius of 1° there is no object with a similar recession velocity, so that IC 1296 can be considered as a largely isolated field galaxy.

The combination of planetary nebula and galaxy is rarely found in images. This is due to the different and almost opposite distribution on the sky. Planetary nebulae are Galactic objects and concentrated in the plane of the disc, mainly in the direction of the Galactic centre. More than 80 per cent of planetary nebulae have Galactic latitudes of <10° and thus lie in a region which, due to strong absorption by great amounts of gas and dust within the Milky Way, is almost completely opaque to the light of extragalactic objects – such as galaxies.

Four Italian observatories collaborate within the Italian Supernovae Search Project (ISSP). Telescopes of diameter 24–53 cm are used to automatically monitor large star fields. From June to December 2011, in the first six months of the project, almost 70 000 observations were made. During this time, the group discovered 19 supernovae. In August 2013 the ISSP also found the supernova SN2013ev (PSN J18531845+3303527) in IC 1296, 9.2″ from the centre, in the lower, southwestern spiral arm. Independently of the ISSP team, a German amateur astronomer also reported the discovery of the supernova. This was accomplished with a 43-cm telescope in his observatory in Bavaria.

OBJECT	IC 1296
CONSTELLATION	Lyra
RIGHT ASCENSION	$18^{h}\ 53^{m}\ 19^{s}$
DECLINATION	+33° 04′ 00″
BRIGHTNESS	14.8 mag
TYPE	SBbc
PHOTOGRAPHERS	Volker Wendel, Josef Pöpsel
TELESCOPE	600-mm reflector
CAMERA	SBIG STL-11000
EXPOSURE TIME	900 min
LOCATION	Skinakas Observatory, Crete, Greece

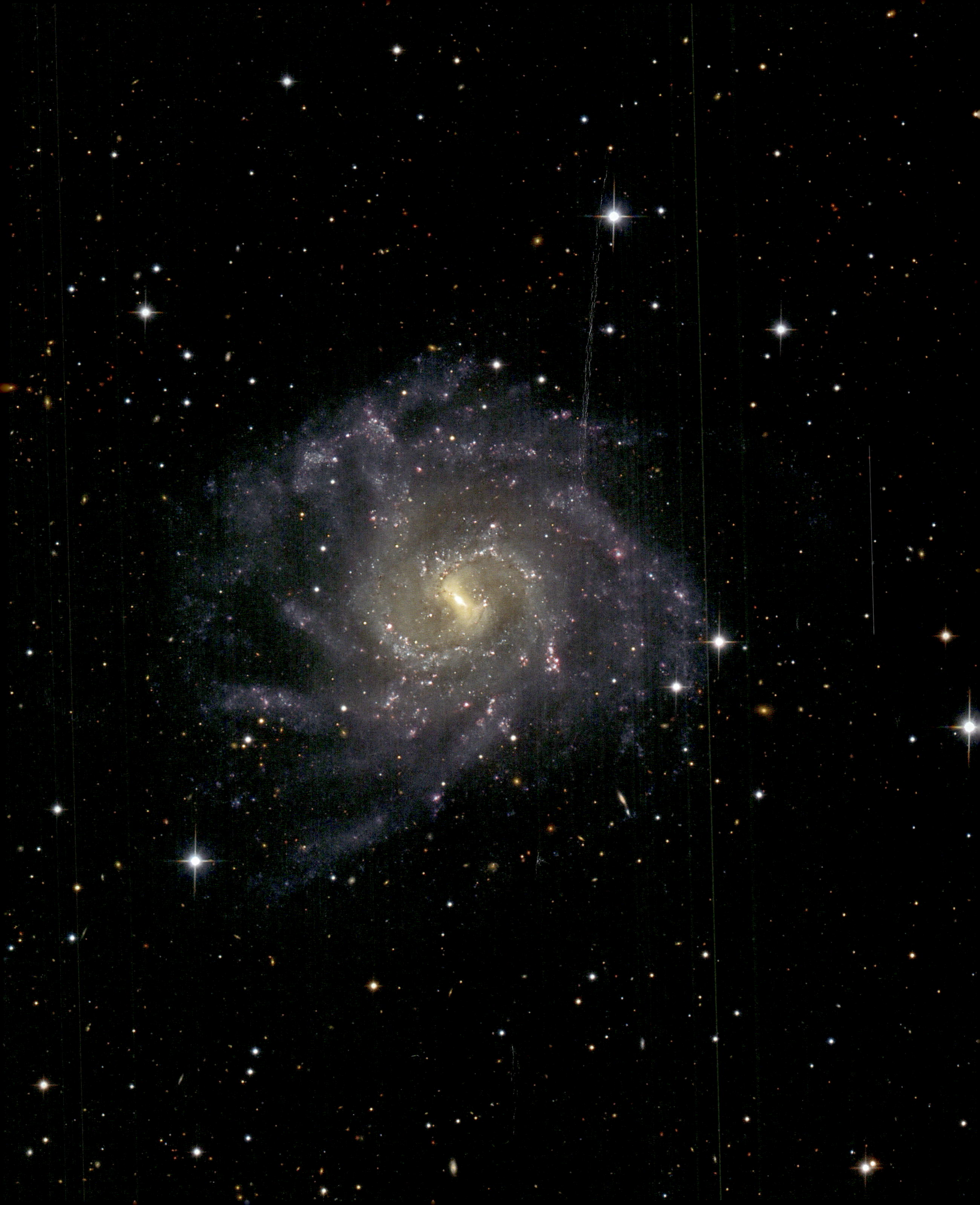

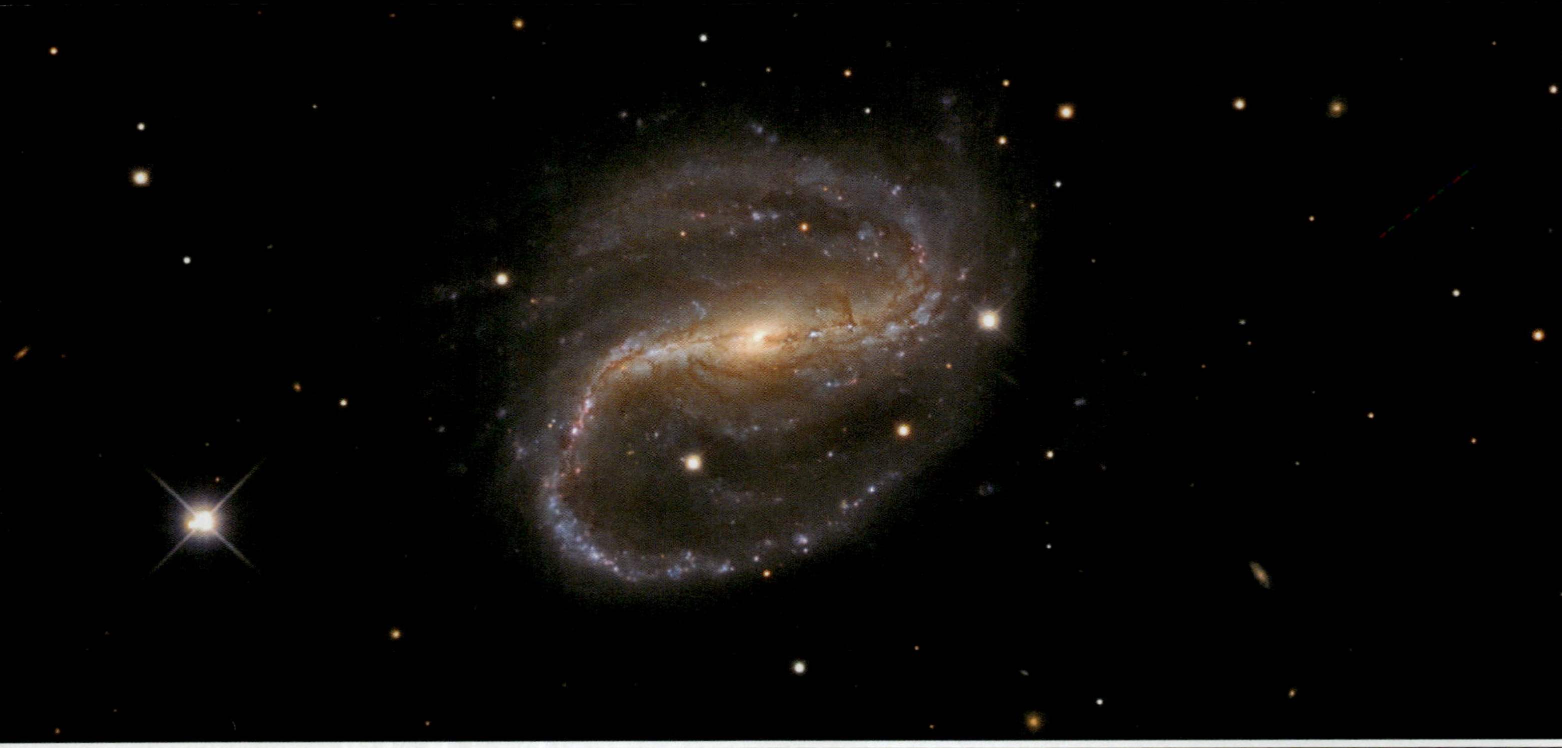

NGC 7479

The barred spiral galaxy NGC 7479 is 115 million light years from the Milky Way. The angular size is 4.1′ × 3.1′ and the projected disc has a diameter of 137 000 light years. The compact centre lies in a bright bar which determines the surface brightness of the entire disc. The small, elliptical bulge is tilted by 45° to the horizontal, delicate bar. One notices that dust lanes running from the inner regions of the disc end at each side of the bar. The spiral arms join the bar at a large angle and the resulting X-shaped form is responsible for the SB(s)c classification of NGC 7479. The spiral arm in the lower, western half of the disc contains more active regions and is better defined than the corresponding one in the east. At the right edge of the image one can see the coloured track left by asteroid 297751 (2001 XL_{63}) due to its motion during the exposure.

The H I radio map of NGC 7479 confirms the visible asymmetry of the spiral form of the outer disc. This refers to the western spiral arm which in the image is seen to change into a displaced tidal tail at the right edge. The H I data also indicate a large velocity gradient in the bar. The corresponding transport of matter into the centre explains the active Seyfer-2 nucleus of NGC 7479. The distinctive feature of this galaxy is the polarization of its radio emission. Astronomers found a jet-like emission structure of length 50 000 light years which emerges from the centre. This continues near the plane of the disc and the resulting compression of the interstellar magnetic fields acts like a particle accelerator, resulting in a relatively high radio luminosity. The cause of this activity is presumably a merger event about 300 million years ago. Simulations of this minor merger event indicate a mass for the destroyed companion of about 1/10 the mass of the disc of NGC 7479. The orbital plane of the companion was tilted by only 10° relative to the plane of the precursor of NGC 7479. The two-arm spiral form of the galaxy is due to the passage of the companion into the disc. The remains of the companion then collected at the centre and formed the bar. Letting the simulations continue into the future reveals that the bar will dissolve after another 150 million years and in this model NGC 7479 will develop into an early, barless type of spiral.

OBJECT	NGC 7479
CONSTELLATION	Pegasus
RIGHT ASCENSION	$23^{h}\ 04^{m}\ 57^{s}$
DECLINATION	+12° 19′ 22″
BRIGHTNESS	11.6 mag
TYPE	SB(s)c
PHOTOGRAPHERS	Makis Palaiologou, Josef Pöpsel
TELESCOPE	1.3-m reflector
CAMERA	SBIG STL-6303
EXPOSURE TIME	240 min
LOCATION	Skinakas Observatory, Crete, Greece

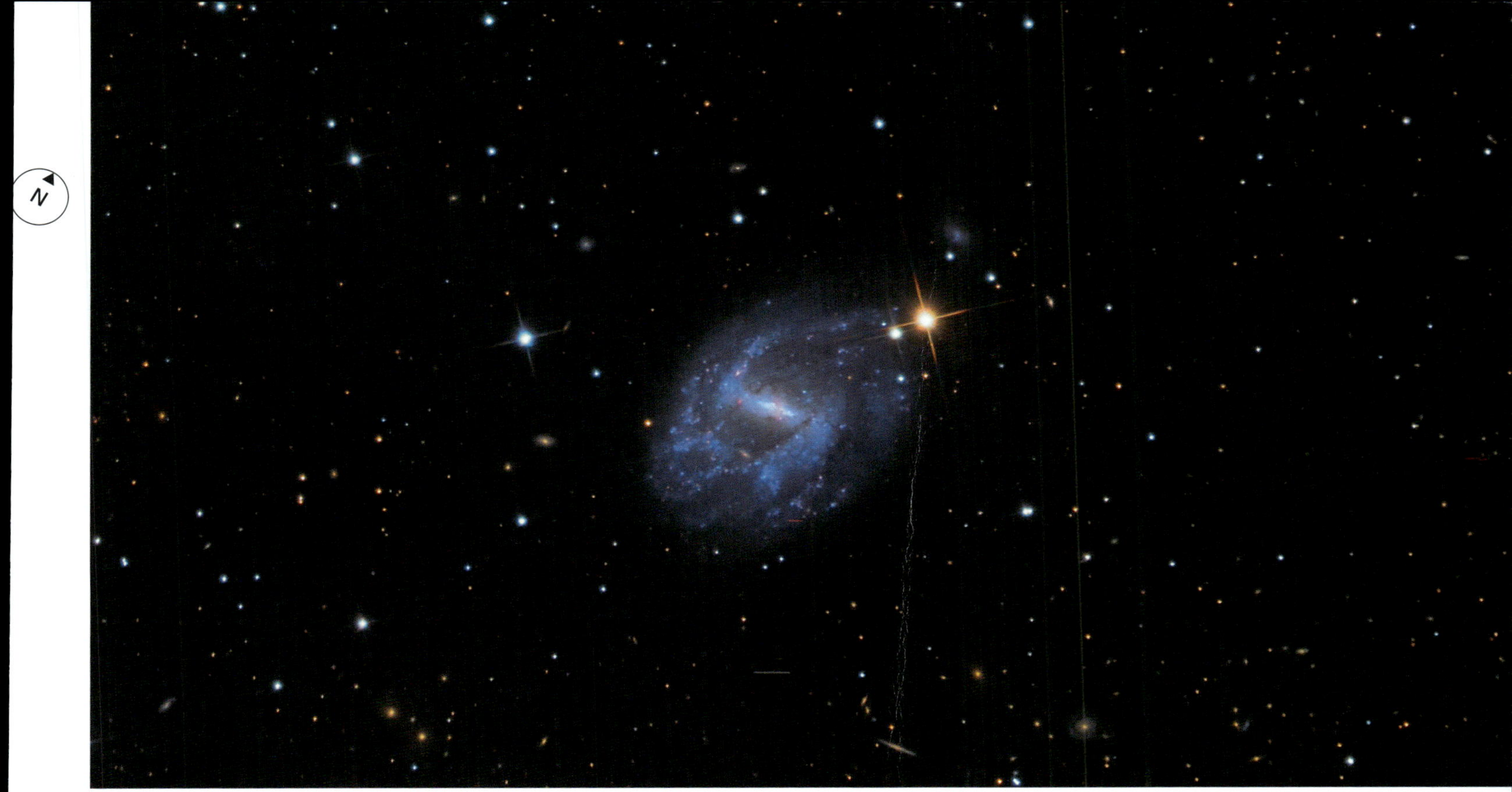

NGC 7741

Next to a 10th-magnitude orange K5 star (SAO 91455) is the SB(s)cd galaxy NGC 7741. The angular size of 4.4′ × 3.0′ corresponds for a distance of 47 million light years to a transverse diameter of 60 000 light years. Thus, NGC 7741 is a comparatively small galaxy with a noticeable contrast between the edge of the disc and the background sky. Short spiral arms connect to the ends of the bright bar. The spiral arm in the lower, southern half of the disc contains many more active regions and a larger number of arm fragments. Across the centre of NGC 7741 is a delicate vertical dust lane which appears to curve around behind the bar. In this dust lane are some of the brightest H II regions in the galaxy. North of NGC 7741 one can see the galaxy LEDA 214984. No distance is recorded in various databases, so it is difficult to ascribe its relation to NGC 7741. However, the diffuse appearance and similar blue colour suggest that LEDA 214984 is a dwarf galaxy interacting with the larger companion.

Optical spectroscopy was used to study the radial velocities of ionized hydrogen in the region of the inner disc of NGC 7741. Taking the inclination of 45° to the line of sight into account, the Hα rotation curve reaches the velocity of ±100 km/s at a distance of ±1′. Close in, the rotation curve shows significant differences between the southeastern and northwestern sides. This is especially pronounced at a distance of 30–40″ where the velocities are up to 40 km/s lower than the local rotation. These irregularities in the velocity field in the zone of the bar can be seen as an indication that NGC 7741 could have captured a small, low-mass companion galaxy. The central dust lane in NGC 7741 would then be a result of this interaction.

OBJECT	NGC 7741
CONSTELLATION	Pegasus
RIGHT ASCENSION	$23^h\ 43^m\ 54^s$
DECLINATION	+26° 04′ 32″
BRIGHTNESS	11.8 mag
TYPE	SB(s)cd
PHOTOGRAPHERS	Adam Block
TELESCOPE	800-mm reflector
CAMERA	SBIG STL-11000
EXPOSURE TIME	300 min
LOCATION	Mount Lemmon SkyCenter/ University of Arizona, USA

NGC 7497

The SB(s)d galaxy NGC 7497 is located in a loose group of galaxies consisting primarily of dwarfs which, however, are not seen in the image. NGC 7497 lies at a distance of 64 million light years and its angular size of 4.9′ corresponds to a disc size of 91 000 light years. The bright core is not in the middle of the disc but rather displaced to the right. The asymmetric form is also visible in the irregular positioning of the H II regions in the left, large half of the disc. The prominent warp in the disc is a further indication of interactions within the galaxy group.

The brown-grey nebular filaments in which the galaxy seems to be embedded are clouds of dust and gas in our Milky Way which are located along the line of sight to NGC 7497 and are known as Galactic cirrus. The elongated structure visible in the image is part of a larger complex which is about 900 light years from the Sun and covers a degree on the sky. The structure is known as MBM 54 and is part of the catalogue of high-latitude molecular clouds by L. Magnani, L. Blitz, and L. Mundi (1985). The UV light of the stars in the region of the disc of the Milky Way is scattered by the dust particles in the cloud and cause this to glow. In addition to this blue scattered light, there is also a red component due to fluorescence of the gas excited by the UV light. Some of the stars visible in the image are farther away than the cloud so that absorption lines of the molecular gas in the cloud are visible in their spectra. The most common molecule in the clouds is hydrogen; however, astronomers have also found carbon monoxide, cyanide, and sodium in MBM 54. The molecules are more varied at higher gas densities and one could even detect complex amino acids in some clouds.

OBJECT	NGC 7497
CONSTELLATION	Pegasus
RIGHT ASCENSION	$23^{h}\ 09^{m}\ 03^{s}$
DECLINATION	+18° 10′ 38″
BRIGHTNESS	12.0 mag
TYPE	SB(s)d
PHOTOGRAPHERS	Adam Block
TELESCOPE	800-mm reflector
CAMERA	SBIG STX-16803
EXPOSURE TIME	840 min
LOCATION	Mount Lemmon SkyCenter/ University of Arizona, USA

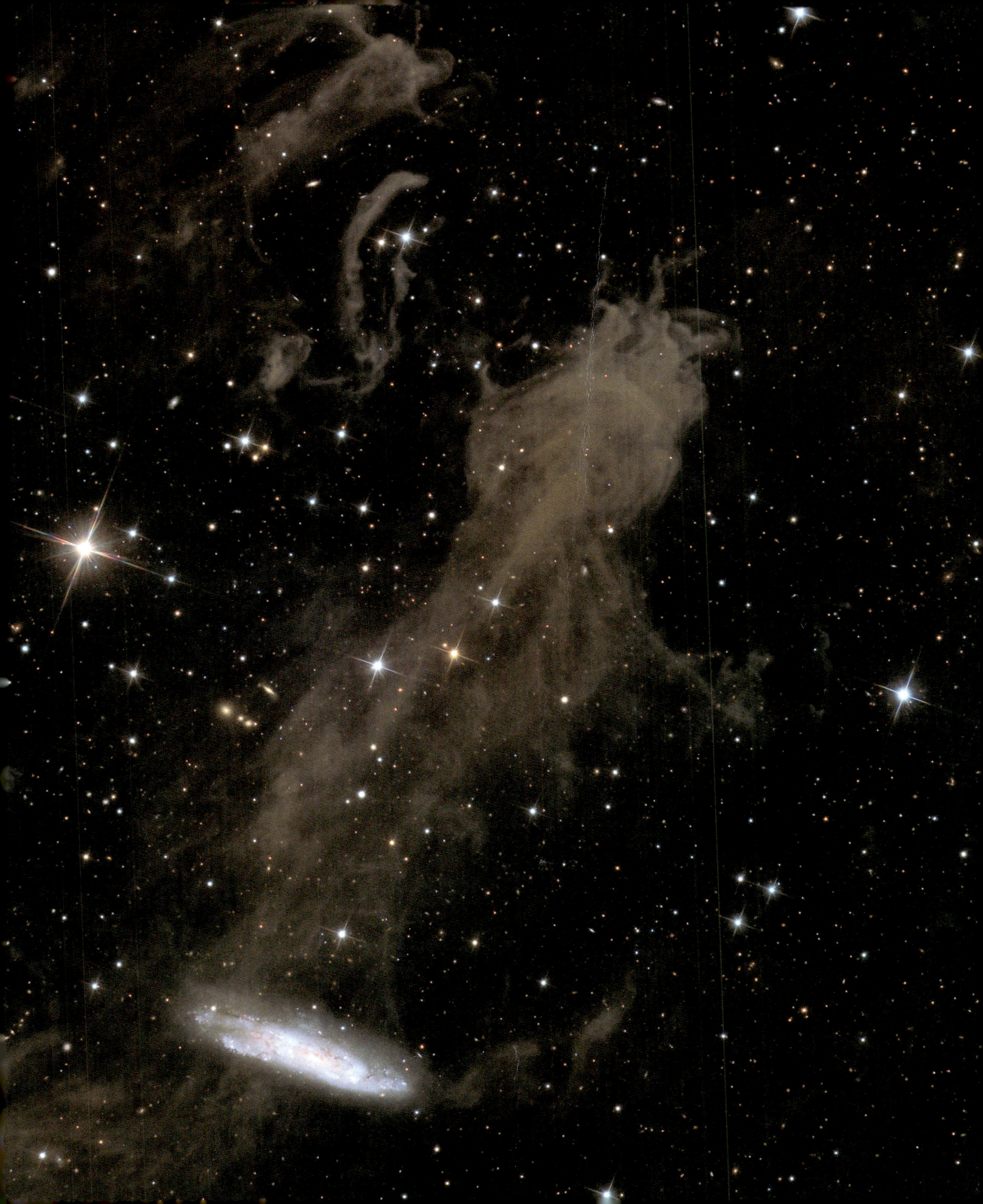

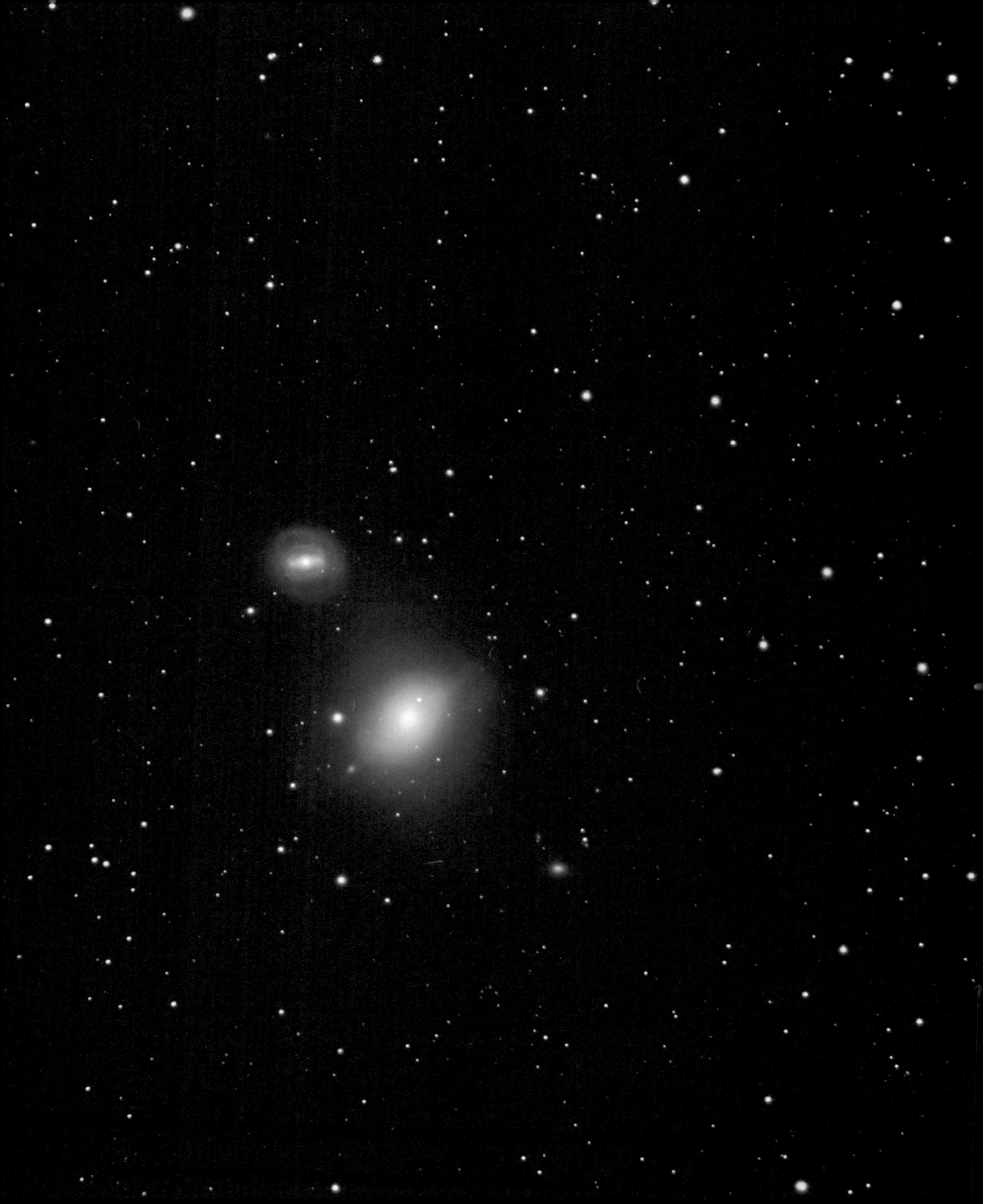

ELLIPTICAL GALAXIES

Elliptical galaxies cover a wide range in size. Thus some of the largest galaxies belong to the galaxy family discussed in this chapter.

THE CLASSIFICATION OF ELLIPTICAL GALAXIES

In the last few years, amateur astrophotography has become more professional. This allows interesting features to be seen even in the apparently uninteresting elliptical galaxies, making visible the results of current research. The high efficiency of sensitive CCDs, their large dynamic range and above all the possibilities of electronic image processing have played an enormous role in making these popular subjects for many photographers.

The classification of elliptical galaxies from E0 to E7 in the Hubble diagram describes their form, from the circular E0 to strongly flattened E7 types. The numbers from 0 to 7 following the E are computed according to the formula $10 \times (a - b)/a$, where a is the semimajor and b the semiminor axis of an ellipse. This allows one to quantify the ellipticity of galaxies. An ellipse also has two foci whose distance from the centre is known as the linear eccentricity, *e*. In the case of circular E0 types, the two foci coincide at the centre and the eccentricity is thus zero. With increasing distance of the foci and thus increasing flattening, the value of the linear eccentricity increases. The classification E0 to E7, however, has no physical background, since depending on the orientation the general triaxial form of a spindle-shaped elliptical galaxy could appear as a round E0 type. The almost circular E2 and E3 types are the most common in surveys.

Dwarf elliptical galaxies with low surface brightness are classified as dE. Some of the dwarf satellite galaxies of the Milky Way are classified as dwarf spheroidal, dSph, which are even dimmer than the dE. The Local Group containing the Milky Way also has irregular satellites. Relatively new are blue compact dwarfs, which are usually classified as irregular dwarf galaxies. At 15–20 per cent, their gas fraction is relatively large and explains their conspicuously large star-formation rates.

The largest elliptical galaxies are the cD types which are found at the cores of galaxy clusters. With up to 10^{14} solar masses, they are extremely massive and contain several tens of thousands of globular clusters in their halos. These cD types can contain up to a trillion stars and thus exceed a typical spiral galaxy by a factor of 10 to 100. Their mass-to-light ratio $M_\odot/L_\odot$ reaches values as large as 750 and is an indication of the presence of dark matter in galaxy clusters. The rest of elliptical galaxies range from large gE through normal E to compact cE types. The corresponding masses are 10^8–10^{13} solar masses, the mass-to-light ratios 7–100.

There are no elliptical galaxies in the Local Group. The closest is Maffei 1 at 11 million light years. It is not visible to the naked eye – in contrast to types S (M 31), SB (Milky Way), and the Magellanic Clouds as representatives of irregulars, all of which we can see unaided at a good location with little light pollution.

THE MORPHOLOGY OF ELLIPTICAL GALAXIES

Elliptical galaxies are relatively bright at the centre and drop at first rapidly then more slowly in brightness towards the outside. The intensity profile I(r) can be well modelled by a de Vaucouleurs profile, in which the logarithm of the intensity I(r) is proportional to the distance from the centre. Since the magnitude scale is logarithmic, the observed profile can be described by an $r^{1/4}$ law.

An important parameter in the description of the profile is the effective radius R_e which contains the central region with 50 per cent of the total luminosity of the galaxy. In cD galaxies, however, the $r^{1/4}$ rule cannot be used, since it cannot describe the extended bright halo of these massive elliptical galaxies. In dwarf galaxies the de Vaucouleurs profile is also

The pair M 85 and NGC 4394 belong to the northern part of the Virgo Cluster. The prominent structure on the halo of M 85 is an indication of its turbulent past (see page 145). Image credit: Stefan Binnewies (105-mm refractor).

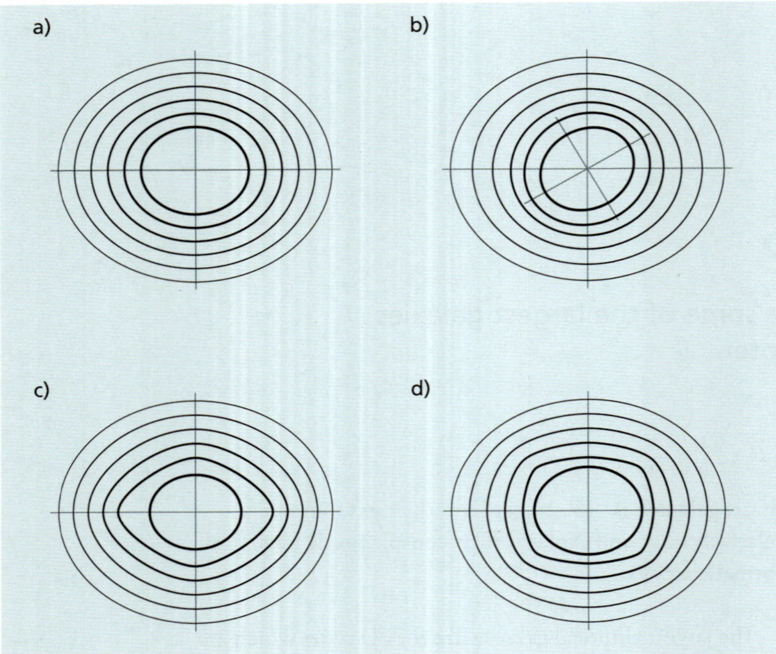

Figure 3.1: a) Schematic luminosity distribution of an elliptical galaxy (isophotes), b) Rotation of the major axis (isophotal twist), c) Disky type, d) Boxy type.

not applicable; their low surface brightness is better described by an exponentially falling King profile.

The surface brightness of elliptical galaxies is described by lines of equal brightness (so-called isophotes) which surround the centre and mathematically can be described quite well by concentric ellipses. The midpoint of these ellipses differs only slightly from the brightest point of the elliptical galaxy; the measured difference is usually less than one per cent of the diameter of the ellipse. It has also been found that the orientation of the major axis of the isophotes changes with radius. This tilting of the symmetry axis of the ellipses is known as isophotal twist and is an indication that elliptical galaxies are not spheroidal, but rather have a triaxial form.

These descriptions indicate that the relatively simple classification of elliptical galaxies in the Hubble diagram overlooks some details. For example, in some galaxies the form of the isophotes is not elliptical. This discrepancy is described by the boxiness parameter which corresponds to the coefficient a_4 of the Fourier expansion of the distance of an isophote from the centre. If this coefficient $a_4 = 0$, then the form is an ideal ellipse. The discrepancies which one sees in elliptical galaxies are in the range of a few per cent. For values $a_4 > 0$, the isophote is disky; for $a_4 < 0$, the form is compressed in the longitudinal direction and is thus classified as boxy. Examining the frequency of these two types, disky dominates with 90 per cent compared to ten per cent boxy types, whereby the transition from disky to S0 types is continuous.

In addition to these two types of distortions there are also physical differences. Disky types are flattened by rotation which is not the case in boxy types. They have an elliptical form, whereby the cause of this is the boxy velocity distribution of stellar orbits. The rotation within disky types is more regular. This fits with the fact that about a third of the boxy types have counter-rotating cores. These oppositely rotating core components almost never occur among disky types. In the environment of boxy types, shell structures and filaments can also be found which are arranged symmetrically about the centre and indicate a past merger with another galaxy. This merger scenario is confirmed by a broad distribution of radio and X-ray luminosities. The very dim, large-scale and concentric scale structures found in the mid-1980s were a big surprise. They were discovered by David F. Malin with the help of classical chemical photography. Modern methods of electronic image processing make this discovery available to the larger audience of amateur astronomers as well.

In a similar manner, unsharp masking of images in the inner regions can also reveal concentric structures with very low contrast, so-called ripples. These were not possible to see in traditional chemically produced images. A nice example of such a shell galaxy is NGC 1344. More detailed studies of merger candidates such as NGC 1316 or NGC 7252 demonstrated not only the existence of tidal tails, but also that the inner ripples belong to the halo ellipsoid. A good source of subjects for amateur astrophotographers is the "Catalog of elliptical galaxies with shells" by the astronomers D. F. Malin and D. Carter.

THE ASTROPHYSICS OF ELLIPTICAL GALAXIES

Comparing the spectra of elliptical galaxies with other galaxy types, one notices differences in the form of their spectral lines. The line width in elliptical galaxies is due to the motion of individual stars. Due to the Doppler effect, the position of the spectral line is shifted; since one cannot observe individual stars in spectroscopic studies, one always observes millions of stars at the same time. Their spectra overlap, causing an additional broadening of the spectral lines. The larger the velocity dispersion of stellar motion in the galaxy, the broader the spectral lines appear.

This dispersion, σ_c, is a statistical measure of the scatter in the measured velocities and depends strongly on the anisotropy of the velocity field of the stars. In elliptical galaxies one measures velocity dispersions at the centre which are strongly correlated with the luminosity of the elliptical galaxy. Doubling the value of σ_c corresponds to increasing the luminosity by a factor of 16. The rotation velocities of elliptical galaxies are smaller than their velocity dispersion, which are a few 100 km/s. Spiral galaxies, by contrast, have much smaller values for velocity dispersion (a few 10 km/s); the orbital velocities of the stars differ only slightly from the rotation velocity.

Direct collisions of stars have been ruled out as a mechanism for interaction within elliptical galaxies. The dynamics of an individual star is determined mainly by the large-scale gravitational field of the galaxy. Since under these conditions stars behave like collisionless gas particles, one can conclude that elliptical galaxies are stabilized by pressure. This "stellar gas" itself also exerts a pressure which is directed oppositely to the gravitation. Ellipsoids result from anisotropies in the distribution of stars in phase space, i.e. in the three-dimensional velocity distribution; this is known as the anisotropic-pressure model of an elliptical galaxy. With the help of the model of such a hydrostatic equilibrium in an isothermal sphere of gas, one can derive a so-called King–Michie intensity profile from the density distribution. It results in better values than the de Vaucouleurs profile and is able to describe the luminosity distribution of other objects. Apart from elliptical galaxies, one can apply it to globular clusters and galaxy clusters in order to model the drop in luminosity.

Elliptical galaxies are sometimes surrounded by very extended halos whose luminosity profile in the outer regions drops so low that it is below the value of the sky background. One assumes that the halos of elliptical galaxies are also filled with dark matter. However, this is not so clearly detectable, since elliptical galaxies have no extended, rotating spiral planes. In massive elliptical galaxies, dark matter could be detected via measurements making use of the gravitational-lensing effect. This must be taken into account, especially in the calculation of stellar orbits. Note that orbital periods are larger the farther the stars are from the centre, so that stars in these outer regions could have completed only a few orbits.

Spectra can also provide information about the abundance of metals in the stars. In astronomy, all elements apart from hydrogen and helium are known as metals. The depth of a spectral line provides information about the relative abundance of the corresponding element whose characteristic line is due to emission or absorption. The more common an element, the stronger the spectral line. More precise information about the metal abundance, which changes during the lifetime of a star, must also take other parameters (e.g. temperature) into account. One thus uses stellar-evolution models and applies them to integrated galaxy spectra. This allowed astronomers to find a relation

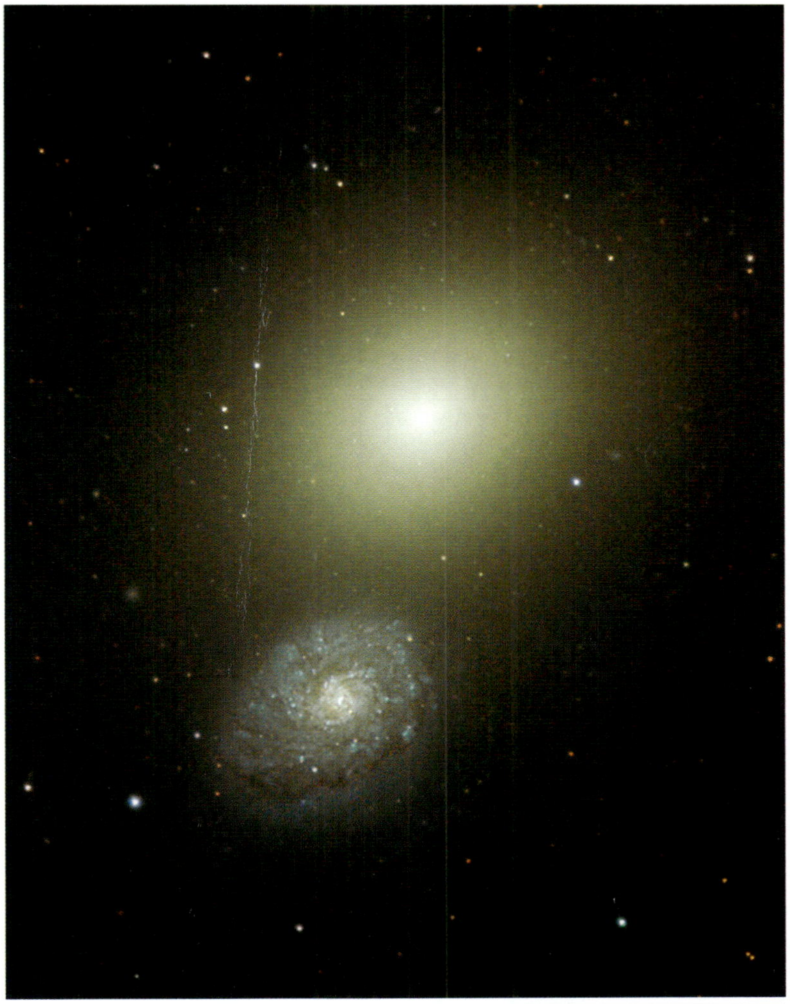

Figure 3.2: M 60 is a large galaxy in the Virgo Cluster. It appears that the spiral galaxy NGC 4647 lies within the diffuse halo of the large elliptical galaxy (see page 148).

between the metallicity and luminosity of elliptical galaxies. This hints at a possible origin of elliptical galaxies in mergers of dust- and gas-rich spiral galaxies containing older stellar populations. Mergers caused a new epoch of star formation, visible in the increased luminosity of the new, metal-rich elliptical galaxy.

In the optical, elliptical galaxies are yellow or, in extreme cases, red objects. Their light is dominated by long-lived, older stellar populations, mainly K stars. In the last few decades, satellite observations have shown the existence of a thin, hot gas of temperature 10^7 K in E-type galaxies. In addition, 10^4 K warm Hα and 10^2 K cold molecular lines have been found. Since the earliest investigations, it has been noted that there is a lack of gas and dust in elliptical galaxies. This missing mass shows up in the form of gas and dust in the surrounding halos. Today it is clear that about half of elliptical galaxies have a small but substantial dust fraction which has a layer-like structure. The relative tilt of this dust plane to

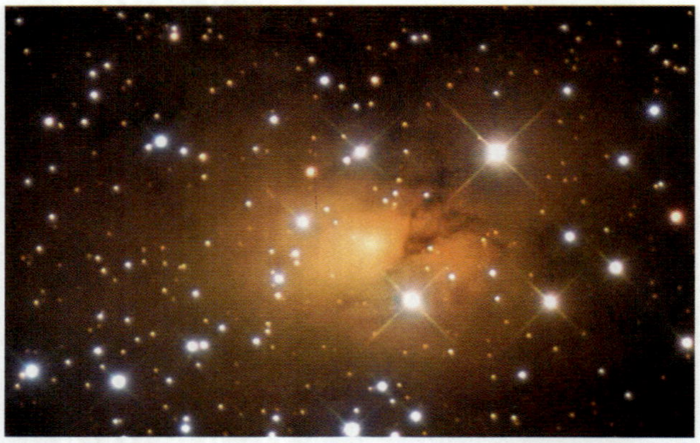

Figure 3.3: A view onto the centre of the elliptical galaxy Maffei 1 shows the yellow colour, which has many causes (see page 140).

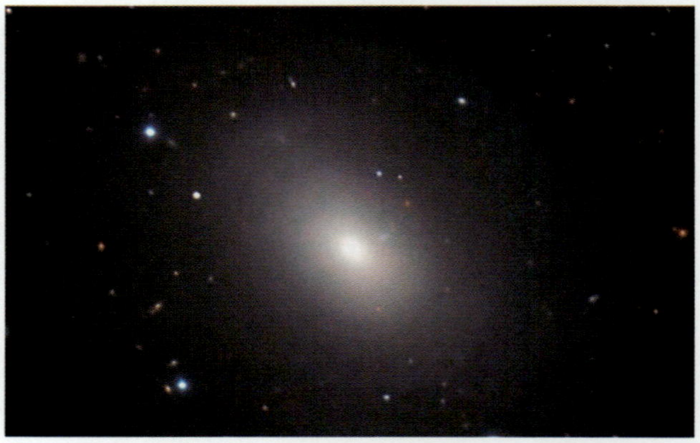

Figure 3.4: The blue colour of the halo of NGC 5982 is an exception and is due to external influences (see page 150).

THE COLOUR OF GALAXIES

The spectrum of a normal, non-active galaxy is determined mainly by the superposition of many stellar spectra. Apart from the closest galaxies, where various populations can be observed individually, astronomers always measure integrated spectra. Usually, these spectra are dominated by the bright central region of the galaxy studied.

The spectrum is dominated by hot stars in the short-wave blue region and by cool stars of late spectral type in the red. Earlier spectral-classification systems for galaxies were oriented on the spectral classes of main-sequence stars. In order to quantify the colour, astronomers measure the luminosity in various parts of the spectrum. Standard filter systems with fixed spectral bands are used, such as the commonly used Johnson filter system (U, B, V, R, I), which defines blue (B) at 440 nm and visual (V) at 550 nm. The term "visual" is due to the fact that the human eye is most sensitive to stars in the green region. The difference in luminosities defines the so-called colour indices such as (B – V). Due to the definition of the astronomical magnitude scale, the larger the colour index, the redder the star. A red star has a low luminosity in the blue, which corresponds to a large magnitude value and hence the difference (B – V) is large. For the Sun, a white G star, one measures the value (B – V) = 0.65 mag, the blue-white star Vega (A0V) has by definition (B – V) = 0 mag, and the red star Betelgeuse has as an M star the value (B – V) = 1.85 mag.

The colour impression of galaxies in images is determined by the optical continuum. In starburst or irregular galaxies, the blue portion exceeds that of other spectral colours. In Sb galaxies, on the other hand, one finds a flat distribution of colours and thus they appear white.

By definition, the colour of a G2 star is white, whereby (B – V) = 0.64 mag. Sa types already have a blue deficit and correspond to the transition between white and yellow. In S0 and elliptical galaxies, the red portion dominates the spectrum. As is the case for stars, the colour index of a galaxy is defined as the difference in luminosity in the blue and visual spectral bands and is correlated with the Hubble type.

The table indicates the approximate (B – V) values for various Hubble types of galaxies. There are types of galaxies which deviate from this correlation. For example, for large elliptical galaxies, the (B – V) value is a few tenths of a magnitude redder than for smaller or even dwarf ellipticals. An exception are blue compact dwarf galaxies whose (B – V) lies between 0.0 and 0.3 mag, which is characteristic of young stars of spectral type A. The amount of interstellar gas is appreciably larger in these dwarf galaxies than in other elliptical galaxies. Plotting the (B – V) values in a histogram, one notices a bimodal distribution. The broad maxima are at the values (B – V) = 0.4 mag and (B – V) = 0.8 mag, whereby the blue-white maximum is due to Sb/Sc/Irr types and the second, yellowish maximum to E/S0/Sa types. This bimodality demonstrates the difference in star-formation rates in the two groups. The galaxy types in the blue-white group have several spectral indications of star formation, such as young stars and OB associations.

Note that these indicators are time-dependent and a galaxy always goes through spectral evolution. For example, if an interaction causes star formation in a spiral galaxy, then the active regions, if there are enough of them in the plane of the disc, contribute a blue component to the colour of the galaxy and reduce its (B – V) value.

(B–V)	HUBBLE TYPE	COLOUR
1.00 … 0.90	for large E	dark yellow
0.95 … 0.80	for E	yellow
0.90 … 0.75	for S0	bright yellow
0.80 … 0.65	for Sa	yellow-white
0.70 … 0.55	for Sb	white
0.52 … 0.35	for Sc	blue-white
0.45 … 0.30	for Irr	bluish

The colour of a galaxy provides an indication of its Hubble type

the axis of rotation is a further indication of a merger history of elliptical galaxies and the spiral galaxies participating in it.

In some elliptical galaxies, the Hubble Space Telescope was able to detect an outer ring containing star-forming regions. Since such galaxies usually occur in groups or clusters, one suspects that they accrete gas from their surroundings and thus form new stars. But this is an exceptional situation which is seen mainly in large elliptical galaxies. As a rule, galaxies in an environment of high galaxy density lose their gas with time via gas stripping. This refers to the motion of a galaxy relative to the cluster medium of hot, thin gas and experiencing a braking force which depends on the density of the medium and causes the atomic gas to flow out of the galaxy. In the last few years, the simple assumption that the loss of gas always means the end of star formation as well has been revised. Hydrostatic models show that this gas loss can lead to turbulent mixing processes between the gas in the disc and the cluster medium. Thus, more metal-rich gas remains in the galaxies than previously assumed, which can positively influence the star-formation rates.

BIBLIOGRAPHY

van den Bergh, S.: *Galaxy Morphology and Classification*, 1998, http://ned.ipac.caltech.edu/level5/VDBergh2/VDB1_4.html

Berry, R. and Burnell, J.: *The Handbook of Astronomical Image Processing*, Willmann-Bell Inc., 2005

Extragalaktische Arbeitsgruppe der Universitäts-Sternwarte München, *Spektrallinien in elliptischen Galaxien*, 2013, www.usm.uni-muenchen.de/people/saglia/dm/galaxien/alldt/node21.html

Fabello, S. et al.: *ALFALFA HI data stacking – I. Does the bulge quench ongoing star formation in early-type galaxies?*, Monthly Notices of the Royal Astronomical Society, **411**, 2011

Fritze, U.: *Spectral Evolution of a Galaxy*, 2013, http://members.galev.org/ufritze/teaching/PostGrad7.pdf

Gerhard, O.: *Diffuse stellare Halos in elliptischen Galaxien*, 2009, www.mpg.de/355706/forschungsSchwerpunkt

Malin, D. F. and Carter D.: *Catalog of elliptical galaxies with shells*, The Astrophysical Journal, **274**, 1983

Meusinger, H.: *Vorlesung Extragalaktik – Eigenschaften von Galaxien*, 2012, www.tls-tautenburg.de/research/meus/vorlesung/ppt/EG10/EG_3.pdf

Meusinger, H.: *Vorlesung Extragalaktik-Klassifikation*, 2012, www.tls-tautenburg.de/research/meus/vorlesung/ppt/EG10/EG_2.pdf

Schneider, P.: *Galaxien – Lectures*, 2006, www.astro.uni-bonn.de/~peter/Lectures/intro3.pdf

Strateva, I. et al.: *Color Separation of Galaxy Types in the Sloan Digital Sky Survey Imaging Data*, The Astronomical Journal, **122**, 2001

Unsöld, A. and Baschek, B.: *Der neue Kosmos – Einführung in die Astronomie und Astrophysik*, Springer Verlag, 2002

Whittle, M.: *Extragalactic Astronomy – Elliptical galaxies, 2014*, www.astro.virginia.edu/class/whittle/astr553/Topic07/Lecture_7.html

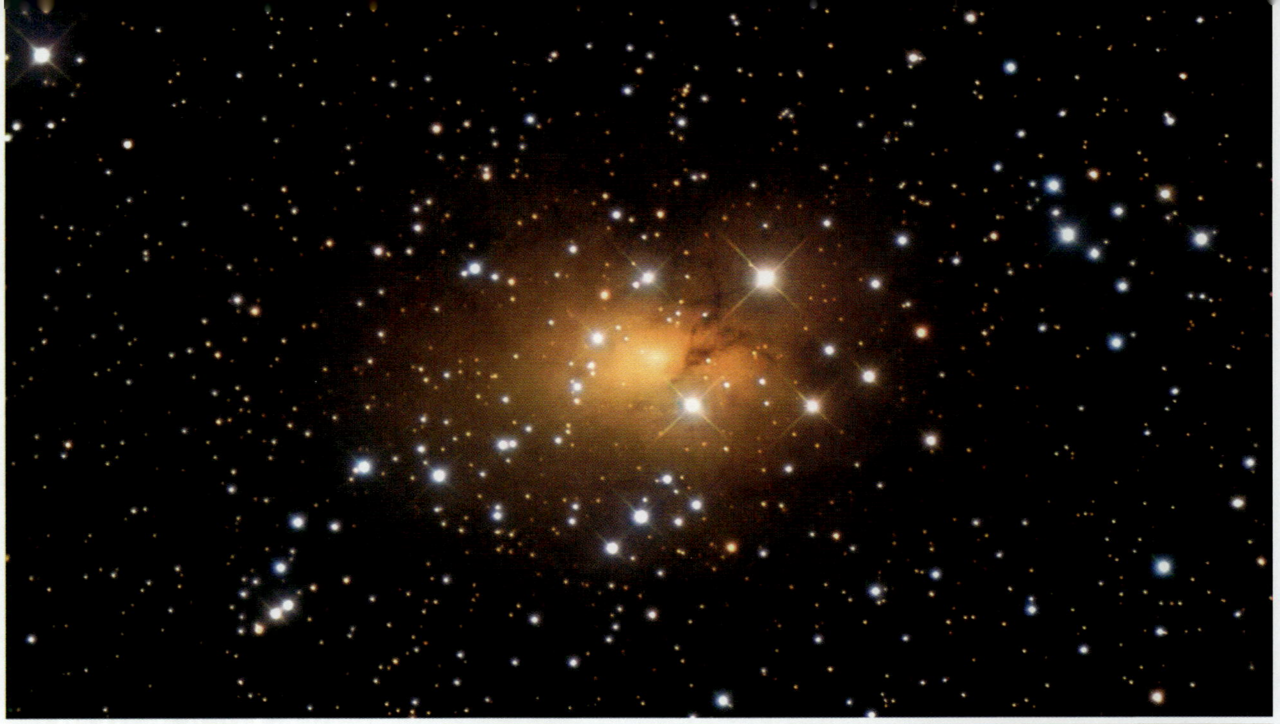

Maffei 1

The E3 galaxy Maffei 1 belongs to the IC 342/Maffei galaxy group. This is a direct neighbour of the Local Group and is bordered by the more distant M 81 Group. At a distance of about ten million light years, it is the closest galaxy group to the Milky Way and contains about a dozen members which surround the two large galaxies IC 342 and Maffei 1. In the NASA Extragalactic Database, the visual extent of Maffei 1 is given as 3.4′ × 1.7′, which at a distance of 11 million light years implies a diameter of 11 000 light years for the visible part of Maffei 1.

A peculiarity of Maffei 1 is its position on the sky, only 33′ from the Galactic plane. Due to the strong extinction by dust and stars in the Milky Way, one finds almost no galaxies in this so-called zone of avoidance. This explains its late discovery in the year 1967 by the Italian astronomer Paolo Maffei, who wanted to investigate the dusty regions around T Tauri stars via infrared-sensitive film on the Schmidt telescope of the Asiago Observatory (University of Padua, Italy). The dark dust lanes seen in the image above the centre of Maffei 1 do not belong to the galaxy, but are dust structures in the Milky Way, perhaps even in the environment of the Sun, which appear dark in front of the bright surface of Maffei 1 and thus become visible. Due to the intervening interstellar matter, distance determination is problematic for Maffei 1.

In high-resolution spectroscopic measurements of the central regions of Maffei 1, astronomers have used magnesium absorption lines to determine the so-called MG_2 index. In elliptical galaxies, this MG_2 index correlates with the colour of the galaxy, whereby typical MG_2 values lie in the range 0.2–0.35 and correspond to (B – V) indices of 0.85–1.0 mag. Since Galactic dust causes reddening of the galaxy light from Maffei 1, this relation can be used to measure the extinction. The MG_2 index of Maffei 1 is 0.31 and the comparison results in an extinction of 4.7 magnitudes in the visual range. The visual magnitude of Maffei 1 is 11.1 mag; without extinction, Maffei 1 would be one of the brightest galaxies in the northern hemisphere. If one corrects the diameter based on the Galactic extinction, one obtains an angular size of 46.8′ which corresponds to a diameter of 134 000 light years and exceeds the value in the NASA database by a factor of almost 14.

OBJECT	Maffei 1
CONSTELLATION	Cassiopeia
RIGHT ASCENSION	$02^h\ 36^m\ 35^s$
DECLINATION	+59° 39′ 18″
BRIGHTNESS	11.4 mag
TYPE	$S0^-$ pec
PHOTOGRAPHERS	Makis Palaiologou, Stefan Binnewies
TELESCOPE	1.3-m reflector
CAMERA	Andor DZ 436
EXPOSURE TIME	197.5 min
LOCATION	Skinakas Observatory, Crete, Greece

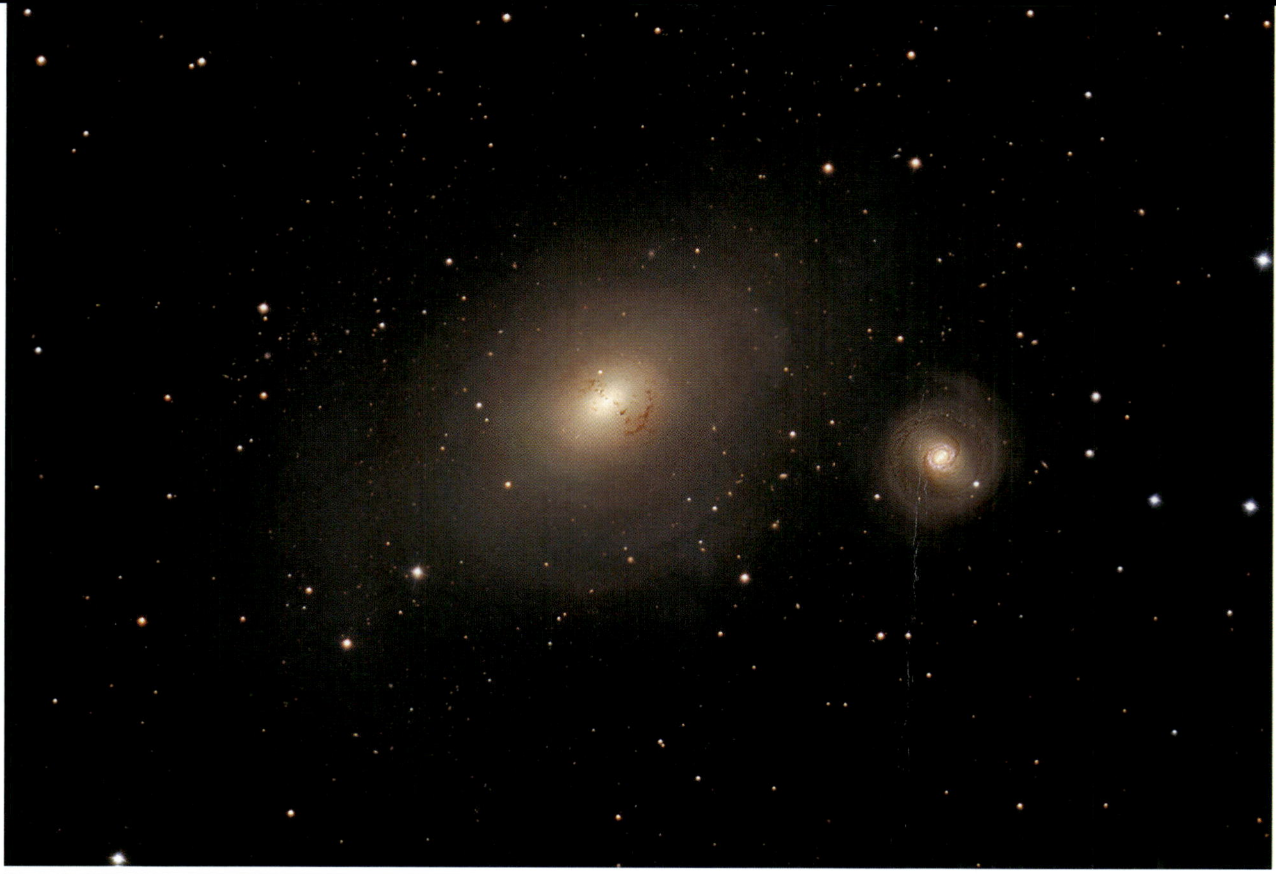

NGC 1316

The image shows the large galaxy NGC 1316 and, 6.5′ to the right, towards the north, the spiral galaxy NGC 1317, both of which belong to the Fornax Cluster. The recession velocity of NGC 1317 is only 181 km/s larger than that of NGC 1316 so that one can assume that the two galaxies form a gravitationally interacting system which lies at a mean distance of 77 million light years. The enormous angular size of 12.0′ × 8.5′ of NGC 1316 becomes clear in comparison to NGC 1317 and the projected separation: the axes correspond to about 270 000 × 190 000 light years. NGC 1316 was classified as E/SAB0, and also as a ring galaxy, but usually with the attribute "peculiar". Like the halo structure of tidal tails and shells, this term indicates a violent past of the galaxy in which it probably merged with two or more compact, gas-rich galaxies.

In the image, one can recognize near the centre of NGC 1316 two dust lanes which can be seen as the remains of a merged spiral galaxy. Measurements of the rotation velocity confirm the existence of a disc component. This has the increase up to a plateau at 160 km/s which is typical of spiral galaxies, but is well defined only on one side and extends only about an arcminute. Farther out, the rotation is no longer regular; here, NGC 1316 is at the transition to an elliptical galaxy. Due to its large luminosity, NGC 1316 is also considered to be an active galaxy. In the radio, it is the strongest source in the Fornax Cluster and is thus known as Fornax A. NGC 1316/Fornax A has two huge radio lobes which are situated symmetrically about the centre. Each of these lobes has a diameter of 650 000 light years. The lobes are powered by jets which emerge diametrically from the centre and whose high-energy particles are accelerated in the environment of the supermassive black hole.

OBJECT	NGC 1316, Fornax A
CONSTELLATION	Fornax
RIGHT ASCENSION	$03^h\ 22^m\ 42^s$
DECLINATION	–37° 12′ 30″
BRIGHTNESS	9.4 mag
TYPE	SAB0(s) pec
PHOTOGRAPHERS	Josef Pöpsel
TELESCOPE	600-mm reflector
CAMERA	SBIG ST-10XME
EXPOSURE TIME	190 min
LOCATION	Amani Lodge, Namibia

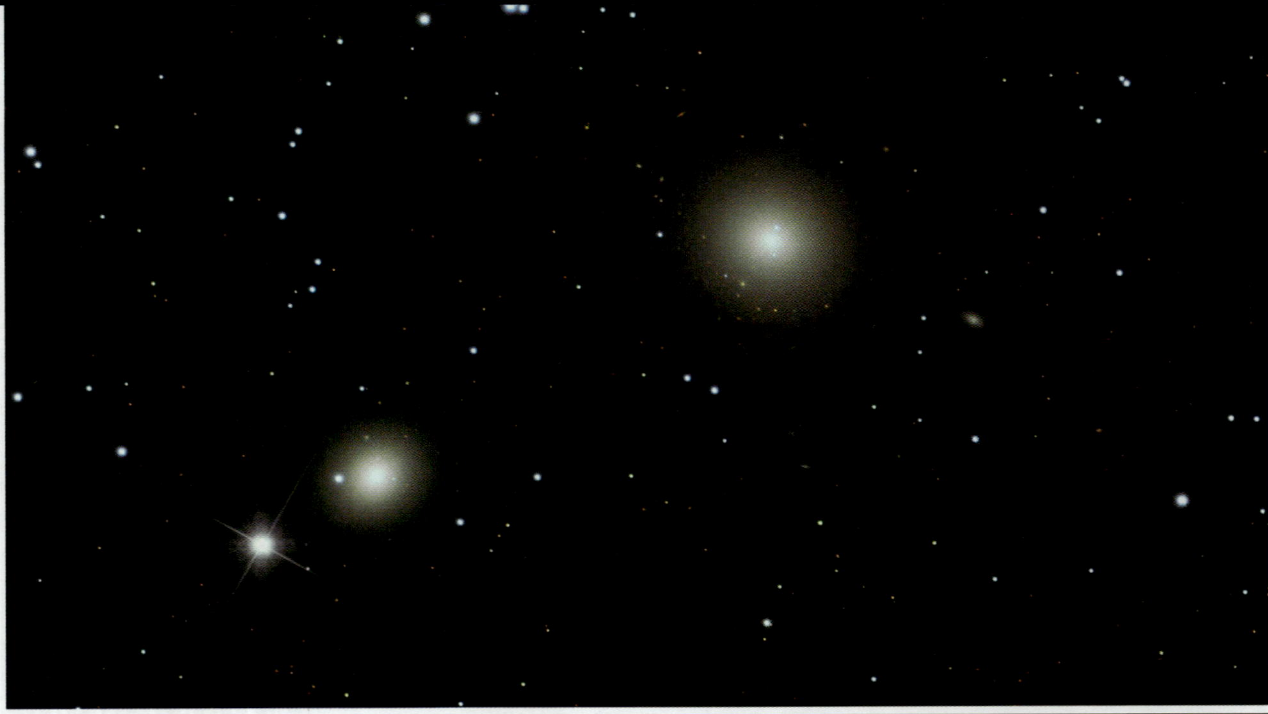

NGC 1399

The galaxy NGC 1399 is the central, dominant elliptical galaxy in the Fornax Cluster. This cD galaxy, at a distance of 58 million light years from the Milky Way, has a true diameter of 116 000 light years. Several background galaxies are visible through its extended, diffuse halo. The two bright blue point sources above and below the centre are foreground stars. NGC 1399 also contains about 5700 globular clusters (GCs), which exceeds the typical number in elliptical galaxies by a factor of five and that in spiral galaxies by a factor of 20. But the number of GCs in NGC 1399 can also be exceeded: one estimates their number in M 87 at 14 000 while the Milky Way contains only 160 GCs. In the image, one can see below and to the left of NGC 1399 at a distance of 10.2′, next to an 8th-magnitude star, the E2 galaxy NGC 1404, which is also part of the Fornax Cluster.

In elliptical galaxies, there are often two populations of GCs which differ in colour. One the one hand are blue GCs which contain young, metal-poor stars. On the other hand are red GCs with metal-rich, older stars. One sees in the distribution of GCs in elliptical galaxies that blue GCs are more extended and the red more strongly concentrated towards the centre. The cause of the formation of these two types of GCs is still being debated among astronomers. A possible scenario is the merger of a gas-rich galaxy with the elliptical galaxy. It brings GCs into the halo, whereby gas in the accreting galaxy induces new star formation during the mixing processes taking place in the interaction. This idea is supported by the fact that the mean metallicity measured in stars in red GCs corresponds to that in giant stars in the halo. This is in indication that the red GCs could have formed together with the giant stars. In NGC 1399, studies of the GCs, however, do not clearly differentiate the GC populations. If one plots the number of GCs as a function of colour, there is a concentration in the blue range and a broad red tail, but no clear separation. The radial distribution of GCs in NGC 1399 also shows that the metal-poor, blue GCs occur more often at radii > 2′ than the metal-rich, red GCs. This confirms the described scenario, whereby the lack of bimodality can be seen as an indication that NGC 1399 experienced a multistage collapse, i.e. a series of mergers of several galaxies led to the formation of the elliptical galaxy.

OBJECT	NGC 1399
CONSTELLATION	Fornax
RIGHT ASCENSION	$03^h\ 38^m\ 29^s$
DECLINATION	−35° 27′ 02″
BRIGHTNESS	10.6 mag
TYPE	E1 pec
PHOTOGRAPHERS	Bernd Flach-Wilken
TELESCOPE	400-mm reflector
CAMERA	SBIG STL-6303
EXPOSURE TIME	175 min
LOCATION	Tivoli Farm, Namibia

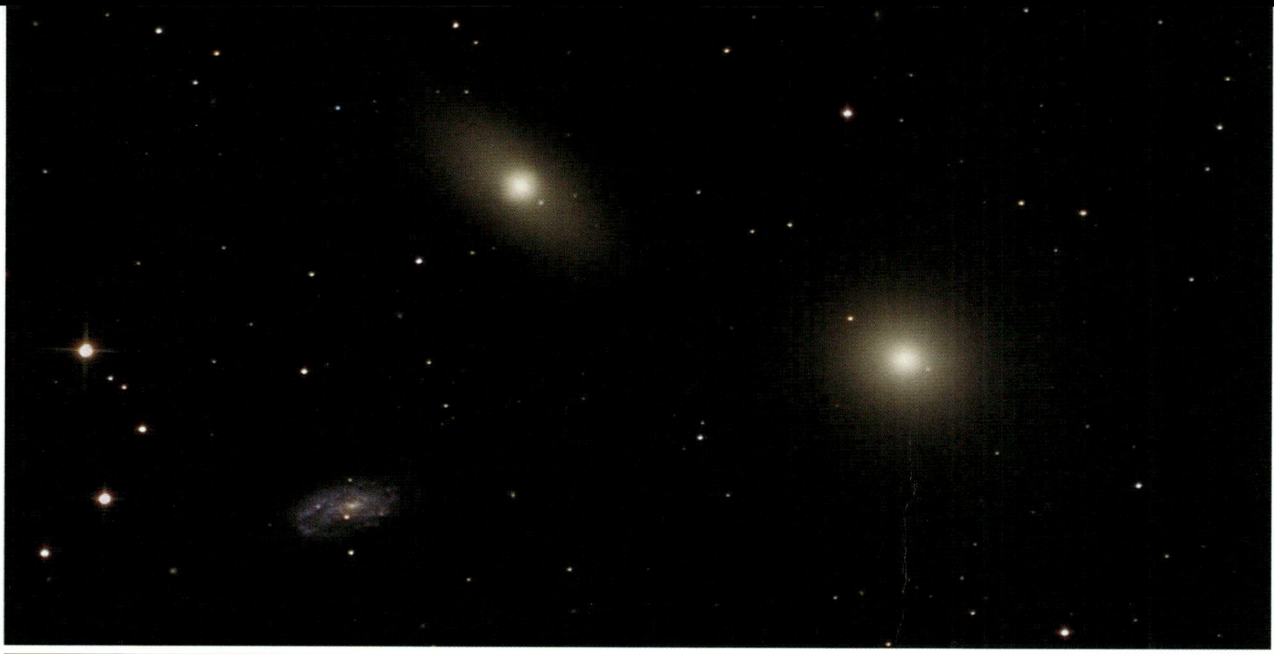

M 105

The E1 galaxy M 105 (NGC 3379) (right), together with the SB0 galaxy NGC 3371 (centre left) and the Sc galaxy NGC 3373 (below left), forms the central visual triplet of the Leo I Group. This galaxy group lies in the direction of the Virgo Cluster at a distance of 32 million light years from the Milky Way. The angular size of M 105 is 5.4′ × 4.8′ and corresponds to a projected size of 53 000 × 47 000 light years. The distance between M 105 and NGC 3371 is 7.1′. According to catalogue distances, NGC 3371 is farther away than M 105, but is moving relative to it at about 200 km/s towards the Milky Way. The small Sc galaxy NGC 3373 (below left) has a recession velocity of 1308 km/s and thus about 500 km/s more than the mean recession velocity of the two other galaxies. Alternative methods of distance determination give a distance of 69 million light years for NGC 3373 and an absolute transverse size of 59 000 light years. The triplet is thus presumably not a real physical triple: NGC 3373 only coincidentally lies in the direction of the other two galaxies.

On the William Herschel Telescope (4.2 m, La Palma, Spain), as part of the SAURON Project (Spectrographic Areal Unit for Research on Optical Nebulae, started 1999), an integral-field spectrograph was used that is capable of recording 1577 spectra simultaneously. M 105 was part of a SAURON observational campaign which used the analysis of the emission lines in spectra to study separately the kinematics of the stars and gas in a galaxy. Astronomers detected rotation in M 105 and discovered that the stellar and gas components are oriented differently, tilted by about 45° with respect to each other. The position of the gas disc fits well with that of a dust ring which could be detected in the innermost arcsecond with the Hubble Space Telescope. Both disc components are an indication of a merger history during which M 105 could have formed by the collision of at least two gas-rich spiral galaxies. With the help of HST data, some stellar objects in M 105 could be studied individually as well. Their spectra have narrow O III emission lines and no detectable Hβ emission. The spectral O III/Hβ signature is typical for planetary nebulae. The velocities of the planetary nebulae in M 105 suggest that the inner halo is rotating – albeit slowly – around the semiminor axis of the isophotes. Such peculiar halo kinematics is also an indication of a possible merger in which angular momentum of the precursor galaxies was transferred to the halo.

OBJECT	M 105, NGC 3379
CONSTELLATION	Leo
RIGHT ASCENSION	$10^h\ 47^m\ 50^s$
DECLINATION	+12° 34′ 54″
BRIGHTNESS	10.2 mag
TYPE	E1
PHOTOGRAPHERS	Volker Wendel, Bernd Flach-Wilken
TELESCOPE	380- and 300-mm reflector
CAMERA	SBIG ST-10XME
EXPOSURE TIME	110 min
LOCATION	Wirges and Weisenheim am Berg, Germany

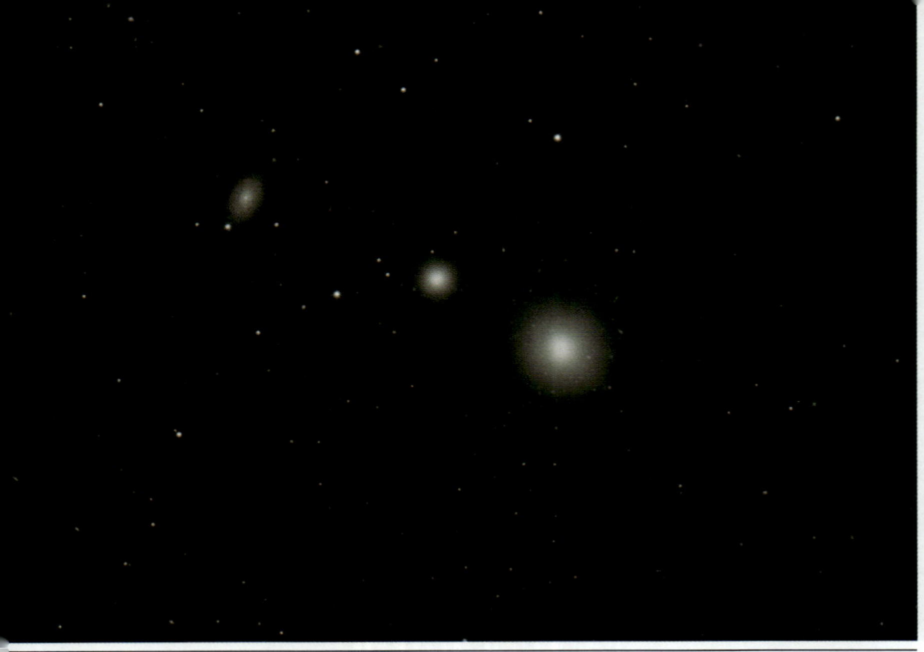

In 1977, NGC 4278 was one of the first elliptical galaxies in which 21-cm radio emission of atomic hydrogen was detected. In many further studies the galaxy was investigated at various other wavelengths, including in X-rays with the Chandra X-ray Observatory. The high angular resolution of this satellite allows astronomers to directly observe extragalactic low-mass X-ray binaries (LMXB). These double stars consist of a compact object, such as a neutron star or black hole, and an evolved star. The latter loses material which flows to and accretes on the compact object, releasing X-rays. NGC 4278 is well suited for this LMXB study since its small distance allows a spatial resolution of 120 light years. Also, the galaxy has no X-ray halo of hot gas which would hinder the detection of weak LMXBs. With a diameter of 4′, NGC 4278 fits completely onto the chip of the X-ray satellite (type ACIS-S3). The Chandra X-ray Observatory had already discovered LMXB populations in some galaxies, and in a deep observation of NGC 4278, 236 X-ray sources were identified. In the total observation time of more than 127 hours, X-ray luminosities of 10^{36}–10^{40} erg/s were measured for the sources. The brightest X-ray source was the galactic centre; all other X-ray sources had luminosities below 10^{39} erg/s, most of them at around 10^{37} erg/s, which is a typical value for an LMXB. Based on the visual halo (see image), about 180 of the 236 X-ray sources lie within the galaxy. The halo of NGC 4278 was also studied with the Hubble Space Telescope, whereby 266 GCs were found. Comparing the sources in X-rays and visible light, there are 39 GC-LMXBs. The LMBXs residing in GCs have an especially high luminosity; the less luminous LMXBs lie scattered about in the halo of NGC 4278.

NGC 4278

The image shows the E1 galaxy NGC 4278 (right), 3.5′ to the left the smaller E0 galaxy NGC 4283, and at the far left, 9′ away from NGC 4278, NGC 4286. With an angular size of 4.1′ × 3.8′, NGC 4278 is much larger than NGC 4283 and NGC 4286, whose maximum size is only about 1.5′. It is difficult to conclude anything about the distance from the size, since two of the three galaxies differ greatly from the standard size. The recession velocity of NGC 4278 at 649 km/s is too small to be used as a distance measure. Alternative methods, however, indicate a distance of 51 million light years and a diameter of 61 000 light years for this elliptical galaxy. Its E0 neighbour has a recession velocity 351 km/s larger, and distance indicators give a compatible distance of 48 million light years. Within the uncertainties, the two galaxies could be close neighbours. NGC 4283 is also classified as a dwarf galaxy due to its small diameter. The third galaxy in the image – NGC 4286 – has a recession velocity of 641 km/s. Since there are no other distance indicators, this results in a distance of 29 million light years and a small projected diameter of 12 000 light years. The dim disc which surrounds the bright core of NGC 4286 is characteristic of this dE,N type, a nucleated dwarf elliptical galaxy. One suspects that dE,N galaxies are so-called UCD types – ultrabright, compact dwarf galaxies, which are still surrounded by a gas disc.

OBJECT	NGC 4278
CONSTELLATION	Coma Berenices
RIGHT ASCENSION	$12^h\ 20^m\ 07^s$
DECLINATION	+29° 16′ 51″
BRIGHTNESS	11.2 mag
TYPE	E1-2
PHOTOGRAPHERS	Michael König
TELESCOPE	350-mm reflector
CAMERA	SBIG STL-6303
EXPOSURE TIME	255 min
LOCATION	Rimbach, Germany

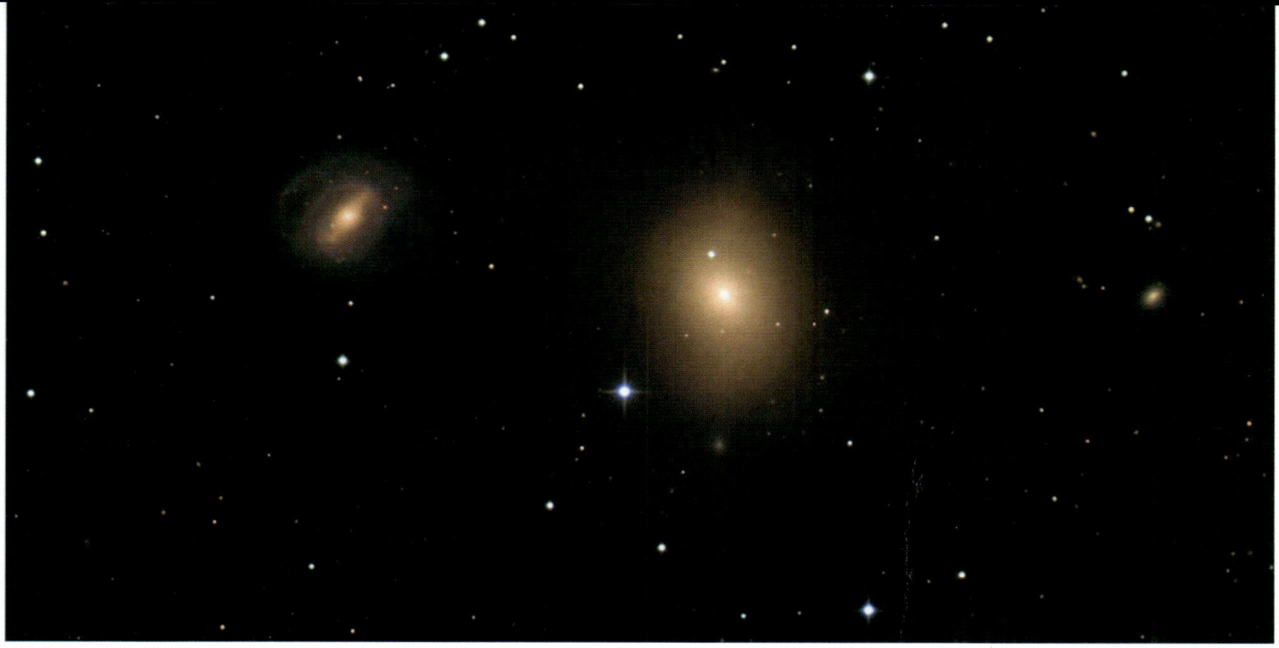

M 85

To the right in the image one can see the core of the galaxy M 85 (NGC 4382) and, 7.8′ to the left, to the east, NGC 4394. Both galaxies belong to the Virgo Cluster and their recession velocities indicate that NGC 4394 is receding from the Milky Way at 193 km/s relative to M 85. Since local velocities within the Virgo Cluster make cosmological distance measurements much more difficult, the recession velocity is not a good indicator. Using distance measures independent of the recession velocity, however, results in a distance of 56 million light years for M 85 and 55 million light years for NGC 4394. Since even the combination of these alternative distance measures doesn't reduce the uncertainties below ten per cent, it is probable that M 85 and NGC 4394 form a physical pair. The transverse distance between the two galaxies in this case is 115 000 light years. This is thus only slightly less than the diameter of M 85, which is given as 116 000 light years or 7.1′.

In NGC 4394, at about half the size, the bar with ansae at the ends has the same yellow colour as the one which dominates in M 85 and is due to old stellar populations. The spiral arms, forming a ring where they join the bar, are blue in NGC 4394: this is especially clear in the upper, northeastern arm. The (R)SB(r)b classification of this galaxy is plausible; this is more difficult for M 85. One finds both E and S0 classifications, whereby the structures in the halo point to a history of interaction, and the term "peculiar" is added as well. The isophotes of M 85 have a boxy intensity profile, indicating a (boxy) elliptical galaxy type. These isophotes extend out to radii of 1′; farther out, the form deviates into more pointed intensity curves in the north–south direction. Many tidal tails and shells are visible in the image, defining a vertical spindle-shaped structure. Astronomers thus describe M 85 as a recent merger, which could have formed out of an elliptical galaxy and a spiral galaxy. The relaxation processes after the merger are known as virialization and the corresponding timescales in the centre of the newly formed galaxy are shorter than in the outer regions. With time, a dynamic equilibrium sets in with a fixed relation between kinetic and potential energy, known among astronomers as the virial theorem. The virialization thus moves from inside to outside, which would also explain why traces of the merger are still visible in the halo while one can already observe an elliptical galaxy in the inner regions.

OBJECT	M 85, NGC 4382
CONSTELLATION	Coma Berenices
RIGHT ASCENSION	$12^h\ 25^m\ 24^s$
DECLINATION	+18° 11′ 29″
BRIGHTNESS	10 mag
TYPE	$SA0^+$(s) pec
PHOTOGRAPHERS	Wolfgang Ries, Stefan Heutz
TELESCOPE	450-mm reflector
CAMERA	SBIG ST-10XME
EXPOSURE TIME	180 min
LOCATION	Altschwendt, Austria

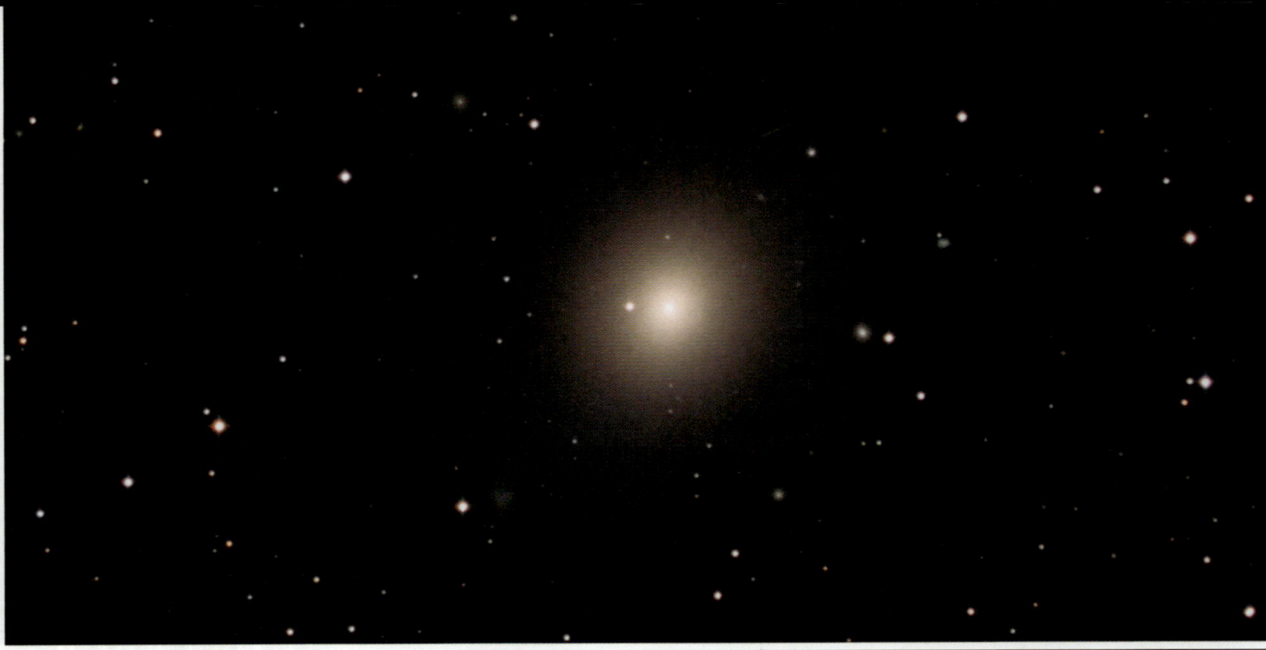

M 49

M 49 (NGC 4472) is a large elliptical galaxy in the Virgo Cluster. It is located in the southern concentration of the cluster and dominates that region. The visual luminosity of M 49, which is also classified as an active galaxy (Seyfert-1), exceeds that of M 87 by 0.2 mag. M 49 is also one of the brightest sources in the Virgo cluster in X-rays, but in this case M 87 is brighter. The extent of the galaxy is given as 10.2′ × 8.3′, which at a distance of 52 million light years corresponds to a diameter of 155 000 light years. In the radio, one can detect a diffuse tidal tail 330 000 light years long. Astronomers suspect that this tail is the trail of gas lost by M 49 due to the ram pressure caused by the motion of the galaxy relative to the cluster medium. This motion is described by astronomers as Virgo infall, with velocities of more than 2000 km/s being measured. The projected distance to M 87 in subcluster A, which is 4° farther north, is 3.88 million light years. In the right edge of the halo one can see tracks of two minor planets. During the exposure, 16.6-magnitude (6983) Komatsusakyo and 17.4-magnitude (31741) 1999 JG_{78} passed by M 49.

At the lower, southeastern halo edge one can recognize a dim, blue fleck of light which is identified as VCC 1249 (Virgo Cluster Catalogue). It is a dIrr galaxy which forms a dissimilar interacting pair with M 49. The dwarf galaxy has bright H II regions, some of which lie outside the inner region visible in the image. They are located along a strip pointing to the companion 5.6′ away. The analysis of stellar populations in VCC 1249 reveals that both effects, ram pressure and tidal forces, have removed hydrogen gas from the galaxy. This switched off star formation in VCC 1249 200 million years ago. Since the difference in radial velocity between VCC 1249 and M 49 at 611 km/s is very high, researchers believe that the orbital motion of VCC 1249 is towards the Milky Way. If one subtracts a fitted profile from the image of M 49, dim residuals appear in the halo. For example, one can see shells extending across 12′ to the southeast and northwest. There is also an inner arc facing VCC 1249. This could have formed during the first approach when M 49 stripped low-density gas out of the outer regions of the dwarf galaxy. During the second passage, higher-density material was pulled out of the inner part of VCC 1249. According to the "in situ" interaction model, the H II regions were formed in the tidal tail out of escaping gas.

OBJECT	M 49, NGC 4472
CONSTELLATION	Virgo
RIGHT ASCENSION	$12^h\ 29^m\ 47^s$
DECLINATION	+08° 00′ 02″
BRIGHTNESS	9.4 mag
TYPE	E2
PHOTOGRAPHERS	Wolfgang Ries, Stefan Heutz
TELESCOPE	450-mm reflector
CAMERA	SBIG ST-10XME
EXPOSURE TIME	126 min
LOCATION	Altschwendt, Austria

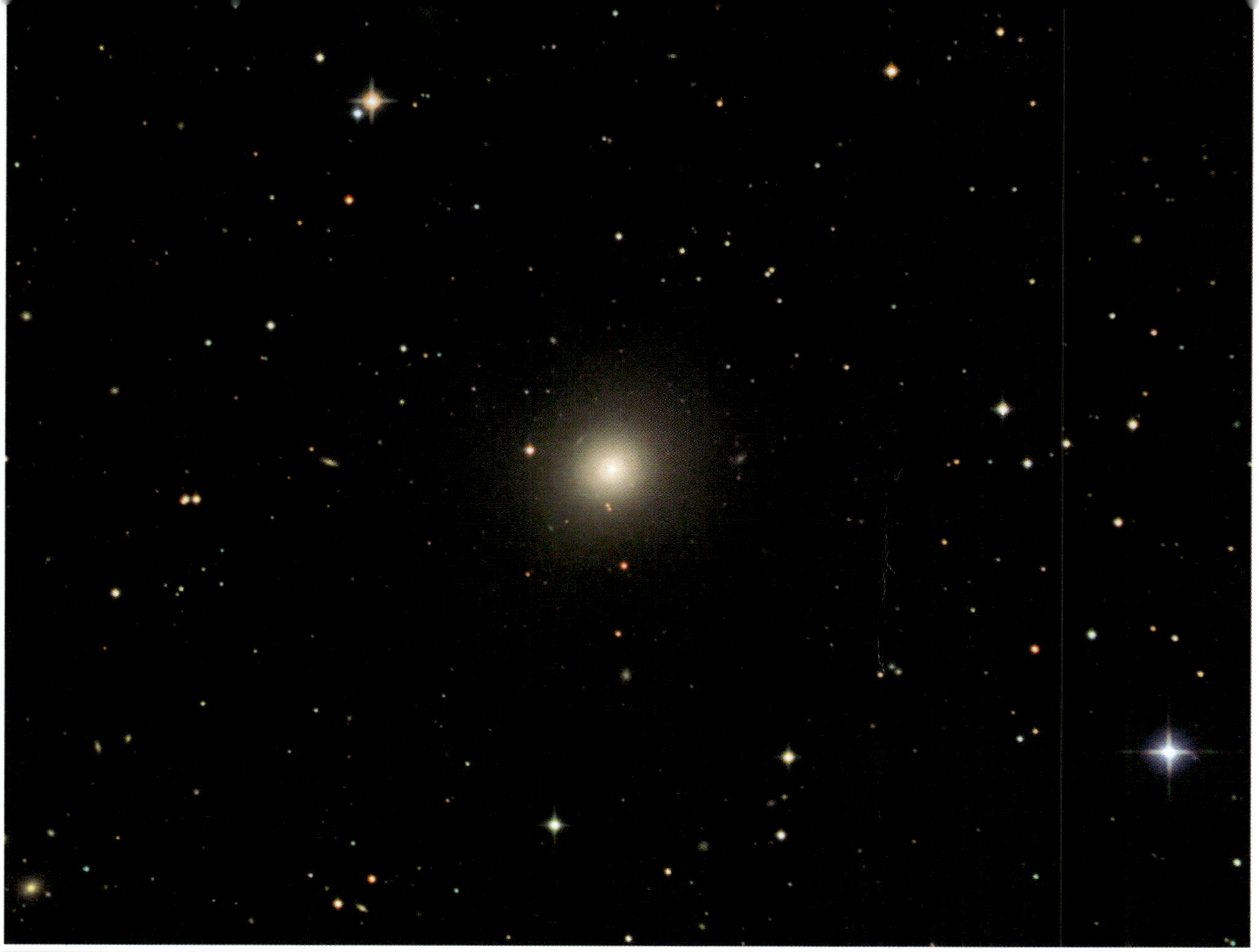

M 89

M 89 (NGC 4552) is an almost circular elliptical galaxy in the Virgo Cluster which has no large neighbour within a radius of 20′. Its angular size is 5.1′ × 4.7′; at its distance of 51 million light years this corresponds to a projected diameter of 76 000 light years.

NGC 4252 was the first galaxy in which an envelope was detected. This is also visible in the image and extends almost twice as far as the bright inner region. Another peculiarity of this galaxy was mentioned in an article in 1979: a jet which seems to emerge from M 89 and is visible as a dim ray in the image, pointing towards the 9th-magnitude near the lower right corner (see also the contrast-enhanced image).

The size of the jet was measured at 5.0′ × 0.5′; its surface brightness is only two per cent above the brightness of the sky background. Spectroscopic observations were also made in order to allow a comparison with the jet of M 87. CCDs were used rather than hypersensitized film which was still the rule at the time and was used to discover the jet. Astronomers found a continuum, but no emission lines which would indicate accelerated, excited gas. The jet also appears significantly bluer than M 89 itself. The jet hypothesis was thus rejected and the structure was interpreted as a tidal tail which could have been formed by the close approach of a satellite galaxy which later merged with the precursor of M 89. The line of sight from the Milky Way to M 89 lies in the former plane of the orbit of this companion galaxy so that the tidal tail is observed as an almost straight line.

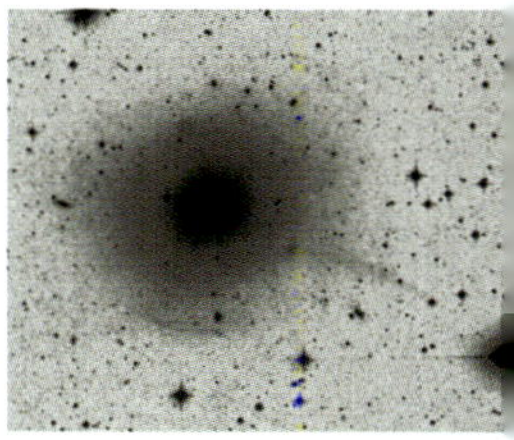

OBJECT	M 89, NGC 4552
CONSTELLATION	Virgo
RIGHT ASCENSION	$12^h\ 35^m\ 40^s$
DECLINATION	+12° 33′ 23″
BRIGHTNESS	10.7 mag
TYPE	E0-1
PHOTOGRAPHERS	Wolfgang Ries, Stefan Heutz
TELESCOPE	450-mm reflector
CAMERA	SBIG ST-10XME
EXPOSURE TIME	84 min
LOCATION	Altschwendt, Austria

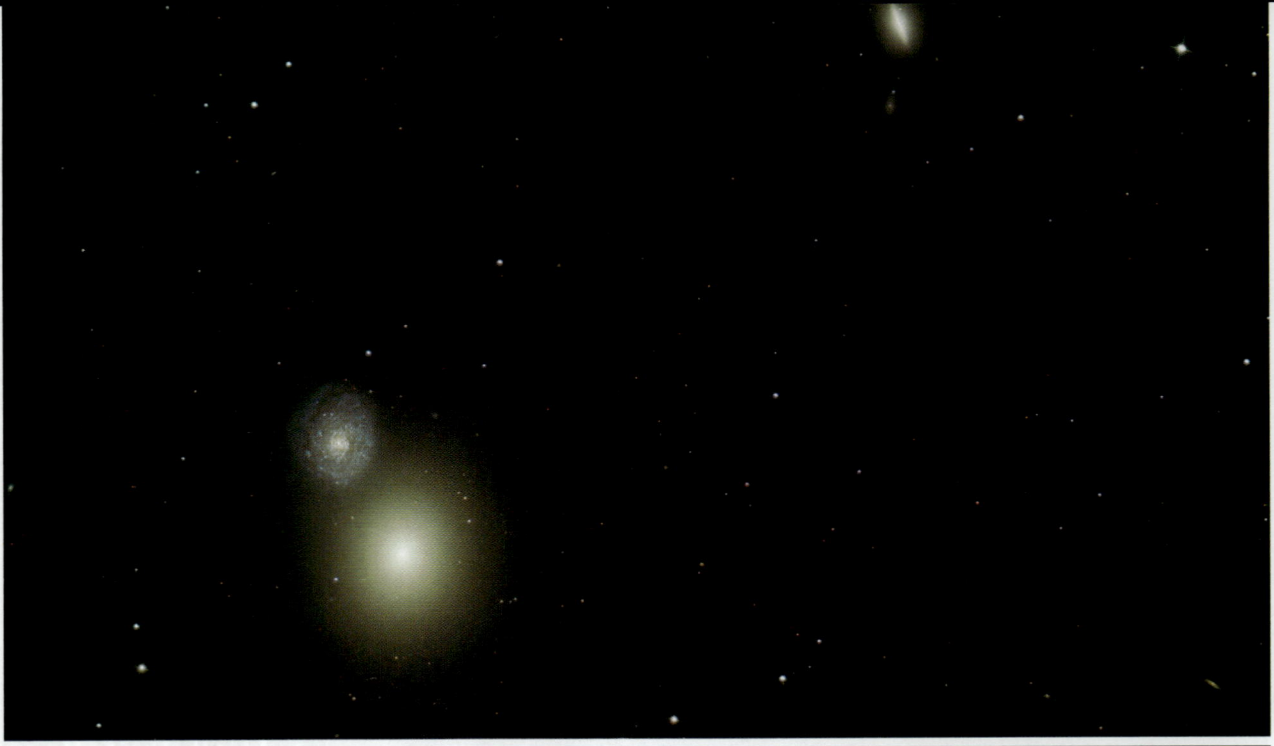

M 60

The elliptical galaxy M 60 (NGC 4649) and the SAB(rs)c galaxy NGC 4647 make up the galaxy pair known as Arp 116 in the *Atlas of Peculiar Galaxies*. At the upper edge of the image one can also see the spindle-shaped S0 galaxy NGC 4638 and 1.5′ below it the diffuse spiral galaxy NGC 4637. The distance from the E2 galaxy M 60 to the spiral galaxy is 2.6′ or 42 000 light years in projection. M 60 at 7.4′ × 6.0′ appears much larger than the spiral galaxy with diameter 2.9′. The recession velocities of the two galaxies are similar, but not identical, differing by 292 km/s. Their mean distance is 55 million light years and they are located in the eastern part of the Virgo Cluster. The question arises how the relative position is to be interpreted, since there are no perturbations in their morphologies. NGC 4647 displays a pronounced, flocculent spiral pattern with a prominent dust lane running across the northern side.

In a detailed kinematic study of NGC 4647, astronomers noted an asymmetry in the rotation curve. The measured rotation velocities on the northern side (with the dust lane) show the flat form with values around 110 km/s typical of spiral galaxies. On the eastern side there is no plateau, but rather a smooth increase up to 150 km/s. This difference can be due to the ram pressure experienced by NGC 4647 as it moves through the cluster medium. It is also possible that this ram pressure is further increased by the passage through the halo of M 60. Applying colour–magnitude diagrams (CMD) to a pixel-by-pixel analysis of images taken by the Hubble Space Telescope, it could be shown that M 60 is well described by a uniform stellar population. Exceptions, however, are dusty pockets as well as blue and red globular clusters. Compared with other pairs of galaxies which are obviously interacting, the analysis showed no prominent blue pixels which would indicate star-forming regions. The only hint of a possible future interaction is just a weak trend of blue pixels in the direction of the companion NGC 4647. Thus a definitive statement about the relative position of the two galaxies is difficult; in spite of their proximity within the galaxy cluster they do not form a close pair.

OBJECT	M 60, NGC 4649
CONSTELLATION	Virgo
RIGHT ASCENSION	$12^h\ 43^m\ 40^s$
DECLINATION	+11° 33′ 10″
BRIGHTNESS	9.8 mag
TYPE	E2
PHOTOGRAPHERS	Stefan Binnewies, Josef Pöpsel
TELESCOPE	600-mm reflector
CAMERA	SBIG STL-11000
EXPOSURE TIME	170 min
LOCATION	Skinakas Observatory, Crete, Greece

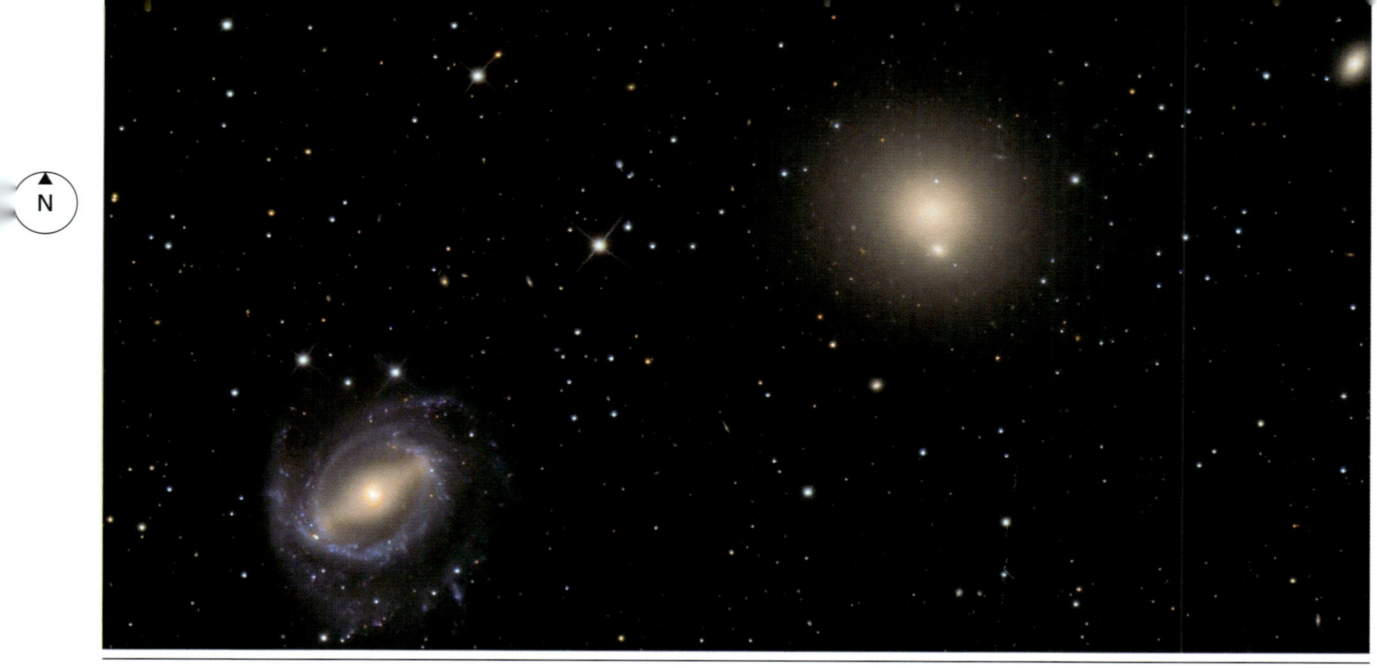

NGC 5846

The E0-1 galaxy NGC 5846 is at the centre of a group of galaxies named after it. NGC 5846 lies at a distance of 77 million light years and its angular size of 4.1′ × 3.8′ corresponds to a transverse diameter of 92 000 light years. Databases record a recession velocity 842 km/s larger for the SB(r)b galaxy NGC 5850 which is visible in the image 10′ to the lower left of NGC 5846. Near the core of NGC 5846 one can also recognize the compact E2 galaxy NGC 5846A, 0.7′ to the south. Various methods of distance determination place it behind NGC 5846, which fits well with its 487 km/s larger recession velocity. Alternatively, NGC 5846A could be located in front of NGC 5846 and its high recession velocity would then indicate rapid motion towards the large elliptical galaxy. However, since no tidal tails are visible around NGC 5846, NGC 5846A is very probably located behind NGC 5846 and belongs to the more distant part of the group.

NGC 5846 is an example of an elliptical galaxy with low but detectable rotation and high velocity dispersion in the core, up to 250 km/s. Studies of the core in both optical and X-ray bands show a few filaments but no connected structures such as a dust disc. By contrast, in the elliptical galaxy NGC 5813, which is part of the NGC 5846 group but not visible in the image, such a circumnuclear disc could be detected. The filaments in NGC 5846 are interpreted as relics of interactions with smaller galaxies. The fact that NGC 5846 is perturbed while NGC 5813 has an unperturbed disc is explained by NGC 5846 having experienced several mergers. These perturbed and finally stopped the uniform rotation. In deep observations of the NGC 5846 group, astronomers have detected the large number of about 250 members. 80 per cent of these galaxies are dwarf galaxies, mostly of type dE. Some of these galaxies can be recognized in the image as diffuse spots of light around NGC 5846. The large number of dwarf galaxies compared to other groups makes it clear that NGC 5846 lies in a dense, dynamically evolved group and could evolve into a cD galaxy. The fraction of dwarf galaxies is so high because the galaxies around which they originally orbited were accreted. The remaining galaxies continue orbiting the centre of mass, their orbital angular momentum preventing them from being cannibalized.

OBJECT	NGC 5846
CONSTELLATION	Virgo
RIGHT ASCENSION	$15^h\ 06^m\ 29^s$
DECLINATION	+01° 36′ 10″
BRIGHTNESS	11.0 mag
TYPE	E0-1
PHOTOGRAPHERS	Adam Block
TELESCOPE	600-mm reflector
CAMERA	SBIG STL-11000
EXPOSURE TIME	410 min
LOCATION	Mount Lemmon SkyCenter/ University of Arizona, USA

NGC 5982

The image shows the well known Draco galaxy triplet, consisting of the SAB(r)b galaxy NGC 5985 (upper right), the E3 galaxy NGC 5982 (centre) and the Sc galaxy NGC 5981 (lower left). Both the recession velocity and other distance measures agree on a distance of 135 million light years for NGC 5982. 7.8′ eastward is the SAB(r)b galaxy NGC 5985 whose spiral plane has a diameter of 5.5′. Relative to NGC 5982, it is approaching the Milky Way at 500 km/s and would thus lie in front of the elliptical galaxy. Disregarding the recession velocity of NGC 5985, databases list a distance of 137 million light years. This would mean that the spiral galaxy would be farther away than NGC 5982. Since these quantities have an uncertainty of ten per cent, it is currently not possible to make a definitive statement about the distances. The angular size of 5.5′ corresponds to a disc diameter of 219 000 light years. NGC 5982 has a shell structure visible in its halo as well as a diffuse outer edge, which also extends about 5′ and thus has a size similar to that of NGC 5985. Much smaller, in contrast, is the third galaxy NGC 5981. Its recession velocity is not, as one might suspect based on its smaller size, larger than that of the two large galaxies. This edge-on galaxy's recession velocity is only 58 per cent of that of NGC 5982, leading to a distance of 86 million light years and a corresponding diameter of 78 000 light years. The well known Draco motif thus presumably shows two unusually large galaxies and a much closer galaxy of normal size.

Some elliptical galaxies are known as shell galaxies because of the shell structure visible in their halos. These shells are dim, usually don't completely surround the core, and are found in 20–40 per cent of all elliptical field galaxies. According to the theory, they are formed in mergers with small, low-mass galaxies with low velocity dispersion, whereby the mass of the small galaxies is only a few per cent of that of the large galaxy. It is known from simulations that they move along radial orbits where they lose more and more stars. The shells are thus not projections of tidal tails, but rather three-dimensional structures formed from accreted stars by density waves in the halo. However, in contrast to spiral galaxies with circulating density waves, no spiral arms form in a plane, but rather concentric shells of stars in the halo of the elliptical galaxy. Stars with a lower binding energy have a longer orbital period and thus a maximum distance which is far out in the halo. These stars are the building blocks of the shells. In principle, stars which have completed an integer number of orbits accumulate along the bright edges of the shells. The large number of shells which one observes in the halo of NGC 5982 is thus seen as an indication of a high age. From the radii of the outer shells and the mass of NGC 5982 of 4.57×10^{11} solar masses, astronomers calculate a dynamic age of 1.1 billion years for the halo structures. It is notable that the shell of NGC 5982 which crosses the semiminor axis above the nucleus is not red as is usual for shells, but rather blue. This could be in indication of an associated young stellar population accreted by NGC 5982 in a manner similar to that of tidal dwarf galaxies. The difference in colour to the other shells could then be explained by the fact that the stars in the shells are due to different galaxies which have merged with NGC 5982, contributing stellar populations of different ages to the halo.

OBJECT	NGC 5982
CONSTELLATION	Draco
RIGHT ASCENSION	$15^h\ 38^m\ 40^s$
DECLINATION	+59° 21′ 21″
BRIGHTNESS	12.1 mag
TYPE	E3
PHOTOGRAPHERS	Josef Pöpsel, Stefan Binnewies
TELESCOPE	600-mm reflector
CAMERA	SBIG STL-11000
EXPOSURE TIME	335 min
LOCATION	Skinakas Observatory, Crete, Greece

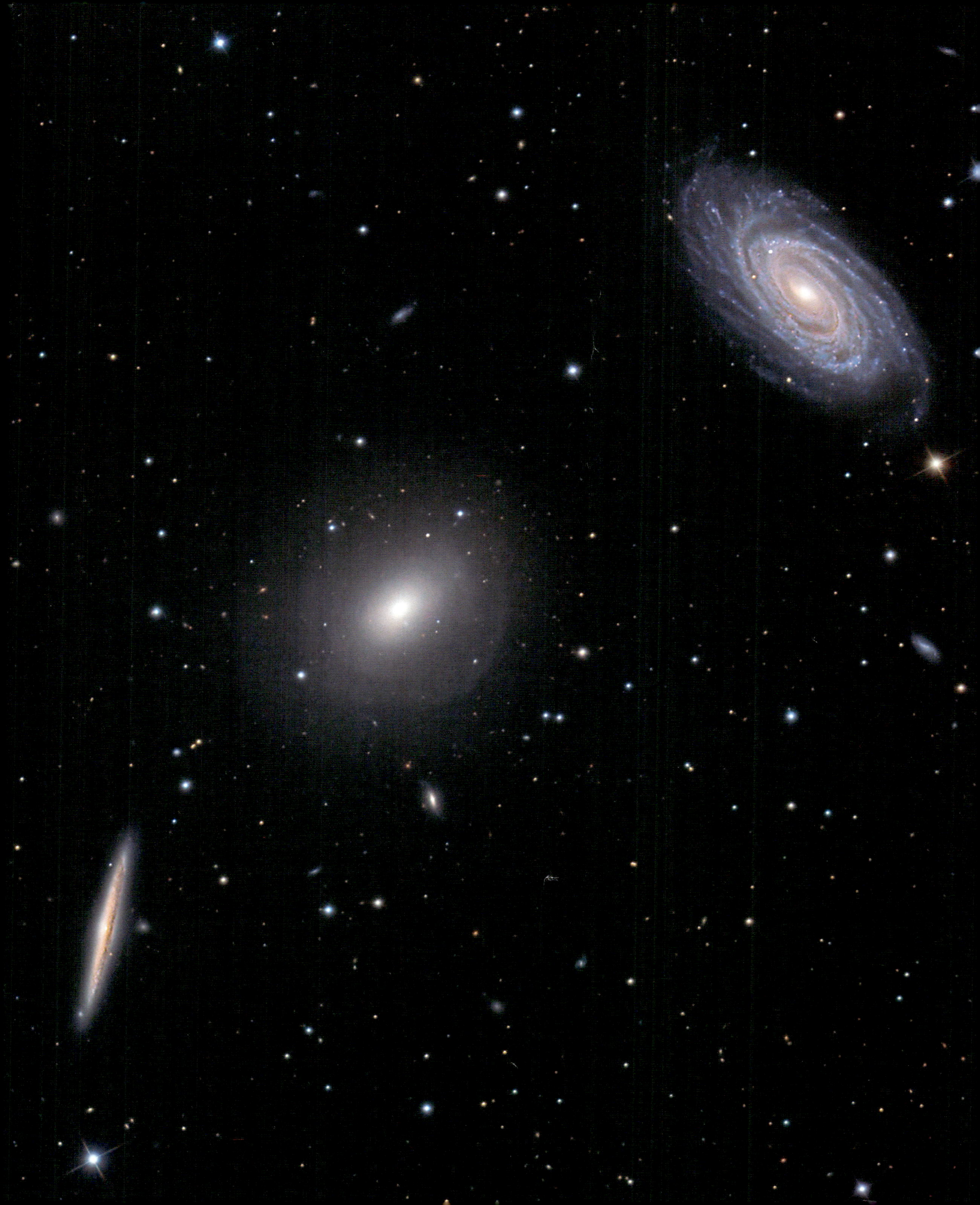

Arp 99

This isolated galaxy trio, known as Arp 99 or HCG 93 (Hickson Compact Group), contains the three large members NGC 7547, NGC 7549 and NGC 7550. The small SAB galaxy NGC 7547 to the right and the barred spiral galaxy NGC 7549 about 4′ away to the upper left display perturbations such as tidal tails or deformations of their spiral arms. These effects point towards NGC 7550, which is classified as E or SA0 in the literature. The largest galaxy in the group is 230 million light years from the Milky Way. The relative velocities of the two spiral galaxies are −300 km/s and −350 km/s. Compared to NGC 7550, they are approaching the Milky Way and this suggests that NGC 7547 and NGC 7549 are orbiting around the larger NGC 7550. This elliptical galaxy dominates the triplet. The V-shaped structure in the western, right part of the halo is an indication that NGC 7547 must have once passed close to NGC 7550. The diameter of the halo of NGC 7550 is 3.5′ and corresponds to a transverse distance of 240 000 light years. The SA(r) galaxy NGC 7558 below and to the left of the trio is at a distance of 340 million light years and is located behind the small group.

If one compares the velocity dispersion of the stars in NGC 7550 with the values of other elliptical galaxies in small groups, one finds that the σ_c value of 220 km/s is at the top of the range. Comparative studies indicate mean values between 100 km/s and 150 km/s. Astronomers explain the discrepancy in NGC 7550 by the special dynamics of its halo, which extends to the core. A prominent emission line, due to excited nitrogen atoms, N II, is visible at 658.3 nm in the spectrum of NGC 7550. Together with other line strengths, this allows one to estimate the mean metallicity of the stars and thus the age of a galaxy. For NGC 7550 this results in a comparatively high metallicity and an age of about four billion years. This high fraction of heavy elements could be the result of a merger of a spiral galaxy with the precursor of NGC 7550.

OBJECT	Arp 99, NGC 7550
CONSTELLATION	Pegasus
RIGHT ASCENSION	$23^h\ 15^m\ 16^s$
DECLINATION	+18° 57′ 42″
BRIGHTNESS	13.2 mag
TYPE	$SA0^-$
PHOTOGRAPHERS	Josef Pöpsel, Stefan Binnewies
TELESCOPE	600-mm reflector
CAMERA	SBIG STL-11000 and STX-16803
EXPOSURE TIME	885 min
LOCATION	Skinakas Observatory, Crete, Greece

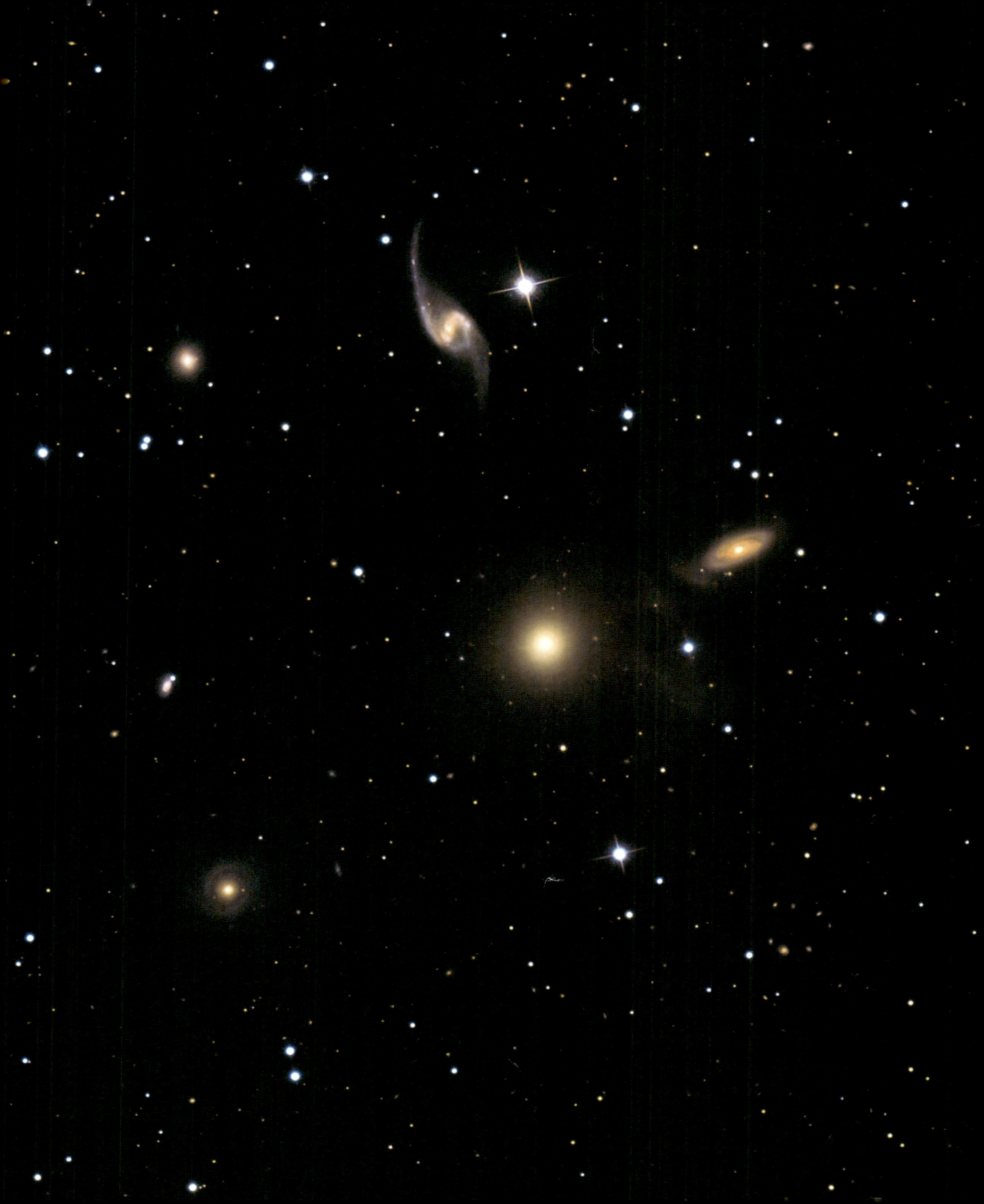

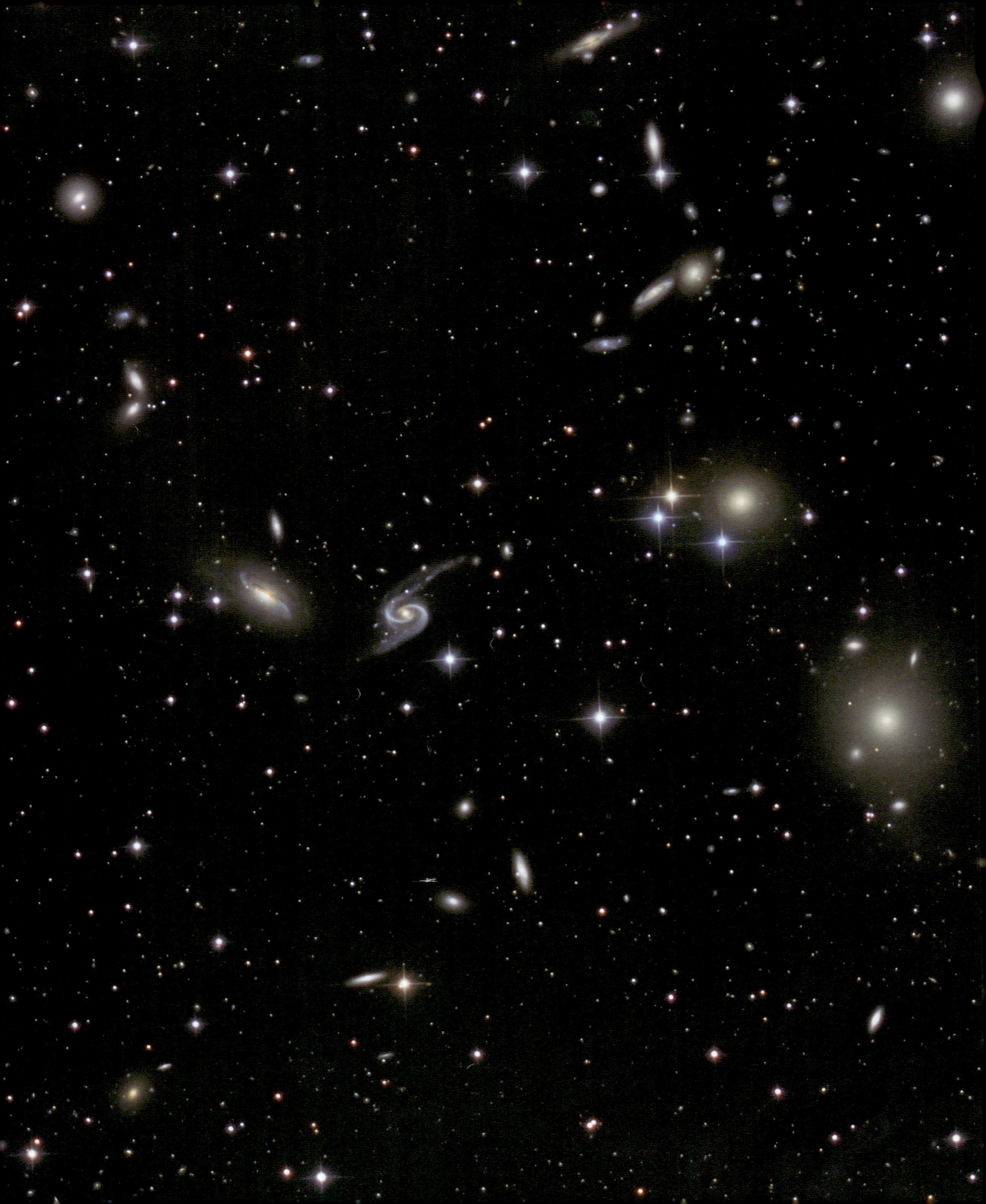

IRREGULAR GALAXIES AND INTERACTING SYSTEMS

It is difficult to determine the original galaxy forms of the systems discussed in this chapter. Compared to other galaxy types, they exhibit a much larger range of possible forms.

THE MORPHOLOGY OF IRREGULAR GALAXIES AND INTERACTING SYSTEMS

In this chapter, irregular galaxies and interacting galaxy systems are considered together. Based on their morphology, which in both cases is characterized by asymmetry and irregularities, the two cases are usually distinguished in astronomy. The examples discussed here, however, demonstrate that one often cannot sharply separate irregular and interacting galaxies. In fact, one often has to take into account combinations of events and interdependencies in the astrophysical interpretation of the visible forms.

About three to four per cent of galaxies cannot be classified as elliptical or spiral since they are lacking the basic structures in their appearance. For example, often a plane of symmetry or clearly defined centre is lacking, which sometimes leads to the observation of collections of large-scale star-forming regions, loose H II regions and individual dust filaments. Their masses are in the range 10^8–10^{10} solar masses and at a few thousand to 30 000 light years, their diameters are very small. The term "irregular galaxy" should not be confused with "peculiar". The latter is used as a qualifier to the standard galaxy types, for example to indicate that a spiral structure exists but is perturbed, but still allows the original classification to be recognized. In the case of irregular galaxies, however, there is no sharp dividing line between perturbation and typical structure, especially since the irregular forms can also be interpreted with respect to the diverseness of their histories.

Interactions with other galaxies are mainly responsible for the appearance of irregular galaxies. There is a broad spectrum of possible interactions. These range from weak perturbations in the gravitational field of two approaching galaxies to stronger tidal interactions which can cause a flow of material to a collision of galaxies which can even end in a merger and produce a new galaxy. The products of such a merger are distinguished by the mass ratio of the galaxies: for similar masses, this is termed a major merger. If the mass ratio is between 1 : 5 and 1 : 10, then it is considered an intermediate merger, and minor mergers have a ratio of less than 1 : 10. In such a merger process, the small, compact partner has only a small influence on the morphology of the larger partner in the interaction. In large spiral galaxies, this type of interaction is often visible only as a warped disc.

The identification of mergers is made difficult by their duration of a few hundred million years, short in cosmological terms. We thus observe only rarely the active phase of the merger, but usually the forerunners of this process; for example, pairs of galaxies which have approached each other to within a few galactic diameters and whose structure already displays perturbations.

Even less violent interactions change the appearance of galaxies. For example, star-formation regions are produced which are visible in images as prominent bright blue regions. If the interacting partners approach each other to within a few diameters, then the systems can orbit each other. During these orbits, long tidal tails with lengths of up to 100 000 light years can be formed out of ripped out gas, dust, and stars, in which new star-forming regions can even occur. These interactions can be well described in simulations, and thus one knows that the collision of two spiral galaxies can form a large elliptical galaxy in which the tidal tails have formed delicate, dim shells which surround the galaxy. In the early universe, the density of galaxies was appreciably larger, which implies a higher probability of interaction with neighbouring galaxies. This is confirmed by the extremely deep view, which thus reaches far into the past, made possible by the Hubble Space Telescope's exposure of the Hubble Deep

The galaxy group around NGC 80 shows a momentary image of galactic interactions (cf. page 161). Image credit: Makis Palaiologou, Stefan Binnewies (600-mm reflector).

Field. Photons were collected over 278 hours from a small area of sky in the constellation of Fornax. The faintest detectable galaxies are seen as they were only 600 million years after the Big Bang. The farther one looks into the past, the more noticeable is the increasing number of irregular systems.

THE ASTROPHYSICS OF IRREGULAR GALAXIES AND INTERACTING SYSTEMS

Studies of the nature of irregular galaxies usually take place on spectacular systems which are in the process of merging or closely approaching and show particularly prominent changes in appearance or long tidal tails.

The probability of an interaction depends on several factors, including the relative velocity of the two galaxies. In the case of a slow approach, tidal interactions can take place over a longer period of time. This increases the probability that material is exchanged or even that a merger takes place. Given a fixed velocity distribution and number of galaxies within a volume of space, it can be calculated how often two interaction partners will merge. The number of mergers increases with time, which reduces the galaxy density and hence the interaction rate. With time, a large initial number of galaxies changes to a small number via mergers, which is often called the merging-tree history. The tree-like structure becomes visible when one plots the number of galaxies as a function of time.

An important interaction quantity – the relative velocity – is partially determined by dark matter. The larger its fraction compared to luminous matter, the higher are the velocities and the shorter the interaction times. Additionally, large halo masses cause stronger gravitational bonding of dust and gas. This in turn prevents the formation of tidal tails in the disc, so that halo-dominated galaxies display fewer perturbation effects despite the presence of interactions.

One knows from analyses of the cosmic microwave background and the distribution of galaxies at high redshifts that most matter is non-baryonic. The length scale of the distribution of fluctuations in the intensity of the microwave background also lies in the sub-galactic range, which means that density fluctuations in dark matter are found in our Galactic back garden. This behaviour is predicted by so-called cold-dark-matter-models (CDM). The idea is that halos fill a volume which contains a galaxy and its luminous components. Dark matter also surrounds groups and larger conglomerations and also the halos of large galaxy clusters. The current understanding of the distribution of dark matter on sub-galactic length scales (lengths < 100 000 light years) is based on simulations of structure formation. One follows the evolution of CDM structures and attempts to make predictions. Models with high spatial resolution have their smallest structures in the mass range 10^{-4}–10^{-5} of the halo mass, which makes it possible to study the effects of the small-scale structures within a galaxy during gravitational interactions.

A model popular among astronomers is that of a large central galaxy which is orbited by companions. One can study the effects of dynamic friction, which occur when smaller galaxies approach the central system too closely and lose material over the course of several orbits, which then collects into tidal tails. The step-by-step merging with these satellite galaxies is the basic principle of hierarchical growth in which galaxies, galaxy groups, and galaxy clusters arise in several steps. All models are based on the idea that the interaction is defined by the presence of dark matter. Structures visible in images – the galaxies, their spiral arms, or tidal tails – are the smaller part consisting of luminous matter which is gravitationally bound to the structures of dark matter. They follow the dynamics of the dark matter and are thus indicators of its existence. Comparing simulations to observations is not always easy. Up until about 1998, there was the problem of too few details: If one compared the substructures of dark matter in the models with those which are actually observed, one noticed that the models had too little small-scale structure. Astronomers traced this to the large number of mergers in this model, so-called overmerging, dissolving the structure at small scales. The model thus lacked galaxies at small scales and instead the distribution appeared more or less smooth. This is now known to have been a problem of insufficient numerical resolution in the simulations. Interestingly, a current problem is the opposite: too much small-scale structure. This obviously can't be caused by insufficient resolution. The cause is still unclear; perhaps it is due to not incorporating enough physics into the simulations, though it is also possible that at least some dwarf galaxies have been overlooked observationally.

It is important to investigate what other effects play a role in mergers of galaxies. Apart from the effects of dark matter, these are classical physical parameters which also play a role in star formation. For example, stars are born in gas clouds, which in turn move and heat gas via their supernova winds. This leads to bubbles of cold or hot gas. They distribute their material not only in the galaxy, but can also influence the accretion of material in the interaction partner.

Stellar mass in a galaxy can increase either by accretion or directly via star formation in cool gas clouds. Star formation plays an important role only in low-mass galaxies; massive galaxies can form only through mergers, since otherwise one cannot explain the amount of luminous matter. In mergers

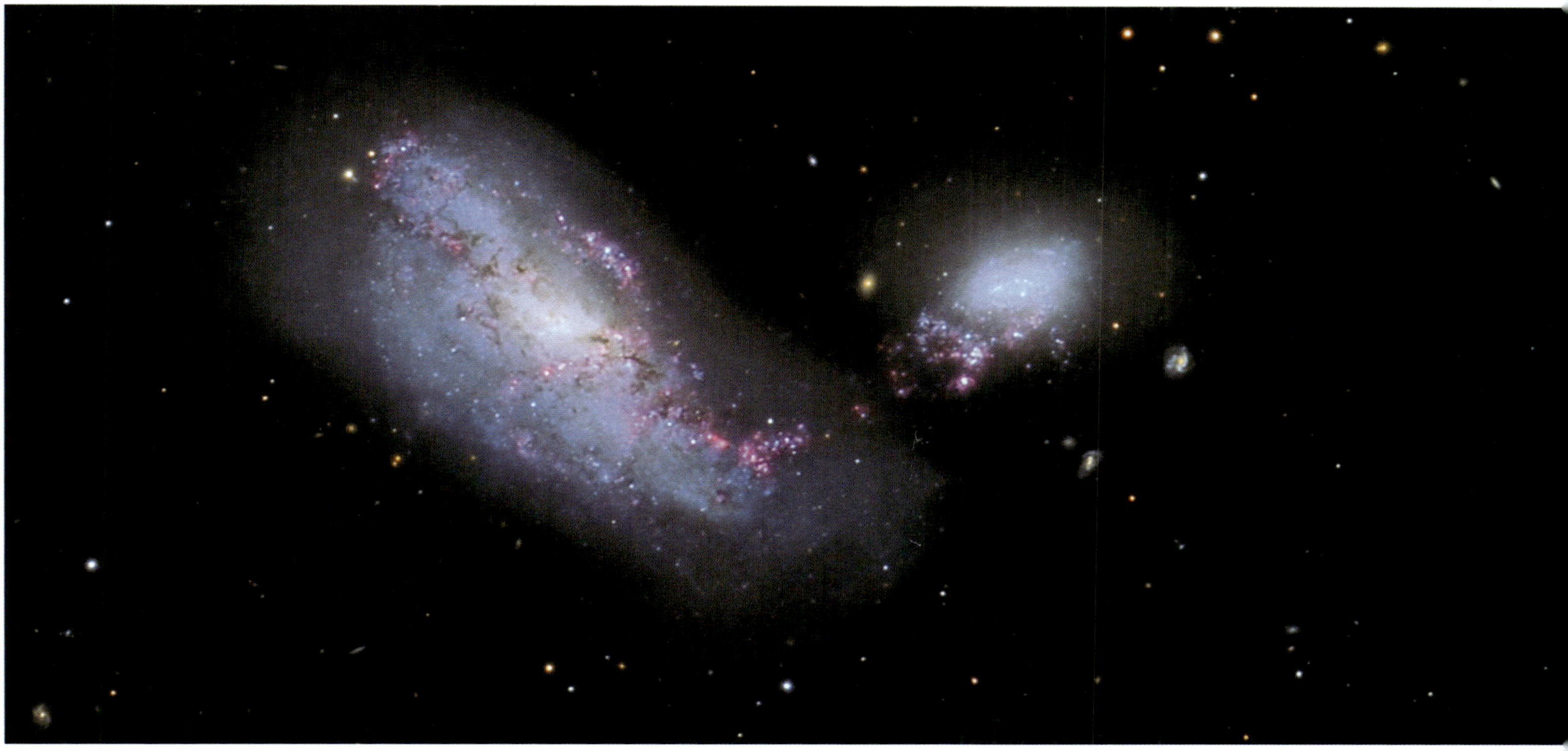

Figure 4.1: The active regions of the system Arp 269 are indicators of tidal interaction and the associated transport of material (see page 187).

there is an important distinction which takes into account whether the galaxies involved are gas-rich or gas-arm; the latter case is known as a dry merger. Since more material is involved in gas-rich galaxies, one usually sees more structure in the surroundings, such as tidal tails.

Galaxy formation began about 13 billion years ago. Today, there are two principally different models to explain this process. In the literature they are described as "top-down" and "bottom-up". The top-down model assumes that galaxies are formed from the collapse of gas clouds whose size is larger than a galaxy due to the fact that the self-gravitation is larger than the gas pressure in the gas cloud. If the cloud formed by the collapse has enough rotational energy, first a disc and then later a spiral galaxy forms. Slow rotation leads to the forerunner of an elliptical galaxy. In this structure-formation scenario, however, there is a feedback effect, since on the one hand gas is used up by the initial star formation but on the other hand stellar winds and supernovae in the first, very massive, stellar population work against the collapse of the gas clouds. A separation between cause and effect is hence difficult, especially when trying to estimate the timescale in this model. An important aspect of this model is the possibility that large gas clouds form smaller cloud fragments when collapsing, out of which galaxies were formed. This is a plausible explanation for the fact that many galaxies can be observed today in groups and clusters.

The problem of the "top-down" model is the very long time which is necessary for the collapse and fragmentation. According to mathematical models, this must still be happening today, which contradicts the fact that today one also observes many clusters and even larger objects, so-called superclusters, which have already reached dynamic equilibrium.

Especially with regard to these large collections of galaxies, the bottom-up model offers many advantages in explaining structure formation. This model assumes a growth process which takes place in several steps, so-called hierarchical clustering. This process begins with simple galaxies which merge with one another and with time form larger galaxies and structures. An important role in these interactions is played by dark matter which provides the gravitation for the luminous matter which we can see and thus must be taken into account in both models. One must also be able to explain exotic effects, such as I Zwicky 18, a nearby dwarf galaxy whose star formation started only about 500 million years ago. It is puzzling how this galaxy could exist for more than 12 billion years in a pristine state while all around it galaxy evolution was taking place. A possible, but just as exotic, explanation is that I Zwicky 18 is a so-called dark galaxy, consisting mainly of dark (but at least partially baryonic) matter. It would be located in a pocket of dark matter which preserves the original state of the early universe. This original gas of hydrogen and helium would not have formed any stars and was only later caused to light up. There are also bottom-up models which predict that there should be many of these dark galaxies. However, they would be difficult to observe since they have a low luminosity. I Zwicky 18 is a dwarf galaxy and thus agrees well with this dark-galaxy hypothesis.

In general, in bottom-up scenarios, galaxies are formed step-by-step via the merger of smaller building blocks. These are gas clouds of a few million solar masses which collapsed in an early phase of the universe. Due to gravitational interactions, larger structures such as galaxy clusters and then superclusters formed. Mergers of galaxies always lead to larger galaxies and thus the number of galaxies always decreases with increasing mass. The consequence of bottom-up models that one should see more small than large galaxies in the past is confirmed by astronomical observations. If one looks out into the universe, then one also looks back in time, and can thus observe for example a galaxy ten billion light years away at an age of about three billion years. There are many examples of interacting galaxies in which one can observe such formation processes "live" and which thus confirm the model.

DARK MATTER AND DARK ENERGY

Dark matter reveals itself only via its gravitational action. The term "dark" describes the fact that it doesn't glow, i.e. neither does it interact with electromagnetic radiation. It is also often referred to as non-baryonic matter. This explains what dark matter is not, but its composition is unknown. The baryons mentioned are strongly interacting elementary particles which consists of three quarks. The most common representatives are protons and neutrons, out of which atomic nuclei are made. Baryonic matter glows across the entire electromagnetic spectrum and can be observed in the form of dust, gas, and stars.

The mass of the universe can be estimated by comparing predictions of the standard cosmological model with data from satellites. From the map of the temperature fluctuations in the cosmic microwave background one can determine whether the distribution is isotropic or anisotropic. This is done by measuring the intensity of the three-kelvin-radiation in the microwave region at wavelengths of 0.1–10 mm. The very small differences in intensity (about 10^{-5}) of two points on the sky can be related to their angular separation and one finds a significant departure from instrumental noise. The results of the satellite missions indicate a mass fraction in the universe of four to five per cent for baryons, 21–27 per cent for non-baryonic matter and something more than 70 per cent for the mass contribution of dark energy.

Dark energy is formally the cosmological constant introduced into the theory of relativity by Albert Einstein and can be described as the energy of the vacuum, which acts in an opposite manner to gravitation. Its gravitational interaction is thus different from that of the other mass contributions. Dark energy thus acts as a repulsive force at very large distances and leads to the acceleration of the expansion of the universe. This discovery superseded the previous assumption that the expansion of the universe must be slowing down because of the action of gravity. Although one can formally describe the behaviour of dark energy mathematically, it is not yet clear what drives this expansion. Astronomers hope to learn more from new satellite missions such as ESA's Euclid telescope (planned for 2020), which will map the three-dimensional distribution of billions of galaxies and thereby look back ten billion years into the past.

What could a combination of these two opposite descriptions look like? First, one assumes that dark matter was distributed homogeneously at the beginning of the universe. Large clumps led to structures which in turn led to clusters and superclusters of dark matter, often connected by a network of filaments. At the locations where these filaments interacted with one another, luminous baryonic matter became concentrated and galaxies formed. If the concentration was especially large, so astronomers believe today, elliptical galaxies were formed. Regions of lower density, on the other hand, are seen as the birthplace of spiral and dwarf galaxies. In 2007, a map of dark matter was constructed which confirms this scenario: the farther away the concentration of dark matter, i.e. the farther back we look into the past, the more uniform and less fragmented is the distribution. One also sees that the visible matter is an indicator of the density concentrations of dark matter, and that its clumpiness has increased with time.

Even though the models explain well the formation, interactions, and diversity of irregular galaxies, there is still a problem which has concerned science in the last few years. If one adds up the number of known galaxies, for example extrapolating from the 10 000 galaxies detected in the Hubble Ultra Deep Field, then one arrives at the number of 200 billion galaxies distributed across the sky. Although this number seems astronomically large, it corresponds to only a tenth of the luminous baryonic matter which was produced in the Big Bang. This baryonic matter should not be confused with dark matter or dark energy, which together make up almost 96 per cent of the mass of the universe. The problem is that we see only about a tenth of the four per cent of the universe which is baryonic. The origin and evolution of galaxies uses and processes only this ten per cent and leaves the rest alone. Astronomers suspect that a possible solution of the riddle of this missing matter are the large spaces between galaxies in galaxy clusters. Their cluster volumes are filled with hot plasma. Taken together, the plasma contribution of clusters, groups and intergalactic space make up about 50 per cent of the total amount of baryonic matter. The dilemma is that the remaining 50 per cent of the baryons are not observed. This could be due to the fact that these baryons are bound to large-scale structures, perhaps gravitationally to a network of dark matter, allowing small groups and clusters of baryons to form. Although this gas is has a temperature of 100 000 K, in contrast to the hot plasma in galaxy clusters it is still too cold to emit X-rays and thus cannot be directly detected. Another hypothesis assumes that the missing 50 per cent of the baryons are to be found in stars which have been ejected from galaxies via interaction processes. Current observations already indicate hints that this idea might be correct. For example, studies of the extragalactic background light have shown that a significant fraction of the diffuse extragalactic background light in the optical and near-infrared bands stems from stars which are located outside of galaxies.

In addition to this fraction of baryonic matter which is not bound to galaxies there is a small part which is part of a cycle in which galaxies are an intermediate stage. Galaxies do not consist of a fixed amount of matter, but are the result of a constant circulation of baryons. The physical processes of star formation and flows of gas into galaxies which is then ejected as hot gas maintain a constant exchange with the intergalactic medium. If one observes an irregular galaxy, then one observes the normal state of this cosmic exchange process; a comparatively inactive galaxy such as our Milky Way is then more of an exception.

BIBLIOGRAPHY

Beygu, B. et al.: *An Interacting Galaxy System Along a Filament in a Void*, 2013, http://arxiv.org/pdf/ 1303.0538v1.pdf

Galagher, J. S. and Hunter, D. A.: *Structure and Evolution of Irregular Galaxies*, Annual Review of Astronomy and Astrophysics, **22**, 1984

Kennicutt, R. C.: *Star formation in galaxies along the Hubble Sequence*, 1998, http://ned.ipac.caltech.edu/level5/Sept02/Kennicutt/frames.html

Konitzer, F.: *Dunkler Kosmos – Die anderen 96 Prozent*, 2013, www.spektrum.de/alias/dunkler-kosmos/die-anderen-96-prozent/1180532

Naab, T.: *Struktur und Dynamik wechselwirkender Galaxien*, 2000, http://archiv.ub.uni-heidelberg.de/volltextserver/1412/1/thesis.pdf

Palma, C.: *Galaxy Interactions*, 2011, www.e-education.psu.edu/astro801/content/l9_p6.html

Space Telescope Science Institute: *Hubble Finds Ring of Dark Matter*, 2007, http://hubblesite.org/newscenter/archive/releases/2007/17/full

Space Telescope Science Institute: *Hubble Maps the Cosmic Web of "Clumpy" Dark Matter in 3-D*, 2007, http://hubblesite.org/newscenter/archive/releases/2007/01/full

Struck, C.: *Galaxy Collisions*, 1999, www.public.iastate.edu/~curt/cg/homepage.html

Theis, C.: *Measuring Dark Matter Halos by Modeling Interacting Galaxies*, Proceedings of the International Astronomical Union, **220**, 2004

Vorontsov-Velyaminov, B. A.: *The Atlas and Catalogue of Interacting Galaxies*, 1959, http://ned.ipac.caltech.edu/level5/VV_Cat/frames.html

Whittle, M.: *Galaxy Interactions and Mergers*, 2011, www.astro.virginia.edu/class/whittle/astr553/Topic12/Lecture_12.html

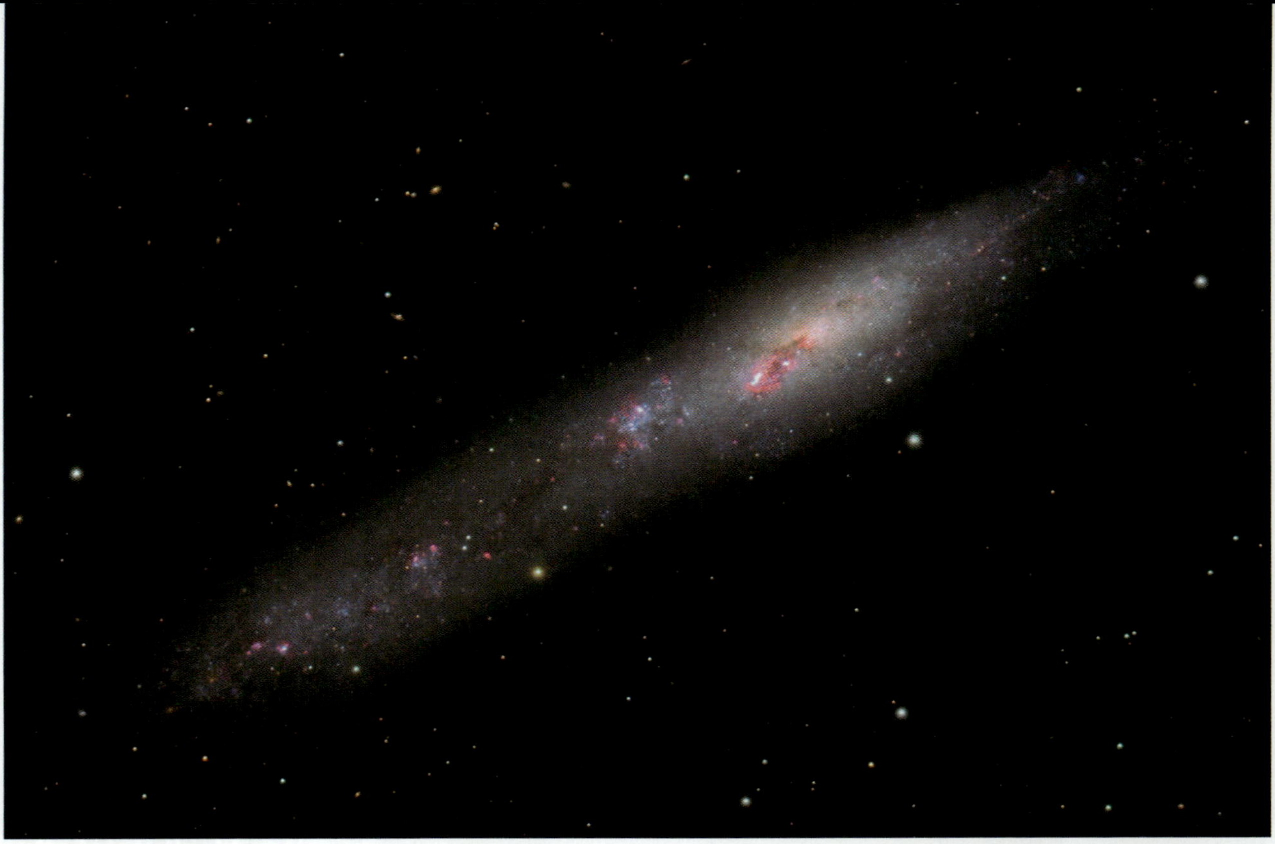

NGC 55

This galaxy, lying far to the south at a declination of –39°, is an SB galaxy seen edge-on, with an impressive angular size of 32.4′ × 5.6′. The diameter of NGC 55 is thus 64 000 light years. Its luminosity distribution is not symmetric and shows a brightening in the upper regions which points towards the hidden centre of the galaxy. The larger part of the disc plane of the galaxy, to the left side of the centre, has a rectangular shape at the boundary to the halo and suggests that if seen face-on, NGC 55 would display perturbations which due to the unequal distribution of matter relative to the centre would support its classification as an irregular galaxy.

The distance to NGC 55 can be determined with standard methods only imprecisely as five to eight million light years since the recession velocity of 130 km/s is too small to allow a clear measurement. NGC 55 has eight large H II regions which are visible in the image as bright red nebulae and lie in the region of bright stars and star clusters. Astronomers have also found Wolf-Rayet stars in this area. These are highly evolved, hot stars which have lost their outer envelopes. Their highly energetic radiation ionizes the outflowing stellar winds and makes itself known via broad He II emission lines in the spectrum. Apart from Wolf-Rayet stars, one can also use the luminosity distribution of planetary nebulae as so-called standard candles for distance determination. This makes use of the fact that particularly bright planetary nebulae are rare and the distribution function has a typical drop, which in NGC 55 lies at a magnitude of 22.5.

For a long time, astronomers assumed that NGC 55 was part of the Sculptor galaxy cluster, located at a distance of nine million light years. However, the methods of distance determination mentioned above using standard candles leads to the conclusion of a probable distance of 7.2 million light years. Thus NGC 55 is in front of the Sculptor Group and, together with its satellite galaxy NGC 300, forms a cosmic link between the Local Group and the closest neighbouring galaxy group.

OBJECT	NGC 55
CONSTELLATION	Sculptor
RIGHT ASCENSION	$00^h\ 14^m\ 54^s$
DECLINATION	–39° 11′ 48″
BRIGHTNESS	8.8 mag
TYPE	SB(s)m: edge-on
PHOTOGRAPHERS	Bernd Flach-Wilken, Volker Wendel
TELESCOPE	400-mm reflector
CAMERA	SBIG STL-6303
EXPOSURE TIME	142.5 min
LOCATION	Tivoli Farm, Namibia

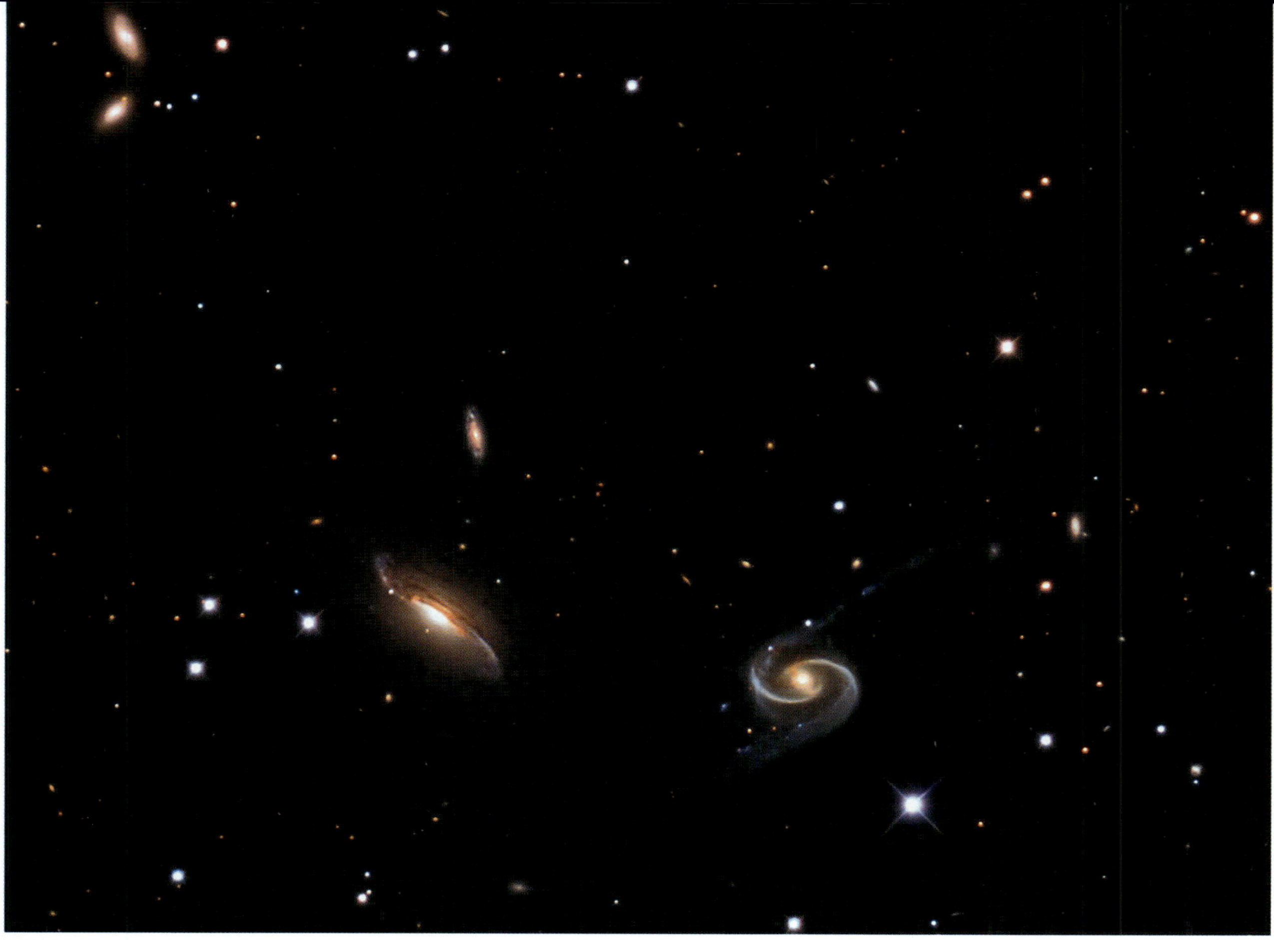

Arp 65

This galaxy pair, consisting of NGC 90 (right) and NGC 93 (left) is also known as Arp 65 and is part of the NGC 80 galaxy group. The SAB(s) galaxy NGC 90 is notable for its spiral arms which are almost linear over a distance of 2′. At its distance of 246 million light years, this corresponds to 143 000 light years. The classification of NGC 93 is more difficult since this galaxy is seen edge-on and in addition to two spiral arms a surrounding dust lane can also be seen. It also has a long, dim structure, presumably a tidal tail, pointing away from the common centre.

The redshifts of the two galaxies have been measured very precisely and show a small relative velocity of only 27 km/s. Both galaxies are thus at the same distance from the Milky Way. The angular separation is 2.77′ and the projected distance between the galaxy centres thus corresponds to the physical distance of 198 000 light years. Since the spiral arms of the two galaxies have a separation of only 115 000 light years from each other, one can assume that tidal forces occur in the galaxies. However, one observes no bridge of material nor tidal tails in the image, either of which would suggest an even closer approach or orbital motion. Data from the ultraviolet satellite GALEX (Galaxy Evolution Explorer) are not very significant for NGC 93 but indicate detailed structures in NGC 90. This UV radiation stems from young, hot stars which lie in the spiral arms. The GALEX images show that star-forming regions are present in the outer regions of the spiral arms of NGC 90, which are beginning to show dissolution and are fanning out. It is thus plausible that there is an outflow of gas here and that further interaction would lead to the formation of tidal tails.

OBJECT	Arp 65, NGC 90 / NGC 93
CONSTELLATION	Andromeda
RIGHT ASCENSION	$00^h\ 21^m\ 51^s$
DECLINATION	+22° 24′ 00″
BRIGHTNESS	14.5 mag
TYPE	SAB(s)c pec
PHOTOGRAPHERS	Makis Palaiologou, Stefan Binnewies
TELESCOPE	1.3-m reflector
CAMERA	Andor DZ 436
EXPOSURE TIME	315 min
LOCATION	Skinakas Observatory, Crete, Greece

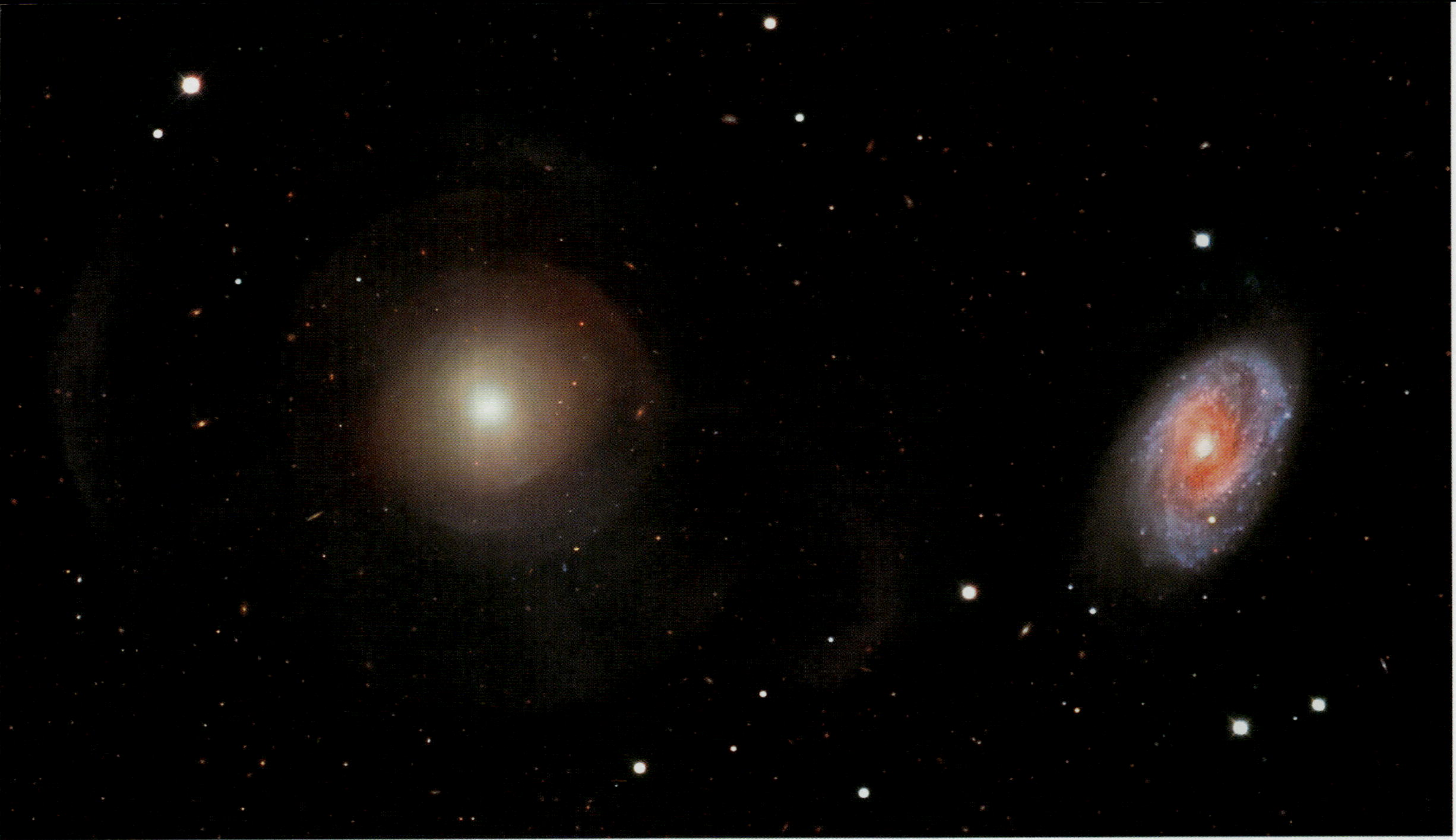

Arp 227

The SA0 galaxy NGC 474, seen at the left of the image next to the SA(rs)b galaxy NGC 470, is surrounded by bright shell structures. Farther to the west and east of these shells are additional arc-shaped patches of low luminosity. These could be tidal tails which indicate a complex history for this galaxy pair. Both galaxies are at a distance of 110 million light years. Since their relative velocity is only 49 km/s, astronomers assume that they are an interacting pair (Arp 227). The projected distance between the galaxy cores is 5.2′ which corresponds to 166 000 light years. Due to the shell structure of NGC 474, it is suspected that a merger with a compact partner took place.

Analyses of the stellar colours in the outer regions of NGC 474 have shown in many studies that the two outer portions have the same colour as NGC 470. The ratio of blue to red luminosity is here between 0.8 and 1.0. NGC 470 orbits about NGC 474 and regularly loses material. The outer regions are thus presumably tidal tails formed out of matter lost by NGC 470. The inner shell structures differ significantly in colour from the outer ones. The ratio here is 1.3; they are thus less blue, which means that they are not formed out of material from NGC 470. A study of the inner 5″ of the nuclear region of NGC 474 ($R < 2600$ light years) indicates that the nucleus has unusual kinematics and orbits around the geometric centre. NGC 474 thus doesn't have a core typical of an S0 galaxy, but is rather a triaxial object, which could be an indication of a minor merger. This merger would have also formed the inner shells in NGC 474.

OBJECT	Arp 227, NGC 474 / NGC 470
CONSTELLATION	Pisces
RIGHT ASCENSION	$01^{h}\ 20^{m}\ 07^{s}$
DECLINATION	+03° 24′ 55″
BRIGHTNESS	12.4 mag
TYPE	SA0(s) / SA(rs)b
PHOTOGRAPHERS	Makis Palaiologou, Stefan Binnewies
TELESCOPE	1.3-m reflector
CAMERA	Andor DZ 436
EXPOSURE TIME	410 min
LOCATION	Skinakas Observatory, Crete, Greece

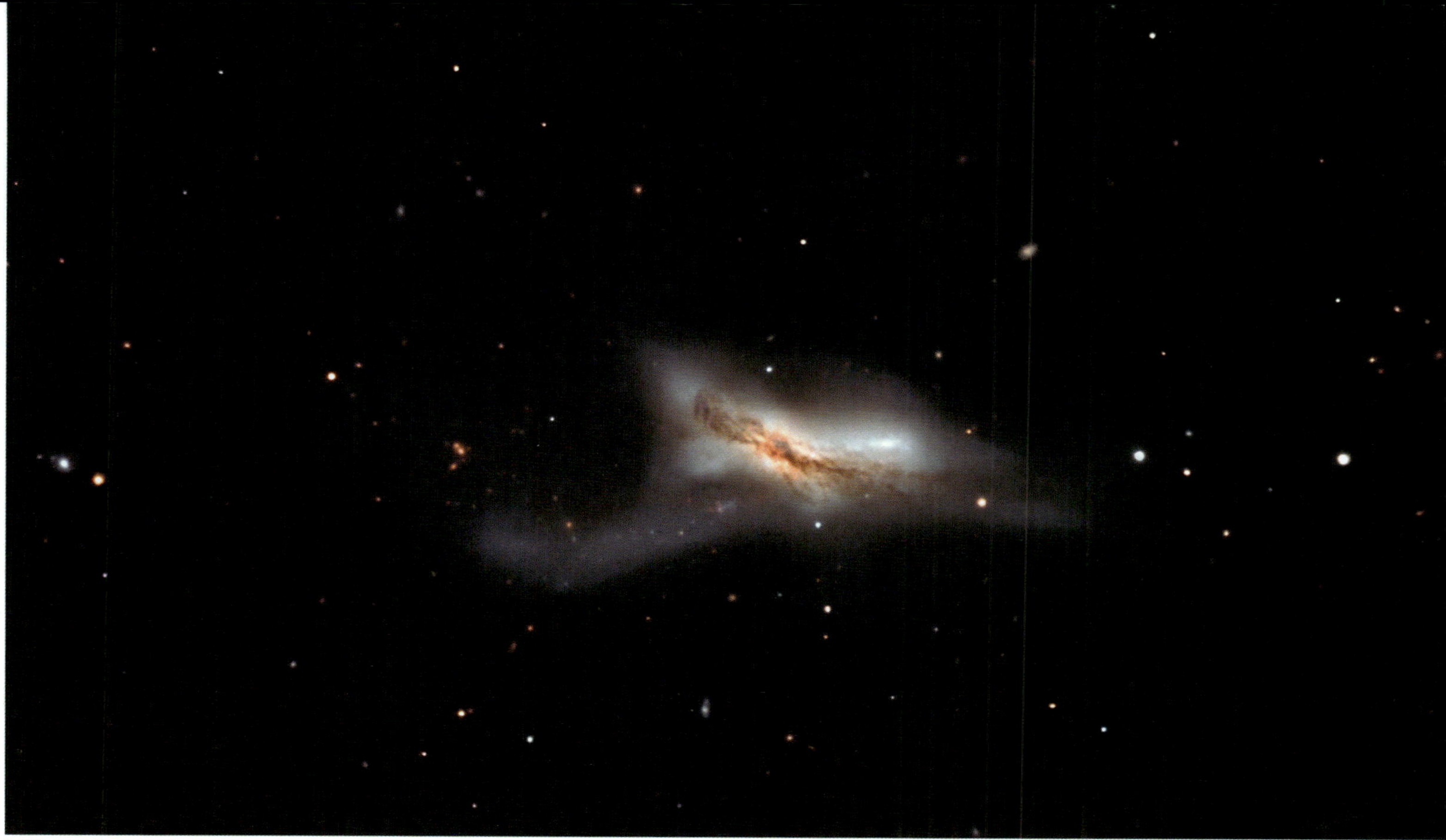

Arp 157

NGC 520 is an irregular galaxy which is catalogued as Arp 157 and often described as a system of two colliding galaxies. In the image, the horizontal direction corresponds to southeast–northwest. At both ends of the galaxy one can recognize tidal tails, whereby the tail at the left, southeastern end is more extended and fades into a dim appendage. The spiral planes of the two merging galaxies are tilted with respect to each other, reflected in the literature by the name NGC 520a for the northwestern part and NGC 520b for the southeastern part with the dust lanes. The system is at a distance of 105 million light years and has an angular size of 4.5′ × 1.7′ from which one can derive a diameter of 137 000 light years.

Using radio observations of atomic hydrogen, astronomers detect two disc planes which are almost perpendicular to the orbital plane in which the two galaxies moved before the collision. This configuration facilitates the exchange of gas, dust and stars and thus aids in the formation of the irregular morphology. The centre of rotation of the entire system is near NGC 520b which indicates a larger central mass in this precursor galaxy. The radio data also show a trail of atomic hydrogen which emerges from the left side of NGC 520b and leads to the right side in a wide spiral arc. The endpoint of this arc is outside of the image and contains a condensation of material with a size of about 1′. This is a tidal dwarf galaxy. The arc is interpreted as the remains of the last orbit of the precursor galaxy. This orbit ended 300 million years ago with the collision and penetration of the two galaxies and finally the formation of the parts visible today with the names NGC 520a and NGC 520b.

OBJECT	Arp 157, NGC 520
CONSTELLATION	Pisces
RIGHT ASCENSION	$01^h\ 24^m\ 35^s$
DECLINATION	+03° 47′ 33″
BRIGHTNESS	11.3 mag
TYPE	pec
PHOTOGRAPHERS	Josef Pöpsel
TELESCOPE	600-mm reflector
CAMERA	SBIG ST-10XME
EXPOSURE TIME	80 min
LOCATION	Amani Lodge, Namibia

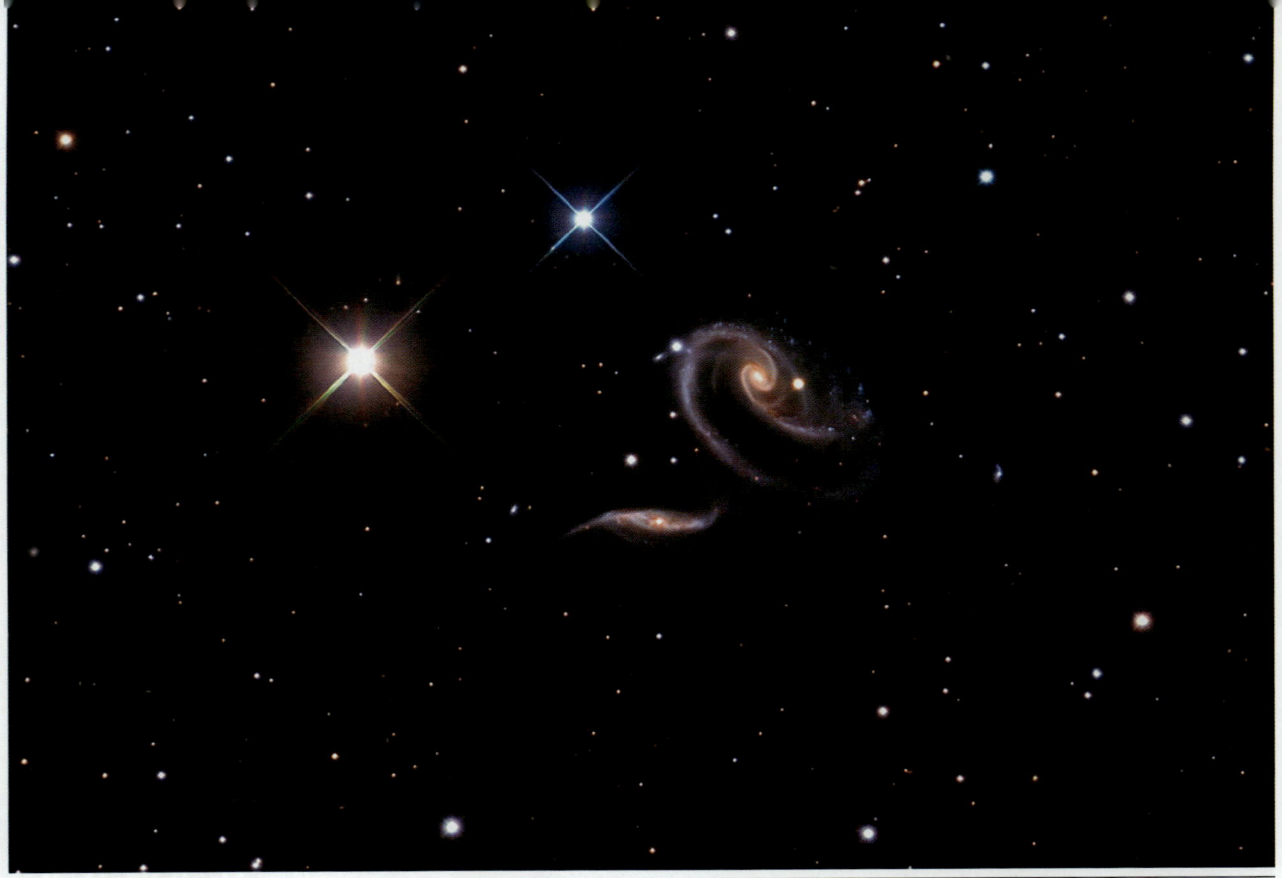

Arp 273

Close to a pair of stars (8.8 mag and 10.3 mag; the brighter of the two stars is a binary star, which is visible in the double diffraction spikes) in the constellation of Andromeda are the spiral galaxies UGC 1810 and UGC 1813. They were catalogued by Halton Arp as number 273 in his *Atlas of Peculiar Galaxies* (1966). UGC 1810 is the northwest face-on partner, classified as SA(s)b pec; the smaller galaxy UGC 1813, of type SB(s)a, is seen more edge-on. The spiral arms in UGC 1810 are very narrow and turn into tidal tails, with the outer ring-shaped structure making almost a complete revolution. Dust lanes are visible within it; the side facing UGC 1810 shows many active regions with young stars. The angular sizes of the two interacting galaxies are only 1.2′ and 1.8′; their separation is 1′. At a distance from the Milky Way of 330 million light years, this corresponds to 96 000 light years.

The deciding role in the interactions is played by dynamical impact parameters such as the direction and position of impact and the relative velocity. Astronomers have used N-body simulations to study these effects, describing the stars as collisionless bodies and the gas and dust masses as a hydrodynamic frictional component. It could be shown that UGC 1810 and UGC 1813 are involved in a high-speed interaction in which a compact partner (UGC 1810) with low mass (15 per cent of the total mass) passes close to the larger galaxy and causes asymmetries. A result is a shift in the centre of mass of the disc in UGC 1810 relative to its halo. The smaller UGC 1813 thus causes, over the course of many orbits, the centre of mass of UGC 1810 to wander along a random orbit with respect to the centre of its halo. This orbital motion of the centre of UGC 1810 reminds one of Brownian motion and attains distances of a few hundred light years from the centre.

OBJECT	Arp 273, UGC 1810 / UGC 1813
CONSTELLATION	Andromeda
RIGHT ASCENSION	$02^h\ 21^m\ 31^s$
DECLINATION	+39° 21′ 58″
BRIGHTNESS	13.7 mag
TYPE	SA(s)b pec / SB(s)a pec
PHOTOGRAPHERS	Makis Palaiologou, Stefan Binnewies
TELESCOPE	1.3-m reflector
CAMERA	Andor DZ 436
EXPOSURE TIME	172.5 min
LOCATION	Skinakas Observatory, Crete, Greece

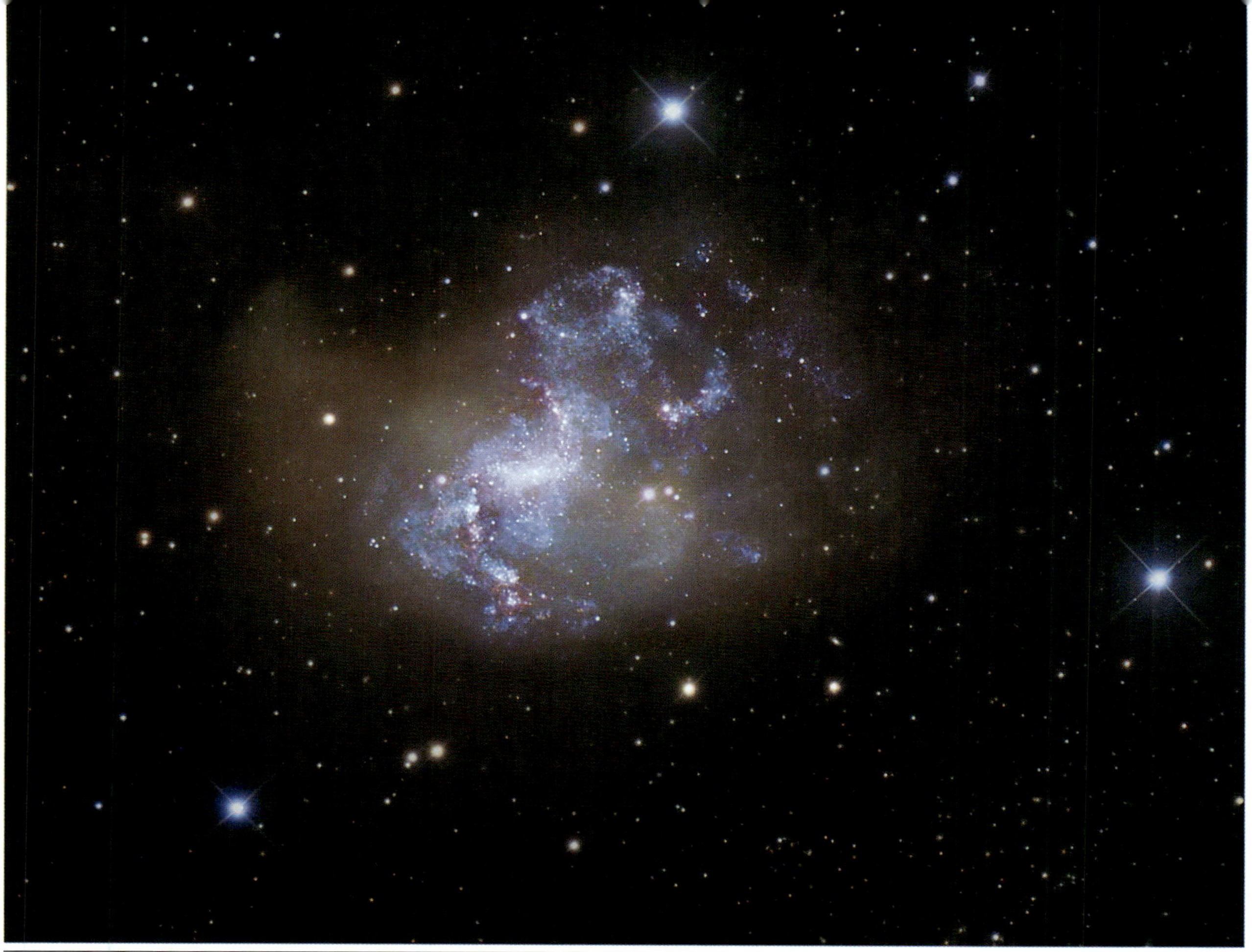

NGC 1313

At the southern end of the constellation of Reticulum is the SB galaxy NGC 1313 with a highly complex morphology. The bar in the centre runs in the north–south direction, corresponding to the horizontal in the image (north left, south right). NGC 1313 is at a distance of 12.8 million light years in the outer edges of the Local Group. Catalogues list a size of 9.1′ × 6.9′, whereby these dimensions also include the dim extensions of NGC 1313, and result in a diameter of 34 000 light years. The bar has a length of 1.6′, which corresponds to 16 000 light years. The interrupted course of the two spiral arms display many hot, bright stars and numerous H II regions whose star-formation rate exceeds that of the Magellanic Clouds. Here are also a few well known ultraluminous X-ray sources. Observations of the bright stars in NGC 1313 were able to identify 70 Wolf-Rayet stars based on their emission-line spectra.

The outer halo of NGC 1313 has a very irregular form. There are perturbations in the brightness profile with slight colour differences, such as at the left edge. These structures are presumably due to tidal tails which formed during an interaction in the past. Since there is no indication of a second nucleus in NGC 1313, the structure of NGC 1313 could have formed during several revolutions of a now completely dissolved dwarf galaxy. The outer regions of the spiral arms were fragmented in the process and since only little gas was transferred, the induced star formation is concentrated in the existing spiral arms.

OBJECT	NGC 1313
CONSTELLATION	Reticulum
RIGHT ASCENSION	$03^h\ 18^m\ 16^s$
DECLINATION	−66° 29′ 54″
BRIGHTNESS	9.2 mag
TYPE	SB(s)d
PHOTOGRAPHERS	Robert Gendler
TELESCOPE	368-mm reflector
CAMERA	SBIG STL-11000
EXPOSURE TIME	900 min
LOCATION	Fair Dinkum Skies, Australia

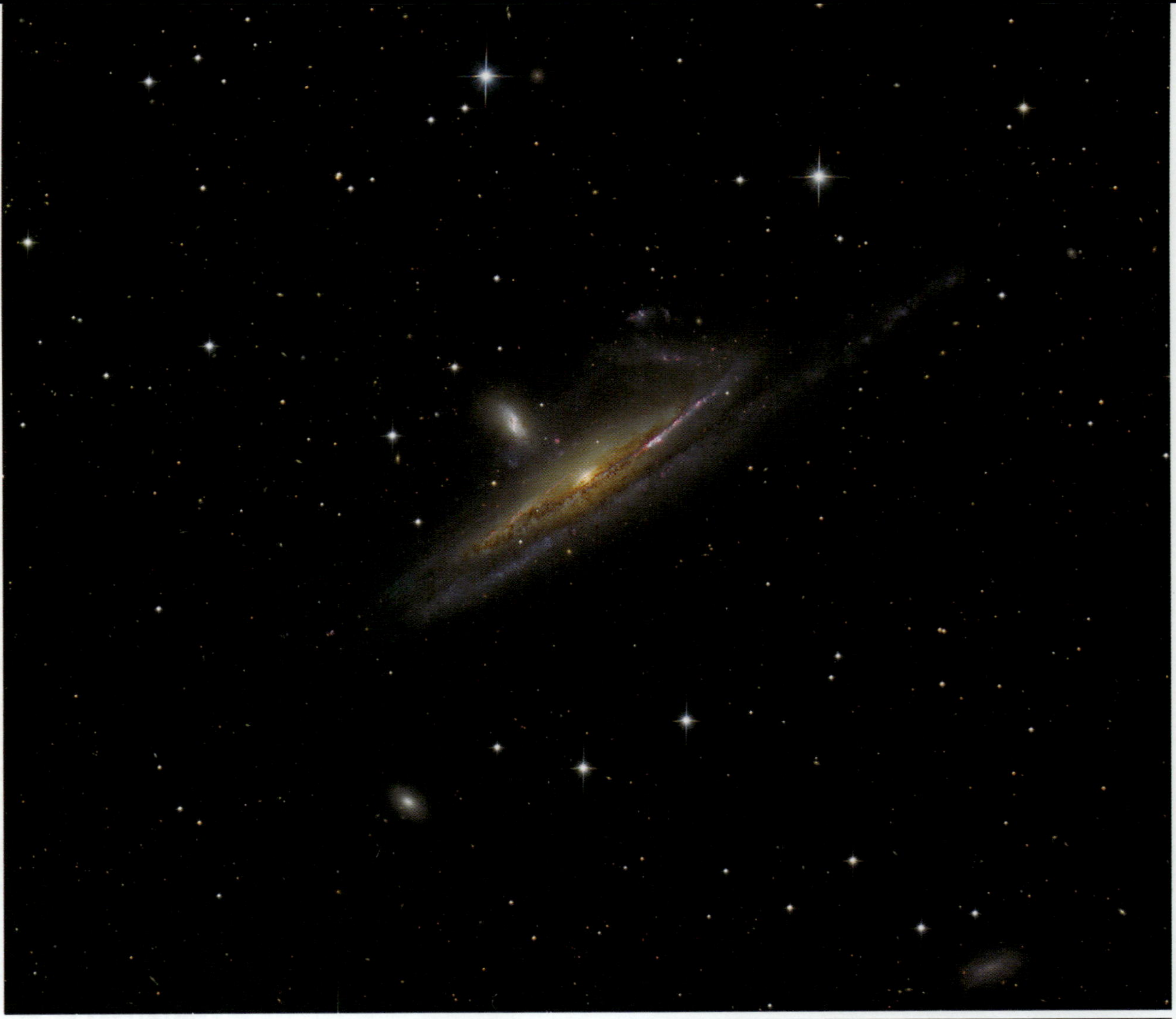

NGC 1532

The SB(s)b galaxy NGC 1532 is one of the largest spiral galaxies. It is at a distance of 55 million light years from the Milky Way, and from its angular size of 12.6′ × 3.3′ follows the enormous diameter of more than 200 000 light years. Northwest of the centre at a distance of 1.7′ is the compact S0 galaxy NGC 1531. Along the dust lane in NGC 1532 are many active, purple-red regions, whereby the prominent region at the northwestern end of the dust lane has a luminosity which, due to the edge-on view, exceeds even that of the nucleus of the galaxy.

The difference in recession velocity between the two galaxies is only 71 km/s so that one can assume that the galaxies form an interacting pair. This is supported by perturbation effects which one can see in NGC 1532. The front spiral arm appears to be stretched; the southwestern arm, enclosing the centre, also displays a longitudinal structure pointing out of the plane. These structures are Vorontsov-Velyaminov rows which are due to a perturbation in the propagation of density waves in the spiral plane and are caused by the companion NGC 1531.

OBJECT	NGC 1532
CONSTELLATION	Eridanus
RIGHT ASCENSION	$04^h\ 12^m\ 04^s$
DECLINATION	–32° 52′ 27″
BRIGHTNESS	10.7 mag
TYPE	SB(s)b pec edge-on
PHOTOGRAPHERS	Philipp Keller, Konstantin Buchhold, Bernd Flach-Wilken, Johannes Schedler, Volker Wendel (Chart 32-Team)
TELESCOPE	800-mm reflector
CAMERA	FLI Proline 16803
EXPOSURE TIME	1062 min
LOCATION	CTIO, Chile

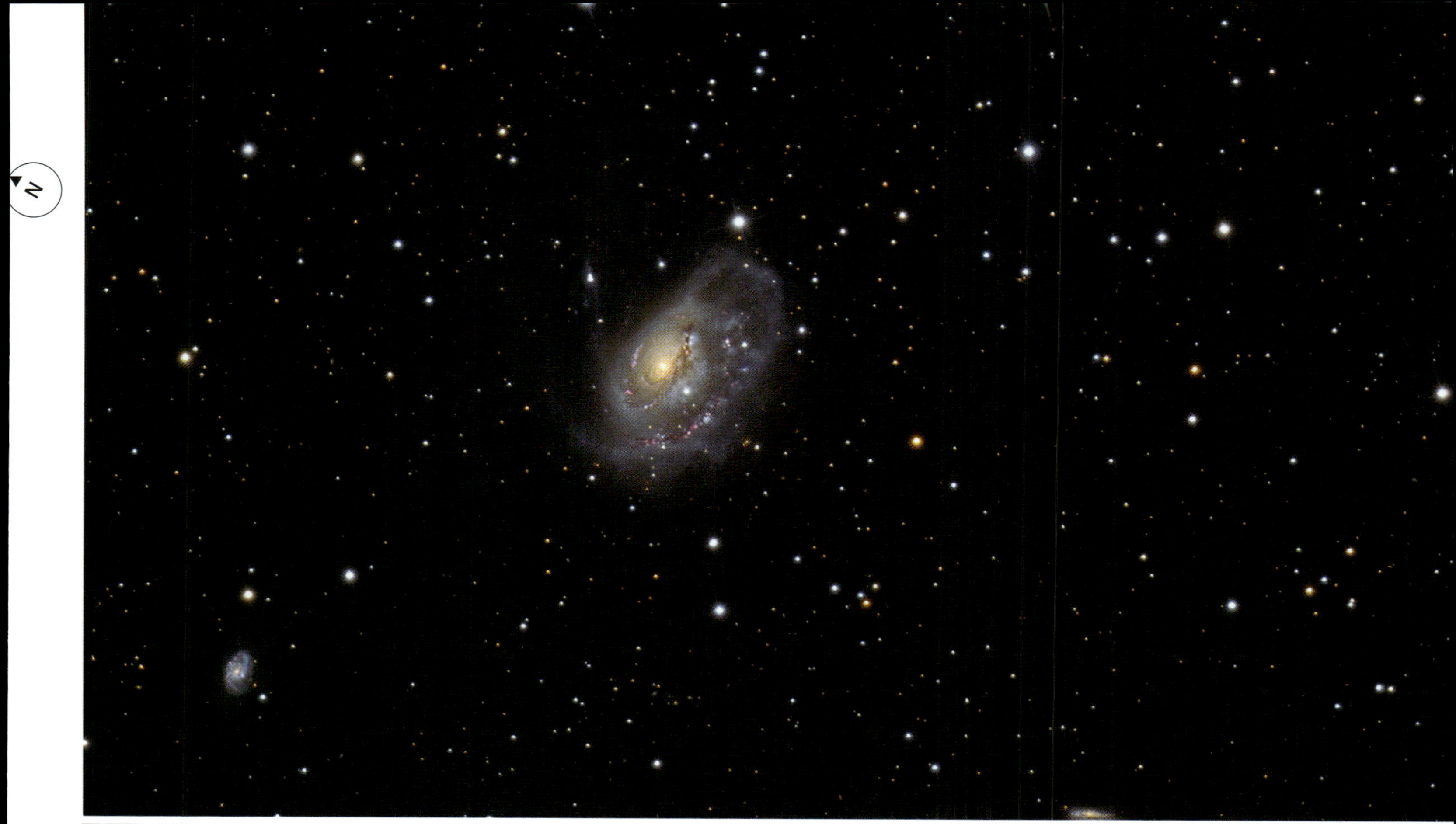

Arp 184

The galaxy NGC 1961 is the largest member of a small group of ten galaxies. Its distance is 180 million light years and its large angular size of 4.6′ corresponds to 240 000 light years. Particularly prominent is the southern spiral arm on the right, which has many areas containing hot, blue stars and large H II regions. These are indicators of a large amount of gas and dust, and also of star formation in this part of the spiral arm, caused by shock waves and condensations. Thus NGC 1961 became known as a bright starburst galaxy. On the opposite, northern side of the galaxy such active regions are lacking and instead one can recognize three faint, outward-pointing arm structures, each with the same angle of about 20° at the point of origin.

NGC 1961 displays a perturbed form, which led to its classification as Arp 184. There is, however, no indication of an interaction. Radio observations show no traces of past orbits of a companion which could have collided and merged with NGC 1961. Also, the thin form and long extent of the spiral arms are incompatible with a past merger. However, the appearance of NGC 1961 in the radio does contain a hint of a possible cause: overlaying the radio contours and the optical image shows a head-tail structure, an elliptic emission region with a bright centre and faint extension. The centre of the radio emission is within the bright, southern spiral arm and the faint side on the northern side of the galaxy. It thus appears probable that NGC 1961 is an intruder into the small galaxy group. Upon entering the group, NGC 1961 piles up the gas in the group medium, compressing it on the southern side, leading to the observed perturbations in its spiral structure.

OBJECT	Arp 184, NGC 1961
CONSTELLATION	Camelopardalis
RIGHT ASCENSION	$05^h\ 42^m\ 05^s$
DECLINATION	+69° 22′ 42″
BRIGHTNESS	10.9 mag
TYPE	SAB(rs)c
PHOTOGRAPHERS	Adam Block
TELESCOPE	600-mm reflector
CAMERA	SBIG STL-11000
EXPOSURE TIME	285 min
LOCATION	Mount Lemmon SkyCenter/ University of Arizona, USA

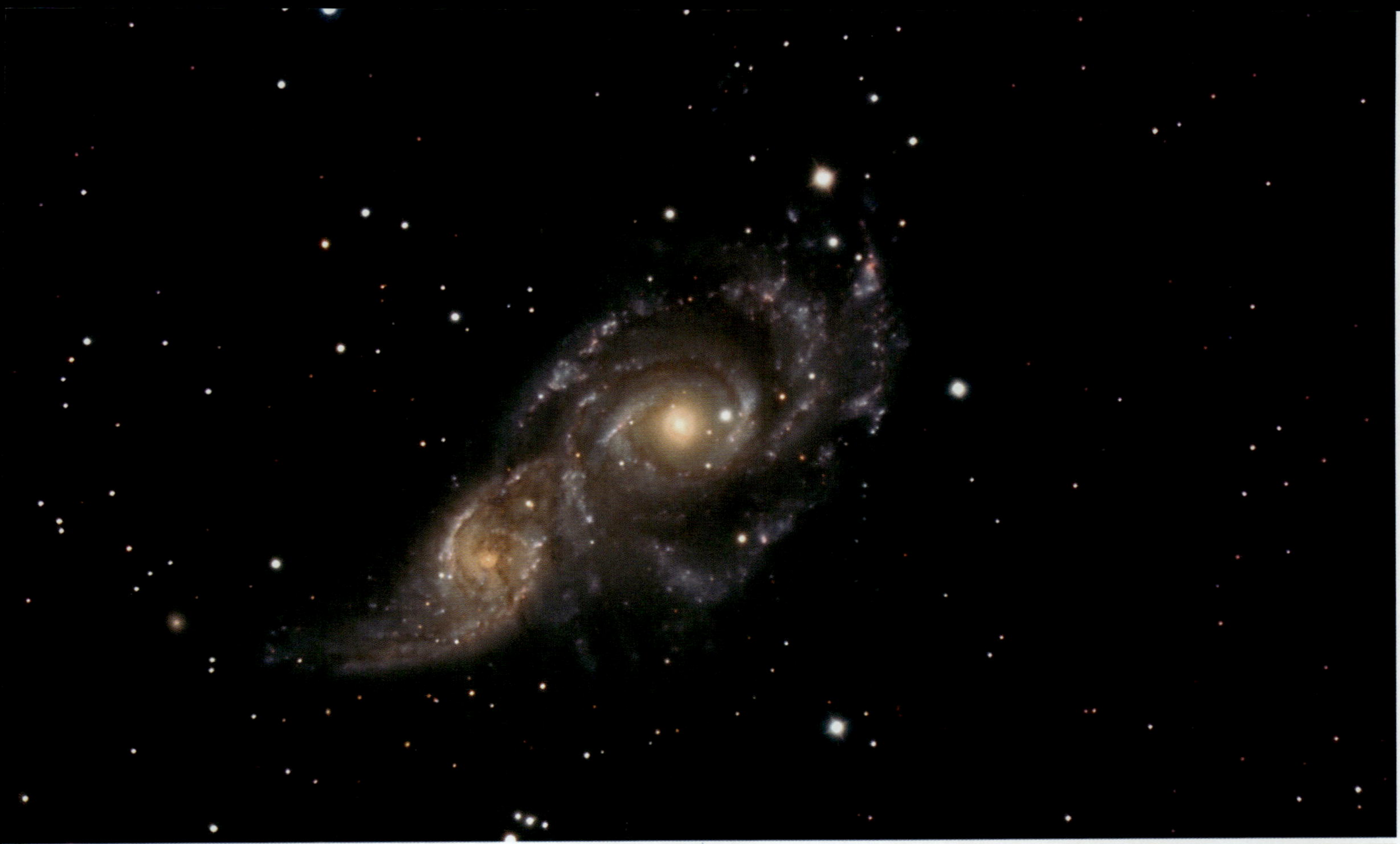

NGC 2207

This interacting pair is in the southern part of the constellation Canis Major at a declination of −21° and contains the galaxies NGC 2207 and IC 2163. The larger partner is of SAB(rs)bc type and is to the west of the SB(rs)c galaxy IC 2163. The galaxy cores are separated by only 1.5′, which at 100 million light years corresponds to a distance of only 50 000 light years in projection. The recession velocity of IC 2163 is 24 km/s larger, which can indicate that IC 2163 is located behind the spiral plane of NGC 2207. This plane shows no perturbations and due to the face-on perspective no possible warp due to tidal forces can be detected. Only in the outer reaches of the spiral arms can hints of Vorontsov-Velyaminov rows be seen, which indicate perturbations of density waves. The dust structures in the eastern spiral arm of NGC 2207 cross the centre of IC 2163 and prove that NGC 2207 lies in front of IC 2163.

Perturbations in the form of IC 2163 are easily seen: the spiral arm at the eastern side fans out at the end and displays a diffuse structure containing dust filaments. Since the relative velocity of the two galaxies is small, one can assume that the less massive partner IC 2163 orbits NGC 2207 and is at the eastern edge of its orbit. Tidal forces from NGC 2207 have already caused perturbations on the far side of the partner. The companion IC 2163 will soon cross through the spiral plane of NGC 2207 and models of this system suggest that the two galaxies will merge in 100–150 million years.

OBJECT	NGC 2207
CONSTELLATION	Canis Major
RIGHT ASCENSION	$06^h\ 16^m\ 22^s$
DECLINATION	−21° 22′ 22″
BRIGHTNESS	11.3 mag
TYPE	SAB(rs)bc pec
PHOTOGRAPHERS	Dietmar Böcker, Ernst von Voigt
TELESCOPE	600-mm reflector
CAMERA	SBIG ST-10XME
EXPOSURE TIME	160 min
LOCATION	Amani Lodge, Namibia

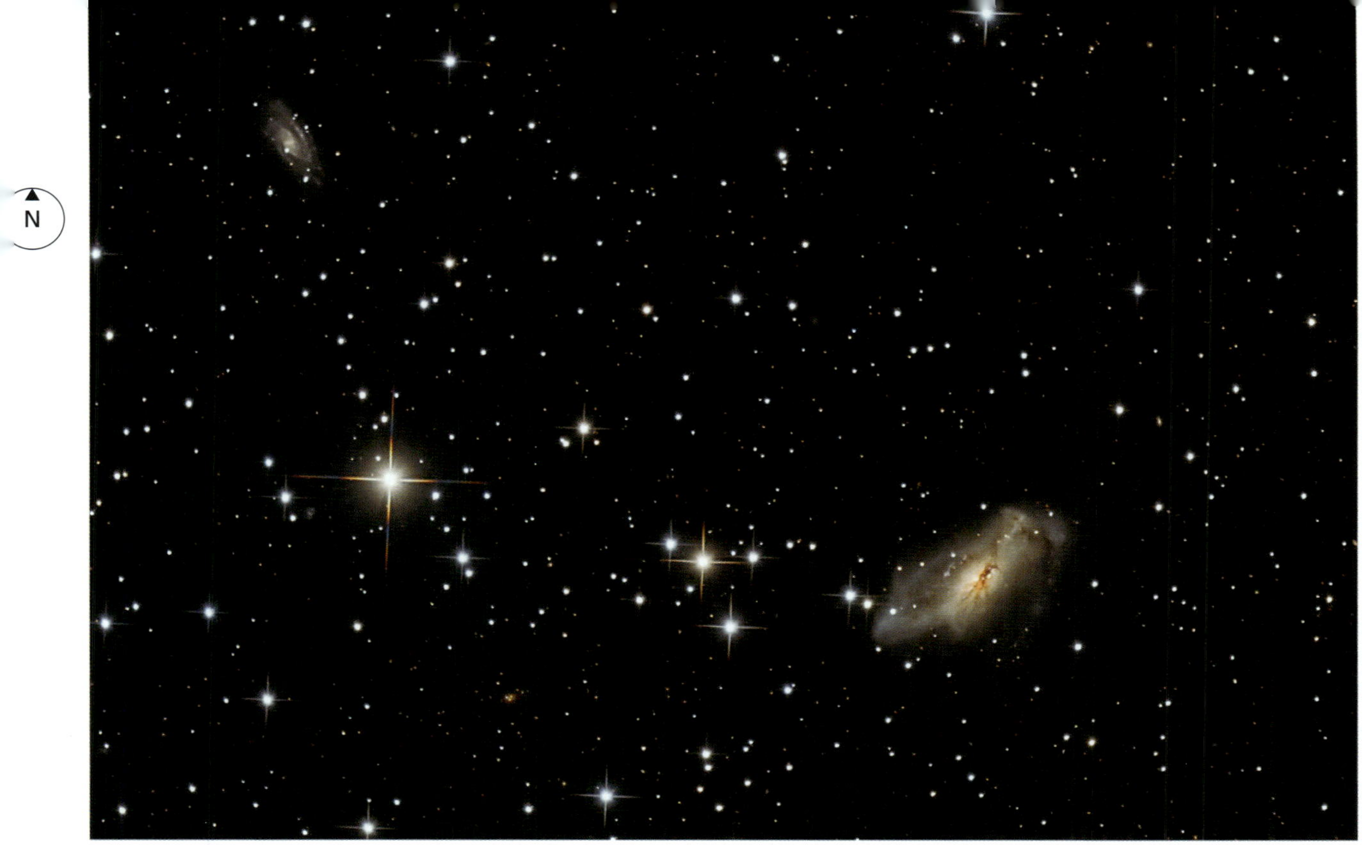

NGC 2146

To the right of a small group of stars in the image is the SB(s) ab galaxy NGC 2146. It has a prominent dust lane which runs in a north–south direction across the bright central region and fans out into dusty strips at the southern end. It is difficult to determine whether NGC 2146 is seen edge-on, as suggested by the dust lane, since the form of the outer region displays strong perturbations. NGC 2146 is thus seen as a possible post-merger. Its angular size is 6.0′ × 4.5′ and based on the imprecise distance values, which lie in the range 50–70 million light years, this corresponds in projection to a diameter in the range 87 000–122 000 light years. To the east of the centre is a blue spiral arm which departs from the plane of the dust lane, kinks farther outside, then points upward again. In this active region are bright OB associations and H II regions, making NGC 2146 a starburst galaxy.

The cause of such activity is usually tidal interaction with another galaxy which affects gas flows, leading to local condensations and hence star formation, In the case of NGC 2146 this scenario is a plausible explanation for the perturbed morphology, although no interaction partner is to be seen. Radio observations at the wavelength of atomic hydrogen indicate that hydrogen gas is distributed in a disc in NGC 2146, its form and measured velocities typical of an unperturbed spiral galaxy. There is no indication at all of a past merger; neither could a bar be detected. Thus one can assume in the case of NGC 2146 that the spiral galaxy is not the product of a merger. The perturbations in the spiral arms are presumably due to a close approach of and tidal interaction with another galaxy – a candidate for this based on the image could be NGC 2146A, visible at the top. However, the fact that the difference in radial velocities of 597 km/s is too large speaks against the interaction hypothesis.

OBJECT	NGC 2146
CONSTELLATION	Camelopardalis
RIGHT ASCENSION	$06^h\ 18^m\ 38^s$
DECLINATION	+78° 21′ 25″
BRIGHTNESS	11.4 mag
TYPE	SB(s)ab pec
PHOTOGRAPHERS	Richard Müller
TELESCOPE	310-mm reflector
CAMERA	Atik 4000
EXPOSURE TIME	1110 min
LOCATION	Lohmar, Germany

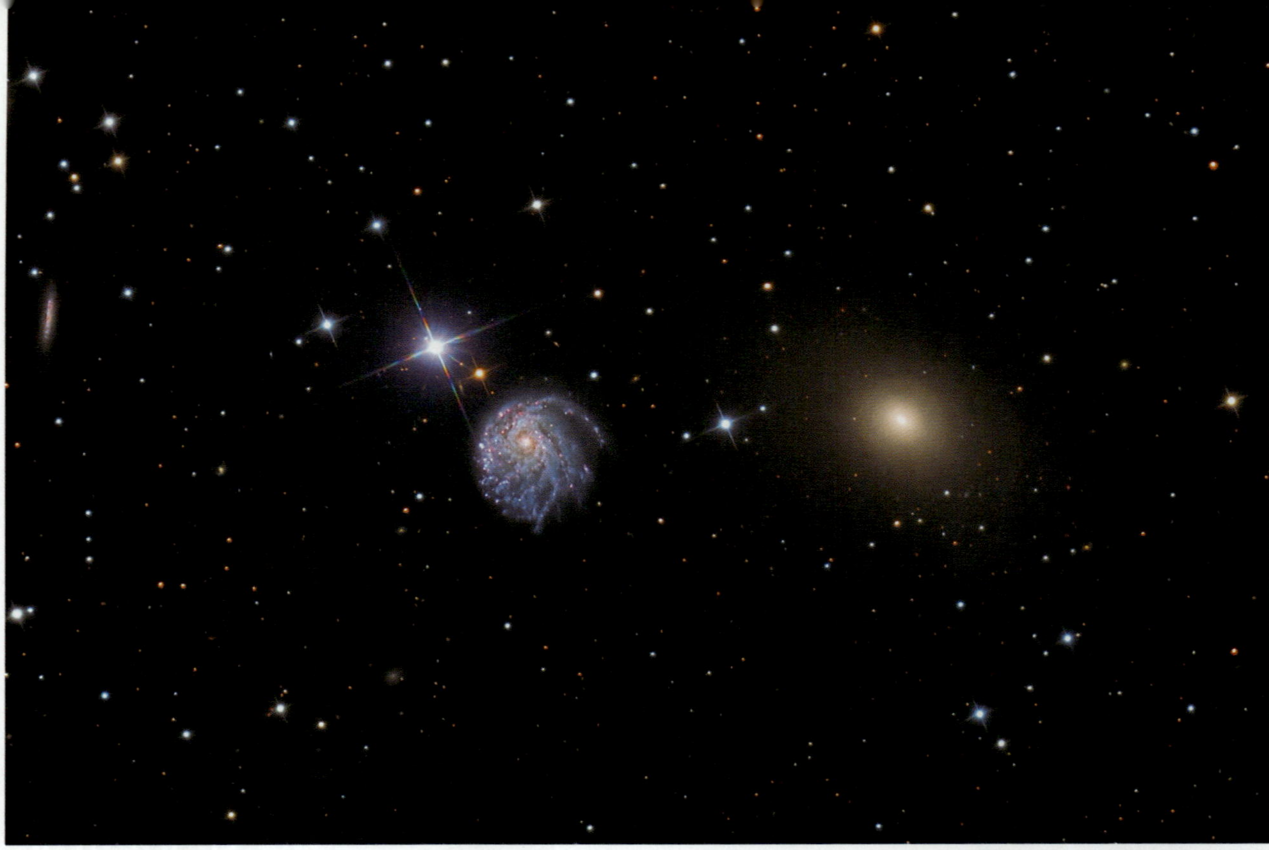

Arp 114

Only 5° away from the north celestial pole, the Sc galaxy NGC 2276 is located next to the a bright 8th-magnitude F star (SAO 1148), which is conspicuous in the image due to its coloured diffraction spikes. To the right of NGC 2276, 6.3′ farther to the east, is the elliptical galaxy NGC 2300. Both galaxies are catalogued as Arp 114; the relative velocity of NGC 2276 exceeds that of NGC 2300 by 511 km/s. Using distance indicators which are independent of the recession velocity, one obtains similar values for both galaxies of 115–120 million light years distance from the Milky Way. From the angular size of 2.8′ × 2.7′, the diameter of its disc is thus 96 000 light years. The spiral pattern in NGC 2276 is highly asymmetric and the core of the galaxy is significantly offset from the centre of the disc. The multiple arms are characterized by many active regions which are especially numerous on the southwestern side. On this side of the disc, the spiral arms seem to be pushed together; along the edge of the disc are located the largest H II regions in NGC 2276.

The possible interaction between NGC 2276 and NGC 2300 was investigated in some studies where astronomers assumed that both galaxies are part of a group and that NGC 2300 has already experienced some mergers with satellite galaxies. During these interactions, the galaxies lost gas which then formed an intragroup medium (IGM) between the galaxies. Observations with the Chandra X-ray Telescope detect this thin gas as a glowing envelope around NGC 2300 which also contains NGC 2276. The X-ray isophotes of NGC 2276 are closer together at the southwest side of the disc plane and a cooler, more diffuse gas tail on the other side shows that NGC 2276 is moving through the IGM at high velocity. This velocity has been calculated to be 850 km/s and leads to ram-pressure stripping with a shock wave at one side and gas loss at the other. Within 1–2 × 10^9 years this will cause the loss of all the neutral hydrogen. NGC 2276 would then further evolve to something like an S0 galaxy.

OBJECT	Arp 114, NGC 2276 / NGC 2300
CONSTELLATION	Cepheus
RIGHT ASCENSION	$07^h\ 27^m\ 14^s$
DECLINATION	+85° 45′ 16″
BRIGHTNESS	11.9 mag
TYPE	SAB(rs)c
PHOTOGRAPHERS	Adam Block
TELESCOPE	800-mm reflector
CAMERA	SBIG STX-16803
EXPOSURE TIME	1200 min
LOCATION	Mount Lemmon SkyCenter/ University of Arizona, USA

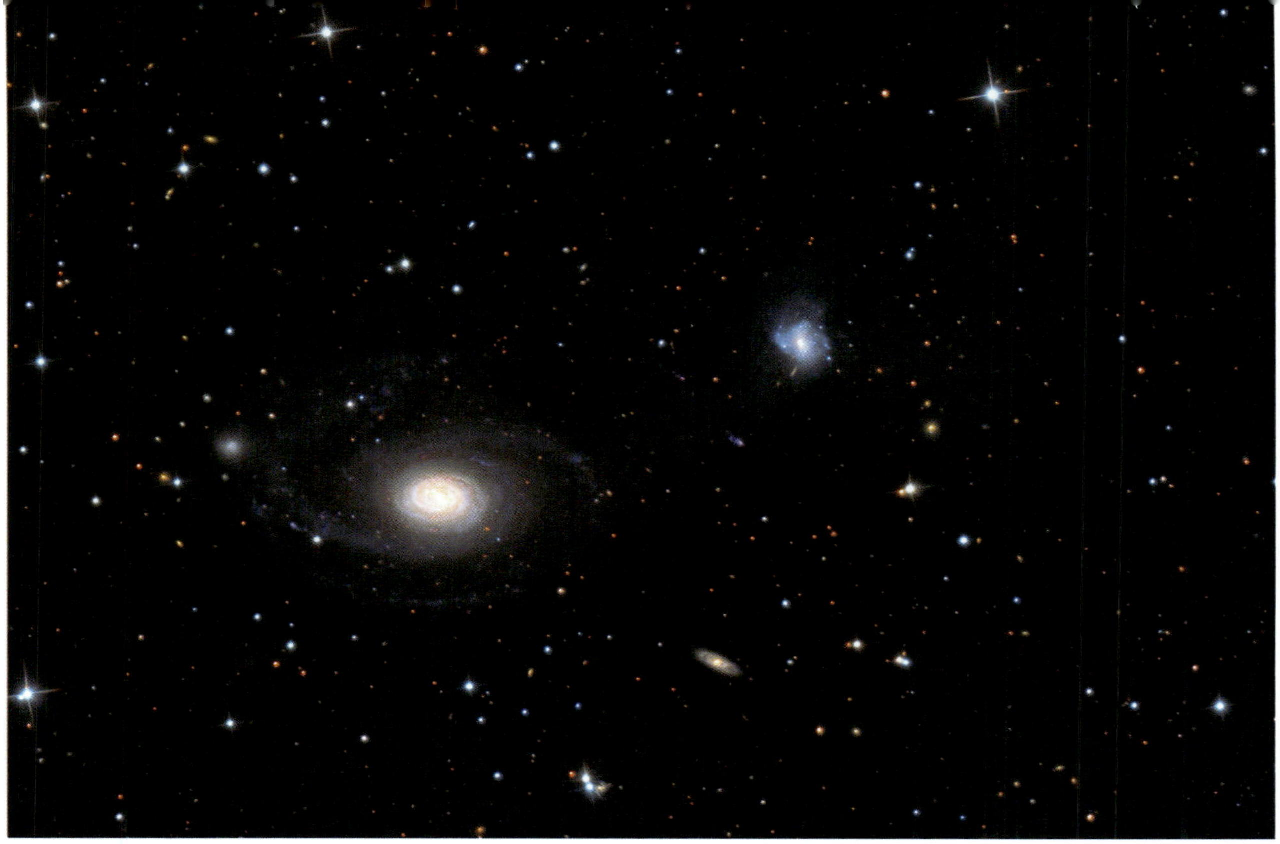

NGC 2460

The image shows the SA(s)a galaxy NGC 2460 which is part of a pair also containing the SB(rs)b galaxy IC 2209 to the right. The distance of the pair is 68 million light years and their separation of 5.4′ corresponds to a distance of about 110 000 light years. The observed diameter of NGC 2460 is of the same order of magnitude – including the dim, outer spiral arms. The companion IC 2209 is much smaller and its angular size of 1.2′ × 0.9′ corresponds to a diameter of its galactic disc of only 24 000 light years. This size, small for an SB galaxy, is due to tidal interaction with NGC 2460 during which the smaller companion has lost gas, dust and stars. Proof of this is given by the tidal tails which surround IC 2209 as a diffuse glow.

Increasing the contrast of the image reveals different colours in the tails which is an indication of their formation at different phases of the interaction. This system has experienced many orbits, and during its last approach NGC 2460 lost material from its outer spiral arms. This is visible as two rows of H II regions and bright blue, active areas which are perpendicular to the line connecting the two galaxies.

The inner spiral structure in NGC 2460 is characterized by a spiral arm covering many revolutions, which also has dozens of large H II regions along its dust filaments. Although NGC 2460 is not catalogued as a barred spiral, the centre has a weak bar structure which is tilted by 30° relative to the horizontal in the image. A two-dimensional analysis of the core region revealed not only hints of a weak bar but also a significant asymmetry in the bulge and disc component of the model. Since the galaxy has large amounts of dust at its centre, it is difficult to make firm statements about the existence of the components. Astronomers assume that interactions can cause perturbations which reach to the core of a galaxy, resulting in an inner structure of gas and dust formations whose complexity exceeds that of the outer regions.

OBJECT	NGC 2460
CONSTELLATION	Camelopardalis
RIGHT ASCENSION	$07^h\ 56^m\ 52^s$
DECLINATION	+60° 20′ 58″
BRIGHTNESS	12.7 mag
TYPE	SA(s)a
PHOTOGRAPHERS	Adam Block
TELESCOPE	800-mm reflector
CAMERA	SBIG STX-16803
EXPOSURE TIME	660 min
LOCATION	Mount Lemmon SkyCenter/ University of Arizona, USA

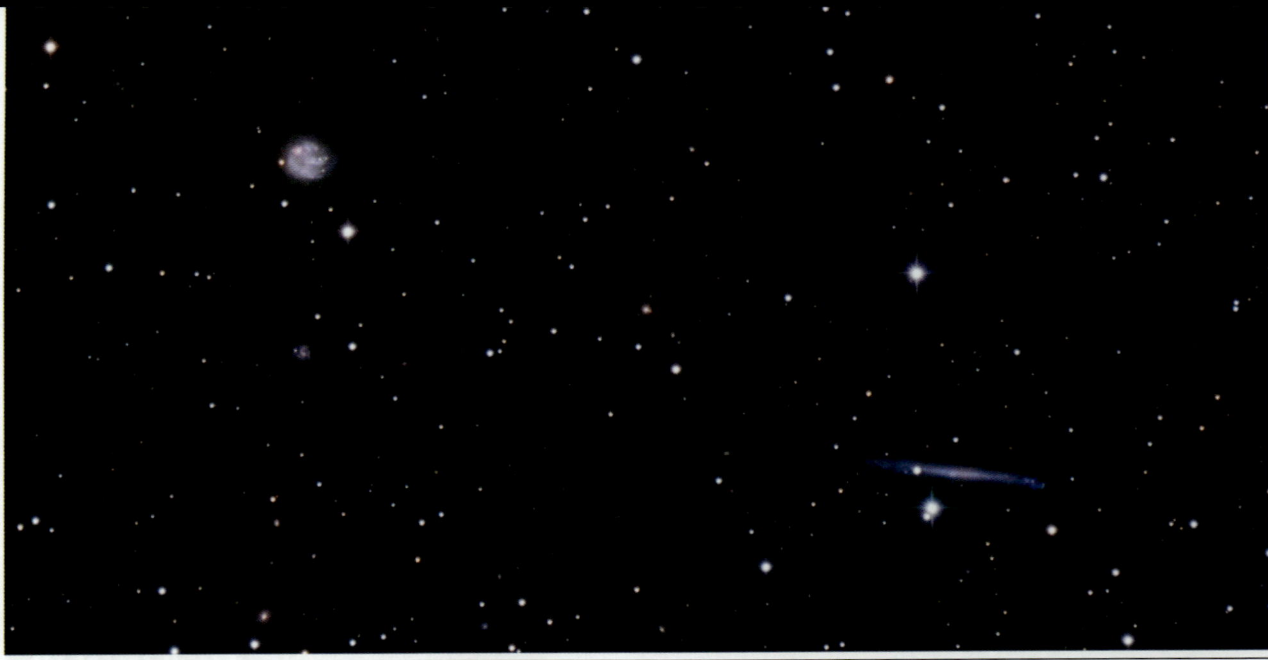

NGC 2537, The Bear Paw Galaxy

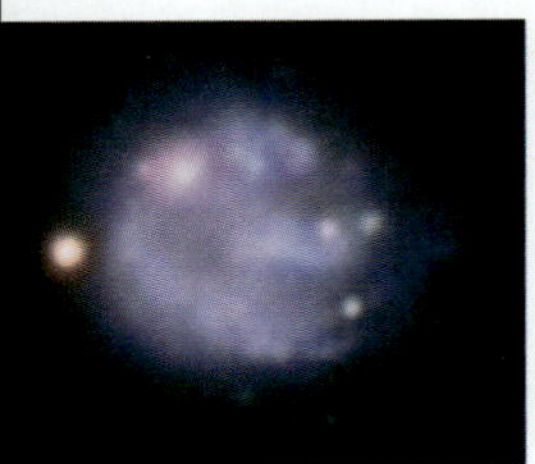

The galaxy NGC 2537 is known as the Bear Paw Galaxy due to its striking morphology. It is listed in catalogues as type SBm and is thus a compact and at 12.3 mag relatively bright type of spiral which is also known as a Magellanic spiral. These galaxies have an asymmetric shape with a stellar bar which is connected to a single spiral arm (see enlargement in the small image). NGC 2537 is located at a distance of 26 million light years and it size of 13 000 light years is less than the extent of the Large Magellanic Clound (LMC).

The spindle-like galaxy IC 2233 is visible in the image to the southeast at a distance of 15′. This SB(s)cd galaxy is at 5.2′ larger and with a relative velocity of +112 km/s also farther away than NGC 2537. The blue colour of IC 2233 is an indication of young stars and a high star-formation rate compared to the Bear Paw Galaxy.

Astronomers have used the VLA radio telescope (Very Large Array, National Radio Astronomy Observatory, New Mexico, USA) to study the distribution of atomic hydrogen gas in both galaxies. Both have an H I envelope, whereby the radio emission of NGC 2537 shows a farther extended, diffuse halo. A bridge of material or tidal tails of atomic hydrogen are not detected between the galaxies nor in their surroundings. The maximum of the radio emission in NGC 2537 is, relative to the optical centre of the galaxy, shifted slightly to the south. This asymmetry in the distribution of atomic hydrogen gas in the galactic plane is not unusual and is observed in 50 per cent of spiral galaxies. Since a neighbouring galaxy is seen near most of these galaxies, the asymmetry is seen as a result of mutual interaction. Simulations also indicate that this is often a temporary effect. The hydrogen gas moves within the disc, reaching its greatest extent at the closest approach. During the next orbit, this inequilibrium of the gas is reduced.

Since NGC 2537 shows no sign of a close approach by IC 2233, it can be assumed that the asymmetry has another cause. This special case is also confirmed by the velocity profile of atomic hydrogen. According to the SBm classification, the disc of NGC 2537 is seen face-on and thus one expects no rotational motion to be seen in the radio. However, relative velocities indicating rotation are nevertheless detected. The southern radio halo (at the right in the image) is moving towards, the northern halo away from the Milky Way. From this one can conclude that NGC 2537 is not a perturbed SBm galaxy, but rather the product of a merger of two compact galaxies.

OBJECT	NGC 2537
CONSTELLATION	Lynx
RIGHT ASCENSION	$08^h\ 13^m\ 15^s$
DECLINATION	+45° 59′ 23″
BRIGHTNESS	12.3 mag
TYPE	SB(s)m pec
PHOTOGRAPHERS	Michael König
TELESCOPE	350-mm reflector
CAMERA	SBIG STX-16803
EXPOSURE TIME	360 min
LOCATION	Rimbach, Germany

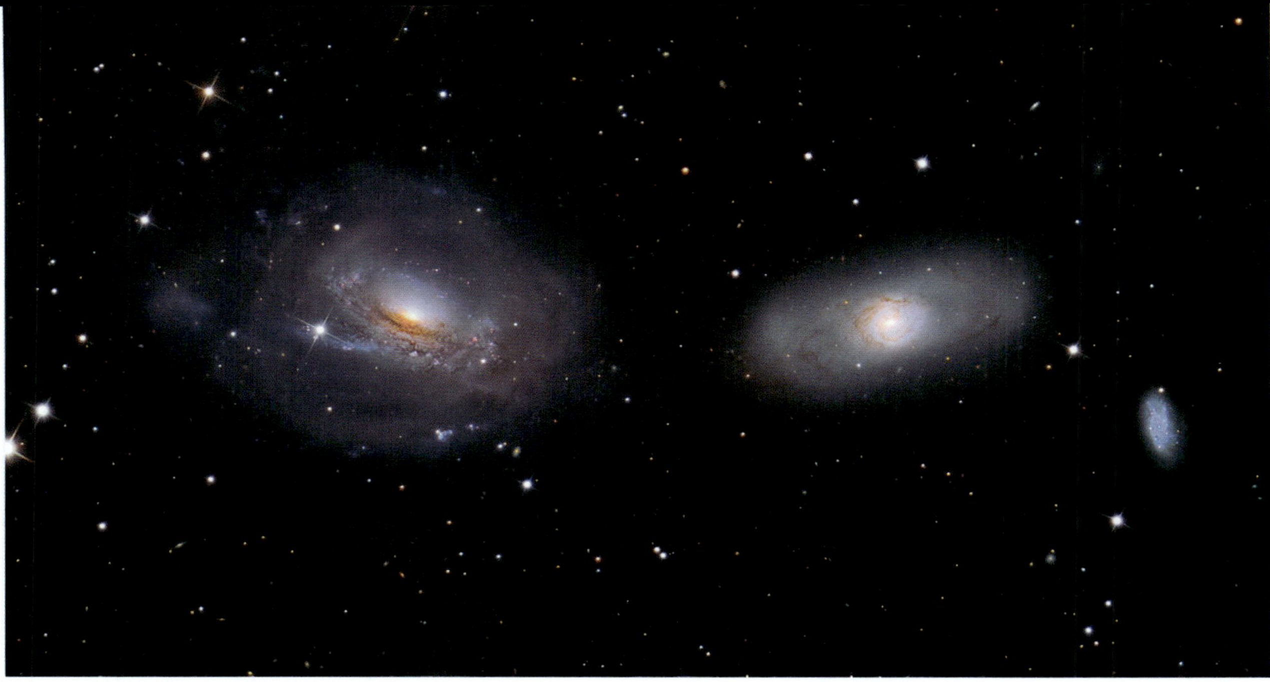

NGC 3169

The interacting galaxies NGC 3169 and NGC 3166 are both located at a distance of 50 million light years. This results in a scaling of 15 000 light years per 1′ and the visible separation of the two galaxy cores of 7.7′ corresponds to 112 000 light years. At the left, eastern side is NGC 3169, an SA(s) galaxy; its companion NGC 3166 is catalogued as SA(rs)0. The galaxies each have a diameter in the range 100 000–110 000 light years. The small SA(s) galaxy NGC 3165, below and to the right of NGC 3166, is not behind the galaxy pair, as one might expect from its small size, but rather at 1340 km/s has almost the same recession velocity as NGC 3166 at 1345 km/s. At 4.8′, NGC 3165 is closer to NGC 3166 than to its larger partner NGC 3169, which is receding at 1238 km/s from the Milky Way. Its angular size of 1.5′ and diameter of 22 000 light years make NGC 3165 the dwarf of a small group of four galaxies, which is completed by NGC 3156. NGC 3156 is 12′ to the southwest and not visible in the image.

NGC 3166 has a diffuse, neutral-coloured halo with brownish dust formations to the right and left of the bright nucleus which extend to the edge of the halo towards NGC 3169. However, even in contrast-enhanced images, no bridge of material to the larger companion is visible. In such images, above and to the right of NGC 3166 one can see a small, dim condensation which could be a tidal tail. In the surroundings of NGC 3169, on the other hand, many more structures and differences in colour are to be seen, indicating a complex history for this galaxy. In the inner halo which surrounds the SA spiral plane, reddish-brown colours dominate, and one sees arc-like structures which seem to be perpendicular to the plane of the spiral. This inner halo is surrounded by a few thick shells which are visible in the light brown filaments of the spiral arms. These could stem from the destroyed disc of a spiral galaxy which interacted with NGC 3169. Since no second nucleus is visible in NGC 3169, one can assume that the galaxy was completely destroyed and formed the halo structures in NGC 3169. The halo shells are surrounded by a third structure. These dim strips and arcs are above and below the galaxy and extend over more than 200 000 light years. Presumably this is the orbital plane of the former companion observed edge-on. The tidal tails which formed during the orbits are presumably visible as relics.

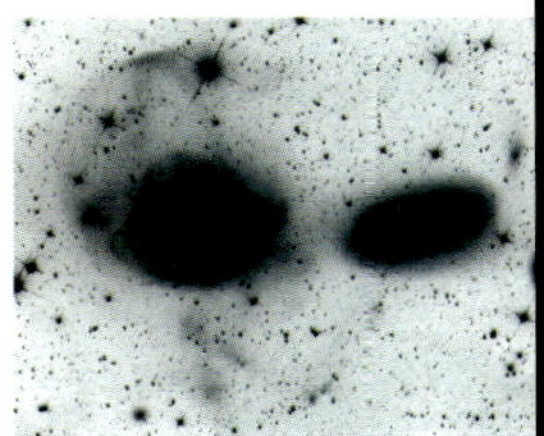

OBJECT	NGC 3169
CONSTELLATION	Sextans
RIGHT ASCENSION	$10^h\ 14^m\ 15^s$
DECLINATION	+03° 27′ 58″
BRIGHTNESS	10.8 mag
TYPE	SA(s)a pec
PHOTOGRAPHERS	Adam Block
TELESCOPE	800-mm reflector
CAMERA	SBIG STX-16803
EXPOSURE TIME	1620 min
LOCATION	Mount Lemmon SkyCenter/ University of Arizona, USA

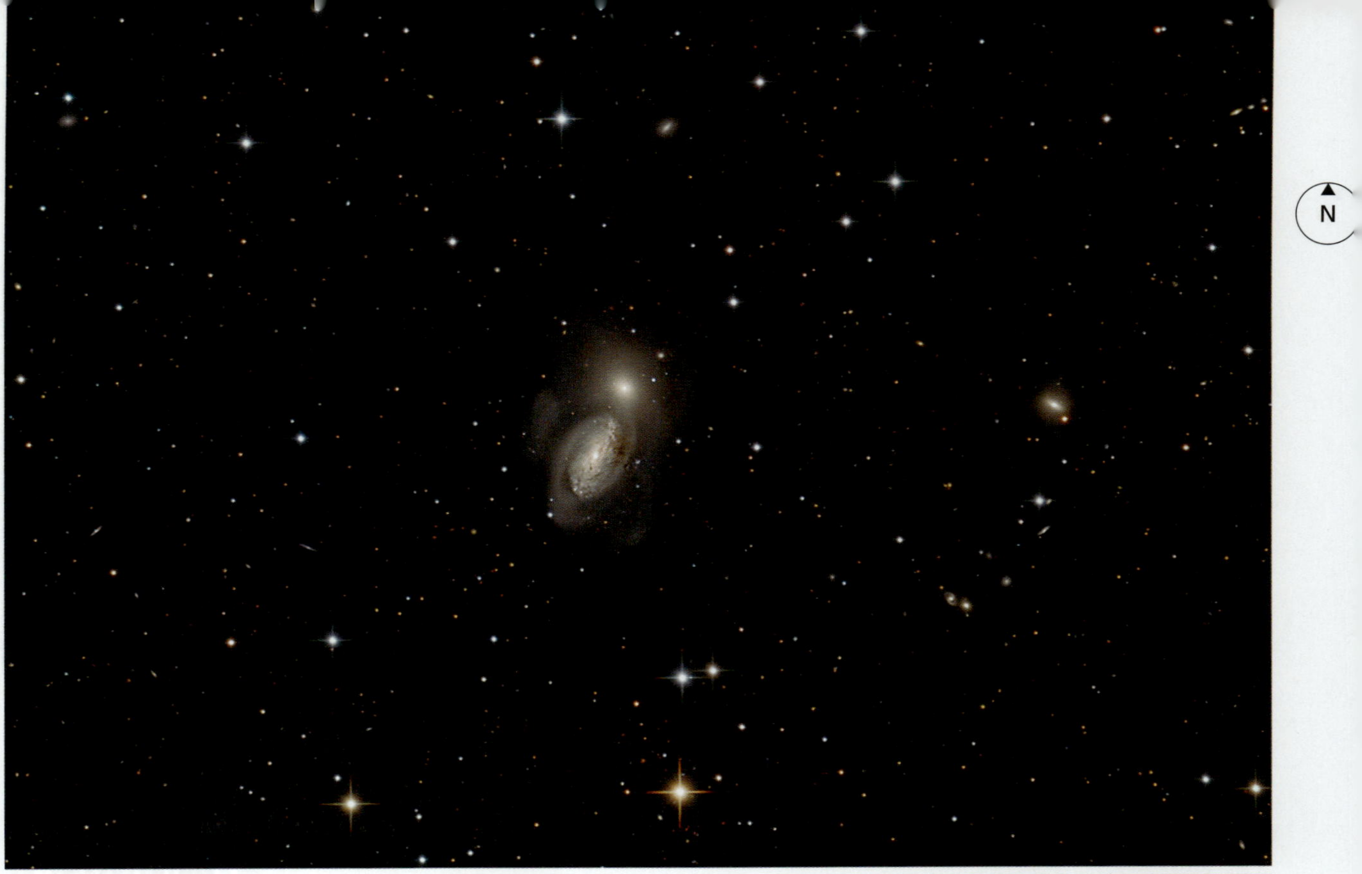

NGC 3227

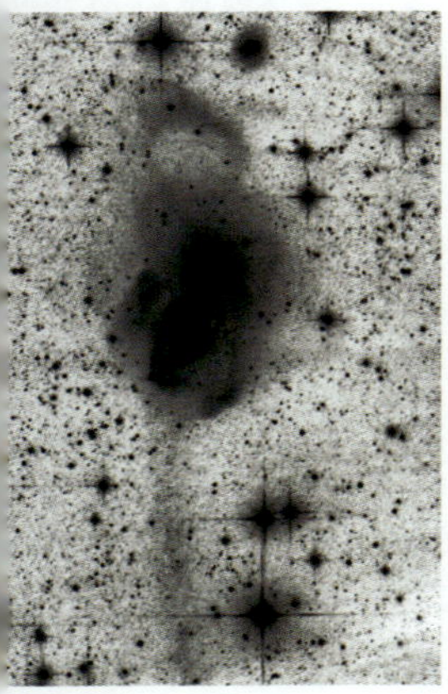

In the middle of the image is the SAB galaxy NGC 3227 which forms an interacting pair with the dwarf elliptical galaxy NGC 3226 to the northwest. The distance from the Milky Way is only 48 million light years, which explains the large number of observations of this galaxy pair in all observable wavelengths. These indicate that NGC 3227 is a Seyfert-1.5 galaxy with its active nucleus shifted relative to the central axis of the bar. This is caused by the interaction with NGC 3226 whose centre is only 2.3′ or 32 000 light years away. Radio observations also indicate that NGC 3226 contains little gas, which is an indication that the tidal tail which surrounds the galaxy pair is composed of material lost by NGC 3226. In one of these tidal tails, at the western edge of NGC 3227 next to a prominent dust filament, a tidal dwarf galaxy was detected. This condensation of gas, dust and stars with a size of 9000 light years is surrounded by a dim halo and contains a few star-forming regions. Its stars indicate an age of less than 100 million years.

If one increases the contrast in the image (see negative image), dim structures appear. For example, the scattered light of Algieba (Gamma Leonis), a 2nd-magnitude star in the constellation of Leo, is visible as rays emerging from the right side of the picture. Below the galaxy pair there is also a dim track of light leading from the lower edge of the image to the pair and continuing to the northern end of the tidal tail. It appears plausible that this track of material was lost by NGC 3226 when it approached NGC 3227 and after passing by decelerated and changed direction. The two arcs which are visible to the right and left of NGC 3226 have, based on their luminosity, a younger dynamic age and are explained by the close orbit of the two galaxies.

OBJECT	NGC 3227
CONSTELLATION	Leo
RIGHT ASCENSION	$10^h\ 23^m\ 31^s$
DECLINATION	+19° 51′ 54″
BRIGHTNESS	11.1 mag
TYPE	SAB(s)a pec
PHOTOGRAPHERS	Bernhard Hubl
TELESCOPE	300-mm reflector
CAMERA	SBIG ST-2000XM
EXPOSURE TIME	1908 min
LOCATION	Nussbach, Austria

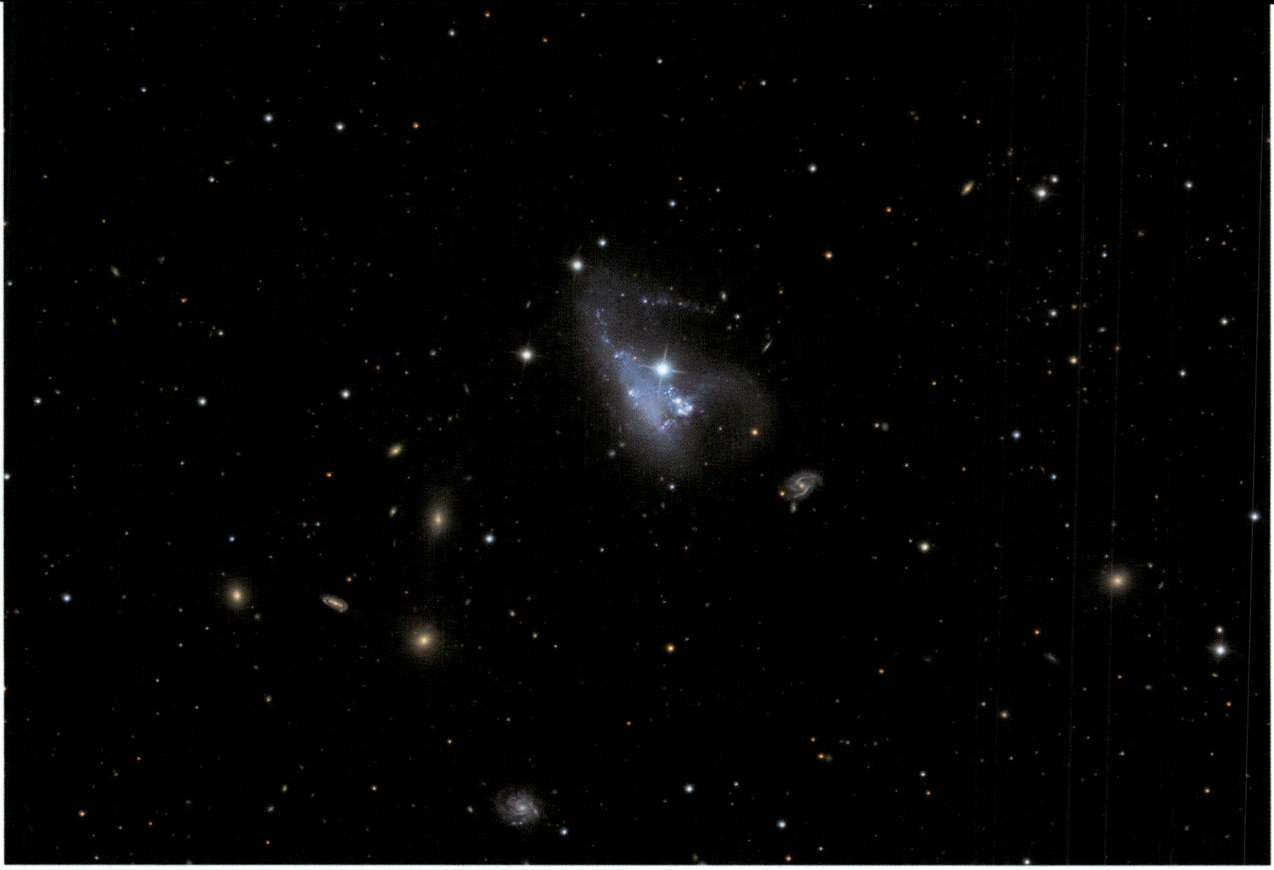

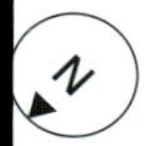

NGC 3239

The irregular galaxy NGC 3239 was classified by de Vaucouleurs as SBm. The image shows that the form of NGC 3239 has several axes of symmetry, which could imply different classifications. The north–south direction is diagonal in the image, with north below left. At the upper, southern end of NGC 3239 one can recognize a row of active regions which can be interpreted as a spiral arm joined to a bar running from upper left to lower right.

NGC 3239 is at a distance of 30 million light years and its angular size of 5.0′ × 3.3′ corresponds to a physical size of 33 000 × 29 000 light years. The optical centre of NGC 3239 is 0.5′ below the 10th-magnitude foreground star. This core region has three bright regions in the image of which the upper left point source is the supernova SN2012A.

NGC 3239 lies within a field of galaxies; there is, however, no companion galaxy nearby. There are also no galaxies with similar recession velocity within a radius of 10′. Radio observations of NGC 3239, however, show a two-dimensional velocity map of atomic hydrogen which corresponds to a rotating gas disc. The corresponding axis of rotation is almost horizontal and is tilted by 30° relative to the bar. This asymmetry of the components can be explained by the merger of two galaxies. This hypothesis is supported by the bright H II regions and OB associations in the galaxy which are due to induced star formation during the interaction. Since no second nucleus can be observed in NGC 3239 and the radio data indicate that the galaxy doesn't have a uniform distribution of neutral hydrogen, it is probable that the smaller interaction partner was destroyed and its material completely transported to the larger partner during just one passage.

OBJECT	NGC 3239
CONSTELLATION	Leo
RIGHT ASCENSION	$10^h\ 25^m\ 05^s$
DECLINATION	+17° 09′ 49″
BRIGHTNESS	11.7 mag
TYPE	IB(s)m pec
PHOTOGRAPHERS	Adam Block
TELESCOPE	800-mm reflector
CAMERA	SBIG STX-16803
EXPOSURE TIME	470 min
LOCATION	Mount Lemmon SkyCenter/ University of Arizona, USA

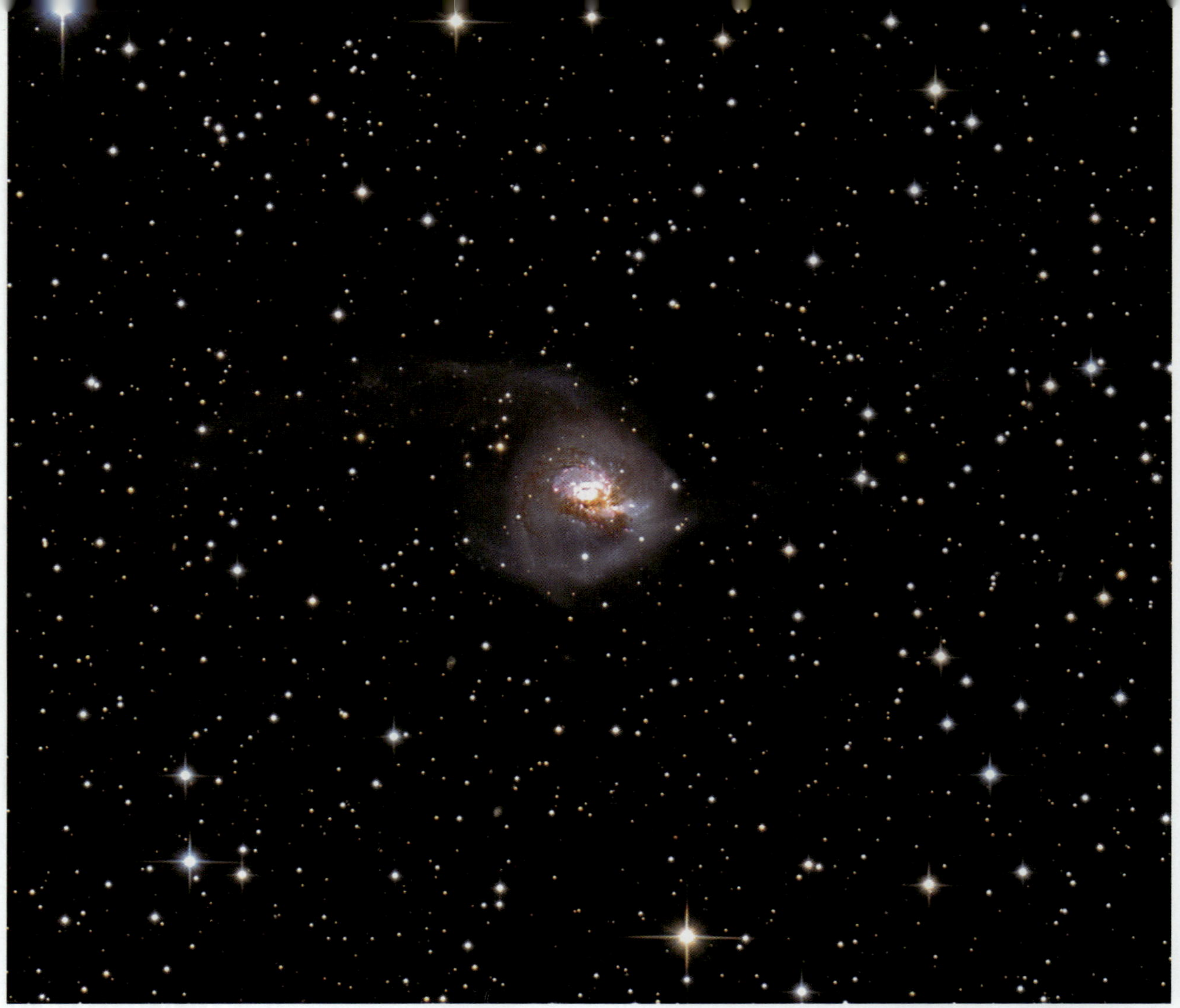

NGC 3256

The galaxy NGC 3256 is located in a galaxy group in the southern constellation of Vela. Near the centre one recognizes star-forming regions which, however, do not follow a spiral structure. This region has a visible diameter of 18 000 light years and is surrounded by a diffuse, triangular structure of angular size 1.5′, which corresponds to about 50 000 light years. NGC 3256 is at a distance of 115 million light years from the Milky Way.

At the upper and lower sides of the triangular structure of the galaxy are dim tidal tails. These point in opposite directions, east and west, and cover an angle of 7′ which in projection corresponds to a distance of more than 230 000 light years.

The galaxy group to which NGC 3256 belongs is part of the Hydra–Centaurus supercluster. Due to the increased probability of interaction within a supercluster, astronomers suspect that NGC 3256 is the relic of a merger of at least two galaxies. This is supported by infrared observations which studied the distribution of dust at the centre. This dust consists of two components which differ in size and temperature and are probably the remains of the precursor galaxies.

OBJECT	NGC 3256
CONSTELLATION	Vela
RIGHT ASCENSION	$10^h\ 27^m\ 51^s$
DECLINATION	−43° 54′ 14″
BRIGHTNESS	12.2 mag
TYPE	pec
PHOTOGRAPHERS	Philipp Keller, Konstantin Buchhold, Bernd Flach-Wilken, Johannes Schedler, Volker Wendel (Chart 32-Team)
TELESCOPE	800-mm reflector
CAMERA	FLI Proline 16803
EXPOSURE TIME	820 min
LOCATION	CTIO, Chile

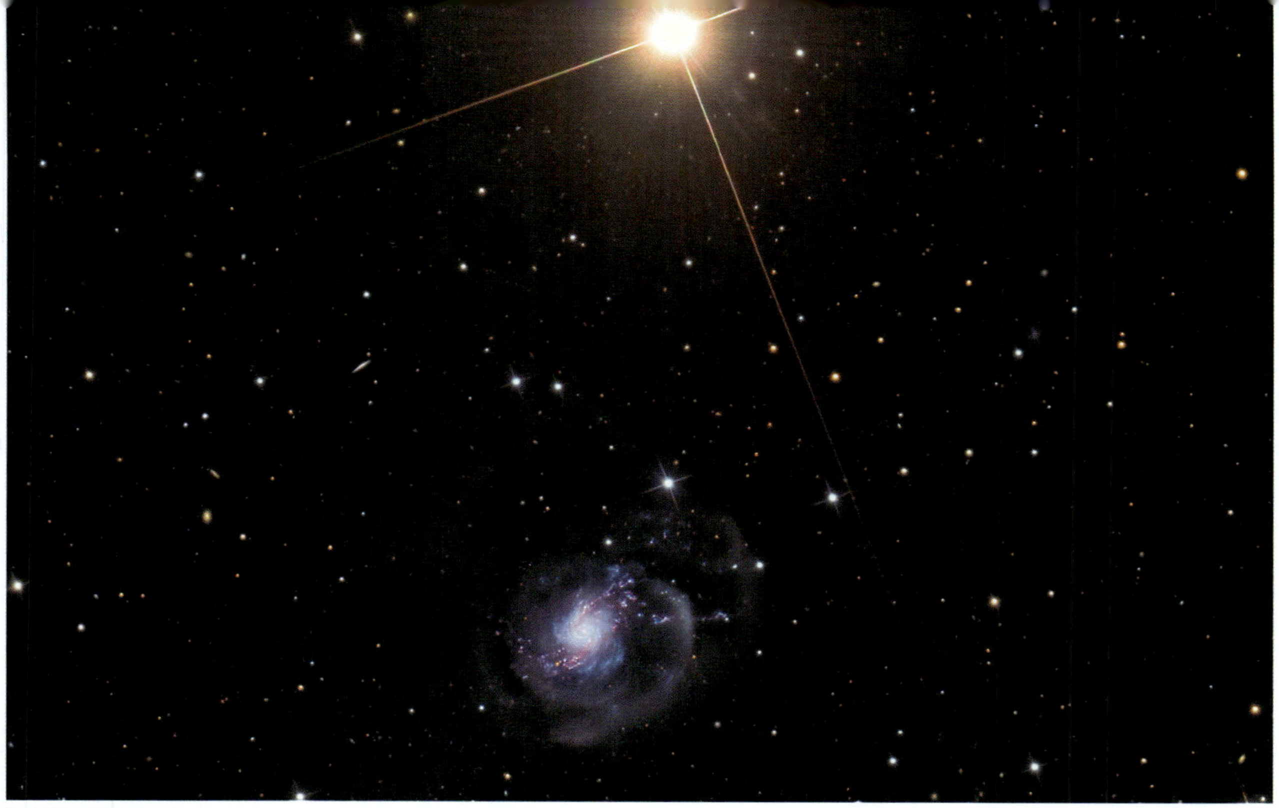

NGC 3310

The 5.5-mag star SAO 27724 at the top of the image is only 10′ north of the SAB(r)bc galaxy NGC 3310. The recession velocity of this galaxy is given as 993 km/s and is of limited use as a distance indicator. Determining the distance by other means results in a value of 59 million light years with an uncertainty of 2.7 million light years. For NGC 3310 the scale is thus 17 000 light years per arcminute. NGC 3310 is a starburst galaxy and a large number of star-forming regions are concentrated in an inner ring. A symmetric spiral pattern begins here with two spiral arms also containing active regions. The nucleus, however, is not at the centre of the inner ring. Astronomers interpret this asymmetry as an indication that NGC 3310 is the product of a merger of two or more galaxies. NGC 3310 is one of the bluest spiral galaxies in the de Vaucouleurs catalogue with (B – V) = 0.3 mag. Its infrared luminosity of 1.1×10^{10} in solar units is comparable to that of M 82, the prototypical starburst galaxy.

The spiral plane of NGC 3310 is surrounded by a complex structure of tidal tails. At the right, western side are two arcs which differ in size and colour. The inner, brighter arc has a radius of 1.8′ and is 30 000 light years from the nucleus. The colour of these structures range from (B – V) = 0.42 mag to values of (B – V) = 0.63 mag and suggest that the material in the tail does not originate in the disc of NGC 3310.

A possible origin is a red precursor galaxy which was completely destroyed in the merger. Its old stellar population survives in the matter in the tidal tail. Astronomers also see NGC 3310 as the possible result of a multiple merger in which several small galaxies could have merged almost simultaneously. This is supported by the high gas fraction and the extended loops and bands in the tidal tails which one can see at the southeastern side of the NGC 3310 system.

OBJECT	NGC 3310
CONSTELLATION	Ursa Major
RIGHT ASCENSION	$10^h\ 38^m\ 46^s$
DECLINATION	+53° 30′ 12″
BRIGHTNESS	11.2 mag
TYPE	SAB(r)bc pec
PHOTOGRAPHERS	Adam Block
TELESCOPE	800-mm reflector
CAMERA	SBIG STX-16803
EXPOSURE TIME	460 min
LOCATION	Mount Lemmon SkyCenter/ University of Arizona, USA

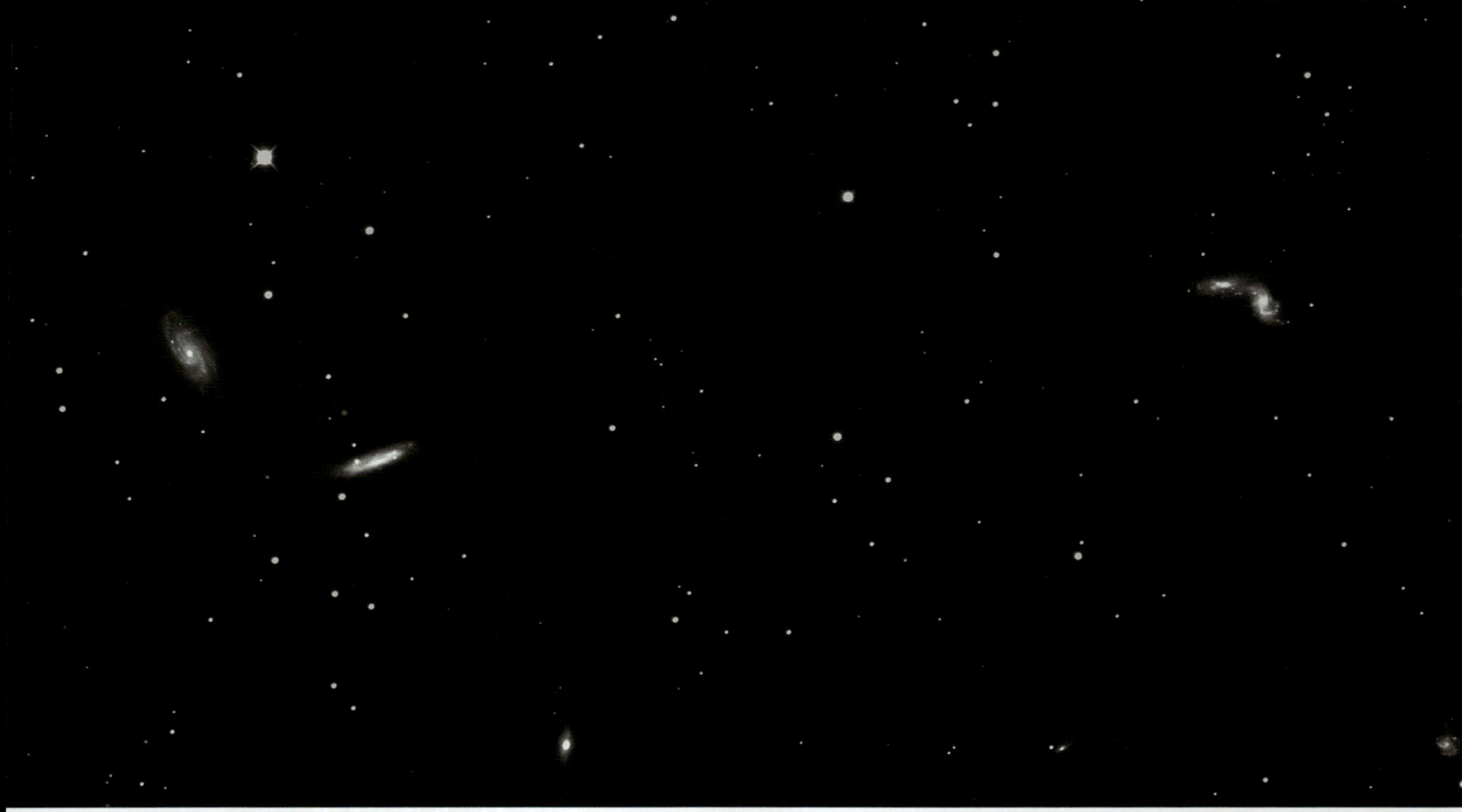

Arp 270

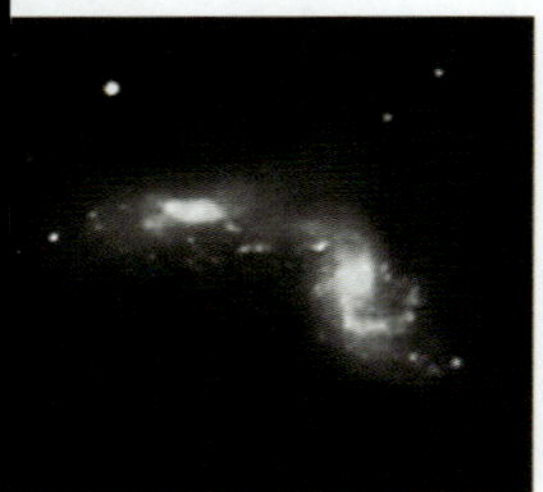

At the eastern edge of the image is the SAB(rs)c galaxy NGC 3430 and 4.5′ to the southwest one can see NGC 3424, an edge-on SB galaxy. Both galaxies are at a distance of 64 million light years from the Milky Way; almost no direct interaction is detected. There is only a dim appendage of what could be a tidal tail in NGC 3624. In contrast, the galaxy pair at the right of the image displays significant interaction (see enlargement in the small image). This pair, consisting of NGC 3395 and NGC 3396, is beginning to merge and has already formed a common halo. The distance between the galactic nuclei is 68″ which corresponds to a transverse separation of 21 000 light years. The system is also known as Arp 270 and forms a small group together with NGC 3430 and NGC 3424.

The original morphology of NGC 3395 is still recognizable; it is an SAB(rs) type on the south side of which a spiral arm with two Vorontsov-Velyaminov rows can be seen. At the opposite, upper side, the spiral structure is no longer intact. The galaxy pair was part of an observational campaign in which astronomers studied the radial velocities in spiral planes via Hα images. The resulting two-dimensional velocity fields highlight the differences between normal and irregular types of galaxies. The inclination of NGC 3396 is 70° and is higher than that of NGC 3395 at 50°. At 198 km/s, NGC 3395 has a much higher rotation velocity than NGC 3396 (66 km/s). The velocity field of NGC 3396 already shows a significant asymmetry and explains its IBm classification. The data for NGC 3395 still show the typical rotation curve of a spiral galaxy, though with perturbations. They are due to material which is in the process of separating from the disc and will probably form a tidal tail.

OBJECT	Arp 270, NGC 3395 / NGC 3396
CONSTELLATION	Leo Minor
RIGHT ASCENSION	$10^{h}\ 49^{m}\ 55^{s}$
DECLINATION	+32° 59′ 27″
BRIGHTNESS	13 mag
TYPE	SAB(rs) / IBm pec
PHOTOGRAPHERS	Michael König
TELESCOPE	350-mm reflector
CAMERA	SBIG STL-11000
EXPOSURE TIME	190 min
LOCATION	Rimbach, Germany

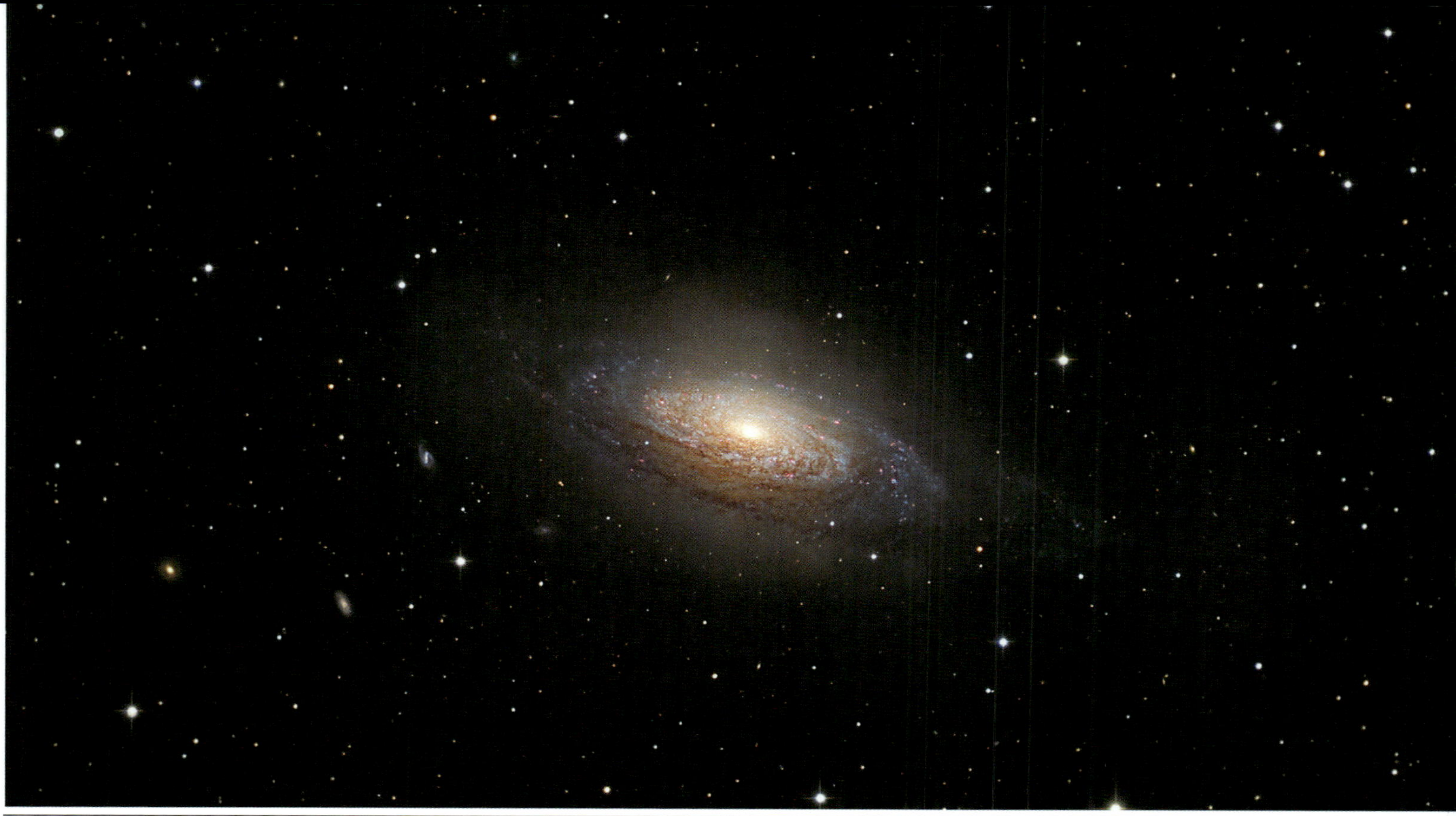

NGC 3521

The SAB(rs)bc galaxy NGC 3521 at a distance of 33 million light years from the Milky Way is a distant neighbour of the Local Group. Its size is given as 11.0′ × 9.8′, whereby the galaxy is embedded in a diffuse halo which in some places displays a delicate shell structure (see contrast-enhanced image). The north–south direction is horizontal in the image. At the right, northern side of the galactic plane one can recognize a dim spiral arm 6.8' from the centre. This corresponds to a distance of 65 000 light years. The part of the disc containing this arm appears to be shifted to the right relative to the core. This outer disc with its flocculent arm structure is tilted by a few degrees with respect to the inner disc.

The drop-like form of the halo of NGC 3521 is more strongly condensed on the left, southern side. The narrow end of the halo coincides with the outer spiral arm mentioned above. Farther to the north is a concentration of material whose projected distance from the centre of the galaxy is more than 120 000 light years. The structures in the halo are the remains of former, small satellite galaxies which have been destroyed by the effects of tidal forces. These stellar streams flow through space free of dynamic friction which is why they remain visible as relics in the halo. Due to the relatively close distance and high inclination of 70°, the rotation curve of the disc can be studied at scales up to a few hundred light years. The H I radio profile within 5000 light years displays the typical steep rise to a plateau of 230 km/s which then falls to 180 km/s at a distance of 75 000 light years before it rises again to 200 km/s at the outer edge. This complex velocity distribution of neutral hydrogen suggests a model for the disc consisting of several components and taking into account the zones displaying various interaction effects in NGC 3521.

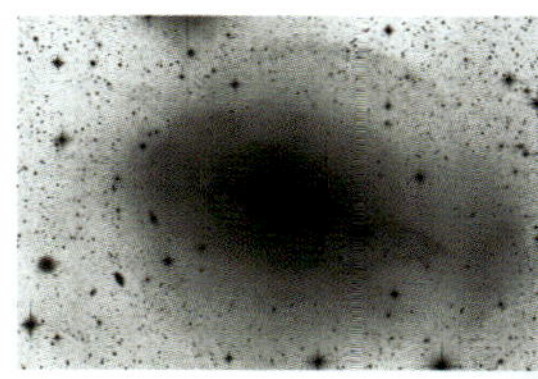

OBJECT	NGC 3521
CONSTELLATION	Leo
RIGHT ASCENSION	$11^{h}\ 05^{m}\ 49^{s}$
DECLINATION	−00° 02′ 09″
BRIGHTNESS	9.8 mag
TYPE	SAB(rs)bc
PHOTOGRAPHERS	Josef Pöpsel, Beate Behle
TELESCOPE	600-mm reflector
CAMERA	SBIG ST-10XME
EXPOSURE TIME	210 min
LOCATION	Amani Lodge, Namibia

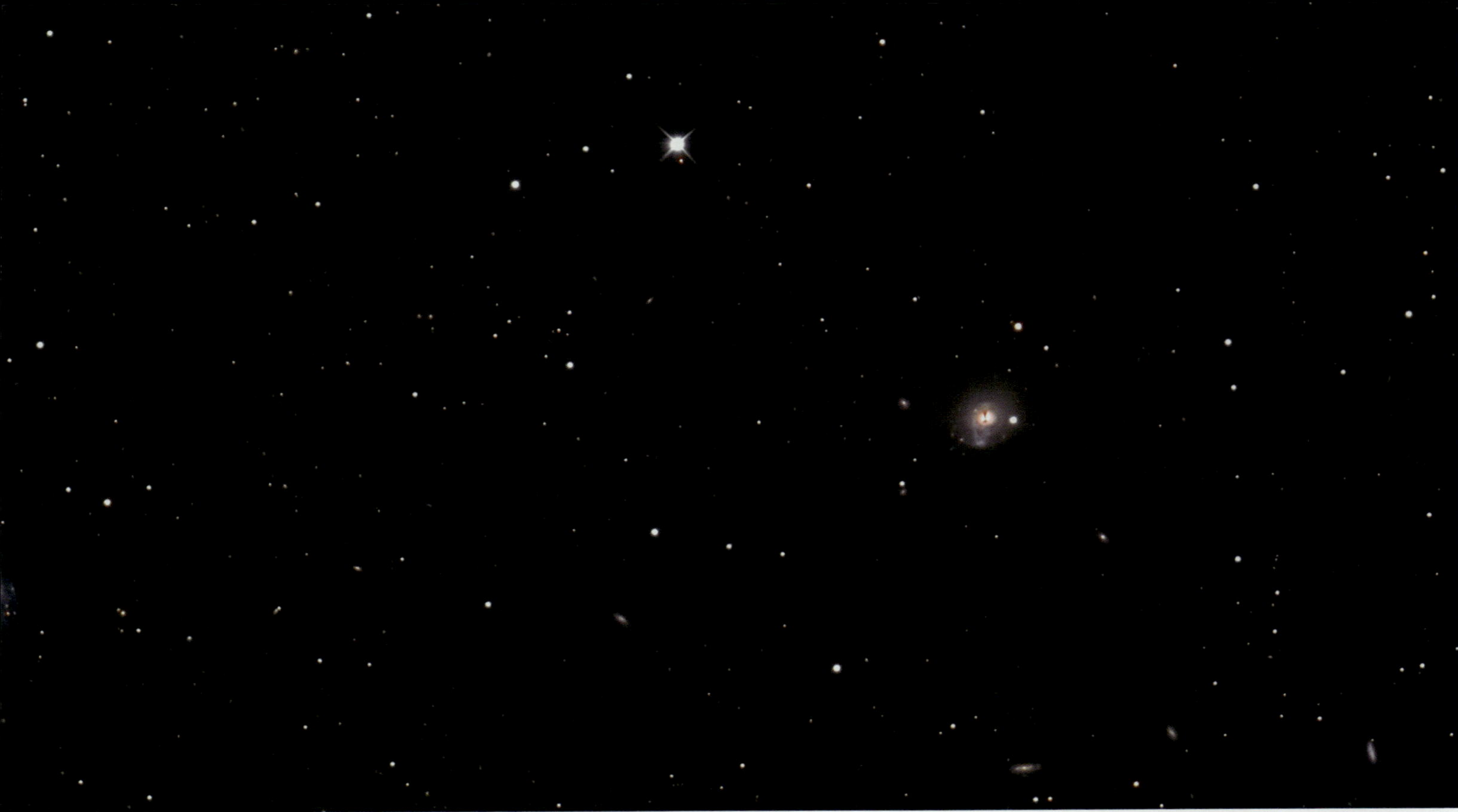

NGC 3656

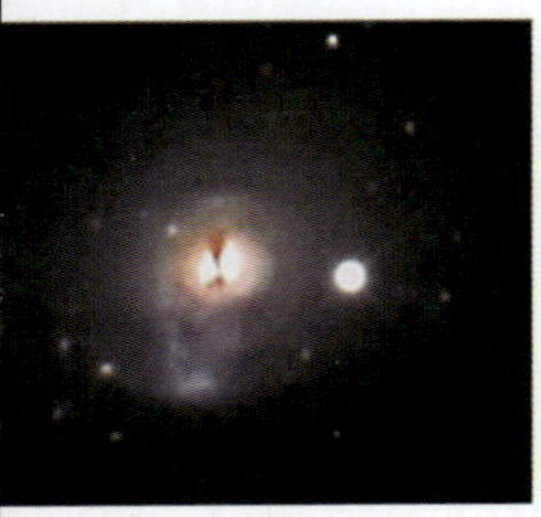

NGC 3656 is a perturbed elliptical galaxy located 130 million light years from the Milky Way. The bright core of the galaxy has an angular size of 0.6′ and is surrounded by two shells (see enlargement in the small image). These in turn are at the middle of a 3.5′ halo whose visible diameter is thus 132 000 light years. A broad dust lane is visible above the centre of NGC 3656 and the remains of a spiral arm run from the southern side of the core to the edge of the halo, ending in a bright condensation of material. At the left edge of the image one recognizes the spiral galaxy UGC 6446. This dim SAd type lies at a distance of 32 million light years and hence only a quarter as far away from the Milky Way.

In interacting or even merging galaxies, gas flows play an important role in determining the morphology. These gas flows induce concentrations of matter and are the basis for starburst regions and the corresponding high infrared luminosity. The newly formed stars carry the angular momentum of the previously ejected galactic gas. Models indicate that in typical cases 60 per cent of the original disc gas in the pre-merger galaxy is compressed into a starburst cloud only a few hundred light years in size. The gas of the precursor which wasn't processed in its central regions is to be found in an extended disc of atomic hydrogen in the environment of the merger. The relation between the tidal tail and the shell structure in the halo can be well studied in the case of NGC 3656. The delicate shells in the galaxy are interpreted by astronomers as the result of a merger of two spiral galaxies with thin discs. The two merged precursor galaxies result in a centrally concentrated disc geometry, whereby 50 per cent of the total gas has been moved into the halo. This gas rotates and thus has a higher angular momentum which prevents the halo from collapsing and is responsible for the well defined shell structure.

OBJECT	NGC 3656
CONSTELLATION	Ursa Major
RIGHT ASCENSION	$11^h\ 23^m\ 39^s$
DECLINATION	+53° 50′ 32″
BRIGHTNESS	13.3 mag
TYPE	(R')I0/a pec
PHOTOGRAPHERS	Michael König
TELESCOPE	350-mm reflector
CAMERA	SBIG STL-11000M
EXPOSURE TIME	520 min
LOCATION	Rimbach, Germany

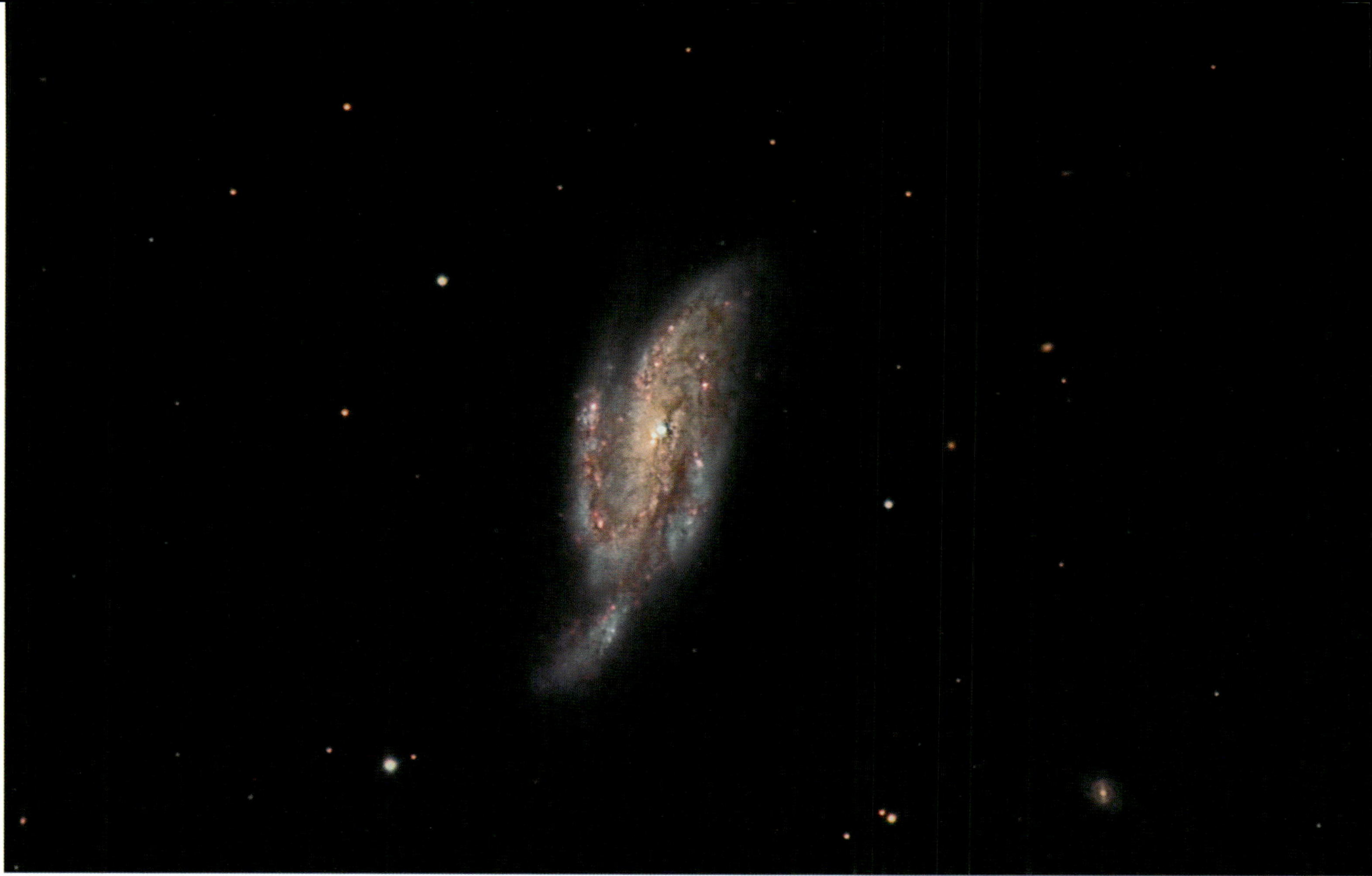

NGC 4088

The SAB(rs)bc galaxy NGC 4088 belongs, together with its neighbour NGC 4085 at a distance of 12′ (not in the image), to the Ursa Major galaxy cluster and is 37 million light years away. From the angular size of 5.8′ × 2.2′ one calculates a projected diameter of 62 000 light years. At the lower, northeastern end of the galaxy a spiral arm appears to be separating from the edge of the disc; at the left, western edge of the disc a spiral arm fades into a tidal tail while moving away from the disc.

The spiral arms in NGC 4088 contain many active areas which are accompanied by H II regions. Since the galaxy is seen edge-on at an inclination of 70°, some of these regions are covered by dust filaments which can be seen along the spiral arms. In the radio, NGC 4088 displays only three further, brighter H I emission regions in addition to the compact nucleus. They coincide with the star-forming regions in the left spiral arm which are also visible in the optical. In the image one can also see the supernova SN2009dd. It is the 13.5-mag star near the centre of the galaxy, dominating its central regions. A separation of 6″ from the centre was measured, which in projection corresponds to about 1000 light years. Two Italian teams of amateur astronomers discovered the supernova independently on 13 April 2009. Although it had already been imaged by others nine days before, it was overlooked because of its proximity to the galactic nucleus.

OBJECT	NGC 4088
CONSTELLATION	Ursa Major
RIGHT ASCENSION	$12^h\ 05^m\ 34^s$
DECLINATION	+50° 32′ 21″
BRIGHTNESS	11.2 mag
TYPE	SAB(rs)bc
PHOTOGRAPHERS	Volker Wendel
TELESCOPE	380-mm reflector
CAMERA	SBIG ST-10XME
EXPOSURE TIME	260 min
LOCATION	Weisenheim am Berg, Germany

Arp 244, The Antennae Galaxies

The Antennae Galaxies NGC 4038 and NGC 4039 are one of the best known interacting pairs. This is also known as Arp 244 and together with five other galaxies belongs to a galaxy group at a distance of 65 million light years. At this distance, an arcminute corresponds to a projected length of 19 000 light years. The system owes its name to two long tidal tails which have a total length of 360 000 light years. The lower, southern system in the Antennae is the SA galaxy NGC 4039 while at the northern end is the SB(s)m galaxy NGC 4038. It has a prominent blue colour which extends into the corresponding tidal tail. This blue tidal tail extends to the right edge of the image and at the end are dim knots which are known as tidal dwarfs. The radio map of the Antennae Galaxies confirms the different nature of the two tidal tails. The reddish-brown tidal tail of NGC 4039 has only weak H I emission, while the blue, southern tidal tail of NGC 4038 appears significantly brighter because it contains more atomic hydrogen. Due to the relatively close distance of this galaxy pair, astronomers were able use high-resolution radio images to show that the end of the tail has a strong velocity gradient and is curved towards the back.

The tidal tails in this system are the textbook example of how tidal interaction causes material to flow out of the outer regions of the disc when two galaxies of equal mass approach each other. The shape of the tidal tails is due to kinematics and it was shown in simulations in 1970 that this morphology can arise in close encounters. The large-scale appearance depends on initial parameters, one of the most important being the relative orientation of the disc rotation of the colliding galaxies. This problem was investigated in many studies and astronomers today assume that in the Antennae Galaxies the direction of rotation of the discs and the orbital motion of the two galaxies were almost aligned during the approach. Models show that in such cases 30 per cent of the disc mass escapes and long tidal tails can form.

At the first pericentre of the mutual orbit, i.e. at the first closest approach of the precursor galaxies, there was no tidal tail. Only 150 million years later did material start to separate from the disc along its edge. After 300 million years there was a second approach of the galaxy cores during which active areas formed in the overlapping regions of the two galaxies. The tidal tails were not influenced by the processes at the centre. Simulations indicate that 350 million years have passed since the first pericentre.

OBJECT	Arp 244, NGC 4038 / NGC 4039
CONSTELLATION	Corvus
RIGHT ASCENSION	$12^h\ 01^m\ 53^s$
DECLINATION	–18° 52′ 03″
BRIGHTNESS	11.2 mag
TYPE	SB(s)m pec
PHOTOGRAPHERS	Adam Block
TELESCOPE	800-mm reflector
CAMERA	SBIG STX-16803
EXPOSURE TIME	370 min
LOCATION	Mount Lemmon SkyCenter/ University of Arizona, USA

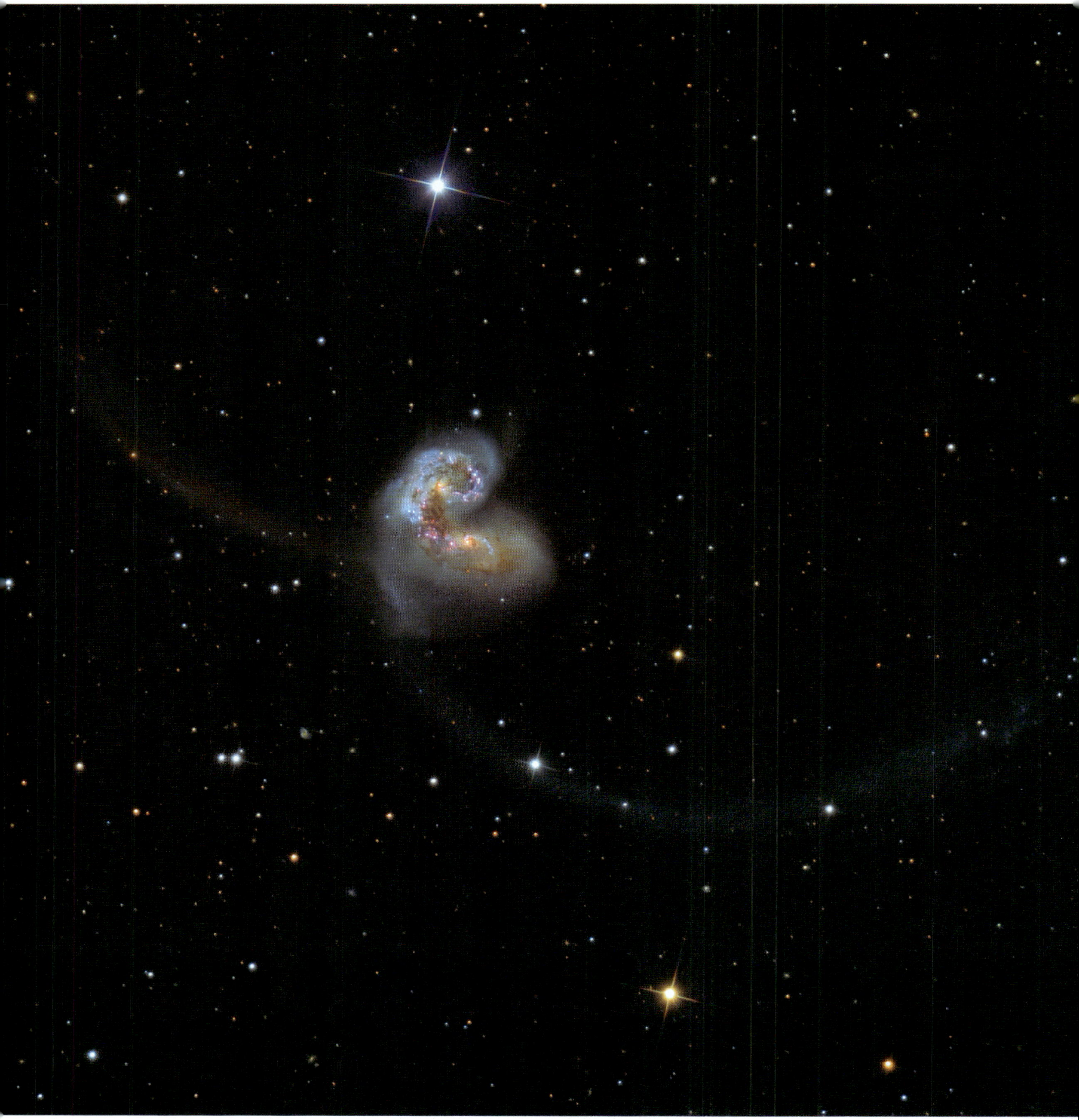

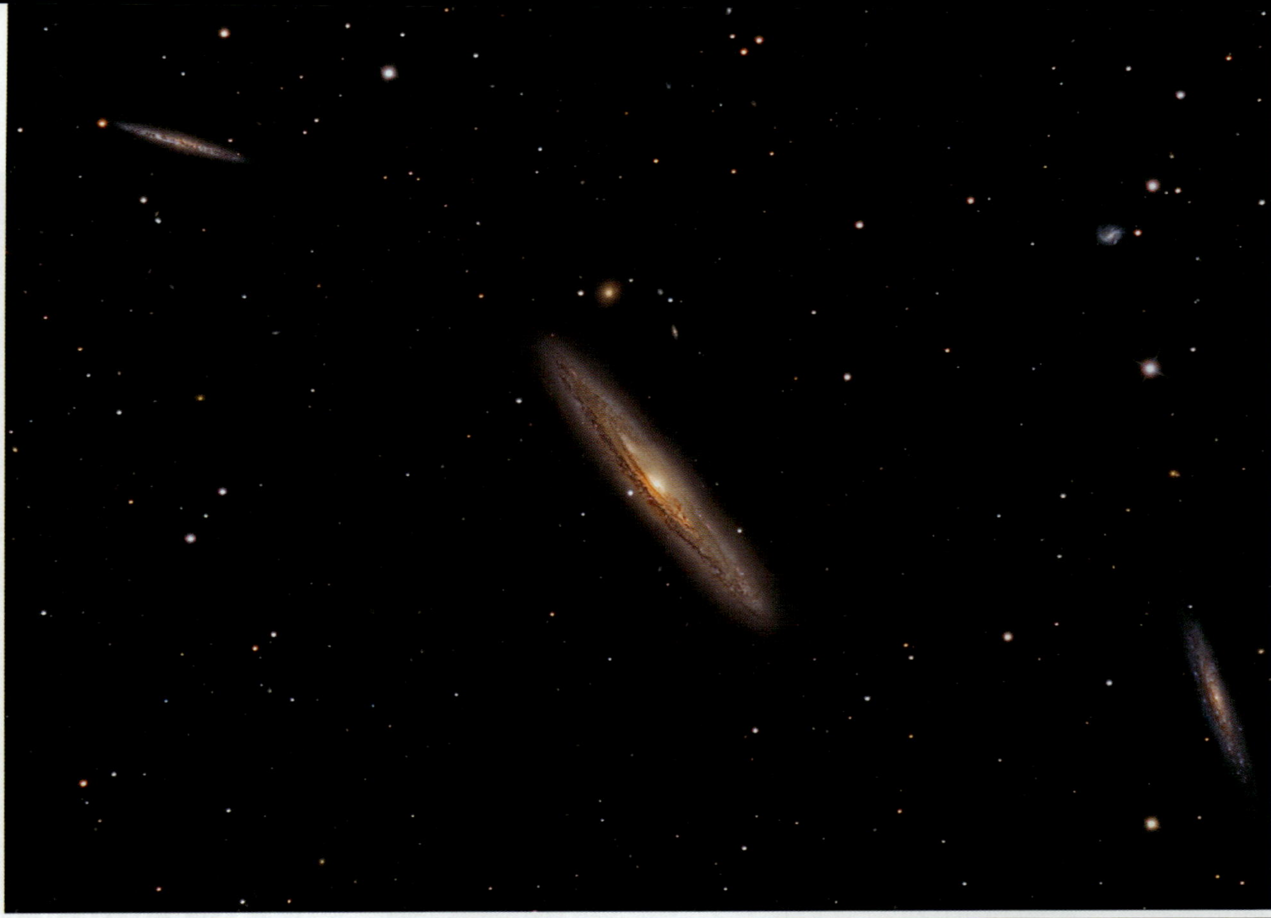

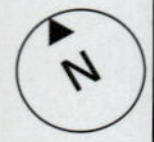

NGC 4216

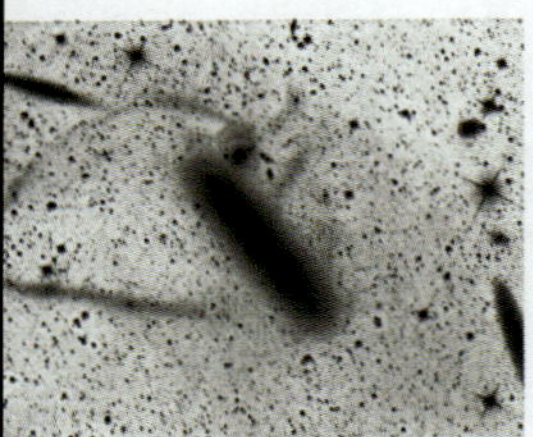

The SAB(rs)b galaxy NGC 4216 belongs to the Virgo Cluster and is located at a distance of 55 million light years from the Milky Way, as are its two companions to the right (NGC 4206) and left (NGC 4222). All three galaxies are seen edge-on. The inclination of NGC 4216 is 89°, whereby dust structures in the near spiral arm cover parts of the bright, compact core. The angular size of 8.1′ × 1.8′ corresponds to dimensions of 130 000 × 29 000 light years – the diameter of the disc is thus a bit larger than that of the Milky Way. One notices that the dust structures near the core are visible only in the southwestern direction while the opposite side instead shows a diffuse brightening which reminds one of the end of a bar. Since the form of the isophotes in the central regions are long ellipses, astronomers suspect that NGC 4216 could have a bar which is also seen edge-on.

NGC 4216 was observed together with seven other galaxies in an observational campaign with the goal of detecting traces of interaction in the environment. The emphasis was on minor mergers which, in contrast to major mergers, don't destroy the stellar disc of the large galaxy. Following the model of cosmological galaxy formation, such minor mergers occur much more often than mergers of galaxies of equal mass. The image of NGC 4216 shows a tidal tail which forms an arc to the east of the galaxy and ends at the small elliptical galaxy PGC 39247. This arc is the projected orbit of the dissolving small companion and the longitudinal structures in the halo of NGC 4216 are indications of previous mergers (see also the contrast-enhanced negative image).

OBJECT	NGC 4216
CONSTELLATION	Virgo
RIGHT ASCENSION	$12^h\ 15^m\ 54^s$
DECLINATION	+13° 08′ 58″
BRIGHTNESS	11 mag
TYPE	SAB(s)b
PHOTOGRAPHERS	Johannes Schedler
TELESCOPE	500-mm reflector
CAMERA	SBIG STL-11000
EXPOSURE TIME	390 min
LOCATION	Blackbird Observatory, New Mexico, USA

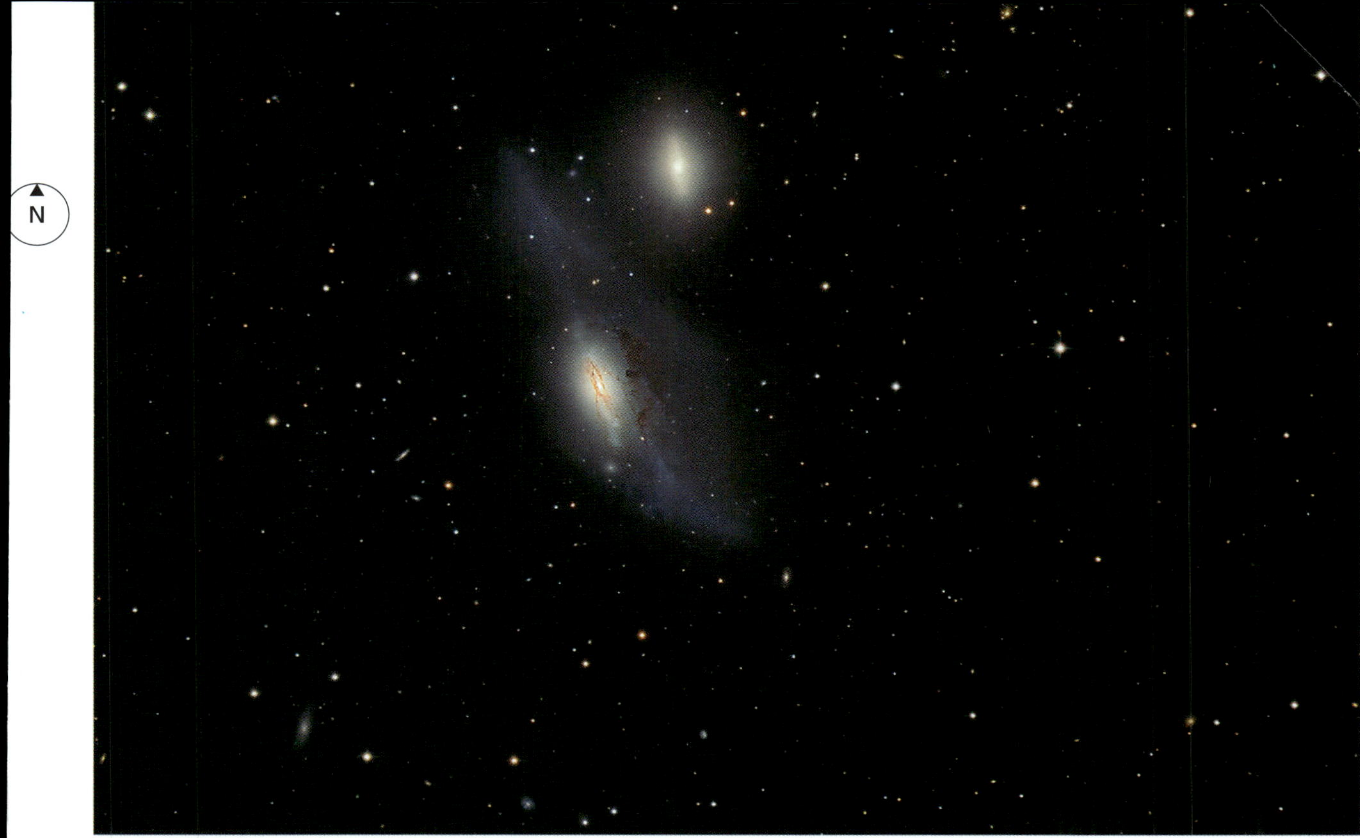

Arp 120

The SA galaxy NGC 4438 and its neighbouring galaxy NGC 4435, of type SB0, form an interacting pair, also known as the Eyes Galaxies (Arp 120). NGC 4435 at 2.4′ is smaller than NGC 4438 whose perturbed disc has an angular size of 8.5′. The system is near the centre of the Virgo Cluster, at a distance of 48 million light years and with a disc diameter of 120 000 light years. NGC 4438 contains little atomic hydrogen, but large amounts of molecular hydrogen have been detected. This gas is not in the plane of the disc, but, along with stars, has been pulled out by tidal forces due to NGC 4435. This is visible in the dust structures in the image which surround the elliptical, vertical core at its right-hand side. This right, western side of NGC 4438 faces the companion and displays a more extended halo. This distortion could be due to a close approach to NGC 4435.

The core of NGC 4438 is active and, as well as increased radio luminosity, displays the optical emission-line spectrum of a Seyfert-2 galaxy. Infrared observations indicate that a large part of the emission stems not from the nucleus itself but rather from a diffuse region to the right, which also contains dust structures. The complex morphology of NGC 4438 is explained by astronomers as due to the high-speed collision with NGC 4435, which took place with a relative radial velocity of more than 750 km/s. The transverse distance between the two galaxies today is 6000 light years. According to simulations, the closest approach was 100 million years ago at only 15 000 light years. At these high velocities, however, the interaction time is too short to lead to a merger. In such cases, gas and dust are ejected out of the disc or the plane of the disc is so perturbed that a lenticular galaxy forms. This also explains why galaxy clusters have a larger number of lenticular galaxies.

OBJECT	Arp 120, NGC 4435/ NGC 4438
CONSTELLATION	Virgo
RIGHT ASCENSION	$12^h\ 27^m\ 46^s$
DECLINATION	+13° 00′ 32″
BRIGHTNESS	11 mag
TYPE	SA0/a(s) pec
PHOTOGRAPHERS	Johannes Schedler
TELESCOPE	400-mm reflector
CAMERA	SBIG STL-11000M
EXPOSURE TIME	960 min
LOCATION	Wildon, Austria

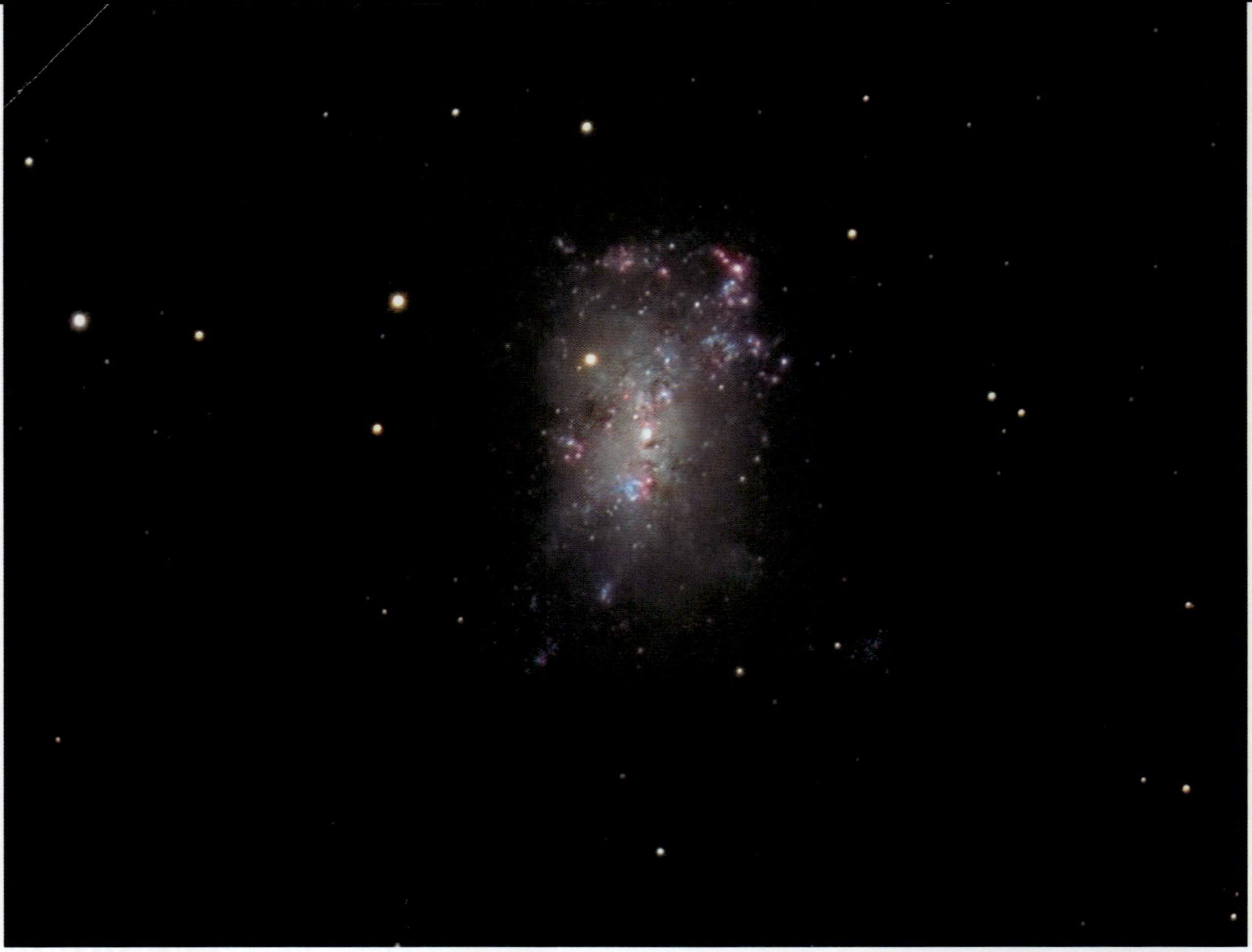

NGC 4449

This irregular galaxy is similar in form to the Magellanic Clouds and is also classified as a dwarf galaxy. Its maximum angular diameter is 6.2′ which at a distance of 12 million light years corresponds to 22 000 light years. The image shows many active regions arranged in bands which contain bright OB associations and large H II regions. This is especially conspicuous at the northern end of the galaxy. Spectral analyses indicate that stars younger than ten million years can be found in these regions. The luminosity of NGC 4449 is 1.4 times that of the Large Magellanic Cloud. The star-formation rate of the galaxy is given as 1.5 solar masses per year and explains its classification as a starburst galaxy.

Radio observations of NGC 4449 show an H I halo which at 250 000 light years exceeds the optical diameter and extends outside the area of the image. The hydrogen gas rotates in the opposite direction to that of the central gas. The halo also displays artefacts which could be the remains of tidal tails surrounding NGC 4449 as a result of previous interactions. Deep optical exposures of NGC 4449 have shown that it has a dim companion. This is catalogued as the dwarf galaxy NGC 4449B and lies outside the image. The distance from NGC 4449B to the core of NGC 4449 is about 30 000 light years and the elongated form of the companion suggests that this galaxy will be pulled apart by the tidal effects of NGC 4449, leading to a stellar stream in its halo.

OBJECT	NGC 4449
CONSTELLATION	Canes Venatici
RIGHT ASCENSION	$12^h 28^m 11^s$
DECLINATION	+44° 05′ 37″
BRIGHTNESS	10 mag
TYPE	IBm
PHOTOGRAPHERS	Volker Wendel
TELESCOPE	380-mm reflector
CAMERA	SBIG ST-10XME
EXPOSURE TIME	190 min
LOCATION	Weisenheim am Berg, Germany

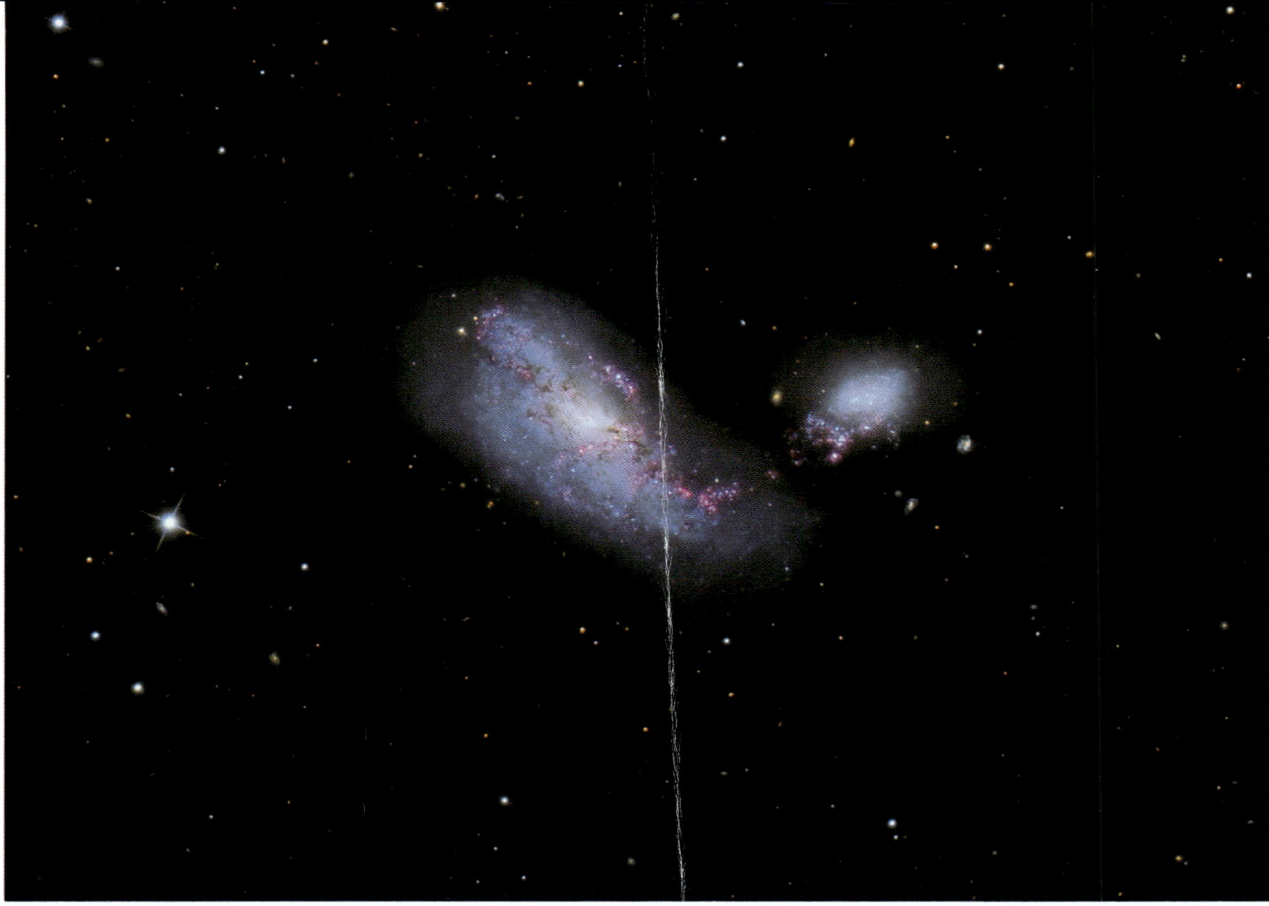

Arp 269, The Cocoon Galaxy

NGC 4485 and NGC 4490 are two neighbouring spiral galaxies which both belong to the NGC 4631 galaxy group. The pair is also known as Arp 269 and due to its diffuse texture is also called the Cocoon Galaxy. It is an SB(s)d type; the small companion NGC 4485, which has only a quarter of the mass of NGC 4490, is an irregular galaxy with a bar. The distance of both galaxies from the Milky Way is 27 million light years and their projected separation is 24 000 light years. The maximum angular diameter of NGC 4490 is 6.3′, of NGC 4485 2.3′. These values correspond to diameters of 49 000 and 18 000 light years. Both spiral galaxies are thus smaller than the standard size of a spiral galaxy, which is about 75 000 light years.

The stars in the space between the galaxies and in the tidal tail of NGC 4485 are younger than the stars in the active regions of NGC 4490. Astronomers believe that this difference in age is due to the fact that the close approach of the two galaxies did not lead to widespread star formation, but rather that stars formed almost only in the neighbouring regions of the galaxies. Simulations of the interaction also indicate that NGC 4490 and NGC 4485 could have almost collided 400 million years ago. This time span corresponds to the age of the youngest stars detectable in the tails. This suggest that both galaxies completed several orbits during this time during which a large part of their gas was distributed in the intergalactic surroundings. The halo of atomic hydrogen gas is one of the largest of these enveloping halos. The diameter of the radio structure exceeds the optical diameter by a factor of nine. The radio halo has a gas mass of 1.8×10^{10} solar masses and thus corresponds to 50 per cent of the mass of the system.

OBJECT	Arp 269, NGC 4485 / NGC 4490
CONSTELLATION	Canes Venatici
RIGHT ASCENSION	$12^h\ 30^m\ 36^s$
DECLINATION	+41° 38′ 38″
BRIGHTNESS	10.2 mag
TYPE	SB(s)d pec
PHOTOGRAPHERS	Adam Block
TELESCOPE	800-mm reflector
CAMERA	SBIG STX-16803
EXPOSURE TIME	780 min
LOCATION	Mount Lemmon SkyCenter/ University of Arizona, USA

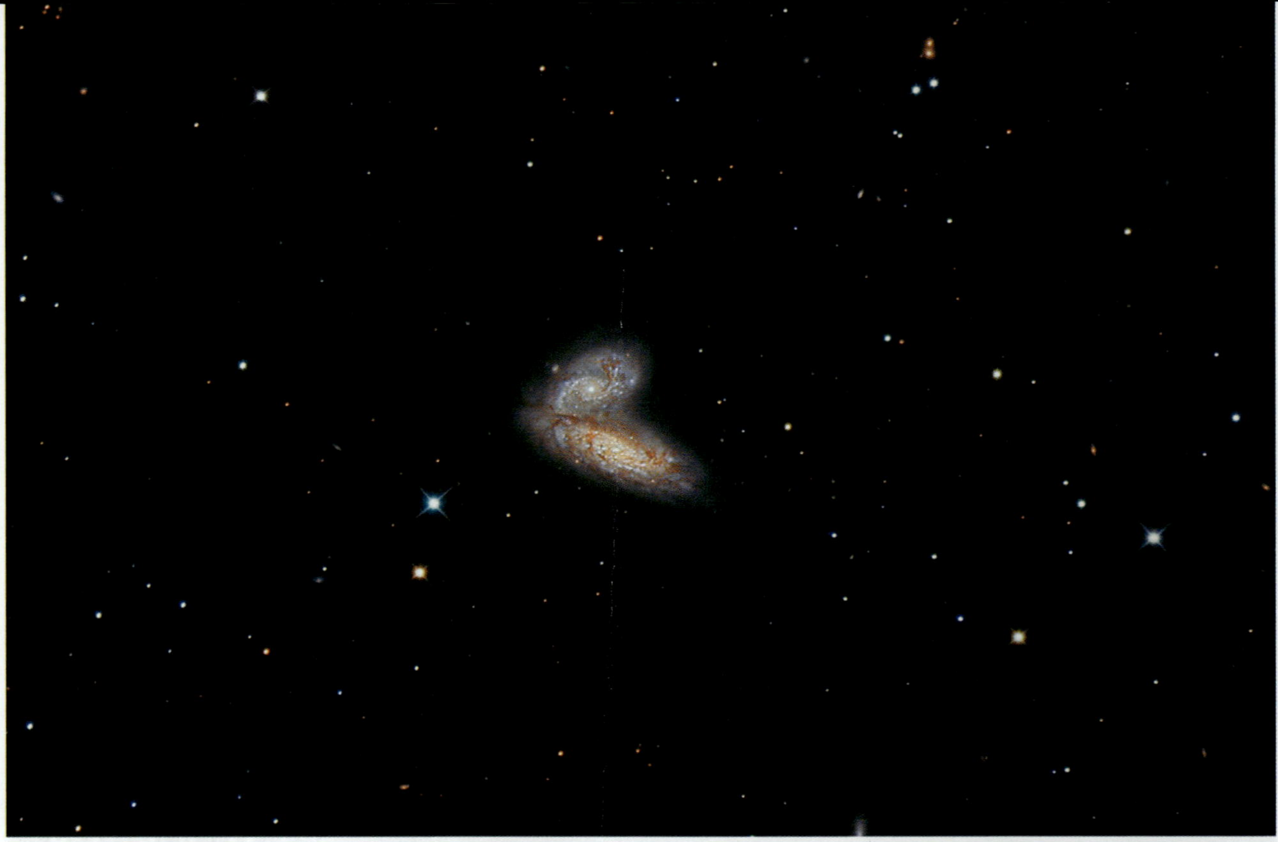

NGC 4567

The image shows the two SA(rs)bc galaxies NGC 4567 and NGC 4568. The smaller NGC 4567 lies above NGC 4568; the transverse separation of the two galaxy cores is 1.3′. The angular size of NGC 4567 is given as 2.9′, that of NGC 4568 as 4.6′. The galaxies are often described as "Siamese twins" and many images indicate that it could be an interacting system. NGC 4567 has a recession velocity of 2247 km/s; a value of 2255 is measured for NGC 4568. Within the uncertainties, the two galaxies seem to be at the same distance. Alternative distance indicators result in a distance of 83 million light years for NGC 4567 and a lesser distance of 69 million light years for NGC 4568. The uncertainties of these methods, however, are significantly larger than those based on the recession velocity and hence don't allow a precise determination of the locations of the galaxies. In contrast to these measurements, however, the visual impression seems to give the correct impression of the relative positions.

The dust structures in the left part of the spiral plane of NGC 4568 lie in the bluish main plane of its partner. One can also see some of the active regions of the lower spiral arm of NGC 4567 here, shining through the plane of NGC 4568. This allows the conclusion that NGC 4567, in relation to the line of sight, is located behind the larger NGC 4568 and is farther away.

The morphology of the galaxies appears unperturbed, and even in deep exposures no tidal tails can be seen. Radio observations indicate intact discs in both galaxies which display no perturbations and thus are not distinguishable from normal spiral galaxies. Thus NGC 4567 and NGC 4568 are probably not an interacting system, but rather two galaxies which lie in almost the same direction from the Milky Way. Since one can assume that both galaxies belong to the Virgo Cluster, it could be that this pair is a momentary effect of perspective due to their paths crossing within the cluster.

OBJECT	NGC 4567
CONSTELLATION	Virgo
RIGHT ASCENSION	$12^h\ 36^m\ 32^s$
DECLINATION	+11° 15′ 29″
BRIGHTNESS	12.6 mag
TYPE	SA(rs)bc
PHOTOGRAPHERS	Wolfgang Ries, Stefan Heutz
TELESCOPE	450-mm reflector
CAMERA	SBIG ST-10XME
EXPOSURE TIME	240 min
LOCATION	Altschwendt, Austria

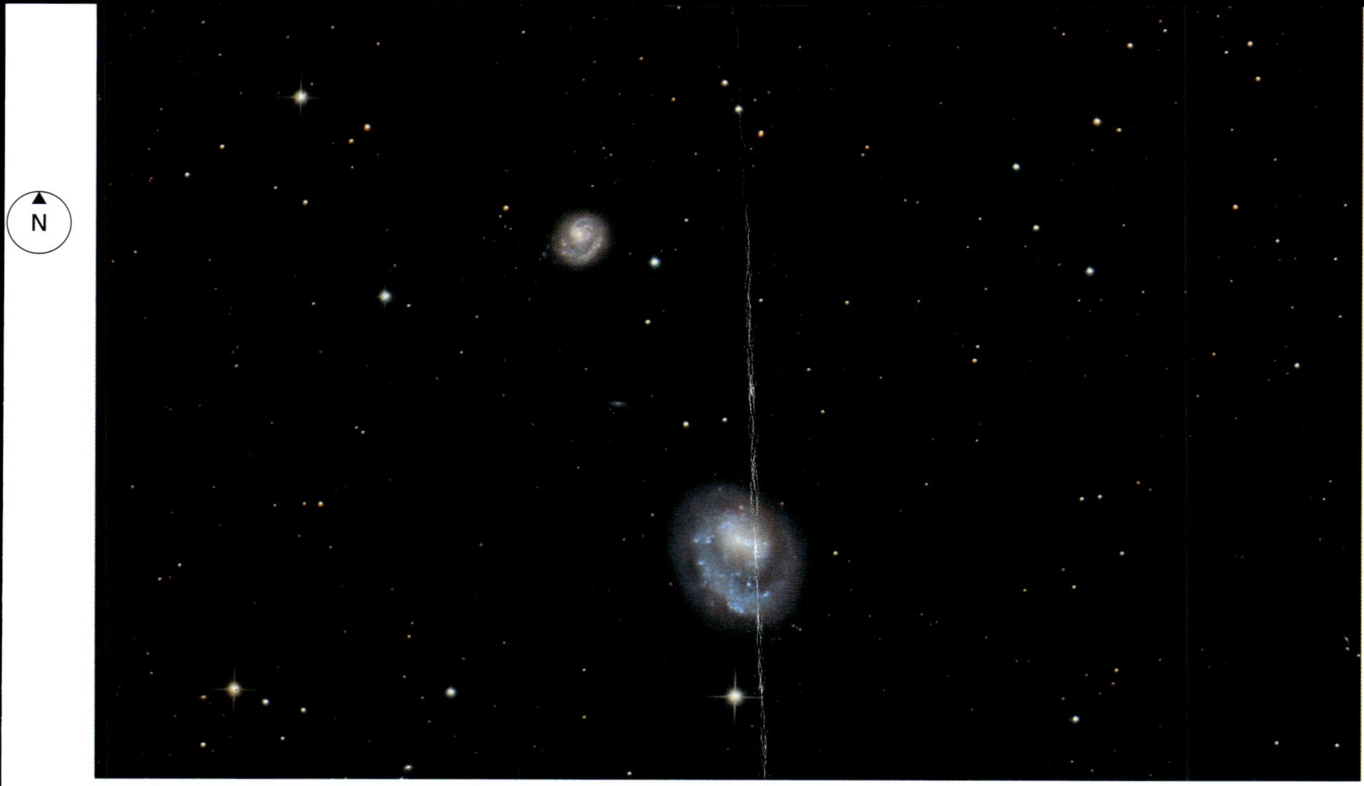

NGC 4618

The SB(rs)m galaxy just below the middle of the image is a Magellanic spiral galaxy. The asymmetric morphology of this type is defined by a massive spiral arm surrounding the core region, which is displaced relative to the centre of the galactic plane. The galactic nucleus is part of a diffuse bar which at the northeastern end is joined to a prominent spiral arm which can be followed over 300°. Only the first half of the spiral arm contains bright blue star-forming regions. NGC 4618 is at a distance of 24 million light years and its angular size of 4.2′ × 3.4′ results in a transverse diameter of the galactic disc of 29 000 light years. At about the same distance is the SAB(rs)m galaxy NGC 4625, which can be seen about 8′ farther to the north of NGC 4618. In the image, NGC 4625 is above and to the left of NGC 4618, and in this galaxy as well one recognizes the dominance of the southern spiral arms. The bright inner region with a size of 24 000 light years is smaller than that of NGC 4618, but is surrounded by a faint spiral pattern which is almost the size of the larger companion.

Both galaxies belong to one of three kinematic subgroups of the Ursa Major galaxy cluster. These subgroups of four to nine members contain galaxies with similar recession velocities. The difference in the recession velocity between NGC 4618 and NGC 4625 is 46 km/s and the projected linear separation is 48 000 light years. H I radio observations were used to study the kinematics of the galactic discs of NGC 4618 and NGC 4625 in detail. It was found that NGC 4618 has a massive disc as compared to the halo, the disc making up 60 per cent of the gas mass. In NGC 4625 this fraction is only eight per cent, since here the halo mass is dominant. Simulations of close encounters of two galaxies with such a large difference in the masses of the discs indicate that a bar should form in the disc-dominated galaxy. The evolution of structures in the halo-dominated galaxy, on the other hand, is much slower. This explains why NGC 4618 already displays a typical Magellanic spiral morphology with a central bar while NGC 4625 displays less asymmetry and the spiral structure of the precursor galaxy is still recognizable.

OBJECT	NGC 4618
CONSTELLATION	Canes Venatici
RIGHT ASCENSION	$12^h\ 41^m\ 33^s$
DECLINATION	+41° 09′ 03″
BRIGHTNESS	11.2 mag
TYPE	SB(rs)m
PHOTOGRAPHERS	Richard Müller
TELESCOPE	310-mm reflector
CAMERA	Atik 4000
EXPOSURE TIME	470 min
LOCATION	Lohmar, Germany

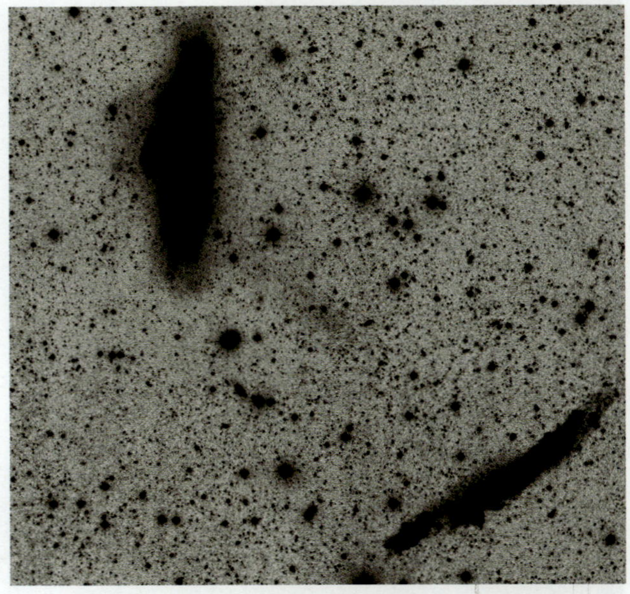

NGC 4631

The images show several galaxies of a group in the constellation of Canes Venatici (CnV II Cloud). Above left one sees the edge-on spiral galaxy NGC 4631 which has a small companion, the small dwarf elliptical galaxy NGC 4627 at a distance of 2.7′. The halo of NGC 4631 is at the northern side, facing NGC 4627, is inhomogeneous, and shows arc-like patches of light which are indications of tidal forces. The halo structures are clearly visible in the negative and contrast-enhanced image. The small companion has induced flows of matter out of and within the plane of the disc and so, in addition to warping the disc, has caused the formation of many H II regions and clusters with young stars across the entire diameter of NGC 4631. The distance of NGC 4631 is 20 million light years and its angular size of 15.5′ corresponds to a visible diameter of 90 000 light years. Due to its form, NGC 4631 is also known as the Herring Galaxy or Whale Galaxy. The galaxy NGC 4656 is visible at the lower right of the image and known as the Hockey Stick Galaxy or Crowbar Galaxy; the translation of the German name "Angelhaken-Galaxie", Fish Hook Galaxy, would also be appropriate. The distance from NGC 4631, which is seen edge on, is 31′, and the corresponding projected distance 180 000 light years. The bright structure forming a kink at the northeastern end of NGC 4656 is catalogued as NGC 4657. It is, however, not an independent galaxy, but rather the relic of a close approach to NGC 4631.

Astronomers have tried to reconstruct the history of interactions within this group of three with the help of radio observations. Several tracks are visible in the radio, indicating atomic hydrogen which was lost along former orbits. Thus, there is a semicircular arc north of NGC 4631 with NGC 4627 at the end. There are also three additional straight tracks unrelated to the companion. One of these tracks points straight at the interaction relic NGC 4657 and radio emission covers 70 per cent of the area between the two large galaxies. It thus seems probable that there was a close approach of NGC 4631 and NGC 4656. During this approach, the disc of NGC 4631 was warped and a tidal tail formed at the northeastern end. Detailed analyses of this tidal tail have shown that it contains 3.8×10^8 solar masses of hydrogen and thus could lead to the formation of a gravitationally independent tidal dwarf galaxy.

OBJECT	NGC 4631
CONSTELLATION	Canes Venatici
RIGHT ASCENSION	$12^h\ 42^m\ 08^s$
DECLINATION	+32° 32′ 29″
BRIGHTNESS	9.8 mag
TYPE	SB(s)d edge-on
PHOTOGRAPHERS	Josef Pöpsel, Stefan Binnewies
TELESCOPE	600-mm reflector
CAMERA	SBIG STL-11000M
EXPOSURE TIME	405 min
LOCATION	Skinakas Observatory, Crete, Greece

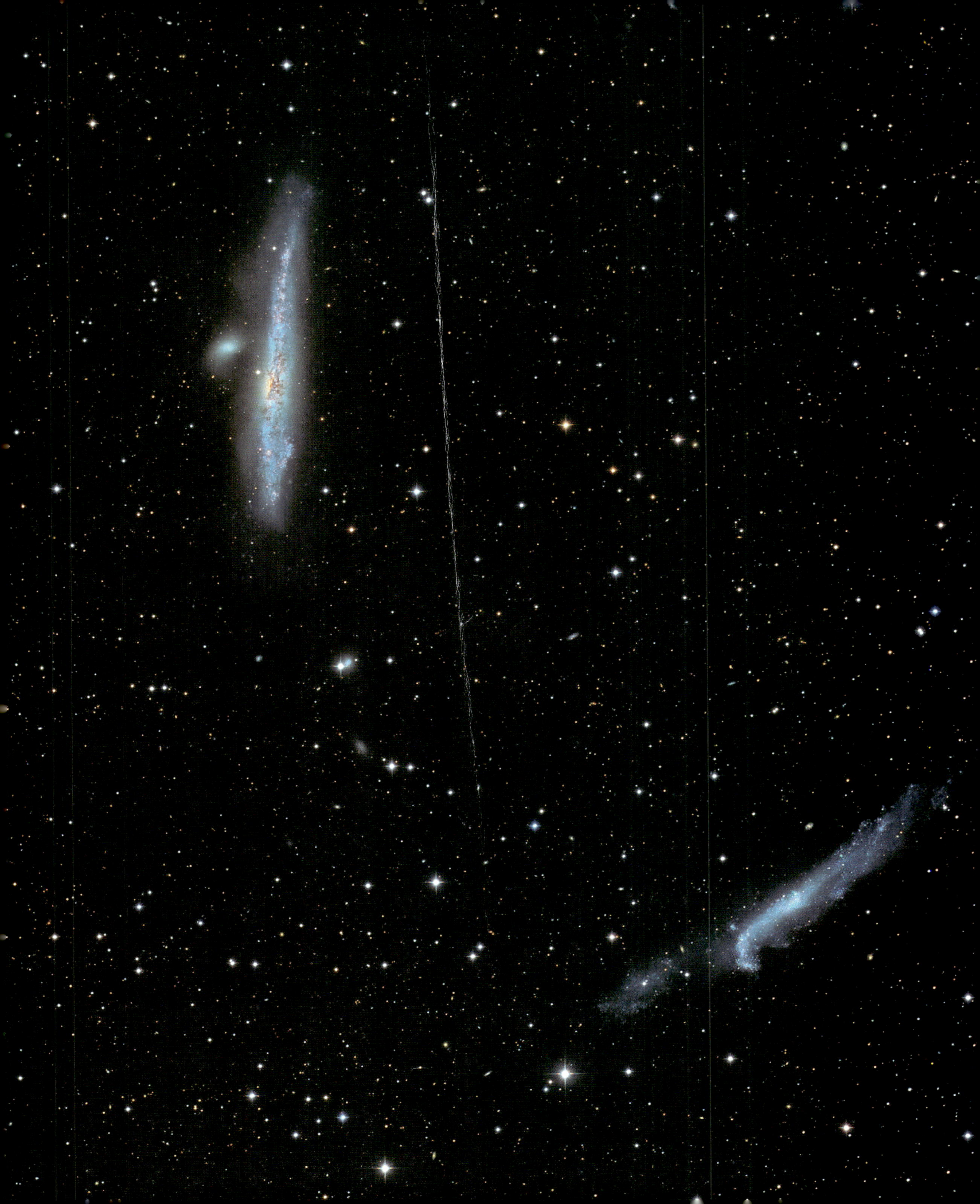

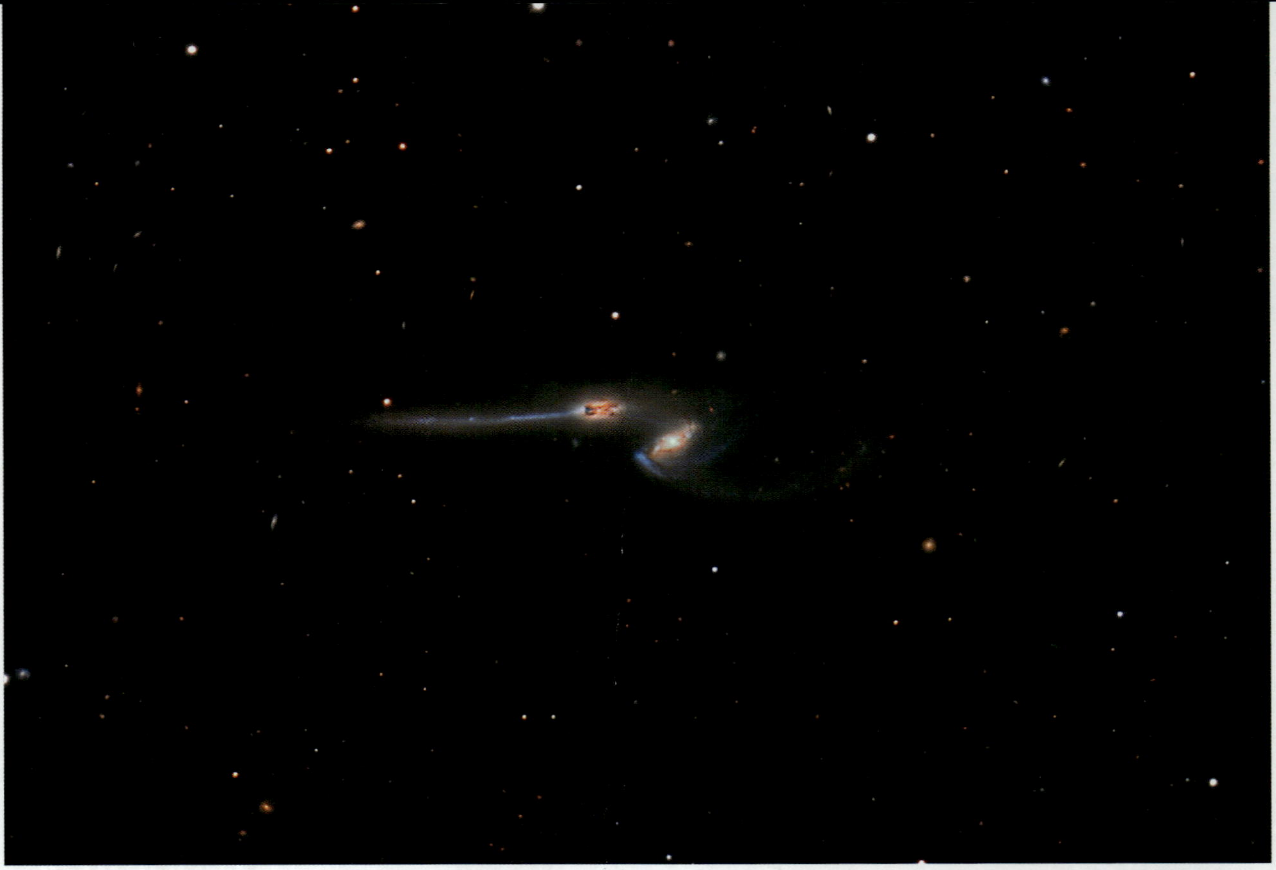

Arp 242

The two galaxies NGC 4676A and NGC 4676B form an interacting pair in which the galactic nuclei are separated by only 0.6′. At the northern end of the system is NGC 4676A whose thin tidal tail extends over 1.9′. Since the southern galaxy also has a tidal tail, the similarity to the tails of mice has led to the name "the Mice" for this system, which is catalogued as Arp 242. The distance from the Milky Way is 300 million light years and the projected separation of the cores corresponds to 52 000 light years. NGC 4676A is classified in catalogues as an irregular galaxy, although the dust lane crossing the core suggests an edge-on spiral galaxy. The SB(s)0 description of NGC 4676B is due to the diagonal bar. At the northeastern end of the bar, a bright blue region can be seen, from which extend a tidal tail pointing to the south and a short spiral arm.

The length and linearity of the northern tidal tail points to a collision or a close encounter (both linear in projection) and to the edge-on orientation of NGC 4676A. The dimmer and bent southern tail is an indication of the face-on view onto NGC 4676B. It can thus be assumed that the spiral planes of the galaxies are perpendicular to each other. The fact that the two tidal tails both have the same length indicates the equal masses of the interacting galaxies. In the last few decades, this system was often used for simulations, enabling astronomers to study precisely the dynamical details of this system. Thus it was shown that both galaxies approached each other on parabolic orbits and that the pericentre was reached 170 million years ago. The orbital plane of the motion of the galaxies is seen almost edge-on from the Milky Way. If one could see the system from the side, however, then this would reveal the conspicuous symmetry of the two tidal tails which form large, diametrically opposed arcs.

OBJECT	Arp 242, NGC 4676
CONSTELLATION	Coma Berenices
RIGHT ASCENSION	$12^h\ 46^m\ 11^s$
DECLINATION	+30° 43′ 38″
BRIGHTNESS	13.6 mag
TYPE	Galaxy pair
PHOTOGRAPHERS	Makis Palaiologou, Stefan Binnewies
TELESCOPE	1.3-m reflector
CAMERA	Andor DZ 436
EXPOSURE TIME	240 min
LOCATION	Skinakas Observatory, Crete, Greece

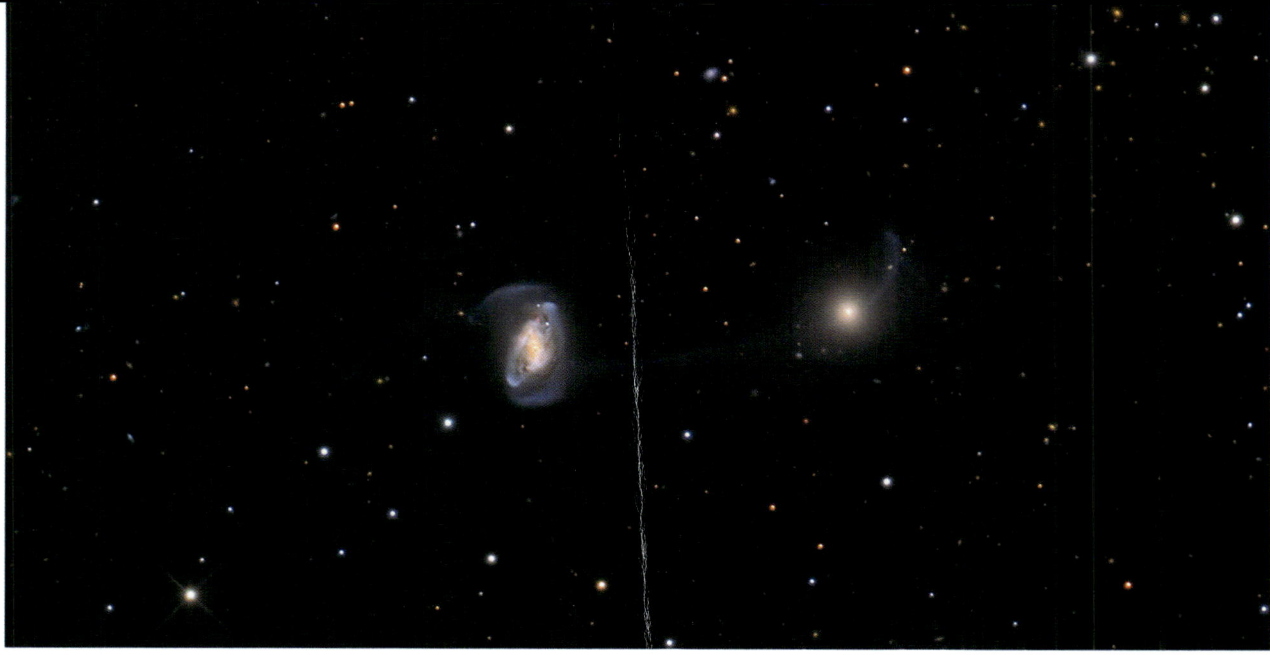

Arp 104

The system Arp 104, consisting of the galaxies NGC 5218 and NGC 5216, is part of the small galaxy group LGG 854. The astronomer P. C. Keenan described this conspicuous system in 1935, which is why the label "Keenan's system" is often found in catalogues. At the right, southern end is the galaxy NGC 5216; 4.2′ farther to the north of NGC 5218. Both galaxies lie at a distance of 137 million light years from the Milky Way and are connected by a thin tidal tail with a length of almost 170 000 light years. In addition, they both have disturbed morphologies which together with the tidal tails at the northern and southern ends of the system indicate a long history of interaction. NGC 5218 is surrounded by bright, prominently blue tidal tails. At the centre of this galaxy is a wide bar surrounded by dust structures with active areas containing H II regions visible at the northeastern end. The E0-type galaxy NGC 5216, on the other hand, is surrounded by a double shell structure, the outer of which has a diameter of 2.8′ and thus exceeds the size of NGC 5218. Both galaxies have tidal tails pointing away from the partner, so-called countertides, which differ in form and colour. In NGC 5216, a short and weakly structured tidal tail continues the curved path of the connecting streamer, whereby this extension has the form of a diffuse and dim fanning out of the tail. The tidal tail emerging from the upper end of NGC 5218 appears similarly diffuse but much bluer. At its concave east side the material is dense and thus brighter. The blue colour is due to young stars and radio observations confirm that most of the neutral hydrogen of this system is located at this northern end.

In detailed studies, astronomers could also detect a two-part structure in NGC 5218 which can be described by two discs tilted with respect to each other. Since these two discs are counter-rotating, one suspects that NGC 5218 is the product of one or more mergers. The described countertides are typically opposite the centre of gravity of the system and indicate a close approach in the past. Simulations of possible scenarios under the assumption that both galaxies are on parabolic orbits results in a mass ratio of 2 : 1 (NGC 5218 to NGC 5216). The last approach thus took place 330 million years ago and the galaxies passed within just 40 000 light years of each other. The simulations also show that the compact galaxy NGC 5216 pulled material out of NGC 5218 which then formed the streamer between the two galaxies and the diffuse tidal tail at the southern end.

OBJECT	Arp 104, NGC 5216 / NGC 5218
CONSTELLATION	Ursa Major
RIGHT ASCENSION	$13^h\ 32^m\ 07^s$
DECLINATION	+62° 42′ 03″
BRIGHTNESS	13.6 mag
TYPE	E0 pec / SB(s)b? pec
PHOTOGRAPHERS	Adam Block
TELESCOPE	600-mm reflector
CAMERA	SBIG STL-11000M
EXPOSURE TIME	500 min
LOCATION	Mount Lemmon SkyCenter/ University of Arizona, USA

N

NGC 5153

The constellation of Hydra with 1300 square degrees has the largest area of all 88 constellations recognized by the International Astronomical Union in 1922. The image shows the elliptical galaxy NGC 5153 together with some neighbouring galaxies. These associations, however, are not part of the well known Hydra galaxy cluster around Abell 1060, which lies 35° farther to the west. The distance to NGC 5153 is 180 million light years and the bright halo with an angular size of 2.7′ corresponds to a diameter of 140 000 light years. Based on its distance, NGC 5153 can be seen as an extension of the cluster Abell 3565 which can be found 10° farther to the south in the constellation of Centaurus.

In the halo of NGC 5153 one can recognize the remains of a spiral galaxy which is catalogued as NGC 5152. Its bright core is 0.7′ from the centre of NGC 5153. Below and to the left, towards the northwest, is the SBbc galaxy NGC 5150 at a distance of 5.1′. Its recession velocity is 437 km/s less than that of NGC 5153; distance indicators independent of the recession velocity, however, place this galaxy behind NGC 5153. Since the values have large uncertainties, and both galaxies are surrounded by dim tidal tails, it is probable that NGC 5150 and NGC 5153 are influenced by the tidal forces of the neighbour and form an interactive triplet with NGC 5152.

The nucleus of the almost destroyed galaxy NGC 5152 is no longer located at its centre. It appears that material from the disc has been ejected to the left. This effect could be explained by a close orbit of NGC 5152 around NGC 5153. Interaction with the halo gas would lead to a loss of material as well. This assumption is supported by two arc-like luminous patches in the upper half of the halo which join to the deformed edge of NGC 5152. During the evolution from an interacting system to the product of a merger, star formation typically increases by a factor of two or three. The corresponding difference in colour is visible at the edge of NGC 5152 – the blue colour of the active regions is substantially different from the yellowish population surrounding the core of the galaxy.

OBJECT	NGC 5153
CONSTELLATION	Hydra
RIGHT ASCENSION	$13^h\ 27^m\ 54^s$
DECLINATION	–29° 37′ 05″
BRIGHTNESS	12.8 mag
TYPE	E1 pec
PHOTOGRAPHERS	Philipp Keller, Konstantin Buchhold, Bernd Flach-Wilken, Johannes Schedler, Volker Wendel (Chart 32-Team)
TELESCOPE	800-mm reflector
CAMERA	FLI Proline 16803
EXPOSURE TIME	940 min
LOCATION	CTIO, Chile

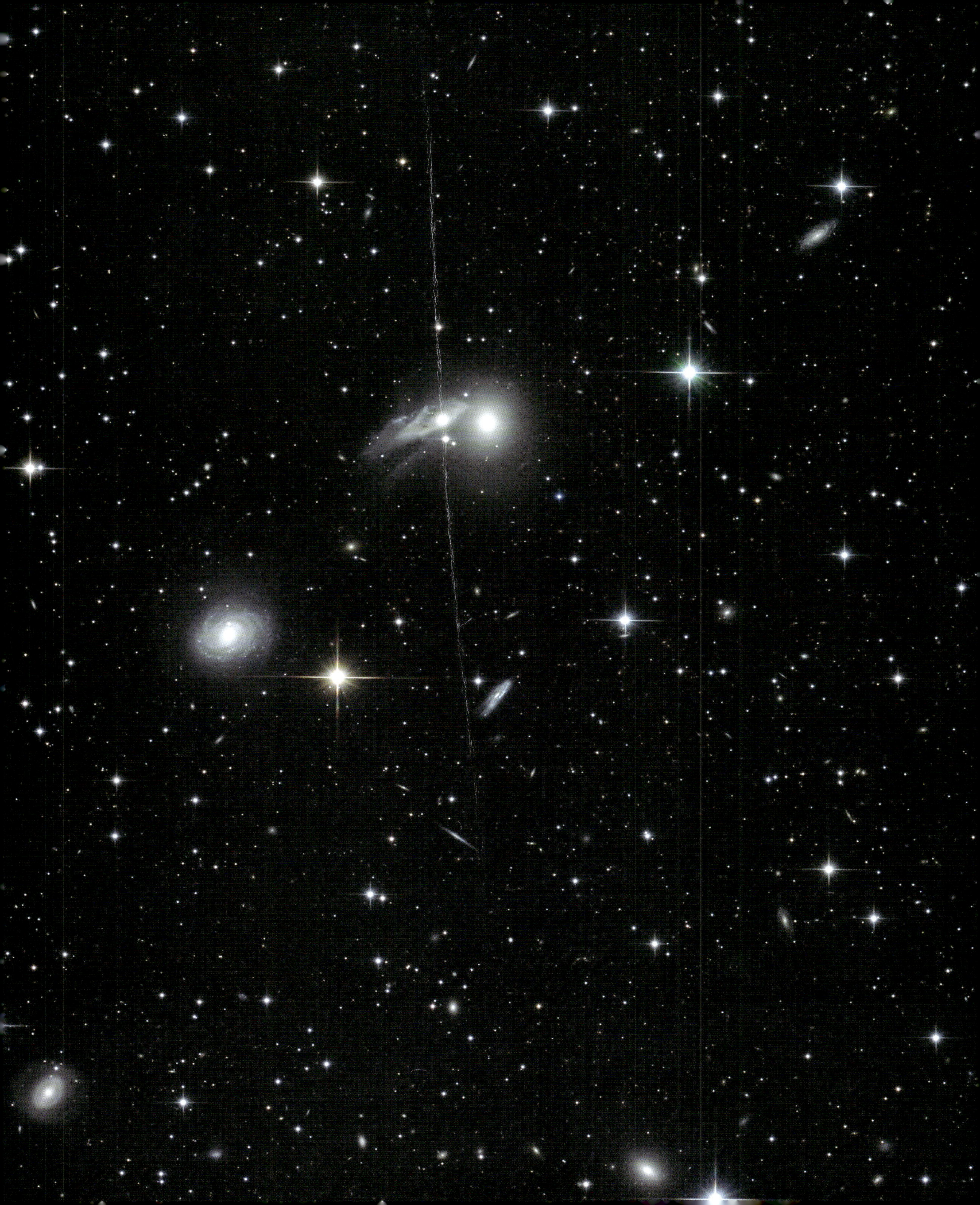

M 51, The Whirlpool Galaxy

M 51 (NGC 5194, the Whirlpool Galaxy) is one of the most photographed galaxies. The term M 51 applies to the pair consisting of the main galaxy NGC 5194 and its companion NGC 5195. The SA(s)b galaxy NGC 5194 is at a distance of 30.7 million light years and its angular size of 11.2′ × 6.9′ corresponds to a projected diameter of 100 000 light years. This is a typical size for spiral systems and thus this galaxy is only somewhat smaller than the Milky Way. NGC 5194 is viewed face-on, revealing a grand-design spiral with two arms which contain delicately structured dust lanes. At the northern end is the companion NGC 5195, an SB0/a galaxy located 4.8' from the core of M 51 and with only a third of its mass. The fanned-out, brownish dust structures are not part of NGC 5195 but rather part of the upper spiral arm of NGC 5194. NGC 5195 illuminates the arm from behind, making the detailed structure of the dust visible.

Astronomers have used simulations to study the history of this galaxy pair in detail and have found that the prominent spiral structure of NGC 5194 is the result of interaction with the companion. The interaction with NGC 5195 began 240 million years ago when they were separated by 80 000 light years. At that time, NGC 5194 had a flocculent spiral-arm pattern which had not formed any large spiral arms. NGC 5195 approached on a highly inclined orbit and, as seen from the Milky Way, moved in front of the plane of NGC 5194. 120 million years ago, it finally penetrated the plane in the southern region, slowing down the companion. It was thus possible for NGC 5195 to orbit the larger galaxy. During the following passage behind, the formation of the spiral arms visible today began in NGC 5194. According to simulations, the system today is in a state shortly before a second penetration of the plane of the larger galaxy by the companion. This forthcoming penetration is indicated by an increase in perturbation effects in NGC 5194. In the zone where they touch and in the neighbouring northern spiral arm of the galaxy one can see a large number of bright, blue stars and large H II regions. These were formed by the flow and compression of gas and dust and by shock waves crossing through this zone. The activity in this region is also visible in the emission of X-rays which is due to hot gas in the immediate surroundings of the stars. The perturbations have also already influenced the density waves responsible for the spiral pattern in NGC 5194. Thus, Vorontsov-Velyaminov rows are visible in both spiral arms, best visible as a 120° kink in the region between the two galaxy cores. According to simulations of the interaction of the two galaxies, NGC 5195 will fall into the inner region of NGC 5194 in 70 million years and the galaxies will merge with each other. In the smaller image of NGC 5194 one can also see the supernova SN2011dh in the outer spiral arm (see arrow). At the time of its discovery in June 2011 SN2011dh had a visual magnitude of 14 mag.

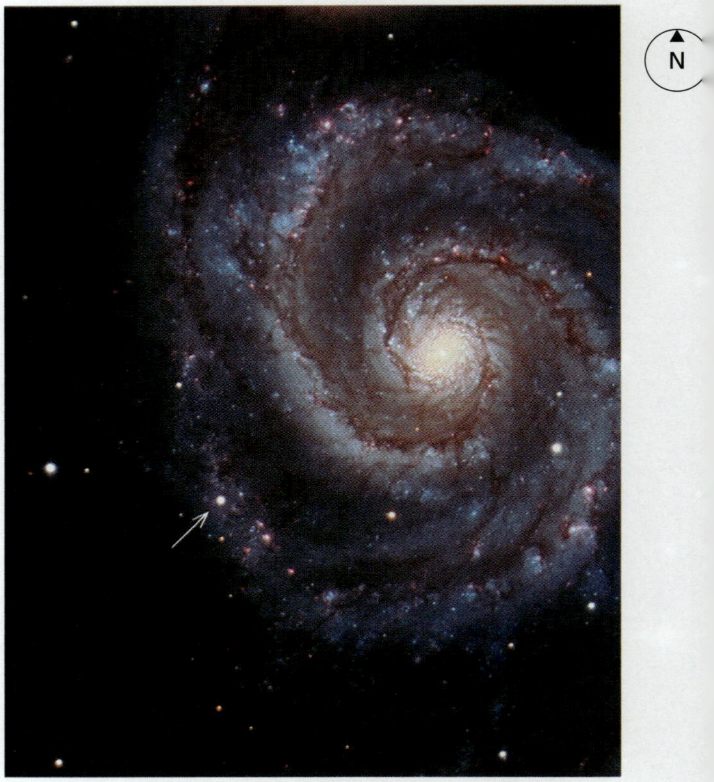

OBJECT	M 51, NGC 5194
CONSTELLATION	Canes Venatici
RIGHT ASCENSION	$13^h\ 29^m\ 56^s$
DECLINATION	+47° 13′ 50″
BRIGHTNESS	8.9 mag
TYPE	SA(s)b and SB0/a
PHOTOGRAPHERS	Josef Pöpsel, Stefan Binnewies
TELESCOPE	600-mm reflector
CAMERA	SBIG STL-11000
EXPOSURE TIME	285 min
LOCATION	Skinakas Observatory, Crete, Greece

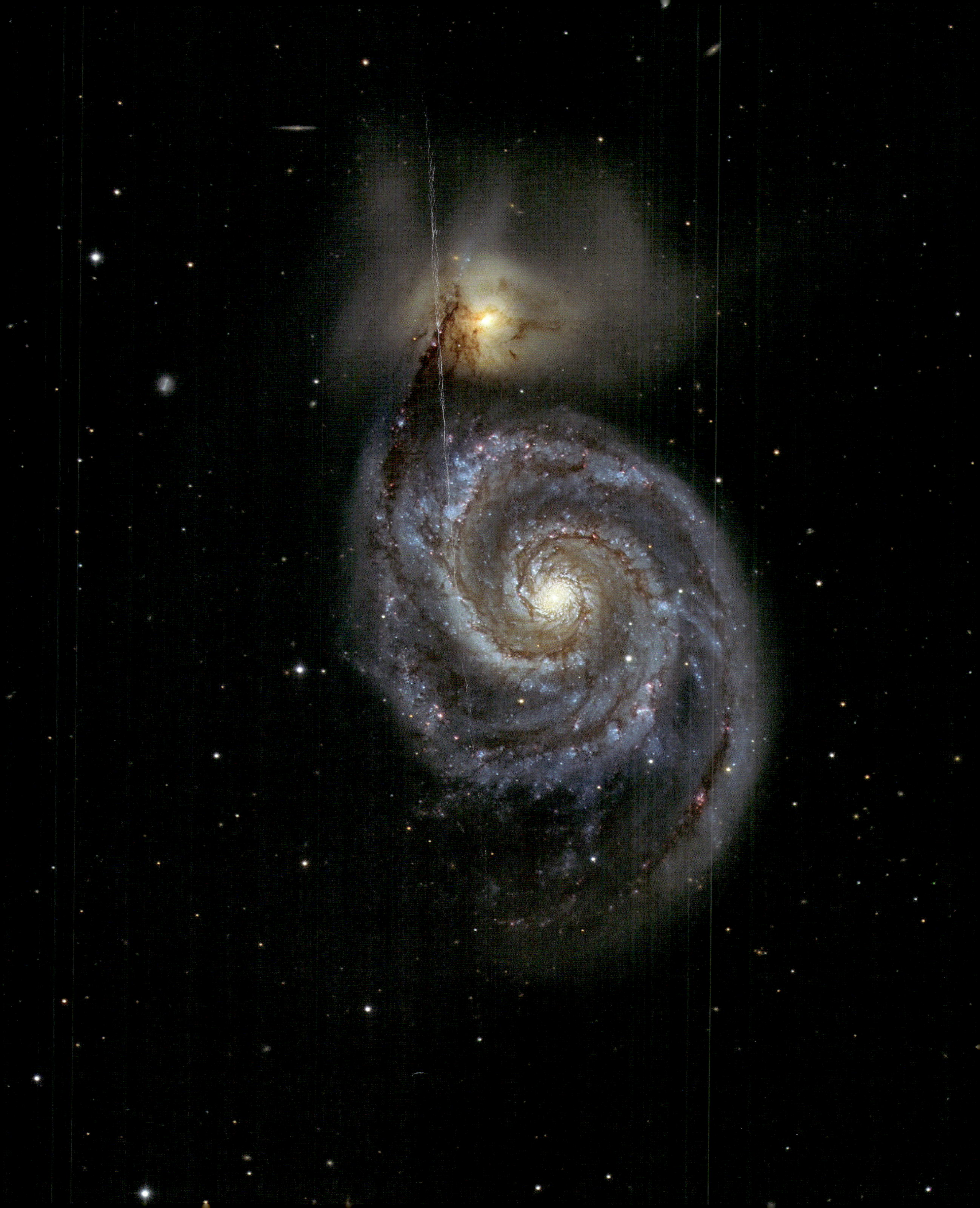

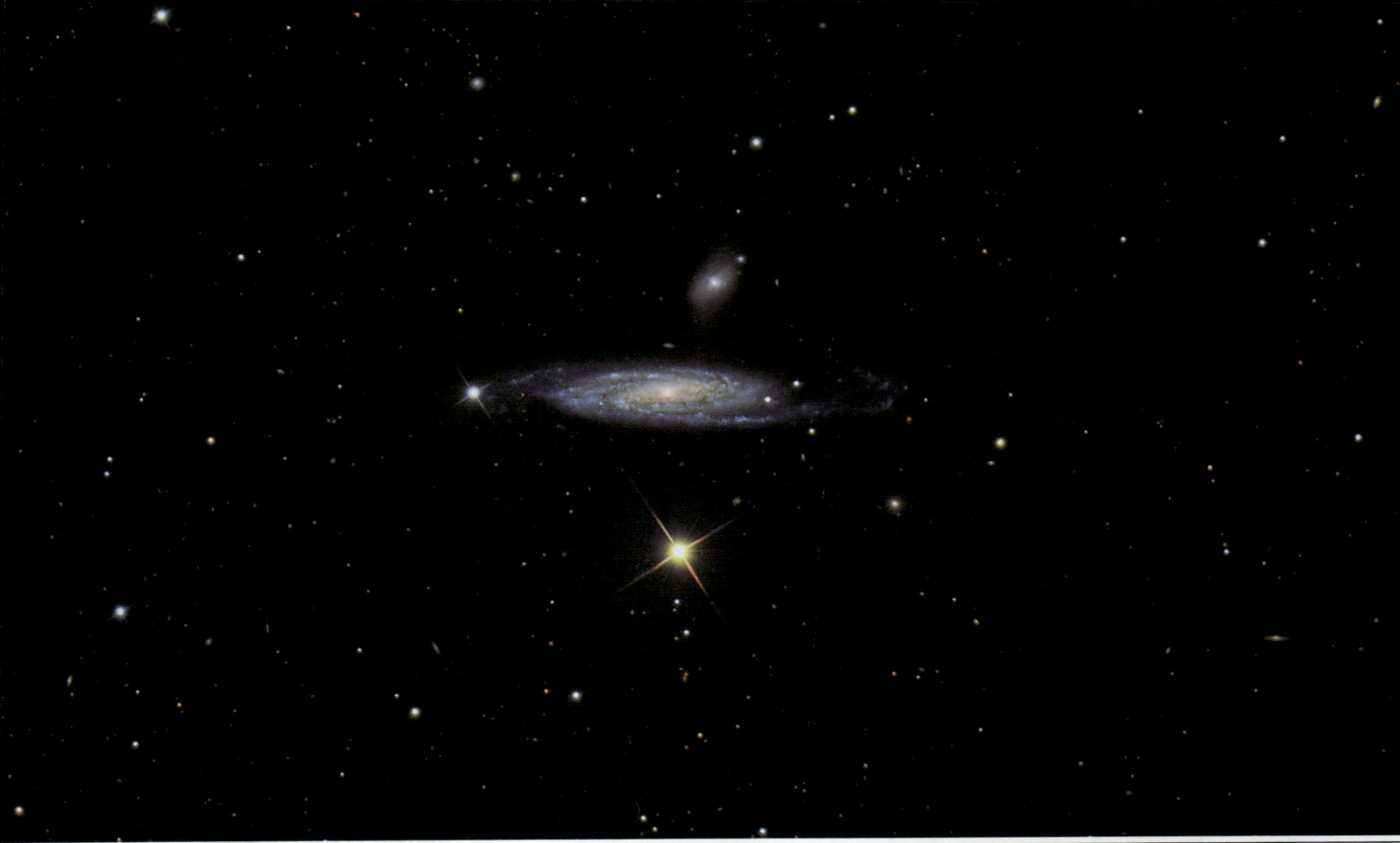

NGC 5297

The SAB(s)c galaxy NGC 5297 and its small companion, the S0 galaxy NGC 5296, have an angular separation of just 1.5′. The recession velocities of the two galaxies differ by only 155 km/s, whereby NGC 5297 is receding faster than its companion. The distance to NGC 5297 is 112 million light years and its angular size of 5.6′ × 1.3′ corresponds to an enormous diameter of 182 000 light years. The transverse separation of the two galactic cores is 48 000 light years. NGC 5296 also has a deformation in its halo in the direction of NGC 5297. The core of NGC 5296 is delimited by a dust lane on its western side which contains bright point sources, which are active regions with OB associations.

The large galaxy NGC 5297 has a less disturbed morphology than its smaller companion NGC 5296. Part of the right, southeastern spiral arm is separating from the galactic disc and its pitch angle here deviates from that of a grand-design spiral. However, neither a tidal tail nor a bridge to NGC 5296 are visible here. Due to the edge-on orientation, the velocity profile of the rotating spiral plane is easy to measure and displays the flat form typical of spiral galaxies. Since the perturbations don't affect the inner regions, astronomers suspect that the galaxies are passing each other very slowly. The relative velocity in such a slow encounter is lower than the internal velocity dispersion of both galaxies, which are 60–150 km/s for NGC 5297 and 300 km/s at the centre of NGC 5296. It thus appears possible that NGC 5296 has collected material from the gas-rich spiral plane of NGC 5297 and this has caused the distortion of the halo in the S0 galaxy.

OBJECT	NGC 5297
CONSTELLATION	Canes Venatici
RIGHT ASCENSION	$13^h\ 46^m\ 24^s$
DECLINATION	+43° 52′ 20″
BRIGHTNESS	12.5 mag
TYPE	SAB(s)c edge-on
PHOTOGRAPHERS	Josef Pöpsel, Stefan Binnewies
TELESCOPE	600-mm reflector
CAMERA	SBIG STX-16803
EXPOSURE TIME	435 min
LOCATION	Skinakas Observatory, Crete, Greece

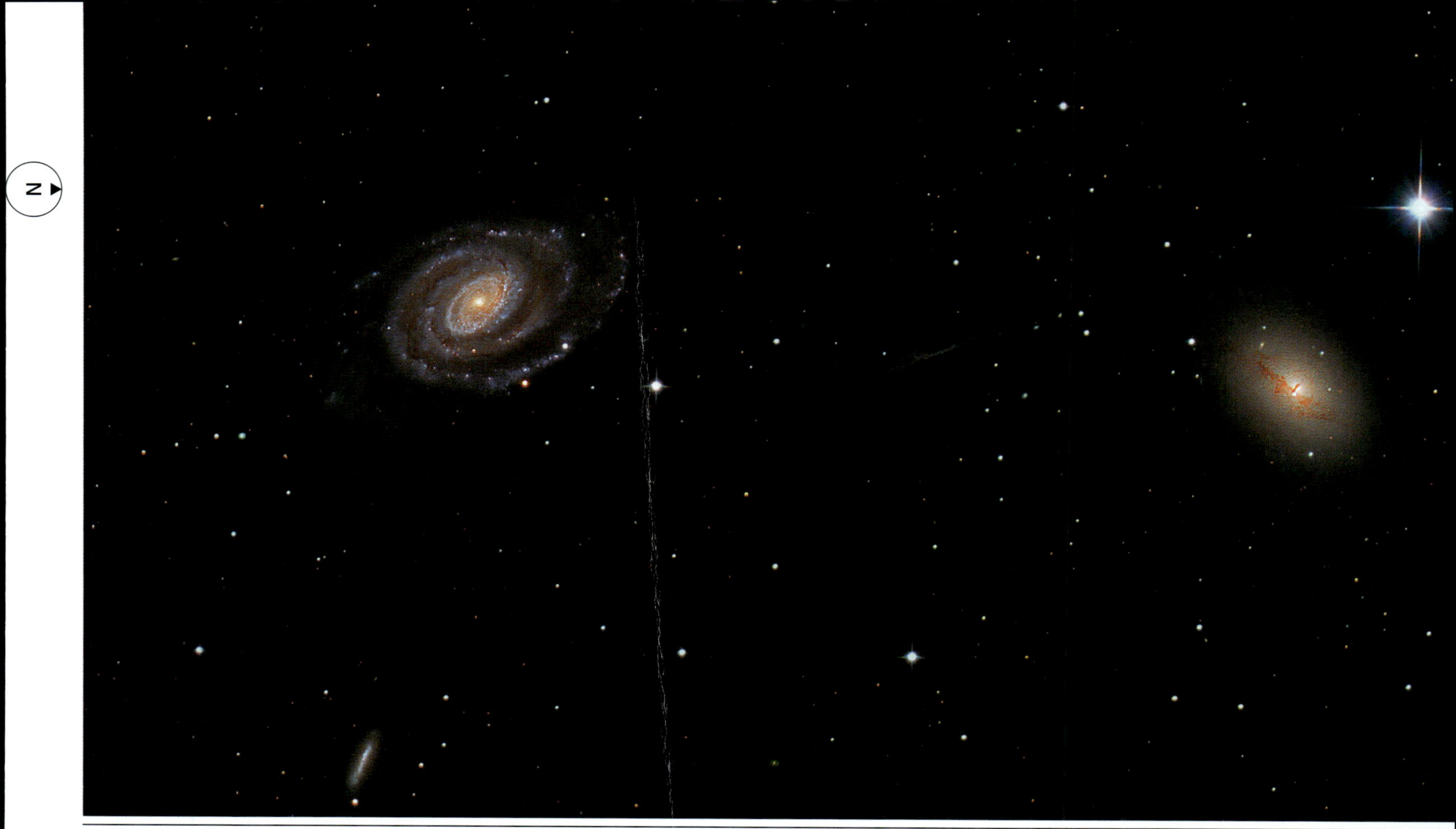

NGC 5364

The three galaxies in the image belong to a filament at the eastern edge of the Virgo Cluster and lie at a distance of about 50 million light years. At the upper left is the grand-design spiral NGC 5364 of type SA(rs)bc and an angular size of 6.8′ × 4.4′. Its two-armed spiral pattern is well defined and the arms can be followed over more than 360°. The inner ring of bright blue stars is complete and is shifted with respect to the centre of the galaxy.

At the right edge, 14′ to the north, is the galaxy NGC 5363 whose diffuse form suggests an elliptical classification. The bright brown dust lane above its compact nucleus, however, shows that NGC 5363 is the product of a merger and that one of the two precursor galaxies must have been a spiral galaxy. The bright point source directly below the centre of NGC 5363 is a foreground star and belongs to the Milky Way.

NGC 5360 is the smallest galaxy of the triplet and lies 8.7′ to the west of NGC 5364 near the lower left edge of the image. With an angular diameter of 2.2′, which corresponds to 32 000 light years, it is only half as large as NGC 5363. Increasing the contrast does not show any traces of tidal tails. Neither do radio observations reveal any emission regions in the environment of the three galaxies. One can thus assume that their arrangement has been stable for a few 100 million years. On the one hand, this is due to the fact that the triplet is at the edge of the Virgo Cluster and not at the core, where the number of interactions is high. On the other hand, it is possible that the cluster filament, with a length of about 5° and to which the galaxies belong, is dominated by dark matter and this dominates the gravitational interaction.

OBJECT	NGC 5364
CONSTELLATION	Virgo
RIGHT ASCENSION	$13^h\ 56^m\ 12^s$
DECLINATION	+05° 00′ 52″
BRIGHTNESS	11.2 mag
TYPE	SA(rs)bc pec
PHOTOGRAPHERS	Stefan Binnewies, Josef Pöpsel
TELESCOPE	600-mm reflector
CAMERA	SBIG STL-11000M
EXPOSURE TIME	525 min
LOCATION	Skinakas Observatory, Crete, Greece

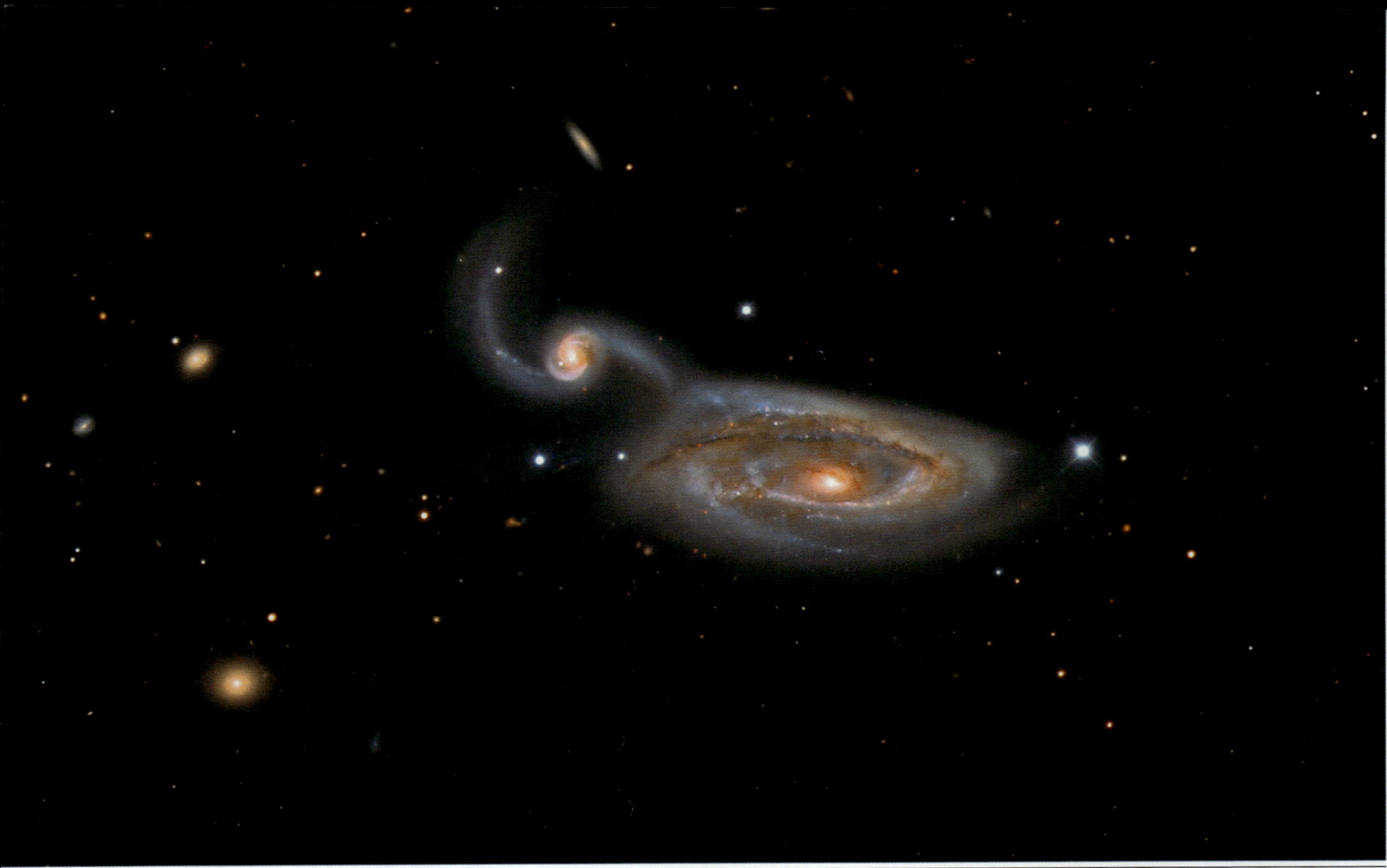

Arp 84, The Heron

The two galaxies NGC 5395 (right) and NGC 5394 (above left) form an interacting pair which is also known as Arp 84 and among amateur astronomers as the Heron. Both galaxies are at a distance of 160 million light years; the separation between their cores is 1.9′, which in projection corresponds to a distance of 88 000 light years. Comparing the recession velocities, one finds that the smaller SB(s) galaxy NGC 5394 is approaching us at 63 km/s. The larger galaxy NGC 5395 is a disturbed SA(s)b type with a spiral arm extending half way around on the southern side and which has almost no dust structures within. The western side, in contrast, has a spiral arm which leads to the beginning of the lower spiral arm in NGC 5394. Here there is also a broad dust lane extending to the northern edge of the disc of NGC 5395 where it turns into a diffuse, dim tidal tail.

Astronomers have learned from simulations that the two galaxies have had a close encounter. The flows of gas and dust which this caused led to radio emission due to an excited state of the CO molecule. About 80 per cent of the CO emission stems from the central starburst region of the galaxy NGC 5394. This is a hint that a large amount of the gas pulled out of NGC 5395 was pulled into this gravitational potential. To be sure, the larger neighbour has a broad CO line in the radio spectrum, which indicates large velocity differences in the gas flows. However, the observed low intensity of the emission means that the tidal forces have not influenced the entire disc of NGC 5395.

OBJECT	Arp 84, NGC 5395 / NGC 5394
CONSTELLATION	Canes Venatici
RIGHT ASCENSION	$13^h 58^m 36^s$
DECLINATION	+37° 26′ 20″
BRIGHTNESS	13.8 mag
TYPE	SA(s)b pec / SB(s)b pec
PHOTOGRAPHERS	Makis Palaiologou, Stefan Binnewies
TELESCOPE	1.3-m reflector
CAMERA	Andor DZ 436
EXPOSURE TIME	300 min
LOCATION	Skinakas Observatory, Crete, Greece

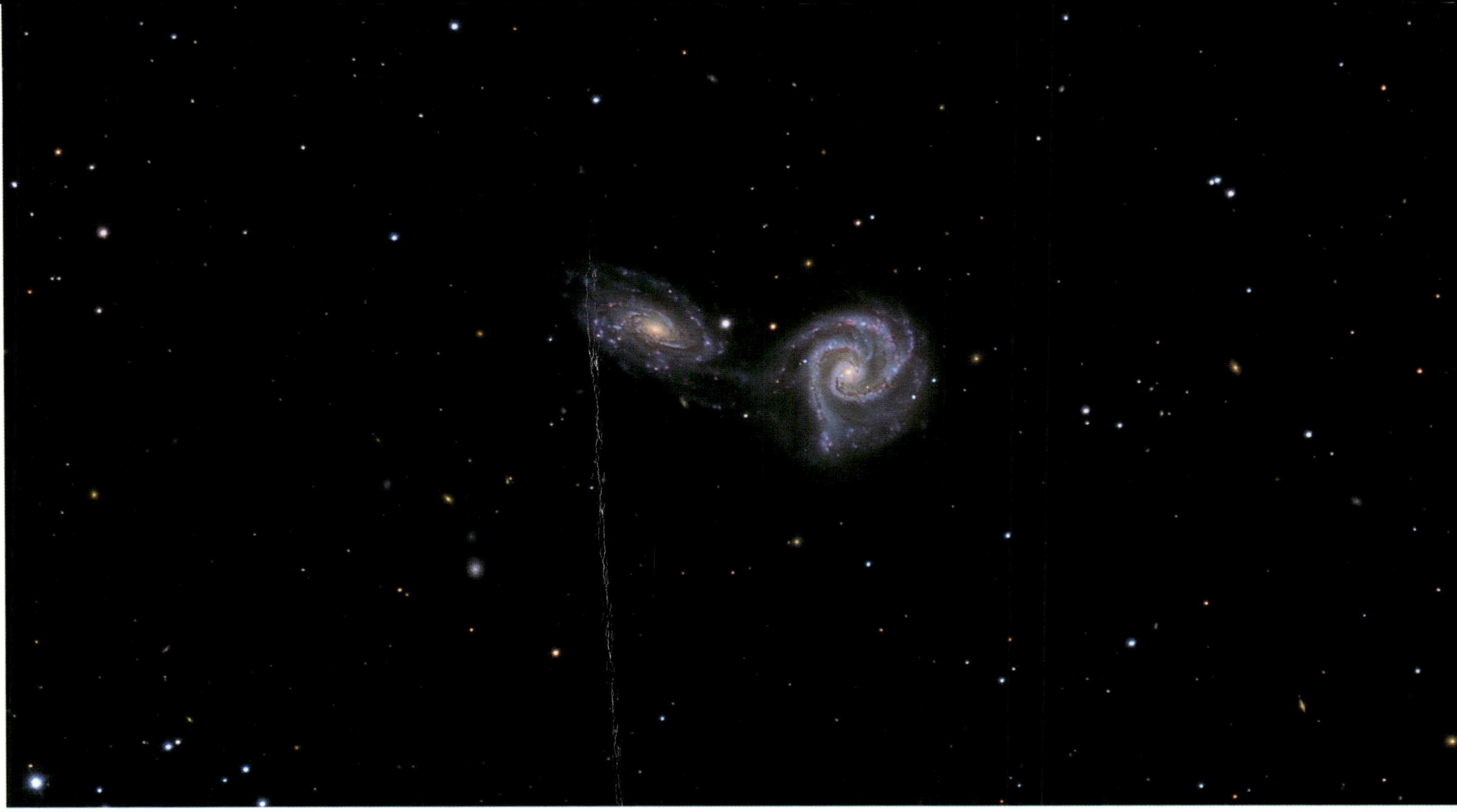

NGC 5426

The image shows the galaxy NGC 5426 at the left and 2.3′ farther north the galaxy NGC 5427. Both galaxies are classified as SA(s)c; they differ only in their orientation to the Milky Way. The diameter of the disc in NGC 5427 is 2.8′; for NGC 5426 one finds a value of 3.0′, but a precise measurement is difficult since the galaxies overlap. Both galaxies thus have a projected disc diameter of about 100 000 light years. Assuming cosmological distances, NGC 5426 lies in front of NGC 5427 and is at a distance of 112 million light years from the Milky Way. Relative to NGC 5426, NGC 5427 is receding 56 km/s more quickly, which corresponds to a distance of about two million light years larger. The uncertainties in the catalogues are about six million light years and thus exceed the calculated difference in the distances.

The morphology of both galaxies is mostly unperturbed except for the Vorontsov-Velyaminov rows visible in the downward-pointing, western spiral arm of NGC 5427. Between the galaxies one observes tidal flows which contain bright, active regions. Using detailed studies of the gas motions, astronomers were able to develop a three-dimensional model and show that NGC 5427 is moving along a parabolic orbit and will pass around the lower, western side of NGC 5426. Since the disc planes of the two galaxies are almost perpendicular to each other, no tidal tails extending out of the galactic planes can form despite the close approach. However, an exchange of material in the overlapping region could be observed with the Gemini South Telescope (8.1-m reflector, Cerro Pachon, Chile). Streams of stars, dust, and hydrogen gas, as well as new star-forming regions, could be observed between the galaxies. This confirmed NGC 5426 and NGC 5427 as a close interacting galaxy pair.

OBJECT	NGC 5426
CONSTELLATION	Virgo
RIGHT ASCENSION	$14^h\ 03^m\ 25^s$
DECLINATION	−06° 04′ 09″
BRIGHTNESS	12.7 mag
TYPE	SA(s)c pec
PHOTOGRAPHERS	Adam Block
TELESCOPE	800-mm reflector
CAMERA	SBIG STX-16803
EXPOSURE TIME	780 min
LOCATION	Mount Lemmon SkyCenter/ University of Arizona, USA

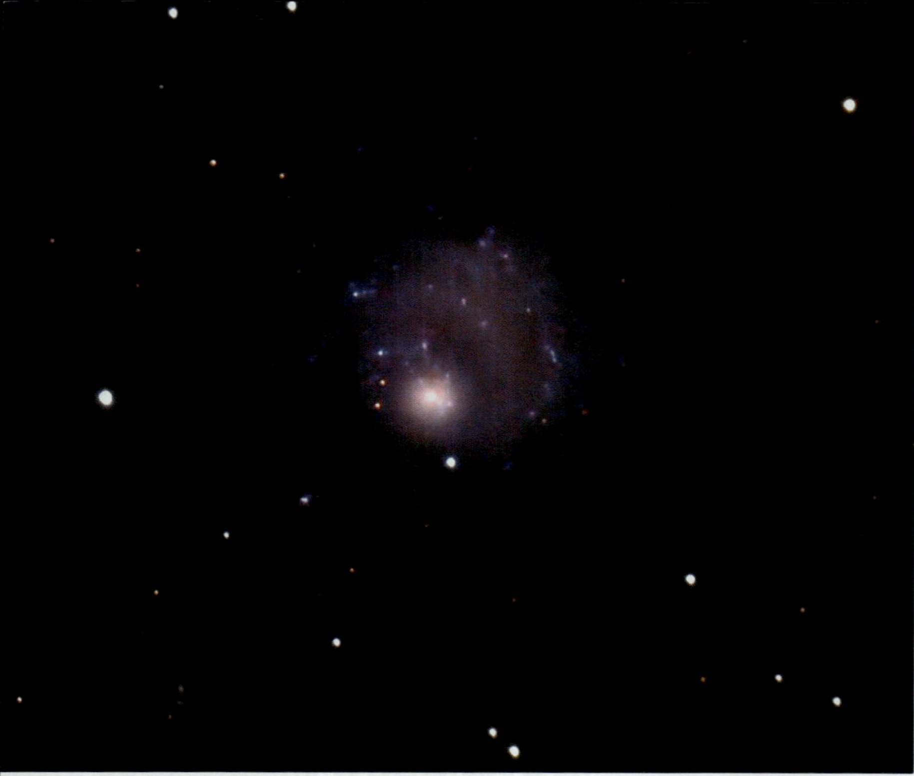

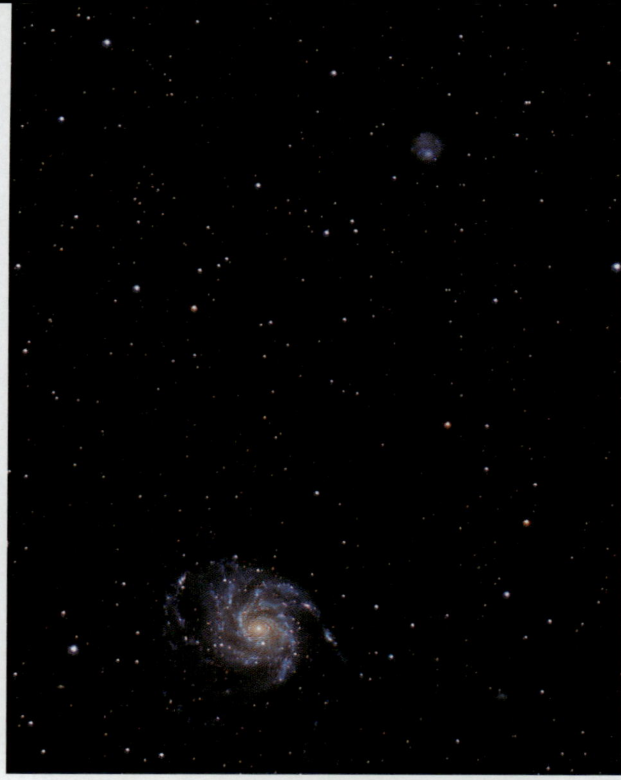

NGC 5474 is part of the group around M 101 (below). Image: Mario Weigand, 105-mm refractor, SBIG STL-11000M, exposure time 330 minutes, Col de Joux Plane, France.

NGC 5474

The galaxy NGC 5474 belongs to the group around M 101. At a distance of 44′, NGC 5474 is the closest neighbour of M 101. The projected transverse separation between the two galaxies is about 275 000 light years. M 101 is below and to the left of the image of NGC 5474 (left), to the northwest (see the image on the right showing both galaxies). NGC 5474 lies at a distance of 21 million light years from the Milky Way and its angular diameter of 4.8′ corresponds to a diameter of 29 000 light years. This particularly small diameter explains why one often finds NGC 5474 described as a dwarf galaxy. The form of NGC 5474 is asymmetric but not completely irregular. The arc visible to the south of the bright core contains the remains of two spiral arms which one can follow based on the H II regions and the bright blue star-forming regions. The brightest and largest H II region is at the edge of the disc, visible in the image below and to the left of the centre as a star-like area of increased bright purple luminosity. It is about 2′ away from the centre. The diameter of this H II region is 1100 light years.

The core of NGC 5474 is displaced compared to the rest of the galaxy. In Hα light, the main plane of the galaxy is a bright, symmetric disc. The velocity field of ionized hydrogen displays normal differential rotation in the disc. The H I radio data on NGC 5474 show that the distribution of atomic hydrogen generally follows that of the optical disc, but in contrast to the latter, the H I disc is significantly warped at the edges. These perturbations are due to tidal forces from M 101, whose total mass of 2.3×10^{11} solar masses is 19 times larger than that of NGC 5474. Since gravitational potentials depend on the ratio of mass to distance, despite its distance M 101 can still be significantly felt. The tidal forces which M 101 induces in NGC 5474 are about the same as the gravitational forces due to its own potential. This relatively strong effect significantly disturbs the spiral structure and is responsible for shifting the galactic core of NGC 5474 from the centre of the disc

OBJECT	NGC 5474
CONSTELLATION	Ursa Major
RIGHT ASCENSION	$14^h\ 05^m\ 02^s$
DECLINATION	+53° 39′ 44″
BRIGHTNESS	11.3 mag
TYPE	SA(s)cd pec
PHOTOGRAPHERS	Michael König
TELESCOPE	280-mm reflector
CAMERA	Starlight Xpress SXV-H9
EXPOSURE TIME	276 min
LOCATION	Berlin, Germany

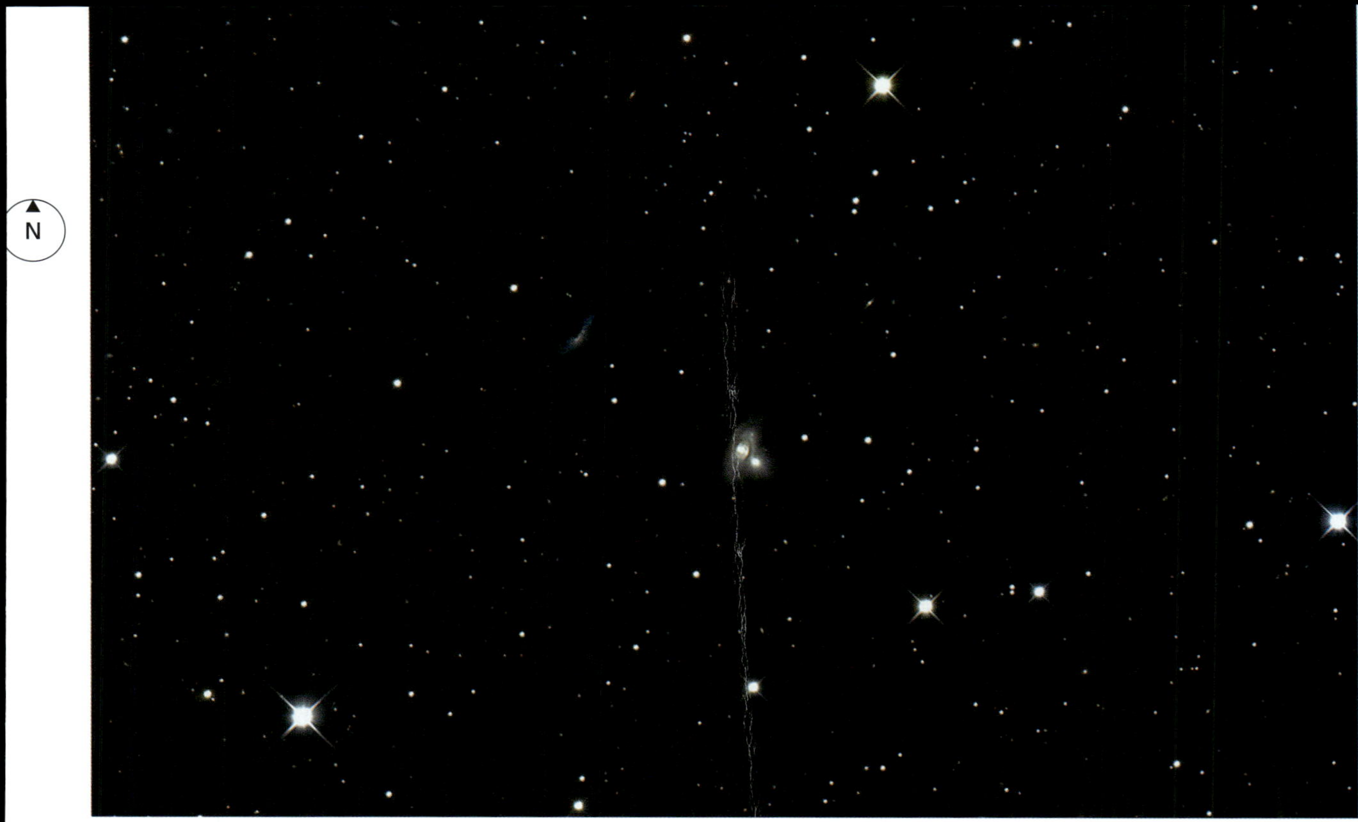

NGC 5929

This interacting pair consists of the larger partner NGC 5929 – an Sab galaxy – and the associated smaller partner galaxy NGC 5930. NGC 5930 is particularly noticeable due to its large, bright core whose large luminosity induces intense star formation in the central regions. The angular separation of the two galaxies is only 0.5′. At a distance of 120 million light years, this corresponds to a transverse separation of 17 000 light years. The diameter of the galactic disc of NGC 5929 is 1.1′, which at 38 000 light years corresponds to a relatively small Sab type. It can thus be assumed that material was removed from the disc of NGC 5929 due to interaction with NGC 5930. In the surroundings of NGC 5929 one can recognize diffuse tidal tails and to the east of the disc a dim brightening which is due to the loss of material. To the northeast of the galaxy pair is the blue, irregular galaxy UGC 9857 whose distance is given as 110 million light years. It is the product of a merger.

UGC 9857, NGC 5929, and NGC 5930 form a loose group together with two other galaxies. Additionally, interactions have led to the flow of material into the core of NGC 5929, resulting in a Seyfert-2 nucleus. This active galaxy type is characterized by highly excited, narrow emission lines in its optical spectrum. The distribution of atomic hydrogen in the central regions of NGC 5929 could be studied in detail using a combination of high-resolution radio observations (MERLIN, the Multi-Element Radio-Linked Interferometer Network) and data from the Hubble Space Telescope. There is a weak, central radio source which is flanked by two brighter radio sources to the left and right. These two sources are separated by 1.26″ or 700 light years. The weak source is the bright region around the supermassive black hole which, however, is obscured by a dust torus and thus appears reduced in luminosity. The dust torus at the core has a high inclination and its edge-on orientation explains the Seyfert-2 spectrum of NGC 5929. It is also obvious that the orientation of the torus has no relation to the face-on disc plane of the galaxy.

OBJECT	NGC 5929
CONSTELLATION	Boötes
RIGHT ASCENSION	$15^h\ 26^m\ 06^s$
DECLINATION	+41° 40′ 14″
BRIGHTNESS	13.9 mag
TYPE	Sab pec
PHOTOGRAPHERS	Michael König
TELESCOPE	350-mm reflector
CAMERA	SBIG STL-11000M
EXPOSURE TIME	220 min
LOCATION	Rimbach, Germany

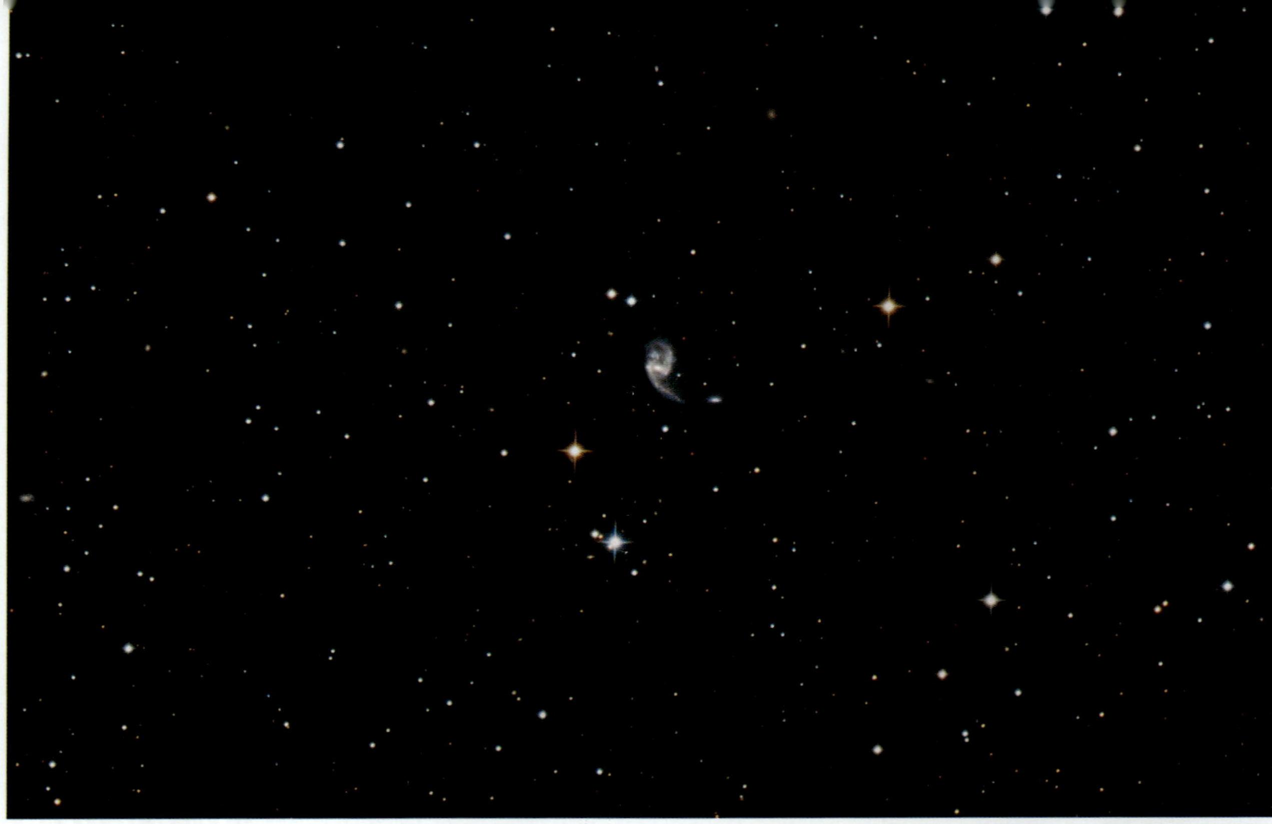

Arp 72

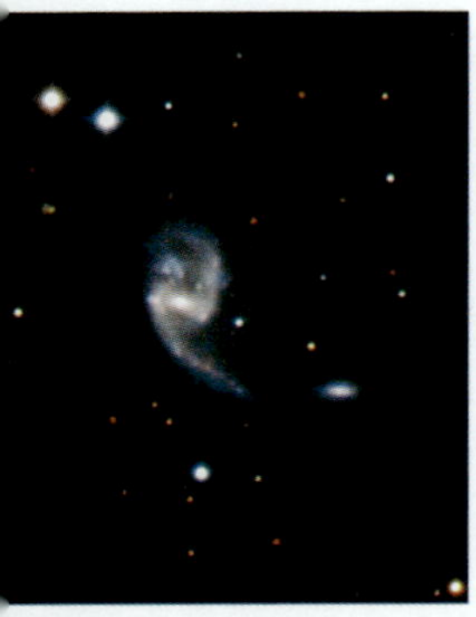

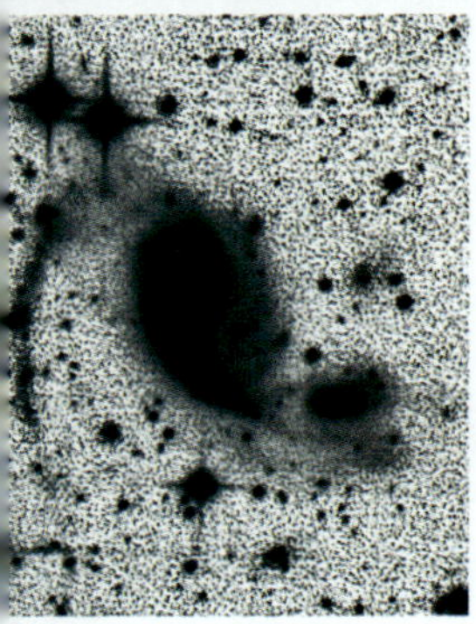

The image shows the system Arp 72 which consists of the interacting galaxies NGC 5996 and NGC 5994. The larger galaxy NGC 5996 has an angular size of 1.6′ × 0.7′ and it is easily recognized as an SBc type due to its two bright spiral arms which emerge from a broad bar. However, the upper, northern spiral arm displays a perturbed form and continues into a slightly blue, dim tidal tail which curves to the left through 180° around the system (see contrast-enhanced image). The southern spiral arm has two parts. After a kink due to a Vorontsov-Velyaminov row it points towards the compact companion NGC 5994. The separation of the two galaxy cores is 1.47′ which corresponds to 64 000 light years. Since the distance of both galaxies from the Milky Way is 150 million light years and the relative recession velocity only 7 km/s, it can be assumed that the orbital plane of their interaction is perpendicular to the line of sight. The tidal interaction during a close approach by the companion led not only to the visible perturbations of the spiral-arm morphology but also triggered the formation of many active regions with hot stars in the inner parts of the two large spiral arms.

NGC 5996 is thus also classified as a starburst galaxy and has a prominent blue contribution in its spectrum which is due to UV radiation from hot O and B stars. A further indication of peculiar activity are the typical absorption profiles due to Wolf-Rayet stars in the spectrum of the galaxy. These massive stars lose their outer envelope via strong stellar winds which are ionized by radiation from the hot core and are responsible for the typical features in the spectrum. Compared to the image above, data from the Sloan Digital Sky Survey (SDSS) indicate a prominent blue colour. This is due to the fact that the spectral data used for the online version of the SDSS cover only the red and blue regions. One can thus assume that the images shown here correspond better to reality.

OBJECT	Arp 72, NGC 5994 / NGC 5996
CONSTELLATION	Serpens
RIGHT ASCENSION	$15^h 46^m 59^s$
DECLINATION	+17° 53′ 03″
BRIGHTNESS	13.4 mag
TYPE	S?
PHOTOGRAPHERS	Wolfgang Ries, Stefan Heutz
TELESCOPE	450-mm reflector
CAMERA	ALccd 9
EXPOSURE TIME	1242 min
LOCATION	Altschwendt, Austria

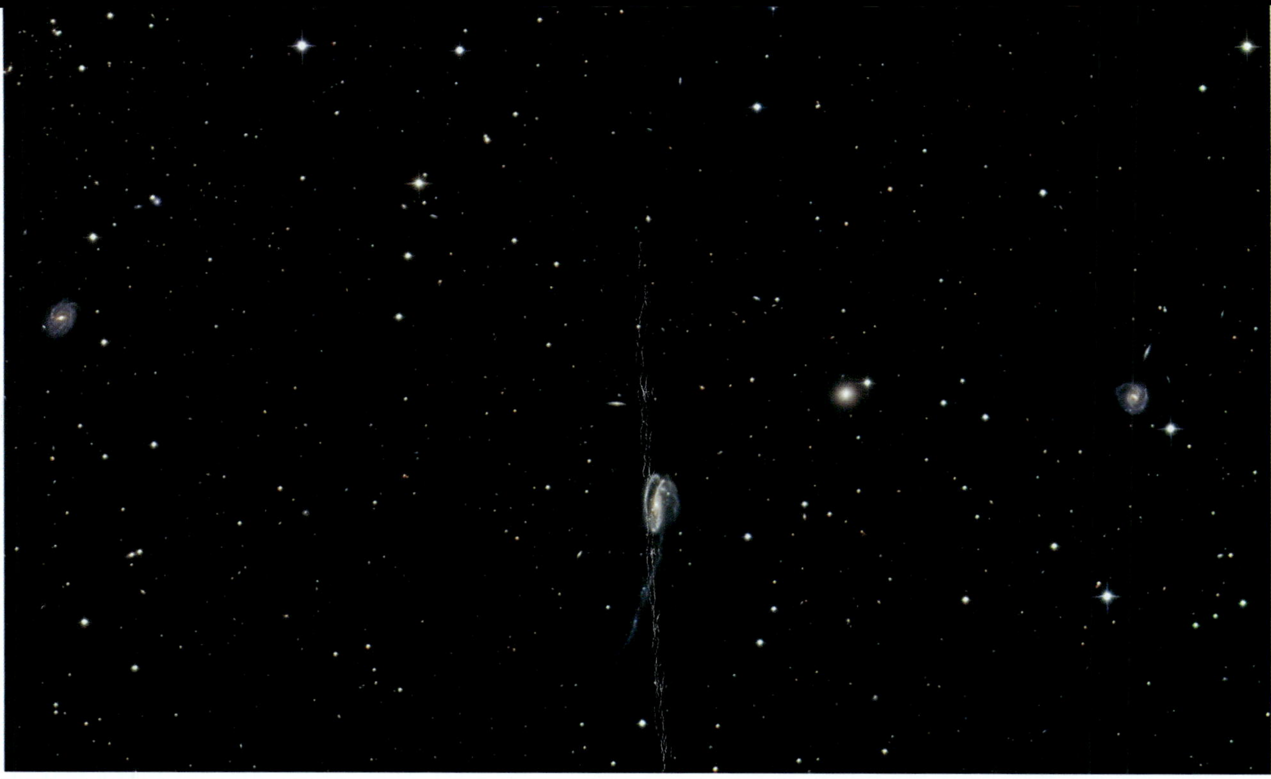

Arp 188, The Tadpole Galaxy

The SB(s)c galaxy UGC 10214 is listed as Arp 188 in the *Atlas of Peculiar Galaxies*; it became famous, however, in 2002 via an image from the Hubble Space Telescope. UGC 10214 is 430 million light years from the Milky Way and the angular size of its spiral plane is 0.9′ × 0.8′. Its most prominent feature is an eastward-pointing tidal tail whose size of 2.3′ exceeds the diameter of the disc. In projection, this corresponds to a distance of 290 000 light years. This tidal tail is responsible for UGC 10214 being known as the Tadpole Galaxy. The inner bar is vertical in the image and two light brown dust lanes can be recognized which continue from the spiral arms into the centre. The position of the arms indicates that the main plane of UGC 10214 has a high inclination of 70°. The elliptical galaxy PGC 57108, visible above and to the right of UGC 10214 close to a foreground star, has the same recession velocity and is thus a direct neighbour. There are no distance measurements for the two spiral galaxies at the right (PGC 57109) and left (PGC 57085) edges of the image. Since the diameters of their discs are only about two-thirds the diameter of UGC 10214 or PGC 57108, the assumption that they lie at a larger distance is justified.

In the downward-pointing tidal tail are prominent bright blue tracks which have been identified as bright star clusters with the help of the Hubble image. In the middle of the tail is the brightest knot, described by astronomers as a super star cluster (SSC). Its visual luminosity is 5×10^8 solar luminosities and its mass 1.3×10^6 solar masses. The stars in the SSC with ages of three to ten million years are still young and since the dynamic age of the tail is 150 million years, the SSC was formed long after the tidal tail. The tail is due to a close approach of UGC 10214 with the small, compact galaxy, the remains of which are visible in the upper right region of the disc of UGC 10214. It can further be assumed that this unnamed companion is already behind the disc of UGC 10214 and is receding from the large galaxy.

OBJECT	Arp 188, UGC 10214
CONSTELLATION	Draco
RIGHT ASCENSION	$16^h\ 06^m\ 04^s$
DECLINATION	+55° 25′ 32″
BRIGHTNESS	14.6 mag
TYPE	SB(s)c pec
PHOTOGRAPHERS	Josef Pöpsel, Volker Wendel
TELESCOPE	600-mm reflector
CAMERA	SBIG STL-11000
EXPOSURE TIME	570 min
LOCATION	Skinakas Observatory, Crete, Greece

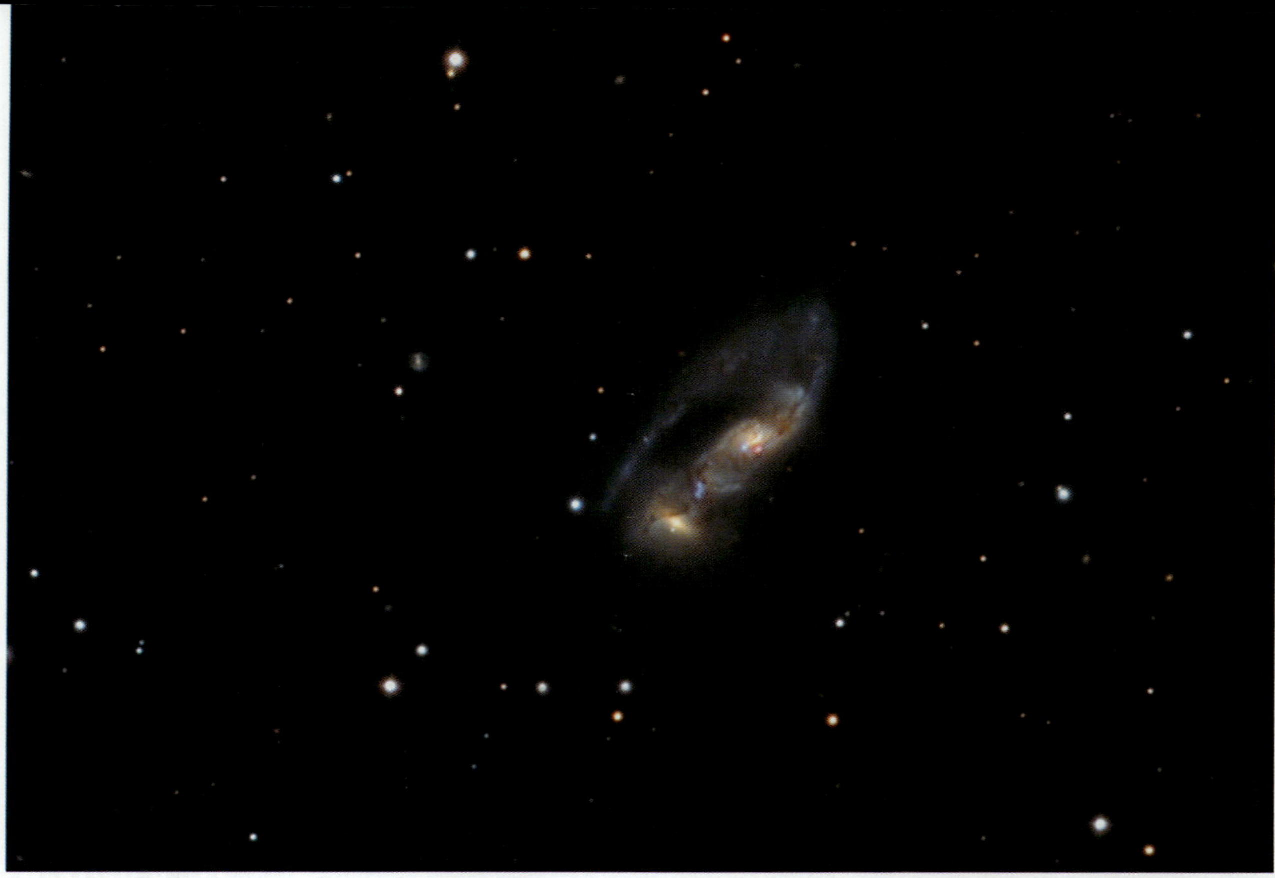

Arp 81

The interacting galaxy pair seen in the image is known as Arp 81 and consists of the galaxies NGC 6621 and NGC 6622. The northwestern partner NGC 6621 is above NGC 6622 and due to its tidal tail, catalogues list an angular size of 1.9′ × 0.7′ for this Sb galaxy; the size of NGC 6622 is smaller at 0.5′ × 0.4′. The mean distance of the two galaxies is 290 million light years; the separation between their cores is 0.7′, which corresponds to a projected distance of 59 000 light years. Both NGC 6621 and NGC 6622 contain dust filaments which are the remains of dust lanes in the spiral arms in the now absent galactic discs. NGC 6621 even has the relic of a spiral arm at the side opposite NGC 6622. Around its nucleus there is also a bright blue star-formation region with similarly bright H II regions nearby. Between the two galaxies are also apparently twisted dust formations and a three-part star-forming region.

The pair is described as an intermediate-state merger. NGC 6621 is orbiting about NGC 6622 and passing it on the right. Its relative velocity is +275 km/s. Simulations suggest that NGC 6621 has been on an arc-shaped orbit since a close approach about 100 million years ago and is now approaching the Milky Way. During the approach, material was removed from the disc plane of NGC 6621 via tidal forces, forming the tidal tail visible along the previous orbital path. The mutual interaction triggered increased star-formation rates in both galaxies and thus the formation of young, massive star clusters along a bridge of material which connects the two galaxies with each other.

OBJECT	Arp 81, NGC 6621 / NGC 6622
CONSTELLATION	Draco
RIGHT ASCENSION	$18^h\ 12^m\ 55^s$
DECLINATION	+68° 21′ 48″
BRIGHTNESS	14 mag
TYPE	Sb: pec
PHOTOGRAPHERS	Stefan Binnewies, Josef Pöpsel
TELESCOPE	600-mm reflector
CAMERA	SBIG STL-11000
EXPOSURE TIME	130 min
LOCATION	Skinakas Observatory, Crete, Greece

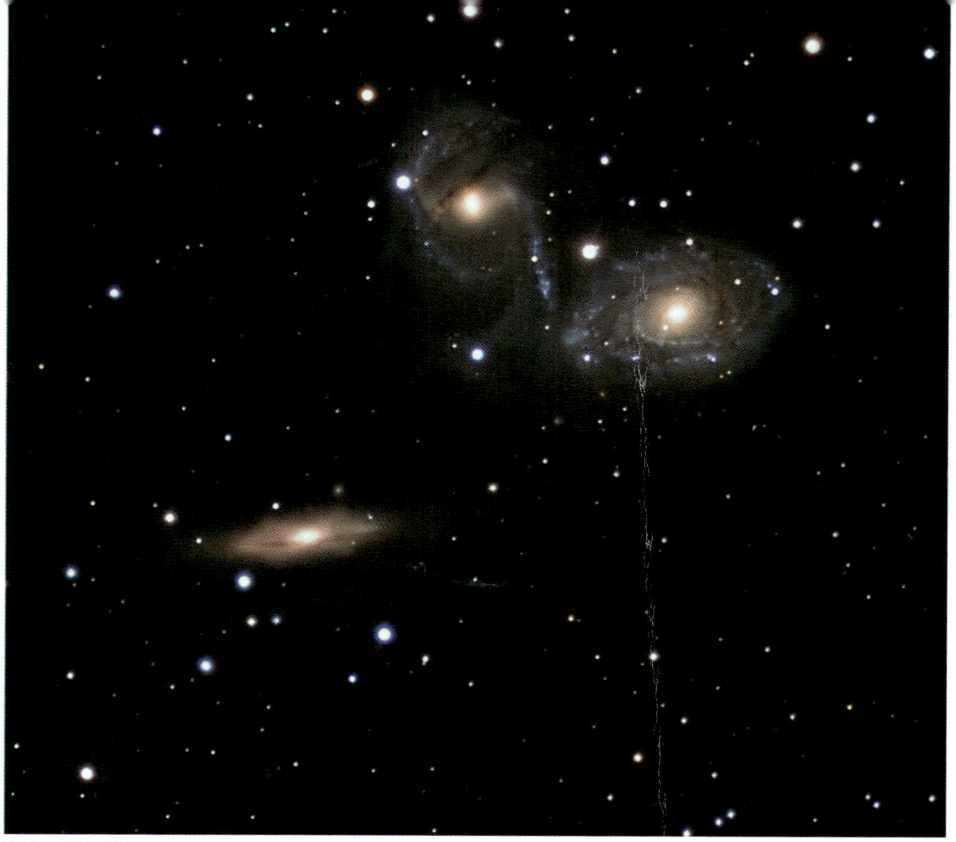

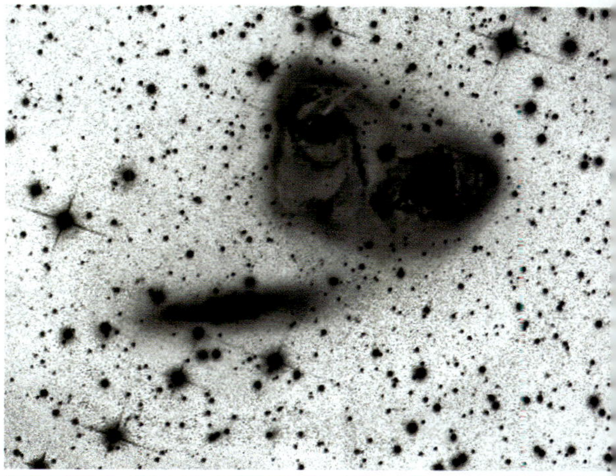

NGC 6771

The galaxy triplet shown in the image consists of the close interacting pair NGC 6770 and NGC 6769 as well as NGC 6771, located about 3′ to the lower left. In the pair, the (R)SA(r)b galaxy NGC 6769 is on the right and 1.7′ away is the SB(rs)b galaxy NGC 6770 with a prominent dust lane in its inner regions. The entire group is 170 million light years from the Milky Way. Since NGC 6769 and NGC 6770 have the same angular diameter, the disc diameter of both galaxies is about 115 000 light years. In NGC 6770, there is a diffuse glow around the bar; the western spiral arm, which conspicuously points to NGC 6769, has many young star clusters. The inner region of NGC 6769, in contrast, has a flocculent spiral pattern consisting of an inner and an outer, ring-shaped, arm structure.

The pair is located in a common envelope; the difference in the recession velocities corresponds to a relative motion of +155 km/s for NGC 6769. This allows the conclusion that the galaxy disrupted the disc structure of its partner during a close approach which also removed the halo gas. This interaction can thus explain the largely unperturbed, elliptic halo of NGC 6769 and the triangular halo of NGC 6770. Relative to the galaxy pair, NGC 6771 has a recession velocity 400 km/s larger. Increasing the contrast in the image (right) reveals hints of a bridge of material between the southern end of the galaxy pair and NGC 6771. The edge-on view of NGC 6771 also reveals the X-shaped luminosity distribution which is due to the vertical motion of stars in the bar. This is often found in galaxies with a small core region and open spiral arms. Astronomers are still debating its possible origins; most models, though, assume a gravitational coupling between the bulge and the bar. Thus tidal forces could have induced an instability in this coupling in NGC 6771, leading to a change in the shape of the bar.

OBJECT	NGC 6771
CONSTELLATION	Pavo
RIGHT ASCENSION	$19^h\ 18^m\ 40^s$
DECLINATION	−60° 32′ 46″
BRIGHTNESS	13.6 mag
TYPE	SB0(r)? edge-on
PHOTOGRAPHERS	Josef Pöpsel, Beate Behle
TELESCOPE	600-mm reflector
CAMERA	SBIG ST-10XME
EXPOSURE TIME	175 min
LOCATION	Amani Lodge, Namibia

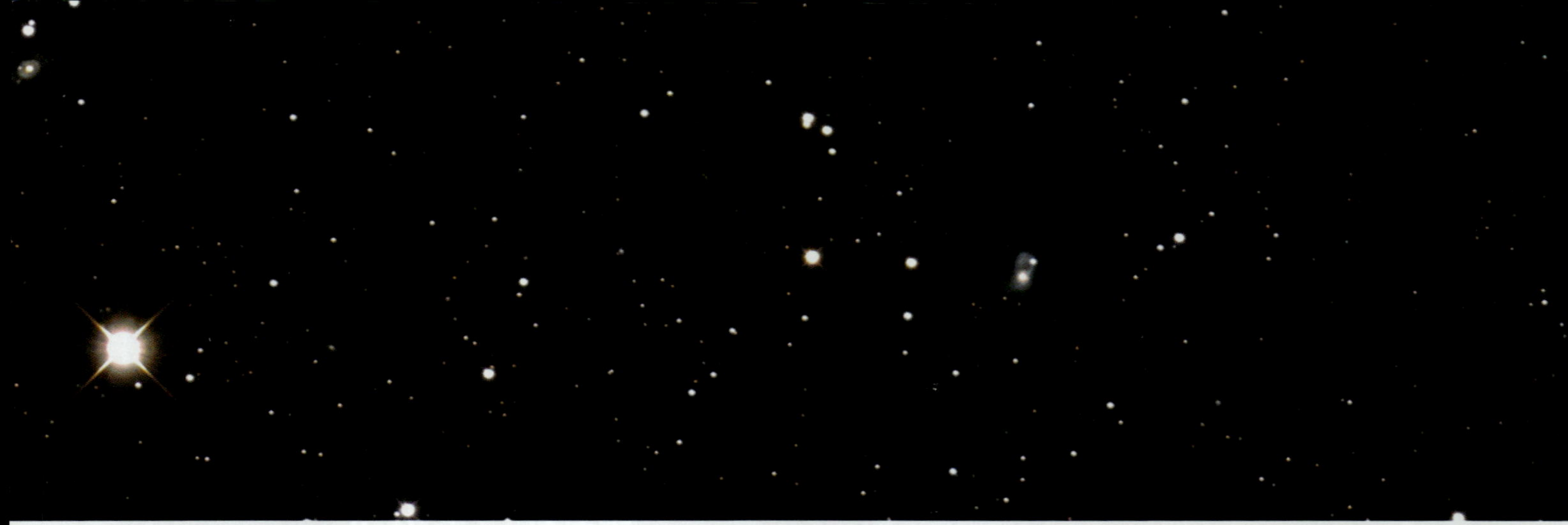

UGC 11871

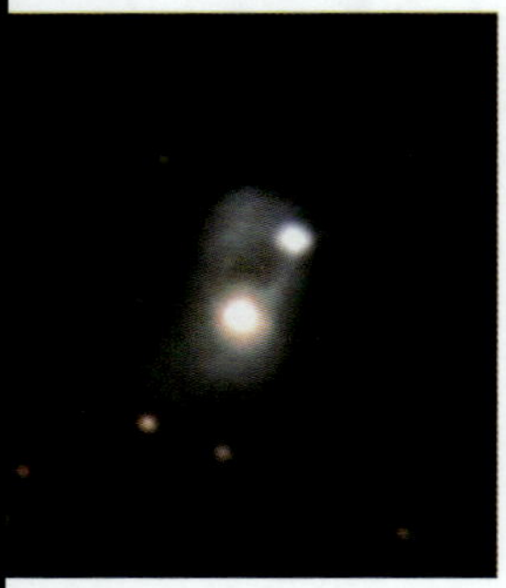

At the left of the image is the 6.8-magnitude bright K star SAO 107567 which is in the southwestern part of the constellation of Pegasus, about 4° from Enif (ε Peg). Farther to the east is a bright blue object. It is the galaxy UGC 11871 whose bright nucleus is almost completely surrounded by an oval arc (see enlargement in the small image). The point of light in the upper right of the arc is a foreground star. The angular size of UGC 11871 is 1.1′ × 0.7′ and at its distance of 365 million light years corresponds to a projected diameter of 117 000 light years. The blue colour is due to interaction-induced star formation and at its lower, northern end is a dim tidal tail in the form of a hockey stick. This extends 1.5′ and indicates that UGC 11871 is the product of a merger of two galaxies. At the left edge of the image is also a prominent ring galaxy which is catalogued as PGC 67786.

The standard theory of cold dark matter (CDM) assumes that the luminous baryonic matter in galaxies, and thus all objects in this image, is embedded in a CDM halo. This omnipresent CDM is also responsible for the constant orbital velocity in the differential rotation of the outer parts of spiral galaxies. In addition, there should be "pure" agglomerations of CDM, so-called sub-halos, with masses in the range 10^8–10^{11} solar masses. CDM theory predicts that the total number of these sub-halos exceeds that of normal galaxies by a factor of one to ten. Based on these cosmological models, astronomers predict about 300 of these sub-halos within the Local Group as compared to the approximately 45 normal members. These models allow the conclusion that physical processes allow baryonic, luminous galaxies to form only in massive sub-halos; light sub-halos, in contrast, remain galaxy-free and thus invisible.

Rather than "galaxy-free", astronomers use the term "completely dark galaxies", which could for example be detected by gravitational-lensing effects. Related to this is the conclusion that some apparently isolated galaxies should display interaction effects, although no luminous neighbour is to be seen. In order to test the hypothesis of CDM sub-halos, astronomers have constructed a *Catalogue of Isolated Galaxies*. This contains eight members which display interaction effects without having a visible interaction partner. These eight galaxies, however, correspond to only 0.8 per cent and are thus fewer than the eight per cent postulated by the theory. The low number thus casts doubt on the existence of CDM sub-halos. This statement is qualified by the fact that in some sub-halo merger products, interaction effects can no longer be seen today. UGC 11871 is one of these eight isolated galaxies with interaction effects listed among the 1050 entries in the *Catalogue of Isolated Galaxies*. In addition to the tidal tails, its active Seyfert nucleus is an indication that it is a candidate which could have formed out of the merger of a normal galaxy with a CDM sub-halo.

OBJECT	UGC 11871
CONSTELLATION	Pegasus
RIGHT ASCENSION	$22^h\ 00^m\ 42^s$
DECLINATION	+10° 32′ 59″
BRIGHTNESS	14.7 mag
TYPE	S?
PHOTOGRAPHERS	Michael König
TELESCOPE	350-mm reflector
CAMERA	SBIG STL-6303
EXPOSURE TIME	355 min
LOCATION	Rimbach, Germany

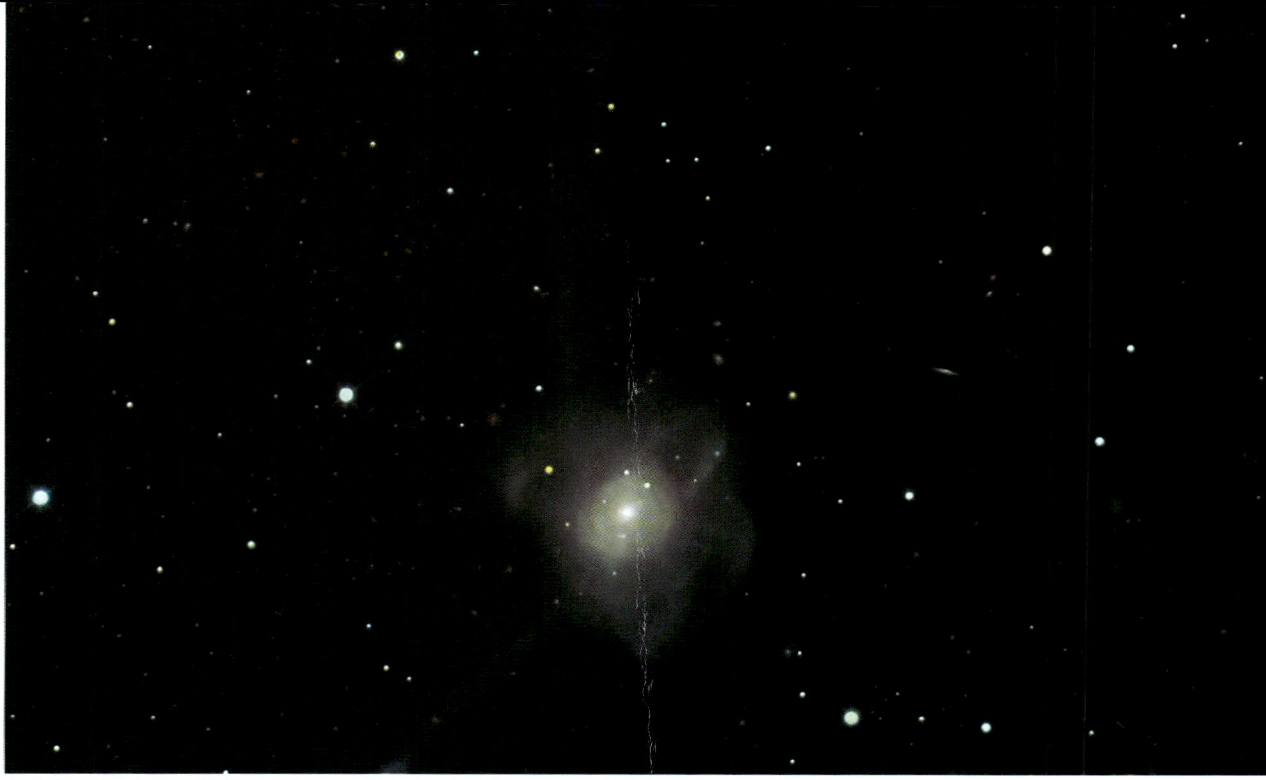

NGC 7252

The galaxy NGC 7252 lies in the southern part of the constellation of Aquarius and is 220 million light years from the Milky Way. Its inner region has an angular size of 1.9′ × 1.6′ which thus corresponds to 120 000 light years. Its outer regions are much larger, however, and reach diameters of more than 400 000 light years. This is due to a large number of tidal tails which surround NGC 7252, allowing a multi-component shell structure to form. This structure reminds one of the electron-shell model of atoms and has led to the nickname the "Atoms for Peace Galaxy".

NGC 7252 was formed in a collision and merger of two gas-rich galaxies. The current state NGC 7252 is a proto-elliptical galaxy. Two almost straight tidal tails emerge from the shells in the halo; the upward, northern tail at 4′ is longer than the eastern one at 3′. At the ends of both tails one can recognize bright blue condensations which are due to star-forming regions. Due to the large distance from the centre of the galaxies, tidal-tail galaxies will probably form.

The best models for the formation of NGC 7252 involve two Sc galaxies with halos about four times as massive as their discs. 620 million years ago, they approached each other within 10 000 light years, orbiting many times. 540 million years ago, induced shock waves led to a first intensive phase of star formation. During this time the two long tidal tails formed, each tail consisting of material from the disc of a precursor galaxy. Dynamical friction led to orbital decay and a further active phase of star formation occurred during a close passage 240 million years ago. 215 million years ago, the galaxies merged completely, though the shell structure remains visible today. The Hubble Space Telescope has studied the core of NGC 7252 in detail and detected a miniature spiral structure. This covers only the innermost 7″, corresponding to 10 000 light years in diameter, and consists of an intact gas disc with an arm structure and dozens of bright, young star clusters. Their size of 60 light years corresponds to the typical size of a medium-size globular cluster. The calculated age of the stars in these clusters corresponds to that of the first close approach.

OBJECT	NGC 7252
CONSTELLATION	Aquarius
RIGHT ASCENSION	$22^h\ 20^m\ 45^s$
DECLINATION	−24° 40′ 42″
BRIGHTNESS	12.7 mag
TYPE	(R)SA0?(r)
PHOTOGRAPHERS	Josef Pöpsel
TELESCOPE	600-mm reflector
CAMERA	SBIG ST-10XME
EXPOSURE TIME	210 min
LOCATION	Amani Lodge, Namibia

Arp 295

This system of two interacting galaxies impressively demonstrates how far tidal tails can extend under certain conditions. The pair was discovered by Fritz Zwicky in 1953 with an 18-inch Schmidt camera (Palomar Observatory, San Diego, USA); it was catalogued as Arp 295 in 1966. The pair consists of the Sc galaxy PGC 72139 (Arp 295A), close to the left edge of the image, and the equally bright Sb galaxy PGC 72155, 4.8′ to the northeast. Between the two galaxies one can recognize a bluish tidal tail which continues as a counter-tail on the other side of PGC 72139. PGC 72155 also shows a similar such small extension, but it appears redder and more diffuse than the tidal tail of the Sc galaxy. Both galaxies are at a distance of 310 million light years, which corresponds to a minimum separation of 430 000 light years. In front of the dust lane in PGC 72139 a 16th-magnitude star can be seen. This is the supernova SN2006dm which was visible at the time of the exposure.

The disc planes of the two galaxies are perpendicular to each other. PGC 72139 is seen edge-on, revealing a dust lane which extends over the diameter of 1.37′, corresponding to 124 000 light years. PGC 72155, on the other hand, is seen face-on and in the inner regions a bright blue ring of star-forming regions is visible with a diameter of 15 000 light years. Further conclusions about spatial motions follow from the recession velocities: the velocities of the two galaxies differ by 360 km/s. Despite the obviously longitudinal interaction, there is also a radial component of the motion. The plane of motion is perpendicular to the plane of the image and also contains the Milky Way. Projection effects thus make the tidal tails appear straight.

OBJECT	Arp 295
CONSTELLATION	Aquarius
RIGHT ASCENSION	$23^h\,41^m\,54^s$
DECLINATION	−03° 38′ 29″
BRIGHTNESS	14.5 mag
TYPE	Sc pec / Sb pec
PHOTOGRAPHERS	Josef Pöpsel, Stefan Binnewies
TELESCOPE	600-mm reflector
CAMERA	SBIG STL-11000
EXPOSURE TIME	260 min
LOCATION	Skinakas Observatory, Crete, Greece

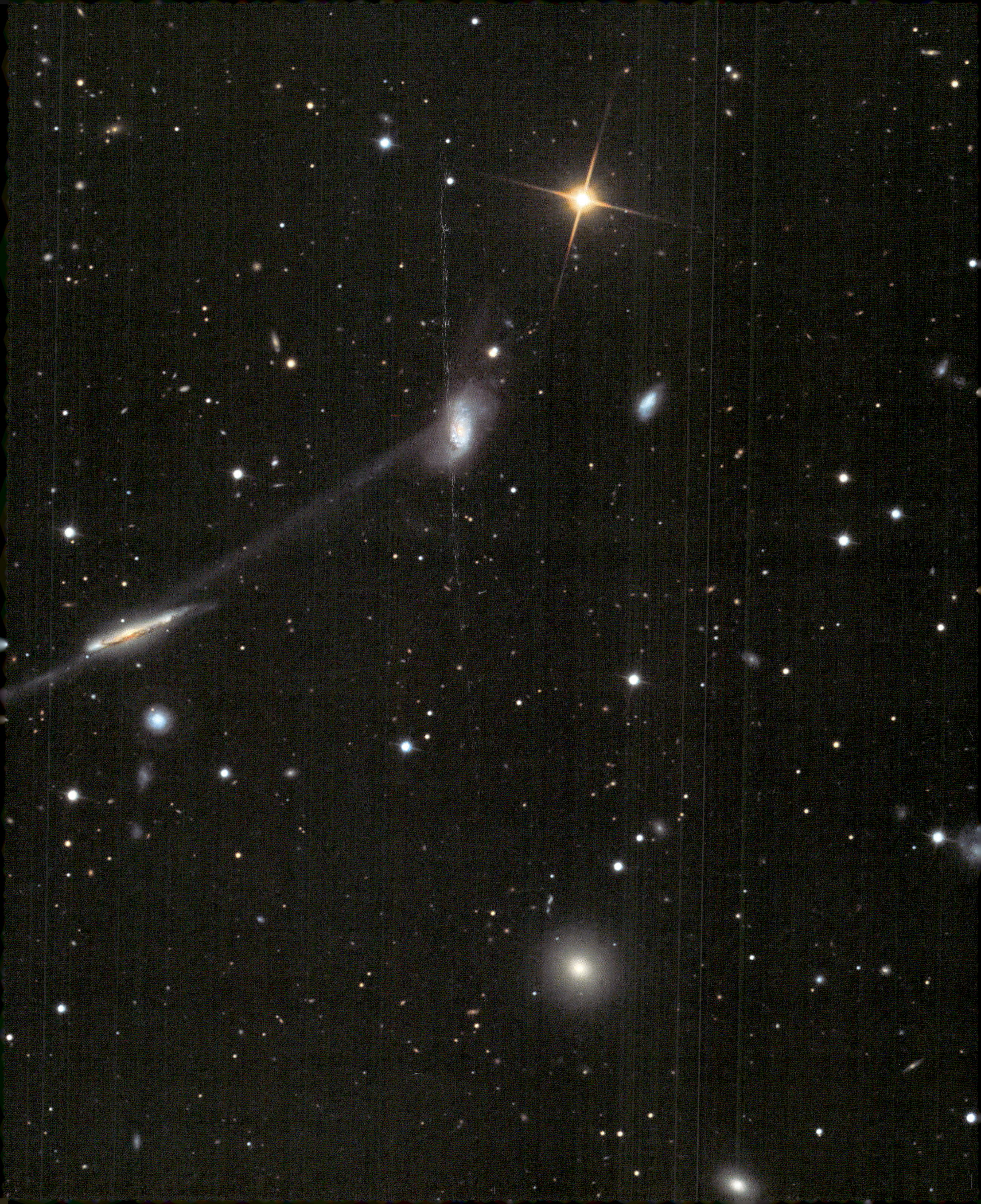

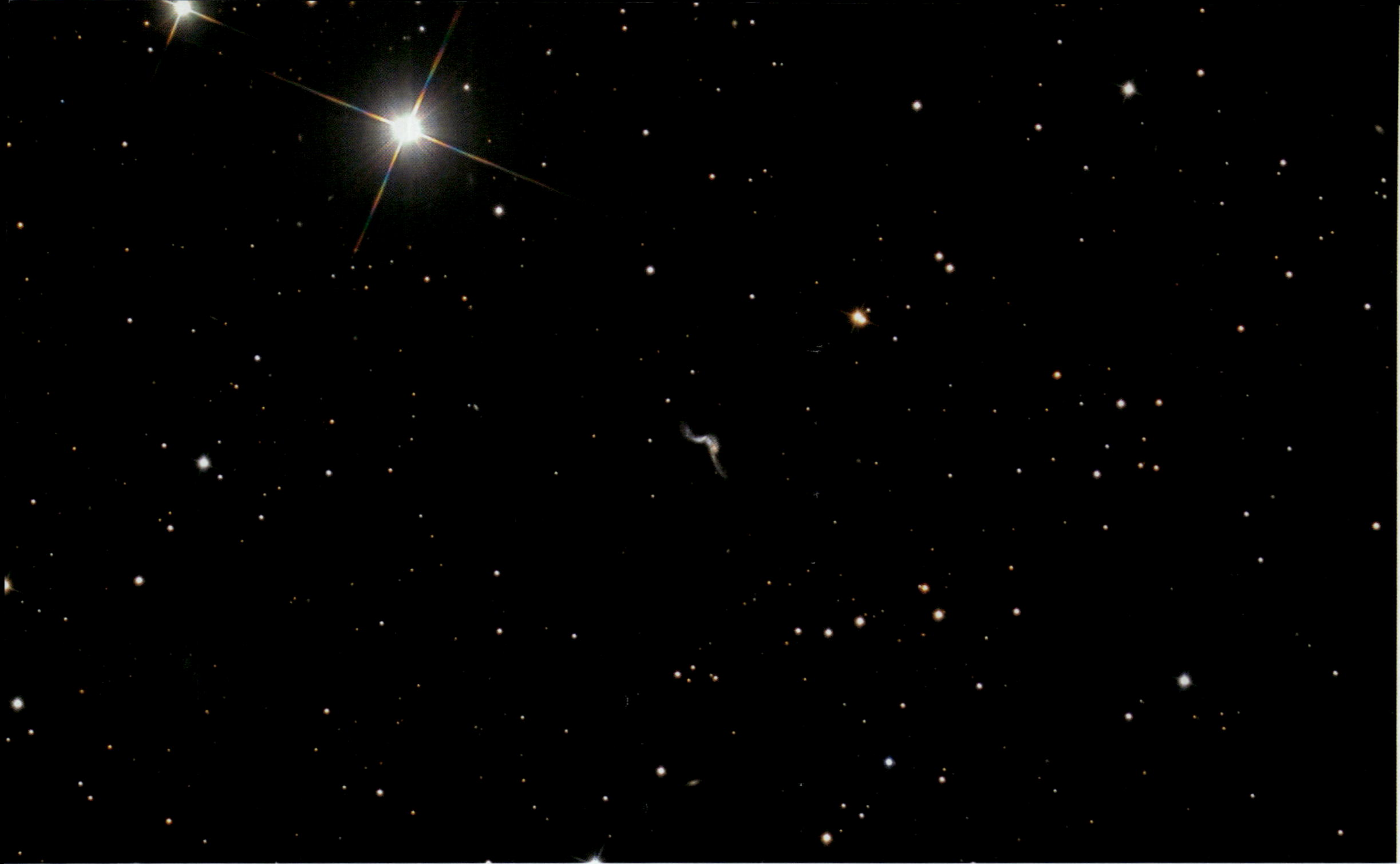

UGC 12342

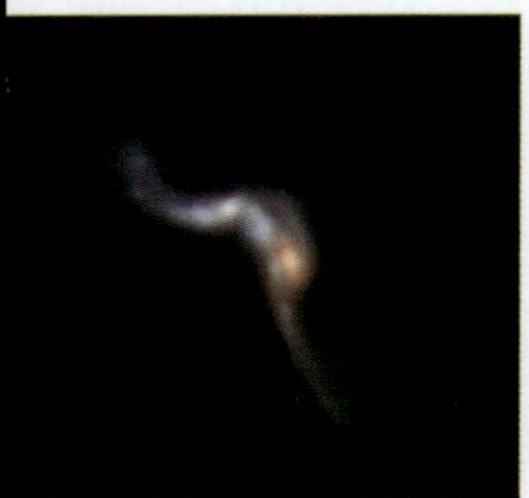

At the centre of the image is UGC 12342, a system consisting of two interacting galaxies which are in the process of merging. The structure, consisting of two connected arcs, has an angular size of 1.3′× 0.5′. Based on the distance of UGC 12342 of 390 million light years, this corresponds to a total length of the system of 147 000 light years. The centre of the upper, southern galaxy is bright blue while the centre of the lower component is yellowish. This difference in colour is also seen in the two oppositely oriented spiral-arm relics (see enlargement in the small image). The shape of these arcs have led to the nickname the Seagull Galaxy. The cores of the colliding galaxies are separated by only 12″ which corresponds to a projected distance of 23 000 light years.

UGC 12342 is at the southwestern end of the Pisces–Perseus Supercluster, which extends over 25° on the sky. Analyses of the galaxy distribution in this cluster show prominent substructures. Recession velocities can be used to study the positions of the subgroup members in more detail. This can be visualized in a wedge diagram showing the recession velocity as a function of angle within a plane through the cluster. This diagram reveals a long filament of galaxies with recession velocities of 9000–13 000 km/s around UGC 12342 whose opening angle is only a few degrees. UGC 12342 with a recession velocity of 8477 km/s is at the apex of this wedge, pointing at the Milky Way.

OBJECT	UGC 12342
CONSTELLATION	Pegasus
RIGHT ASCENSION	$23^h\ 04^m\ 54^s$
DECLINATION	+16° 40′ 42″
BRIGHTNESS	15 mag
TYPE	SB?
PHOTOGRAPHERS	Wolfgang Ries, Stefan Heutz
TELESCOPE	450-mm reflector
CAMERA	SBIG ST-10XME
EXPOSURE TIME	599 min
LOCATION	Altschwendt, Austria

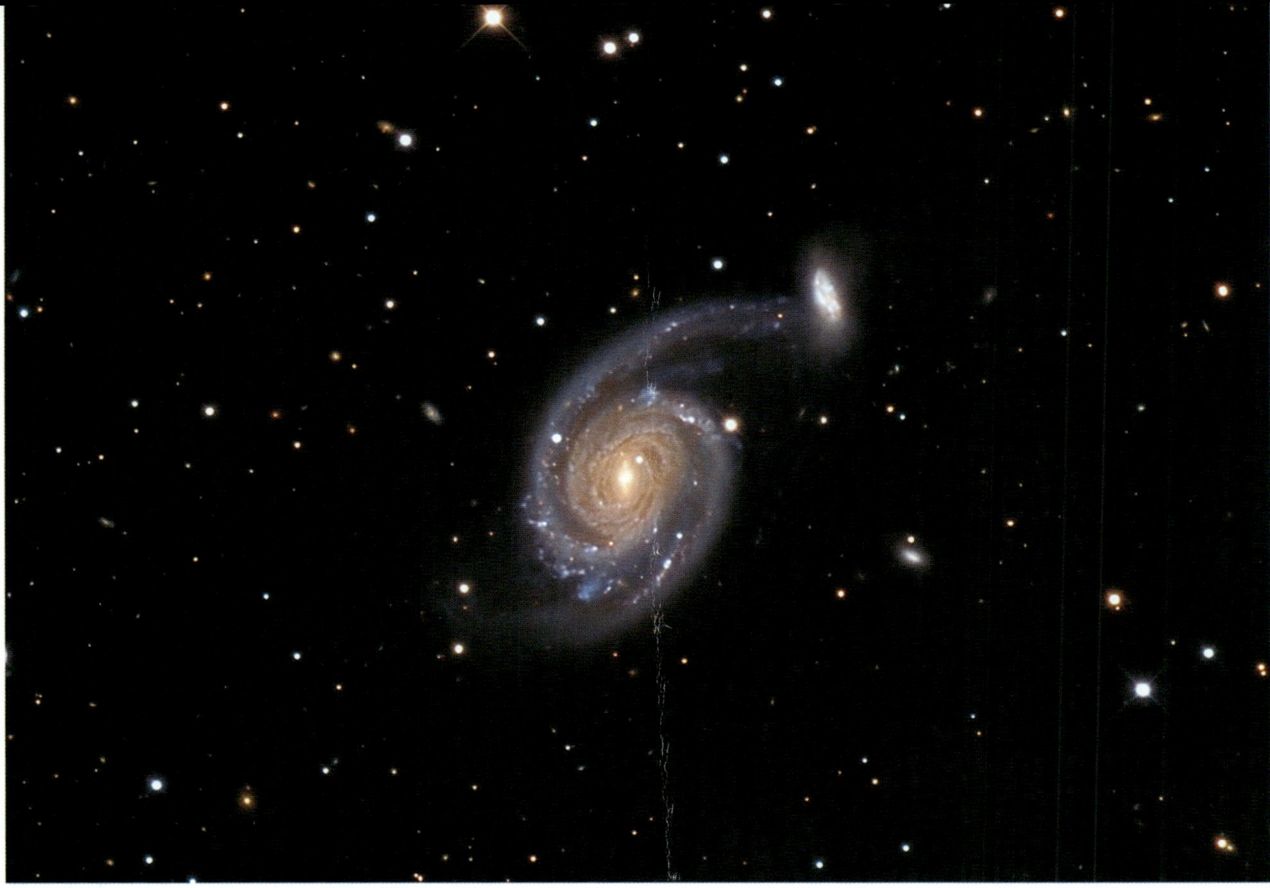

NGC 7753

The SAB(rs)bc galaxy NGC 7753 is 240 million light years from the Milky Way and has an angular size of 3.3′ × 2.1′. One of its outer, bright blue spiral arms points to the southwest at the irregular companion NGC 7752. The recession velocity of NGC 7752 is only 96 km/s less than that of the larger galaxy 2′ away. The corresponding transverse velocity is 140 000 light years. Both spiral arms of NGC 7753 are symmetric and can be followed far into the inner regions where they fade into a loose spiral structure with several arms. Their outer regions are filled with star-forming regions, whereby it is conspicuous that there are especially many of these active areas on the side of NGC 7753 facing the companion. At the core of NGC 7753 is a bright yellow, vertically oriented oval structure which has caused NGC 7753 to be classified as AB(r)bc as well. The colour of the galactic disc is markedly different from that of the two spiral arms. The colour index of NGC 7753 suggests an Sa type, which could be explained by the large number of star-forming regions. The almost white colour of the companion NGC 7752 is typical of irregular galaxies and is in stark contrast to NGC 7753.

Radio observations of atomic hydrogen in this galactic system reveal a tidal bridge between NGC 7753 and NGC 7752. The measured velocities of the H I gas also indicate that the southeastern side of the rotating disc in NGC 7753 is moving towards the Milky Way. The same is true for the hydrogen gas in the tidal bridge, which however points to the southwest side of NGC 7753. The velocity field has a complex structure which could be an indication that the companion has orbited many times. Radio observations confirm this by revealing two arcs which surround the system. They lie at distances of 3.2′ and 2.1′ and suggest that they formed out of gas lost by NGC 7752 and are part of the spiral-shaped orbit of NGC 7752.

OBJECT	NGC 7753
CONSTELLATION	Pegasus
RIGHT ASCENSION	$23^h\ 47^m\ 05^s$
DECLINATION	+29° 29′ 00″
BRIGHTNESS	12.8 mag
TYPE	SAB(rs)bc
PHOTOGRAPHERS	Makis Palaiologou, Stefan Binnewies
TELESCOPE	1.3-m reflector
CAMERA	Andor DZ 436
EXPOSURE TIME	200 min
LOCATION	Skinakas Observatory, Crete, Greece

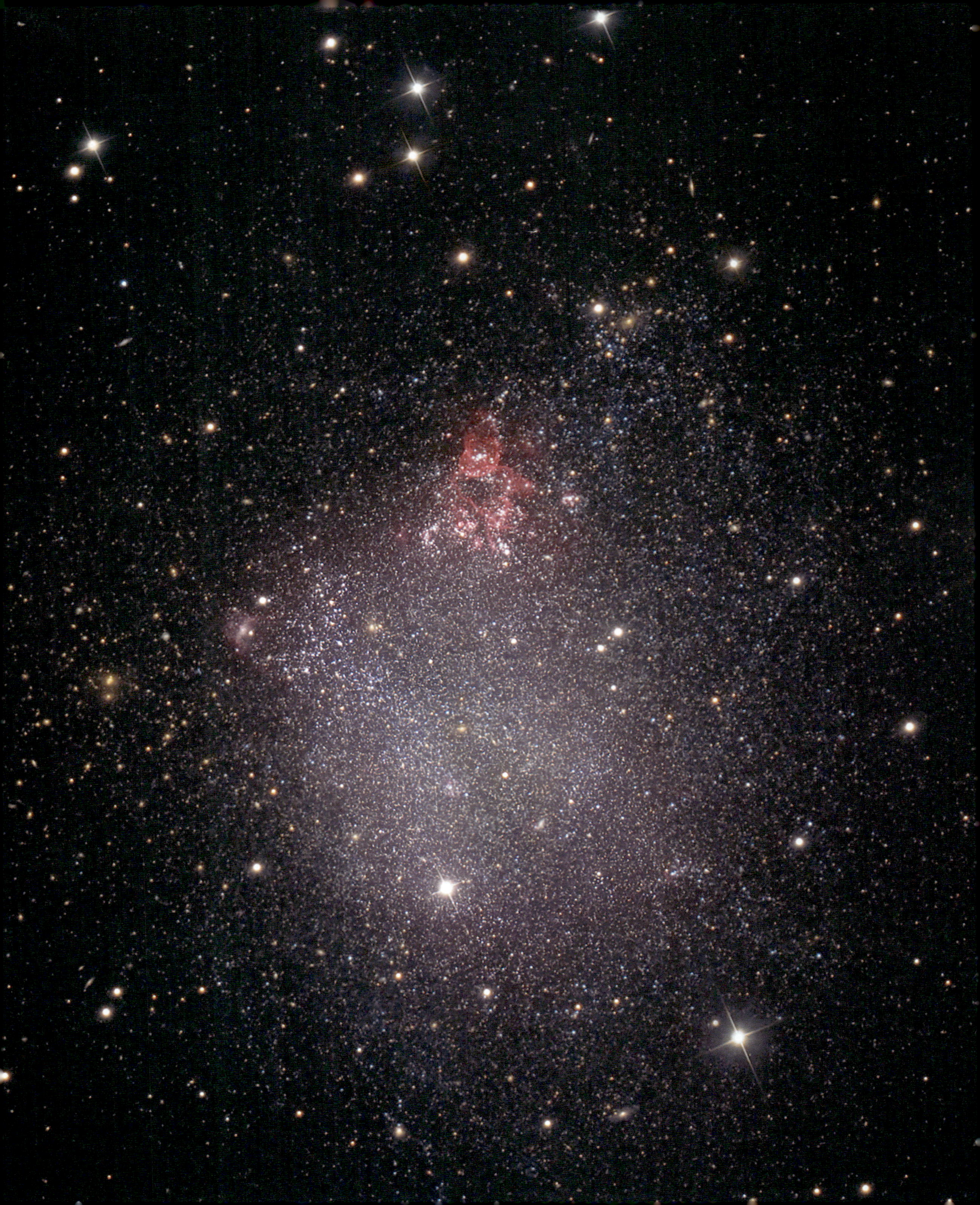

DWARF GALAXIES

This chapter concerns the smallest of all galaxies. Examined more closely, they reveal, apart from their small dimensions, further properties which distinguish them strongly from their larger siblings.

THE MORPHOLOGY OF DWARF GALAXIES

The classification of irregular galaxies often contains a separate group of especially small, dim galaxies. These are dwarf galaxies. Dwarf galaxies are differentiated according to their appearance: blue compact dwarfs (BCDs), dwarf spheroidals (dSph), dwarf ellipticals (dE), and dwarf irregulars (dIrr), which also include tidal dwarfs. The terms are not used uniformly in the literature, which is due to the fact that because of the difficulty of detecting them, studies of these galaxies began in earnest only in the last two decades. A further type is ultra-compact dwarf galaxies (UCDs), which were discovered in nearby galaxy clusters in 1999. The UCDs are more compact than other dwarf galaxies and their appearance is similar to that of bright globular clusters.

The term "d" for dwarfs was introduced to extend the "regular" galaxies in the Hubble sequence to dwarf morphologies. Thus, a "d" was placed before the "E" or "Irr" to denote an elliptical or irregular galaxy as a dwarf. The BCDs have the highest surface brightness among dwarfs. Their prominent blue colour is due to a high star-formation rate at the centre. Much dimmer are dwarf spheroidal galaxies. These have less or even almost no concentrated gas and, along with the UCDs, are the smallest type of dwarf galaxy.

The classification of dwarf galaxies also includes so-called tidal dwarf galaxies which can form from the material of tidal tails which are common in the interaction of large galaxies. These tails extend from a few ten thousand to a few hundred thousand light years into space. As the interacting galaxies orbit each other, twisted tracks of gas, dust, and stars form which surround the pair. A local gravitational potential far enough away from the centre of the galaxy can collect and concentrate material, which can lead to a tidal dwarf galaxy forming in the tidal tail.

The dwarf galaxy IC 1613 is a member of the Local Group (see page 236). Image credit: D. Böcker, E. von Voigt (600-mm reflector).

THE ASTROPHYSICS OF DWARF GALAXIES

The stars in dwarf galaxies are composed mainly of hydrogen and helium; compared to normal galaxies, they have a metallicity from a few millionths to a few per cent of solar metallicity. This is an indication that dwarf galaxies have experienced only a few stellar-population cycles and so there are almost no heavier elements. These small galaxies thus have histories which differ from those of large spiral galaxies. Apart from the special case of tidal dwarf galaxies, dwarf galaxies are not recent products of the interaction of evolved galaxies, but rather are among the oldest objects in the universe and make up an independent class.

Criteria for the classification as a dwarf galaxy are a low surface brightness and an absolute visual magnitude of more than −16 mag. This low surface brightness makes them difficult to detect, and thus only a limited comparison is possible between the fraction of dwarf galaxies among distant galaxies and in the Local Group. For example, the Local Group contains the dwarf galaxy Sextans A, which has star-forming regions with bright, young stars and purple-red H II regions whose gas glows due to excitation by stars in its vicinity. Even closer than Sextans A are the Magellanic Clouds, the two larger companions of the Milky Way, which are also considered to be irregular galaxies. Between these two galaxies (and beyond them) there is a tidal bridge of neutral hydrogen. Recent radio observations suggest that this Magellanic Stream extends much farther and is older than previously assumed and that these two satellite galaxies were interacting even before their approach to the Milky Way. M 31 is also a large spiral galaxy with two nearby dwarf galaxies, NGC 221 and NGC 204. Both of the two large spiral galaxies in the Local Group are surrounded by systems of about 40 irregular, elliptical, and spheroidal galaxies.

An average dwarf galaxy has a mass about one hundredth that of the Milky Way, but usually the masses are much lower. Dwarf galaxies are not, however, miniature forms of spiral or elliptical galaxies, but rather an independent galaxy type whose form and evolution differ significantly from those of large galaxies. The most important difference with regard to dwarf galaxies, except dSph galaxies, concerns star formation. They have a much higher star-formation rate than for example a spiral galaxy.

Dwarf spheroidals have usually lost most of their gas during many passages through the halo of their parent galaxy so that star formation has ceased in them. In dwarf irregulars, by contrast, the number of stars doubles every ten million years – in comparison, the Milky Way would need 1000 times longer to double its stellar population. The Hubble Space Telescope was able to detect such star-formation rates in many dwarf galaxies since the hot, young stars illuminate the hydrogen in the surrounding gas clouds. The discovery of a large number of dwarf galaxies in the last two decades has made it clear that these small galaxies, often barely noticeable in astronomical images, are the most common type of galaxy in the universe.

A few years ago, the CANDELS observational programme (Cosmic Assembly Near-infrared Deep Extragalactic Legacy Survey) carried out a census and classification of distant galaxies. Astronomers used this view back into early epochs of the universe to investigate how the star-formation rate has changed with time. Star formation is basically a slow process, taking many billions of years. The CANDELS results indicate, however, that there are galaxies with much higher rates of star formation in their youth. This discovery cannot be explained with the usual assumptions about the evolution of dwarf galaxies.

In the CANDELS programme, 68 dwarf galaxies were discovered in images taken by the Hubble Space Telescope in the near infrared. The results indicate that dwarf galaxies were very common about nine billion years ago and had large star-formation rates at this phase of their existence. However, dwarf galaxies have too little mass to sustain this rate. The detected star-formation rates thus correspond to evolutionary episodes and vary greatly with time. In principle, compared to the surrounding galaxy star formation takes place on small length scales. Thus the basic process of a cooling and collapsing gas cloud which finally produces stars is the same – whether in a spiral or dwarf galaxy. During the life of a star this gas cloud will be heated again – for example, by supernovae – and thinned out by stellar winds. This can lead to a collapse of the cloud again, describing a cyclical process of star formation. This classical description, however, cannot explain the very intense phases of star formation observed.

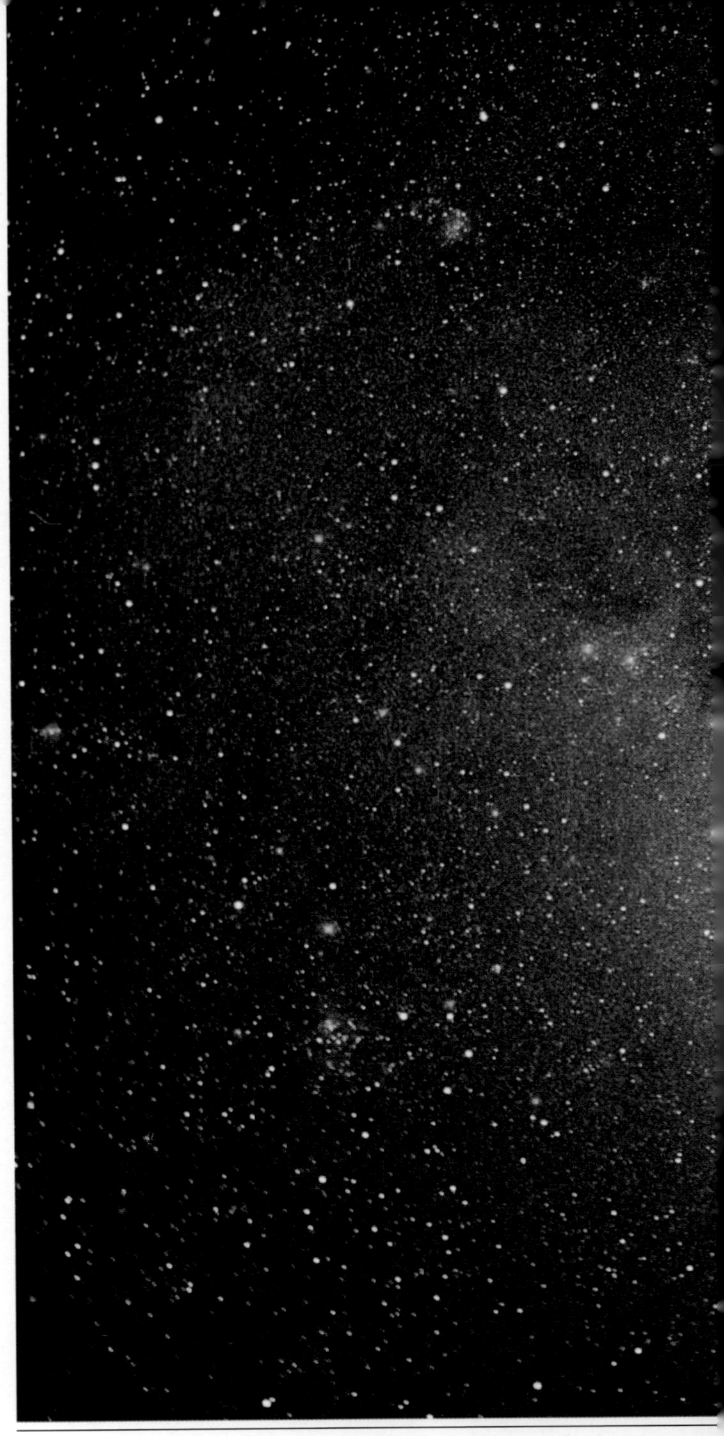

Newer theories thus assume that the idea of small length scales does not apply to dwarf galaxies. Supernova winds expel large amounts of gas from the inner regions of the dwarf galaxy. This amount is so large compared to their total mass that the mass loss significantly reduces the central density. Thus no concentration of stars can form at the centre, such as one observes by contrast in the bulge of a spiral galaxy. Previous models of dwarf galaxies which don't take this important effect into account suggested too dense a core region. Only in the last few years have complex stellar-wind models, assuming a constant density of dark matter at the centre of dwarf galaxies, allowed such galaxies to be simulated that agree with

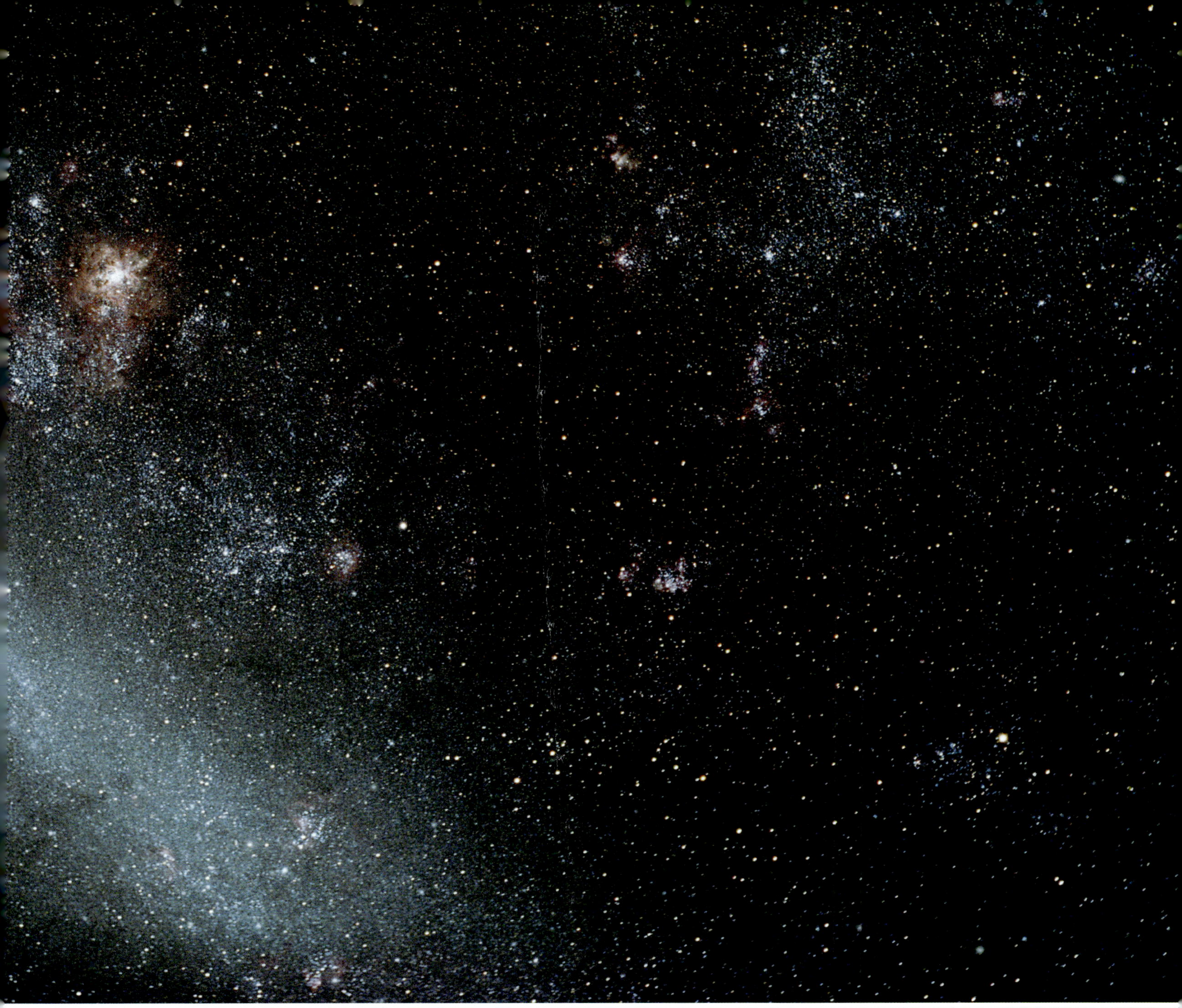

Figure 5.1: The Large Magellanic Cloud is a dwarf galaxy in the Milky Way system (see page 228).

those observed in the cosmos. One must note, however, that there is a subgroup of dwarf galaxies which is characterized by just this enhanced central stellar density. They are known in the literature as nucleated dwarfs.

Star formation in dwarf galaxies displays a further peculiarity as well. Although processes occur which correspond to those known in larger galaxies, the small size itself causes some differences. Due to spatial limitations, certain physical limits are reached which can create a barrier for star formation. Observations indicate that in dwarf galaxies there is usually little variation in the metallicity of the stars and that heavy elements, such as iron, are almost non-existent. This can be reproduced in simulations only by assuming an upper limit to the mass of stars formed in the dwarf galaxy. The lower the mass of a star, the fewer heavy elements can be produced by nuclear fusion. According to these models, there are no large, massive stars in small galaxies and the calculated limit is about 25 solar masses. By comparison, the most massive stars in our Milky Way of about 120 solar masses appear huge. This new result of astronomical research suggests that dwarf galaxies have only a small range of stellar masses at their disposal. Current speculation suggests that the reason could be the low mass density of dwarf galaxies. This implies that the gas

THE MASS–LUMINOSITY RELATION FOR GALAXIES

The luminosity of a galaxy is defined as power and describes the energy per time radiated by all components of the galaxy (nucleus, disc, halo, etc.). The entire radiation power of a galaxy is measured by astronomers in solar luminosities, often distinguished by spectral region and termed, for example, radio luminosity or X-ray luminosity. In spiral galaxies, the maximum rotation velocity is related to the luminosity. This means that brighter spiral galaxies also spin faster. This so-called Tully–Fisher relation can be used to calculate the luminosity of a typical spiral galaxy if the width of the 21-cm radio line is known. This line width is a measure of the rotation velocity. A similar correlation is the Faber–Jackson relation for elliptical galaxies which connects the velocity dispersion and luminosity. The larger the measured velocities of the stars at the centre, the brighter the elliptical galaxy.

By contrast, dynamical methods are used to determine the mass of a galaxy or galaxy cluster. Astronomers usually use the virial theorem, which describes the relation between the time-averaged values of kinetic energy E_{kin} and potential energy E_{pot}: $2 \times E_{kin} = E_{pot}$. The total mass of the system can thus be determined from the radius and the distribution of velocities. A requirement for the use of the virial theorem, however, is a closed system, also known as a relaxed system, which is in equilibrium and not subject to any external influences.

The table shows the mean *M*/*L* values for different systems, whereby there are exceptions which are not taken into account here. The Milky Way has *M*/*L* = 50–80; for dwarf galaxies in the Local Group one finds values in the range 3–60 in the literature.

The physics of stellar structure leads to a mass–luminosity relation for main-sequence stars. For the Sun, *M*/*L* = 1; for a star of four solar masses, *M*/*L* = 0.2 and *M*/*L* decreases with increasing stellar mass. Low-mass stars of 0.5 solar masses have *M*/*L* = 2.8. If a galaxy were composed only of stars, the mass could be determined by adding up the luminous matter and would result in small, single-digit values for *M*/*L*. The computed large *M*/*L* values thus indicate a dominant contribution from non-luminous matter and hence are seen as proof of the existence of dark matter. It is conspicuous that dwarf galaxies have the largest range in *M*/*L* values of all galaxy types. Dwarf elliptical galaxies thus have a dark-matter fraction which corresponds to that of large elliptical galaxies or galaxy clusters.

Since using the virial theorem to determine masses demands a closed system, some researchers suggest that dwarf galaxies do not meet this requirement and that the gravitationally calculated mass values are overestimated. For example, in dwarf spheroidal galaxies there is the trend that those which are closer to the Galactic centre have the largest *M*/*L* values. Studies of the motions of the satellites of M 31 have shown that half of these dwarf galaxies orbit within the same plane. This subgroup also has the same direction of rotation and similar dynamical parameters and forms a co-planar group.

These observations cast doubt on the assumption that these satellite galaxies are really closed, unperturbed systems. Only if this is the case would dark matter be needed to explain their composition.

	$M/M_\odot$	$L/L_\odot$	M/L
GLOBULAR CLUSTERS	10^5–10^6	10^4–10^6	1–2
ULTRACOMPACT DWARF GALAXIES	10^6–10^7	10^6–10^7	2–3
DWARF SPHEROIDAL GALAXIES	10^6–10^8	10^6–10^7	3–250
DWARF ELLIPTICAL GALAXIES	10^7–10^9	10^7–10^8	3–300
SPIRAL GALAXIES	10^9–10^{12}	10^8–10^{11}	5–50
ELLIPTICAL GALAXIES	10^8–10^{13}	10^7–10^{12}	10–100
GALAXY CLUSTERS	10^{12}–10^{15}	10^{11}–10^{14}	100–500

clouds which produce stars also never reach a certain density value. This would define an upper limit on stellar mass.

The different types of dwarf galaxies differ according to their star-formation rates and the amount of stellar evolution which has taken place. One of the most recently discovered groups of UCDs is characterized by old populations and reminds one of globular clusters, which are similar to UCDs. The situation is very similar in dSph galaxies and has been known for a long time. Compared to globular clusters, however, UCDs and dSphs are larger, more massive, and have a higher absolute luminosity. They reach the enormous stellar density of about a million stars per cubic lightyear, with diameters of 100 light years. The observed mass-to-light ratio (M/L) of UCDs is twice as large as that of globular clusters and is thus too high to be explained by stars alone. There are attempts to explain these mass-to-light ratios without dark matter. The assumption is that due to the high densities, stars merge at the centre of UCDs. This leads to massive, short-lived stars which end as supernovae, causing neutron stars and black holes to be formed at the cores of UCDs. They thus add a non-luminous contribution to the dynamic mass of the UCDs and explain the observed mass-to-light ratio without dark matter. In recent studies, astronomers have made statistical studies of the distribution of UCDs in galaxy clusters. For about 180 UCDs in the Fornax Cluster, it could be shown that their luminosity functions are very well modelled as the bright end of the distribution for globular clusters. Studies of other galaxy clusters also suggest that UCDs are extremely large and bright globular clusters rather than a type of dwarf galaxy.

BIBLIOGRAPHY

Coleman, M. G. et al.: *Dwarf Galaxies Under the Magnifying Glass*, 2009, www.mpg.de/345100/forschungsSchwerpunkt1

Da Costa, G.: *Local Group of Galaxies – Research School of Astronomy & Astrophysics*, 2006, www.mso.anu.edu.au/~gdc/talks/LocalGroupLect_17Dec2012.pdf

Faber, S. M.: *Masses and Mass-to-Light Ratios of Galaxies*, 1980, http://ned.ipac.caltech.edu/level5/Faber/Faber3_2.html

Fendt, C.: *Sternentwicklung*, 2009, www.mpia.de/homes/fendt/Lehre/Lecture_ANPh/07_CF_sterne_entwickl-09-06-11.pdf

Frank, M. J. et al.: *Spatially resolved kinematics of an ultra-compact dwarf galaxy*, Monthly Notices of the Royal Astronomical Society, **414**, 2011

Grebe, E. K.: *Dwarf Galaxies as Archaeological Tracers of Galaxy Evolution, Evolution of galaxies, their central black holes and their large-scale environment*, Potsdam, Germany, 2010

Gronwall, C.: *Dwarf Galaxies*, 2012, www2.astro.psu.edu/~caryl/a480/lecture9_10.pdf

Hiroyuki, H. et al.: *Physical Interpretation of the Mass-Luminosity Relation of Dwarf Spheroidal Galaxies*, The Astrophysical Journal, **504**, 1998

Ibata, R. et al.: *Do globular clusters possess dark matter haloes? A case study in NGC 2419*, 2012, http://mnras.oxfordjournals.org/content/early/2012/11/26/mnras.sts302.full

Ibata, R. et al.: *A vast, thin plane of co-rotating dwarf galaxies orbiting the Andromeda galaxy*, Nature, **493**, 2013

Mieske, S.: *Ultra-compact dwarf galaxies (UCDs)*, 2012, www.eso.org/~smieske/ucds.htm

Mieske, S. et al.: *The specific frequencies of ultra-compact dwarf galaxies*, Astronomy and Astrophysics, **537**, 2012

De Rijcke, S. et al.: *The internal dynamics of the Local Group dwarf elliptical galaxies NGC 147, 185 and 205*, Monthly Notes of the Royal Astronomical Society, **369**, 2006

Stanek, R. and Evrard, A.: *Mass-Luminosity Relationship of Galaxy Clusters*, American Physical Society, Ohio Section Spring Meeting 2004, Ohio University, 2004

Tollerud, E. J. et al.: *From Galaxy Clusters to Ultra-Faint Dwarf Spheroidals: A Fundamental Curve Connecting Dispersion-supported Galaxies to Their Dark Matter Halos*, 2010, http://arxiv.org/pdf/1007.5311v2.pdf

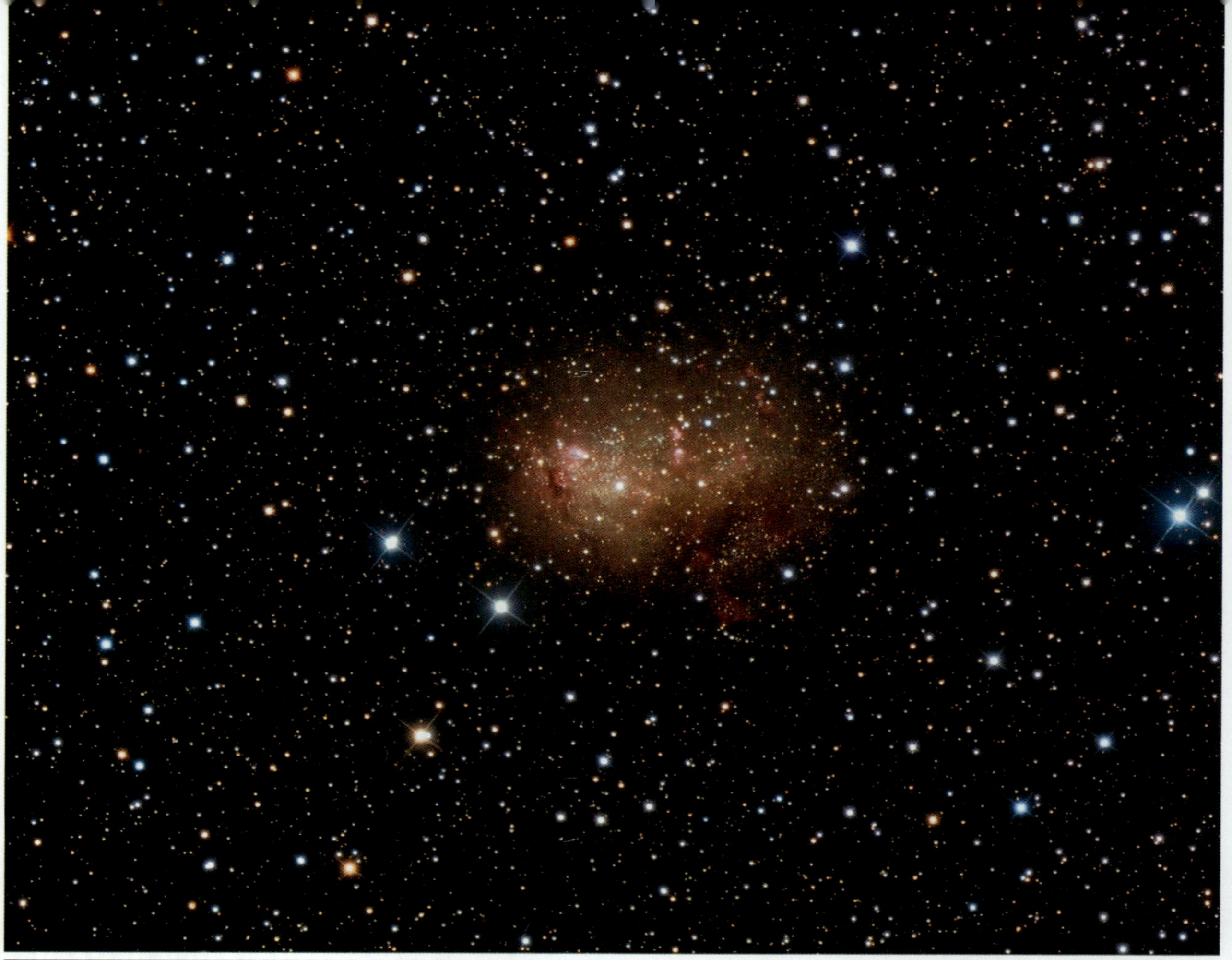

IC 10

In the constellation of Cassiopeia, 12′ to the northeast of a 9th-magnitude A5 star (outside the image) is the irregular galaxy IC 10. Its angular size is 6.8′ × 5.9′ and catalogues list its distance from the Milky Way at 2.8 million light years. In spite of the large uncertainty of 35 per cent in this measurement, one can conclude that IC 10 is part of the Local Group and thus one of the distinct satellites of M 31. As suggested by the diffuse and irregular morphology, IC 10 is a dwarf irregular galaxy with an absolute size of only 5600 light years. The colour–magnitude diagram of stars in IC 10 shows an unusually large number of blue stars for a dwarf galaxy.

IC 10 displays dust structures and a dust lane in the southern part, bordered by active areas with H II regions and OB associations. Dust plays an important role in the process of star formation. It shields molecular gas from the intense radiation of hot stars and thus hinders its dissociation into atoms. It has been shown that in IC 10 molecular hydrogen and CO gas is found in 14 giant molecular clouds (GMC). The individual masses of some of these GMCs are more than 40 000 solar masses and its total mass is more than 400 000 solar masses. Star formation induced by M 31 and other companions attains a rate of 0.2 solar masses per year in IC 10 and thus exceeds the typical values of normal spiral galaxies. In the case of irregular galaxies, however, the comparatively large values can be only momentary snapshots and can change quickly due to the limited amounts of gas and other factors.

OBJECT	IC 10
CONSTELLATION	Cassiopeia
RIGHT ASCENSION	$00^h\ 20^m\ 17^s$
DECLINATION	+59° 18′ 14″
BRIGHTNESS	11.8 mag
TYPE	IBm
PHOTOGRAPHERS	Rainer Sparenberg
TELESCOPE	1.12-m reflector
CAMERA	SBIG STL-11000M
EXPOSURE TIME	240 min
LOCATION	Melle, Germany

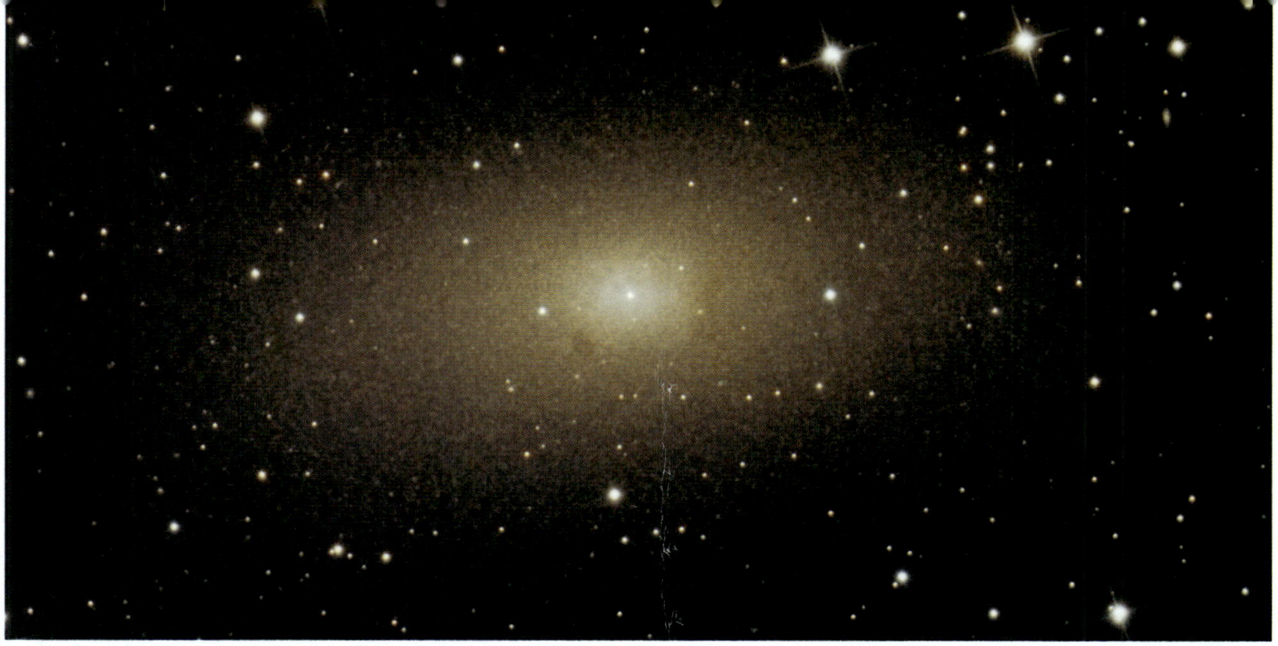

M 110

The dwarf elliptical galaxy M 110 (NGC 205) is one of the larger satellite galaxies of the Andromeda Nebula. The angular size of 21.9′ × 11.0′ corresponds to physical dimensions of 16 500 × 8300 light years, assuming a distance of 2.59 million light years. Compared to M 32, the other close companion of the Andromeda Nebula, M 110, is farther away and, as seen from the Milky Way, presumably directly behind the spiral plane of M 31. M 110 is about 1° to the northwest of the core of M 31 and is, after M 32, the second-brightest dwarf galaxy in the Andromeda system. Compared to M 32, M 110 has almost two-and-a-half times the diameter, but its morphology is more diffuse and at 13.8 mag/arcsec^2 its surface brightness is substantially lower than that of M 32 (12.5 mag/arcsec^2). Together with M 31, the two companions form a triplet also known as Holmberg 17.

The mass of the dwarf galaxy M 110 is estimated at 10–15 billion solar masses and the colour–magnitude diagram of its stars indicates the presence of old and middle-aged stars as well as evolved red giants. In M 110, 12 planetary nebulae and 8 globular clusters have been detected. The inner form of the galaxy is very similar to the dwarf galaxy NGC 85. One can also recognized dust structures at the centre of M 110, and astronomers have also detected young stellar populations. It is notable that not only is this proof that star formation has taken place in the dwarf galaxy, but also that the distribution of giant stars in the colour–magnitude diagram separates into groups. The distances between these groups correspond to spans of time, thus describing a scenario of episodic star formation.

Simulations of the motion of the two satellite galaxies of M 31 – M 110 and M 32 – lead to the same conclusion, namely that their current positions and velocities result from two complete orbits about M 31. The radius of the orbit of M 32 in this model is 38 000 light years with a period of 270 million years. The orbit of M 110 is larger with a radius of 41 000 light years; the galaxy also needs more time for an orbit, namely 290 million years. The researchers' models also specify other orbital parameters: the orbital plane of M 32 is almost polar while that of M 110 is tilted by about 45° with respect to the main plane of the Andromeda Galaxy. M 32 last passed through this plane 130 million years ago, M 110 93 million. The last intense interaction between M 31 and M 110 occurred when the latter passed through the high particle density in the plane of the former. This time corresponds well to the detected episode of star formation in M 110.

OBJECT	M 110, NGC 205
CONSTELLATION	Andromeda
RIGHT ASCENSION	00^{h} 40^{m} 22s
DECLINATION	+41° 41′ 07″
BRIGHTNESS	8.9 mag
TYPE	E5 pec
PHOTOGRAPHERS	Volker Wendel
TELESCOPE	380-mm reflector
CAMERA	SBIG ST-10XME
EXPOSURE TIME	115 min
LOCATION	Weisenheim am Berg, Germany

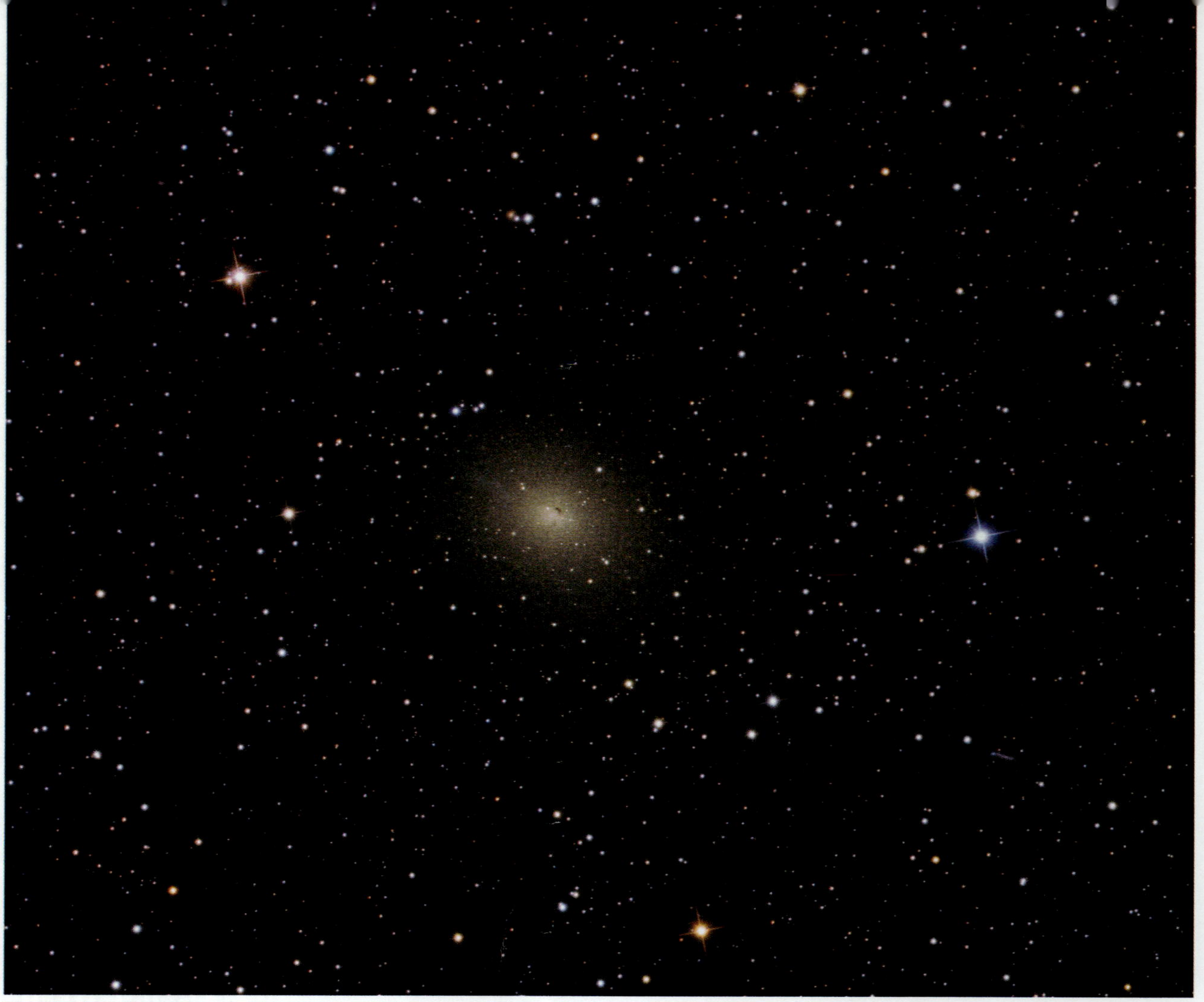

NGC 185 / NGC 147

The dwarf galaxy NGC 185 is among the larger satellite galaxies of the Andromeda Nebula M 31. It is at a distance of 2.20 million light years and its angular size of 11.7′ × 10.0′ corresponds to a maximum diameter of 7500 light years. The Andromeda Nebula is 2.56 light years from the Earth and NGC 185 is located 7.3° to the north – very close to NGC 147 (see image on facing page). The separation of the two galaxies is 55′ which corresponds to a projected distance of 35 200 light years.

In the literature, NGC 185 is listed both as a dwarf spheroidal and as a dwarf elliptical galaxy. Dwarf spheroidal galaxies (dSph) have almost no gas and contain no young stars, making them dimmer than dwarf ellipticals (dE). Due to the low surface brightness, NGC 185 could be a dwarf spheroidal galaxy. This is contradicted by the fact that dust and gas are detected at the centre. Proof is given by two central dust structures visible in the image. In contrast, NGC 147 is free of gas and dust and astronomers have shown that there is a population of young, blue stars in the core of NGC 185. Therefore the two partners NGC 185 and NGC 147 differ significantly in their chemical composition and thus in their evolutionary state. This implies that the two dwarf galaxies do not form an interacting pair, but rather that one is located

OBJECT	NGC 185
CONSTELLATION	Cassiopeia
RIGHT ASCENSION	$00^h\ 38^m\ 58^s$
DECLINATION	+48° 20′ 15″
BRIGHTNESS	10.1 mag
TYPE	E3 pec
PHOTOGRAPHERS	Johannes Schedler
TELESCOPE	400-mm reflector
CAMERA	SBIG STX-16803
EXPOSURE TIME	760 min
LOCATION	Panther Observatory, Wildon, Austria

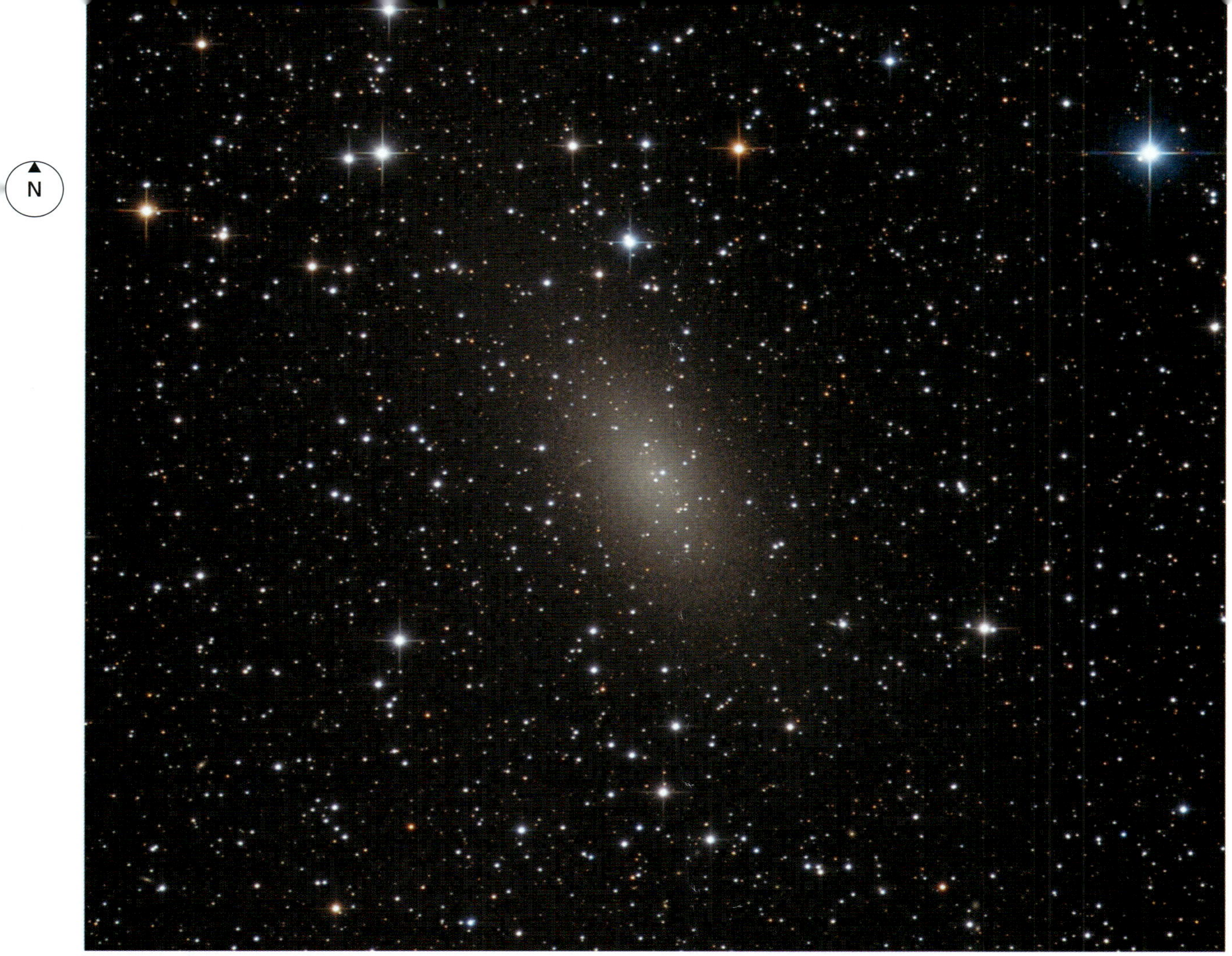

NGC 147, image credit: Bernhard Hubl, 300-mm reflector, SBIGST-2000XM, 606 min, Nussbach, Austria.

behind the other, whereby NGC 147 is about 180 000 light years farther away.

The infrared spectrum of NGC 185 shows emission lines of complex molecules, so-called polycyclic aromatic hydrocarbons, and absorption lines due to silicates (dust). Both spectral signatures indicate that star formation has taken place, over a span of about 100 million years, but with a low star-formation rate. Together with other weak emission lines in its spectrum, NGC 185 thus fulfils the requirements for classification as a Seyfert-2 galaxy – although the galaxy doesn't have an active nucleus containing a black hole. The classification of this galaxy as Seyfert-2 in some catalogues is thus not correct, since the emission lines are not produced in an active nucleus but rather are due to the superposition of stellar emission processes.

Further dwarf galaxies were found in a study of an area of 57 square degrees surrounding M 31 (Megacam CCD Array, Canada–France–Hawaii Telescope, Mauna Kea, USA). In more recent studies such as the Sloan Digital Sky Survey (SDSS), further dwarf galaxies were discovered which, however, lie at a distance of more than 950 000 light years from M 31. They are thus outside of the virial radius of the galaxy and are barely influenced gravitationally by M 31. Astronomers have been able to detect more than two dozen dwarf galaxies in the Andromeda Nebula; the largest companions are M 32, M 110, NGC 185, and NGC 147.
In addition to the many dSph companions, designated Andromeda I to Andromeda XXVIII, further studies in the last few years have discovered several stellar streams both in the halo of the Milky Way and in the M 31 system. The Andromeda Stream is the best known of these. The halo structure has been found to be even more complicated, however. On the one hand, this demonstrates that up until recently many dwarfs have been accreted, but on the other hand there are still many waiting to be cannibalized in the future.

The Small Magellanic Cloud

The Small Magellanic Cloud (SMC, NGC 292) is only half as large as the LMC and at 200 000 light years from the Milky Way farther away than its larger companion. In spite of the uncertainty of ±10 per cent in the distance, one can assume that the SMC is 35 000 light years farther from the Earth than the LMC. The distance between the cores of the two Magellanic Clouds is about 20°; the SMC at −73° is farther south than the LMC at −69°. The angular size of the SMC of 5°20′ × 3°5′ makes it an ideal object for binoculars. However, astronomers from the northern hemisphere must travel to the equator in order to experience the culmination of the SMC 17° above the horizon.

At the upper right edge of the image is the globular cluster 47 Tucanae (NGC 104). It belongs to the halo of the Milky Way and is 18 000 light years away from the Sun in the direction of the SMC. 47 Tucanae is the second-largest globular cluster in the Milky Way after Omega Centauri. It contains many red giants, easily identified as yellow-red stars in the cluster. 47 Tucanae has an angular size of 30′; its diameter is listed in catalogues as 160 light years. Like the SMC itself – whose visual magnitude is 2.7 mag – 47 Tucanae at 4.0 mag is also a naked-eye object. The globular cluster consists of a few million stars and thus has a mass comparable to that of a dwarf galaxy.

Like in the LMC, in the SMC there is a special type of large, dense star cluster which can reach 150 light years in extent. The above-average amount of blue-white stars in these clusters led to their description as blue globular clusters, although the number of stars is closer to that of an open star cluster. The corresponding colour–magnitude diagram of these clusters shows main-sequence stars and blue supergiants; the main-sequence turn-off in the diagram implies an age of 100–130 million years. They are thus appreciably younger than the globular clusters in the Milky Way, which are usually much older at eight to twelve billion years. In addition to the two Magellanic Clouds, blue globular clusters are also observed in M 33; they are not visible in the observable part of the Milky Way. Similar star clusters exist, however, in starburst galaxies or in strongly interacting systems.

The blue giant stars have masses of 10–50 solar masses. Their high luminosity compared to red giants can be explained by the fact that they leave the giant stage after about ten million years. They are also found in binary systems. Due to the size of the giant star in the binary system, it reaches the so-called Roche limit beyond which material can flow onto the smaller companion without using energy. If this companion star is a white dwarf, a neutron star or a black hole, material which leaves the giant star in the form of stellar winds can fall onto the companion. This creates such high temperatures that X-rays are emitted. Such X-ray binaries, XRBs, are often found in globular clusters since the formation of double stars is aided by the high local stellar density. Using observational data from the high-energy satellite HEASARC, a catalogue of high-mass X-ray binaries was constructed. With 92 XRBs, the SMC has even more than the LMC, which has 36.

In addition to the RBG data, the image also contains an exposure in Hα light. This shows clearly the location of the active regions, in which young, hot stars excite the surrounding gas clouds, causing them to glow, and the prominent H II nebulae in the SMC are formed. The image shows an almost vertical, axial bar from which a spiral arm with many bright knots emerges to the left, containing many H II regions and supergiants. A second, dimmer spiral arm leads to the right. It contains fewer bright stars and no giant stars. The asymmetric perturbations visible in the structures point in the direction of the LMC. Both galaxies lie within the Magellanic Stream, a tail of neutral hydrogen gas more than 800 000 light years

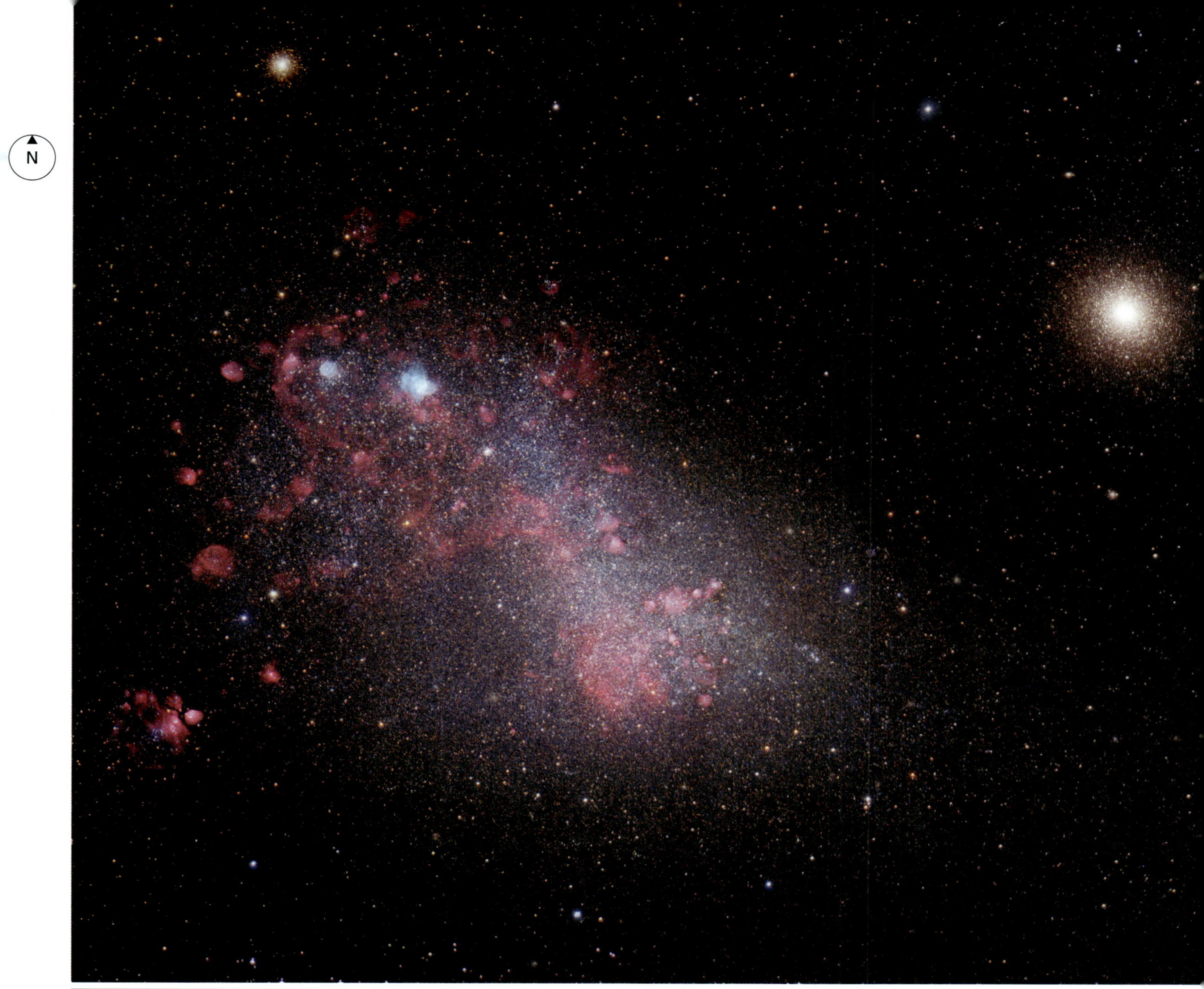

long. In wide-angle radio maps it is seen as a long band which stretches over almost 180° of the sky. The densest region of this emission is the envelope of the two Magellanic clouds, whereby one assumes that the hydrogen gas in the Magellanic Stream stems from the Magellanic Clouds themselves. On the one hand, the tidal forces of their interaction removed gas from their halos. On the other hand, the larger gas loss is caused by the strong stellar winds of the supergiants and by shock waves from supernovae in the many active regions. Under the assumption that the Magellanic Stream corresponds to the previous orbital path of the two Clouds, the length of the Magellanic Stream leads to an age of 2.5 billion years. The motion of the Clouds relative to each other and their orbits about the Milky Way began at this time.

OBJECT	SMC
CONSTELLATION	Tucana
RIGHT ASCENSION	$00^h\ 52^m\ 45^s$
DECLINATION	–72° 49′ 43″
BRIGHTNESS	2.7 mag
TYPE	SB(s)m pec
PHOTOGRAPHERS	Robert Gendler, Ryan Hannahoe
TELESCOPE	106-mm reflector
CAMERA	SBIG STX-16803
EXPOSURE TIME	2100 min
LOCATION	Fair Dinkum Skies Observatory, Moorook, Australia

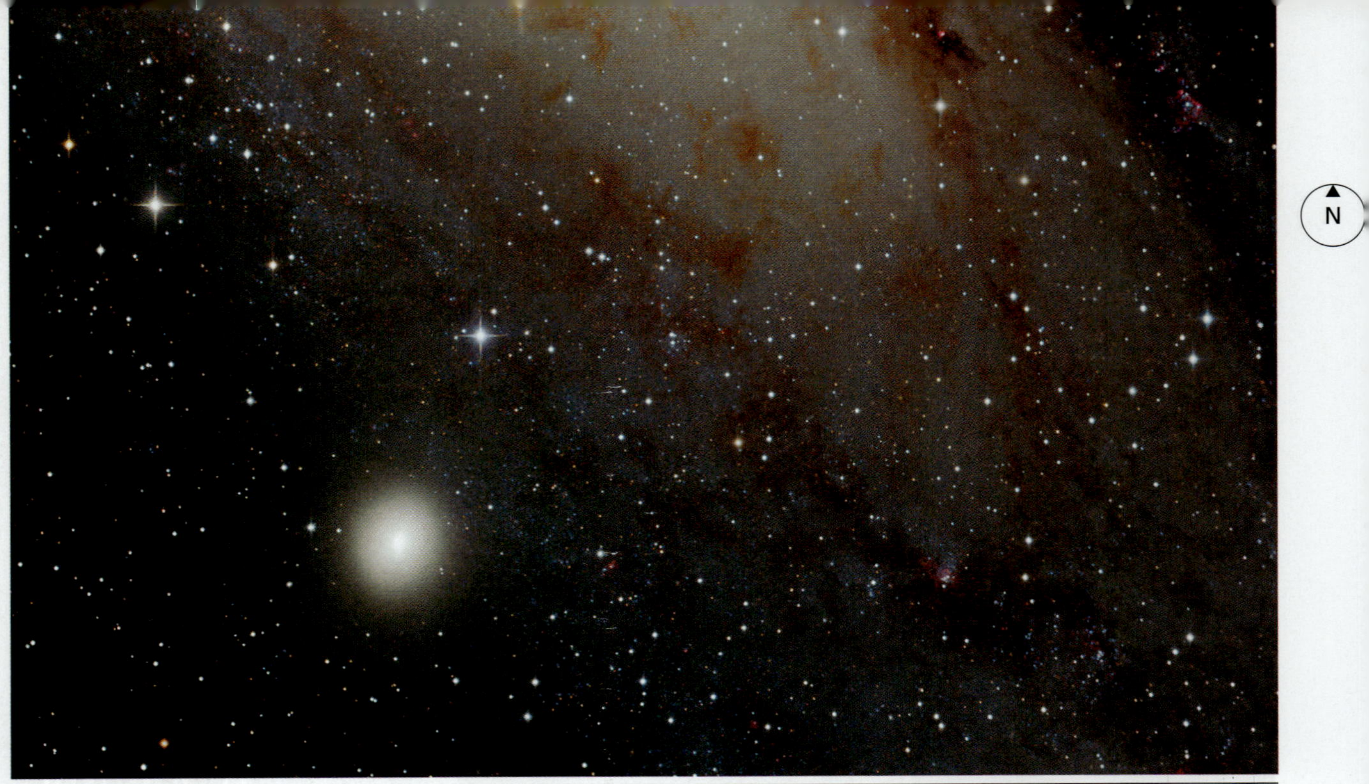

M 32

M 32 (NGC 221) is the closest companion of the Andromeda Nebula. This dwarf galaxy of size 9.1′ × 6.3′ is described in some catalogues as an elliptical galaxy. Due to its high surface density and its compact form, it is also classified as cE, a compact elliptical galaxy. The image also gives the impression that M 32 looks like the core of a spiral galaxy which has lost the surrounding disc. This hypothesis is supported by the size of 6500 light years and also by the almost complete lack of star formation. M 32 is composed mainly of an old stellar population and there are almost no hints of gas and dust in the spectra of its stars. This property is also an indication that M 32, as seen from the Earth, is in front of the spiral plane of M 31. And there is a further hint: if the light of the stars passed through the plane of M 31, astronomers would notice the gas and dust located there on the basis of absorption lines in the stellar spectra.

Using the orbital model (see text on M 110, page 221) for M 32 allows one to conclude that 130 million years have passed since M 31 last traversed the main plane of M 31, 48 per cent of the total orbital period. Thus M 32 will soon pass through again; according to the model, its current distance from the plane is about 5000 light years.

Further indications of the extraordinary nature of M 32 are the photometric and dynamic data on its stars. On the one hand, astronomers measure a steep brightness profile from the edge towards the centre, but on the other hand the rotational velocity peaks near the centre. The combination of results from ground-based telescopes and the Hubble Space Telescope agree with predictions only when there is a compact object of 3.4 million solar masses within 0.85 light years of the centre. The required density can be provided only by a supermassive black hole. A further indication that M 32 could be the core of a former spiral galaxy is that such black holes are normally not found in dwarf galaxies.

OBJECT	M 32, NGC 221
CONSTELLATION	Andromeda
RIGHT ASCENSION	$00^h\ 42^m\ 42^s$
DECLINATION	+40° 51′ 55″
BRIGHTNESS	9 mag
TYPE	compact E2
PHOTOGRAPHERS	Wolfgang Ries, Stefan Heutz
TELESCOPE	460-mm reflector
CAMERA	SBIG ST-10XME
EXPOSURE TIME	165 min
LOCATION	Altschwendt, Austria

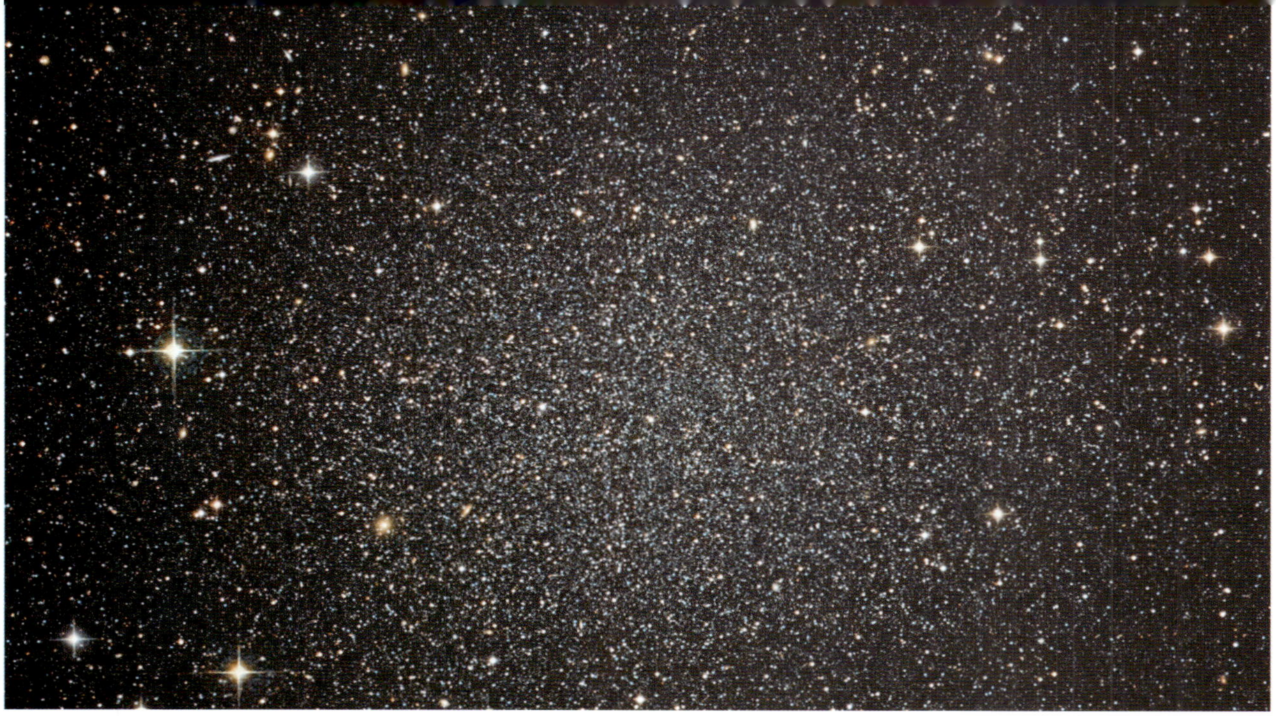

ESO 351-G030, The Sculptor Dwarf Galaxy

The dwarf galaxy ESO 351-G030 (PGC 3589) is also known as the Sculptor Dwarf Galaxy and is considered to be a satellite galaxy of the Milky Way. This dwarf galaxy has about the same galactic longitude as the Magellanic Clouds. The angular distance from ESO 351-G030 to the plane of the Milky Way of −83° is 45° larger and its distance of 260 000 light years exceeds that of the Magellanic Clouds by more than 50 per cent. Its angular diameter is almost 40′ and the physical diameter of 3000 light years is much smaller than that of the Magellanic Clounds (20 per cent of the diameter of the SMC).

In 1937, the astronomer Harlow Shapley discovered the Sculptor Dwarf Galaxy on plates taken by the Harvard 24-Inch Survey Telescope at Harvard Observatory's Boyden Station (Bloemfontein, South Africa). The dwarf galaxy was catalogued in 1982 in the ESO *Uppsala Survey of the ESO(B) Atlas* as ESO 005747-3358.7. Instead of this coordinate-based name, it is often referred to as ESO 351-G030 which contains the ESO field/plate number and the class of object (G = galaxy). ESO 351-G030 is a dwarf spheroidal galaxy. In the image one can clearly see that such dSph types, compared to dwarf elliptical galaxies, have a less compact form and a flatter intensity profile. Astronomers see in this galaxy the relic of hierarchical merger processes which in a series of steps formed large galaxies such as the Milky Way.

Radio observations of ESO 351-G030 show two equally large radio lobes located at 15–20′ to the northeast and southwest of the centre of the dwarf galaxy and which are similar to the morphology of large radio galaxies. The galaxy itself has almost no radio emission and most of the detected hydrogen gas does not correspond to the position of the stars. Since dwarf galaxies have no active nuclear region, they are not able to produce bipolar radio emission by themselves. Thus it is more probable that ESO 351-G030 is part of a tidal stream and that the distribution of neutral hydrogen along this stream causes the radio structure. This effect is strengthened on the one hand by the consumption of hydrogen in star formation and on the other hand stellar winds could have blown gas into the interstellar medium. These mechanisms could have led to condensations of neutral hydrogen gas within the tidal tail in front of and behind the dwarf galaxy. The amount of hydrogen gas detectable in radio observations corresponds to about ten per cent of the mass loss of the giant stars which formed during the strong star-formation phase eight to ten billion years ago.

OBJECT	ESO 351-G030
CONSTELLATION	Sculptor
RIGHT ASCENSION	$01^{h}\ 00^{m}\ 09{,}3^{s}$
DECLINATION	−33° 42′ 33″
BRIGHTNESS	8.6 mag
TYPE	dSph
PHOTOGRAPHERS	Dietmar Böcker, Ernst von Voigt
TELESCOPE	600-mm reflector
CAMERA	SBIG ST-10XME
EXPOSURE TIME	57 min
LOCATION	Amani Lodge, Namibia

The Large Magellanic Cloud

This dwarf irregular galaxy at a distance of 165 000 light years from the Milky Way is the third-nearest satellite galaxy. Closer are only the Sagittarius and Canis Major dwarf galaxies. The name Large Magellanic Cloud (LMC) refers to the Portuguese seaman and explorer Ferdinand Magellan (1480–1521). His circumnavigation of the world 1519–1522 in the service of the Spanish crown was accompanied by the Italian chonicler Antonio Pigafetta who was interested in geography and astronomy. He was responsible for the fact that the two companions of the Milky Way – the Large and Small Magellanic Clouds – were described and thus introduced to European science. Thanks to his access to the southern sky, the Persian astronomer Abd al-Rahman al-Sufi already described the LMC in his main work *Book of the Fixed Stars* in 964.

The LMC has a mass of about 10 billion solar masses and thus 1/20 the mass of the Milky Way. That and its diameter of 31 000 light years make it a large example of a dwarf galaxy. It is the fourth-largest galaxy in the Local Group after the Milky Way, M 31, and M 33. Its extent on the sky including all extensions is 10.7° × 9.2° and its approximately 100 square degrees make the LMC the most extended extragalactic object which can be seen with the naked eye. The longitudinal structure visible in the inner part of the LMC can be interpreted as a bar and the dimly recognizable spiral arms suggest a weakly evolved barred-spiral type. Its low rotational velocity of 40 km/s could also be seen as the sign of a past interaction which destroyed an existing spiral pattern. The dim arcs visible in the negative image which surround the LMC at the east and carry on underneath the LMC are tidal tails formed due to interaction with the Small Magellanic Cloud (SMC). Galactic cirrus makes a detailed description of these difficult.

The Large Magellanic Cloud contains a few dozen NGC objects, whereby NGC 2070 (S Doradus) is a cluster of young, hot giant stars embedded in the well known Tarantula Nebula. This cluster contains R136a1, one of the most massive and brightest stars (265 solar masses). The inner structure of NGC 2070 consists of about 30 subclusters. Each of them holds 10–150 stars and are part filaments of large arcs which stretch over 30–60 light years. The ionized

OBJECT	LMC
CONSTELLATION	Dorado
RIGHT ASCENSION	$05^h\ 23^m\ 35^s$
DECLINATION	–69° 45′ 22″
BRIGHTNESS	0.9 mag
TYPE	SB(s)m
PHOTOGRAPHERS	Stefan Binnewies, Rainer Sparenberg
TELESCOPE	28-mm Weitwinkel Objective
CAMERA	Canon EOS 5D
EXPOSURE TIME	3 min
LOCATION	Atacama Desert, near Cerro Armazones, Chile

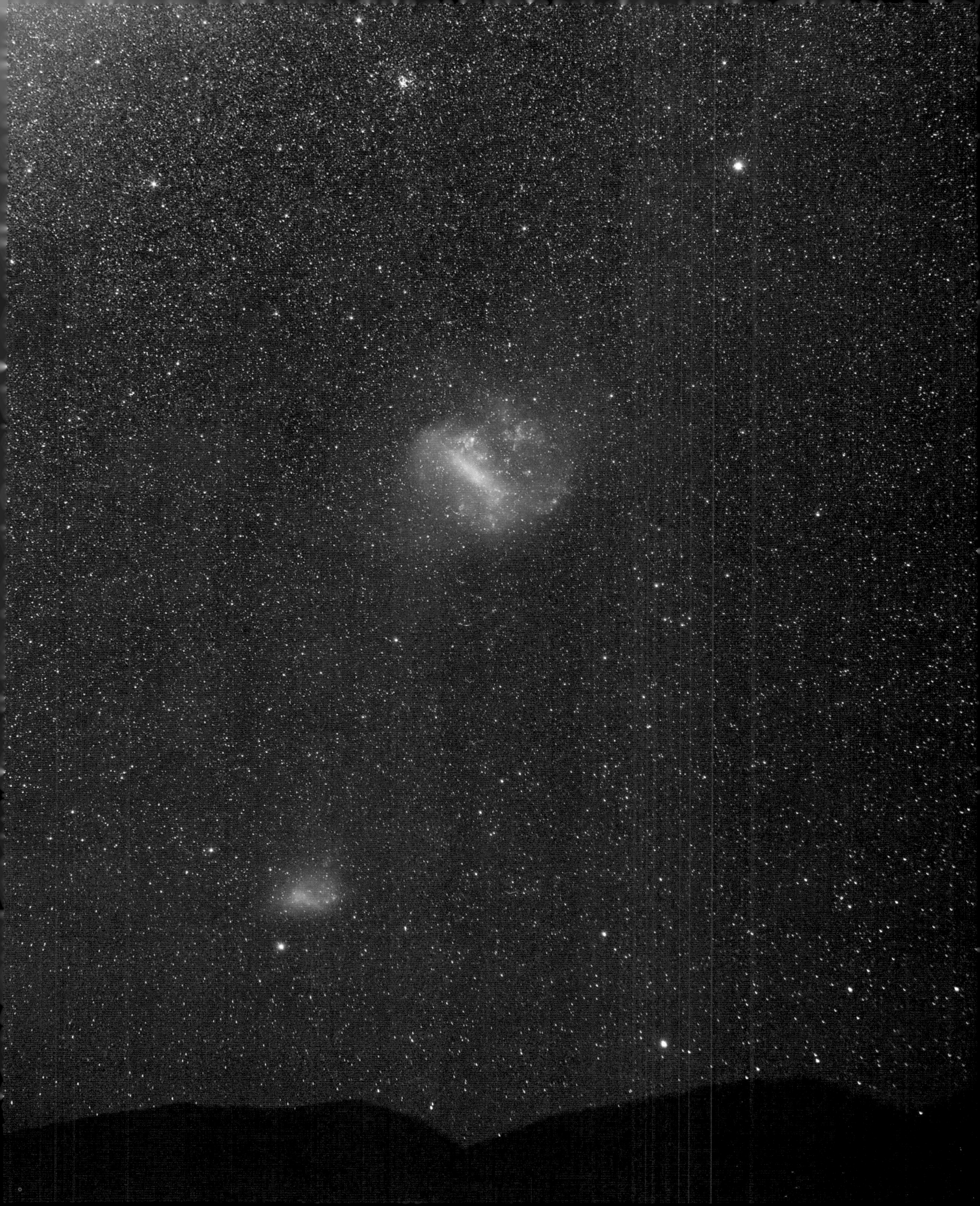

gas in these arcs is responsible for the luminous parts at the centre of the Tarantula Nebula. The analysis of the colour–magnitude diagram of the subclusters in NGC 2070 indicates that those near the centre contain bright, massive stars. The subclusters at the edge of NGC 2070, on the other hand, have fewer bright stars, which can be explained by a higher age or other physical conditions in the zone near the edge. The exceptional behaviour of the LMC with regard to star formation is also supported by the supernova 1987A. It exploded in the LMC, making it the closest and hence brightest supernova since 1604.

View towards the centre of the Large Magellanic Cloud. Image credit: Stefan Binnewies, 105-mm refractor, roll film Kodak Panther 400, 90 min, Réunion, France.

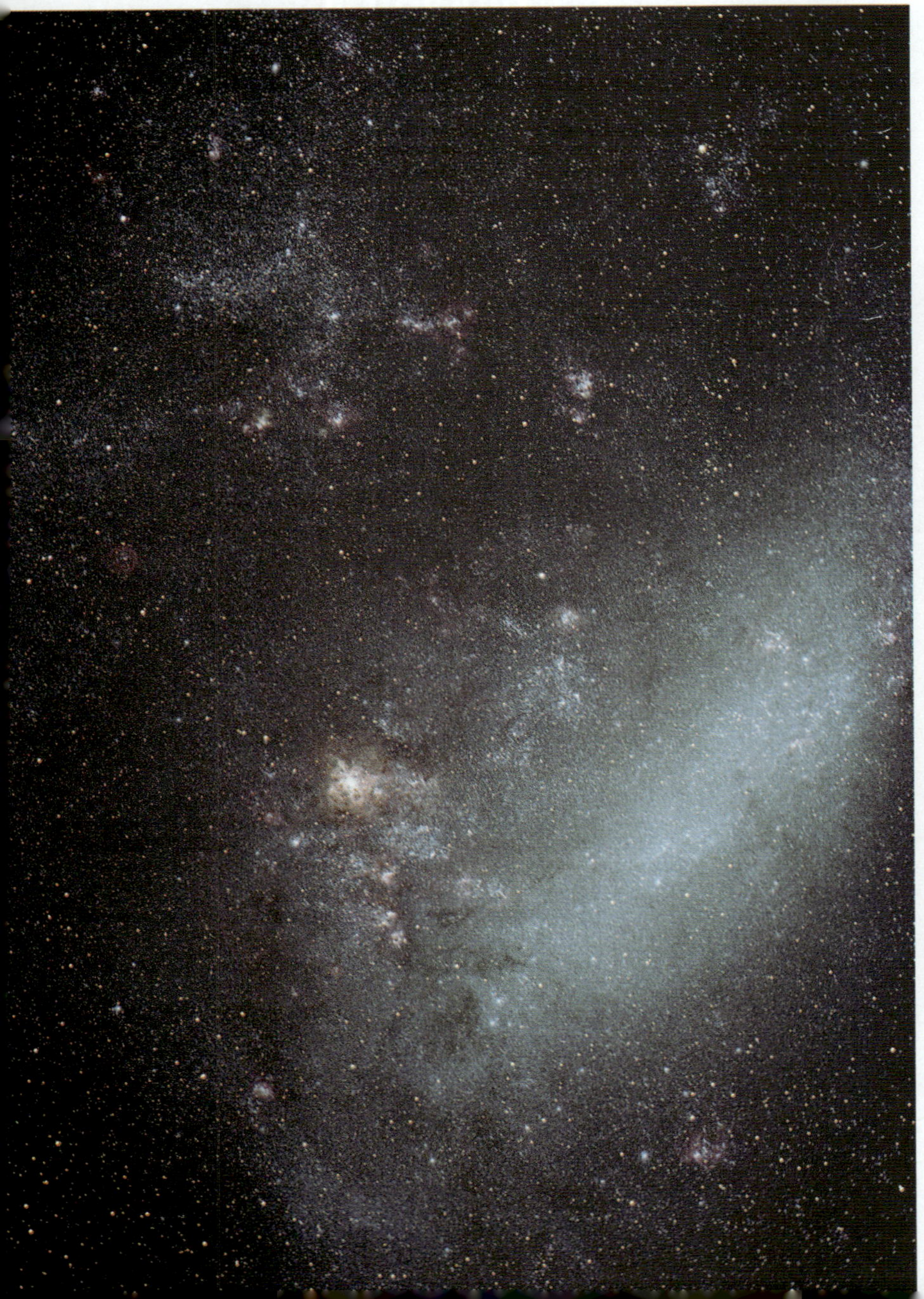

The high star-formation rates in the LMC are due to the high stellar densities, which in some regions can be 100 000 times as high as in the environment of the Sun. The metallicity of the LMC is only about half that of the Milky Way. This favours the formation of large stars. Active regions lead to the formation of large bubbles which can reach sizes of a few hundred light years and are caused by stellar winds and shock waves from hot stars and supernovae. Astronomers have found in the LMC 60 globular clusters, 400 planetary nebulae, and hundreds of thousands of giant stars and supergiants – a huge number of objects which demonstrate the special status of the LMC as one of the most active galaxies in the Local Group.

The determination of distances and velocities is especially difficult in nearby objects such as the LMC since systematic errors can be proportionally large. An example is use of the period–luminosity relation for Cepheids as standard candles, calibrated with Cepheids in the LMC. Current research indicates that the LMC is approaching the Milky Way more rapidly than previously supposed.

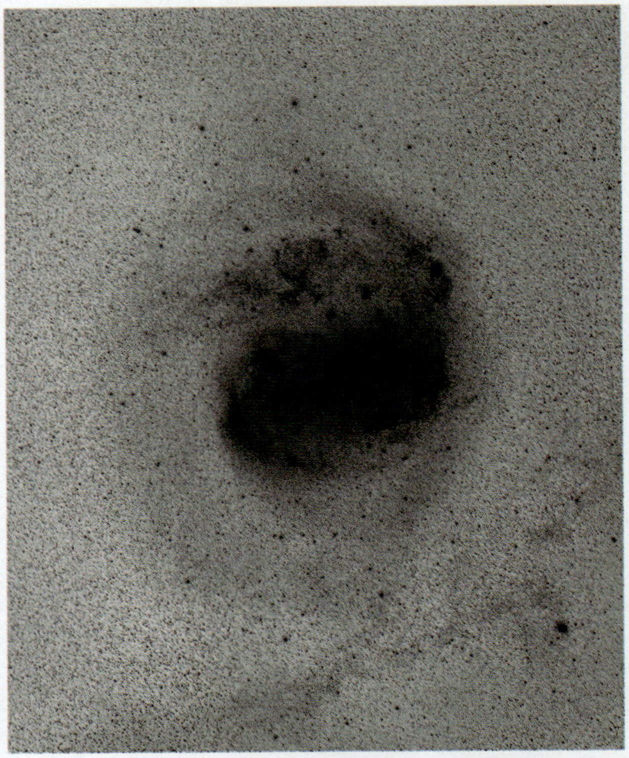

Contrast-enhanced image of the LMC and its surroundings. Image credit: Stefan Binnewies, 165-mm medium-format objective, roll film Kodak Ektachrome Pro Gold 400, 30 min, Tivoli Farm, Namibia.

Astronomers use simulations to analyse the kinematics of the Magellanic Clouds and the Milky Way, whereby the starting point are three-dimensional positions and velocities. These indicate that the LMC orbits the Milky Way in an eccentric orbit with a diameter of two million light years, closely accompanied by the Small Magellanic Cloud (SMC, with 15 per cent of the mass of the LMC). Due to the more rapid motion towards the Milky Way, researchers suspect that this orbit is just a first approximation. This also led to the observed increased star formation, the gas starting to interact with the halo of the Milky Way in the last one to three billion years. The larger velocities, however, also mean a shorter interaction time. Such a short time raises new questions, such as what caused the warp of the disc of the Milky Way, which up until now can be explained only by several orbits of the satellite galaxies.

A further question which astronomers ask about the Magellanic Clouds concerns the uniqueness of such dwarf galaxies as satellites of large spiral galaxies. Data from the Sloan Digital Sky Survey (SDSS Data Release 7) were used to answer this question. Spiral galaxies similar to the Milky Way were extracted from the spectral catalogue and candidates for Magellanic-Cloud-like systems from the photometric catalogue. This was used to calculate the probability that large spiral galaxies have satellites similar to the Magellanic Clouds. On average, 12 per cent have one satellite and only 3.5 per cent have two satellites like the Milky Way does. So it seems that the Magellanic Clouds are not only special among dwarf galaxies due to their high star-formation rates, but also that their very existence is a cosmic exception. This is in contradiction to the Copernican principle, which states that the Earth and the Milky Way do not occupy a privileged position in space. This could be resolved if the Magellanic Clouds are a temporary phenomenon. In a few billion years the Milky Way might lose its satellites – they could either merge with the Milky Way or escape from its gravitational field. Then the Milky Way would be an average galaxy.

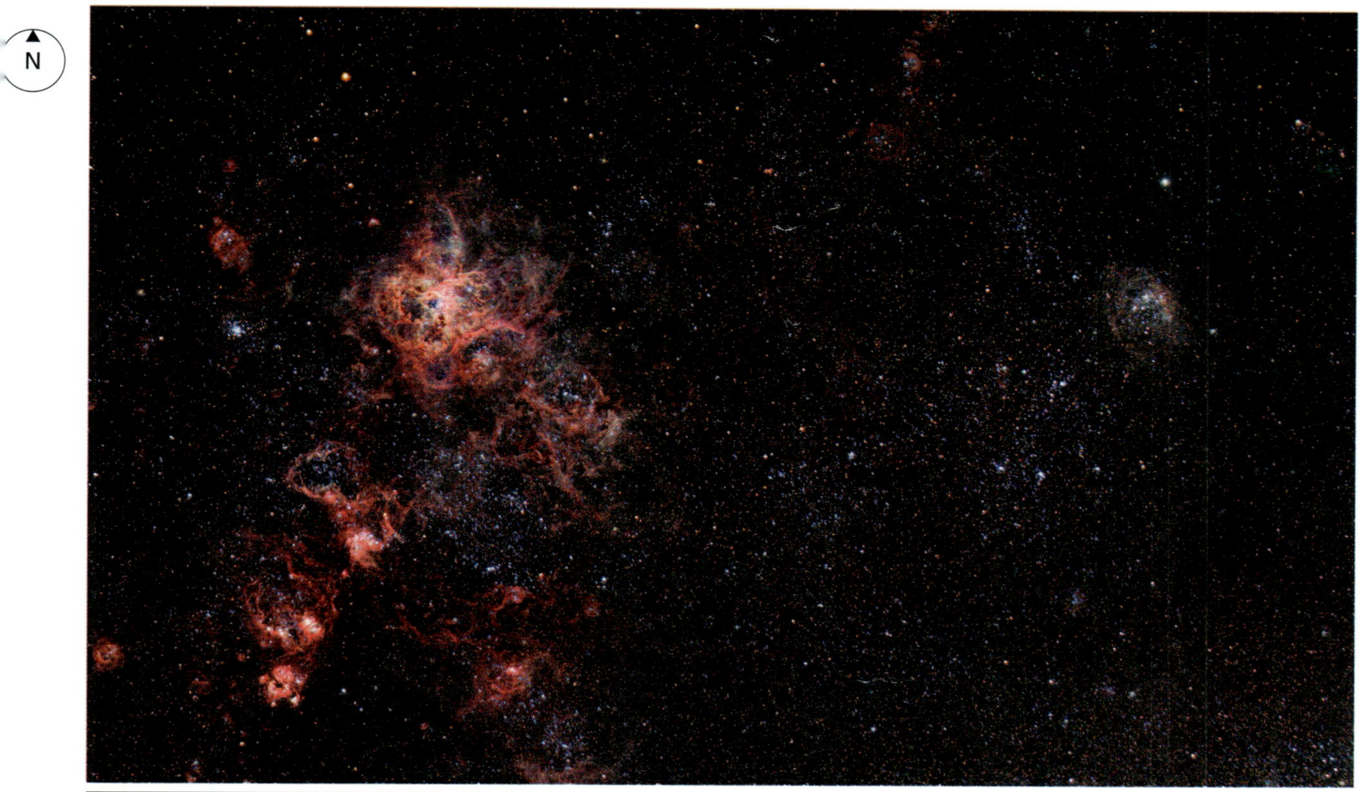

The Tarantula Nebula (above) in the LMC. Image credit: Johannes Schedler, 140-mm refractor, SBIG STL-11000, 210 min, Hakos Farm, Namibia.

Holmberg II

Holmberg II (UGC 4305, Arp 268) is an irregular galaxy which belongs to the group around M 81. Its distance is 11.4 million light years and its angular size of 7.9′ × 6.3′ corresponds to a mean diameter of 26 200 light years. Holmberg II has an irregular, funnel-shaped form which contains many bubbles. These have sizes of a few thousand light years and are surrounded by active regions which contain chains of OB associations and H II regions. Particularly prominent is a U-shaped formation of active regions at the eastern end of Holmberg II. It extends across a minute of arc, so its angular length corresponds to more than 3300 light years. The individual active zones all have almost the same diameter of 400–500 light years. Studies of Holmberg II describe it as a dwarf irregular galaxy whose structure and appearance are similar to the Large Magellanic Cloud, which is why it is also described in the literature as a dwarf Magellanic galaxy. This also means that Holmberg II has prominent signs of star formation and is thus a transitional type between a gas-poor dwarf galaxy and a normal galaxy.

It is notable that irregular galaxies such as Holmberg II have neither a spiral plane with differential rotation nor a central bulge to disturb the bubble structure of these dwarf galaxies. It is thus possible that star-forming regions are stable over a time span which would never be achieved in a spiral galaxy. Thus, several stellar populations can follow one another since new star-forming regions can always be formed via the mixing processes and concentration of gas due to a supernova in the surroundings. Thus in a dwarf galaxy, star formation can take place as a mostly undisturbed chain reaction.

The optical spectrum of the galaxy Holmberg II shows strong emission lines of the elements hydrogen and oxygen. This radiation is due to photoionization, i.e. by the energetic light of young, hot stars which cause the surrounding gas nebula to glow. The two strongest lines are that of hydrogen Hα (656.3 nm) and the oxygen line O III (436.3 nm), which is referred to as a temperature-sensitive line in the scientific literature. The temperature of the emission region can be determined from the line strength via models of physical excitation. Astronomers have studied the temperatures of various active regions in the galaxy by varying the position of the slit during spectrographic studies. The line strengths indicate temperatures in the range 12 500–15 600 K. It could also be shown that although Holmberg II has a similar luminosity to that of other dwarf galaxies, it also has a below-average metallicity. This metallicity, which in astronomy indicates the abundance of elements heavier than helium, was determined by the ratio of oxygen to hydrogen. This ratio in Holmberg II is below the average value one would expect based on the luminosity of the dwarf galaxy. Based on general considerations of the relation between luminosity and metallicity in dwarf galaxies, these observations lead to the conclusion that Holmberg II is an unusually bright dwarf galaxy, for its metallicity. As proof of this, astronomers point to the helium II emission line (468.6 nm) which indicates an additional ionization source. This could be so-called Wolf-Rayet stars, which are 20 to 100 times as massive as the Sun. At the end of their lives, the outer envelope is ejected, revealing the hot core. This emits powerful stellar winds, forming an expanding nebula out of the surrounding material. The high-energy radiation of the core finally causes the gas bubble formed around the Wolf-Rayet star to glow. One assumes that the most massive examples of these stars end as a supernovae only a few ten thousand years after this phase.

The high temperatures of these regions make them interesting for X-ray astronomy as well, in which Holmberg II and other irregular galaxies are studied via satellite observations. Where the X-rays could be spatially resolved, it was possible to associate them with the large bubbles of hot gas.

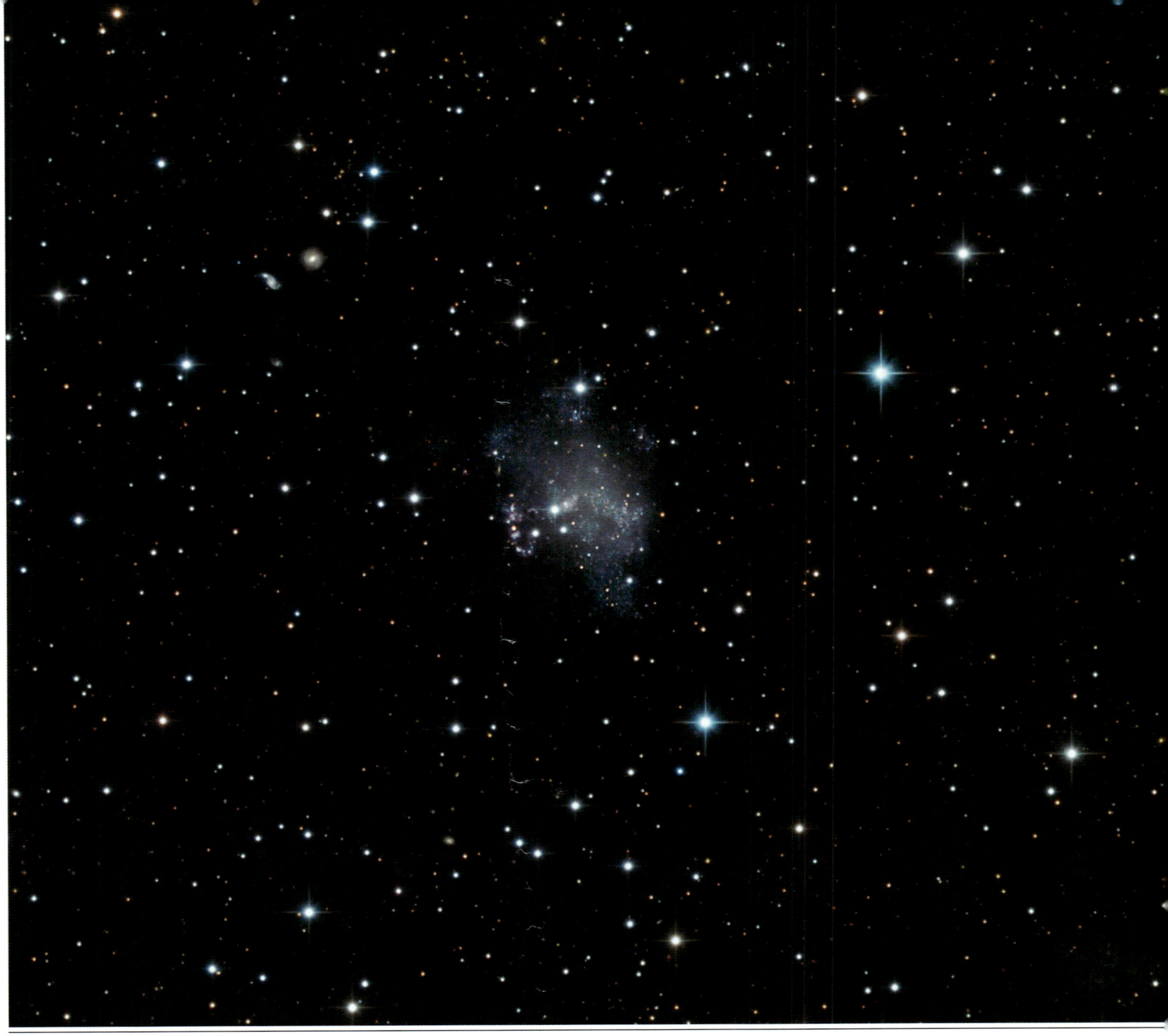

In contrast to other galaxies, the X-ray emission in Holmberg II comes from a point source which exhibits non-periodic variation on timescales of days to a few months. This ultraluminous X-ray source (ULX) received the name Holmberg II X-1 and was studied in detail in many missions. It is difficult to uniquely classify Holmberg II X-1, however, since the source is located in a star-forming region and the measurements could be contaminated by other nearby sources. It is suspected that Holmberg II X-1 is a stellar black hole with a mass of 10–20 solar masses onto which material is flowing and which is surrounded by a gas cloud. The position of this ULX in the image is at the lower part of the U-shaped structure mentioned above.

OBJECT	Holmberg II, UGC 4305, Arp 268
CONSTELLATION	Ursa Major
RIGHT ASCENSION	$08^h\ 19^m\ 05^s$
DECLINATION	+70° 43′ 12″
BRIGHTNESS	11.1 mag
TYPE	Im
PHOTOGRAPHERS	Bernhard Hubl
TELESCOPE	305-mm reflector
CAMERA	SBIG ST-2000XM
EXPOSURE TIME	468 min
LOCATION	Nussbach, Austria

IC 2574

The American astronomer Edwin Foster Coddington, using the Crocker Photographic Telescope at the Lick Observatory (University of California, USA) to search for comets and asteroids, discovered the dwarf irregular galaxy IC 2574 in Ursa Major on a photographic plate in 1898. It subsequently became known as the Coddington Nebula. This dwarf galaxy is part of the M 81 galaxy group and with an angular size of 13.2′ × 8.4′ and a visual magnitude of 10.8 mag is one of the five largest galaxies in this group. This consists of a total of 30 members and the largest galaxies are M 81, NGC 2403, and NGC 4236. The distance of IC 2574 from the Milky Way is 12.5 million light years and its size of 48 000 light years exceeds that of the starburst galaxy M 82 (40 000 light years in diameter).

Compared to M 81, IC 2574 appears to be a dim variant of an irregular galaxy, although it also has many active regions where new stars are formed. This is described in astronomy by the star-formation rate (SFR) and expressed in units of solar masses per year. In IC 2574, this rate is about 0.1 solar mass per year. In the larger Milky Way, the star-formation rate varies between 0.5 and 1.6 solar masses per year. In such a comparison, however, one must keep in mind that the Milky Way has spiral arms as large-scale generators of star formation and that the metallicity of the gas which can be used for star formation is lower in irregular galaxies than in spiral galaxies. There is thus a fundamental relation between the star-formation rate, the metallicity, and the stellar mass of a galaxy. Applied to a dwarf galaxy, this means that they are usually gas-rich systems with low metallicity but with no or almost no dust and with the highest scatter in metallicity of all galaxy types. The reason for this is that clumpy distributions of matter with especially high local SFR values contribute significantly to the total star-formation rate.

In many dwarf galaxies, star-forming regions are located in the periphery of large bubble structures with an H I deficit at the centre. IC 2574 researchers found a special case of these, a supergiant shell (SGS). This has an expansion velocity of 25 km/s and an enormous diameter of 26 000 light years, corresponding to an age of 15.8 million years. The mass of hydrogen which must lie within this bubble is 820 000 solar masses. At the edge of this bubble, gas is pushed out from the inner regions and thus new star clusters are formed, causing the SGS to dominate the infrared luminosity of the galaxy. In the image, the SGS can be easily identified at the left, northern edge of IC 2574 by the dense clusters of bright, blue-white stars. The hot stars also emit ultraviolet radiation which is absorbed by the dusty surroundings and re-radiated as infrared light. There are relations between various indicators of a high star-formation rate such the Hα intensity or the infrared luminosity which are used by astronomers to study in detail the processes at work in a galaxy. IC 2574 was observed in October and November 2004 for a total time of 127 minutes and 36 seconds with the Infrared Array Camera and the Multiband Imaging Photometer of the Spitzer Space Telescope within the Spitzer Infrared Nearby Galaxies Survey (SINGS). The star cluster which is at the centre of the SGS and surrounded by dense zones of dust and gas can be seen in the infrared spectrum in the range 24–160 μm. The infrared measurements show that the temperature of this dust varies by a factor of three, with the warmest regions reaching 30 kelvin.

OBJECT	IC 2574
CONSTELLATION	Ursa Major
RIGHT ASCENSION	$10^h\ 28^m\ 23^s$
DECLINATION	+68° 24′ 44″
BRIGHTNESS	10.8 mag
TYPE	SAB(s)m
PHOTOGRAPHERS	Rainer Sparenberg
TELESCOPE	1.12-m reflector
CAMERA	SBIG STL-11000
EXPOSURE TIME	290 min
LOCATION	Melle, Germany

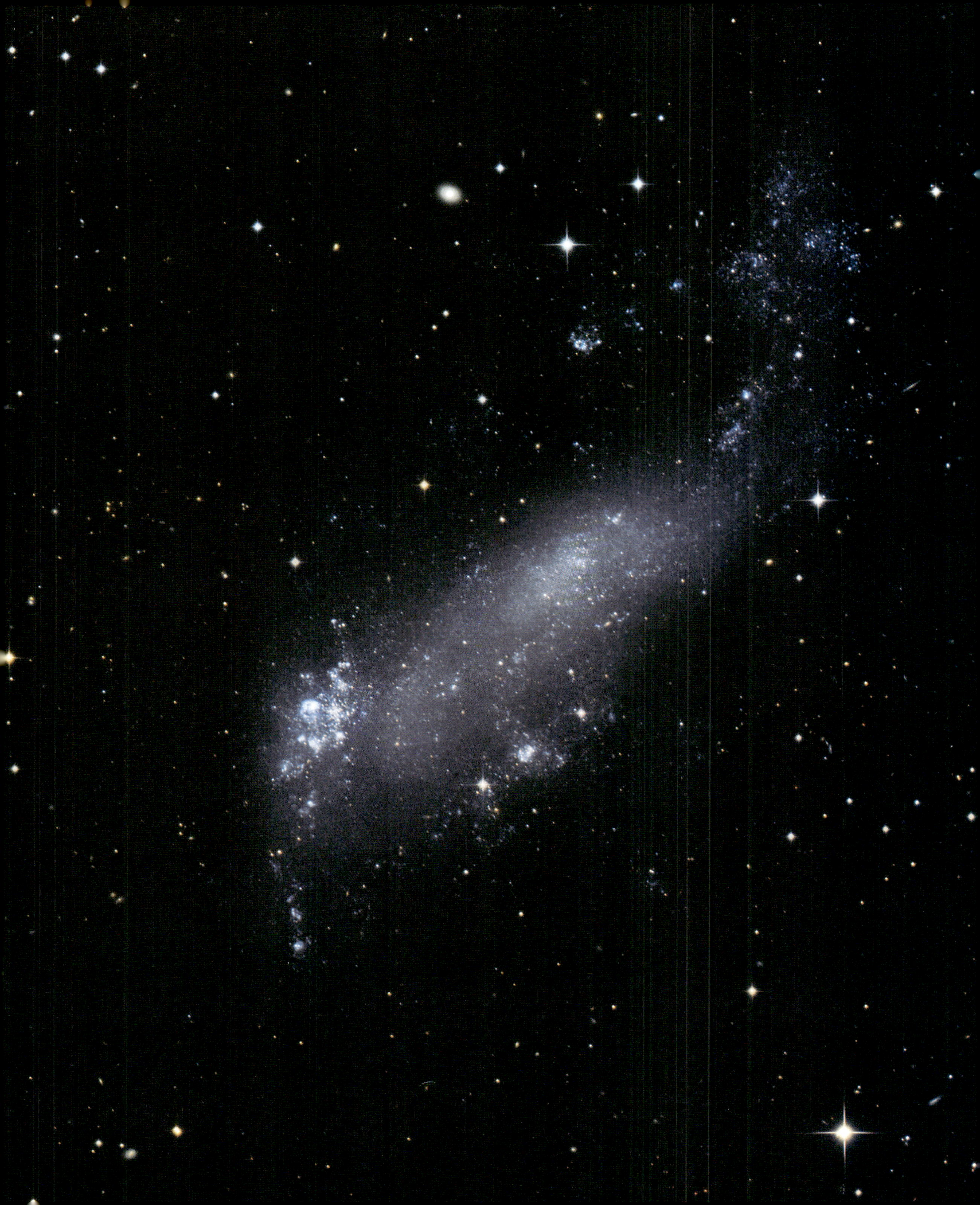

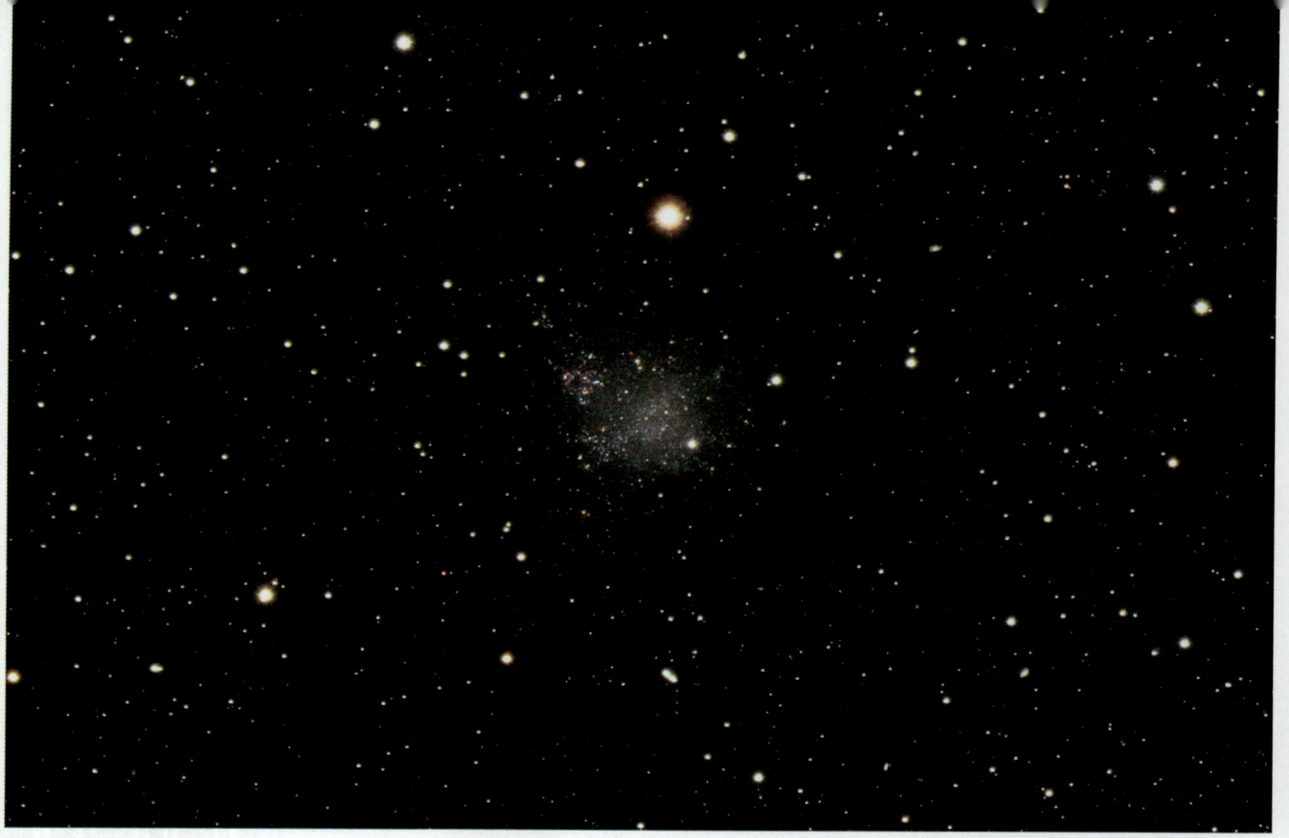

IC 1613

The irregular galaxy IC 1613 is a member of the Local Group located at a distance of 2.4 million light years from the Earth. It is thus about as distant as the Andromeda Nebula. However, it is located 40° away, and thus cannot be seen as a satellite of the Andromeda Nebula. The angular size of IC 1613 is 16.2′ × 14.5′ and its magnitude is less than 10 mag. The maximum physical diameter is 11 300 light years. Based on the size and classification of this galaxy, it is morphologically between the Large and Small Magellanic Clouds.

However, IC 1613 displays significant differences from the Magellanic Clouds. Its form is more compact and the distribution of the stars is concentrated in a central area about 9′ in diameter. IC 1613 has no history of active star formation like the LMC and SMC and there is no indication of several generations of stars. Astronomers measure only a low star-formation rate in IC 1613 which is due to the low metallicity of the irregular galaxy. Thus, the image displays a few active regions only at the left, eastern edge of the galaxy. Young, massive stars in OB associations are also located here. These OB stars are relatively short-lived and their short lifetime is a good, compact indicator for the chemical composition of the gas in the galaxy.

There is thus a typical number of OB stars in these associations, which also results in a typical size. In the case of IC 1613, these associations usually consist of fewer than ten stars; about half of these groups consist of three OB stars. Only ten per cent have as many as ten OB stars The average size of the OB associations is between 140 and 200 light years. Radio observations can also confirm the tendency of neutral hydrogen to be located in the same areas, as in IC 1613. This contains the active star-forming regions and there are also shells of hydrogen gas which are fuelled by the stellar winds of OB stars. They are also present in larger numbers in the Magellanic Clouds.

OBJECT	IC 1613
CONSTELLATION	Cetus
RIGHT ASCENSION	$01^h\ 04^m\ 48^s$
DECLINATION	+02° 07′ 04″
BRIGHTNESS	9.9 mag
TYPE	IB(s)m
PHOTOGRAPHERS	Bernhard Hubl
TELESCOPE	101-mm reflector
CAMERA	SBIG ST-2000XM
EXPOSURE TIME	143 min
LOCATION	Steinbach am Ziehberg, Austria

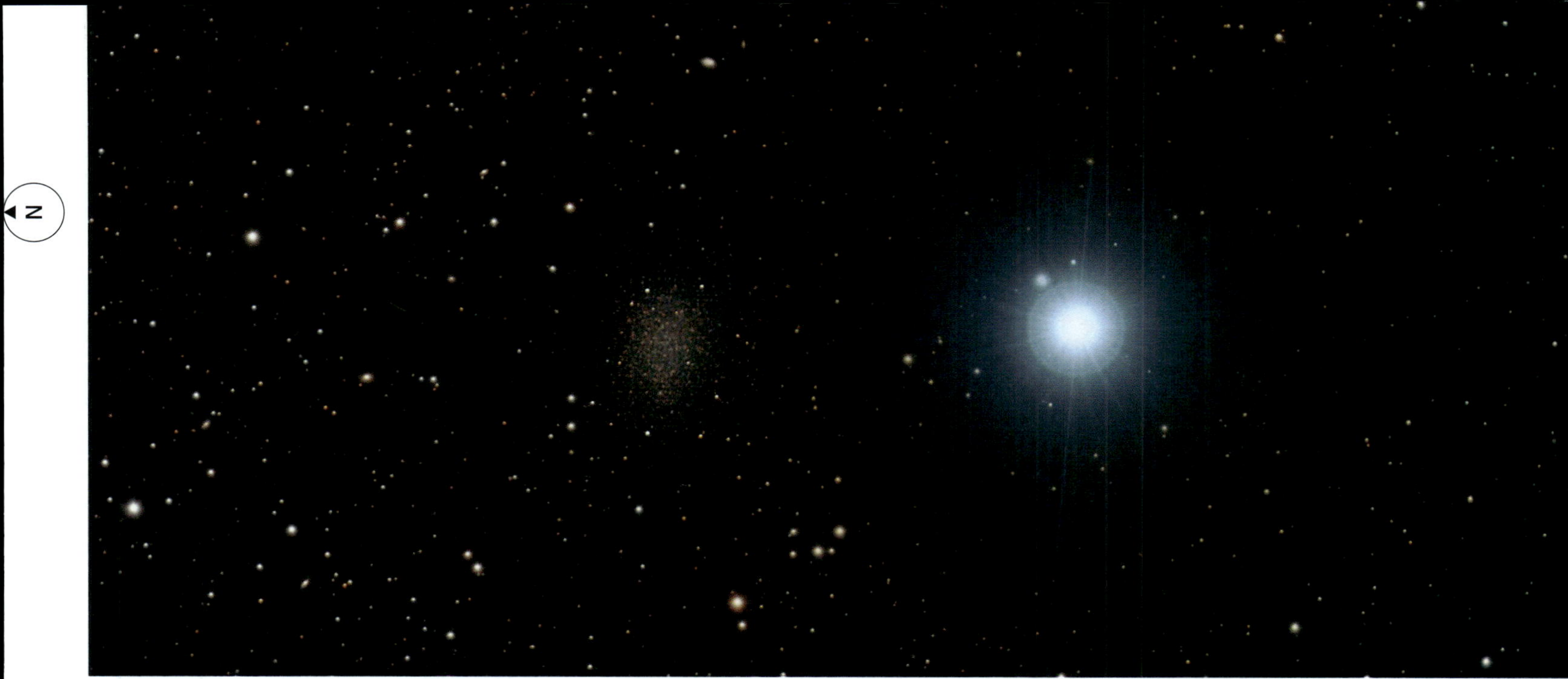

Leo I

The dwarf galaxy Leo I is a dwarf spheroidal galaxy seen in the image only 20′ away from the star Regulus. Compared to the other dwarf satellite galaxies of the Milky Way, at 789 000 light years Leo I is relatively far away and thus has a compact angular size of 9.8′ × 7.4′; its physical size is given as 2300 × 1700 light years. Leo I is receding from the Milky Way at +286 km/s and its distance of 120° from the dwarf galaxies Ursa Mino and Draco, which are approaching the Milky Way, is also a hint that Leo I belongs to a different tidal stream surrounding the Milky Way.

Astronomers studied galaxies in the Local Group in the Leiden/Dwingeloo H I Survey (1997, Dwingeloo Radio Observatory, Netherlands). However, the resolution of 36′ was low compared to the size of the galaxy of almost 10′. The range in velocity covered was −450 km/s to +450 km/s, which is sufficient for the Local Group. H I emission was detected in about half of the dSph galaxies in the Local Group – Leo I displayed a complex intensity distribution. In contrast, no radio emission could be detected in the dwarf galaxies Ursa Minor, Draco, and Leo II. The radio map shows Leo I not as a point source but rather as a structure consisting of several bands of emission. This is known among astronomers as highly fragmented emission, which can cover several degrees, with bands 30–60′ wide. Leo I is at the edge of such a band close to two knots of H I emission separated by 40′.

The textbook description of dwarf galaxies describes them as gas-poor or gas-free systems, which raises the question as to how dwarf galaxies can be radio sources at all. This is explained via so-called high-velocity clouds, which orbit the galactic centre and are associated with streams of material. These clouds consist mainly of atomic hydrogen of low metallicity, only a tenth that of the Sun. The atomic hydrogen observed in some of these galaxies in the Leiden/Dwingeloo survey can be associated with the envelope of the dwarf galaxy, but not with its core. This explains why other radio observations of the cores of dwarf galaxies, using telescopes with higher resolution, could not detect any radio emission.

OBJECT	Leo I
CONSTELLATION	Leo
RIGHT ASCENSION	$10^h\ 08^m\ 28^s$
DECLINATION	+12° 18′ 23″
BRIGHTNESS	11.2 mag
TYPE	dE3
PHOTOGRAPHERS	Bernhard Hubl
TELESCOPE	101-mm reflector
CAMERA	SBIG ST-2000XM
EXPOSURE TIME	504 min
LOCATION	Schlierbach, Austria

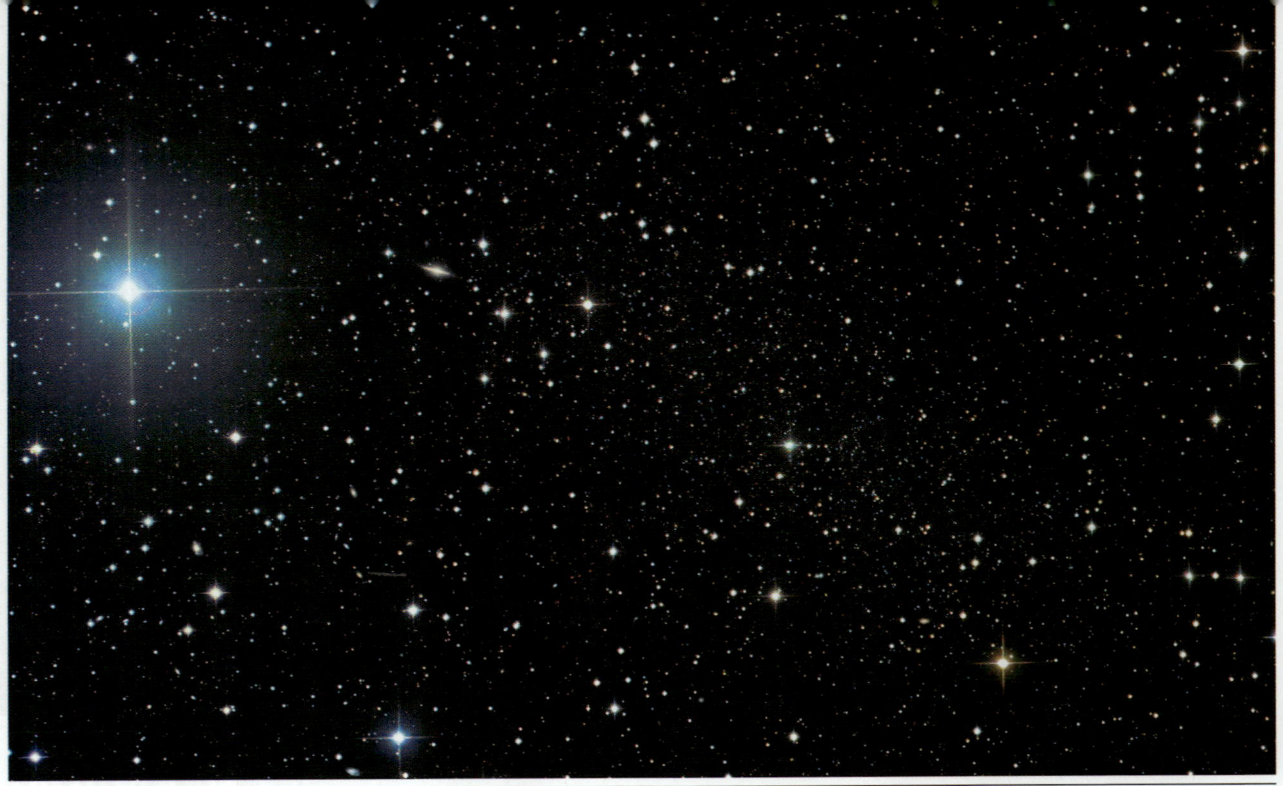

Ursa Minor Dwarf

The dwarf galaxy Ursa Minor Dwarf (UGC 9749, PGC 54074) is a small satellite galaxy of the Milky Way. It was discovered in 1954 by Albert G. Wilson of the Lowell Observatory. In addition to the Ursa Minor Dwarf Galaxy, he also discovered three other dwarf galaxies, which all belong to the Milky Way (Leo I and II, Draco Dwarf). The dwarf galaxies in the Local Group can be divided into irregular, dwarf elliptical, and dwarf spheroidal galaxies. The last are much more diffuse and appear less well defined than dwarf ellipticals, whereby the difference in size makes comparison with normal galaxies difficult. The Ursa Minor Dwarf is classified as dSph, whereby it is visible only as a light brightening in the middle of the right half of the image. The stars in dSph galaxies are of an old population. Most of the stars are several billion years old and since these galaxies contain almost no gas, there is no formation of new stars.

The recession velocity of the Ursa Minor Dwarf Galaxy is −247 km/s – it is thus approaching the Milky Way and demonstrates that there is large scatter among the velocities of galaxies in the Local Group, with some even being negative. Thus, astronomers must use other methods for determining the distance. These result in a distance of 220 000 light years, although this distance measurement has an uncertainty of 22 per cent (according to NED). The angular size of the Ursa Minor Dwarf Galaxy is 30′ × 19′; its physical size is calculated as 1900 × 1200 light years.

The number of stars in a dwarf galaxy is usually larger than that of a Galactic globular cluster; they are, however, spatially orders of magnitude less concentrated. Results from the last few years indicate that more than ten million stars in a region of size 1000 light years exceeds the stability limit for dwarf galaxies. Dwarf galaxies have proportionally much more dark matter than normal spiral galaxies and it appears that dwarf galaxies are the first condensations of luminous matter around conglomerations of dark matter. In the scenario of hierarchical galaxy formation, a large galaxy is formed by a series of mergers of such dwarfs.

OBJECT	Ursa Minor Dwarf, UGC 9749, PGC 54074
CONSTELLATION	Ursa Minor
RIGHT ASCENSION	$15^h\ 09^m\ 08^s$
DECLINATION	+67° 13′ 21″
BRIGHTNESS	11.9 mag
TYPE	E
PHOTOGRAPHERS	Stefan Binnewies, Josef Pöpsel
TELESCOPE	600-mm reflector
CAMERA	SBIG STL-11000
EXPOSURE TIME	470 min
LOCATION	Skinakas Observatory, Crete, Greece

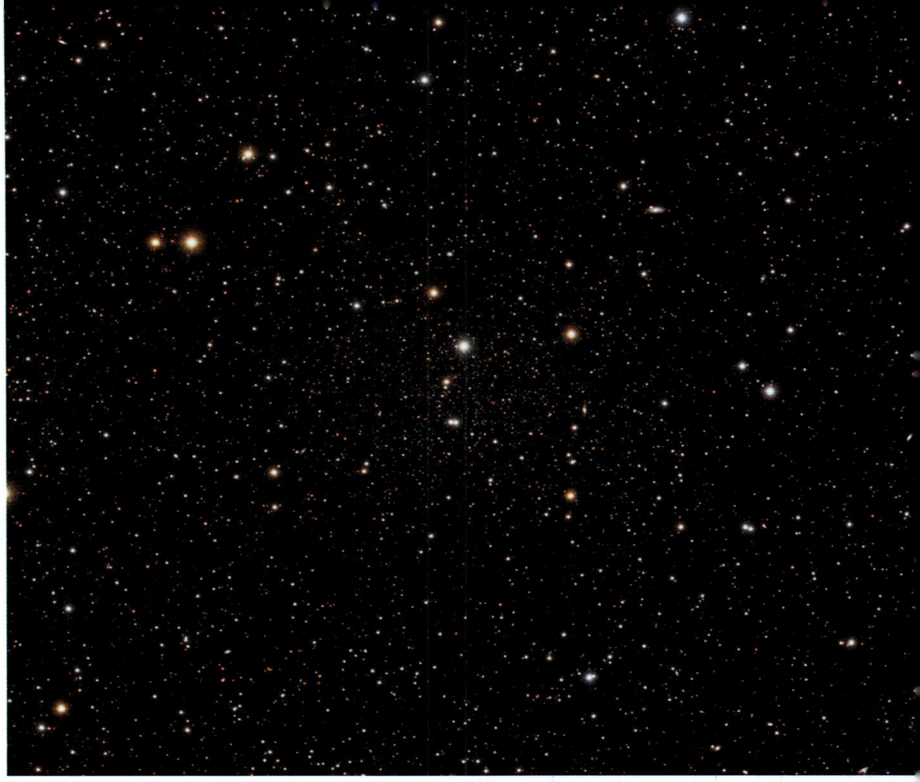

Draco Dwarf

The Milky Way has eleven dwarf galaxies, all gravitationally bound, at distances between 50 000 light years and 800 000 light years. Among these is the dwarf spheroidal galaxy UGC 10822 in the constellation of Draco, visible as a dim cloud in the centre of the image. This galaxy is among the dimmest companions of the Milky Way and is at a distance of 265 000 light years. Its relative velocity of −292 km/s is close to that of the Ursa Minor Dwarf, which is 20° away from the Draco Dwarf and thus lies in a similar direction from the Milky Way. The diameter of the Draco Dwarf is 2800 light years.

Dwarf galaxies usually have a low mass and a size of just several hundred to a few thousand light years. This makes them susceptible to the locally varying tidal effects which the Milky Way exerts on its satellites. At typical orbital radii of up to 60 000 light years, the dwarf galaxies are pulled apart and can even be completely destroyed. Their remains are distributed along the former orbits and thus allow one to directly observe the kinematical state of the surroundings of the Milky Way. Such observations can answer questions about the existence of streams of material which originate in groups of dwarf galaxies and surround large galaxies such as the Milky Way or the Andromeda Nebula. The kinematics of the dwarfs also allows one to study the gravitational potential of the Milky Way in order to test models of the distribution of mass in the Galactic plane.

However, the tidal effects also make it impossible to treat dwarf galaxies as equilibrium systems. Mass estimates are thus difficult and it can be assumed that the derived masses are too large. Thus the unusually large mass-to-light ratio in dwarf galaxies probably has a systematic error. This mass-to-light ratio for dwarf galaxies is in the range 30–100, for the Draco Dwarf galaxy one finds in the literature values of 60–250. In contrast, based on kinematic methods, spiral galaxies have values of about 30, elliptical galaxies between 10 and 20. Thus the especially large mass-to-light ratios derived for some galaxies are probably due to tidal effects and not just to a higher-than-average fraction of dark matter. Future astrometry missions will be able to help to answer this question, since more accurate orbital-velocity measurements will allow the pericentre of the orbit of a dwarf galaxy to be better calculated. The larger the actual mass-to-light ratio, the smaller is the smallest possible distance of this pericentre. A ratio of 30 results in a distance of 156 000 light years, a ratio of 100 implies that a dwarf galaxy can avoid destruction even if the distance of the pericentre is only 81 000 light years.

OBJECT	Draco Dwarf
CONSTELLATION	Draco
RIGHT ASCENSION	$17^h\ 20^m\ 12^s$
DECLINATION	+57° 54′ 55″
BRIGHTNESS	10.9 mag
TYPE	E pec
PHOTOGRAPHERS	Bernhard Hubl
TELESCOPE	101-mm reflector
CAMERA	SBIG ST-2000XM
EXPOSURE TIME	540 min
LOCATION	Schlierbach, Austria

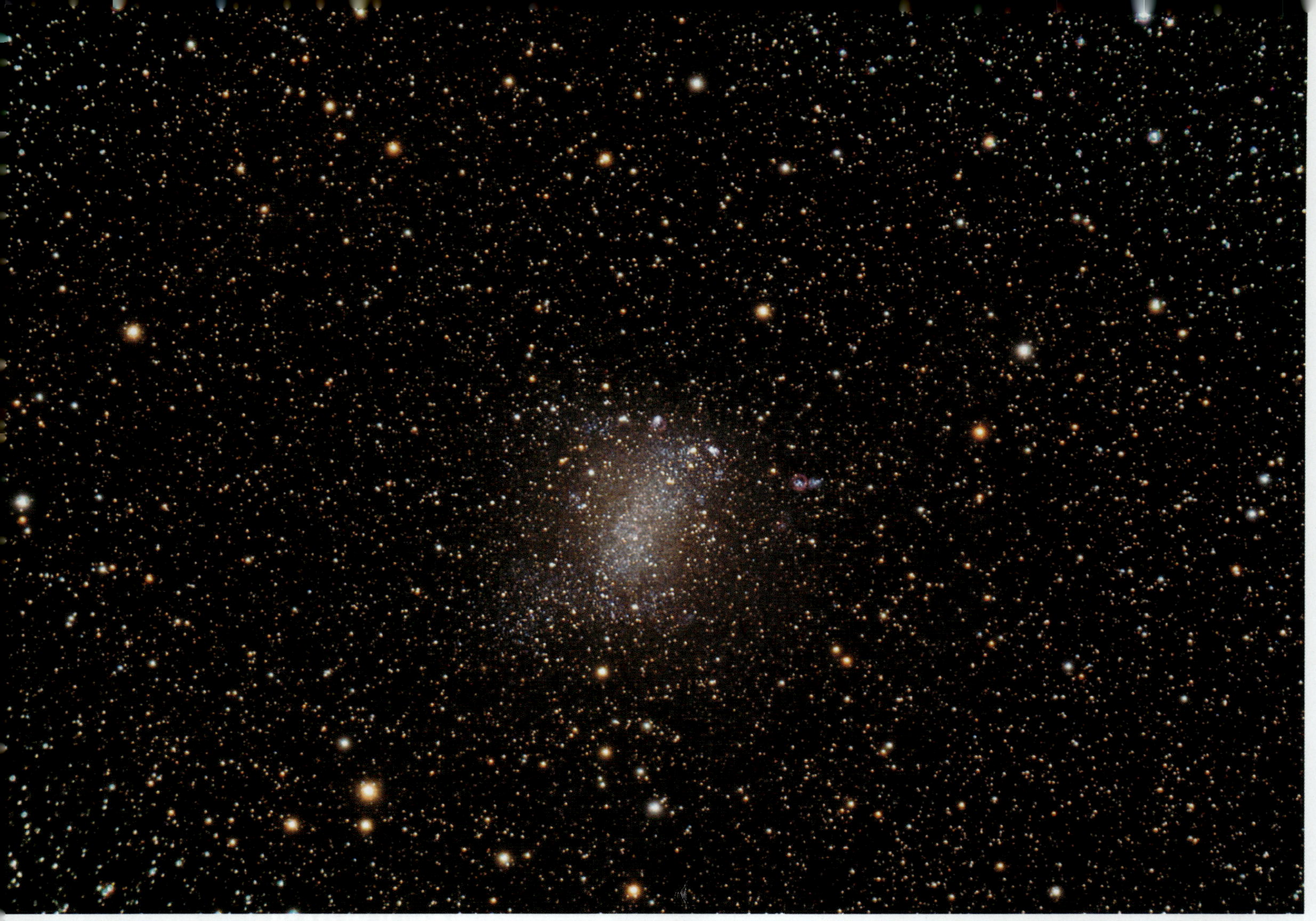

NGC 6822

NGC 6822 is a dwarf irregular galaxy which is also known as Barnard's galaxy. This dwarf galaxy is closer to the Milky Way than to the Andromeda Nebula, but due to its large distance from both large galaxies is often considered to be an isolated member of the Local Group. NGC 6822 was discovered in 1884 by the American astronomer Edward Emerson Barnard, one of the first pioneers of astrophotography. He first observed NGC 6822 with a five-inch refractor and described the object in his notes as an "excessively faint nebula".

The angular size of NGC 6822 is 14′ × 18′; the visible bar structure in the inner regions measures 6′ × 11′. The distance from the Milky Way of 1.62 million light years results in a diameter of 8500 light years. The yellowish colour of NGC 6822 is caused by many middle-aged stars, though there is also a contribution from the extinction of their light by the Galactic plane. The dwarf galaxy at a galactic latitude of only −18° lies behind this dust- and gas-rich zone which in the wide-angle image reveals itself as dim, diagonal veils – galactic cirrus. The detailed image displays a few prominently red gas clouds at the northern edge of the galaxy which are

OBJECT	NGC 6822
CONSTELLATION	Sagittarius
RIGHT ASCENSION	$19^h\ 44^m\ 58^s$
DECLINATION	−14° 48′ 12″
BRIGHTNESS	9.3 mag
TYPE	IB(s)m
PHOTOGRAPHERS	Stefan Binnewies, Rainer Sparenberg
TELESCOPE	105-mm reflector
CAMERA	SBIG ST-2000XM
EXPOSURE TIME	180 min
LOCATION	Amani Lodge, Namibia

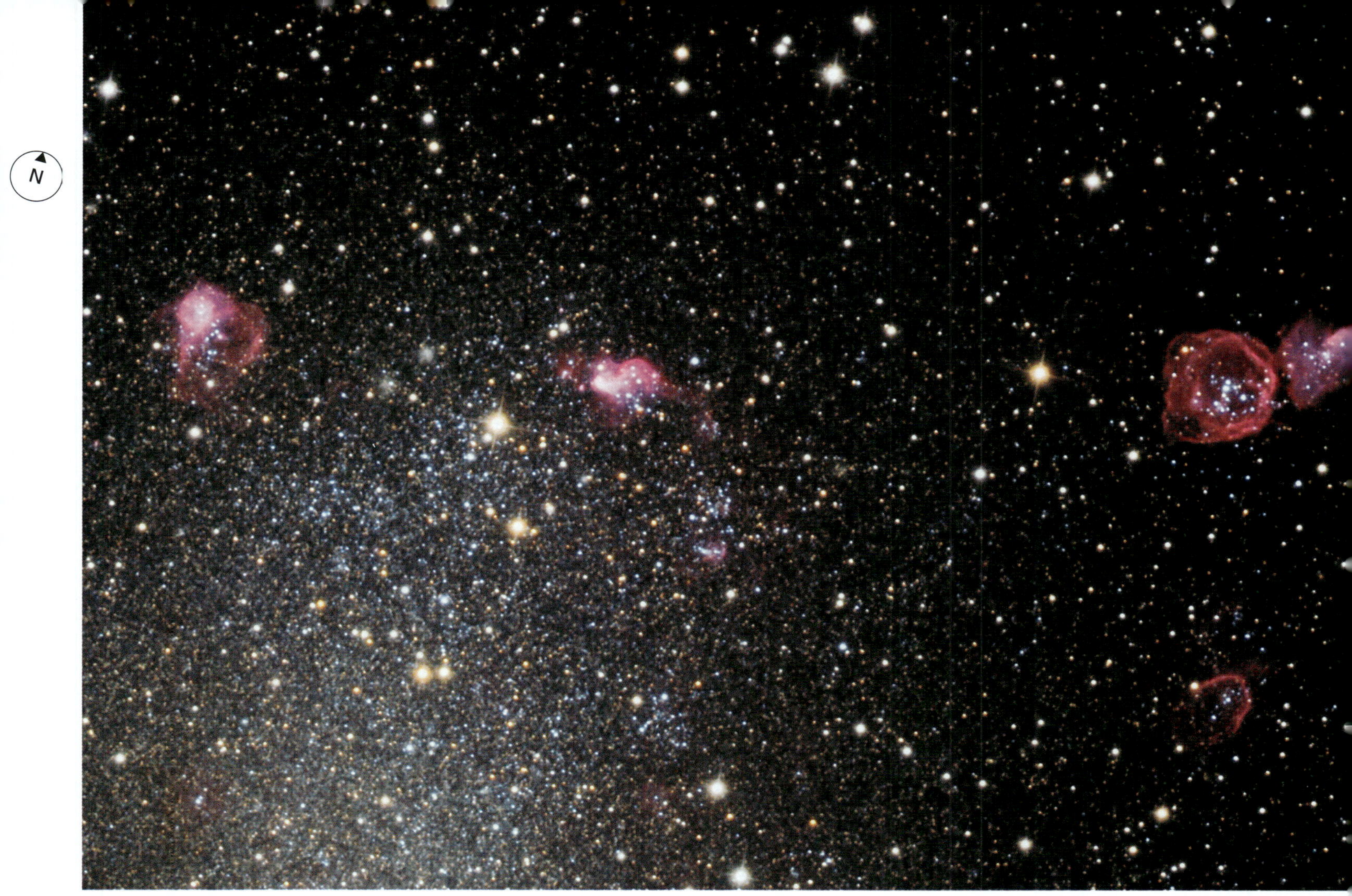

Large H II regions in NGC 6822. Image credit: Rainer Sparenberg, Stefan Binnewies, 600-mm reflector, SBIG ST-10XME, 180 min, Amani Lodge, Namibia.

H II regions and the corresponding OB stars. More than 150 such H II regions, with diameters of up to 600 light years, have now been catalogued in NGC 6822.

The first step on the cosmic distance ladder is the use of Cepheids, a type of pulsating variable star, as standard candles. There is a linear relation between their absolute magnitude and the logarithm of their period. This allows astronomers to derive the absolute magnitude M from the measured period and via the apparent magnitude m the distance modulus $(m - M)$ and thus to calculate the distance. If there are several Cepheids in a galaxy, one obtains a mean distance, which can reduce the statistical uncertainty to a few per cent. Edwin Hubble was the first astronomer who looked for Cepheids in NGC 6822 and was able to identify 11. Hubble made this distance measurement after using the same procedure for the Magellanic Clouds and before he began his studies of the Andromeda Nebula.

In two nights during an observational campaign in 2002 on the William Herschel Telescope (WHT, diameter 4.2 metres, La Palma, Spain), the magnitudes of thousands of stars in different colour bands were derived, allowing the construction of a colour–magnitude diagram for NGC 6822. The photometry data of the stars confirm the distance found from Cepheids. On the other hand, it was revealed that the range of metallicity in NGC 6822 is twice as large as that in the Magellanic Clouds. In addition, the distribution of metallicities has two maxima. This indicates two epochs of star formation in NGC 6822 separated by about two billion years, each producing heavy elements. Based on the velocity of 64 km/s relative to the centre of the Milky Way, the galaxy covered a distance of 430 000 light years between these two star-formation epochs. This indicates an interaction with another member of the Local Group during this time, which caused the first phase of star formation.

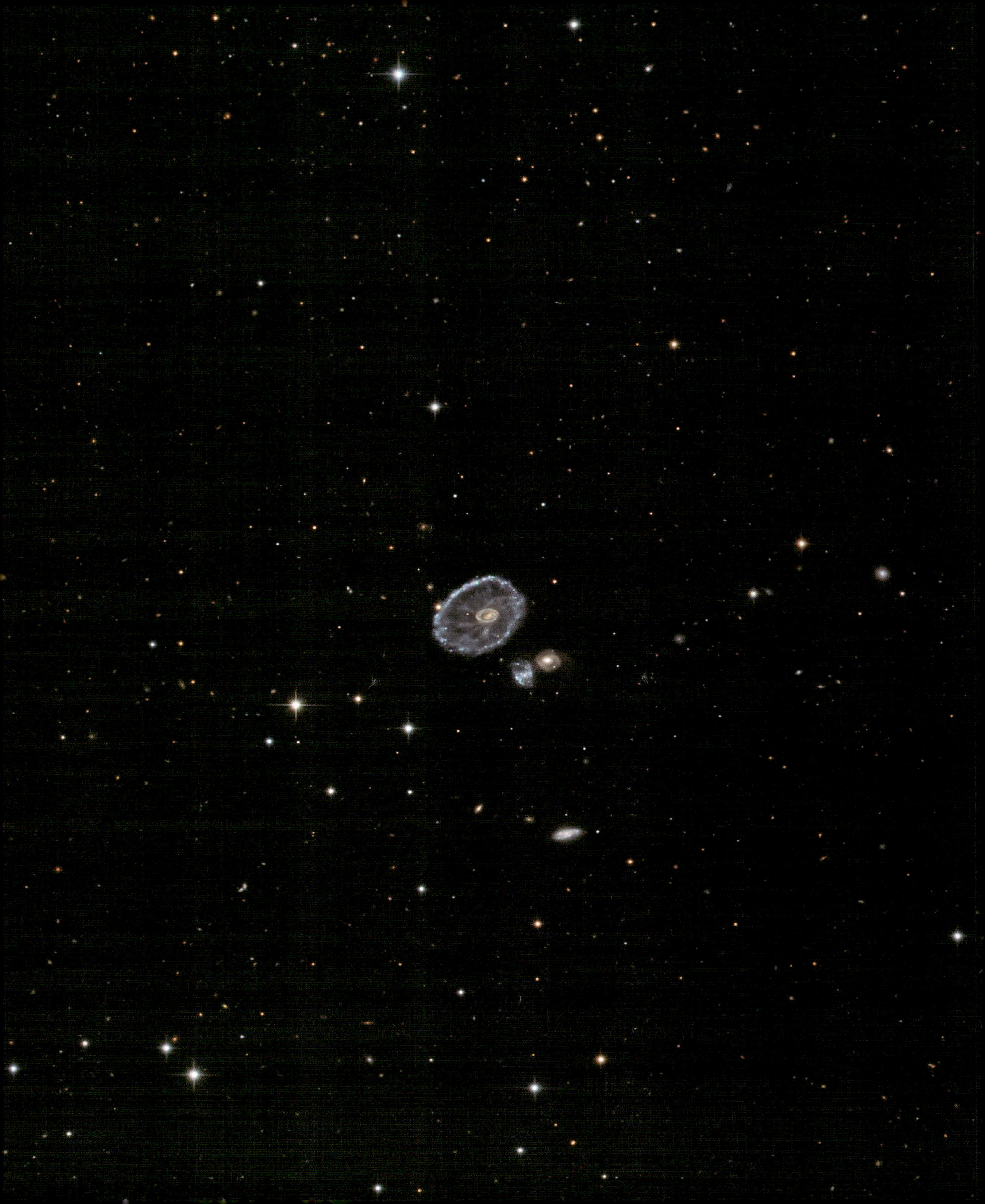

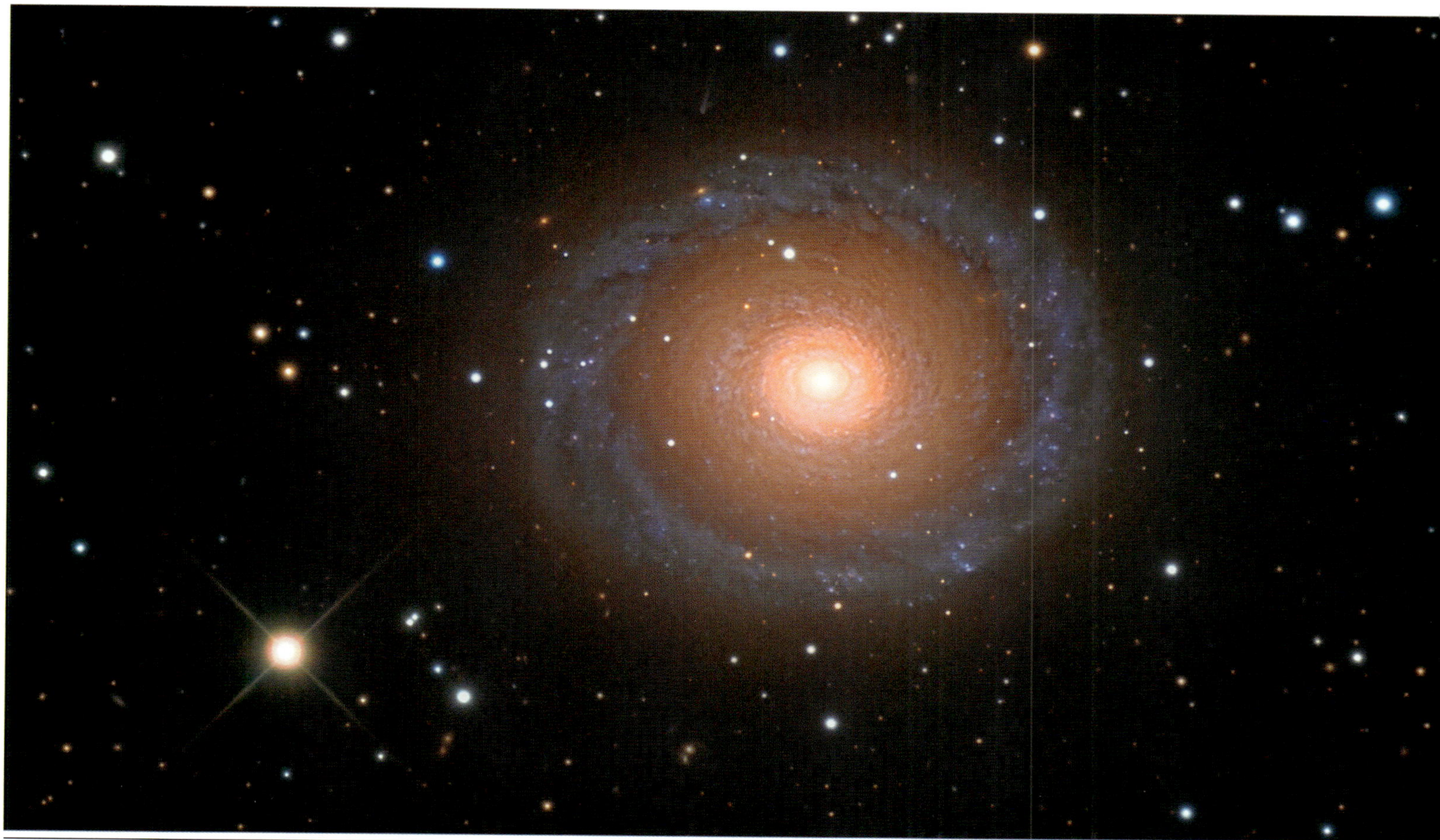

Figure 6.2: In many ring galaxies, as here in NGC 7217, one finds a disturbance of the morphology in the galactic plane – here, the transition from a spiral pattern to a surrounding ring (see page 267).

surface which expands as a wave. If the mass of the compact galaxy is large or its motion very slow, then the impact causes especially large perturbations. These act perpendicular to the plane and lead to a warp of the disc. Thus the theory explains how departures from the ideal circular form come about, for example in the form of banana- or sickle-shaped ring elements. If the compact galaxy doesn't centrally impact the plane of the disc, the result is also an asymmetric ring or only part of a ring. Such a displacement at the collision can also lead to the nucleus of the spiral galaxy being ejected from the plane, leaving behind an empty ring, such as that seen in Arp 147.

The astrophysics of polar-ring galaxies differs from the scenario described above due to the perpendicular angular momenta of ring and central galaxy. These two components can co-exist only under certain boundary conditions. An important role is played by the mass of the ring: in order for this to remain stable over several billion years, it cannot have less than a certain minimum mass. Theoretical studies also suggest that warps can occur in the plane of the ring. In the case of heavy rings, this perturbation by the central galaxy is small and the ring remains stable in spite of the wavy plane. However, a connection with the embedding of the ring in dark matter is not excluded.

The question of the formation of polar rings is explained today by two scenarios: on the one hand a merger of two galaxies into a polar-ring galaxy; the other hand a scenario in which the rings of polar-ring galaxies are gradually formed by accretion processes.

In the merger scenario, a spiral galaxy frontally impacts a second spiral galaxy, where the two galactic planes are perpendicular. For large relative velocities, this also corresponds to the formation of ring galaxies, whereby in this case ring galaxies with cartwheel morphology are formed. Smaller relative velocities, on the other hand, increase the interaction time, leading to a merger of the components. Simulations indicate that the impacting galaxy forms the new central galaxy and the outer regions of the impacted spiral galaxy form the polar ring. Its core is completely dissolved and in

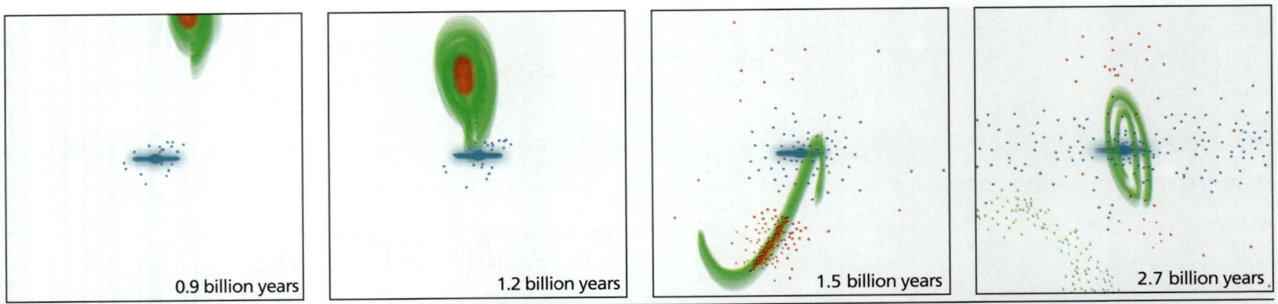

Figure 6.3: Snapshots from a simulation leading to the formation of a ring galaxy (after F. Bournand and F. Combes, 2003). A spiral galaxy (gas is green and stars are red) approaches another spiral galaxy (blue) from above and collides with it. The captions indicate the time of the snapshot in the simulation. A side of a box corresponds to 140 000 light years. The blue galaxy is the centre of the simulation.

SIMULATIONS OF CLOSE APPROACHES AND COLLISIONS OF GALAXIES

In 1970, the mathematician and astronomer Alar Toomre and his brother Juri Toomre, also a mathematician, were the first to successfully use a computer to simulate the collision of two galaxies. Already at the beginning of the 1960s, there were first studies of gravitational effects in close approaches of galaxies, carried out at the Astronomical Institute of the University of Tübingen by Jörg Pfleiderer and Heinrich Siedentopf. The results, however, were published only in German journals and are thus much less known than the works of the Toomre brothers, who cited the work of the Tübingen astronomers.

The calculations were carried out by Alar and Juri Toomre on computers at the Goddard Institute for Space Studies (GISS, Cambridge, Massachusetts, USA) and the results, displayed as ASCII characters on a screen, were filmed with a 16-mm camera. The first galaxy models had simple forms of a few point masses. The galactic planes were represented by point masses arranged in a few concentric circles exhibiting differential rotation. A galaxy consisted of 100–200 point masses and the gravitational effects on the point masses in both galaxies were determined by step-wise N-body calculations. The individual steps covered time spans of just a few million years and thus allowed the tidal interaction between two closely approaching galaxies to be resolved in time.

In spite of the simplicity of the simulation, the results explained how tidal tails form from material pulled out of the disc and how bars and long spiral arms are formed from perturbations. In a head-on collision of a disc galaxy with a compact partner it was possible to follow the formation of ring structures in the galactic plane.

One of the most important results of the simulation was the disproof of the paradigm that spiral structures exist for the entire Hubble time.

Modern simulations no longer take into account just individual point masses but also contain gas components whose hydrodynamic behaviour in the interactions is also modelled. To this end, astronomers make use of the numerical-analysis method of adaptive mesh refinement (AMR) in order to increase the resolution where the gas density is high and hydrodynamic instabilities and shock waves can occur. For ring galaxies, one can use this method to study a range of impact parameters of the target and intruder galaxy and the physical size of the rings formed.

One result was the fact that a heavier impact partner causes more-sharply defined and faster-expanding ring structures than is the case with a light impact partner. The calculations also show that there are standard values for the majority of ring galaxies: a ring with a typical diameter of 40 000–60 000 light years forms about 50 million years after the collision.

Special forms of ring galaxies, such as those without a nucleus, can be well modelled by modern simulations. This special morphology occurs in off-centre collisions where the offset is 15 000–30 000 light years. The empty ring produced has an asymmetric form and consists of a wide arc and a narrow portion. Simulations also indicate that the gas in the ring is compressed, allowing the star-formation rate to rise, exceeding that of the original galaxy.

thinned-out form makes up the new galactic halo. A stable polar ring is found in these models only when the angle of impact is between 50° and 90° and the point of impact lies within half of the radius of the disc. Assuming a uniform distribution of initial conditions, the probability is only four per cent that the collision of two spiral galaxies results in a polar-ring galaxy – this fits very well with the probability actually observed.

In the accretion scenario it is assumed that the central galaxy collects material from another galaxy. There is thus no collision, but rather only a close approach. Due to tidal interactions, gas, dust, and stars are accreted and flow towards the central galaxy due to its stronger gravitational field. Since the two galaxies orbit each other, the material first forms a disc. Simulations indicate that this is stable only when the orbital plane is highly inclined, i.e. when it is already in the polar region. If there is enough mass in the ring, the ring stabilizes itself in the maximally polar position. At the end of this interaction, the galaxy supplying the material is completely dissolved. Polar-ring galaxies are thus two stellar systems whereby each has its own direction of rotation. A polar-ring galaxy is thus a seldom case of the co-existence of two systems which are very close to each other, though stable on cosmic time scales.

The formation of polar rings can, however, also be completely explained without merger or accretion. The theory assumes that infalling gas from cosmic gas flows can form a polar ring. However, this places strong constraints on the angular-momentum vector.

A fourth and rather exotic explanation makes use of a mechanism which needs just one spiral galaxy whose halo is described by a triaxial ellipsoid of dark matter. This halo rotates relative to the galaxy and due to resonance effects stars wander out of the galactic plane into a polar-ring structure. In this model, however, so far no simulation could confirm the theoretical assumptions. The merger and accretion scenarios, however, explain both the morphology and the actual probabilities and are thus accepted by many astronomers as the most probable models.

BIBLIOGRAPHY

Allam, S. S.: *Ring Galaxies*, http://home.fnal.gov/~sallam/MergePair/SDSSRingGalaxies.html

Appleton, P. N. and Struck-Marcell, C.: *Collisional Ring Galaxies*, Fundamentals of Cosmic Physics, **16**, 1996

Bournard, F.: *Polar ring galaxies and dark matter*, 2012, www.obspm.fr/actual/nouvelle/apr03/prg.en.shtml

Casertano, S. and Sackett, P. D.: *Warped Disks and Inclined Rings Around Galaxies*, Cambridge University Press, 1991

Hogg, D. W.: *SDSS images of selected RC3 galaxies*, http://cosmo.nyu.edu/hogg/rc3/index.html

Lauter, M.: *Galactic Bridges and Tails*, A. Toomre and J. Toomre – restored film, 2007, http://kinotonik.net/mindcine/toomre/

Leibinger, H.: *Struktur von Polar Ring Galaxien*, 2009, http://othes.univie.ac.at/6874/1/2009-09-28_0009449.pdf

McElroy, D. B.: *Polar Ring Galaxies*, http://alumnus.caltech.edu/~mcelroy/polar.html

Mihos, C.: *Galaxy Crash JavaLab*, 2012, http://burro.astr.cwru.edu/JavaLab/GalCrashWeb/dynamic.html

Pfleiderer, J.: *Gravitationseffekte bei der Begegnung zweier Galaxien*, Zeitschrift für Astrophysik, **58**, 1963

Pfleiderer, J. and Siedentopf, H.: *Spiralstrukturen durch Gezeiteneffekte bei der Begegnung zweier Galaxien*, Zeitschrift für Astrophysik, **51**, 1961

Toomre, A. and Toomre, J.: *On Intergalactic Bridges*, Bulletin of the American Astronomical Society, **2**, 1970

Toomre, A. and Toomre, J.: *Galactic Bridges and Tails*, The Astrophysical Journal, **178**, 1972

Whitmore, B. C.: *A few statistics from the catalog of polar ring galaxies*, http://ned.ipac.caltech.edu/level5/Whitmore/Whit_contents.html

Whitmore, B. C. et al.: *New observations and a photographic atlas of polar-ring galaxies*, The Astronomical Journal, **100**, 1990

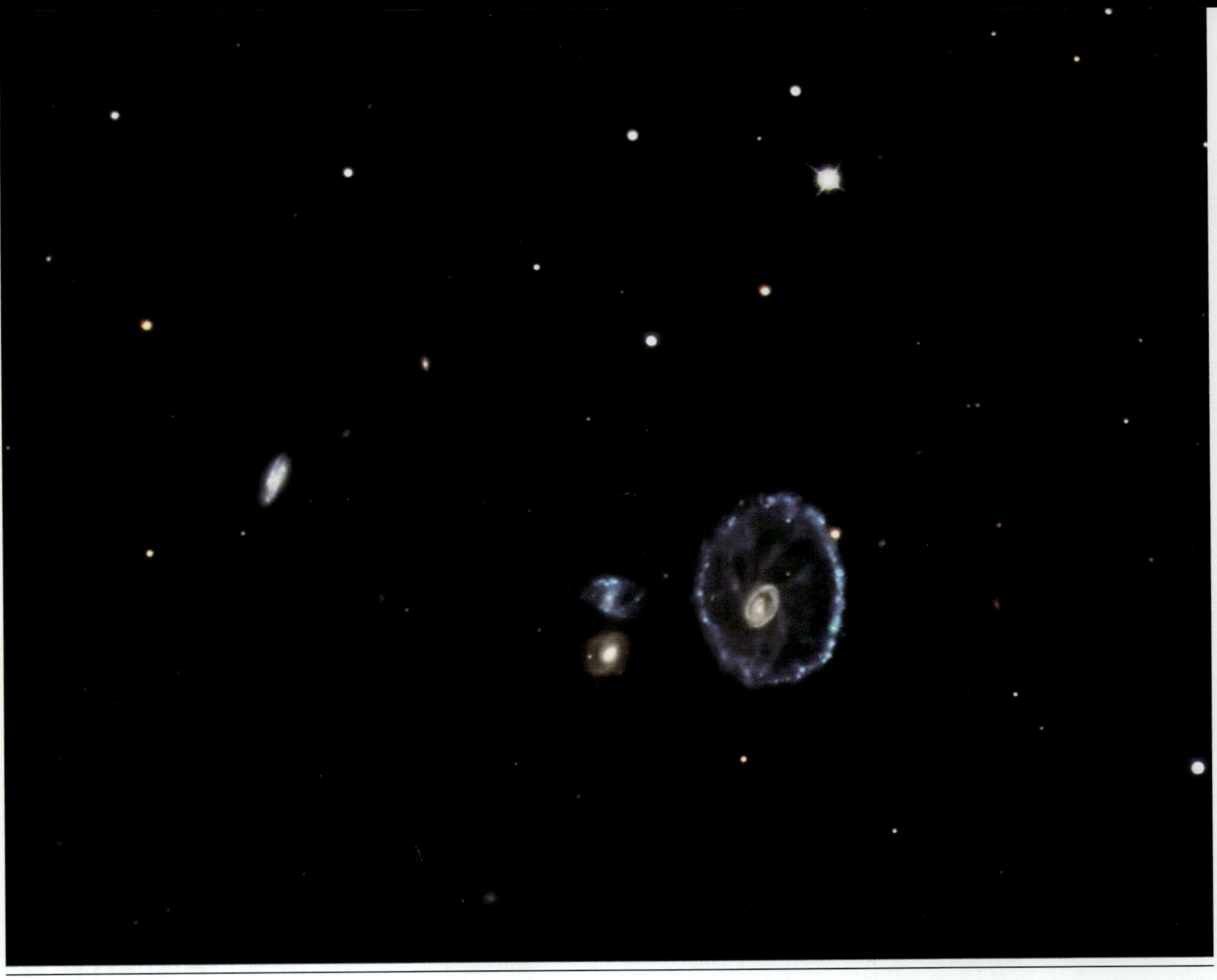

ESO 350-40, The Cartwheel Galaxy

ESO 350-40 is one of the most impressive ring galaxies. Its nickname, the Cartwheel Galaxy, is due to Fritz Zwicky, who discovered it in 1941. Its declination of −33° makes it difficult to access for astrophotographers at mid-northern latitudes and its small angular size of about an arcminute requires a long focal length in order to record details. ESO 350-40 lies at a distance of 400 million light years, which leads to a diameter of about 116 000 light years for the prominently blue ring. More exactly, this is an ellipse. In the inner regions, ESO 350-40 displays a smaller, second ellipse, which directly surrounds the centre and in addition to a prominent yellow colour is also rotated by about 30° with respect to the outer ellipse. Spoke-like filaments are found between the two rings, seeming to join them and reminding one of spiral arms.

Radio observations of the Cartwheel Galaxy and its surroundings with the VLA (New Mexico, USA) show directed emission which connects the centre of the ring galaxy with neighbouring filaments in the plane of the ring. The history of the two galaxies which formed this ring galaxy is still visible, whereby the asymmetry of the H I structures is a direct indication of the direction from which the compact galaxy approached the spiral galaxy. Simulations can model the unique morphology of this ring galaxy very well. These indicate a slightly offset collision with a compact component with 20 per cent the mass of the spiral galaxy.

OBJECT	ESO 350-040
CONSTELLATION	Sculptor
RIGHT ASCENSION	$00^h\ 37^m\ 41^s$
DECLINATION	−33° 42′ 59″
BRIGHTNESS	15.2 mag
TYPE	Ring galaxy
PHOTOGRAPHERS	Josef Pöpsel
TELESCOPE	600-mm reflector
CAMERA	SBIG ST-10XME
EXPOSURE TIME	105 min
LOCATION	Amani Lodge, Namibia

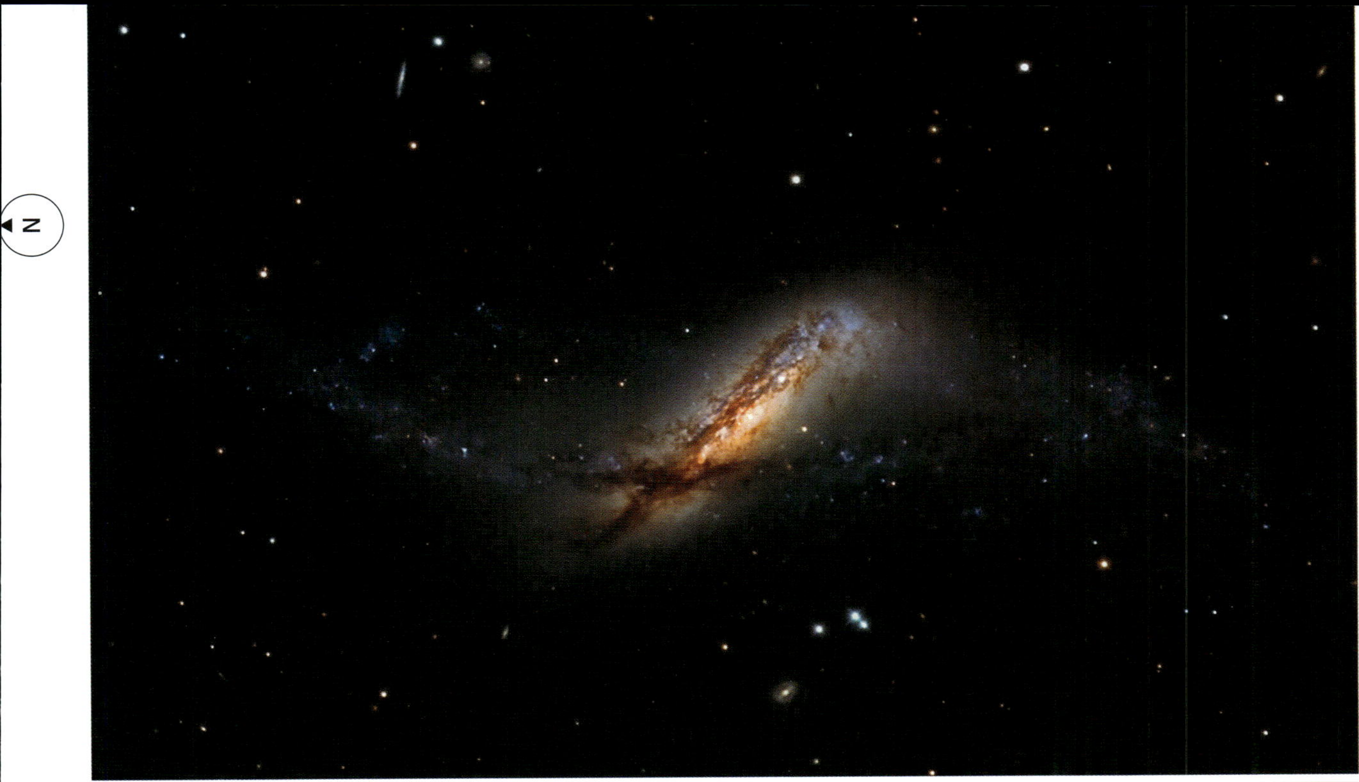

NGC 660

NGC 660 is a ring galaxy whose outer ring is tilted by about 60° relative to the central galaxy. This angle is easily seen at the overlap of the dust lanes, whereby one notices that the plane of the outer ring is warped and has an even steeper inclination in the polar regions. The image shows a detailed view of the two components which make up NGC 660. The entire object has an angular size of 8.5′ × 3.2′. Taking into account the low distance of 42 million light years, the scaling of 12 000 light years per arcminute implies a compact size in the range 33 000–50 000 light years for the inner galaxy. The prominent warp in the plane of the ring is probably due to tidal interaction with UGC 1195. The diameter of the ring is 100 000 light years; the companion UGC 1195 (at a distance of 22′ and not visible in the image) is thus at a distance of 270 000 light years.

The central galaxy in NGC 660 is not a structureless spindle galaxy as in the polar-ring prototype, but rather an SB galaxy with detailed structures visible in the dust lanes and a yellowish, diffuse glow which surrounds the disc and appears to be deformed in the direction of the outer ring where the two overlap. This bluish outer ring has many young stars and large amounts of dust and molecular gas. In the image one can even recognize OB associations and H II emission regions in the particularly active areas.

It is estimated that the ring contains about ten per cent of the total molecular mass of the object, which explains the stability of the ring in this almost polar position. Detailed studies of the population of stars in the ring reveal blue and red giants and the colour–magnitude diagram indicates that the youngest stars in the ring are about seven million years old. Thus the ring must be relatively young and a comparison with models of galactic-colour evolution indicate a possible age of about a billion years. NGC 660 was probably formed in a merger process. One the one hand this is supported by the compactness of the central component, and on the other hand by the yellowish halo of old and middle-aged stars which is typical of this scenario.

OBJECT	NGC 660
CONSTELLATION	Pisces
RIGHT ASCENSION	$01^h\ 43^m\ 02^s$
DECLINATION	+13° 38′ 42″
BRIGHTNESS	12 mag
TYPE	SB(s)a pec
PHOTOGRAPHERS	Makis Palaiologou
TELESCOPE	1.3-m reflector
CAMERA	Andor DZ 436
EXPOSURE TIME	180 min
LOCATION	Skinakas Observatory, Crete, Greece

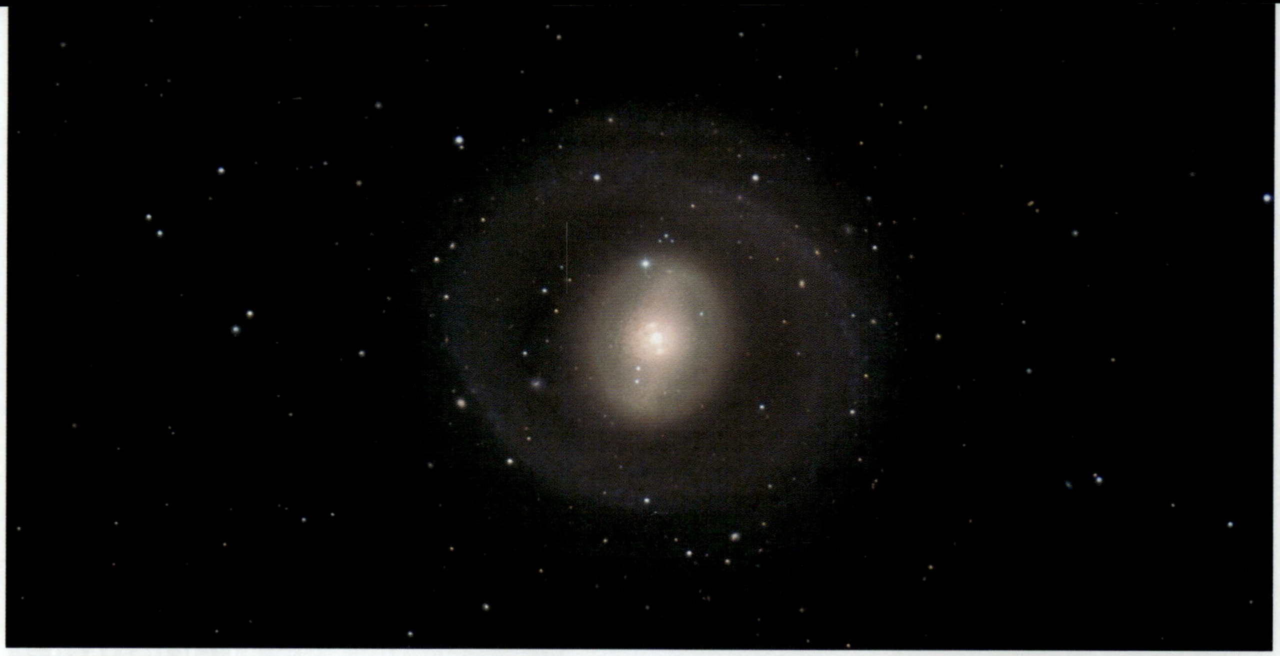

NGC 1291

NGC 1291 is basically an early-type barred spiral galaxy seen face-on. A closer look reveals that this SBa galaxy has no spiral arms in the outer regions, but rather a complete ring containing individual dim structures. At a distance of 32 million light years the diameter of the ring corresponds to a projected distance of about 81 000 light years. Looking closely, one can see a small ring within the bar structure, tilted by about 30° relative to the axis of the bar. This small ring is particularly symmetric and is only a sixth the length of the bar, corresponding to a diameter of 4900 light years.

The lintel-shaped form of the inner regions surrounding the bar corresponds to reality – we are not viewing a ring tilted with respect to the line of sight which appears as an ellipse only due to perspective. The lintel-like shape of this region can be confirmed by spectroscopic measurements of velocities of the gas components it contains. In contrast to the outer ring, in this region there are almost no indications of star formation. Contrast-enhanced images show a complex pattern of gas and dust filaments, which are usually on only one side, thus the diametral symmetry often found in barred-spiral systems is lacking.

In the radio, as in many galaxies which are the product of an interaction process, astronomers find in NGC 1291 many more molecular-hydrogen structures than are visible in the optical image. The measurements indicate a dominant circular concentration of the gas which extends much farther than the visible bar. The extended distribution of hydrogen gas appears diffuse, although a few spiral structures can be seen. The radio image is significantly different from the optical: the spiral arms in the radio lie between the inner and outer rings, just where no structures are visible in the image above. In addition, it is notable that in the radio the central region of NGC 1291 has a hole of width four arcminutes – there is thus almost no hydrogen gas detectable in the core. In X-rays, NGC 1291 behaves in the opposite manner; it has a central maximum. A combination of the two reveals an outer radio ring and a core bright in X-rays. This can be explained by the fact that the hydrogen gas has been used up by star formation. Thus these stars lie in a region of low gas density and their radiation causes very high temperatures, which leads to a prominent X-ray luminosity in which they appear to be embedded.

OBJECT	NGC 1291
CONSTELLATION	Eridanus
RIGHT ASCENSION	$03^h\ 17^m\ 19^s$
DECLINATION	−41° 06′ 29″
BRIGHTNESS	9.4 mag
TYPE	(R)SB0/a(s)
PHOTOGRAPHERS	Josef Pöpsel
TELESCOPE	600-mm reflector
CAMERA	SBIG ST-10XME
EXPOSURE TIME	35 min
LOCATION	Amani Lodge, Namibia

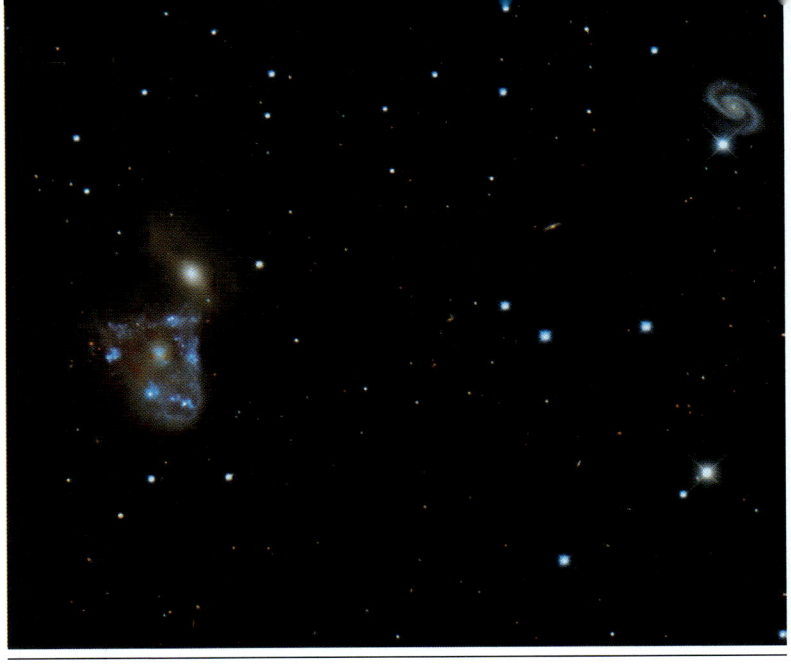

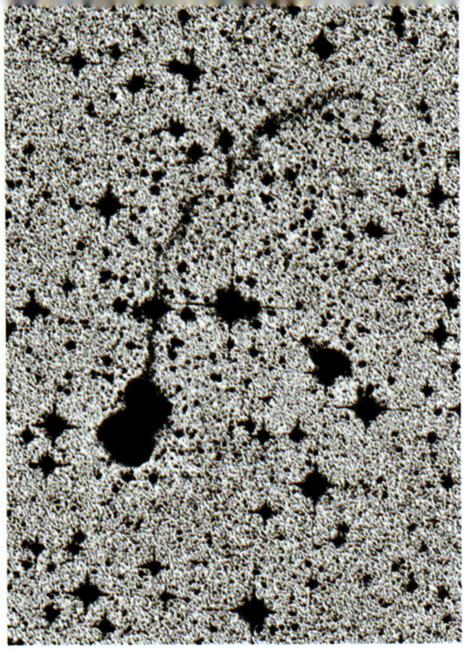

Richard Müller, 320-mm reflector, Artemis 4021, 300 min, Lohmar, Germany.

Arp 143

NGC 2444 and NGC 2445 form an interacting system which is also known as Arp 143 and as the Lynx Ring Galaxy. In this pair, the yellowish, S0-like galaxy NGC 2444 is about an arcminute north of NGC 2445, whose ring has a diameter of 1.2′. Both galaxies are 180 million light years away, an arcminute corresponding to a transverse scale of 52 000 light years.

Deep H I radio observations (VLA, New Mexico) reveal a huge intergalactic gas tail with the galaxy system at the end. This gas tail exceeds the length of the entire system of NGC 2444 and NGC 2445 by a factor of seven or eight, corresponding to an extent of more than 330 000 light years (see negative image). One can see this tail as part of a huge expanding ball of hydrogen gas and observations of the gas velocity indicate that the ball was formed in a collision about 100 million years ago. The gas-rich precursor galaxy of NGC 2444 penetrated the originally spiral galaxy NGC 2445 out of which the ring then formed. The ring galaxy and the gigantic hydrogen bubble are products of the collision, whereby with time the galaxy pair has moved away from the centre of the bubble. It can further be assumed that the two partners orbited each other several times before the collision. During this time, tidal forces transported much hydrogen gas into the intergalactic medium, and during this phase of interaction there could have been several collisions of the two galaxies.

In recent years, extragalactic research has increasingly been concerned with molecular emission lines in the infrared. The spectra of warm molecular hydrogen and polycyclic aromatic hydrocarbons (PAHs) in NGC 2444 were studied with the Spitzer Space Telescope (wavelength range 3–180μ) The "warm" temperature should be taken literally, since the bright knots in the ring structure of NGC 2445 contain hydrogen gas at room temperature, about 300 kelvin. The strong hydrogen lines are due to an expanding shock wave which has compressed gas and dust, leading to condensations and finally knots and star-forming regions. The temperature implies a collision velocity of 120 km/s and thus 85 million years to escape from the nuclear region. This age corresponds well to that obtained from radio observations. The dynamic evolutionary history of Arp 143 is, however, not yet over, since due to the small separation it is probable that NGC 2444 will collide with NGC 2445 yet again.

OBJECT	Arp 143, NGC 2444 / NGC 2445
CONSTELLATION	Lynx
RIGHT ASCENSION	$07^h\ 46^m\ 53^s$
DECLINATION	+39° 01′ 55″
BRIGHTNESS	14.2 mag
TYPE	Ring galaxy
PHOTOGRAPHERS	Makis Palaiologou, Stefan Binnewies
TELESCOPE	1.3-m reflector
CAMERA	Andor DZ 436
EXPOSURE TIME	235 min
LOCATION	Skinakas Observatory, Crete, Greece

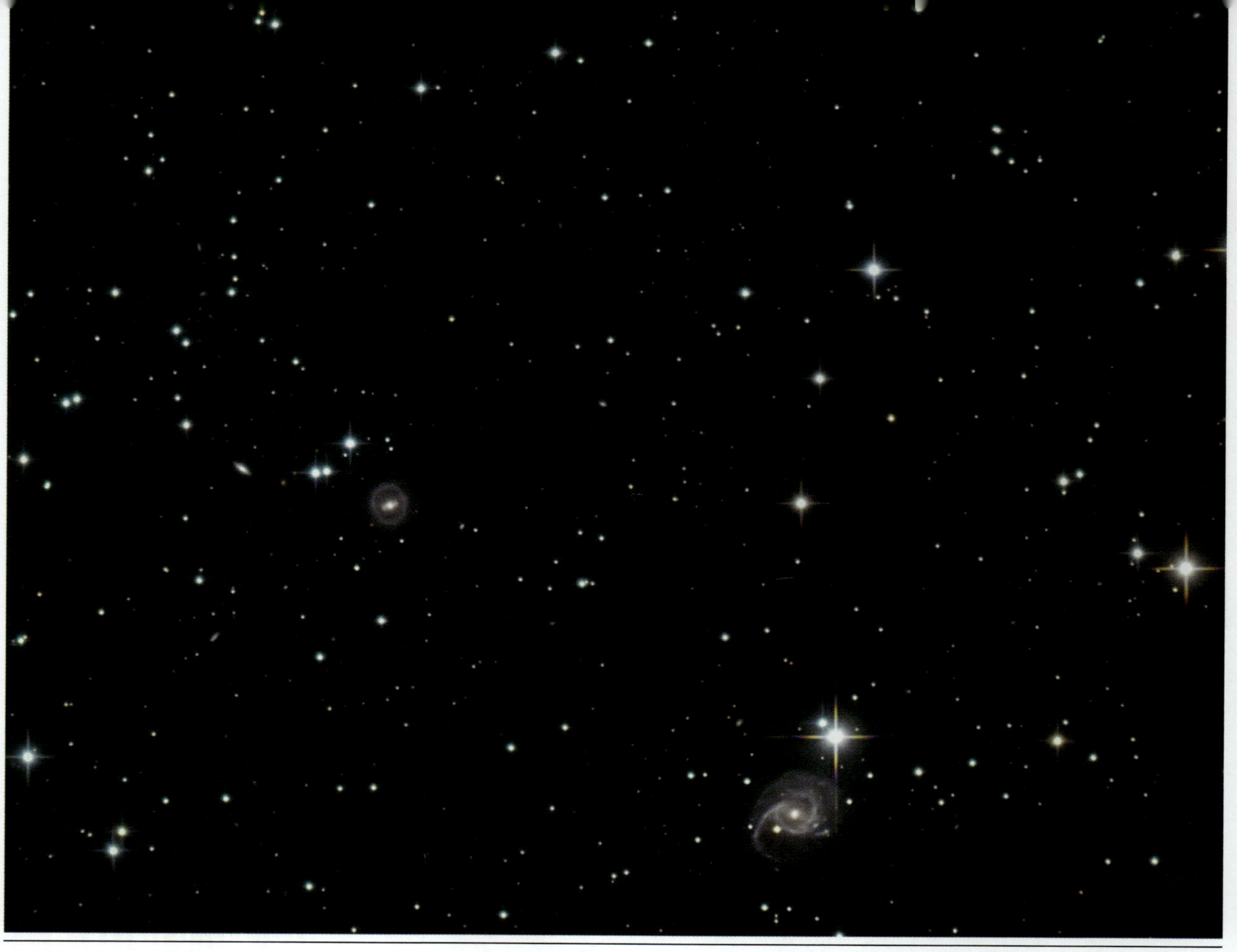

NGC 2595

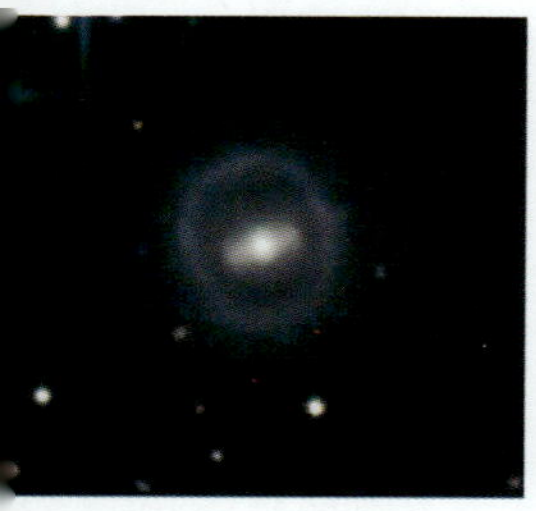

The image shows the SABc galaxy NGC 2595 at the lower edge, close to a ninth-magnitude star. One of its spiral arms is deformed, making the spiral plane appear asymmetric. Without this perturbation there would be a ring-like structure where the spiral arms begin and the classification would then be SAB(rs)c. NGC 2595 lies at a distance of 190 million light years and its angular size of 2.8′ × 3.2′ applies to the extended, diffuse outer regions of the spiral plane. In the inner regions one recognizes a dim yellowish bar structure, tilted by 15° relative to the horizontal in the image.

The image also shows a few field galaxies near a group of three bright blue stars. The most prominent is the ring galaxy UGC 4414, about 12′ away from NGC 2595 (see enlargement in the small image). Despite its great distance of more than 330 million light years one can recognize the typical components: the bluish ring and the contrasting yellowish core region with a bar. The angular size of UGC 4414 is 42″ × 33″, corresponding to a projected extent of 67 000 light years. Compared to this image of UGC 4414, that of the Sloan Digital Sky Survey is deeper, and also shows a yellowish oval bulge. It surrounds the brighter nucleus with the bar, but is not visible in the image above. It is interesting that the sizes of the components of the ring galaxy UGC 4414 are almost the same as those of NGC 5701 (cf. page 263). Both form a morphological pair of twins, supporting the hypothesis for the origin of the closer and better studied NGC 5701, in which conditions are similar.

OBJECT	NGC 2595
CONSTELLATION	Cancer
RIGHT ASCENSION	$08^h\ 27^m\ 42^s$
DECLINATION	+21° 28′ 45″
BRIGHTNESS	13.7 mag
TYPE	SAB(rs)c
PHOTOGRAPHERS	Richard Müller
TELESCOPE	320-mm reflector
CAMERA	Artemis 4021
EXPOSURE TIME	460 min
LOCATION	Lohmar, Germany

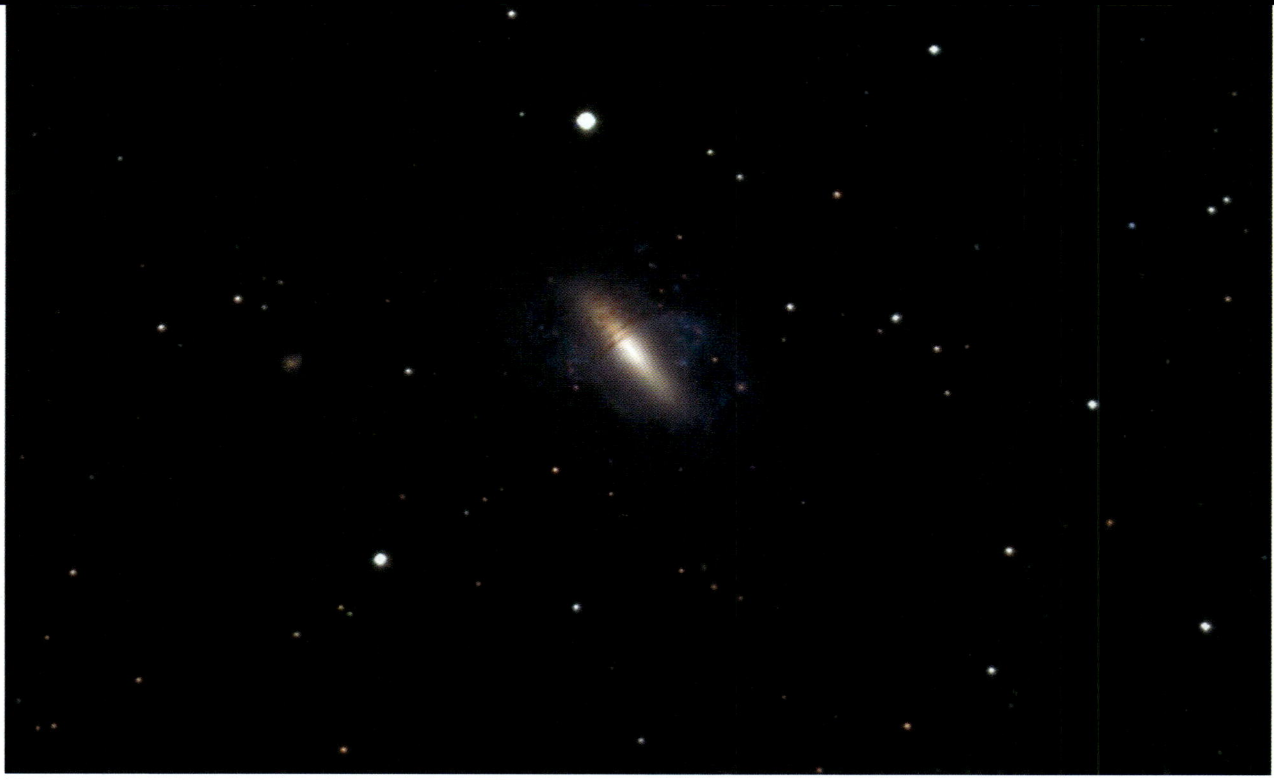

NGC 2685, The Helix Galaxy

The polar-ring galaxy NGC 2685 is also known as the Helix Galaxy since it has several polar rings which appear to be wrapped around the major axis of the central galaxy. The angular diameter of the rings is about 2′, which at a distance of 46 million light years corresponds to a projected diameter of 27 000 light years. A further prominent characteristic in NGC 2685 is an outer equatorial ring which exceeds the polar rings in size and has the same orientation as the spindle-like central galaxy. The latter measures about 4.5′ in diameter and at 13 mag is the brightest part of NGC 2685.

In a polar-ring galaxy, the entire system is in equilibrium, i.e. the central galaxy and the ring co-exist over time. In NGC 2685, the form of the polar rings indicates an age of two to five billion years for the helix structure. The H I ring has a radius three times as large as that of the optical ring of gas, dust, and stars, whereby only the latter is visible in the image.

Only recently, deep H I observations (Westerbork Synthesis Radio Telescope, the Netherlands) have shown that the optical and radio structures correspond well with each other. This is true not only in the region of the central galaxy but also at large radii. In addition, these studies show that NGC 2685 is probably an exceptional case of a ring galaxy – a galaxy which not only has a polar and an equatorial ring but also an extremely warped disc. The ring structures would then be filaments of gas and dust in the common disc. However, the simple model of a disc with two inclined rings is not sufficient: the best fit of this model differs significantly from the measured values. A good description is provided, however, by a disc model consisting of several rings. This disc is decoupled from the spindle-like core and the inclination varies with radius, due to its warp, around a value of about 70°. In this model, the visible helix-like structures would not be explained as accreted rings, but rather as spiral arms of the warped galactic plane. This hypothesis is supported by the measured H I velocity profiles which are very similar to those of spiral arms of low surface brightness.

OBJECT	NGC 2685
CONSTELLATION	Ursa Major
RIGHT ASCENSION	$08^h\ 55^m\ 35^s$
DECLINATION	+58° 44′ 04″
BRIGHTNESS	12.1 mag
TYPE	(R)SB0$^+$ pec
PHOTOGRAPHERS	Michael König
TELESCOPE	280- and 355-mm reflector
CAMERA	SBIG STL-11000 and Starlight Xpress SXV-H9
EXPOSURE TIME	230 min
LOCATION	Rimbach, Germany

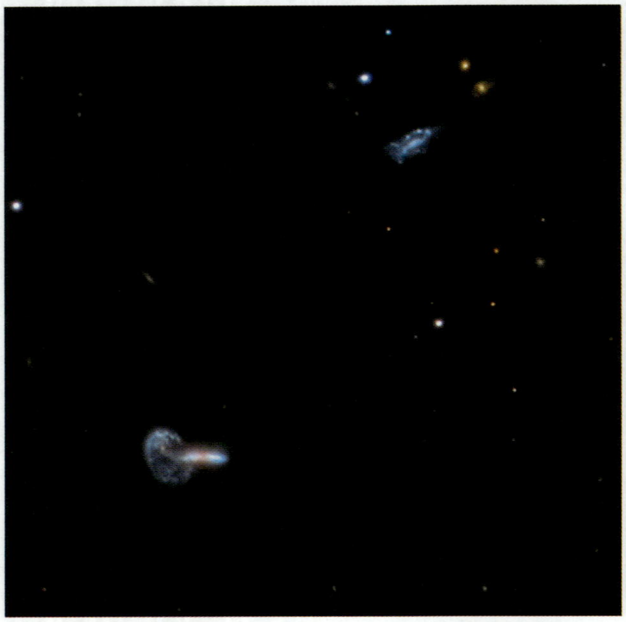

Enlargement

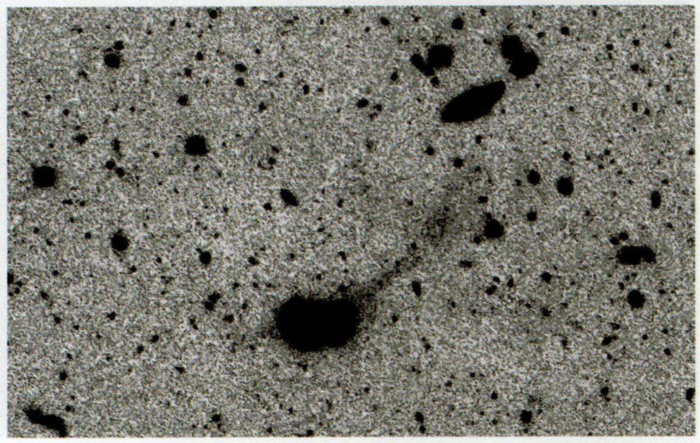

Inverted, contrast-enhanced image

Arp 148

The interacting system Arp 148 consists of two colliding galaxies at a distance of 465 million light years from the Milky Way. The system is also known as KUG 1101+410 (Kiso Ultraviolet Galaxy Catalogue) and GALEXASC J110400.72+404901.7. Both catalogue entries are an indication of the high luminosity of the pair in the blue and ultraviolet spectral bands. The system was discovered in 1940 by the American astronomer Nicholas U. Mayall on plates taken at the 0.91-m Cossley reflector at the Lick Observatory (University of California, USA). In honour of its discoverer, Arp 148 is also known as Mayall's Object.

The enlargement shows Arp 148 at the lower left. It is a conspicuous object, consisting of a spindle-shaped S0-like galaxy and a diffuse blue ring. The spindle galaxy is also known as PGC 33423 and its angular size of 0.32′ corresponds to a diameter of 43 000 light years. The ring in Arp 148 has an angular size of 0.37′ and a transverse diameter of 50 000 light years. Both parts are particularly compact compared to the sizes of standard galaxies, which is why astronomers assume that Arp 148 formed from the collision of two galaxies. The S0 object penetrated the precursor of the galaxy from which the ring structure was mainly formed. Relative to the plane of the ring, the S0 object is approaching the Milky Way at 200 km/s and the angle of about 25° between the major axis of the S0 galaxy and a line perpendicular to the plane of the ring is an indication of a collision which was almost head-on. The image also shows a yellowish bright spot above the centre of the ring which can be interpreted as a relic of the central bulge of the precursor galaxies. Simulations of galaxy collisions show such an offset when the penetrating galaxy doesn't pass through the centre of the penetrated galaxy, thus allowing the strongly modified gravitational field during the penetration to shift the position of the entire bulge.

During some interactions or collisions, the distribution of interstellar matter and thus the star-formation rate in the galaxies involved is massively influenced. The resulting starburst phases usually last a few hundred million years. In Arp 148, this led to a ring structure in the penetrated galaxy. The previously existing density waves in the galactic plane and thus the original spiral pattern were destroyed and instead radial oscillations occurred in the plane. This caused ring-like shock waves which compressed the material, causing a ring of star-forming regions. In addition, a further

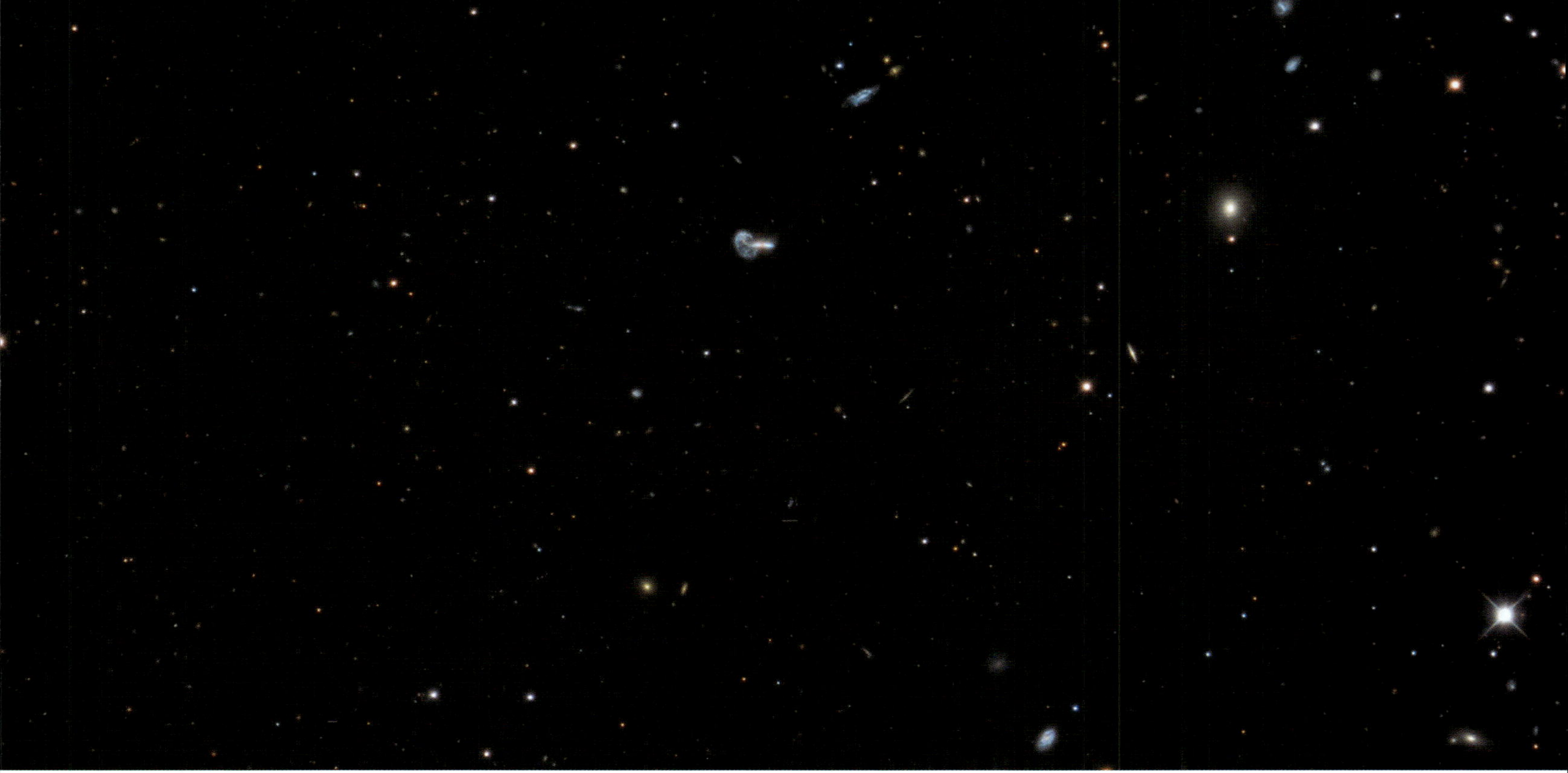

effect of the radial transport of material is observed in Arp 148 – the high concentration of hydrogen in the centre of the ring. In a study of the H I concentration of 16 ring galaxies, Arp 148 had the highest value at 1.5×10^{10} solar masses.

A peculiarity of this galaxy is visible in the contrast-enhanced image: a tidal tail pointing to the southeast. Its formation is presumably due to the dynamic history of this system of colliding galaxies. The direction of the tail indicates the possible participation of other galaxies visible in the image, especially the galaxy SDSS J110400.47+404904.3, which lies 2.6′ to the above and to the right of Arp 148.

OBJECT	Arp 148
CONSTELLATION	Ursa Major
RIGHT ASCENSION	$11^h\ 03^m\ 53^s$
DECLINATION	+40° 50′ 57″
BRIGHTNESS	15 mag
TYPE	Galaxy pair
PHOTOGRAPHERS	Adam Block
TELESCOPE	800-mm reflector
CAMERA	SBIG STX-16803
EXPOSURE TIME	780 min
LOCATION	Mount Lemmon SkyCenter/ University of Arizona, USA

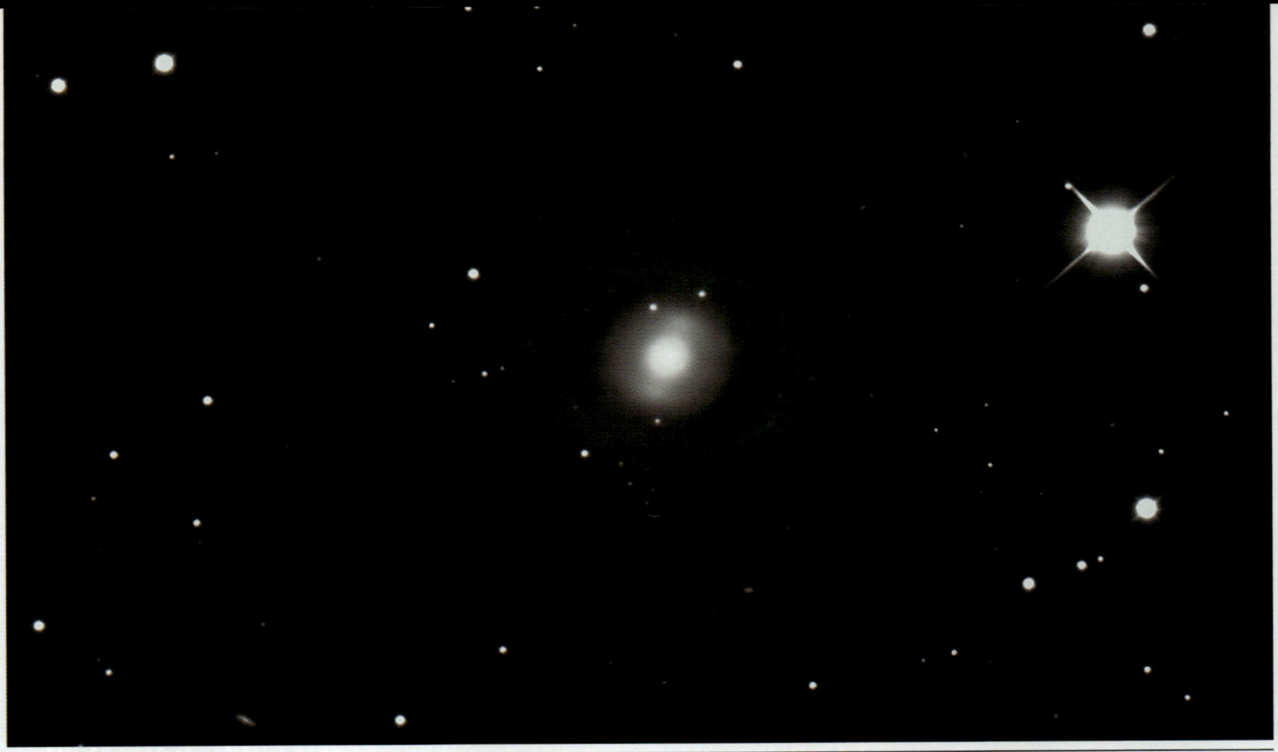

NGC 2859

This ring galaxy is at a distance of 75 million light years and its outer ring is viewed face-on.

The inner region of NGC 2859 contains a bright ring-like disc in which a bright bar is embedded, thus resulting in the classification (R)SB(r). The bar is inclined at an angle of about 70° to the main plane of NGC 2859. Bright ansae typical of SB galaxies are observed at the ends of the bar. Barred-spiral galaxies are rather rare among ring galaxies, though they are common among normal spiral galaxies. About a third of these have weak and a third strong bars in the inner regions. NGC 2859 is also special for another reason, namely the fact that a second, smaller bar has been detected in the inner regions of the prominent bar.

The bulge of NGC 2859 has a form similar to an S0 galaxy. It has the classical, soft appearance and in images taken by large telescopes one can see that the bulge is crossed by a few weak dust lanes. The outer ring measures about 4.3′ × 3.8′. Its mean diameter is 94 000 light years and so it is larger than the diameter of a typical spiral galaxy (65 000–80 000 light years). This large size fits well with the merger scenario described at the introduction to this chapter involving the frontal collision of a compact galaxy with a spiral galaxy.

Both of the rings visible in the image are somewhat blurred, although the inner ring is better defined. There is a dim region between the rings. It has two minima which lie perpendicular to the major axis of the bar. According to theory, this structure is due to a lack of stars, since these regions correspond to two Lagrange points of the bar potential in which no stable stellar orbits are possible.

NGC 2859 has been the target of several observational campaigns which looked for nuclear stellar rings deep in the core regions of galaxies. These rings of gas, dust and young stars with radii of a few hundred to a thousand light years are relatively common in early-type spiral galaxies. NGC 2859 has such a ring with a radius of 2800 light years, though it is visible only in the form of a delicate structure of gas and dust, but no stars. The reason for this has to do with the merger history as well. Presumably this ring must form again along with the galactic plane of the ring galaxy.

OBJECT	NGC 2859
CONSTELLATION	Leo Minor
RIGHT ASCENSION	$09^h\ 24^m\ 19^s$
DECLINATION	+34° 30′ 49″
BRIGHTNESS	11.8 mag
TYPE	$(R)SB0^+(r)$
PHOTOGRAPHERS	Michael König
TELESCOPE	355-mm reflector
CAMERA	SBIG ST-10XME
EXPOSURE TIME	120 min
LOCATION	Rimbach, Germany

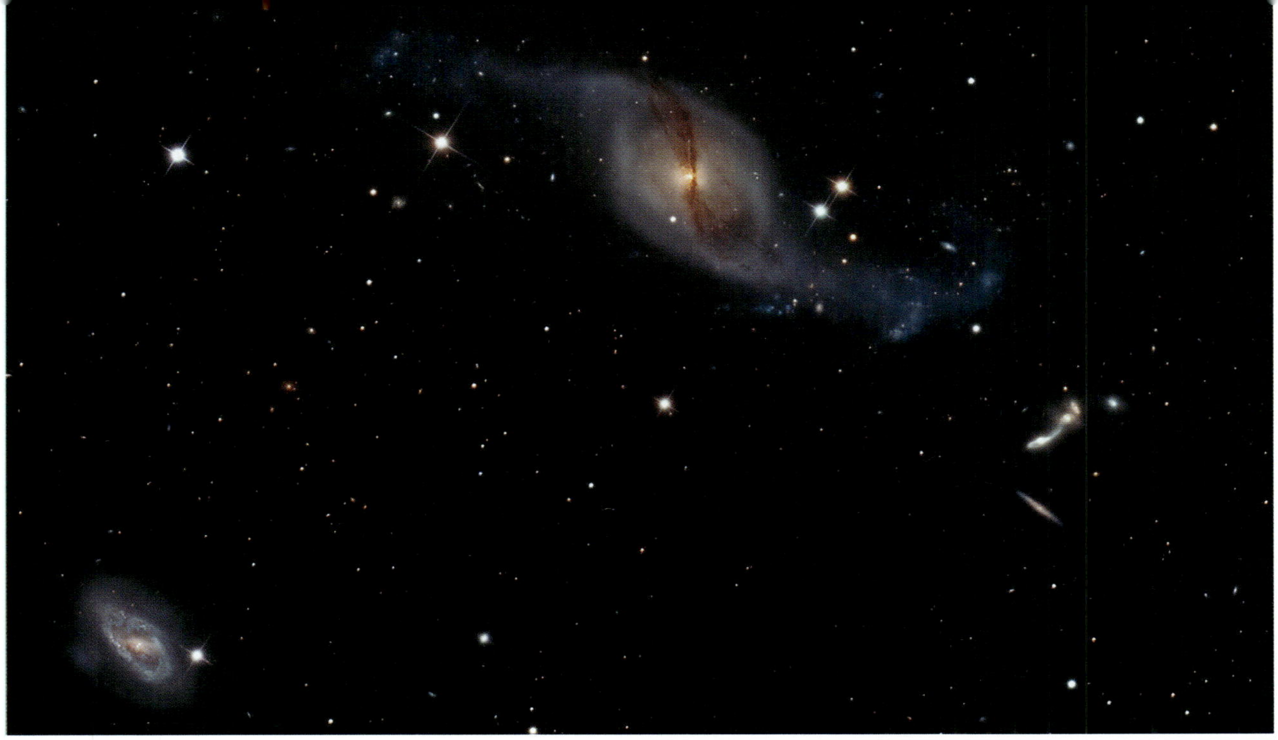

NGC 3718

The image shows NGC 3718 and, 13′ to the east at the lower left corner of the image, its companion NGC 3729. Both galaxies lie at about the same distance of 49 million light years, which implies a separation of 185 000 light years. Both this distance and some radio observations indicate that they are interacting with each other. The smaller galaxy measures 2.8′ × 1.9′ and is classified as SB(r)a, whereby in the outer regions is the rudiments of a countertail, i.e. a tidal tail of gas, dust, and stars which points away from the larger interaction partner.

NGC 3718 is much larger and at 12 mag its core is somewhat brighter than that of the small companion. In the literature, NGC 3718 is described as a polar-ring galaxy, whereby its morphology is not simple and it appears to be made up of several components. Seen edge-on is a sweeping, twisted dust lane crossing the central regions. Perpendicular to this an inner ring plane is visible, from which spiral arms extend far outward.

Radio measurements present astronomers with a so-called H I data cube for NGC 3718 – a spatially resolved distribution of velocities of atomic hydrogen gas. In the inner regions, the currents of hydrogen gas are almost polar with respect to the plane of the disc. The larger the radius, the lower the inclination of the plane of the current, which is only 30° in the outer regions, where the hydrogen gas is 7′ or 100 000 light years from the core and the orbital period of the gas is a billion years. Simulations indicate that the observed distribution of hydrogen in NGC 3718 can be very well modelled via concentric rings which are tilted relative to one another.

Very close to NGC 3718 is a conspicuous group of five galaxies catalogued as Hickson 56, in which there are signs of strong interaction. These galaxies at a distance of about 390 million light years are quite blue despite their great distance, which indicates an increased star-formation rate and that the colour is thus due to the luminosity of young, hot stars.

OBJECT	NGC 3718
CONSTELLATION	Ursa Major
RIGHT ASCENSION	$11^h\,32^m\,35^s$
DECLINATION	+53° 04′ 05″
BRIGHTNESS	11.6 mag
TYPE	SB(s)a pec
PHOTOGRAPHERS	Stefan Binnewies, Josef Pöpsel
TELESCOPE	600-mm reflector
CAMERA	SBIG STL-11000
EXPOSURE TIME	510 min
LOCATION	Skinakas Observatory, Crete, Greece

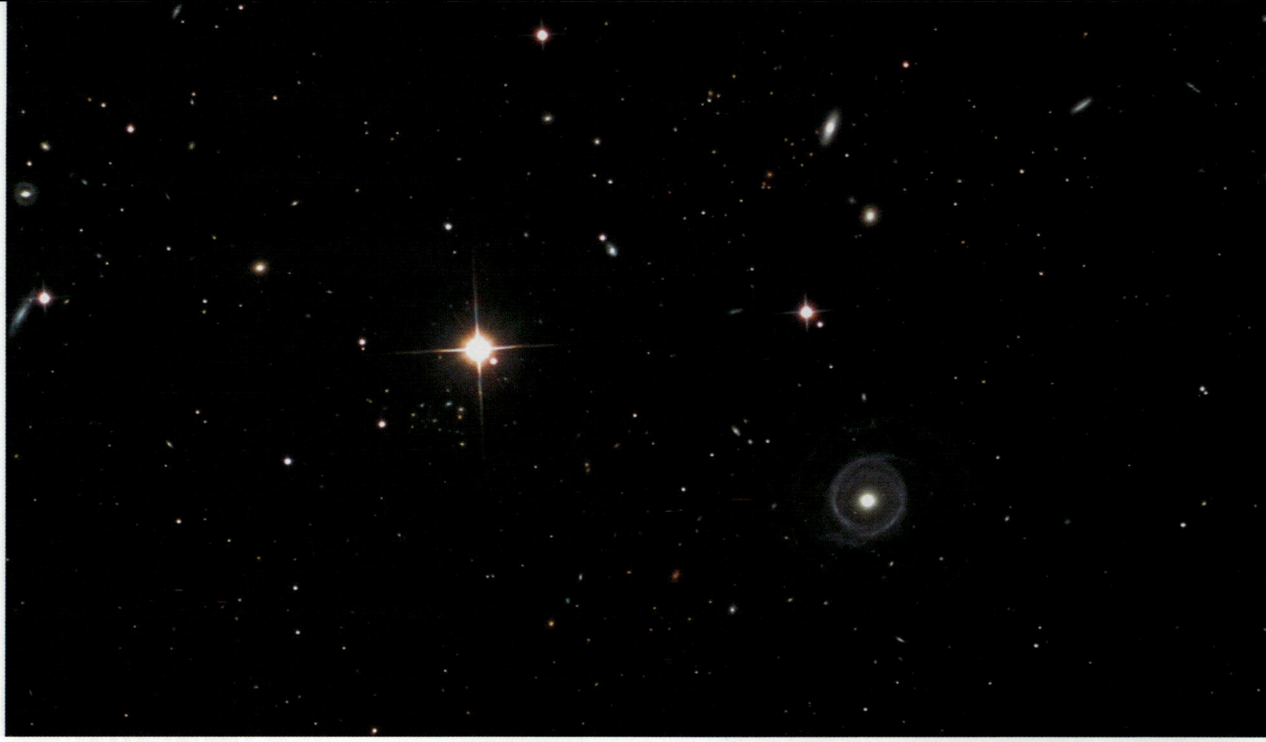

UGC 6614

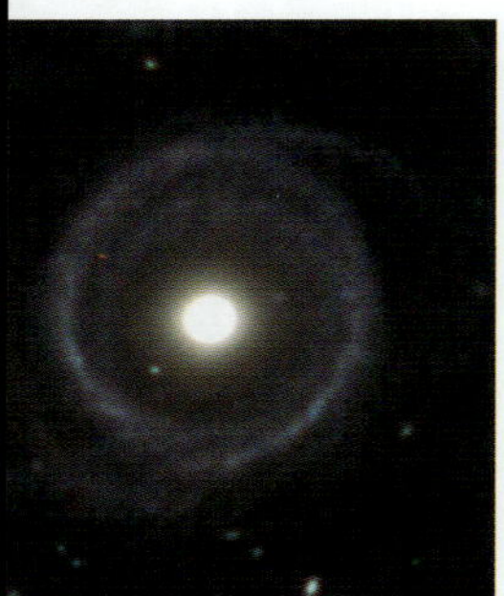

UGC 6614 is seen almost face-on and its classification as a giant low surface brightness galaxy (GLSBG) appears immediately plausible. A bright, structured inner region is separated by a ring from a dim, extended outer region. This ring measures about 1.4′ × 1.7′ in diameter. At a distance of 280 million light years from the Milky Way this corresponds to a mean ring diameter of 120 000 light years. Its luminosity of 14 mag is due almost entirely to the inner region; the much larger but much dimmer outer region makes a negligible contribution to the total luminosity. The large difference in surface brightness is demonstrated by the fact that the outer region is not visible in the POSS catalogue (Palomar Observatory Sky Survey); this shows up only on sensitive CCDs. UGC 6614 is huge at a diameter of more than 330 000 light years – three Milky Ways could fit side-by-side in this spiral plane.

There are no large-scale features in UGC 6614. No star-forming regions with bright, blue stars nor any H II emission regions are visible. This lack of structure is typical of GLSB galaxies. Interestingly, the nucleus of UGC 6614 is active in X-rays as in a Seyfert galaxy. This is an indication of a supermassive black hole at the core (with a mass of 1.2×10^5 solar masses) and of an interaction which drove material into the core and thus set it into an active state. Computer simulations describing merger scenarios which lead to ring galaxies indicate a logical connection to GLSBGs such as UGC 6614 when one follows the evolution after the collision. This indicates that the ring galaxy corresponds to the typical image with a prominent ring after 100–200 million years. Continuing the simulation shows that the ring expands and becomes dimmer and dimmer. This corresponds to an increase in the size of the spiral plane, which after about 500 million years forms a very flat disc with a size of about 300 000 light years – the ring galaxy has thus become a GLSBG. This result is also important for the theory of CDM since for a long time this model could describe the formation of GLSBGs only with difficulty.

OBJECT	UGC 6614
CONSTELLATION	Leo
RIGHT ASCENSION	$11^h\ 39^m\ 15^s$
DECLINATION	+17° 08′ 37″
BRIGHTNESS	14.4 mag
TYPE	(R)SA(r)a
PHOTOGRAPHERS	Stefan Binnewies, Josef Pöpsel
TELESCOPE	600-mm reflector
CAMERA	SBIG STL-11000
EXPOSURE TIME	645 min
LOCATION	Skinakas Observatory, Crete, Greece

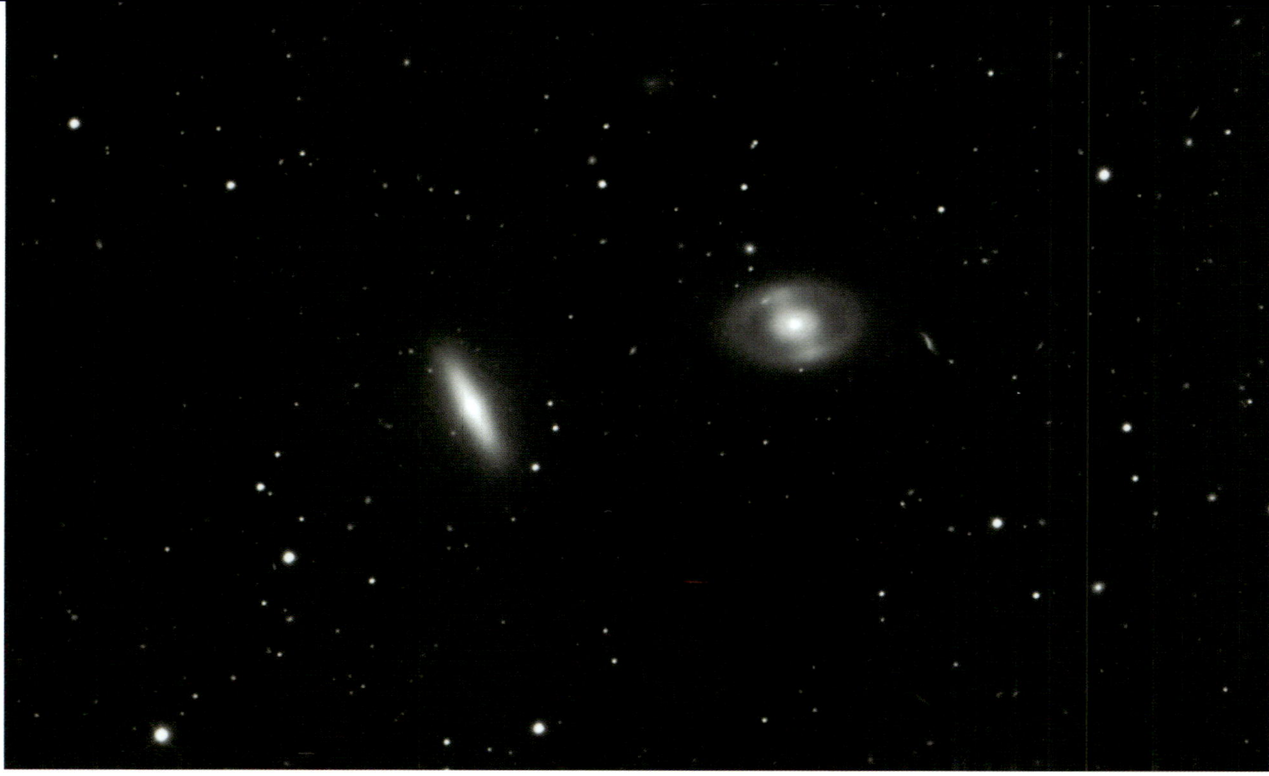

Holmberg 391

Both galaxies visible in the image belong to the Virgo Cluster and make up the system Holmberg 391. On the left is the SA0 galaxy NGC 4350 and 5′ to the west is the SB(r)0 galaxy NGC 4340. Redshift-independent distance measures indicate 51±15 million light years for NGC 4340 and 55±10 million light years for NGC 4350. Redshifts indicate 40±3 million light years for the ring galaxy and 52±4 for NGC 4350. For NGC 4340, the two distance values differ significantly, while for NGC 4350 they are consistent within the errors. It could be that the two galaxies form a pair, but that the ring galaxy NGC 4340 is approaching the Milky Way at 277 km/s relative to NGC 4350. The projected diameter of NGC 4350 is 48 000 light years, of NGC 4340, 52 000. In the image one can see that at 3.5′ its diameter is larger than that of NGC 4350, which is 3′. NGC 4340 displays no spiral structure; the disc is dominated by a closed ring. The visible bar in NGC 4340 extends all the way to this ring and at the ends of the bar bright ansae are visible.

A second bar surrounded by a nuclear ring is visible in the bright central region of NGC 4340, though not in the image above. The radius of this second ring is 12″ and here as well the bar does not extend past the ring. This second ring-bar structure is 15 times smaller than the surrounding one and the axes of the large and small bars are not parallel, but inclined by 12°. This can be an indication that the two bars rotate independently of each other.

Observations of the central region in different colours can provide more information about the inner nuclear ring. This is bright but has no star-forming regions, but rather is composed of relatively old stars which lie in a gas-poor environment. The presence of several dynamic components in NGC 4340 suggests that at least one merger with a compact partner has taken place.

OBJECT	Holmberg 391, NGC 4350 / NGC 4340
CONSTELLATION	Coma Berenices
RIGHT ASCENSION	$12^h\ 23^m\ 58^s$
DECLINATION	+16° 41′ 36″
BRIGHTNESS	11.9 mag
TYPE	SA0 edge-on, SB(r)0
PHOTOGRAPHERS	Michael König
TELESCOPE	200-mm reflector
CAMERA	SBIG ST-10XME
EXPOSURE TIME	10 min
LOCATION	Rimbach, Germany

M 94

The galaxy M 94 (NGC 4736) lies at a distance of only 21.5 million light years and is thus one of the brightest galaxies in the constellation of Canes Venatici. Its luminosity is given as 9 mag which explains why it can also be found in the Messier catalogue. M 94 is a member of the Canes Venatici I Group, which is part of the Virgo Supercluster. The central component of this ring galaxy is an early-type Sab galaxy, which shows hints of two spiral arms, its spiral plane characterized by a flocculent pattern of filaments of gas and dust with a few star-forming regions.

The central component is surrounded by a dim glow, which is an indication of a halo which forms the border to the star-poor ring zone farther out. This halo has a maximum diameter of about 5′, which corresponds to a diameter of only 31 000 light years for the central galaxy. The dimension of the outer, much fainter ring is larger and the major and minor axes have projected lengths of 50 000 × 60 000 light years. The faint outer structure hints at coiled-up spiral arms containing some active areas in which young stars, dust, and gas nebulae are visible. Due to the outer ring and the preserved spiral structure of the central galaxy, complete accretion of a companion galaxy is a plausible formation scenario for the ring galaxy M 94.

Due to the low distance and favourable view of the spiral plane, the rotation velocity can be measured spectroscopically, allowing one to determine the mass density, including dark matter, in the plane. The small error bars of the measurements allow one to fit detailed models which usually consist of the three components core, disc, and spherical halo. The most massive component – the halo – has the least empirical support in the literature. One had assumed that 68 per cent of the mass of M 94 is in the dark-matter halo, assuming a total mass of the galaxy of 5×10^{10} solar masses. However, the measured rotation curves indicate that a spherical halo component is not a good mass model. Much better, in contrast, is a mass distribution with a flattened halo and a lower total mass of 3.4×10^{10} solar masses. This model needs little dark matter and is consistent with the observed luminosity distribution and the measured hydrogen fraction in the outer regions of the galaxy.

OBJECT	M 94, NGC 4736
CONSTELLATION	Canes Venatici
RIGHT ASCENSION	$12^h\ 50^m\ 53^s$
DECLINATION	+41° 07′ 14″
BRIGHTNESS	9 mag
TYPE	(R)SA(r)ab
PHOTOGRAPHERS	Stefan Binnewies, Josef Pöpsel
TELESCOPE	600-mm reflector
CAMERA	SBIG STX-16803
EXPOSURE TIME	480 min
LOCATION	Skinakas Observatory, Crete, Greece

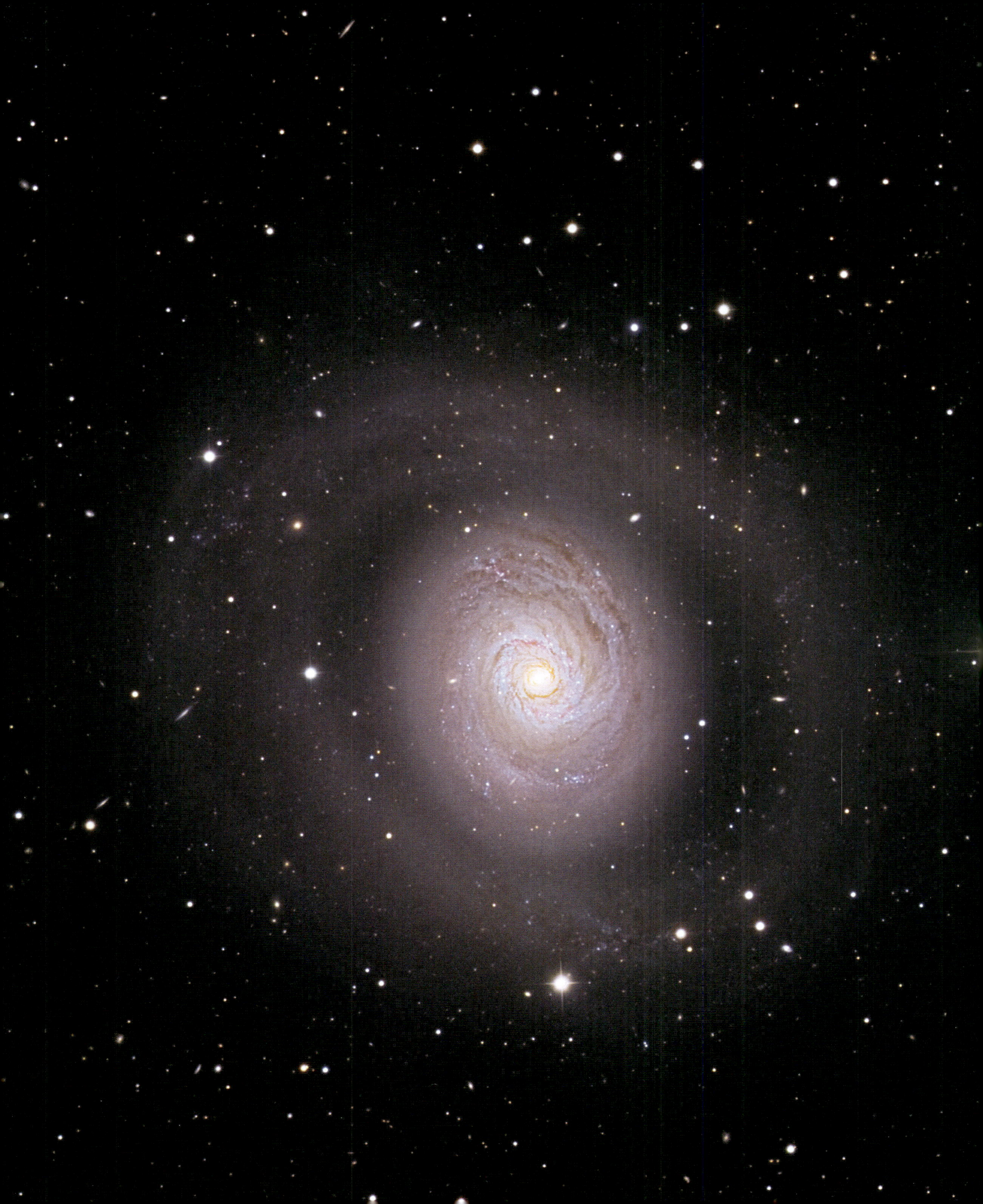

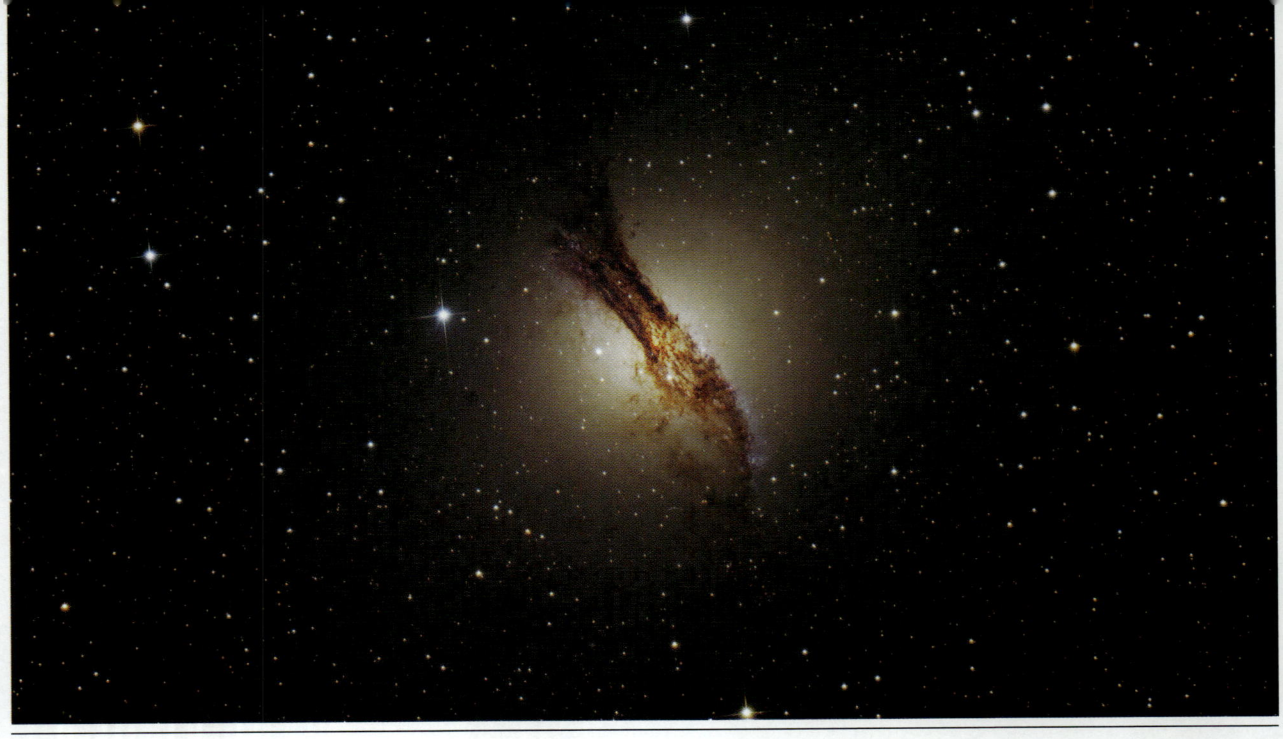

NGC 5128, Centaurus A

Astronomers use the letter A to denote the brightest radio source in a constellation. The centre of our Milky Way lies in the direction of the constellation of Sagittarius, where the brightest radio source Sagittarius A is located. For the galaxy NGC 5128 shown here, the radio designation Centaurus A is much better known than its NGC number or its being catalogued as Arp 153. Centaurus A is the closest extragalactic radio source. Its size of 10° on the radio sky substantially exceeds its optical diameter. This is 25.7′ × 20.0′ and at 7.8 mag is bright for a galaxy – and thus for observers in the southern hemisphere a worthwhile object for binoculars.

Due to its interesting history of formation, the redshift of NGC 5128 has been measured very often and precisely. The NASA Extragalactic Database gives a velocity of 547±5 km/s. With an error of less than one percent, the cosmological distance of NGC 5128 can be calculated at 17.3 million light years. At this distance, an arcminute corresponds to 5000 light years, which results in an extent of about 129 000 × 100 000 light years for the optical part. The two invisible radio lobes extend to the huge distance of 1.5 million light years and are fuelled by two jets emerging in opposite directions from the black hole at the core of NGC 5128.

The galaxy NGC 5128 is often mentioned in the literature as a candidate for a polar-ring galaxy. It is also classified morphologically as an elliptical or S0 type, whereby the broad, warped dust lane and the star-forming regions visible there indicate a collision with another galaxy. One assumes that NGC 5128 is no longer an unperturbed polar-ring galaxy, but rather that the ring structure has been destroyed by its enormous radio luminosity. The huge radio lobes have pushed the material in the outer shells of the galaxy farther and farther to the outside, causing it to thin out and finally turn into halo gas. Interestingly, the jet activity and the formation and disappearance of the outer shell structures are due to the same cause: the merger of two galaxies.

Very deep optical images of NGC 5128 reveal a large, faint halo, which has a broad spindle-like form, and the optically prominent part with conspicuous dust lanes with an almost perpendicular orientation. The spindle-shaped halo has about three times the diameter of the central region and is thus similar to the warped disc of NGC 2685.

OBJECT	NGC 5128
CONSTELLATION	Centaurus
RIGHT ASCENSION	$13^h\ 25^m\ 28^s$
DECLINATION	−43° 01′ 09″
BRIGHTNESS	7.8 mag
TYPE	S0 pec
PHOTOGRAPHERS	Johannes Schedler
TELESCOPE	500-mm reflector
CAMERA	SBIG STL-11000
EXPOSURE TIME	190 min
LOCATION	Hakos Farm, Namibia

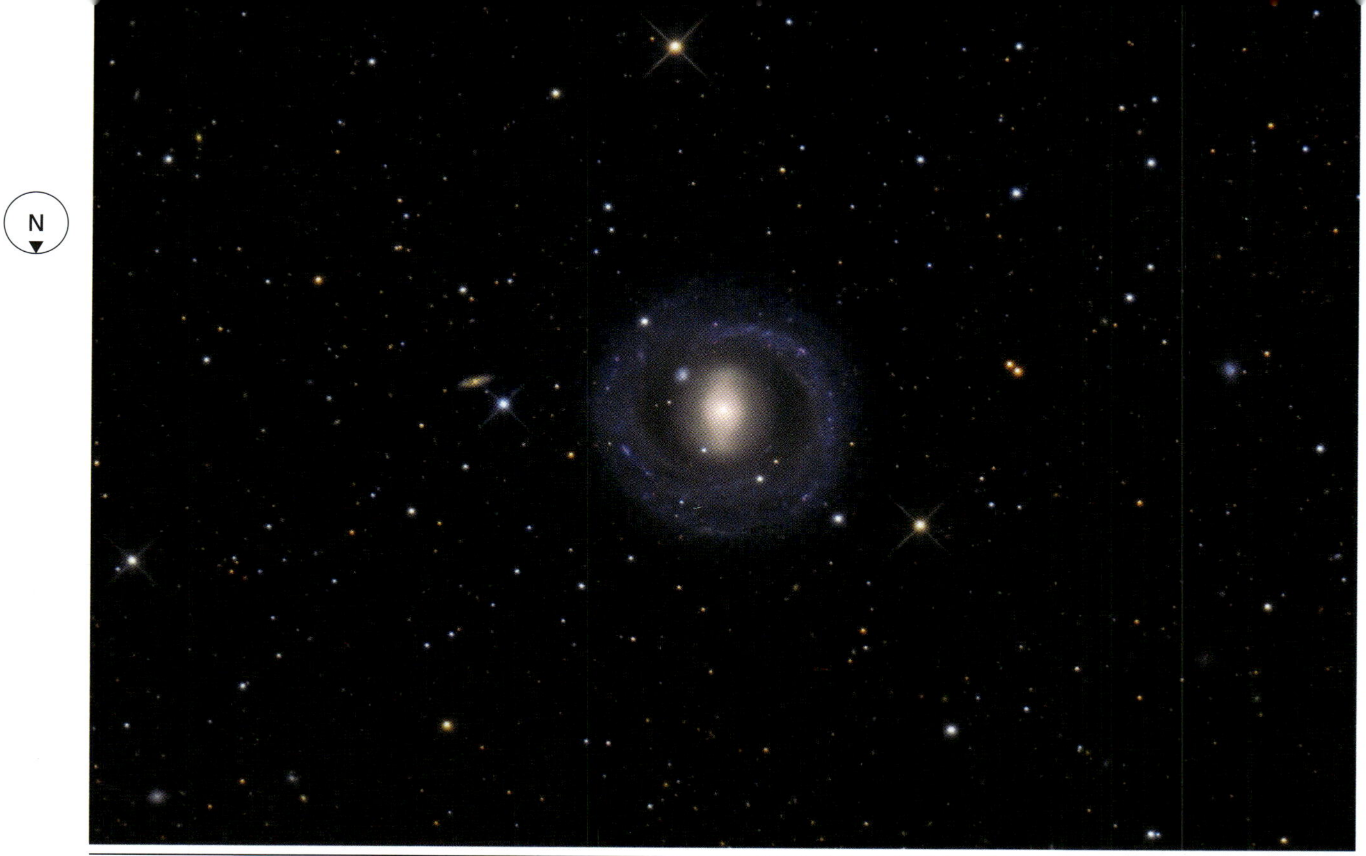

NGC 5701

The ring galaxy NGC 5701, whose two colours are immediately obvious at first glance, is located in the Virgo cluster. The diffuse and lens-shaped inner region appears yellowish, in contrast to the blue spiral arms emerging from it. At the centre a bar is visible which has a higher luminosity than the oval bulge in which it is embedded. In ultraviolet light, the appearance is dominated by star-forming regions in the spiral-arm fragments of the ring. One can still see the nucleus, but the bar and the oval inner region are invisible (*The Galex Ultraviolet Atlas of Nearby Galaxies*). The distance of NGC 5701 is 69 million light years. Since its low inclination of only 15° allows us an almost face-on view of the plane of the ring, one can calculate a mean diameter of 59 000 light years for the ring.

In order to analyse the luminosity distribution in NGC 5701, a model consisting of a bulge and a disc was fit to the pixel values of a high-resolution image of this galaxy. One needs 11 parameters for NGC 5701 in order to achieve a good fit. This model galaxy was then subtracted from the image of the galaxy, resulting in a residual image revealing substructures which were visible only faintly or not at all in the original image. The best fit for NGC 5701 is a model with a combination of bulge and bar with an outer ring. Surprisingly the model needs no disc component, so the question arises as to how the bar could have formed in this discless galaxy. A possible answer is provided by simulations which predict a bar when the galaxy is surrounded by a non-spherical, massive halo of dark matter. This hypothesis is supported by the resulting bar size of about 33 000 light years and by the collision history of NGC 5701 which could have led to such a halo.

OBJECT	NGC 5701
CONSTELLATION	Virgo
RIGHT ASCENSION	$14^h\ 39^m\ 11^s$
DECLINATION	+05° 21′ 49″
BRIGHTNESS	11.8 mag
TYPE	(R)SB0/a(rs)
PHOTOGRAPHERS	Adam Block
TELESCOPE	600-mm reflector
CAMERA	SBIG STL-11000
EXPOSURE TIME	320 min
LOCATION	Mount Lemmon SkyCenter/ University of Arizona, USA

ESO 138-IG29

The ring galaxy ESO 138-IG29 is part of the catalogue of merging galaxies by Halton Arp and Barry Madore and is listed there as AM1724-622. The mushroom-like morphology consisting of a ring component and an elongated companion is also found in a few other galaxies in the AM catalogue. ESO 138-IG29 is at a distance of 200 million light years and its angular size is given as 2.5′ × 1.4′ – an arcminute corresponds to a scale of 58 000 light years.

At first sight, ESO 138-IG29 doesn't offer a bird's-eye view of the plane of the ring, but rather a view from the side of a galactic interaction with a partner and a connecting tidal tail. Since the core of the ring galaxy is offset and the thickness of the ring also varies, ESO 138-IG29 is described as a young ring galaxy. Since the collision partners and impact direction are still directly visible, this ring galaxy is a good model for simulations. In contrast to the standard model of ring galaxies, the ring in ESO 138-IG29 is reddish, i.e. an older stellar population is dominant and the usual increased star-formation rate in the ring is not to be seen. H I observations show that in this object there is little dust and gas in the ring.

Simulations highlight the fact that caustic forms in the plane of the disc of the impacted galaxy result from an off-centre collision. This can also lead to a shell-like form with a gap in the ring. The model which best fits the visible morphology of ESO 138-IG29 has an impact partner with only 15 per cent of the mass of the spiral galaxy which impacts at half of the radius. Interestingly, the paths of the impact partners are not straight, but parabolic, which increases the asymmetry of the induced radial density waves. The lighter interaction partner experiences dynamic friction but at first there is no significant braking. Only after the impact does a tidal tail arise which, antiparallel to the direction of motion, moves towards the spiral galaxy and further influences the structure of its disc.

In the analysis of galaxy mergers, in addition to theoretical studies via simulations, one can look for so-called pre- and post-merger candidates in catalogues of galaxies. Thus one can find galaxies whose appearance corresponds to a snapshot of a merger. The analysis of a sequence which displays the course of the merger frame by frame indicates that the infrared luminosity and hence the efficiency of star formation increases just before the collision. Afterwards, these values decrease and reach the level of a normal elliptical galaxy. This behaviour can be explained by the fact that the previous star formation has used up the hydrogen, which is thus no longer available. One has to keep in mind, though, when studying merger candidates that there is always some scatter due to the individual differences between the galaxies.

The ratio of radio to blue luminosity is also an indicator for developments along the merger sequence. The blue luminosity is directly related to the number of young, hot stars. This ratio is also larger in ESO 138-IG29 than in isolated spiral galaxies, since due to the interaction material flows into the central region, which stimulates the radio emission via increasing activity of the central black hole.

OBJECT	ESO 138-IG29
CONSTELLATION	Ara
RIGHT ASCENSION	$17^h 29^m 10^s$
DECLINATION	−62° 26′ 44″
BRIGHTNESS	12.8 mag
TYPE	(R)SAB0+(s) pec
PHOTOGRAPHERS	Josef Pöpsel
TELESCOPE	600-mm reflector
CAMERA	SBIG ST-10XME
EXPOSURE TIME	140 min
LOCATION	Amani Lodge, Namibia

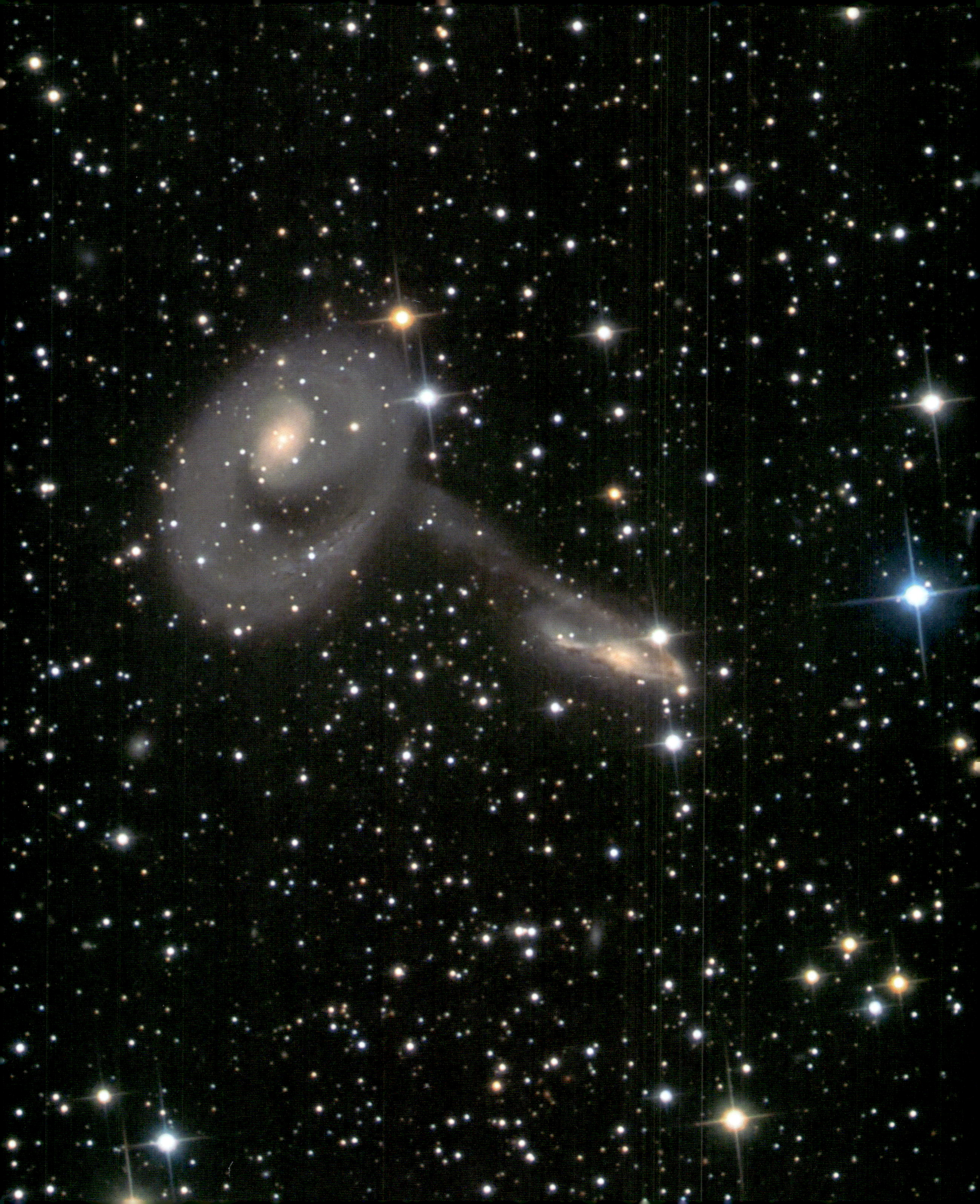

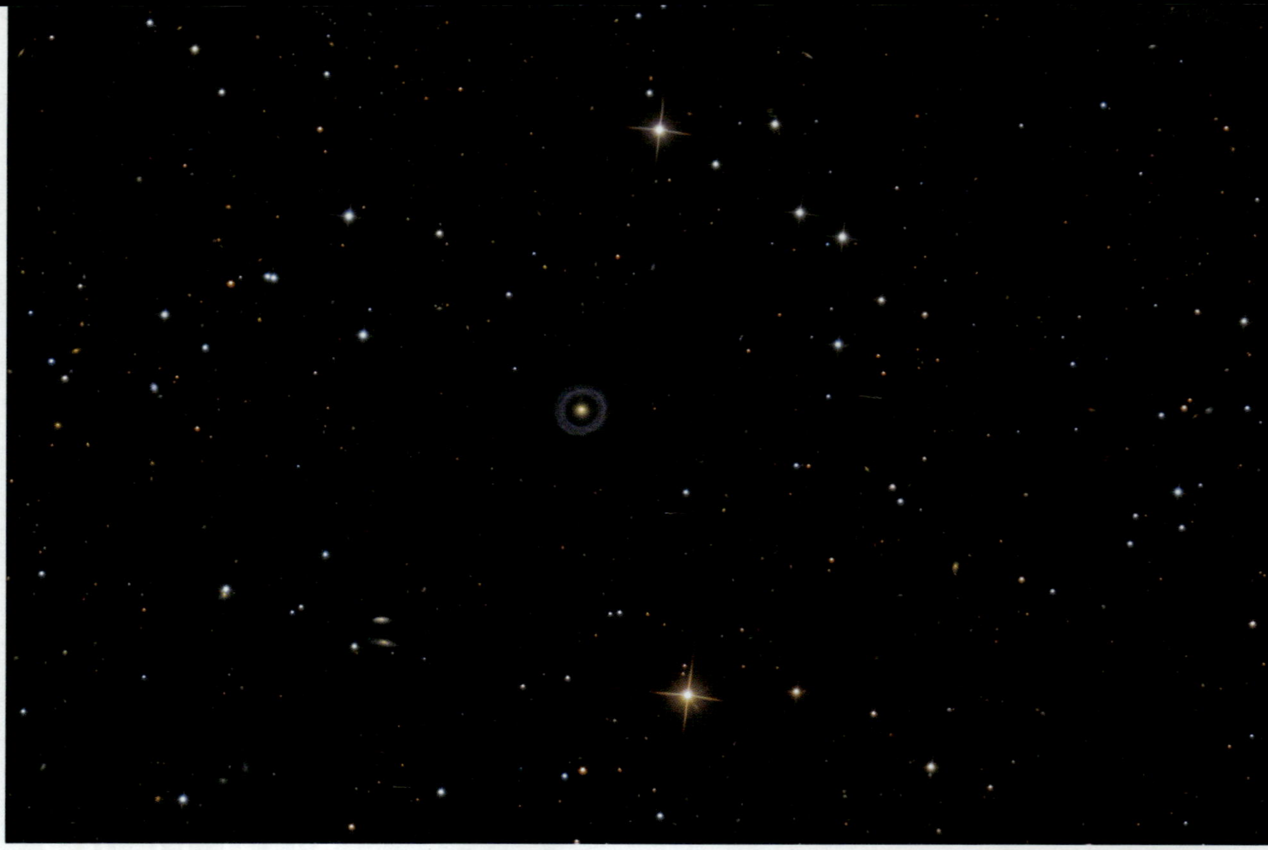

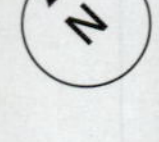

PGC 54559, Hoag's Object

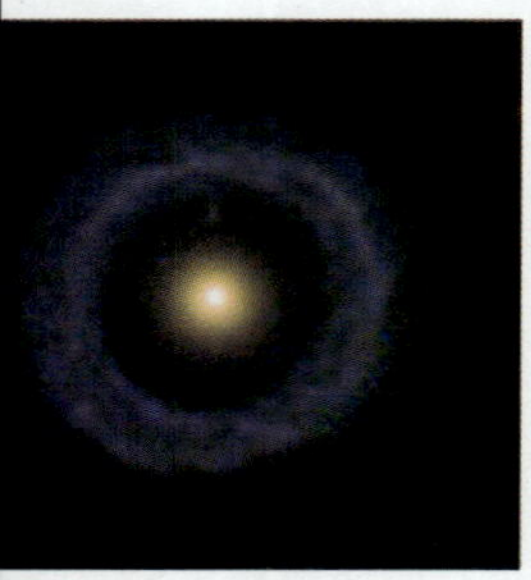

In 1950, the American astronomer Arthur Allen Hoag discovered an unusual object in an image taken by the Jewett Schmidt Telescope at Harvard College Observatory (Cambridge, Massachusetts, USA). In the 75-minute exposure an object was visible which was similar in appearance to a planetary nebula. The core, however, was atypically diffuse and too large for such an object. Thus Hoag suspected that it could be a peculiar galaxy. He also speculated whether the ring could be due to a gravitational-lens effect.

Today it is known that Hoag's Object is a polar-ring galaxy, catalogued as PGC 54559. Its distance from the Milky Way is 550 million light years and the angular size of the ring of 45″ corresponds to 120 000 light years. The circular form of the core is due the special perspective; there are similar ring galaxies with oval cores. In contrast, the yellow core of the galaxy is only 17 000 light years in size. An image of PGC 54559 by the Hubble Space Telescope shows that there are almost no stars in the space between the core and the ring. It is also remarkable that a further, much smaller ring galaxy is present in the background. This is visible as a bright dot in the upper part of the empty ring.

Astronomers now assume that PGC 54559 formed in the head-on collision of a compact galaxy with a spiral galaxy. Parts of the former spiral arms were compressed by an outward-propagating density wave, leading to the spiral-arm ring visible today. The image also shows that the active regions in the ring are not exactly circularly symmetric. Rather, two semicircles are visible in the brightening of the inner part of the ring (upper left) and, not exactly opposite, the outer part (lower right). This dichotomy is probably due to the fact that the precursor galaxy had two spiral arms whose spiral pattern was partially conserved in the collision.

OBJECT	PGC 54559
CONSTELLATION	Serpens
RIGHT ASCENSION	$15^h\ 17^m\ 14^s$
DECLINATION	+21° 35′ 08″
BRIGHTNESS	16.2 mag
TYPE	Ring Galaxy
PHOTOGRAPHERS	Adam Block
TELESCOPE	800-mm reflector
CAMERA	SBIG STX-16803
EXPOSURE TIME	1260 min
LOCATION	Mount Lemmon SkyCenter/ University of Arizona, USA

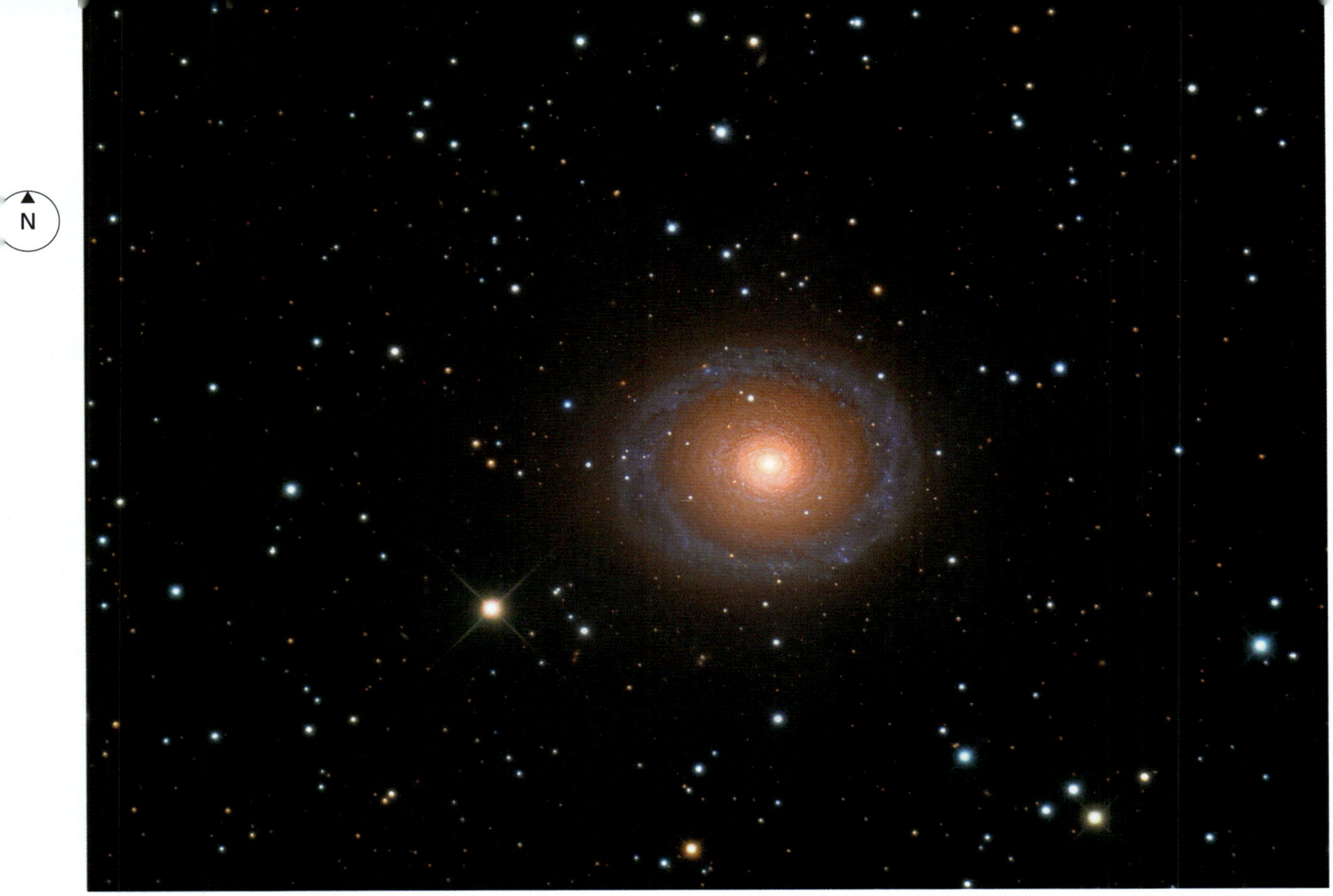

NGC 7217

Conspicuous in NGC 7217 are the spiral-arm rings and dust patterns which repeat several times and extend far into the core region. The outer ring of spiral arms displays areas of active star formation and scattered H II regions, recognizable via their violet-red colour. The low distance of NGC 7217 of 52 million light years allows a detailed view of the spiral plane. The largest H II regions measure about two arcseconds, which corresponds to about 500 light years. The outer ring has a maximum diameter of 3.9′ and thus corresponds to 59 000 light years – a typical value for ring galaxies which have formed via a merger scenario.

Multiple structures are also revealed by deep spectroscopic observations of the inner 6500 light years of the core. Two stellar discs are seen – an inner and an outer disc which differ in luminosity and density. The cause is presumably a merger several billion years ago of a small compact galaxy with a larger spiral galaxy. This redistributed the stars which were previously organized into a disc. The outer gas disc is almost in the main plane of the outer galaxy; the inner disc by contrast is highly inclined and some of the stars in this disc show retrograde motion.

Simulations show that such a minor merger leads to central gas discs which are stable only at high angles of inclination of the disc with respect to the main plane. New stars arise in these gas discs and create star-forming rings. This conspicuous feature is observed in some ring galaxies, which also have a further peculiarity – none of them has a bar, since this would quickly destroy the stellar disc.

OBJECT	NGC 7217
CONSTELLATION	Pegasus
RIGHT ASCENSION	$22^h\ 07^m\ 52^s$
DECLINATION	+31° 21′ 34″
BRIGHTNESS	11 mag
TYPE	(R)SA(r)ab
PHOTOGRAPHERS	Makis Palaiologou, Stefan Binnewies
TELESCOPE	1.3-m reflector
CAMERA	SBIG STX-16803
EXPOSURE TIME	340 min
LOCATION	Skinakas Observatory, Crete, Greece

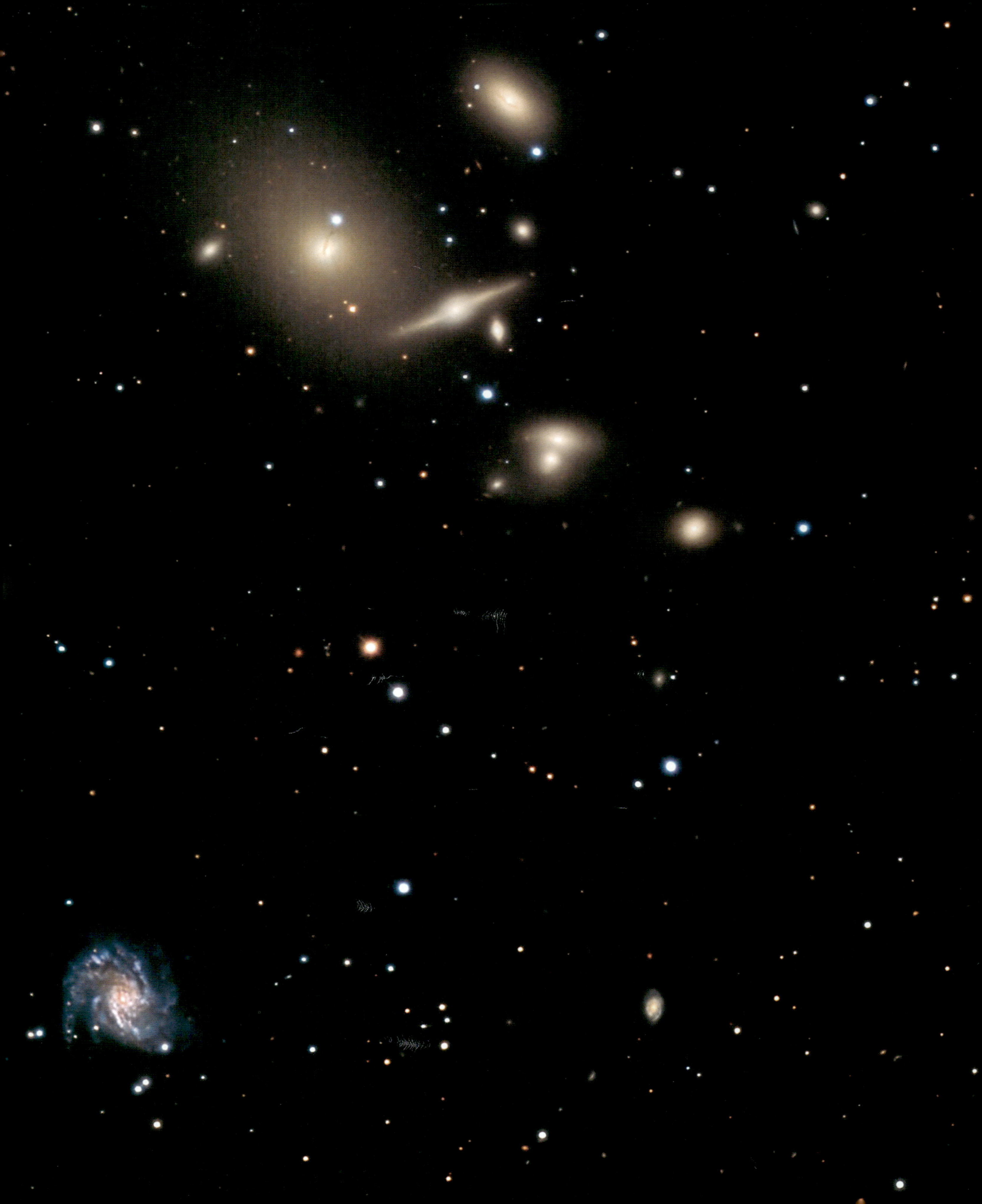

GALAXY GROUPS AND GALAXY CLUSTERS

Most galaxies are not isolated in space, but rather form groups. Galaxy clusters are the largest structures visible in wide-angle images of the universe. These galactic groups are challenging subjects for astrophotographers.

THE CLASSIFICATION AND MORPHOLOGY OF GALAXY GROUPS AND GALAXY CLUSTERS

Independent of the focal length, when studying images of galaxies one notes that galaxies are not solitary objects. This does not necessarily imply spatial proximity, since even if the angular distance of two suspected neighbours is small, their true separation can be much larger. Since the third dimension, the distance of the individual galaxies from the Milky Way, is not obvious from the photograph, there is room for speculation. It is thus a logical assumption that galaxies which appear about the same size and display a similar amount of detail are located at similar distances and are thus neighbours in space. However, sizes are just a rough estimate since the appearances and thus relative sizes of galaxies in a group can be greatly changed by interactions. Redshift-derived galaxy distances in astronomical databases, by contrast, allow firmer statements to be made concerning which galaxies have a high probability of being part of a pair or group. Collections of up to 50 galaxies are deemed to be a group if the diameter is in the range three to five million light years. Due to the common gravitational potential, the members of the group differ only slightly in redshift. The so-called virial theorem implies that the mass of a bound group is always less than $2–3 \times 10^{13}$ solar masses. The Milky Way is itself part of such a group – the Local Group. It consists of about 35 galaxies, although the number is increasing as further members have been discovered in the last few years. These new members are usually dwarf galaxies in the surroundings of large group members. Apart from the Milky Way, M 31 and M 33 are the largest and brightest members of the Local Group. Looking beyond the Local Group, nearby galaxies in our Galactic neighbourhood are also arranged in groups. The Sculptor Group contains more than 20 members. Five of the galaxies are brighter than 10 mag and their distances are in the range 5.9–13.3 million light years. This is followed by the M 81 group with more than 44 individual galaxies. The centre of this group is at a distance of about 12 million light years; its projected extent is 5.9 × 2.9 million light years. Within a large sphere of radius 35 million light years one also finds the Centaurus Group (17 members), the IC 342/Maffei Group (24 members), the M 101 Group (10 members), and finally the NGC 1023 Group (6 members). The Local Group is the richest group in the nearby cosmic environment.

If a galaxy group has more than 50 members, it is then termed a galaxy cluster. In the literature there is a smooth transition from poor to rich clusters, whereby the latter contain thousands of galaxies and can reach a size of more than ten million light years. The distribution of galaxies in a cluster displays structures which are used for further classification. There is a coarse distinction between regular, spherical clusters and irregular clusters. The latter are less compact and often have galaxies distributed on planar or linear structures. At the centre of regular clusters is a large elliptical galaxy, surrounded by an inner core region with high galaxy density. If a pair of these large galaxies dominates the centre, it is known as a B-type cluster. An L-type also has a linear

The galaxy cluster Abell 262 is located in the constellation of Andromeda, close to the border with Triangulum (see page 277). Image credit: Makis Palaiologou, Josef Pöpsel (1.3-m reflector).

arrangement of the dominant galaxies, and a C-type cluster a prominent concentration of bright members and a thin outer region. An F-type has a flat distribution, and finally the I-type describes irregular galaxy clusters, which have fewer galaxies and a uniform distribution. Irregular clusters also indicate that the distribution of galaxies is type-dependent. Spiral systems and irregular galaxies are often found in irregular clusters; in regular clusters, by contrast, they are found only in the outer regions and not at the centre.

The closest galaxy cluster is the Virgo Cluster at a distance of 65 million light years. It consists of about 250 large and more than 2000 small galaxies, filling a volume with a diameter 10 million light years. This corresponds to an angle of about 10° on the sky. Even larger is the Coma Cluster at a distance of 450 billion light years from the Milky Way and a diameter of 20 million light years. This cluster consists of 10 000 galaxies, most of which are dwarf elliptical galaxies. It also has about 1000 bright galaxies which are irregular or spiral types. Compared to the Virgo Cluster, the Coma Cluster is not only larger and has a significantly larger number of members, but is also more regular and thus more symmetric. As impressive as the number of galaxy clusters is, clusters are still the exception. Just along the way to the Virgo Cluster are 20 smaller galaxy groups. It is thus assumed that only 20 per cent of all galaxies are found in the large, rich clusters; most galaxies are found in much smaller galaxy groups.

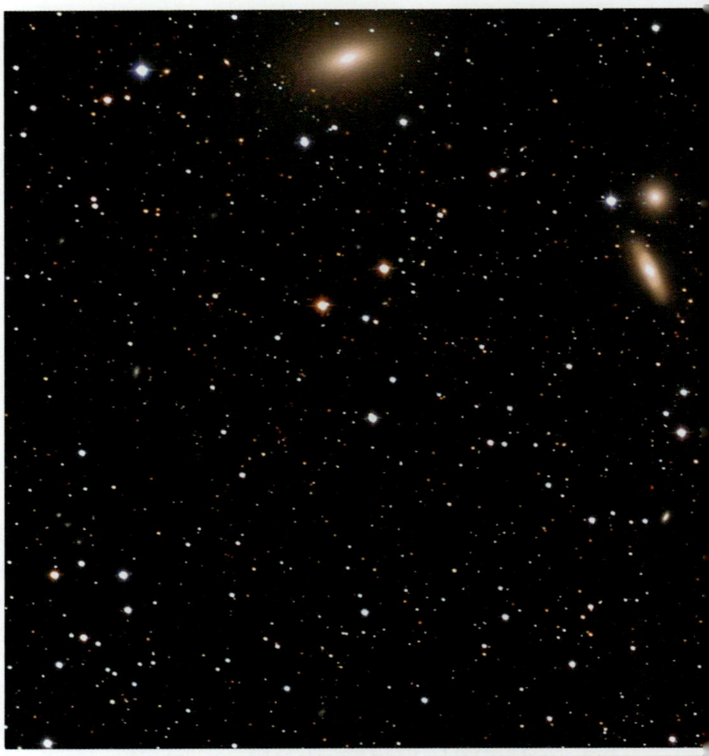

THE ASTROPHYSICS OF GALAXY GROUPS AND GALAXY CLUSTERS

Both in groups and in clusters, gravitation is responsible for the mutual binding of the members. A common potential is formed from the overlap of the gravitational potentials of the individual galaxies, defining a centre of mass. Thus, the velocity differences between the members become uniform with time, stabilizing the structure of the group or cluster. Due to the difference in size, the velocity dispersion of groups is usually less than 200 km/s and 800–1000 km/s in clusters. This value is even higher in very large clusters, since mutually orbiting galaxies can be strongly accelerated in short-range interactions. Thus frequent interactions between the galaxies are mainly responsible for the structure of the galaxy cluster. The centre is almost always dominated by a cD galaxy. This galaxy is a massive elliptical which can form only in rich clusters. Due to the high galaxy density in such a rich cluster, many mergers take place in cascades, allowing the mass of the cD galaxy to steadily increase. Since usually elliptical galaxies are formed in such mergers, in the inner regions of a cluster – where most mergers take place – the E-type galaxies dominate. In many galaxy clusters there are thus no irregular or spiral galaxies at the centre. These are found almost only in the outer regions, since here the frequency of merger processes is lower. In contrast to clusters, elliptical galaxies are rare in most galaxy groups and field galaxies. These observations lead to a general morphology-density relation: the higher the density of galaxies, the more frequently the elliptical population is dominant. Spiral galaxies, by contrast, are usually found in regions with low galaxy density.

The interactions between the galaxies in groups and clusters also have further consequences. Tidal forces due to close approach pull gas and dust out of the galaxies. This material collects in the gravitational potential of the group or cluster, forming the intragroup or intracluster medium which fills out the entire volume of a galaxy cluster. The corresponding gas is strongly rarefied, leading to a temperature of a few million kelvin. This causes the emission of X-rays, which is why galaxy clusters are among the most luminous X-ray sources on the sky. In X-rays, many clusters display substructures with several bright regions, which indicates non-uniform emission. This is due to the fact that the distribution of the hot gas resulting from mergers is not yet complete. Observations of the mass and luminosity of a galaxy cluster at various wavelengths of the electromagnetic

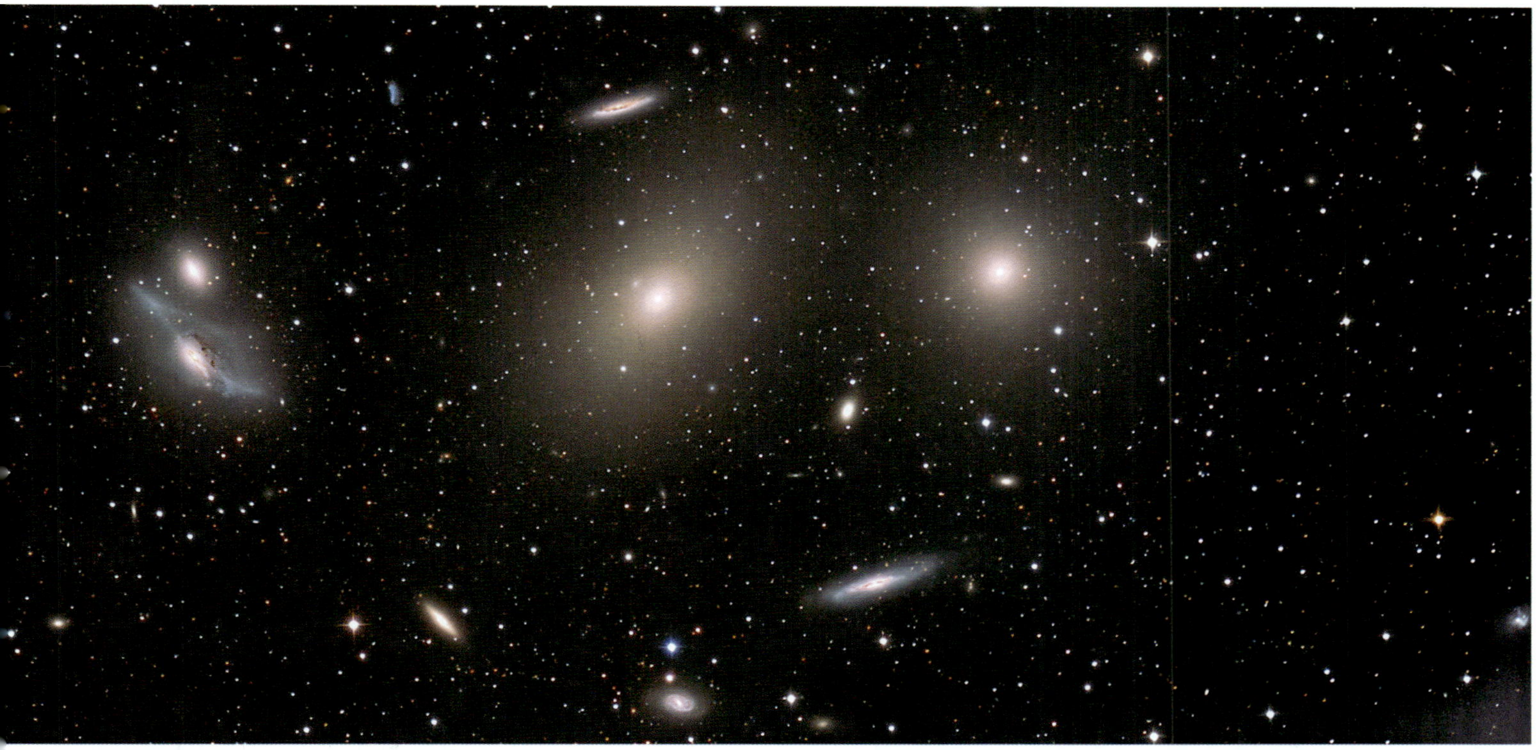

Figure 7.1: The Virgo Cluster, together with the Local Group and other galaxy groups, makes up the Virgo Supercluster (see page 289).

spectrum indicate that the cluster medium typically contains the same amount of mass as the total mass of all visible galaxies. This means that after many interactions and mergers a relaxed galaxy cluster develops, with the number of members decreasing with time. In smaller galaxy groups, this hot and diffuse intragroup medium is the first physical evidence of the existence of a group of neighbouring galaxies.

A special case in the classification of galaxy groups are the fossil groups which consist of a large elliptical galaxy surrounded by several dwarf galaxies. Here as well there is hot gas enveloping the scenery, allowing it to glow in X-rays. An example is NGC 1131, an isolated elliptical field galaxy which revealed the surrounding X-ray halo and a population of dwarf galaxies only when studied in more detail. Following the hierarchical model of galaxy evolution, NGC 1131 can be interpreted as an example of a transition in which a single elliptical galaxy forms via several merger processes involving the members of the galaxy group. This galaxy is the fossil relic of the process and part of the universal evolution which transforms galaxy groups into larger galaxy clusters. This is confirmed by the halo gas between the galaxies. Compared to the gas in groups, the gas in clusters has a higher metallicity. This implies that the cluster gas has already been through several stellar populations and has been ejected from the galaxies in the group into the cluster medium during the evolution of the cluster.

There is a further difference between groups and clusters: their stability with time. The members of a group or cluster lose more and more of their orbital angular momentum due to dynamical friction and slowly fall into the centre, though in a group this takes place much more quickly due to the smaller velocities. The lifetime of a galaxy group – the corresponding so-called dynamic time scale – is only about two per cent of the Hubble time and thus appreciably smaller than the age of the universe. A galaxy group is thus a transitionary form to a larger galactic structure. In other words: one observes only snapshots of mergers, which result in new building blocks used to form larger structures.

In the last few decades, astronomers have detected a three-dimensional pattern: there are regions with a high density of matter which attract galaxies. In contrast, there are large voids which are avoided by galaxies. This results in a foam-like structure. It appears to be constructed of bubbles with most galaxies lying on the surfaces. The typical size of a bubble is 100–300 million light years. This structure is,

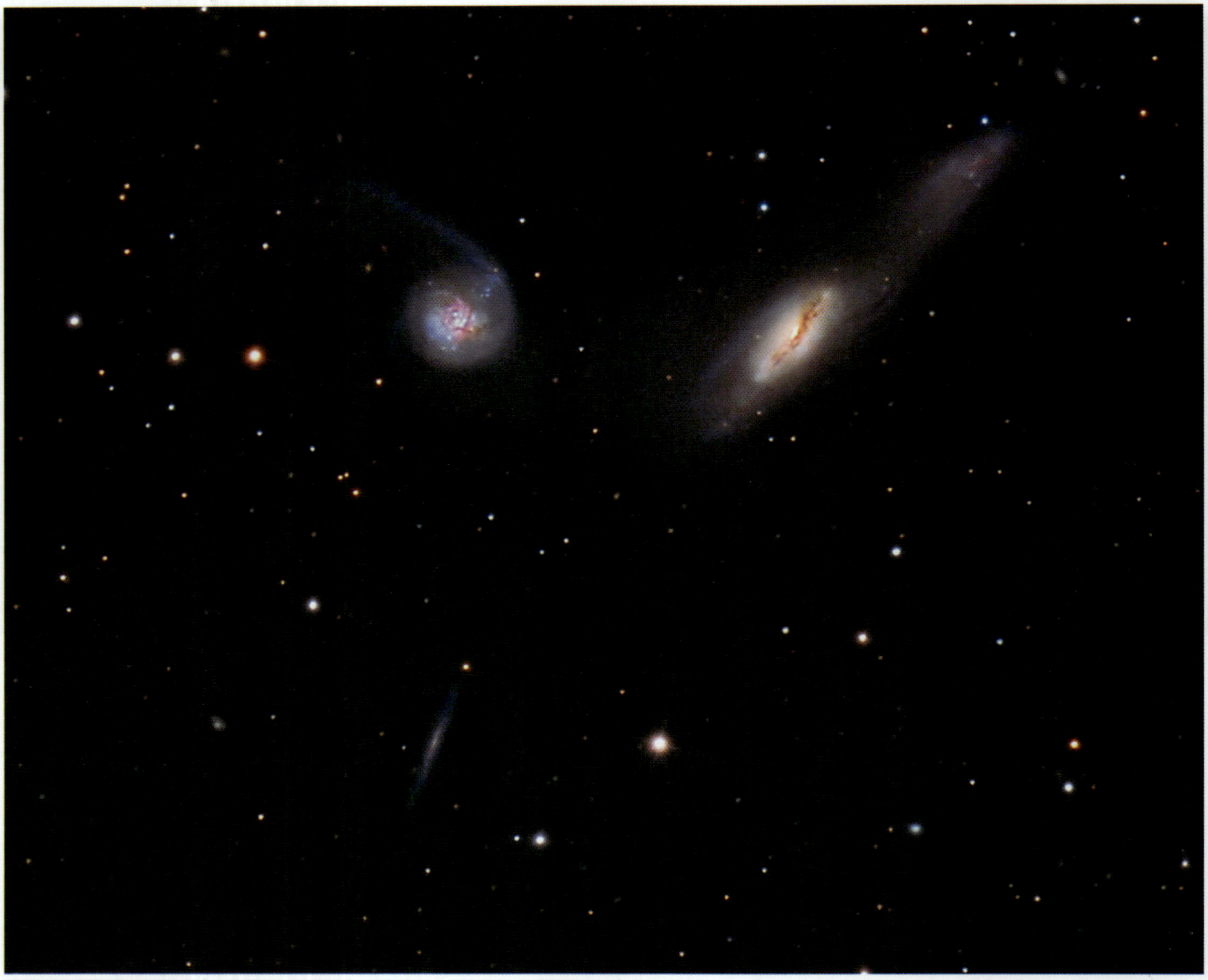

Figure 7.2: The interacting galaxy pair NGC 2992/NGC 2993 awaits the final merger into a new, larger galaxy.

however, not uniform like the atoms in a crystal, but rather reveals organic, filament-like structures such as are often found in nature. Today cosmologists agree that this bubble structure formed in the early phase of the universe. After the Big Bang 13.7 billion year ago, fluctuations occurred in the quantum vacuum of the universe, which then had a size of only 10^{-35}m. In the extremely short span of time from 10^{-35} to 10^{-32} seconds after the Big Bang, from this small scale known as the Planck length, the universe expanded in an inflationary phase. The expansion of space took place faster than the speed of light and covered 30 orders of magnitude. At the end of this phase the universe had a diameter of about a metre and was filled with the foundation of these bubble-like filaments, which were the source of the first stars, galaxies, and, later, galaxy clusters.

LARGE-SCALE STRUCTURES IN THE UNIVERSE, THE SHAPLEY SUPERCLUSTER AND QUASAR GROUPS

In the 1980s, the motions of galaxies within 200 light years were analyzed. Velocities pointed in the direction of the constellation of Centaurus, where it seemed there is a great concentration of mass pulling galaxies towards it. This mass was dubbed the Great Attractor. The total mass was estimated at 10^{16} solar masses and it was assumed that it consists almost totally of dark matter.

In 1992 it was shown that the velocity trends found in the galaxies were not real but rather due to insufficient corrections and assumptions about the distribution of galaxies and the measured distances. Better models no longer showed an infall towards the Great Attractor, neither from the near nor from the far side of the suspected mass concentration. However, there remained a motion towards a mass concentration at a redshift $z > 0.02$. This concentration is now associated with the Shapley Supercluster, which lies in the northern part of the constellation of Centaurus and was discovered in 1930 by Harlow Shapley. In an ESO observational campaign, velocities of 8000 galaxies were measured in order to construct a cone diagram of the distribution of galaxies. This cone, with an opening angle of 15°, plots the distance of galaxies up to a redshift of $z = 0.2$ (= 2.5 billion light years). The distribution indicates a bridge of galaxies between the Centaurus Supercluster in the foreground and the Shapley Supercluster. The Shapley Supercluster contains 5700 galaxies in the survey at a mean redshift of $z = 0.049$, corresponding to 650 million light years. There are also filaments between the Shapley Supercluster and other rich galaxy clusters which, however, do not attain its extent and density.

The Shapley Supercluster has a linear form, stretching over 120 million light years and containing more than 20 large galaxy clusters. Such large mass concentrations often contradict cosmological models, since theoretically such large structures are very improbable. Cosmological models favour smaller scales for mass concentrations, which are related to the energy density and age of the universe. These cosmological models predict that large mass concentrations of the dimension of the Shapley Supercluster are relatively rare.

In 2003, a structure was discovered via data from the Sloan Digital Sky Survey whose extent exceeds that of the Shapley Supercluster. This is a group of galaxies with almost identical redshifts of $z = 0.1$. It stretches over more than a billion light years. This Sloan Great Wall, however, in contrast to a galaxy cluster, is not a connected, gravitationally bound object. A similar situation exists for Large Quasar Groups (LQGs) which since the 1990s have often been found in surveys.

Due to their luminosity, these quasars indicate the pattern of matter distributions which extend to distances a from a few hundred thousand to a billion light years. The quasars are arranged in twisted linear structures containing groups of a few dozen members each. Such a group is also not a gravitationally bound object, but rather displays the spatial pattern of large-scale structure and thus the framework of dark matter, which is responsible for mass distributions on these large scales.

BIBLIOGRAPHY

Bender, R. et al.: *Extragalaktische Astronomie*, 2012, www.usm.uni-muenchen.de/people/saglia/dm/galaxien/alldt/node35.html

Camenzind, M.: *Astronomie und Kosmologie – Teil III: Galaxien und das Universum*, www.lsw.uni-heidelberg.de/users/mcamenzi/APCOSMO_3_tlg.pdf

Clowes, R. G. et al.: *A structure in the early universe at z ~ 1.3 that exceeds the homogeneity scale of the R-W concordance cosmology*, 2012, http://arxiv.org/abs/1211.6256

Landy, S. D. and Szalay, A. S.: *A general analytical solution to the problem of Malmquist bias due to lognormal distance errors*, The Astrophysical Journal, **391**, 1992

Meusinger, H.: *Thüringer Landessternwarte Tautenburg – Vorlesung Extragalaktik*, 2012, www.tls-tautenburg.de/research/meus/vorlesung/ppt/EG10/EG_9.pdf

Proust, D.: *ESO Messenger – Shapley Supercluster*, 2006, www.eso.org/sci/publications/messenger/archive/no.124-jun06/messenger-no124-30-31.pdf

Raychaudhury, S.: *The distribution of galaxies in the direction of the 'Great Attractor'*, Nature, **342**, 1989

Sparke, L. S. and Gallagher, J. S.: *Galaxies in the Universe: An Introduction*, Cambridge University Press, 2007

Weiß, A.: *Einführung in die Astrophysik – Galaxienhaufen*, 2006, www.mpa-garching.mpg.de/lectures/EinfuehrungAW/Galaxienhaufen.pdf

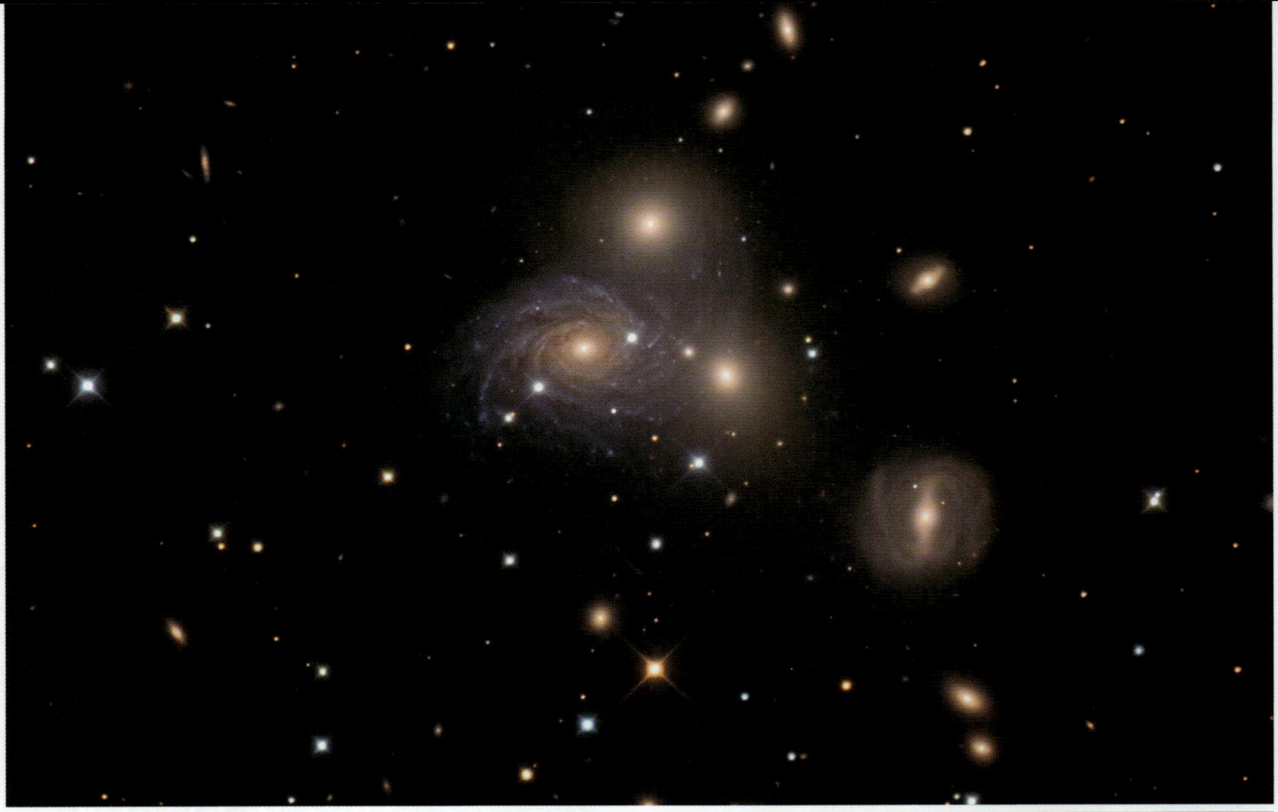

Arp 113

The image shows the galaxy group around NGC 70 in the constellation of Andromeda. This SA(rs)c galaxy forms the northern point of an equilateral triangle with sides of an arcminute. The SA0 galaxies NGC 71 and NGC 86 form the other vertices, whereby one vertex points in the direction of the SB(rs)ab galaxy NGC 72. This group is known as the NGC 70 group or Arp 113, whereby NGC 70 seems to lie in front of the two other galaxies. All galaxies have an angular size of one to two arcminutes, increasing the impression that they belong together. Studies of the distances of the individual galaxies indicate for NGC 70 a distance of 326 million light years, for NGC 68, 264 million light years, and for NGC 71, 307 million light years. The distance to NGC 72 is given as 333 million light years. In order to estimate the true separations, one must take the uncertainties in the distance measurements into account. This implies that the spiral galaxies NGC 70 and NGC 72 are very probably close neighbours. Their separation of 2.3′ corresponds to a projected distance of 212 000 light years. The two lenticular galaxies NGC 68 and NGC 71 are located in front of the spirals, whereby NGC 68 is closer to them. NGC 71 is an isolated field galaxy along the line of sight to the galaxy group.

However, the sizes of the galaxies derived from the distances appear too large. Thus NGC 70 at 183 000 light years, NGC 68 at 153 000 light years, NGC 71 at 130 000 light years, and NGC 72 at 121 000 light years would all be the largest examples of their types. Since no disturbances in their morphologies are visible, physical sizes this large are improbable. More plausible is that the entire galaxy group has too large a recession velocity. This could result from a local velocity component due to the fact that the group is located in a filament in the surroundings of the Perseus–Pisces Supercluster, which is receding from us.

OBJECT	Arp 113 / NGC 70
CONSTELLATION	Andromeda
RIGHT ASCENSION	$00^h\ 18^m\ 23^s$
DECLINATION	+30° 04′ 47″
BRIGHTNESS	14.5 mag
TYPE	SA(rs)c
PHOTOGRAPHERS	Makis Palaiologou, Stefan Binnewies,
TELESCOPE	1.3-m reflector
CAMERA	SBIG STL-6303
EXPOSURE TIME	330 min
LOCATION	Skinakas Observatory, Crete, Greece

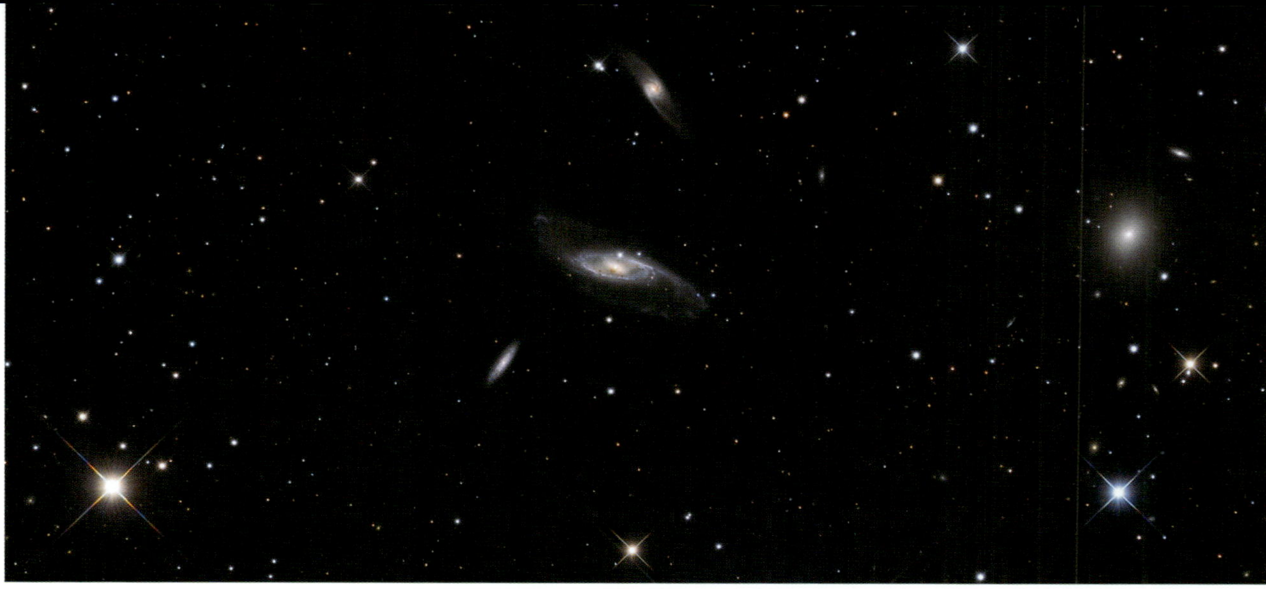

NGC 536 GROUP

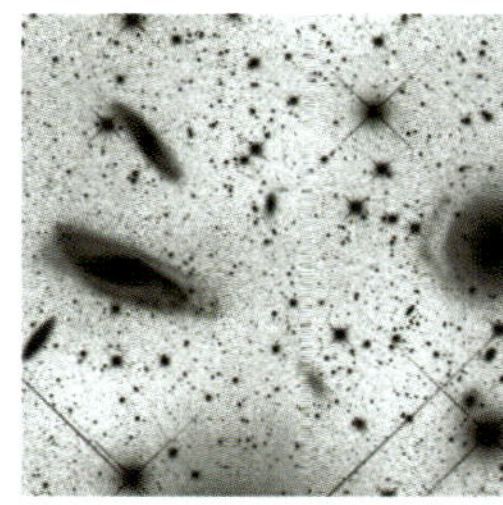

The image shows four obvious galaxies above a sixth-magnitude star (SAO 54695): at the far right NGC 529 and 9′ farther to the left a group of three, the SBa galaxy NGC 531 (top), the SB(r)b galaxy NGC 536 (3.5′ below) and, 2.7′ farther to the left, to the southeast, the Sc galaxy NGC 542. The small group lies at the edge of the large Pisces–Perseus Supercluster. Their mean distance is 223 million light years; at 238 million light years, the relative velocity of NGC 536 is 475 km/s larger than the mean of the group of three. These galaxies, NGC 529, NGC 531, and NGC 542, are direct neighbours. Their radial velocities differ by only ±80 km/s. This 3+1 configuration is supported by the contrast-enhanced image. There are no tidal tails extending from NGC 536 to the two galaxies above and below it. The enhanced contrast also reveals a peculiarity of NGC 529 (see small image). In addition to a faint disc or halo with a strong edge, to the left and right of this circle are two very faint arcs which have a diameter of more than 150 000 light years. These have not yet been described in the literature. There is also no mention of the brightening about 5′ to the southeast of NGC 529. This could be the relic of a tidal tail or of a small, compact galaxy which has merged with NGC 529.

The method of *JHK* photometry is used in astrophysics in order to study the near-infrared spectrum of galaxies. The standard filter names correspond to the IR colours J (1220 nm), H (1630 nm), and K (2190 nm). In an infrared colour diagram, the ordinate displays the short-wave colour index $(J - H)$ and the abscissa the long-wave colour index $(H - K)$. Points in this diagram cluster in a region corresponding to an inactive state. From this point, arrows can indicate the colour changes which take place in a galaxy when the physical conditions change. For example, reddening due to scattering, emissions due hot dust or gas, or starburst regions are directly visible in the colour space. Studies of the distribution of galaxy colour indices indicate that most of the data points at the edge of the distribution are due to interacting galaxies. The reason is that there is a transfer of gas and dust as a result of the interaction. The induced star formation also causes a contribution from the thermal emission of dust heated by stars. Thus high $(H - K)$ values are due to dust at a temperature of 600–1000 K in the galaxies. This peculiar colour is visible in interacting galaxies before perturbations become stronger and a merger takes place.

OBJECT	NGC 536
CONSTELLATION	Andromeda
RIGHT ASCENSION	$01^h\ 26^m\ 22^s$
DECLINATION	+34° 42′ 11″
BRIGHTNESS	13.2 mag
TYPE	SB(r)b
PHOTOGRAPHERS	Adam Block
TELESCOPE	800-mm reflector
CAMERA	SBIG STX-16803
EXPOSURE TIME	520 min
LOCATION	Mount Lemmon SkyCenter/ University of Arizona, USA

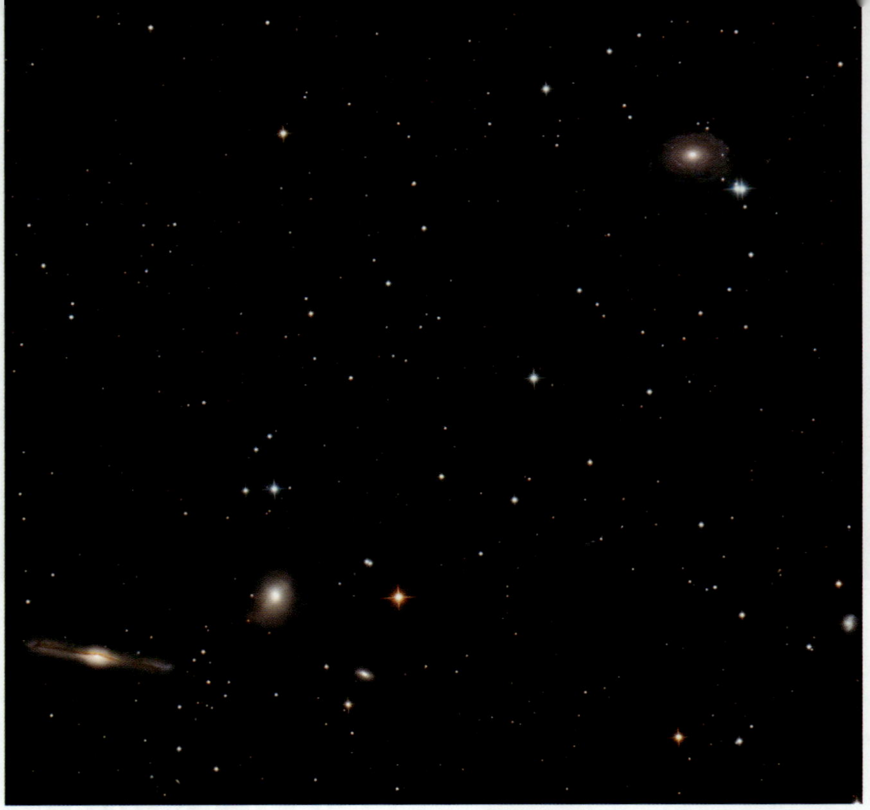

NGC 678 / NGC 680 / NGC 691

The image shows (from bottom to top) the three galaxies NGC 678, NGC 680, and NGC 691. They are part of a larger galaxy group which also contains NGC 697, NGC 693, IC 163, and IC 167. The brightest galaxy of this group is NGC 697, which is why it is known as the NGC 697 Group. The image above, however, shows just a small portion of the group.

The SB galaxy NGC 678 is seen edge-on and displays a dust lane which covers its entire length of 4.5′. The distance of 130 million light years leads to a diameter of the spiral plane of NGC 678 of 170 000 light years. Its orientation also makes it clear that the outer regions of the spiral plane are extended, with their width even exceeding that of the bulge. At a distance of 3.7′ southeast of NGC 678 is the elliptical galaxy NGC 680. Its size of 1.9′ corresponds to a projected diameter of 75 000 light years. The redshifts of both galaxies differ by only 93 km/s, so that they probably form an interacting pair with a minimum distance of 140 000 light years. Following the diagonal of the image leads to the galaxy NGC 691 at a distance of 20′. This SA(rs)bc galaxy is very close to a double star and it appears to have a softer form than a typical late-type spiral. Closer inspection reveals a multipart structure in the spiral arms. They form a diffuse ring at the outer edge of the plane.

Radio observations of the galaxy group show the distribution of emission due to atomic hydrogen in the surroundings of the galaxies and thus indicate current or past interactions. The pair NGC 678 and NGC 680 is surrounded by a common H I envelope which extends far beyond the two galaxies and points in the direction of NGC 691. Since NGC 691 itself also has unperturbed H I emission, the extension of the envelope of NGC 678 and NGC 680 is probably only coincidentally oriented in the direction of NGC 691 and is an indication of a third, smaller galaxy which has interacted with NGC 680 in a minor merger. The result is the perturbed elliptical galaxy NGC 680, which in addition to a dust lane in the inner regions also displays shell-like structures in its halo.

OBJECT	NGC 678
CONSTELLATION	Aries
RIGHT ASCENSION	$01^h\ 49^m\ 25^s$
DECLINATION	+21° 59′ 50″
BRIGHTNESS	13.3 mag
TYPE	SB(s)b edge-on
PHOTOGRAPHERS	Makis Palaiologou, Stefan Binnewies
TELESCOPE	600-mm reflector
CAMERA	SBIG STX-16803
EXPOSURE TIME	405 min
LOCATION	Skinakas Observatory, Crete, Greece

N

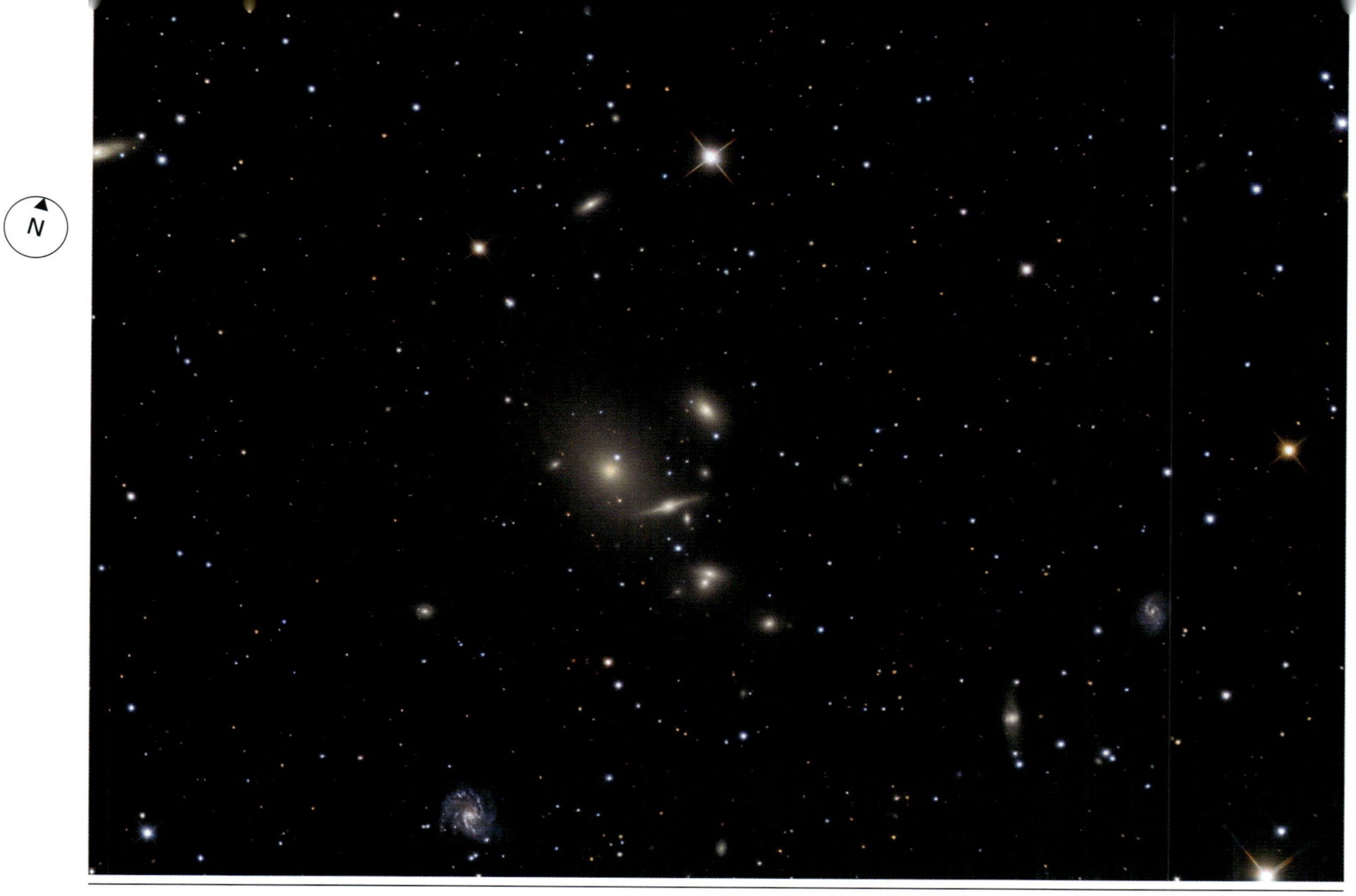

Abell 262

The image shows the centre of the galaxy cluster Abell 262. It is dominated by the cD galaxy NGC 708, which is surrounded by several cluster members. The image shows, however, only some of the approximately 200 cluster members. These make Abell 262 a comparatively small cluster within the large filament of the Pisces–Perseus Supercluster.

NGC 708 is closely surrounded by about a dozen galaxies whose redshifts lie at about 4800 km/s and thus indicate that the distance of the cluster is 222 million light years. NGC 708 has a size of 3.0′ × 2.5′; its halo is elongated in the north–south direction and extends as far as a few cluster members. Prominent is the S0 galaxy NGC 705 1.5′ to the southwest whose distance is given as 209 million light years. It thus lies in front of the cD galaxy. It is also possible that it is approaching us with a relative velocity of about 250 km/s. Assuming that NGC 705 and NGC 708 are orbiting each other, the projected velocity corresponds to an orbital radius of 100 000 light years and is an indication of the dynamics of the galaxies at the centre. In the bright core of NGC 708, there is also a light brown dust lane, which is almost perpendicular to the major axis of the galaxy. This is a relic of the merger of a spiral galaxy with NGC 708, which has caused a flow of material into the centre. This also explains the active state of the galaxy, making NGC 708 a Seyfert-2 galaxy. This is also supported by deviations from a flat luminosity distribution of the core, measured within a radius of two arcseconds and thus only a few hundred light years from the black hole at the centre.

The group of galaxies around NGC 708 is also known in the literature as the Fath Cluster. The origin of this name is unclear. It might relate to the astronomer Edward Arthur Fath, who studied spiral nebulae, as they were then known, with James E. Keeler at the Lick Observatory (Mount Hamilton, California, USA), but there seems to be no direct connection between him and NGC 708.

OBJECT	Abell 262
CONSTELLATION	Andromeda
RIGHT ASCENSION	$01^h\ 52^m\ 47^s$
DECLINATION	+36° 09′ 05″
BRIGHTNESS	–
TYPE	Galaxy cluster
PHOTOGRAPHERS	Adam Block
TELESCOPE	800-mm reflector
CAMERA	SBIG STL-11000
EXPOSURE TIME	220 min
LOCATION	Mount Lemmon SkyCenter/ University of Arizona, USA

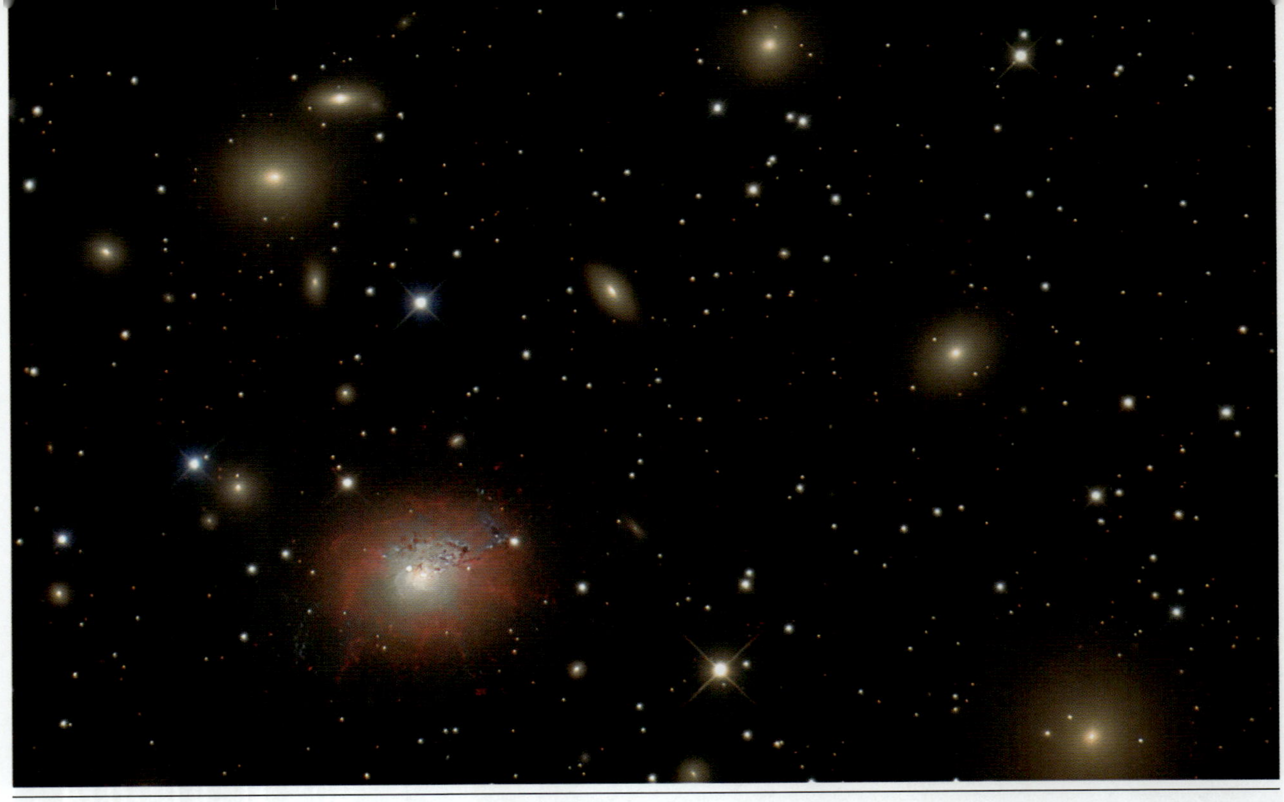

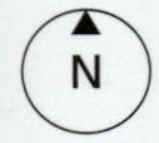

Abell 426

Abell 426, also known as the Perseus Cluster, is a rich galaxy cluster with almost 1000 members. It belongs to the Pisces–Perseus Supercluster and, due to its richness classification of 2, occupies an outstanding position. The richness parameter was introduced in 1958 by George O. Abell in his galaxy catalogue in order to have a statistical description of the galaxies. To this end, he first counted the number of galaxies in a cluster within a range of two magnitudes and later defined six classes. Abell 426 with class 2 is a medium-size cluster. In contrast, Abell 1835 in the constellation of Virgo has class 4 and more than twice as many galaxies within the Abell criterion.

The large elliptical galaxy NGC 1275 at the centre of Abell 426 lies at a distance of 235 million light years. It is at the lower left of the image and has an angular size of 2.3′ × 1.7′, corresponding to 157 000 × 116 000 light years. NGC 1275 was already identified as a radio source (Perseus A) in the 1950s. It thus displays a typical property of a cD galaxy within a galaxy cluster and indicates that material is accreted onto the supermassive black hole at the centre. NGC 1275 is an active merger machine and the product of many cannibalized cluster members. Its halo thus gives a direct view into such a merger process. The image shows blue and red filaments in the halo of NGC 1275 whereby the red H II emission regions form a net about the centre. The band-like blue structures are the remains of the spiral arms of a destroyed galaxy and display regions of active star formation as well as clouds of dust and gas. Near the centre one can also recognize spiral-arm fragments which belonged to this merging galaxy. In the infrared, one observes the emission lines of molecular gas (carbon monoxide, CO) in the H II filaments. At 20–500 K, this gas is relatively warm and dense and could form from the cooling of hotter gas which flows out of the centre in large lobes. The filaments of molecular gas have a mass of 10^9 solar masses, which corresponds to about ten per cent of the total mass of molecular gas in NGC 1275.

OBJECT	NGC 1275
CONSTELLATION	Perseus
RIGHT ASCENSION	$03^h\ 19^m\ 47^s$
DECLINATION	+41° 30′ 47″
BRIGHTNESS	11.8 mag
TYPE	E pec
PHOTOGRAPHERS	Makis Palaiologou, Stefan Binnewies
TELESCOPE	1.3-m reflector
CAMERA	Andor DZ 436
EXPOSURE TIME	122.5 min
LOCATION	Skinakas Observatory, Crete, Greece

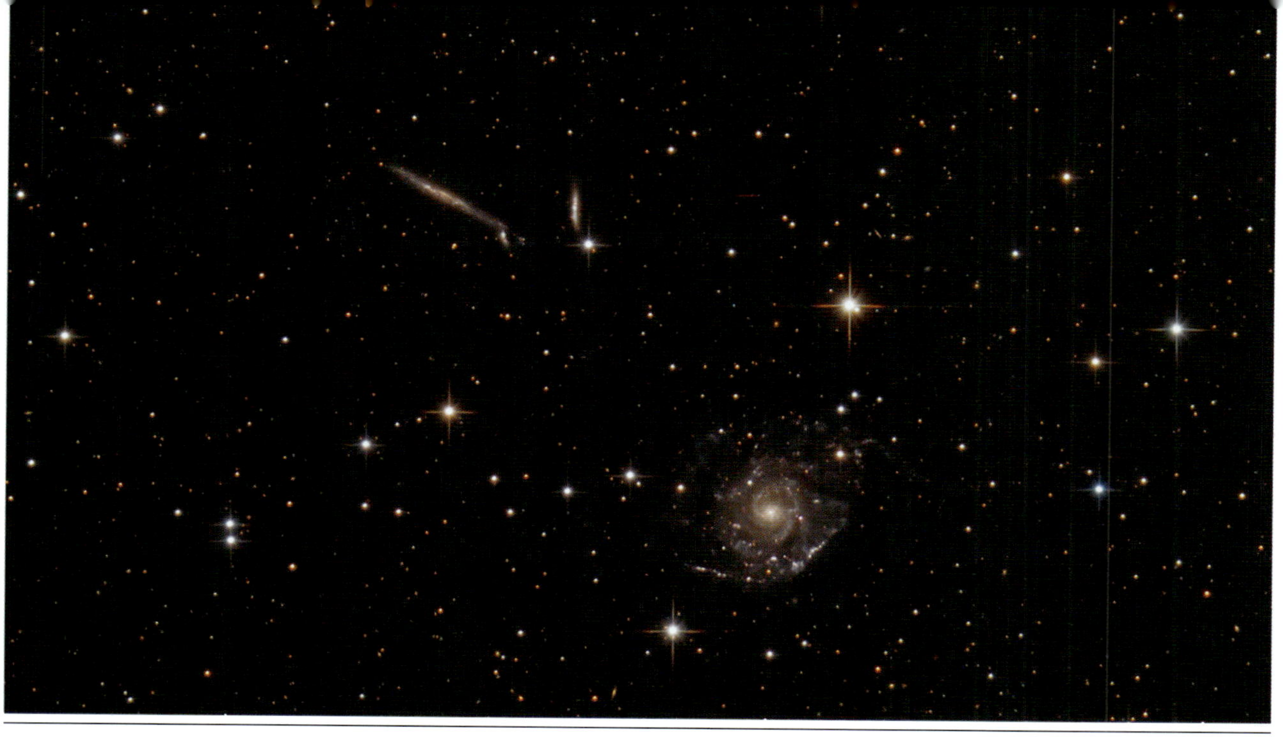

Holmberg 124

Holmberg 124 is a compact group consisting of the galaxy NGC 2805 (below right) and the three closely neighbouring galaxies NGC 2820, NGC 2814, and Mrk 108. NGC 2805 is a late-typ SAB at a distance of 82 million light years and has the enormous diameter of more than 196 000 light years. Presumably there is an interaction with the three galaxies 10′ to the northeast since the spiral plane of NGC 2805 displays asymmetric perturbations which lie either in the direction of the group of three or in the opposite direction. Assuming that all galaxies lie at the same distance, their minimum distance from NGC 2805 is about 230 000 light years. The relative velocities of the three galaxies differ by less than 60 km/s and, in relation to NGC 2805, all are approaching us at about 150 km/s. The largest galaxy in the group of three is NGC 2820 which is seen edge-on. It is a barred spiral since the inner region displays a rectangular luminosity distribution rather than a bright core. Only 2.1′ away from NGC 2820 is the irregular galaxy Mrk 108 with a size of only 0.5′. Also seen edge-on is the somewhat larger Sb galaxy NGC 2814, two arcminutes farther away.

In 1983, a common halo of these three galaxies was first detected in observations with the Very Large Array (VLA, National Radio Astronomy Observatory, New Mexico, USA). The radio emission is due to the radiation of atomic hydrogen, which forms bridges between the three galaxies which are not seen in the optical. The intensity of the radio emission indicates a hydrogen density of about one atom per ten cubic centimetres. This low density is not sufficient for star formation, which explains why the bridges are invisible in the image above.

Radio observations made 20 years later confirmed this discovery and provided further details. For example, they indicate that the bridge between NGC 2820 and NGC 2814 contains three knots, with Mrk 108 located at one of them. A detailed radio map was made for NGC 2820 as well. A narrow arc-shaped emission region extends over a distance of 57 000 light years parallel to the spiral plane. This atomic gas which lies within the spiral plane was pulled out of the disc by tidal interactions in close approaches of Mrk 108 and NGC 2814 and is now flowing into the intragroup medium which fills the space between the galaxies. The ram pressure released by the gases is the energy source of the observed radio emission in NGC 2820.

OBJECT	Holmberg 124
CONSTELLATION	Ursa Major
RIGHT ASCENSION	$09^h\ 21^m\ 12^s$
DECLINATION	+64° 12′ 47″
BRIGHTNESS	–
TYPE	Galaxy group
PHOTOGRAPHERS	Bernhard Hubl
TELESCOPE	305-mm reflector
CAMERA	SBIG ST-2000XM
EXPOSURE TIME	540 min
LOCATION	Nussbach, Austria

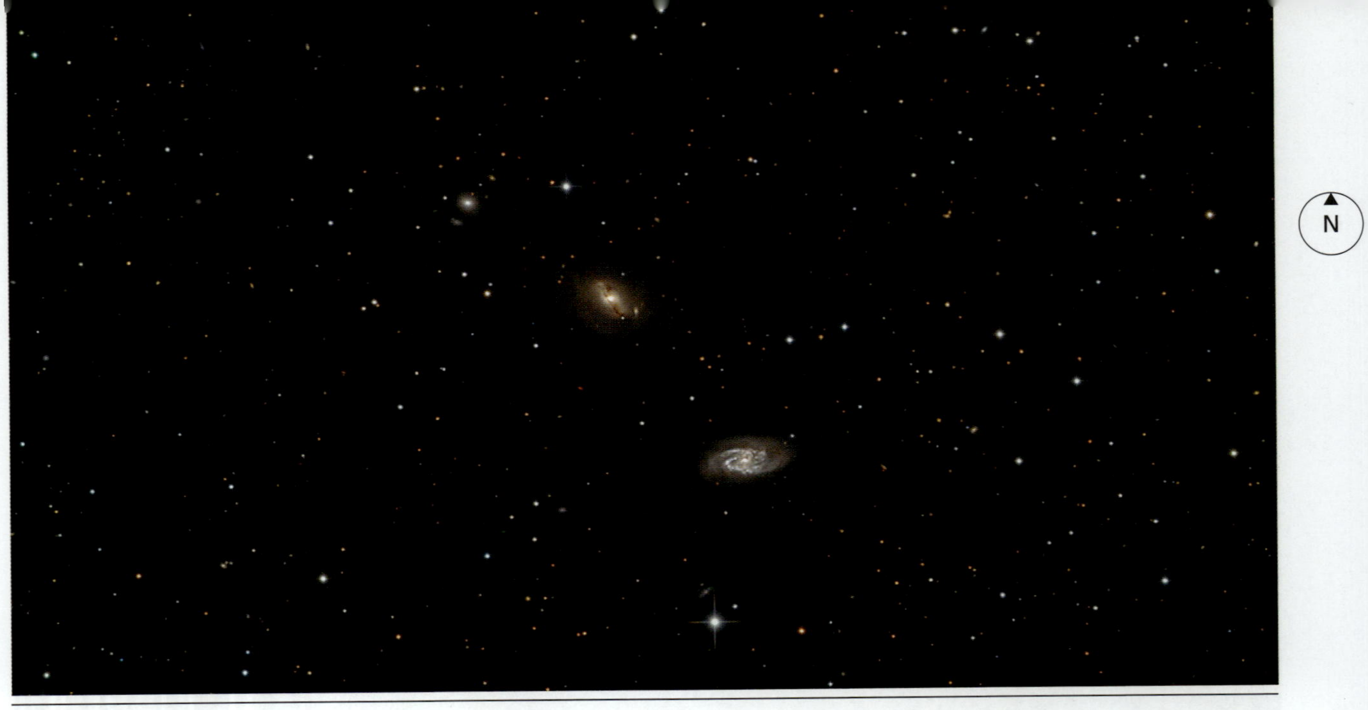

NGC 2964 Group

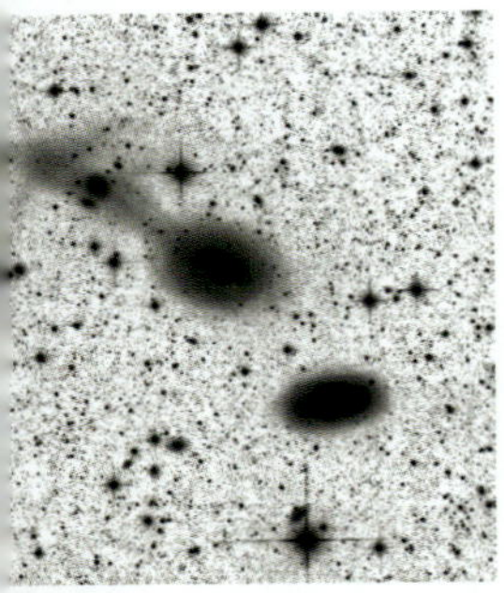

This group is located about 6° to the north of the head of the lion in the constellation of Leo. The images show the galaxy triplet NGC 2964, NGC 2968, and NGC 2970 (from lower right to upper left). All three galaxies lie at a distance of about 62 million light years and are mutually interacting because their transverse separations are only 150 000 light years (NGC 2964 to NGC 2968) and 100 000 light years (NGC 2968 to NGC 2970). NGC 2964 is a late-type SAB(r). The outer ends of its spiral arms merge into a diffuse halo whose luminosity distribution appears rectangular, an indication of tidal interaction with the galaxy NGC 2968 to its upper left. The form of NGC 2968 reminds one of an E/S0 type, whereby it is not clear if the dust lane crossing the galaxy belongs to NGC 2968 itself. It could also be the product of a merger, so that in reality the remains of the destroyed precursor galaxy are observed. However, since no rotation velocity can be measured for this dust lane, which would be expected in a spiral galaxy, it can be assumed that the dust lane in NGC 2968 stems from a spiral galaxy which has fallen in.

The contrast-enhanced image reveals a structure consisting of two faint tidal tails. One tail points from NGC 2968 towards NGC 2970 and extends beyond it, ending in a concentration after 114 000 light years. The width of this tail is less than 16 000 light years across its entire length. The second tidal tail is circular and passes through the concentration in the first, straight tail. The arc covers about 50° and thus also more than 100 000 light years. It appears that its centre lies in NGC 2968. No connection is seen between NGC 2968 and NGC 2964.

Stellar populations are born and die within the tidal tail. An indication of this could be the supernova SN1970L which was discovered in the material bridge between NGC 2968 and NGC 2970. However, since no active star-forming regions are observed in this bridge, the question arises as to how a supernova could take place there. An alternative explanation is hyper-velocity stars. These are high-speed stars which are formed in the galaxy but accelerated so strongly by tidal forces that they are ejected from the galaxy. In the case of SN1970L, one calculates a mass of 15 solar masses for the progenitor star and a velocity of more than 1000 km/s. During its stellar lifetime, which lasted only 13 million years, the star thus travelled 50 000 light years through half of the bridge of material before it exploded.

OBJECT	NGC 2964
CONSTELLATION	Leo
RIGHT ASCENSION	$09^h\ 42^m\ 54^s$
DECLINATION	+31° 50′ 51″
BRIGHTNESS	12 mag
TYPE	SAB(r)bc
PHOTOGRAPHERS	Bernhard Hubl
TELESCOPE	305-mm reflector
CAMERA	SBIG ST-2000XM
EXPOSURE TIME	1344 min
LOCATION	Nussbach, Austria

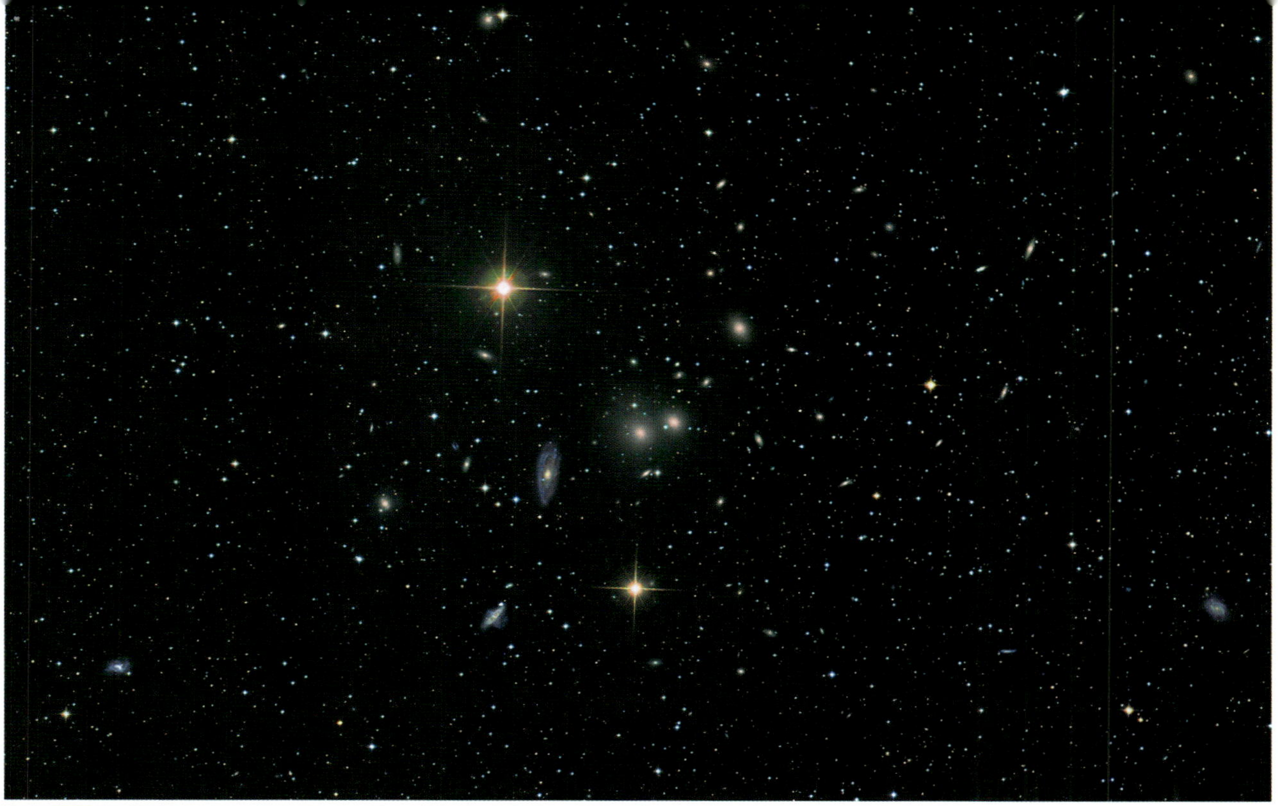

Abell 1060

Along with the Virgo Cluster and the Centaurus Cluster (Abell 3526), the Hydra Cluster Abell 1060 is one of the three large clusters near the Milky Way. It consists of 157 members and covers about 2° on the sky. The fifth-magnitude foreground star is a red giant at a distance of 490 light years. The distance to Abell 1060 is listed at 176 million light years. The Hydra Cluster has a spherical form and is not perturbed by other clusters or supercluster filaments which makes it a preferred object of study. Its isolation also helps in deciding which individual galaxies are members.

There are three large galaxies at the core. These are the two ellipticals NGC 3309 (E3) and NGC 3311 (cD) and the Sab spiral galaxy NGC 3312. The two E-type galaxies are close to each other, separated by only 1.7′, the larger NGC 3311 to the left of NGC 3309. At 3.9′, its halo is larger than that of its neighbour. Both galaxies show no obvious signs of interaction and their nominal difference in distance of 11 million light years seems realistic, also taking into account the relative velocity (250 km/s). NGC 3312 is the brightest galaxy in the Hydra region, and its angular size of 3.3′ corresponds to a huge physical diameter of almost 170 000 light years.

Although Abell 1060 is described in studies as a well isolated cluster, the relative velocities of its members deviate significantly from a Gaussian normal distribution. The histogram of velocities has no clear maximum, but rather a flat plateau at a mean velocity of 3400 km/s. The tails of the distribution reach as far as ±1500 km/s. There is thus a substructure within the Hydra Cluster, which is explained by the merger of two clusters. More exactly, the cluster is at the end of this merger process. The direction of motion of the two precursor clusters was along the line of sight, which explains the observed distribution of velocities. NGC 3312 is part of the closer subcluster, NGC 3309 to the right part of the farther subcluster, and the cD galaxy NGC 3311 is in between.

OBJECT	Abell 1060
CONSTELLATION	Hydra
RIGHT ASCENSION	$10^h\ 36^m\ 42^s$
DECLINATION	−27° 31′ 28″
BRIGHTNESS	–
TYPE	Galaxy cluster
PHOTOGRAPHERS	Josef Pöpsel, Stefan Binnewies
TELESCOPE	600-mm reflector
CAMERA	SBIG STL-11000
EXPOSURE TIME	120 min
LOCATION	Amani Lodge, Namibia

NGC 3190 Group

The image shows shows all four large members of Hickson Compact Group 44 in the constellation of Leo. At the upper edge is the E2 galaxy NGC 3193 and 5.5′ below lies NGC 3190; 4′ to the west is NGC 3187. A further 10′ below is NGC 3185. All four galaxies lie within a circle 16.4′ in diameter, which corresponds to 280 000 light years. This galaxy group is the closest Hickson group and is at a mean distance of 59 million light years. Its velocity dispersion, i.e. the distribution of the measured recession velocities, of 360 km/s exceeds the Hickson isolation criterion, which assumes a maximum of 200 km/s, whereby NGC 3187 is the outlier. If this galaxy is ignored, the velocity dispersion decreases to 182 km/s and thus half of the original value of all four galaxies. The galaxy group HCG 44 is thus mainly composed of the three galaxies NGC 3193, NGC 3190, and NGC 3185.

The elliptical galaxy NGC 3193 has a regular form and a few globular clusters can be recognized in its halo. Its diameter of 85 000 is around the average value. Much larger is the Sa galaxy NGC 3190, which has a diameter of 104 000 light years and an irregular dust lane. The lane forks below the core of the galaxy and follows a spiral arm outwards and then upwards. Since there is no corresponding spiral arm on the other side, this asymmetric perturbation can be interpreted as part of a strongly warped disc. This is surrounded by a torus of diffuse tidal tails. In addition, the spiral arm has dim extensions which point upward and downward and thus reach into the polar regions of NGC 3190. This gives the galaxy an Ω shape. The galaxy NGC 3187 located at the side is often described as a barred spiral. This classification is questionable, though, because although there is a detailed structure in the central regions where some OB associations with young stars and H II regions are visible, there is no bar structure going all the way to the nucleus. The impression of the bar is due solely to the form of the outer regions. They can be interpreted as two tidal tails pointing in opposite direction out of the spiral plane. The southern-most galaxy NGC 3185 is a ring galaxy formed in a merger.

Various interaction indicators are used to describe the evolution of a galaxy group. The most prominent are the tidal tails, envelopes of atomic hydrogen visible in the radio, and the hot gas between the galaxies visible in X-rays. There are also many stars in this space which have been pulled out of the galaxies and contribute a dim, diffuse glow to the optical emission of the galaxy. An indication of the existence of such diffuse light are planetary nebulae (PN), which due to their emission-line spectrum are easy to find in the group. 12 such PN have been found in HCG 33 which are not part of the galaxies but rather belong to the group itself. The contribution of the diffuse light to the surface brightness in HCG 44 is about 30 mag/arcsec^2. This rather low value is an indication that HCG 44 is only at the beginning of its evolution as a group.

OBJECT	NGC 3190
CONSTELLATION	Leo
RIGHT ASCENSION	$10^h\ 18^m\ 06^s$
DECLINATION	+21° 49′ 56″
BRIGHTNESS	11.3 mag
TYPE	SA(s)a pec edge-on
PHOTOGRAPHERS	Adam Block
TELESCOPE	800-mm reflector
CAMERA	SBIG STX-16803
EXPOSURE TIME	320 min
LOCATION	Mount Lemmon SkyCenter/ University of Arizona, USA

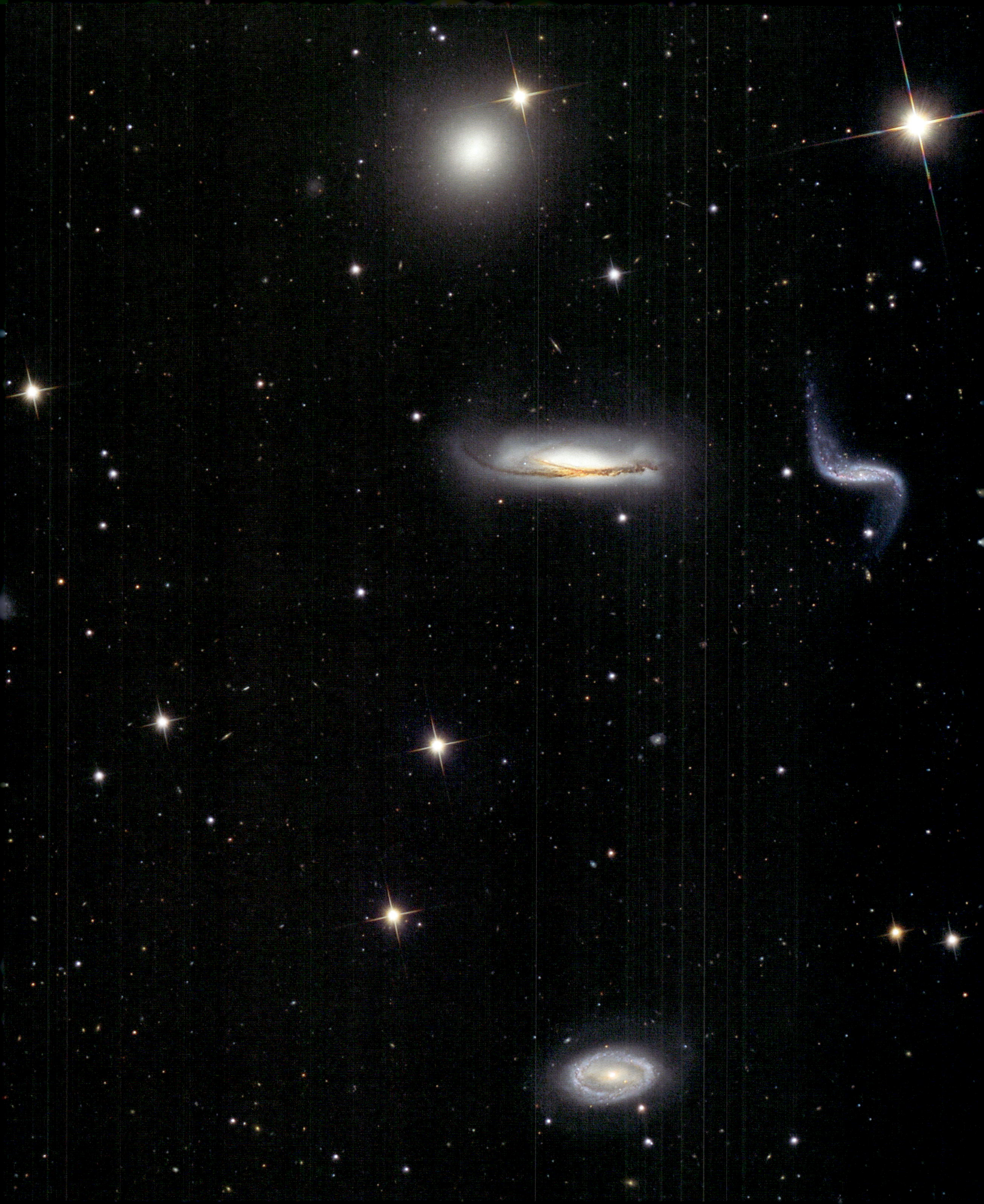

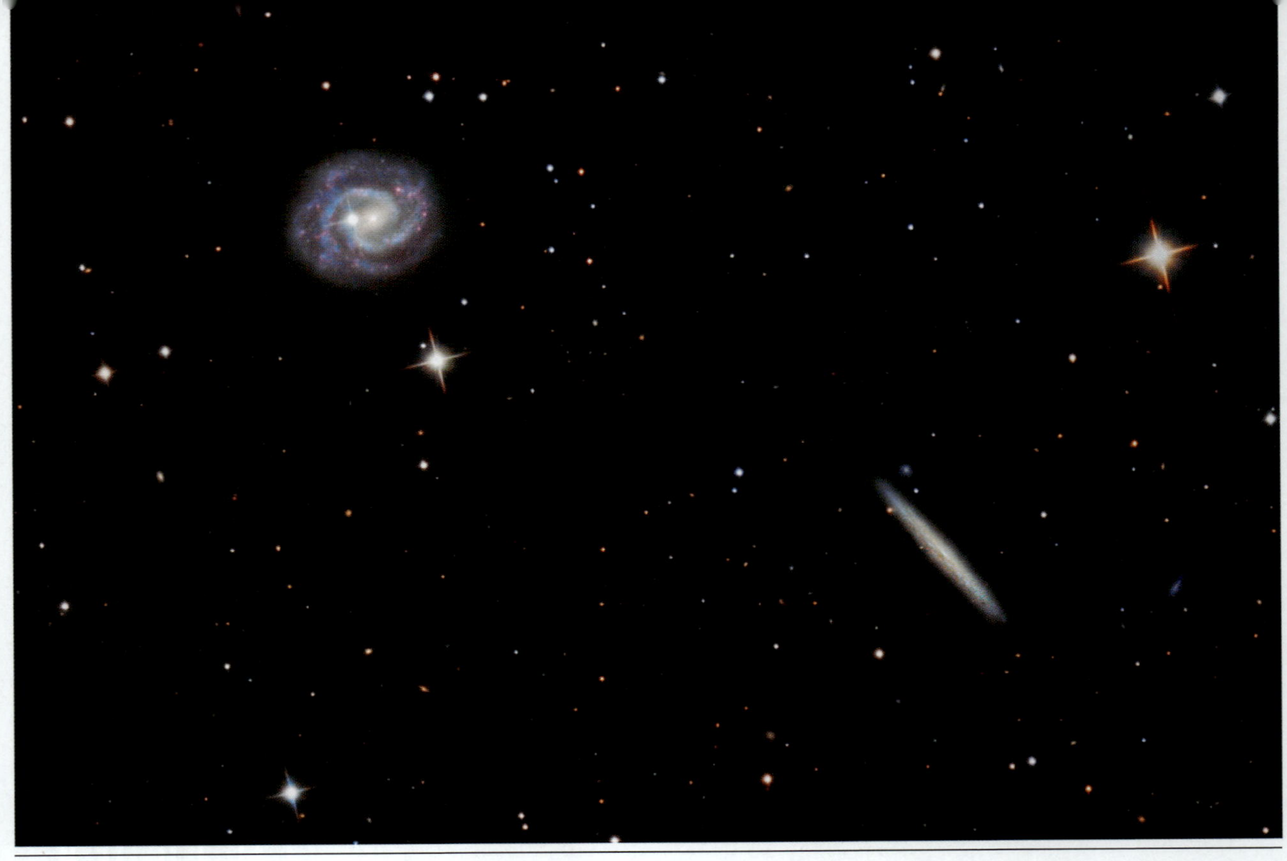

Holmberg 224

The galaxy pair consisting of NGC 3507 and NGC 3501 is known as Holmberg 224. The SB(s)b galaxy NGC 3507 is at the upper left in the image and as a grand-design spiral one can follow both spiral arms from the ends of the bar over 1.5 revolutions. The bar is easily seen and its yellowish nucleus contrasts with the colour of the arms. In the outer regions of the left, eastern spiral arm one sees the hint of a 120° kink and the corresponding Vorontsov-Velyaminov row. The luminosity distribution of NGC 3507 indicates an inclination of about 25°. 12′ to the southwest of NGC 3507 is the Scd galaxy NGC 3501 which is seen almost edge-on. While the distance to NGC 3507 is given as 40 million light years, one finds a value of 47 million light years for NGC 3501. The difference between the radial velocities of the two galaxies of 140 km/s could be interpreted as relative motion, but in that case their minimum distance of 150 000 light years would cause a tidal interaction. Since no corresponding perturbation of the morphology of the pair is visible, one galaxy is probably located behind the other in space.

At the end of the 1930s in a study of galaxy pairs, Erik Holmberg discovered that the colours of galaxies in a physical pair of the same type (EE or SS) are very similar. He made use of statistical methods to study the relation between pairs regarding their type and nearby surroundings. This effect of similar colours in similar types in close pairs is explained today by a similar evolutionary history within almost the same environment. Although Erik Holmberg used just 32 galaxy pairs to test his theory, modern campaigns such as the Sloan Digital Sky Survey (SDSS) impressively support the effect. The SDSS Data Release 4 has 1100 close pairs. The pair Holmberg 224 confirms the Holmberg effect indirectly: the colours of NGC 3507 and NGC 3501 are different, and NGC 3507 is much bluer. This would seem to contradict Holmberg's theory, but would not be in conflict if the galaxies are not a physical pair.

OBJECT	Holmberg 224
CONSTELLATION	Leo
RIGHT ASCENSION	$11^h\ 03^m\ 06^s$
DECLINATION	+18° 03′ 44″
BRIGHTNESS	–
TYPE	Galaxy group
PHOTOGRAPHERS	Wolfgang Ries, Stefan Heutz
TELESCOPE	460-mm reflector
CAMERA	SBIG ST-10XME
EXPOSURE TIME	768 min
LOCATION	Altschwendt, Austria

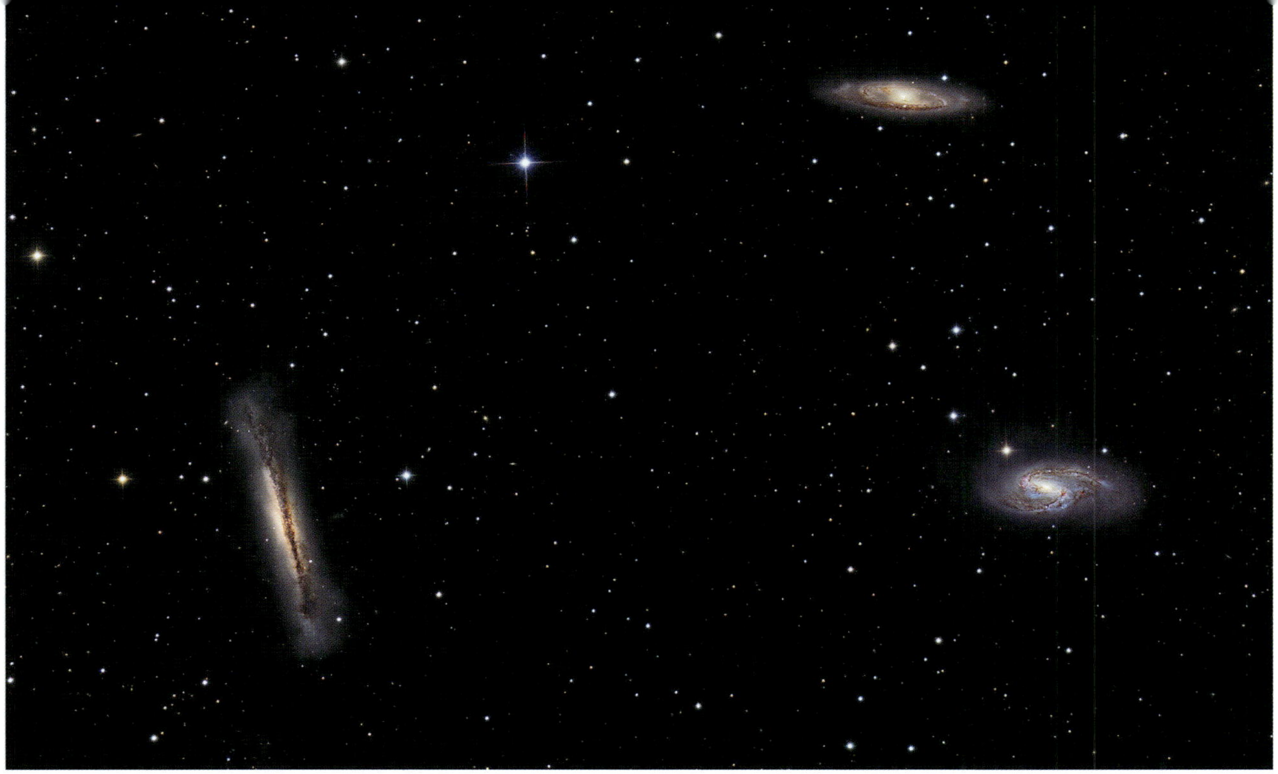

M 65 / M 66 / NGC 3628

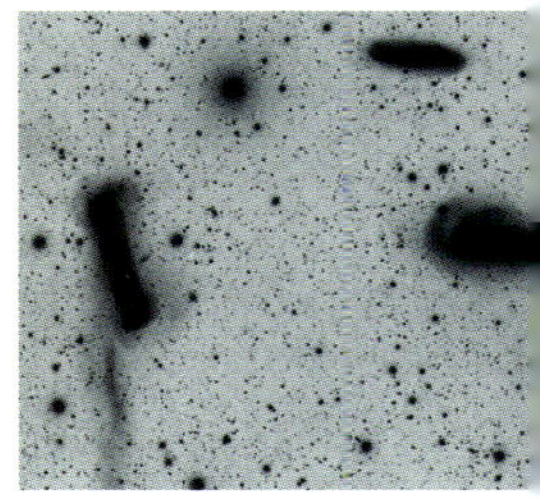

The three galaxies M 65 and M 66 (upper and below right) and NGC 3628 (left) make up the Leo Triplet. The SAB(s)b galaxy M 66 (NGC 3627) is closest at only 29 million light years. M 65 (NGC 3623) is next at 32 million light years. M 65 is of type SAB(rs) and thus similar to M 66. The inner regions of the spiral arms form an almost complete circle. The difference in the recession velocities of M 65 and M 66 is only 80 km/s, which is why it is assumed that their direct separation is much smaller than three million light years. The minimum direct separation follows from the projection of their angular separation of 20′ and is about 177 000 light years.

The angular separation between M 65 and M 66 and NGC 3628 to the north is about 35′, which corresponds to more than 310 000 light years and a relative velocity of 116 km/s to M 66. The SA galaxy has a much more unusual form than the Messier pair. The morphology of the halo reminds one of a dog's bone. The image also shows rudiments of a tidal tail extending from the northeastern end of the halo and continuing far beyond the edge of the image. Deep images reveal the length of this slightly curved tail of gas, dust, and stars as 330 000 light years, which is more than twice the diameter of NGC 3628. This tidal tail contains 15 per cent of the total mass of atomic hydrogen in the galaxy.

The length of the tidal tail also suggests an interaction with one of the other neighbouring galaxies. Since, in contrast to M 65, M 66 is surrounded by diffuse structures in the region of the outer spiral arms, an approach to M 66 is seen as the most probable. The bluer colour of both galaxies also indicates more star-formation regions and thus young stars than in M 65. Radio observations of the triplet also show that 21-cm emission of atomic hydrogen surrounds the galaxy NGC 3628, follows the tidal tail, and contains three condensations. Their radio intensity is about 1/30 the radio luminosity of the galactic centre. Particularly conspicuous is that in comparison M 66 emits less in the radio, although this radio source is not concentrated at the centre but rather is extended to the east. Since M 65 doesn't show any such peculiarities, this supports an approach between M 66 and NGC 3628.

OBJECT	M 65 / M 66
CONSTELLATION	Leo
RIGHT ASCENSION	$11^h\ 18^m\ 56^s$ / $11^h\ 20^m\ 15^s$
DECLINATION	+13° 05′ 33″ / +12° 59′ 30″
BRIGHTNESS	10.3 mag / 9.7 mag
TYPE	Galaxy group
PHOTOGRAPHERS	Johannes Schedler
TELESCOPE	400-mm reflector
CAMERA	SBIG STL-16803
EXPOSURE TIME	1102 min
LOCATION	Wildon, Austria

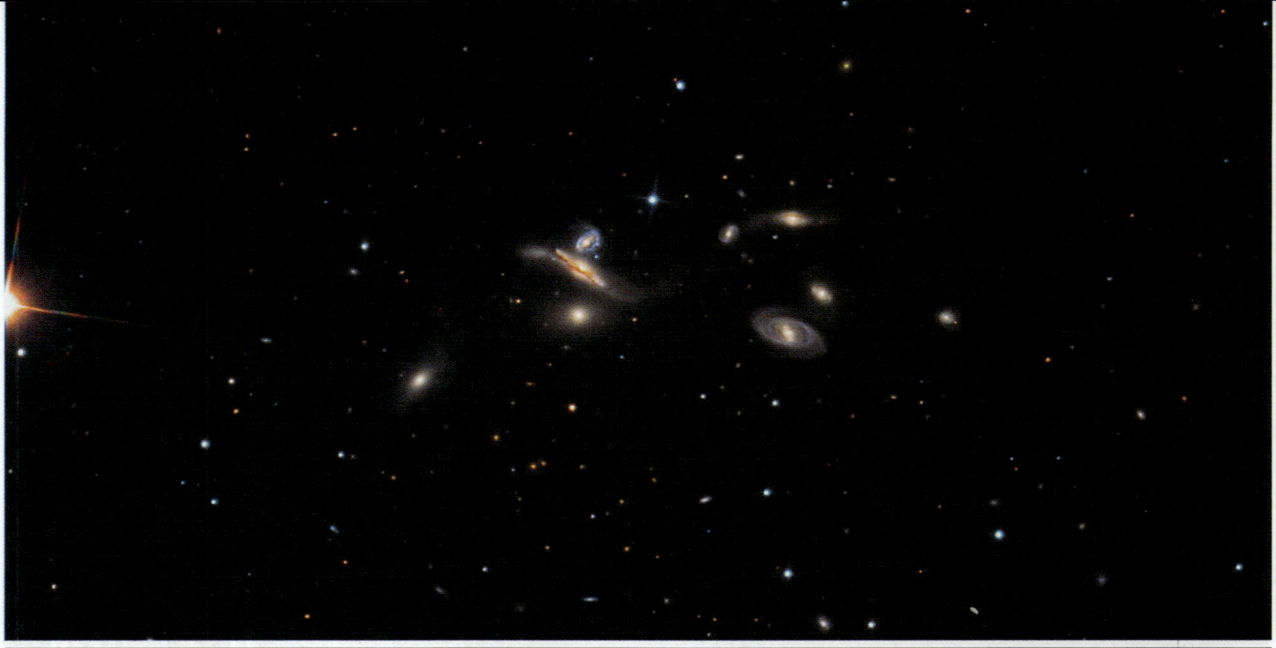

Copeland's Septet

The image shows Copeland's Septet, a compact galaxy group in Leo lying about 8° northwest of Denebola. The term was first used by Gérard-Henri de Vaucouleurs in his *Second Reference Catalogue of Bright Galaxies* (RC2, 1976) and refers to Ralph Copeland, who had discovered the group in 1874. His observational data were also used by the astronomer Johann Louis Emil Dreyer in constructing the *New General Catalogue*.

All galaxies in the septet have NGC numbers and their luminosities lie in the range 13.6–15.2 mag. Members of this group of seven are NGC 3571 (E4) at the far left, the trio in the middle, consisting of NGC 3754 (SBc, above), NGC 3753 (SB, centre), and NGC 3750 (SAB0, below), and the second trio 2.4′ away. This consists of NGC 3748 (SB0, above), NGC 3754 (SB(s)0, centre), and NGC 3746 (SB(r)b, below).

The association is completed by a further, eighth galaxy. This is the small BBb galaxy with an angular size of 0.33′ × 0.28′ which is seen to the left of NGC 3748. Its redshift clearly indicates that it is a member of the galaxy group. However, with a luminosity of just 17.4 mag it was not visible to Ralph Copeland.

The entire galaxy field measures 5′ × 2′. The two largest galaxies are the spirals NGC 3753 at 1.7′ × 0.5′ and NGC 3746 at 1.1′ × 0.5′. Due to its compactness, the group is also catalogued as HCG 57. The redshifts of all eight galaxies deviate by only ±370 km/s from the mean of 9087 km/s; the distance to the group is 400 million light years.

NGC 3753 shows strong signs of a series of interactions which have led to a deformation of the galactic plane. Several tidal tails were formed which now surround the galaxy. In addition, there are extremely faint bridges which at high contrast reveal the connections between the members. This is especially prominent in NGC 3753 and NGC 3754 above it in which one spiral arm leads into a tidal tail. The radial velocity of NGC 3753 deviates most strongly from the value of the group and explains its exceptional dynamics. It is the high-velocity member of the group and tears out gas and dust from the halos of other nearby group members. This leads to star formation in the central regions of the corresponding galaxies and in the dense parts of the tidal tails. The UV radiation of young stars increases the temperature of the dust there to about 40 K, causing the emission of infrared radiation. In the 60–100 μm band, the galaxy group has only one infrared source, about two arcminutes in size, in the centre of which are the galaxies NGC 3754 and NGC 3753. The other galaxies are not seen in this part of the spectrum.

OBJECT	NGC 3750
CONSTELLATION	Leo
RIGHT ASCENSION	$11^h 37^m 52^s$
DECLINATION	+21° 58′ 27″
BRIGHTNESS	14.9 mag
TYPE	SAB0-
PHOTOGRAPHERS	Adam Block
TELESCOPE	800-mm reflector
CAMERA	SBIG STX-16803
EXPOSURE TIME	330 min
LOCATION	Mount Lemmon SkyCenter/ University of Arizona, USA

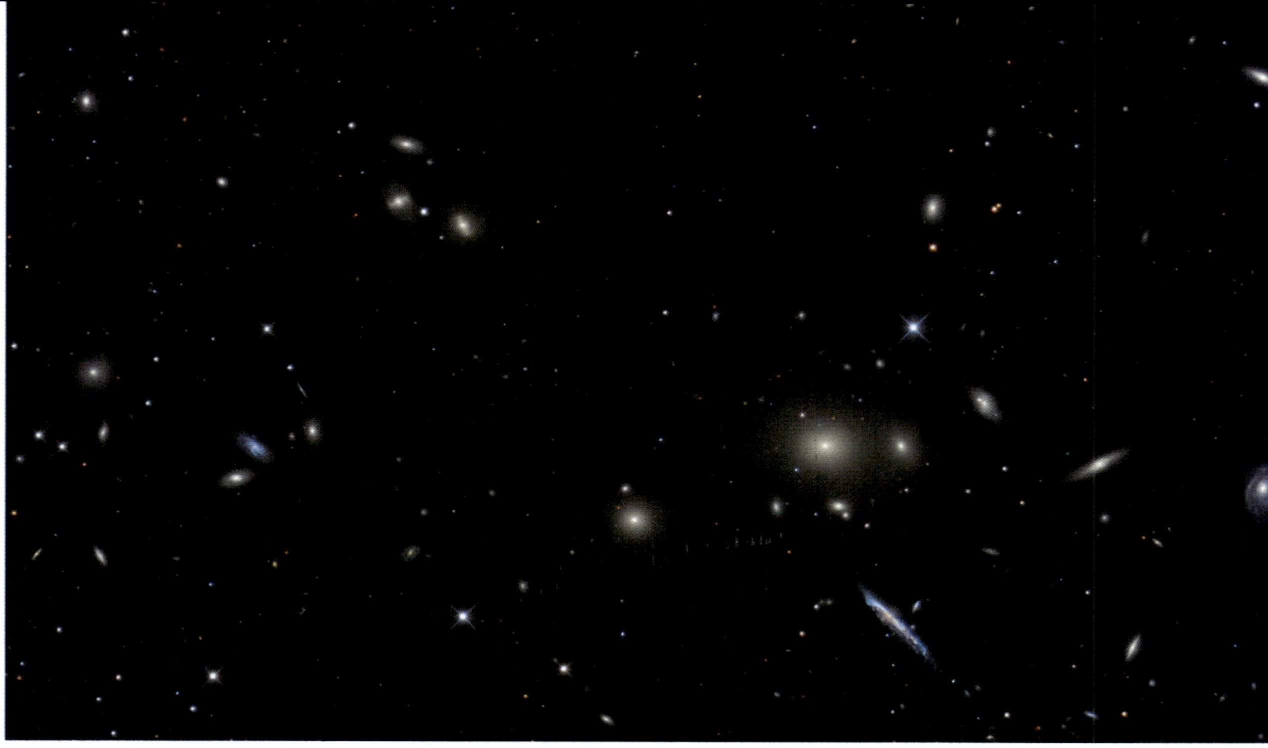

Abell 1367

The large elliptical galaxy NGC 3842 is in the dense part of the galaxy cluster Abell 1367, also known as the Leo Cluster. NGC 3842 is the brightest galaxy in this cluster, which is 300 million light years from the Milky Way and together with the Coma Cluster one of the two largest clusters in the Coma Supercluster. Within a radius of 1° around NGC 3842 lie 166 galaxies whose recession velocities deviate by only ±10 per cent from that of NGC 3842. Thus all these galaxies can be considered part of the cluster. The visual luminosities of the faintest members, which are dwarf galaxies, are dimmer than 19 mag. In contrast, NGC 3842 is more than 300 times as bright as these faint cluster members.

The size of the bright inner shell of NGC 3842 is given as 1.4′ × 1.0′ and corresponds to 122 000 light years. The diffuse halo, however, extends more than twice as far in the image. Below NGC 3842 one notices the spiral galaxy UGC 6697 whose blue colour contrasts strongly with that of the other galaxies. Despite the low angle at which its spiral plane is seen, it can be seen that tidal forces have destroyed its spiral structure and a tidal tail in the direction of NGC 3842 has already formed. The forthcoming destruction of UGC 6697 is also noticeable in high star-formation activity and the enormous length of the disc, which reaches almost 200 000 light years.

The comparatively large number of spiral galaxies near the centre and the observed double structure of the cluster are indications that Abell 1367 formed by the merger of several small clusters. It is expected that NGC 3842 will form a cD galaxy as the cluster further evolves. Although NGC 3842 is thus only at the beginning of its growth, the supermassive black hole at the centre of the galaxy already attains a record value. Astronomers have studied the spectra of stars in the immediate vicinity of the black hole and from the Doppler shift of their spectral lines calculated a mass of 9.7 billion solar masses. Thus the black hole in NGC 3842 is about 2500 times as massive as the black hole at the centre of the Milky Way. This record-breaking galactic central mass also supports the theory that supermassive black holes are not formed by steady accretion of mass but rather by the mergers of galaxies.

OBJECT	Abell 1367
CONSTELLATION	Leo
RIGHT ASCENSION	$11^h\ 44^m\ 37^s$
DECLINATION	+19° 45′ 32″
BRIGHTNESS	–
TYPE	Galaxy cluster
PHOTOGRAPHERS	Adam Block
TELESCOPE	800-mm reflector
CAMERA	SBIG STX-16803
EXPOSURE TIME	1140 min
LOCATION	Mount Lemmon SkyCenter/ University of Arizona, USA

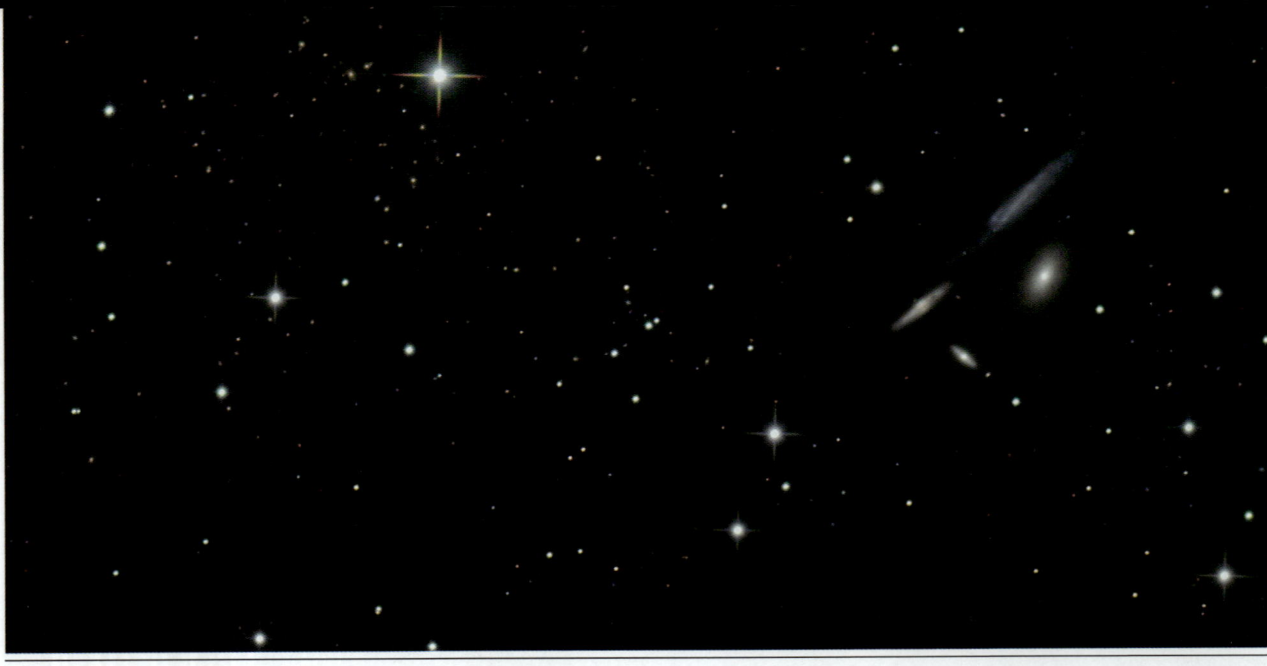

Hickson Compact Group 61

The Coma Cluster and the cluster Abell 1367 are the largest structures in the Coma Supercluster. Both clusters lie at a distance of about 300 million light years from the Milky Way and their separation of 18° corresponds to a projected distance of 95 million light years. Between the two clusters are several galaxy groups and smaller clusters. Also located there is the group of four galaxies visible in the right half of the image. At the top is the spiral galaxy NGC 4173 which due to its size of 5′ × 0.7′ and its blue colour contrasts strongly with the other three galaxies. Two arcminutes below and to the right is the S0 galaxy NGC 4169, sideways from it the SBc galaxy NGC 4175, the edge-on orientation revealing a dust lane. At the bottom of the group is the smallest galaxy, NGC 4174, of type S0 and size of only 0.82′ × 0.50′.

All four galaxies are classified as HCG 61. Due to its form, the group is also known as the Box. Analyses of the redshifts of the group members indicate that only three of them are close neighbours. The galaxy NGC 4173 is not a member of the group and is at a distance of 50 million light years. The other three galaxies lie at a mean distance of 170 million light years with relative velocities of ±130 km/s.

The near galaxy NGC 4173 appears larger, but its diameter is only 73 000 light years. It is thus smaller physically than NGC 4169 and NGC 4175. Their absolute diameters are 88 000 light years and 94 000 light years. The morphology of NGC 4173 is most similar to a spiral galaxy, although only a diffuse, dim plane without nucleus or spiral arms is visible. At its lower, southeastern end are conspicuous active regions which extend into the tidal tail which is separating from the galactic plane. The bottom-most and smallest galaxy NGC 4174 has a further peculiarity: it is surrounded by a faint polar structure, making it a candidate for a polar-ring galaxy. Typical of these rings is a radio signature due to large amounts of atomic hydrogen gas. Radio observations of NGC 4174 detect such emission, though this is not confined to the galaxies but rather indicates hydrogen gas filling out the space between the members of the Box.

At the upper left of the image is the 8.9-magnitude star SAO 82192. Within a radius of 5′, more than 100 galaxies are visible, most with luminosities in the range 18–20 mag. They belong to the galaxy cluster Abell 1495. This lies at a distance of 1.95 billion light years and is thus almost 12 times as far from the Milky Way as the HCG 61 quartet.

OBJECT	NGC 4173
CONSTELLATION	Coma Berenices
RIGHT ASCENSION	$12^h\ 12^m\ 21^s$
DECLINATION	+29° 12′ 25″
BRIGHTNESS	13.6 mag
TYPE	SBd
PHOTOGRAPHERS	Richard Müller
TELESCOPE	320-mm reflector
CAMERA	Artemis 4021
EXPOSURE TIME	760 min
LOCATION	Lohmar, Germany

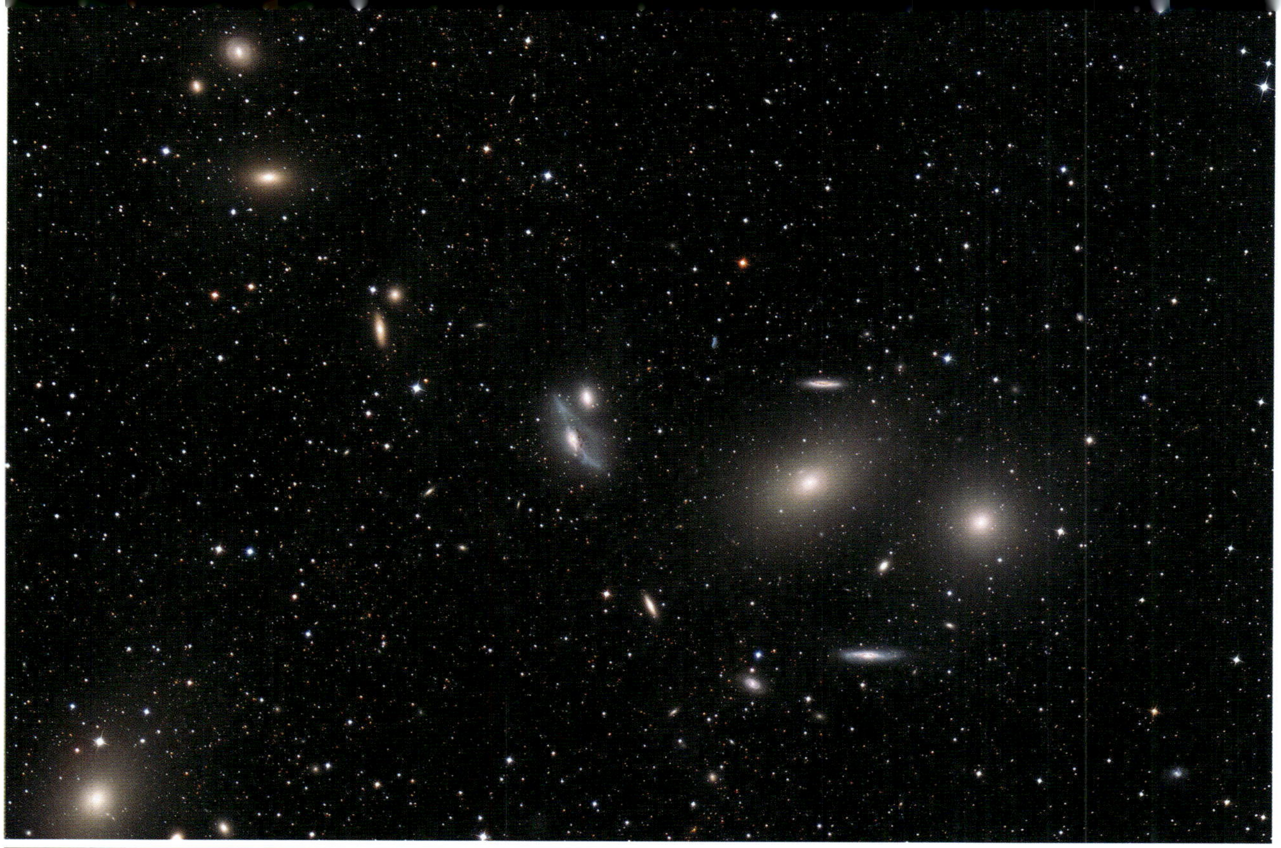

The Virgo Cluster

The image shows a view of the northern part of the Virgo Cluster. Of the 16 Messier galaxies which belong to it, visible in the image are M 84 (NGC 4374), M 86 (NGC 4406), and M 87 (NGC 4486). M 84 is the right end of a chain which leads in almost equal steps to M 86, Arp 120 (NGC 4438), and NGC 4435, then to IC 4461 and IC 4458 and finally to IC 4473. This chain, which extends a bit farther, is known as Markarian's Chain. The distance between M 84 and M 86 is 17′ and the two E/S0 galaxies form an equilateral triangle with the dimmer SAb galaxy NGC 4388, the small elliptical galaxy NGC 4387 at the centre. About 8′ above M 86 is NGC 4402, in which a concise dust lane is visible despite the small scale of the image. At the lower left edge of the image the well known cD galaxy in the Virgo Cluster, M 87, is visible. This large elliptical galaxy is about 1° to the southeast of the row of galaxies mentioned above and is at the centre of the Virgo Cluster.

The Virgo Cluster is the nearest large cluster to the Local Group and dynamically connected with it. Its distance is about 54 million light years and the cluster radius is more than 6.5 million light years. The Virgo Cluster consists of more than 2000 galaxies and dominates our intergalactic neighbourhood. It is the physical centre of the Local Supercluster (Virgo Supercluster or Coma–Virgo Supercluster) to which the Milky Way also belongs. The enormous mass concentration of the Virgo Cluster affects the recession velocities of galaxies surrounding us so strongly that a flow of galaxies into the Virgo Cluster is detectable. In the case of the Local Group, this Virgo-centric flow is responsible for a velocity component of 100–400 km/s in the direction of Virgo. Taking into account the expected orbit of the Local Group towards the Virgo Cluster, it can be calculated that it will first pass by the Virgo Cluster, but then be decelerated by and finally fall into the cluster.

OBJECT	Virgo Cluster
CONSTELLATION	Virgo
RIGHT ASCENSION	$12^h\ 30^m\ 47^s$
DECLINATION	+12° 20′ 13″
BRIGHTNESS	–
TYPE	Galaxy cluster
PHOTOGRAPHERS	Makis Palaiologou, Stefan Binnewies
TELESCOPE	305-mm reflector
CAMERA	SBIG STL-11000
EXPOSURE TIME	465 min
LOCATION	Skinakas Observatory, Crete, Greece

NGC 4650 Group

The Centaurus Supercluster is our nearest supercluster. It occupies an elongated area of the sky with a length of 30° which covers the declination range between −30° and −45° and is thus near the plane of the Milky Way. In addition to several galaxy groups, the supercluster also contains the galaxy clusters Abell 3526, Abell 3565, and Abell 3581. The Centaurus Supercluster also has an almost linear form. The four Abell clusters lie above one another in the order above and the corresponding distances range from 145 million light years (Abell 3526) to 300 million light years (Abell 3581). It is, however, questionable whether this supercluster filament really extends so far into space. It could also be that Abell 3581 only coincidentally lies beyond the other three Abell clusters.

With about 1000 cluster members, there are a similar number of galaxies in Abell 3526 as in the Virgo Cluster. Since the Virgo Cluster doesn't have an Abell designation, the Centaurus Cluster is the nearest Abell cluster. Also note that behind the Centaurus Cluster is a second cluster at a distance of 200 million light years, which contains the galaxy NGC 4709. This is the second-brightest galaxy in the region, after the cD galaxy NGC 4696 in Abell 3526. Images of the Centaurus Supercluster thus always show a mixture of two galaxy clusters. A detailed examination of the Centaurus Supercluster shows that this picture is even more complex, since near the clusters Abell 3565 and Abell 3574 a further, massive concentration of about 20 galaxy clusters was found at a distance of about 650 million light years. This Supercluster, named after Harlow Shapley, is a popular object of study. The image, however, shows no galaxies in this cluster, since it shows only the northern area of Abell 3526, which is about 1° from the centre of the supercluster.

The image shows two large spiral galaxies. One is the SB(rs)a galaxy NGC 4650 in the lower half of the picture. In the upper right is NGC 4603, an SA(rs)bc galaxy whose angular diameter of 3.4′ is somewhat larger than that of NGC 4650. The angular separation of the two galaxies is 34′ and at their mean distance of 116 million light years this corresponds to a transverse separation of 1.1 million light years. The barred spiral galaxy NGC 4650 has two tidal tails which fade into a diffuse halo. Cloe by it lie galaxies with further clear signs of past interaction. 4′ below NGC 4650 is the polar-ring galaxy NGC 4650A. Its blue ring is seen edge-on and thus appears as a blue line, passing through the middle of the lenticular galaxy. Above NGC 4650 is the compact galaxy pair NGC 4622A which has already started to merge. Particularly prominent due to the human ability to recognize patterns is NGC 4622 which is located near the middle of the picture. This is due to two narrow, unusually well defined spiral arms which are connected to a complete inner ring. NGC 4622 is a ring galaxy which has formed from the merger of a spiral galaxy with a compact companion. This also explains the peculiar spiral-arm orientation of the galaxy. The outer spiral arms do not rotate more slowly with increasing radius as in most spiral galaxies. The measured velocity field of the spiral arms indicates that the outer regions of NGC 4622 rotate faster than the inner regions.

OBJECT	NGC 4650
CONSTELLATION	Centaurus
RIGHT ASCENSION	$12^h\ 44^m\ 20^s$
DECLINATION	−40° 43′ 55″
BRIGHTNESS	12.6 mag
TYPE	SB0/a(s) pec
PHOTOGRAPHERS	Stefan Binnewies, Josef Pöpsel
TELESCOPE	600-mm reflector
CAMERA	SBIG STL-11000
EXPOSURE TIME	110 min
LOCATION	Amani Lodge, Namibia

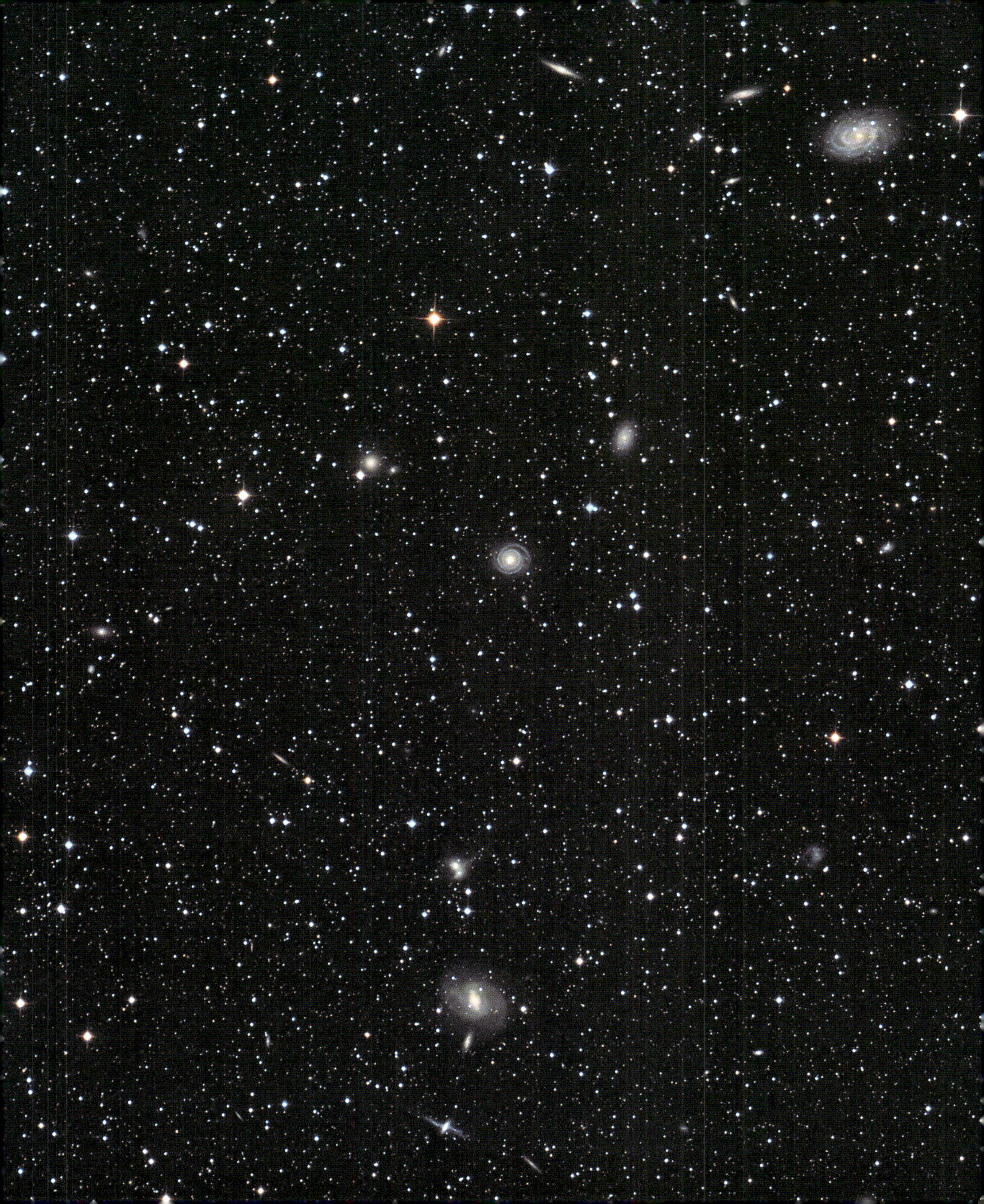

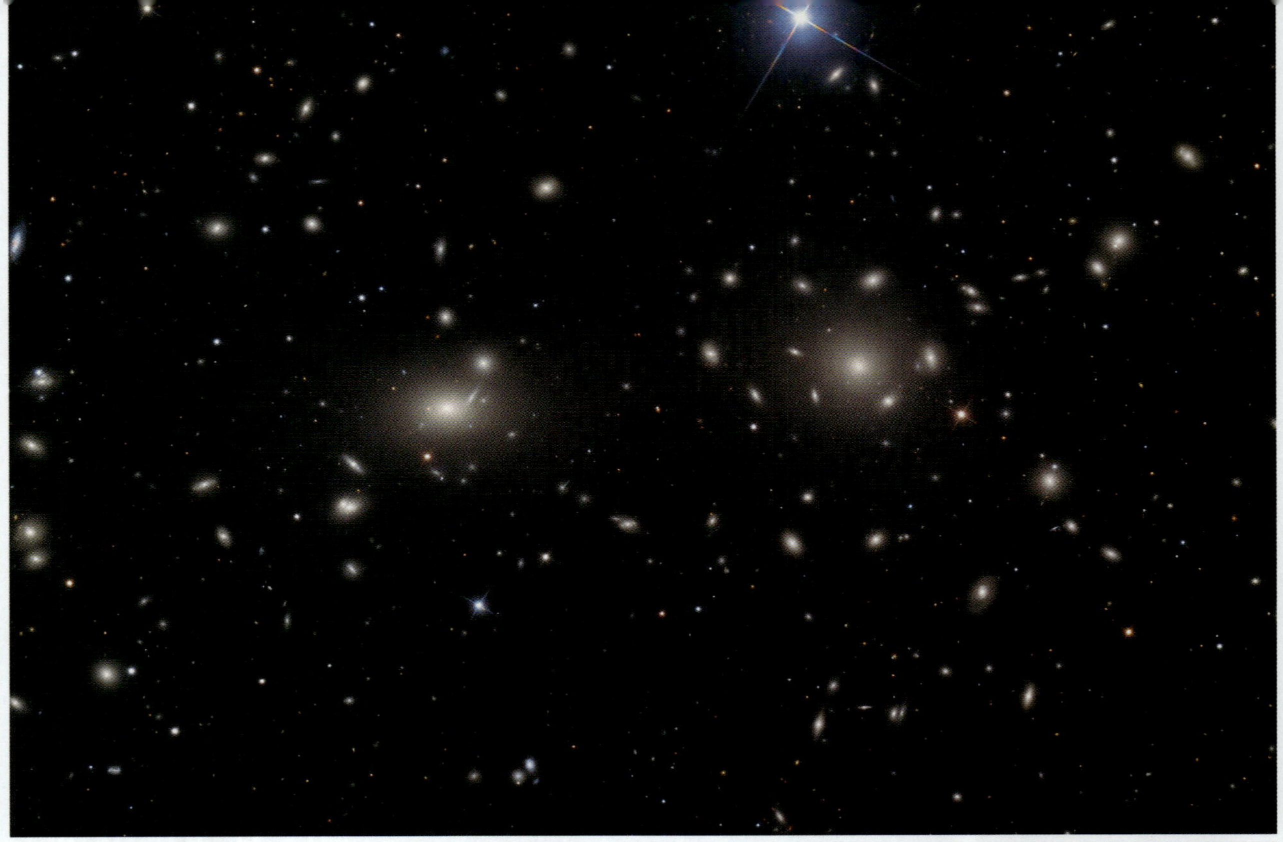

Abell 1656

The Coma Cluster (Abell 1656) consists of more than 3000 galaxies and is considered a rich galaxy cluster. It lies at a distance of 310 million light years and has an almost spherical form with a diameter of 20 million light years. The centre is dominated by the two large cD galaxies, NGC 4874 and NGC 4889, which are separated by 7′. In the image, NGC 4874 is at the right, to the west of NGC 4889. The size of NGC 4874 is given as 1.9′ × 1.9′ and the halo has the form of an ellipse with a uniform luminosity distribution. The neighbouring galaxy NGC 4889 is somewhat brighter and at 2.9′ × 1.9′ appreciably larger. Its halo has a flattened, almost boxy structure and stretches across a diameter of about 250 000 light years.

The field of the Coma Cluster contains about 26 active galaxies. This large number is, however, not a special property of the cluster itself, but is due to the fact that in the last few years it has been studied in detail via X-ray and radio observations.

The fame of the Coma Cluster is related to the history of studies of it. In 1933 the astronomer Fritz Zwicky published an article concerning the mass of this cluster. His paper began by examining the radial and relative velocities of the cluster members. He first determined the distribution of the relative velocities – the velocity dispersion. Since he also assumed that the system is gravitationally bound, he could assume the validity of the virial theorem. This describes a simple relation between the mean kinetic energy and the mean potential energy and allows the gravitationally active mass to be derived from the velocity dispersion. This mass is appreciably larger than the total mass of the luminous material in the cluster, which is just the sum of the visible galaxies. Fritz Zwicky termed this the missing mass in the cluster. It is necessary in order to explain the high velocities of the cluster galaxies. The term missing mass has been superseded by dark matter, which fills and stabilizes the Coma Cluster.

OBJECT	Abell 1656
CONSTELLATION	Coma Berenices
RIGHT ASCENSION	$12^h\ 59^m\ 49^s$
DECLINATION	+27° 58′ 51″
BRIGHTNESS	13.5 mag
TYPE	Galaxy cluster
PHOTOGRAPHERS	Adam Block
TELESCOPE	800-mm reflector
CAMERA	SBIG STX-16803
EXPOSURE TIME	700 min
LOCATION	Mount Lemmon SkyCenter/ University of Arizona, USA

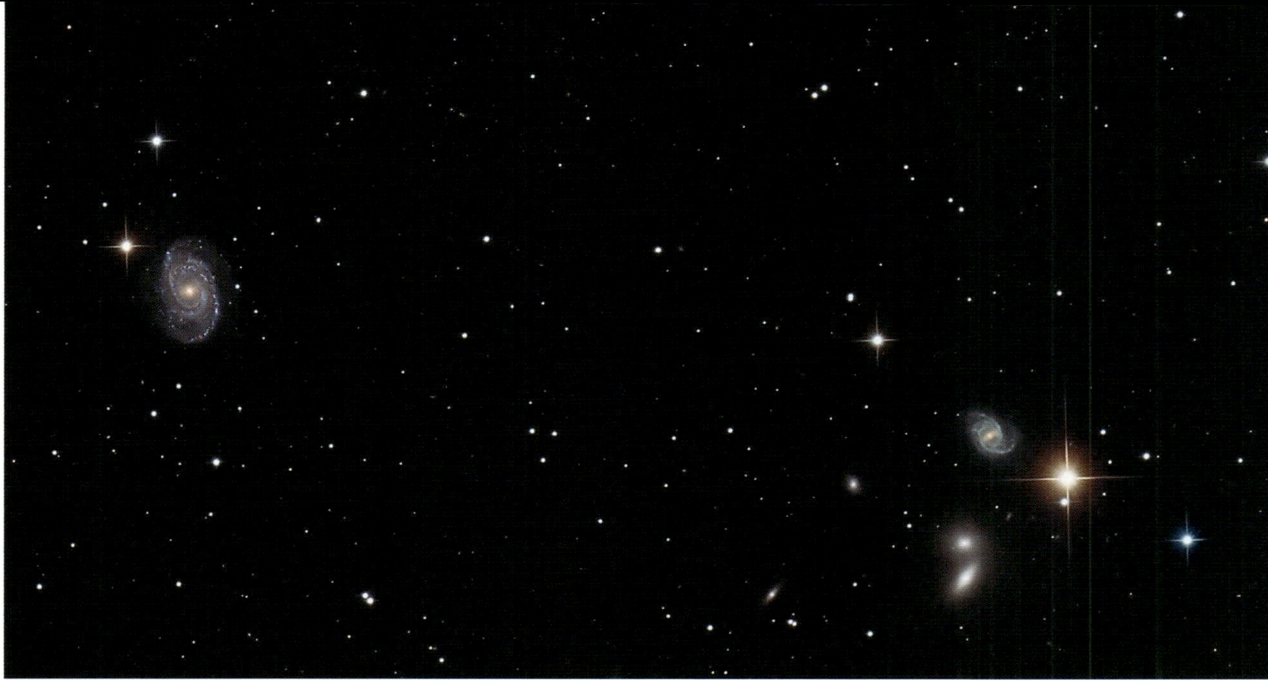

Hickson Compact Group 68

In 1982, Paul Hickson published a list of 100 compact galaxy groups which he had found by studying the red prints of the Palomar Observatory Sky Survey. The requirement was that at least four galaxies with luminosities within three magnitudes meet the isolation criterion for the group selection. This criterion is that no other bright galaxy is visible between radii corresponding to one and three times the diameter of the group.

The group HCG 68 consists of five galaxies which lie within a circle of radius of 9.2′. In the image these galaxies are near a 6.5-magnitude K5 star at the lower right. The SBb galaxy NGC 5350 is only two arcminutes away and the E/S0 systems NGC 5354 and NGC 5353 are somewhat below. To the left along the horizontal line separating the spiral galaxy from the other two is the fainter S0 galaxy NGC 5355; farther below is NGC 5358. Ignoring NGC 5354, the redshifts lie in the range 2320–2400 km/s and thus close to one another. At this distance, an arcminute corresponds to about 32 000 light years. Thus one can assume that tidal actions are taking place. The most conspicuous pair, consisting of NGC 5353 and NGC 5354, is probably not part of this process, since NGC 5354 at 120 million light years is 10 million light years farther away.

The left half of the image is dominated by NGC 5371, an SAB-type galaxy with many H II regions and areas of active star formation. At 118 million light years, this galaxy is at the same distance as NGC 5354 and indicates that HCG 68 is a subset of a linear association of galaxies pointing away from the Coma Cluster.

The H I map of the entire HCG 68 group is interesting. It displays a sharp H I spectral line between 2150 km/s and 2450 km/s. However, between the steep tails is a central depression. This indicates an H I deficit in the compact group, indicating that they contain less neutral hydrogen than comparable galaxies in the field. The reasons for this deficit are, on the one hand, gas loss due to tidal interactions and, on the other hand, star formation triggered by them, which uses up the gas. However, perhaps the most important reason for the H I deficit is the evolution of the H I distribution in the group. The luminosity of all galaxies within such a small space completely ionizes the neutral hydrogen, converting it into hot intragroup medium, which means that it does not radiate in H I.

OBJECT	Hickson Compact Group 68
CONSTELLATION	Canes Venatici
RIGHT ASCENSION	$13^h\ 53^m\ 41^s$
DECLINATION	+40° 19′ 07″
BRIGHTNESS	–
TYPE	Galaxy group
PHOTOGRAPHERS	Michael König
TELESCOPE	355-mm reflector
CAMERA	SBIG STL-6303
EXPOSURE TIME	290 min
LOCATION	Rimbach, Germany

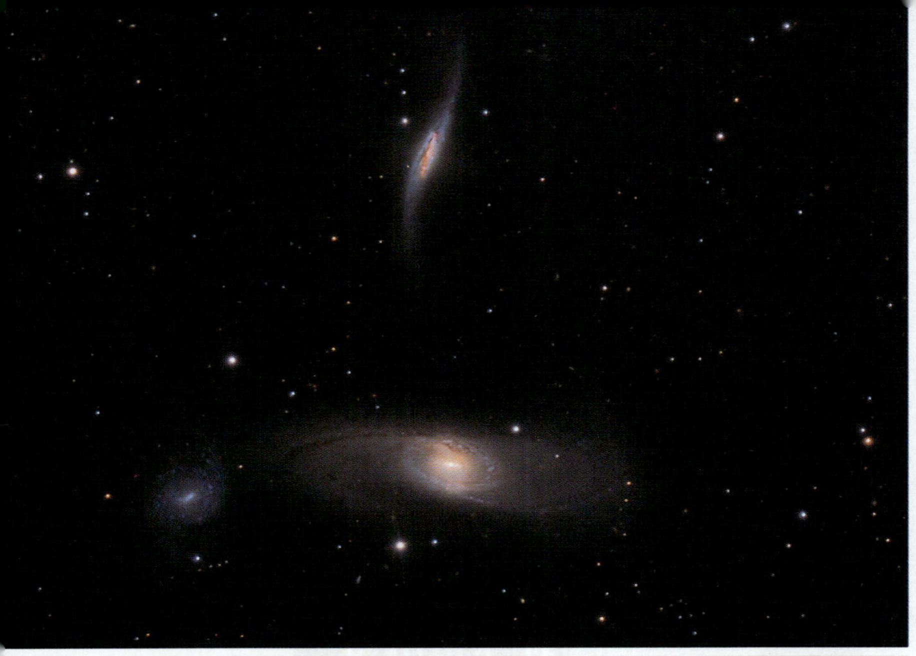

Arp 286

The galaxies NGC 5566, NGC 5569, and NGC 5560 form a triplet known as Arp 286 or Holmberg 360. The SAB(rs)cd galaxy NGC 5569 (left) is located 4.2′ from NGC 5566 while the SB(s)b galaxy NGC 5560 (above) is 5′ away from NGC 5566. Due to its size, the triplet in an extension of the Virgo Cluster is a popular target for visual observers. The orientation of these three group members can be studied via data from astronomical catalogues and via the visible luminosity distribution in deep exposures. The redshifts of the two smaller galaxies differ by only 50 km/s – they are thus very probably at the same distance of about 78 million light years. In contrast, the larger galaxy NGC 5566 is approaching us at 240 km/s. It is thus possible that the three galaxies with the corresponding velocities form a physical triplet. The relative cosmic separation based on the difference in the velocities, however, would imply that NGC 5566 is 10.7 million light years closer to the Milky Way.

An indication of a direct interaction between NGC 5566 and NGC 5560, though, is the faint bridge of material extending from the edge of the disc in NGC 5560 towards NGC 5566. This could also be a tidal tail, however, which is behind the larger galaxy NGC 5566 and is seen in projection. The latter seems plausible since although the SB(r)ab morphology of NGC 5566 is perturbed, no disturbance of the inner ring and associated dust-lane structures is seen where the bridge would be attached. The elongated galaxy NGC 5560 is also characterized by the fact that its halo has an almost horizontal extent, making it polar and thus perpendicular to the spiral plane. This could be due to tidal tails formed by orbiting satellite galaxies, which could have led to these galaxies losing material and merging with NGC 5560. A further indication that NGC 5566 is located in front of the other two galaxies is its close neighbour NGC 5569. Since no traces of interaction are visible here and especially since the nearby spiral arm of NGC 5566 has a regular form, the direct separation between the two galaxies is probably more than 300 000 light years. This distance is mentioned in the literature as the typical limit below which mutual tidal interactions lead to perturbations in form. In NGC 5566 one has the impression that the inner ring consists of two overlapping, bright spiral arms which are connected to the end of the bar. The outer spiral arm appears warped relative to the inner ring.

Studies of the luminosity distribution of the core region of NGC 5566 show a small feature which could be interpreted as a possible second, smaller bar. Since NGC 5566 is also an active galaxy (LINER), such a feature would not be unusual. It is known that the existence of a bar is related to the transport of material into the core. This transport mechanism extends, however, to within only a few thousand light years of the central black hole and accretes material there which one can observe in a so-called nuclear starburst ring. Only the scenario of a second bar allows the fuelling of the black hole via the further flow of material into its immediate vicinity, i.e. within 30 light years. Up until 1997, the consensus was that there is such a second bar in NGC 5566. However, an image taken by the Hubble Space Telescope suggests that the structure within the large bar is not a second bar, but presumably a pair of spiral arms. If this holds up, NGC 5566 would be the product of a galaxy merger.

OBJECT	Arp 286 / Homberg 360
CONSTELLATION	Virgo
RIGHT ASCENSION	$14^h\ 20^m\ 19^s$
DECLINATION	+03° 58′ 11″
BRIGHTNESS	–
TYPE	Galaxy group
PHOTOGRAPHERS	Adam Block
TELESCOPE	600-mm reflector
CAMERA	SBIG STL-11000
EXPOSURE TIME	255 min
LOCATION	Mount Lemmon SkyCenter/ University of Arizona, USA

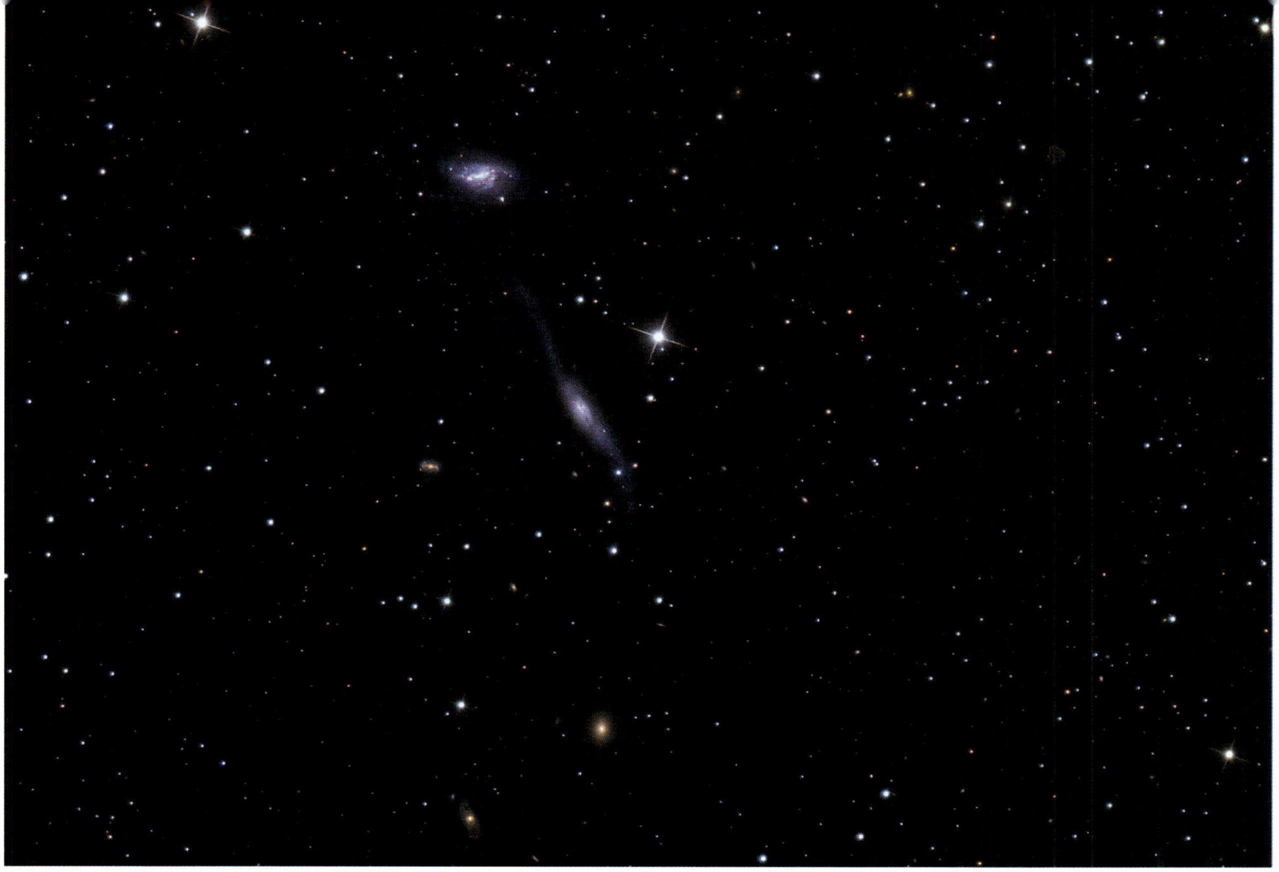

Arp 254

The image shows the galaxy pair NGC 5917 (Arp 254) and PGC 54817. They have an angular separation of 4.2′. To the left is the third galaxy PGC 1018861 (2MAXX J15214381–0726440), forming an isosceles triangle with the other two. At the upper, northern corner of the triangle in NGC 5917 a bright, elongated central region is visible which could be interpreted as a bar. The galaxy with size 1.5′ × 0.9′ shows no core differing in colour; the active areas with large OB associations and H II emission regions are arranged irregularly, so only rudimentary spiral arms are visible. It is thus probable that NGC 5917 is a transitionary type between spiral and irregular. This is caused by the interaction with PGC 54817, from which tidal tails extend to the halo of NGC 5917. At the distance of 85 million light years at which both galaxies are located, this corresponds to a length of more than 100 000 light years for the tidal tail. At the opposite end of PGC 54817 is also a short tail. This countertail, on the side opposite the interaction partner, is a typical effect due to mutual loss of material in closely orbiting galaxies.

The third and smaller galaxy PGC 1018861 is an unperturbed SB(r)c type and differs not only in colour and form from the other two galaxies. Its redshift is ten times larger and it lies at a distance of 810 million light years. Thus the three galaxies do not form a physical triplet, and only NGC 5917 and PGC 54817 are physically close to each other. The two galaxies 2MASX J15214119–0731520 (E) and 2MASX J15215155–0732210 (SAB) 6′ to the south are at the same distance as PGC 1018861. However, the separations between them are too large to cause any visible tidal effects. The transverse distance between 2MASX J15214119–0731520 and 2MASX J15215155–0732210 is more than 600 000 light years.

OBJECT	Arp 254 / NGC 5917
CONSTELLATION	Libra
RIGHT ASCENSION	$15^h\ 21^m\ 33^s$
DECLINATION	−07° 22′ 38″
BRIGHTNESS	14.5 mag
TYPE	Sb pec
PHOTOGRAPHERS	Adam Block
TELESCOPE	800-mm reflector
CAMERA	SBIG STX-16803
EXPOSURE TIME	850 min
LOCATION	Mount Lemmon SkyCenter/ University of Arizona, USA

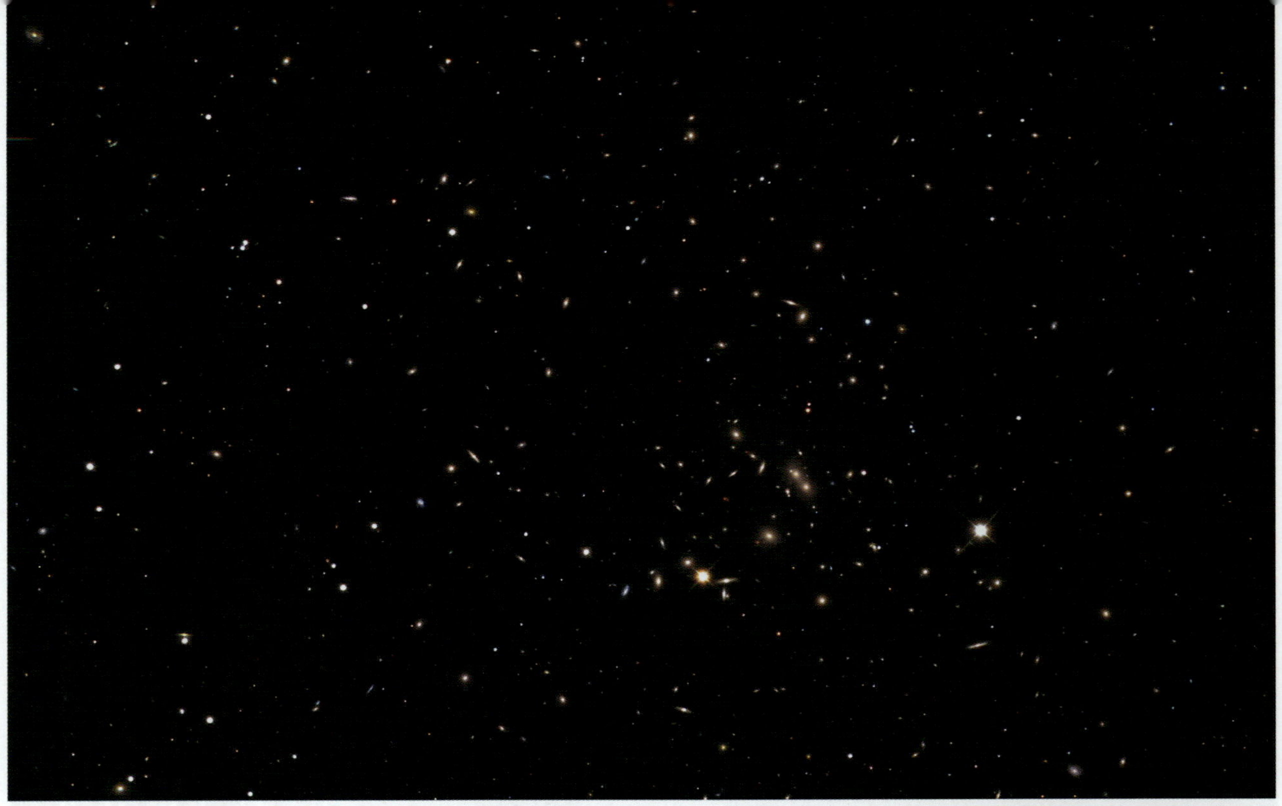

Abell 2065

The galaxy cluster Abell 2065 seen in this image is close to two foreground stars with luminosities of 10 mag and 11 mag separated by 5′. Most of the galaxies in Abell 2065 are fainter than 16 mag and their angular sizes are usually less than 30 arcseconds. The brightest galaxy is PGC 54883 with a size of about 0.72′ × 0.52′. Its physical size can be calculated from the large distance to the galaxy cluster of 980 million light years, resulting in a diameter of 200 000 light years which is typical of a cD galaxy in a middle-size galaxy cluster. The diameter of the cluster itself is eight to ten million light years.

The image shown here also allows an estimation of the possibilities available to the modern amateur astronomer. However, this needs a comparison image, such as that of the POSS (Palomar Observatory Sky Survey, 120-cm Schmidt telescope with photographic plates). On the POSS plate, the detailed arrangement of the galaxies in Abell 2065 can hardly be recognized and the four specks of light are only with difficulty identifiable as galaxies. Like many other galaxy clusters, Abell 2065 displays structures in the arrangement of its galaxies.

When considering in detail the spatial structure of a galaxy cluster, it must be kept in mind that only a two-dimensional projection of the cluster is visible to us. The measurement of the redshift of the galaxies, however, offers some information about the structure of the cluster along the line of sight, so that radial velocities can be used to estimate the relative motions of the galaxies in the third dimension. However, we have no information about transverse velocities, i.e. motions perpendicular to the line of sight. Statistical studies of the two-dimensional structure of Abell clusters indicate that there is no relation between the structures observed and the morphological form of the Abell cluster, such as the presence of two cluster centres or rows of galaxies. According to the analyses, about 34 per cent of the studied Abell clusters display a strong substructure and in the case of Abell 2065 a dominant, central structure with a conspicuous group of a few galaxies about 4′ above the centre was identified.

OBJECT	Abell 2065
CONSTELLATION	Corona Borealis
RIGHT ASCENSION	$15^h\ 22^m\ 43^s$
DECLINATION	+27° 43′ 22″
BRIGHTNESS	15.6 mag
TYPE	Galaxy cluster
PHOTOGRAPHERS	Stefan Binnewies, Volker Wendel
TELESCOPE	600-mm reflector
CAMERA	SBIG STL-11000
EXPOSURE TIME	450 min
LOCATION	Skinakas Observatory, Crete, Greece

Seyfert's Sextet

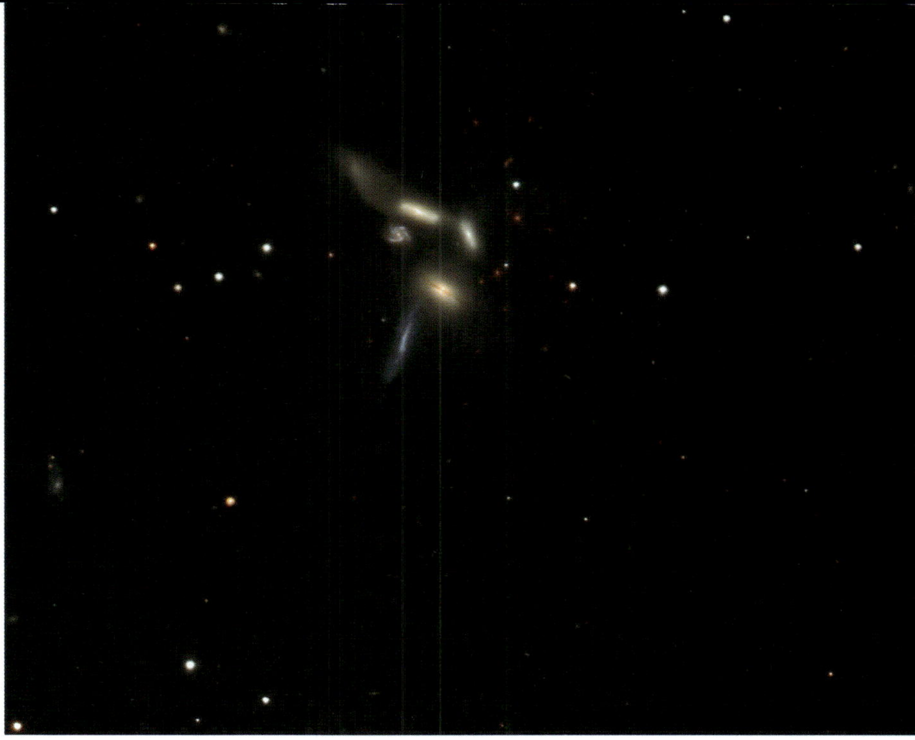

This galaxy group was discovered in 1951 by the astronomer Carl Keenan Seyfert (Barnard Observatory of Vanderbilt University, Nashville, Tennessee, USA) during a study of faint galaxies which also used plates from Harvard College Observatory (Cambridge, Massachusetts, USA). On one of these plates, Seyfert found the compact galaxy group with the brightest galaxy NGC 6027 and five further objects which were denoted NGC 6027a/b/c/d/e. In the image, the S0 galaxy NGC 6027 (PGC 56575) is at the upper left of the group. A broad tidal tail surrounds it and points to the upper left. It first fans out then narrows again, ending in a bright condensation. At the discovery, this was identified as NGC 6027e and hence wrongly as an independent galaxy. Seyfert already mentioned in his paper that further observations by Walter Baade at the 100-inch and 200-inch reflectors at the Mount Wilson and Palomar Observatories (California, USA) had determined that this object is only a filament of the tidal tail of NGC 6027 which exceeds its parent galaxy in extent.

The term NGC 6027, which Carl Seyfert used for the bright S0 galaxy, was later replaced with NGC 6027E (NGC, RC3, *Third Reference Catalogue of Bright Galaxies*). In the modern notation, 14″ below NGC 6027E is the SB(s) bc galaxy NGC 6027D. This galaxy with a size of only 0.2′ is at the enormous distance of 890 million light years and is thus almost five times as far away as NGC 6027E, whose distance from the Milky Way is only 188 million light years. 20″ to the right of NGC 6027D is NGC 6027B, seen edge-on and almost vertical in the image. It is a neighbour of NGC 6027E but is moving relative to it at −350 km/s and thus approaching the Milky Way compared to others in the group. Below NGC 6027E 24″ farther to the south is the Sa galaxy NGC 6027A with a light brown dust lane running across the spherical halo. At the southern tip of the group is finally NGC 6027C. The relative velocity of this SB galaxy is +170 km/s so that this galaxy is moving away from the Milky Way compared to the others in the group. It is interesting to combine and sort the recessional velocities in order to derive their radial positioning. The sequence would be NGC 6027C, NGC 6027B, NGC 6027A, and NGC 6027E, which is the nearest of the four galaxies. The galaxies' relative velocities, when plotted, lie on a line, one after the other. This could be explained by rotational motions around a common centre. In the sequence of galaxies, this centre would be located between NGC 6027A and NGC 6027B. The galaxy NGC 6027C would be farther from the centre; this galaxy has a higher radial velocity. All four neighbouring galaxies are catalogued as HCG 79. Since all lie within only two arcminutes, this is one of the most compact galaxy groups known.

N

At the distance of NGC 6027E an arcminute corresponds to a distance of 55 000 light years, thus tidal forces massively distort the morphology of the members and tidal tails are formed in the course of the orbits. Studies have shown that the tidal tails in the group consist of older stellar populations whose metallicity is about 40 per cent of the solar metallicity and thus fits well with the typical characteristics of tidal tails and the tidal dwarf galaxies formed from them. Since these tidal tails all point away from the centre of the group, it is suspected that the group has already reached a stage in which the mutual orbits are becoming smaller and smaller and a merger to a large elliptical galaxy is in store.

OBJECT	NGC 6027
CONSTELLATION	Serpens
RIGHT ASCENSION	$15^h\ 59^m\ 13^s$
DECLINATION	+20° 45′ 48″
BRIGHTNESS	14.7 mag
TYPE	S0 pec
PHOTOGRAPHERS	Stefan Binnewies, Rainer Sparenberg
TELESCOPE	600-mm reflector
CAMERA	SBIG STL-11000
EXPOSURE TIME	160 min
LOCATION	Skinakas Observatory, Crete, Greece

Abell 2151

At the southern edge of the Hercules Supercluster is Abell 2151, its richest Abell cluster. This is also known as the Hercules Cluster, is at a distance of 490 million light years, and contains 150 members. An arcminute corresponds to 143 000 light years. The peculiarity in Abell 2151 is the lack of a cD galaxy at its dense centre. Instead, there one can see several ongoing mergers of cluster galaxies, which will lead to a large elliptical galaxy only in 100 million years. Due to this property, Abell 2151 is also known as a merging cluster, which is also the reason for the unusually large number of irregular and spiral galaxies at its core.

In the upper half of the image are the large elliptical galaxy NGC 6041 and the somewhat smaller E-type galaxy NGC 6039 only two arcminutes away. Above and to the left, to the northwest, is the conspicuous pair of spiral galaxies NGC 6040A and NGC 6040B which are about to merge. To the left of the centre of the picture are the galaxies NGC 6043A, NGC 6043B, the SBc-type NGC 6045A and at the right NGC 6047. It is an E/S0 type and has the unusual morphology of a large, oval halo and a bright, circular core. This is an indication of a past merger in which the discs of the precursor galaxies were destroyed. As a result of this merger, no uniform velocity profile is measurable today.

Conspicuous at the lower left corner is IC 1182, which is classified as S0 but is also surrounded by tidal tails. This indicates that a merger of two spiral galaxies has taken place. Two bright knots separated by 3″ are visible at the core, which could be the merging nuclei. The striking linear structure which appears to emerge from the core extends over 1.45′ and contains several condensations in which star formation is taking place and out of which dwarf galaxies could develop. Such tidal dwarfs support the idea that the structure is a tidal tail and thus the result of a merger. An optical jet, in which material is flowing out of the core, is improbable.

At the right edge is the pair IC 1178 and IC 1181, separated by only 0.5′. The two galaxies have over many orbits produced a common halo, some of which is outside of the image. IC 1178 at 1.13′ × 0.60′ is larger as the SAB(rs) galaxy IC 1181. Its size, which corresponds to 161 000 light years, is due to the diffuse tidal tail – the precursor galaxies were certainly smaller.

OBJECT	Abell 2151
CONSTELLATION	Hercules
RIGHT ASCENSION	$16^h\ 05^m\ 15^s$
DECLINATION	+17° 44′ 55″
BRIGHTNESS	13.8 mag
TYPE	Galaxy cluster
PHOTOGRAPHERS	Stefan Binnewies, Rainer Sparenberg
TELESCOPE	600-mm reflector
CAMERA	SBIG STL-11000
EXPOSURE TIME	380 min
LOCATION	Skinakas Observatory, Crete, Greece

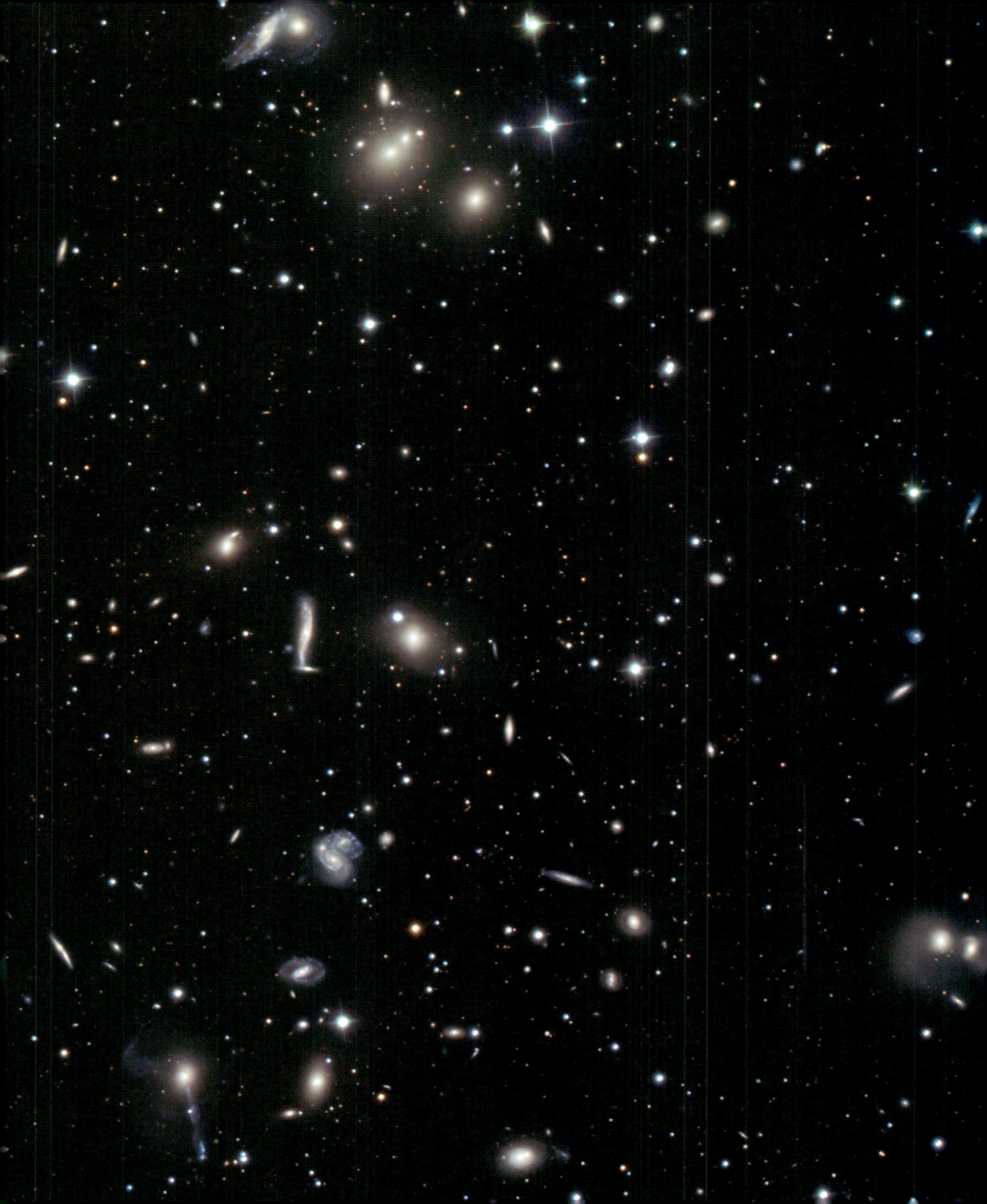

NGC 6340 Group

NGC 6340 is the largest galaxy of a triplet of dissimilar galaxies in the constellation of Draco, visible at the lower right of the image next to the orange-white pair of stars. 6′ above it to the north lies the Scd galaxy IC 1251 whose radial velocity is only 12 km/s smaller than that of NGC 6340. Both galaxies are at a distance of 62 million light years; their spatial separation is 110 000 light years. The diameter of NGC 6340 of 71 000 light years is typical for a spiral galaxy; its companion IC 1251 is a smaller example with a diameter of only 36 000 light years.

NGC 6340 has a bright, compact core. It is surrounded by a disc from which two diffuse spiral arms lead to an outer ring. The edge of the disc is dark, but the idea that it is a ring-like dust lane was revised by an image taken by the Hubble Space Telescope. This shows the zone at the edge as a transparent ring through which background galaxies can be seen. The inner region of NGC 6340 is that of an S0 galaxy; its outer region is similar to an SAab type. In the case of NGC 6340, such a mix of types is denoted in catalogues by SA(s)0/a.

To the upper left northeast of NGC 6340 is the galaxy IC 1254. It is a perturbed Sb type which also belongs to the group around NGC 6340. The relative velocity to NGC 6340 is +73 km/s; the distance between the two galaxies is 150 000 light years and thus within the range in which significant tidal interactions are possible. As is the case with IC 1251, the companion galaxy IC 1254 is notable for its small size. The calculated 40 000 light years refer to the diameter of the halo; the core region is only half as large. At the northeastern end of the halo are tidal tails which extend to a possible, even smaller interaction partner, which however has not been catalogued. Near IC 1254, further to the northeast, is the spiral galaxy PGC 2749453. It is not part of the group. There are hints that the supernova 1999bt exploded here. Through the type of the supernova and its corresponding absolute luminosity, the distance was estimated at 760 million light years, making the small spiral galaxy 12 times as far away as IC 1254.

Analyses of the motions of the stars in NGC 6340 also show that its inner and outer components are two separate kinematic components which are offset relative to each other. Simulations show that the cause could have been a major merger, i.e. the merger of two galaxies. Possible precursors are a non-rotating large elliptical galaxy and a spiral galaxy. Three billion years after the merger, they have formed a large lenticular galaxy which is surrounded by a highly extended, rotating disc. According to simulations, other peculiarities in NGC 6340 could be due to a further merger with a smaller, gas-rich companion galaxy. This presumably took place a few hundred million years ago.

OBJECT	NGC 6340
CONSTELLATION	Draco
RIGHT ASCENSION	$17^h\ 10^m\ 25^s$
DECLINATION	+72° 18′ 16″
BRIGHTNESS	11.9 mag
TYPE	SA(s)0/a
PHOTOGRAPHERS	Josef Pöpsel, Stefan Binnewies
TELESCOPE	600-mm reflector
CAMERA	SBIG STL-11000
EXPOSURE TIME	450 min
LOCATION	Skinakas Observatory, Crete, Greece

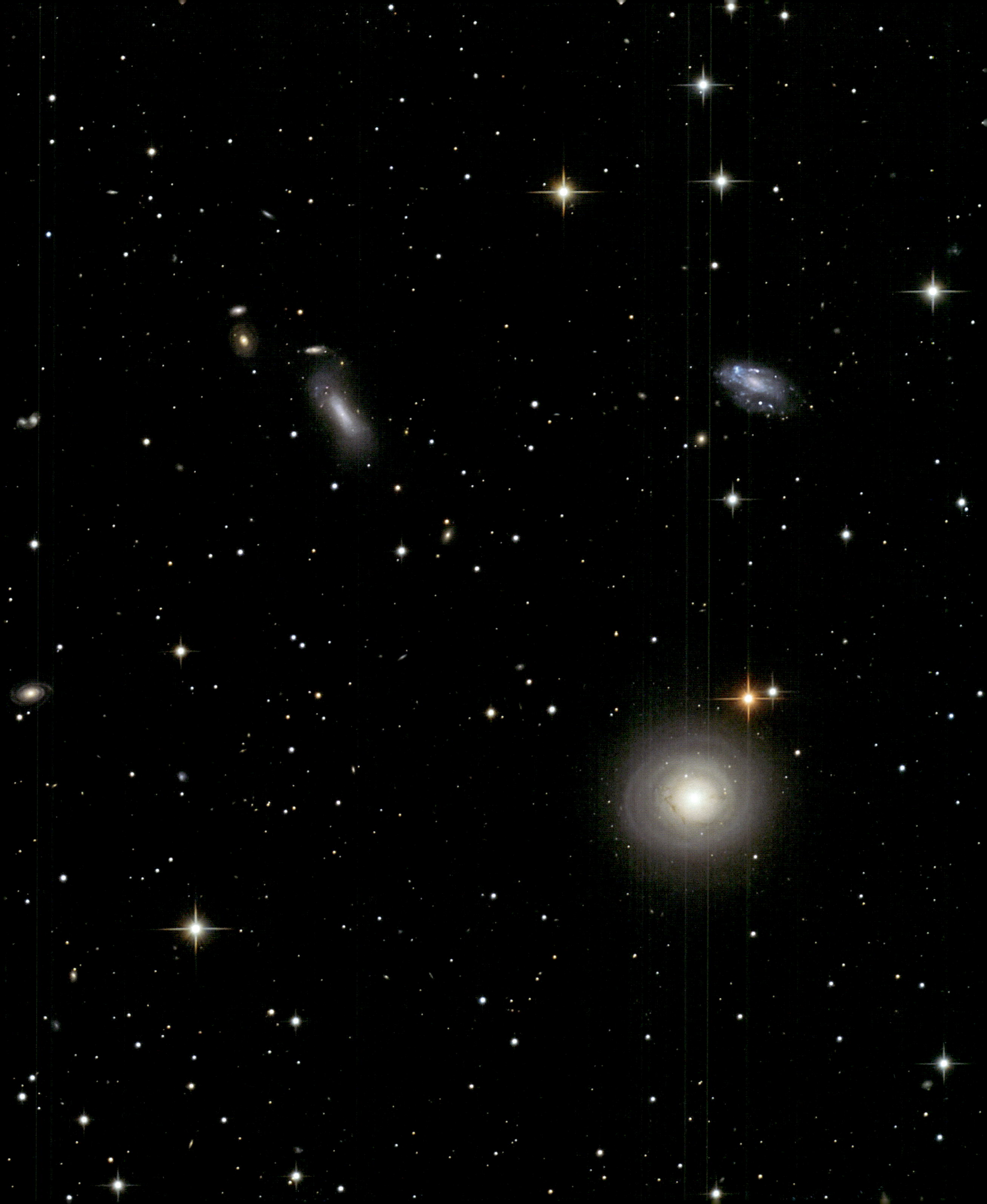

NGC 6580 Group

Near the centre of the image are the galaxy NGC 6580 and its companion NGC 6579 (separation 0.55′) which have a common halo, recognizable as a faint, diffuse glow. The larger galaxy also has a barely defined spiral arm to the right of the nucleus and a semicircular tidal tail at the lower left end. 0.63′ away from NGC 6580 is the edge-on spiral galaxy UGC 11153. There is no radial velocity listed for this galaxy, so no direct test as to whether it forms a physical triplet with the other two. The two galaxies NGC 6580 and NGC 6579 lie at a distance of 249 million light years and their relative velocities differ by the large amount of 860 km/s. The visible common halo does not, however, indicate any direct interaction, so that the large velocity difference suggests that NGC 6579 is about to fall into NGC 6580. Comparing the physical sizes of the two galaxies, one notices that at 80 000 light years NGC 6580 corresponds to the standard size of a spiral galaxy. NGC 6579, by contrast, is appreciably smaller with a diameter of only 28 300 light years. Its morphology is similar to that of a small elliptical galaxy and can be explained by past close approaches resulting in heavy gas loss. The result is a relic core region without a spiral plane.

Within a radius of 15′ around NGC 6580 are three further galaxies: NGC 6577 (7.8′ separation), NGC 6576 (11′ separation), and CGCG 142-016 (15′ separation), as well as three other galaxies which belong to this small galaxy group. Most conspicuous of these is NGC 6577 in the right third of the image. It exceeds NGC 6580 both in size and in luminosity and has a diameter of 130 000 light years. NGC 6577 is known as a radio source and the radio map of this galaxy can be described as a fan-like structure about 4′ × 2′ in size with a bright point source at the end. The position of this core corresponds to the nucleus of NGC 6577 and the weaker structures point roughly in the direction of the pair consisting of NGC 6580 and NGC 6579. An interaction, however, seems improbable, due to the projected distance of 500 000 light years.

OBJECT	NGC 6580
CONSTELLATION	Hercules
RIGHT ASCENSION	$18^h\ 12^m\ 34^s$
DECLINATION	+21° 25′ 34″
BRIGHTNESS	14.5 mag
TYPE	E
PHOTOGRAPHERS	Josef Pöpsel, Stefan Binnewies
TELESCOPE	600-mm reflector
CAMERA	SBIG STX-16803
EXPOSURE TIME	360 min
LOCATION	Skinakas Observatory, Crete, Greece

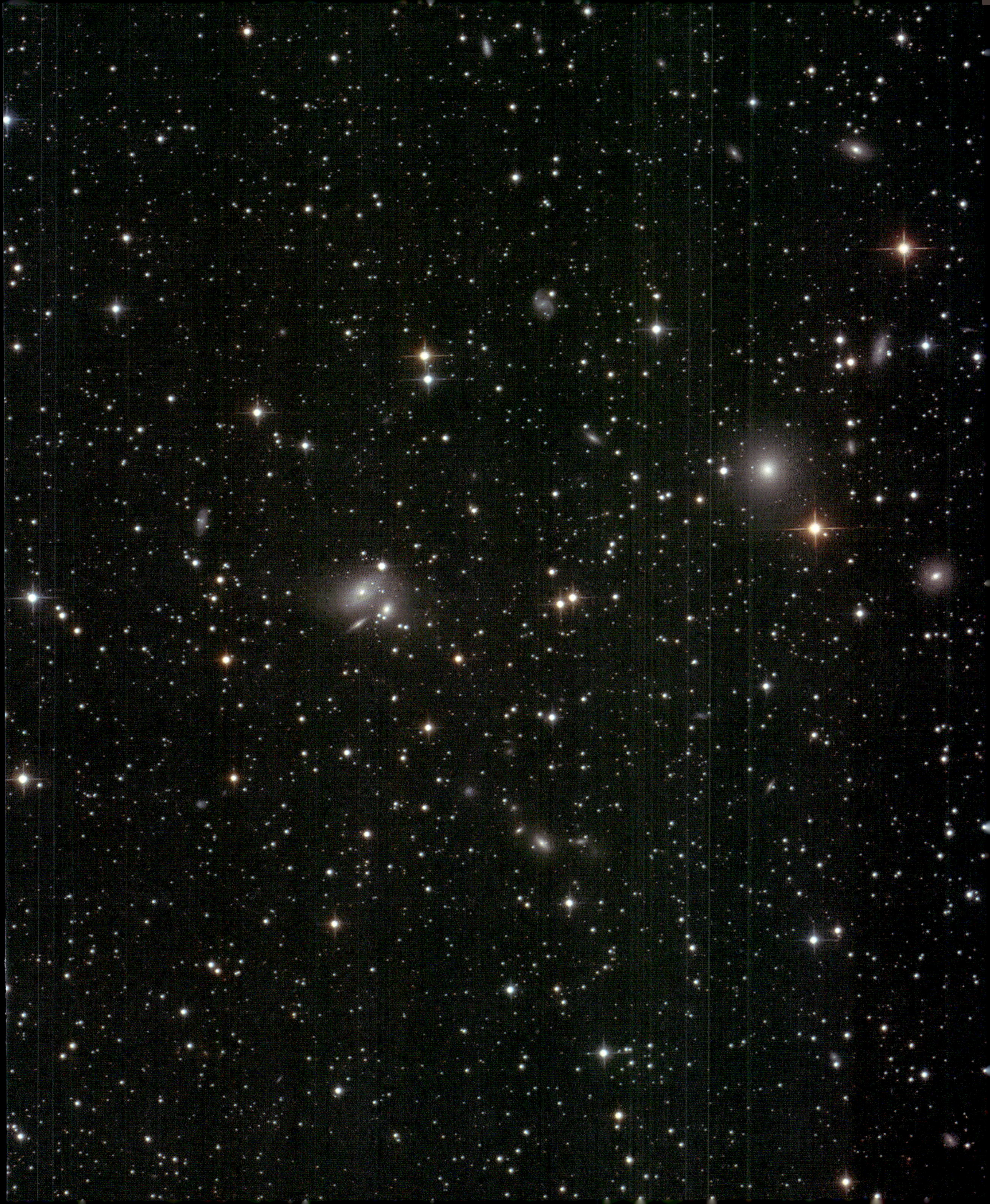

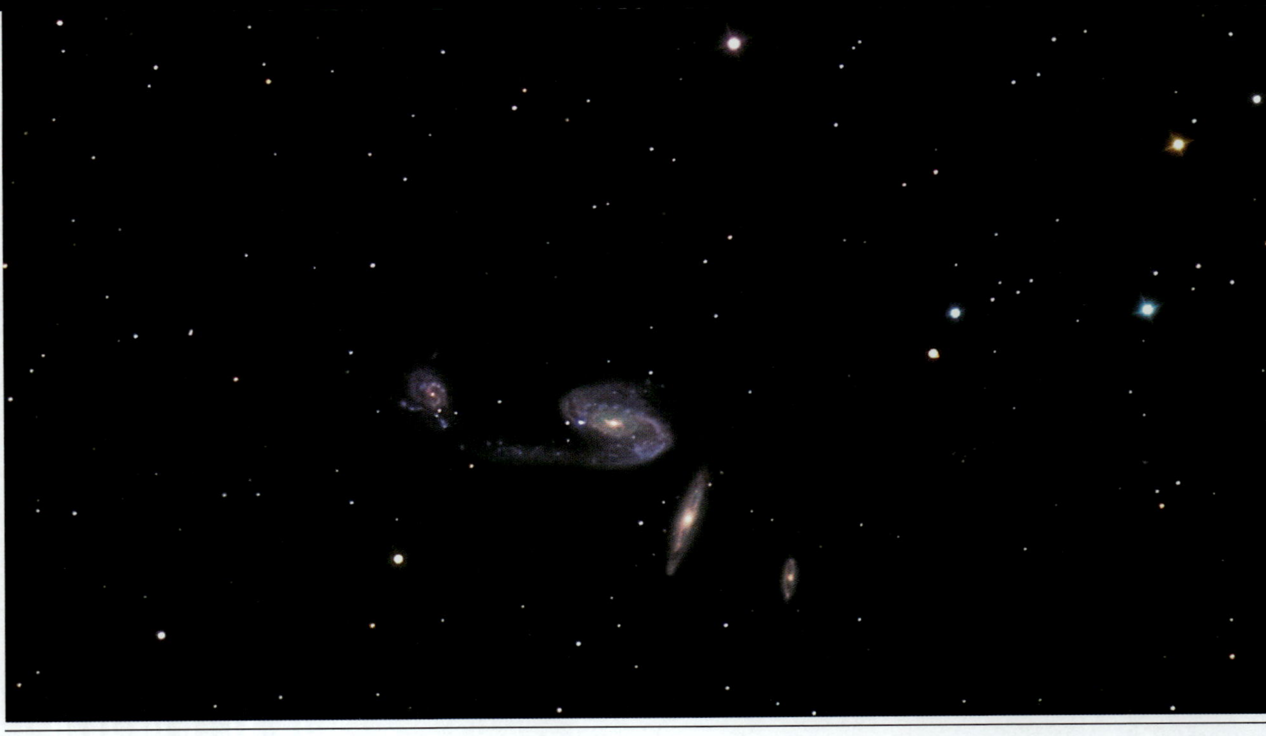

NGC 6845 Group

NGC 6845A is the dominant spiral galaxy in the compact group Klemola 30. Its four members lie along a line running from northeast to southwest, covering about 3′. At the left, 1.35′ away from NGC 6845A, is the small, strongly perturbed spiral galaxy NGC 6845B (PGC 63986). It is connected directly to the larger companion by a tidal tail which continues into the lower spiral arm of NGC 6845A and extends over at least 120 000 light years. Below NGC 6845A are the spiral galaxies NGC 6845C (S0/a, PGC 63979), 0.88′ away from NGC 6845A, and NGC 6845D (S0, PGC 63978), 1.71′ away. The distribution of the radial velocities of these four galaxies has a width of ±355 km/s and thus exceeds the mean value for small galaxy groups of 100–200 km/s.

The mean distance of the group is 294 light years, whereby an arcminute corresponds to a projected length of 86 000 light years. For NGC 6845A this results in the enormous dimensions of 322 000 × 132 000 light years. Thus the diameter of this galaxy corresponds to about three Milky Ways side by side. The radio emission of atomic hydrogen also shows a clear bimodal distribution within the group. The interacting galaxies NGC 6845A and NGC 6845B correspond to maxima in the emission; the two other galaxies in the group lie below the sensitivity of the radio telescope used.

NGC 6845 is classified as SB(s)b, but doesn't have a prominent bar. Instead, one sees a horizontal, linear structure within an inner disc onto which the spiral arms are attached. Studies show that the colour of this bar is not uniform. The southwest part of the bar is redder, which for astronomers is an indication of dust along the line of sight which scatters blue light, causing the red impression. The fact that this is seen on only one side is explained by asymmetric shock waves which occur in this part of the bar and cause material (gas and dust) to pile up at the end. In the spiral arm and tidal tail there are bright knots whose blue colour indicates star-forming regions. There are also areas of H II emission of which many can be kinematically associated with the tail. Analyses of the stars in these active regions of the tidal tail result in an age of 100 million years. This was triggered by the large perturbation in the morphology during the last close approach of the two partners in the interaction.

OBJECT	NGC 6845
CONSTELLATION	Telescopium
RIGHT ASCENSION	$20^h\ 00^m\ 58^s$
DECLINATION	−47° 04′ 12″
BRIGHTNESS	14 mag
TYPE	Galaxy group
PHOTOGRAPHERS	Josef Pöpsel, Beate Behle
TELESCOPE	600-mm reflector
CAMERA	SBIG ST-10XME
EXPOSURE TIME	140 min
LOCATION	Amani Lodge, Namibia

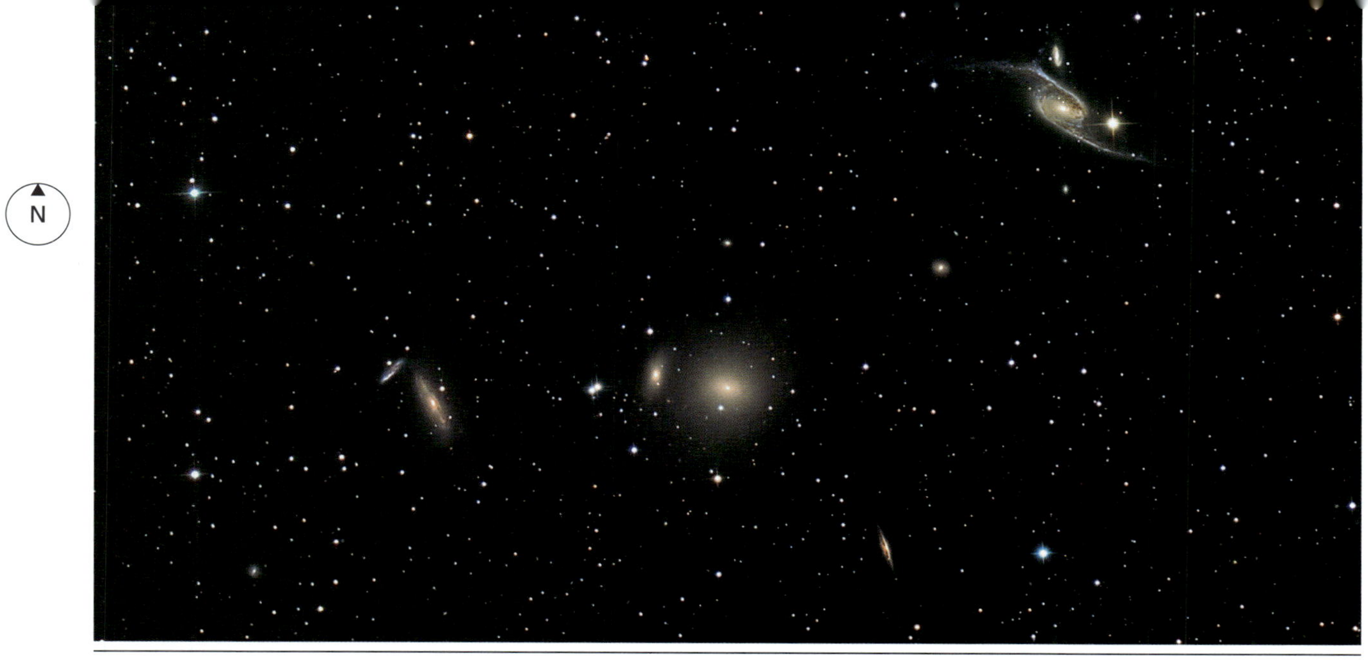

NGC 6872 Group

The image shows the inner regions of the Pavo galaxy group with three interacting pairs of galaxies: NGC 6872 and IC 4970, NGC 6876 and NGC 6877, and NGC 6880 and IC 4981 (from upper right to lower left). The perspective under which the SAB(rs)c galaxy NGC 6872 is observed is difficult to describe, since its horizontal extent suggests that it is seen edge-on while but the inner ring onto which the two large spiral arms are attached is seen almost face-on. The galaxy measures 6.0′ × 1.7′ which at the distance of 200 million light years corresponds to 349 000 × 99 000 light years. These enormous dimensions of the SAB galaxy are due to perturbations within the group, caused mainly by the companion IC 4970. It is separated by 65 000 light years and orbits NGC 6872 with a velocity of 160 km/s. The core of NGC 6872 is bright, its bar weak, and the upper, northern spiral arm shows traces of a 120° angle due to Vorontsov-Velyaminov rows.

At the centre of the image is a pair of ellipticals with the larger E3 galaxy NGC 6876 and, 1.4′ away, the E5-type NGC 6877. Their separation is at least 71 000 light years and the size of 142 000 light years for NGC 6876 means that NGC 6877 is within the halo of NGC 6876. 5′ farther east is a pair of spiral galaxies consisting of the larger SB galaxy NGC 6880 and the small irregular IC 4981. The latter could also be classified as a spiral whose dust lane darkens the central region.

The relative velocities of the members of the Pavo group display a large scatter of several hundred kilometres per second. The dynamic evolution of the group is thus still at the beginning and the high velocities demonstrate that there are not only interactions within the close pairs but also between them. In X-rays there is also a structure 330 000 light years in length between NGC 6872 and NGC 6876. This track of gas at 12 million kelvin is hotter than the intergalactic medium whose temperature is about five million kelvin.

The fact that NGC 6872 is one of the largest known spiral galaxies is explained by a series of perturbations. Simulations of these interactions deliver good results if a companion which has 20 per cent of the mass of the large galaxy is on a parabolic orbit only slightly inclined with respect to the spiral plane of the latter. The loss of gas and stars takes place in the direction of the spiral arms and pulls these far to the outside along the tidal tail formed. After 130 million years, the simulations indicate not only the observed spiral-arm structure but also the measured relative velocities and thus explain the formation of the peculiar morphology of NGC 6872.

OBJECT	NGC 6872
CONSTELLATION	Pavo
RIGHT ASCENSION	$20^h\ 16^m\ 57^s$
DECLINATION	–70° 46′ 05″
BRIGHTNESS	12.7 mag
TYPE	SB(s)b pec
PHOTOGRAPHERS	Josef Pöpsel, Beate Behle
TELESCOPE	600-mm reflector
CAMERA	SBIG ST-10XME
EXPOSURE TIME	170 min
LOCATION	Amani Lodge, Namibia

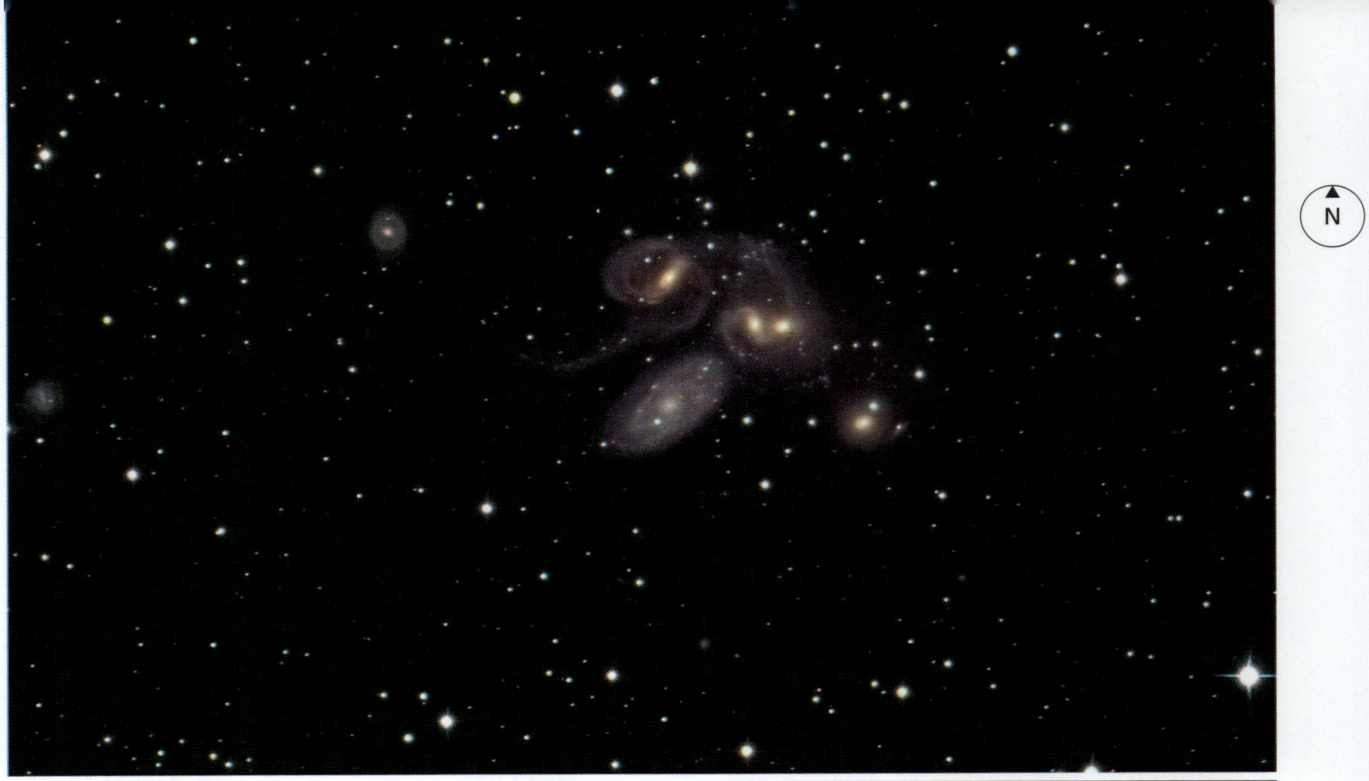

Stephan's Quintet

In 1876, Edouard Stephan, Director of the Observatory at Marseille, discovered a small galaxy group in the constellation of Pegasus using the then-modern 80-cm reflector designed by Léon Foucault. This group contains the five closely neighbouring galaxies NGC 7319, NGC 7320, NGC 7317, the close pair NGC 7318A/B, and the galaxy NGC 7320C somewhat farther away. Edouard Stephan had discovered the main group as a quartet since he saw 7318A/B as a single galaxy. Only with time did the name Stephan's Quintet arise for this group (also known as Arp 113 and Hickson Compact Group 92).

The distances of NGC 7313 and NGC 7319 are almost the same at 303 million light years and 310 million light years; the pair NGC 7318A/B is at 287 million light years. It soon became clear, however, that the SA galaxy NGC 7320 has a lower redshift and is at a distance of 45 million light years. In the image, NGC 7320 is at the lower left and compared to the other four galaxies, its core is less prominent and not as yellow. Its distance is similar to that of the large neighbouring galaxy NGC 7331 which lies 35′ to the northeast at a distance of 46 million light years. However, even in deep exposures of 26.7 mag/arcsec^2 no sign of interaction between NGC 7320 and NGC 7331 is visible. Today, some exclude NGC 7320 from the group, since it is a foreground galaxy, but include NGC 7320C instead (also discovered by Stephan, one day later), now known to be at the same distance as the other four and physically associated with them.

Information from more and better observations of the galaxies in the group have allowed for refinement of the models of Stephan's Quintet. For example, condensations have been observed in the tidal tails of the galaxies. Simulations also confirm that dwarf galaxies can form in the tails between the galaxies. Also, a diffuse, faint band of light was discovered which connects the area around the four close members and NGC 7320C. This band is interpreted as an indication of a close approach with NGC 7319 about 260 million years ago, which in addition to the observed tidal tail also formed a second, weaker tail. Since NGC 7320C also has a prominent ring, it is suspected that this approach led to a minor merger in which NGC 7320C collided with a compact dwarf galaxy, forming the ring.

OBJECT	NGC 7320
CONSTELLATION	Pegasus
RIGHT ASCENSION	$22^h\ 36^m\ 03^s$
DECLINATION	+33° 56′ 53″
BRIGHTNESS	13.2 mag
TYPE	SA(s)d
PHOTOGRAPHERS	Josef Pöpsel, Stefan Binnewies
TELESCOPE	600-mm reflector
CAMERA	SBIG STL-11000
EXPOSURE TIME	250 min
LOCATION	Skinakas Observatory, Crete, Greece

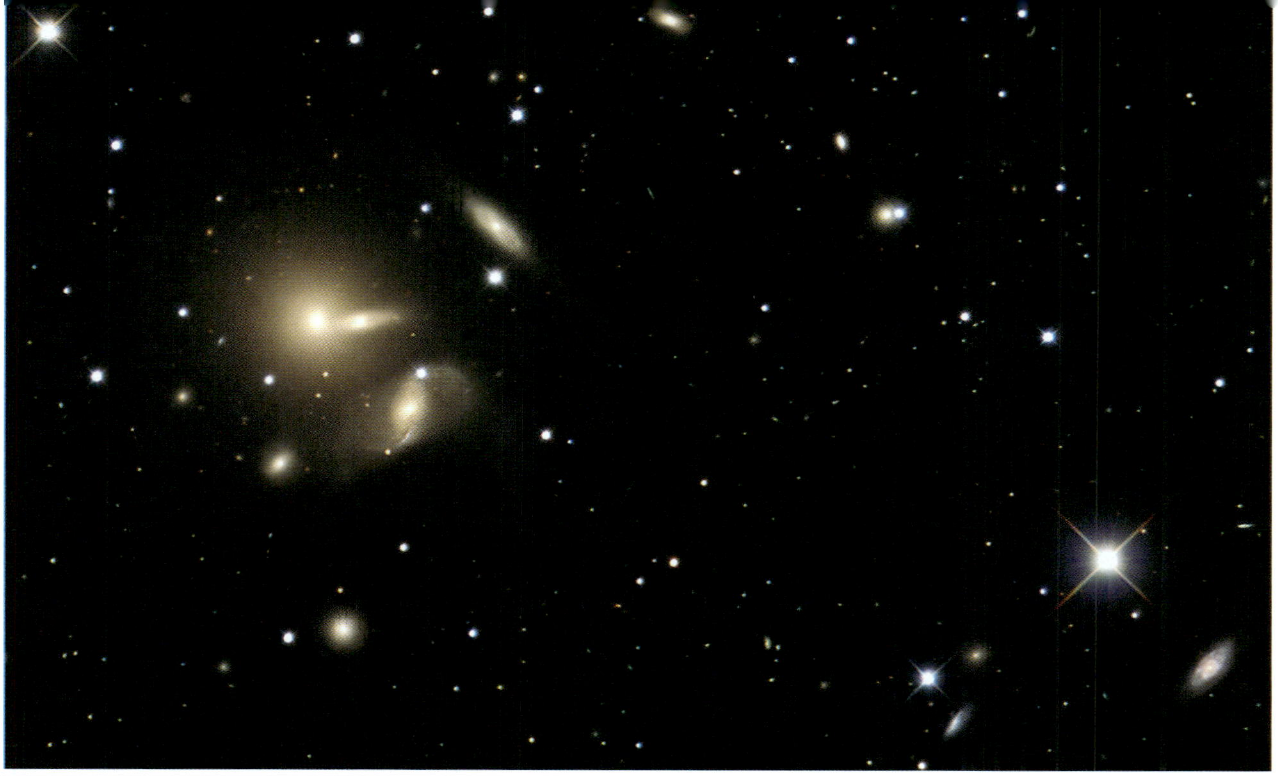

Holmberg 800

The compact group of galaxies surrounding the E0 galaxy NGC 7436 was discovered in 1784 by William Herschel. This group is sometimes described as a multiple galaxy, and is catalogued as Holmberg 800. This group belongs to the outskirts of the galaxy cluster Abell 2513, which is about 20′ to the west. The distance of NGC 7436 from the Milky Way is 340 million light years. The field containing the four bright galaxies measures 2′ × 2′ and has a projected diameter of about 200 000 light years.

0.33′ away and directly to the right of NGC 7436 is the spiral galaxy PGC 70123 (NGC 7436A). The velocity relative to its large elliptical companion is 220 km/s. Since neither tidal tails nor distortions in the halo are visible, it is probable, but not verified, that PGC 70123 is receding from NGC 7436, thus lying behind it.

In contrast, the galaxy NGC 7435, an arcminute below and to the right of NGC 7436, shows clear signs of interactions. This SB(s)a galaxy has tidal tails attached to the edge of the disc, pointing away from NGC 7436. The right, western tail has a sharp kink downward which at a length of 156 000 light years is comparable to the diameter of the galaxy. The velocity relative to NGC 7436 is 800 km/s and thus appreciably more than the mean velocity dispersion for small groups. NGC 7435 thus probably also lies behind NGC 7436 and is interacting with PGC 70123. 1.5′ above and to the right of NGC 7436 lies NGC 7433, an elongated S-type galaxy whose spiral plane is almost 2 arcminutes in diameter if the dim extensions are included. The radial velocity of the galaxy is near that of PGC 70123, which makes it plausible that NGC 7433 and PGC 70123 form a pair moderately interacting with NGC 7435. The qualifier "moderate" is needed because all of these galaxies are reddish and thus they have neither a large number of star-forming regions nor active nuclei. The dynamics of the members of this group also indicate that the large galaxy NGC 7436 is only marginally part of it.

OBJECT	Holmberg 800
CONSTELLATION	Pegasus
RIGHT ASCENSION	$22^h\ 57^m\ 54^s$
DECLINATION	+26° 08′ 29″
BRIGHTNESS	–
TYPE	Galaxy group
PHOTOGRAPHERS	Makis Palaiologou, Stefan Heutz
TELESCOPE	1.3-m reflector
CAMERA	Andor DZ 436
EXPOSURE TIME	285 min
LOCATION	Skinakas Observatory, Crete, Greece

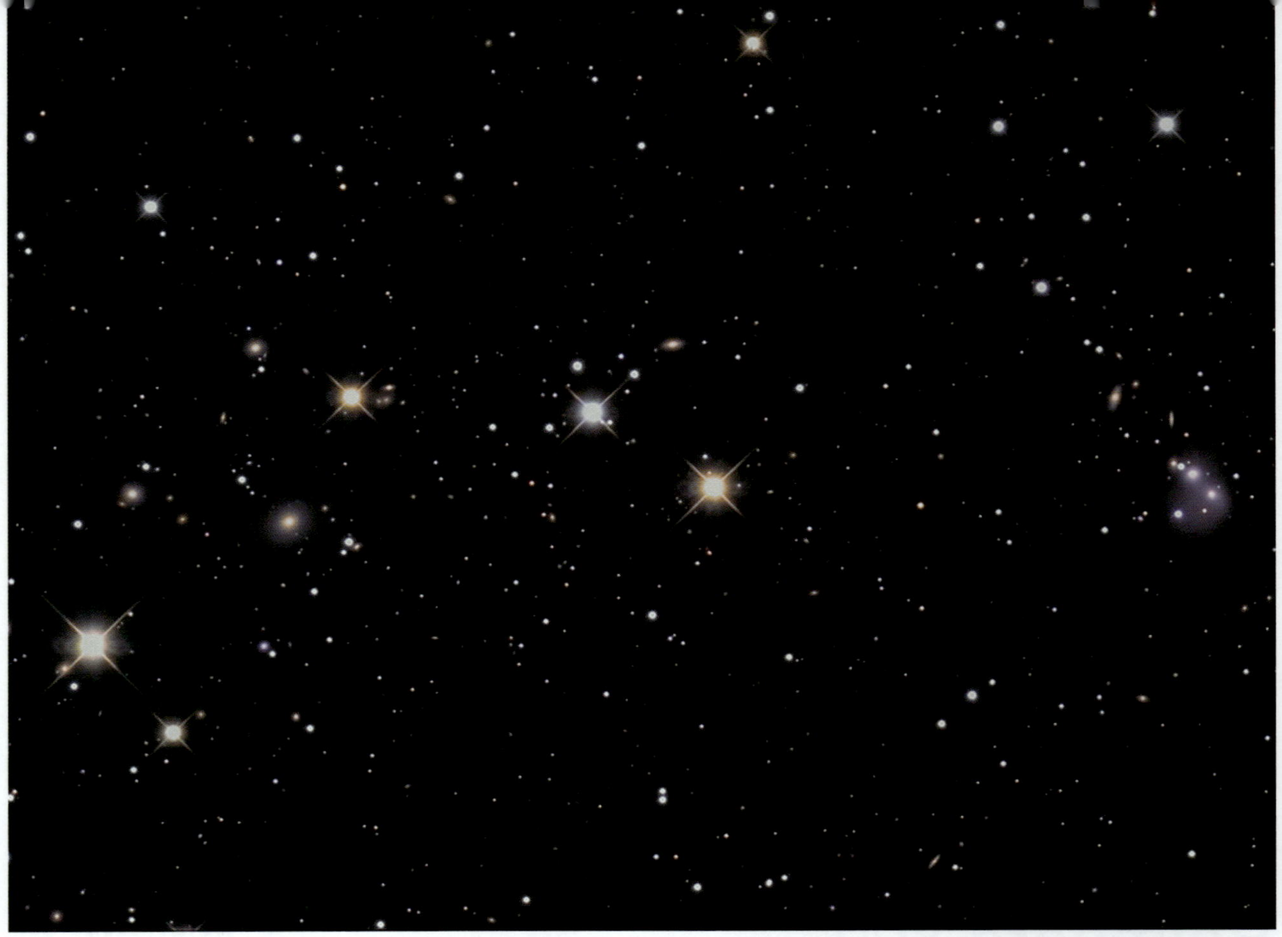

Abell 2572

Various entries in the catalogue *Hickson Compact Groups of Galaxies* were studied with the X-ray satellite ROSAT. Of particular interest is the cluster Abell 2572 with HCG 94, also catalogued as Arp 170. It was found that HCG 94 is not an isolated group of galaxies but rather the core of a galaxy cluster which exceeds the X-ray luminosity of the core of the Abell cluster. This fits well with the fact that HCG 94 also has the highest density of all HCGs. Both systems, the Abell cluster and HCG 94, have the same redshift of $z \approx 0.04$ and thus lie at a distance of 555 million light years. Their separation on the sky is 17′.

The image shows Abell 2572 in the left half and HCG 94 on the right. In HCG 94, almost a dozen galaxies could be found within a radius of a few arcminutes. The centre of the X-ray emission in HCG 94 is near the galaxies NGC 7578A and NGC 7578B, which are somewhat below. Due to the spatial energy distribution and absorption signatures, the structure of the emission is more that of an autonomous cluster than that of a galaxy group. HCG 94 would then be a medium-luminosity cluster which appears relaxed and shows no sign of an impending merger with Abell 2572.

The mean velocities of Abell 2572 and HCG 94 differ by about 935 km/s, which is roughly the velocity dispersion of the cluster itself. The projected distance is about three to seven million light years. In the literature one can find several scenarios with regard to the possible construction and interaction – among others a variant in which both systems, decoupled from the cosmological expansion, are falling into each other. In this case, HCG 94 would be closer despite the higher redshift.

OBJECT	Abell 2572
CONSTELLATION	Pegasus
RIGHT ASCENSION	$23^h\ 18^m\ 24^s$
DECLINATION	+18° 44′ 25″
BRIGHTNESS	–
TYPE	Galaxy cluster
PHOTOGRAPHERS	Michael König
TELESCOPE	355-mm reflector
CAMERA	STL-11000
EXPOSURE TIME	610 min
LOCATION	Rimbach, Germany

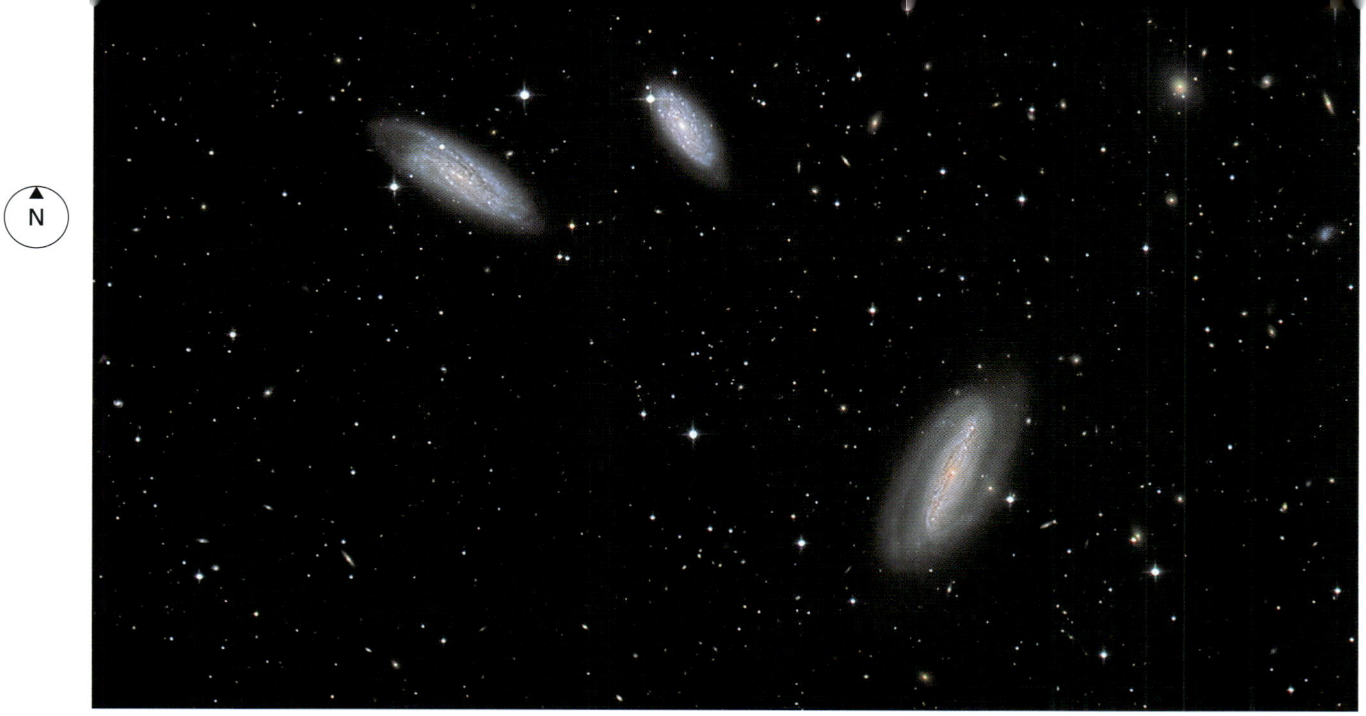

NGC 7582 Group

The three galaxies shown lie in the constellation Grus, a constellation relatively unknown to amateur astronomers at mid-northern latitudes, and they have a declination of 42° south. They form the so-called Grus Triplet. The brightest member of this triplet is the SB(a)rs galaxy NGC 7582, an early-type barred spiral. Its angular size of 5.0′ × 2.1′ and its distance of 69 million light years result in a diameter of 100 000 light years. It is thus a large example of a spiral galaxy. The two other members of the triplet lie to the northeast, NGC 7590 at 9.4′ and NGC 7599 at 12.2′ from NGC 7582. The relative velocity of NGC 7582 and NGC 7590 is almost zero and NGC 7599 deviates only slightly from this and thus forms a physical triplet with the other two, though receding from them at 190 km/s. The separation between the two larger members is 238 000 light years, which means that the outer areas of each galaxy feel the gravitational potential of the neighbour, influencing the flow of material in the spiral planes. Due to the high inclination at which we observe NGC 7582, an analysis of the spiral-arm pattern is difficult. However, one can recognize that the two spiral arms in the outer regions differ in that the arm on the side next to the two neighbours winds at a sharper angle than that on the other side.

The galaxy NGC 7582 is known as a Seyfert-2 and thus displays the narrow emission lines typical of active galaxies. This galaxy has been intensively studied due to its complex spectrum, which consists of several components. Observations with the Chandra X-ray satellite and the Hubble Space Telescope have shown where the X-rays in NGC 7582 originate and what conclusions can be drawn about the internal form of the core. Overlaying the two high-resolution images allows a common coordinate system to be constructed, showing the locations of star formation near the core and the point sources of X-ray emission. This covers the innermost 10″ and corresponds to an extent of 3300 light years. The X-ray-emitting gas is not homogeneous but rather has two soft-X-ray hotspots which indicate absorption by dust and gas along the line of sight. This absorption is due to a central flow of gas and dust and confirms NGC 7582 as a galaxy with an active nucleus.

OBJECT	NGC 7582
CONSTELLATION	Grus
RIGHT ASCENSION	$23^h\ 18^m\ 24^s$
DECLINATION	−42° 22′ 14″
BRIGHTNESS	11.4 mag
TYPE	(R′)SB(s)ab
PHOTOGRAPHERS	Dietmar Böcker
TELESCOPE	600-mm reflector
CAMERA	SBIG ST-10XME
EXPOSURE TIME	72 min
LOCATION	Amani Lodge, Namibia

NGC 7771 Group

The image shows the two neighbouring and interacting galaxies NGC 7771 and NGC 7769. Only 1′ away from NGC 7771 (near the centre of the picture) is the small Sa galaxy NGC 7770. With the larger barred spiral NGC 7771 it shares a halo, which is marked by many tidal tails. The three galaxies form a close triplet located at a distance of almost 196 million light years, whereby the largest relative velocity of 60 km/s is between NGC 7770 and NGC 7771. This is an indication that there have been several close orbits of the two galaxies in the last 200–400 million years. During each orbit, dust, gas and stars were distributed along it, and in the spiral arms of the satellite galaxy NGC 7770 new stars were formed. This post-starburst population has been detected. The ratio of the stellar masses of NGC 7770 and NGC 7771 is 10 : 1. For this reason, the system is described as a minor merger. In addition to the tidal tails, other effects are observed which are induced by the close approaches of the orbiting galaxies. For example, there are filaments of glowing hydrogen gas which lie between the galaxies near the tidal tails and extend over 10 000–20 000 light years. This emission is stimulated by shock waves in the gas which run through the dense parts of the tidal tails.

NGC 7769, which lies a few minutes below (about 300 000 light years in projection), displays active regions. It is an SA(rs) type whose spiral arms form an almost complete, flocculent ring in the central regions. In the outer regions, the spiral arms extend into a pair of tidal tails which, due to their symmetric arrangement, appear like spiral arms as well. With a length of about 230 000 years, they contain, however, no young stellar populations, which makes it clear that they are tidal tails.

Apart from the galaxies described above, this very deep image also shows Galactic cirrus. This is light which is radiated by interstellar dust particles in our Milky Way and – like cirrus clouds in the atmosphere – in deep images is visible as cloud-like structures running across the field. Galactic cirrus was first detected on the photographic plates of the Palomar Observatory Sky Survey, the cataloguing of which began in the 1960s. The surface brightness of these structures is very low and is in the range 23–25 mag/arcsec^2.

Physically, this dust consists mostly of simple carbon compounds which originate in stellar winds or supernovae in the Galaxy and is concentrated mainly in the Galactic plane. The dust layer is about 330 light years thick, whereby the temperature of the dust particles is 20–30 K. Heated by starlight, these particles radiate in the far infrared with a maximum at about 100–300 μm. In the image, a veil of Galactic cirrus lies across NGC 7771 and NGC 7769. The wavy dust cloud is ochre in colour, in contrast with the blue spiral arms of the galaxies, since blue light is more strongly scattered than red. Thus the light which has passed through this Galactic dust curtain is reddened. The change in colour is enhanced by the fact that the illumination is due mainly to the bright galactic bulges consisting of population II stars.

OBJECT	NGC 7771
CONSTELLATION	Pegasus
RIGHT ASCENSION	$23^h\ 51^m\ 25^s$
DECLINATION	+20° 06′ 42″
BRIGHTNESS	12.2 mag
TYPE	SB(s)a
PHOTOGRAPHERS	Makis Palaiologou, Stefan Binnewies
TELESCOPE	600-mm reflector
CAMERA	SBIG STX-16803
EXPOSURE TIME	585 min
LOCATION	Skinakas Observatory, Crete, Greece

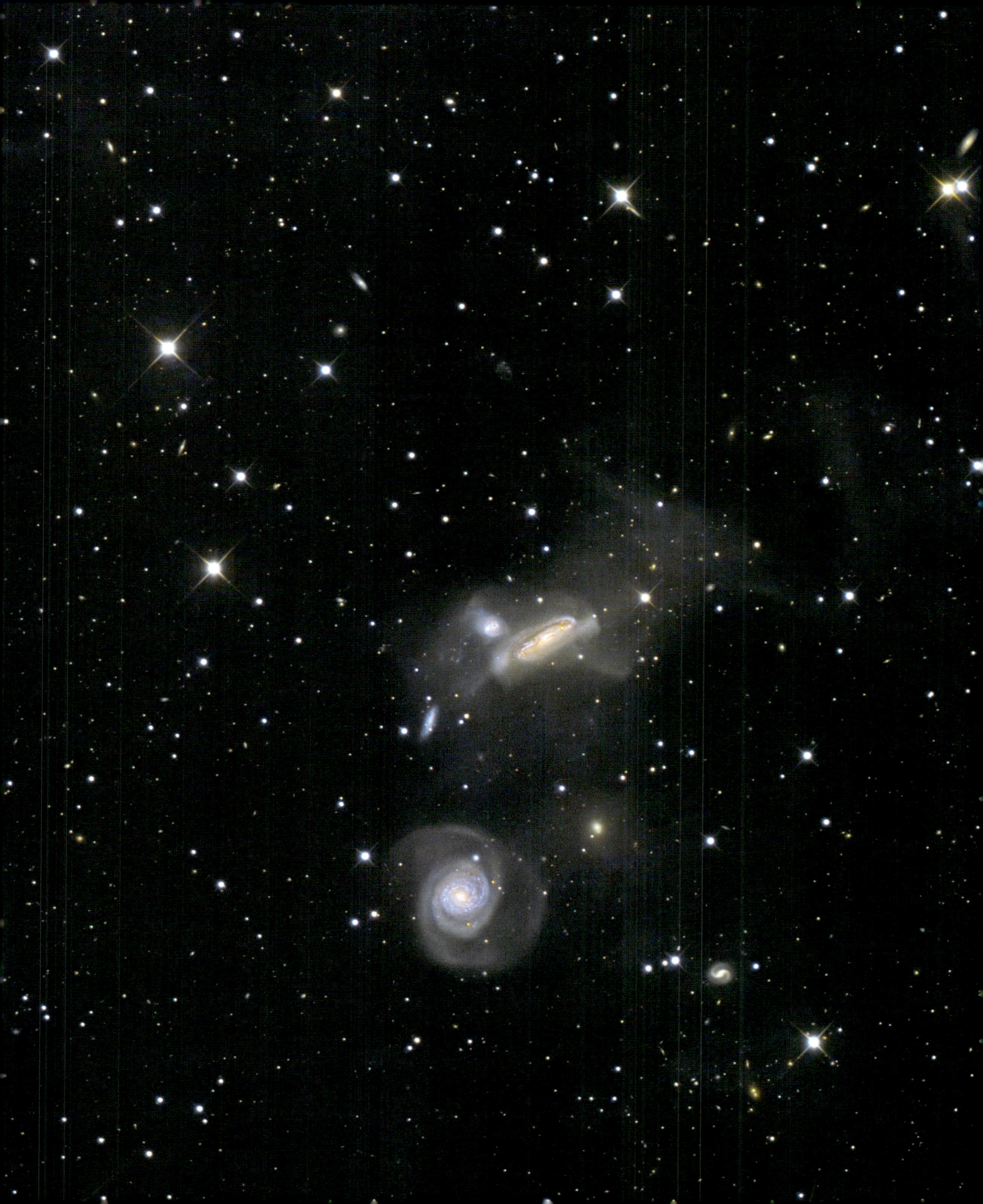

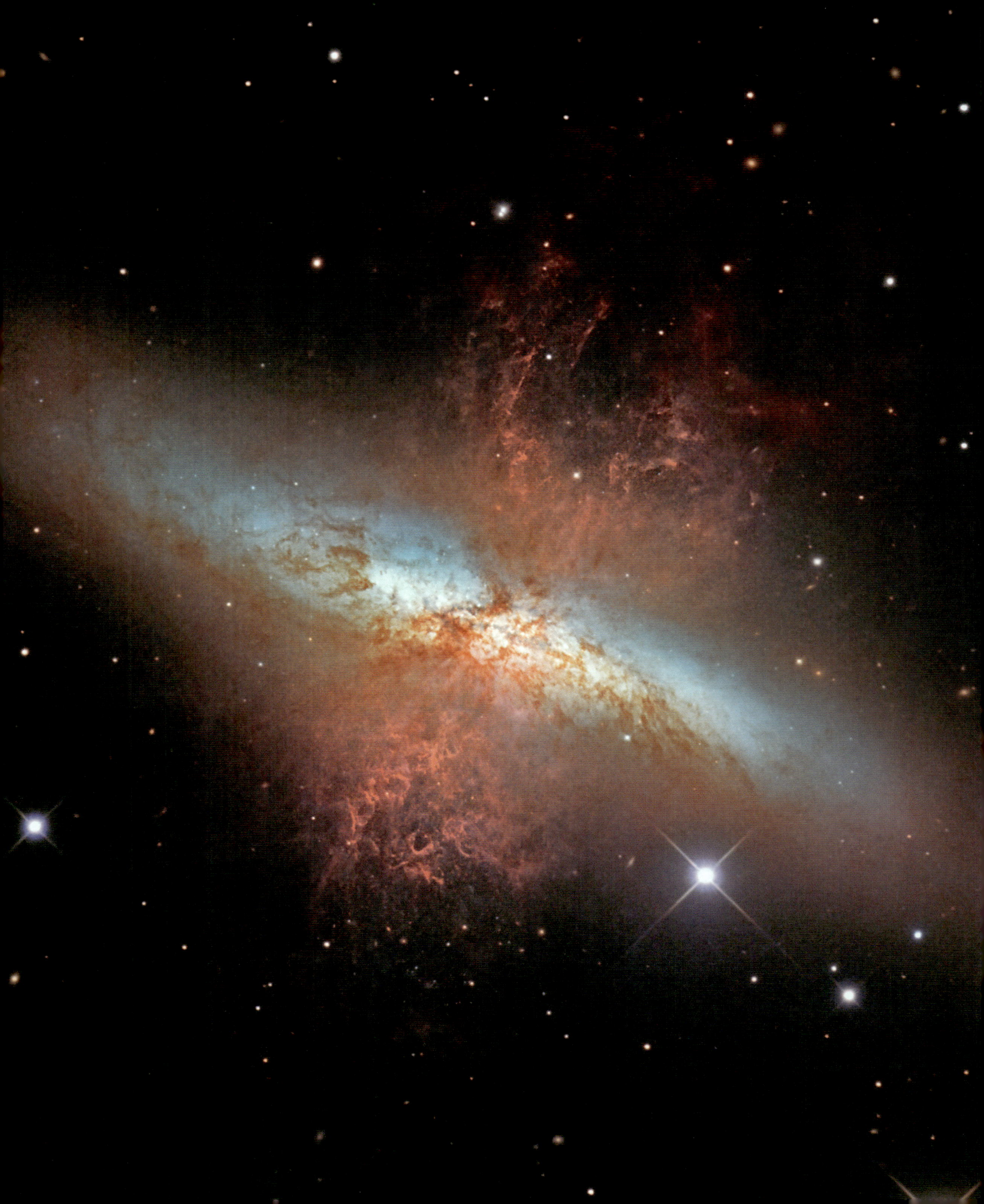

ACTIVE GALAXIES, QUASARS, AND GRAVITATIONAL LENSES

This chapter is concerned with objects which are only rarely observed by amateur astronomers. This is due in part to the fact that observing them is difficult and requires almost professional equipment.

THE MORPHOLOGY OF ACTIVE GALAXIES, QUASARS, AND GRAVITATIONAL LENSES

The objects in this chapter have a typical angular size of a few arcseconds. Apart from a few exceptions, long exposure times and optics with long focal length are needed in order to recognize details in the objects. A few quasars are achievable goals for today's astrophotographers, but gravitational lenses are a challenge even for dedicated amateurs.

Quasars, whose name stems from the abbreviation of "quasi-stellar" objects, are a prominent class in the family of active galaxies. They are highly luminous in the radio part of the electromagnetic spectrum, and today it is known that they are the very bright, active nuclei of distant galaxies. The high central luminosity is characteristic of active galaxies and will be described later in this chapter from the astrophysical point of view. More exactly, it is thus only the quasi-stellar, bright core region of the galaxy which is visible, since due to the great distance the dim outer regions are very faint or not visible at all. The morphology of quasars is much simpler than that of the various types of elliptical or spiral galaxies.

Only the nearest quasars allow one to recognize structures in the galaxies which contain them. However, first the bright core region must be removed via special image processing. When this is done, about half of the host galaxies of quasars show fragment-like rudiments of spiral arms or indications of past interactions. Such a host galaxy is two to three magnitudes fainter than the quasar and is located in the wings of its point-spread function, which has a width of only 1–2″. Under good seeing conditions and with elaborate post-processing it is possible to separate the spectrum of the quasar from that of the host galaxy. The isolated spectrum is similar to that of an active galaxy: the continuum contains prominent emission lines which indicate a strong ionization source. A few quasars have been discovered in non-active host galaxies as well. However, the brightest quasars are always found in active galaxies. Thus there must be a direct connection between the local processes in the galaxy, which transport gas and dust into the centre, and the high luminosity of the quasar.

The galaxy M 82 has an active nucleus whose luminosity has a maximum in the infrared (see page 323). Image credit: Makis Palaiologou, Stefan Binnewies (1.3-m reflector).

The brightest quasar is 3C 273. It has a luminosity of about 13 mag and is in the constellation of Virgo. Discovered in 1962 by Maarten Schmidt in a study of quasars in Ryle's *Third Cambridge Catalogue* (3C), the quasar has a redshift of $z = 0.156$, which corresponds to a distance of 2.12 billion light years. At the time, photographic plates showed only an inconspicuous star at the position of the radio source; no typical galaxy morphology was recognizable. Two years before the discovery of 3C 273, Allan Sandage and Thomas Matthews had already used radio interferometry to identify an optical counterpart of the radio source 3C 48, with a redshift of 0.367. It was not until the discovery of 3C 273, however, that the term "quasar" for these objects was introduced. Close examination of the corresponding Palomar Observatory Sky Survey plate reveals a small, elongated speck of light next to 3C 273, later associated with a visible jet. This jet does not originate directly at the quasar and is visible only at a distance of 11–20″ from the centre, corresponding to a projected distance of 100 000 light years. The peculiarity of the jet of 3C 273 is its visibility at both radio and optical wavelengths. It has a row of bright knots of emission whose brightness is variable. Spectroscopy has also revealed the existence of magnetic fields in the jet which surround it and bundle the plasma within it. An optical jet is also easily seen in M 87, whereby the core of this galaxy does not reach the luminosity of a quasar. This is an indication that the inner structure of galaxies always follows the same plan, causing the same phenomena, with the only variation that in the activity of the core.

Over the years variations have been observed at many wavelengths, indicating a particular activity of these objects, which are among the most luminous in the universe. Light curves show variations of up to 30 per cent within one to two years. The rise and fall in luminosity is not periodic as in a variable star, but appears to be random. Analyses of the temporal structure of the variation show long-term trends which are interrupted by short-term, rapid variation. Sometimes, very rapid flares are observed, with time scales of only a few hours in X-rays. These variations on various time scales make up the total observed variability of 3C 273.

Quasar variability makes it difficult to compile a list of the brightest, which could be observed by amateur astronomers. Based on the mean visual magnitude, 3C 273 and two other objects are mentioned in the literature as being brighter than 13 mag. However, more exactly these are not quasars but BL Lac objects. These are members of the family of active galaxies which, like quasars, are radio-loud and thus have above-average radio luminosity. Their designation refers to the prototype BL Lacertae – a variable active galactic nucleus. BL Lac objects are conspicuous due to the special perspective under which they are observed, in which the observer sees the galaxy along the radio jet pointed at us. This special geometry leads to amplification via relativistic effects, giving the jet a high radio luminosity. BL Lac objects also show even more rapid optical variation compared to quasars and their emission is up to 20 per cent polarized.

Based on the distribution of mean magnitudes of all known quasars, five quasars and a BL Lac object can be identified in the luminosity range 13–14 mag. The numbers increase rapidly as the quasars become fainter. There are thus 27 entries in the literature in the range 14–15 mag. Among these are two active galactic nuclei (AGNs) and four BL Lac objects. At even larger magnitudes the number of quasars increases even more, reaching its maximum at 19–21 mag. The corresponding redshifts lie in the range $z = 0.5–2.2$. After this maximum the number of fainter quasars decreases strongly. Only a few per cent of quasars known today are dimmer than 24 mag (corresponding to about $z = 3.5$). Large surveys have provided astronomers with an appreciable amount of data for statistical analyses. For example, the Sloan Digital Sky Survey (SDSS) has identified more than 40 000 quasars. Astronomers expect this to rise to 100 000 before the end of the survey.

As the most luminous examples, quasars have an exceptional status among active galaxies. They are the brightest and thus the longest-range measuring instruments in astronomical research and allow one to study in detail the mass distribution in the universe along the line of sight. It is not a coincidence that the first multiply-imaged source discovered was a quasar. This is the quasar QSO 0957+561A/B, found via radio observations in 1979 in the constellation of Ursa Major, also known as the twin quasar. The search for an optical counterpart found two: at a separation of 6″, two sources were found with very similar spectra typical of quasars and identical redshifts ($z = 1.41$). The double image is explained by astronomers as a result of the gravitational-lensing effect. In the decades since the discovery, the components A and B with luminosities of 16.7 and 16.5 mag have been studied in detail. This has shown that variations in the luminosity in the two light curves do not reach the Earth simultaneously. Instead, 417 days pass before a change in brightness seen in component A is seen in component B. Among the known multiple quasars are more exotic forms with triple or quadruple images; even a complete ring has been observed. In honour of Albert Einstein, who had predicted the gravitational-lensing effect, this ring is also known as an Einstein ring.

While photographing multiply-imaged quasars is not a difficult challenge for amateur astronomers, another form of the gravitational-lensing effect is more demanding. These are small arcs or arclets which were first found in 1986 in deep images of galaxy clusters. These arcs are oriented tangentially to the centre of the cluster and they have an extent from a few tenths of an arcsecond to several tens of arcseconds. The redshift of the arcs is appreciably higher than that of the cluster galaxies. Thus these objects lie at greater distances from the Milky Way. More detailed studies reveal that these are the elongated, strongly distorted images of distant galaxies.

THE ASTROPHYSICS OF ACTIVE GALAXIES AND QUASARS

Active galaxies are known as AGNs, for active galactic nuclei. This term is used for galaxies which are conspicuous due to intense radiation from their cores. The spectra of active galaxies and their time-variable luminosity from the radio to X-rays are the main differences to normal, non-active galaxies. Thus the luminosity of an active galaxy can be up to 10 000 times as large as that of a non-active galaxy with similar morphology. Statistical analyses also show that only a few per cent of galaxies are active. It is also conspicuous that many active galaxies appear relatively normal in the optical and in images are difficult to distinguish from non-active galaxies.

In addition to quasars, there are also fainter members of the family of active galaxies, notable more for their special emission-line spectra than for their high luminosities. These are the spiral systems with compact core known as Seyfert galaxies. They radiate a large part of their energy in the ultraviolet and infrared spectral ranges. The emission lines of the Seyfert galaxies can be separated into two groups based on their line width: the narrow-line and broad-line types.

This distinction allows conclusions about the velocities of ionized gas clouds which are in motion at the centre of the galaxy and responsible for the formation of the spectral lines.

These galaxies were named after the astronomer Carl Keenan Seyfert, who studied their spectra in the 1940s and discovered the peculiarity of their spectra. The brightest lines belong to the elements oxygen, hydrogen, and helium. In Seyfert galaxies of type 1 are also narrow forbidden lines such as O III with a spectral width of about 400 km/s but also lines with widths of up to 10 000 km/s which are emitted from the so-called broad-line region with higher gas densities. Seyfert-2 galaxies, by contrast, contain only narrow forbidden emission lines which are due to the so-called narrow-line region of low gas density. Over the years, astronomers have learned that the distinction of type 1 and type 2 is not sharp, but rather that the two types are just the ends of a scale. A Seyfert galaxy is perhaps always a mixture of the two types, classified by the relative strengths of the narrow and broad emission lines. This mixture plays an important role in the description of the generally accepted standard model of active galaxies, where a more precise classification with decimal points is used, such as Seyfert 1.5.

The appearance of emission lines is a general indicator of activity. However, there is a wide range of forms of such activity. In addition to Seyfert galaxies, starburst galaxies are often observed. They are the first form of an active galaxy and more common than Seyfert galaxies. Their emission lines stem from star-forming regions which lie at the centre of the galaxy and are composed of clusters of young, hot stars. Especially prominent in their spectra are the emission lines of the elements hydrogen, nitrogen, and oxygen, which are superimposed on the continuum. This spectral continuum of the galaxy nucleus is in normal galaxies dominated by stars of spectral type K such as Aldebaran. These display characteristic absorption lines of the elements magnesium and sodium. In starburst galaxies, the light of the active cluster is stronger than this absorption signature. Spectroscopic methods, which can also be used by amateur astronomers, can be used to learn about the dynamics of the core from the position of the emission lines. This results in the picture of circumnuclear star clusters which form a disc or ring structure orbiting the centre (see Figure 8.1).

A further subgroup of active galaxies are the Markarian galaxies, first described in the 1960s by the Armenian astronomer Benjamin Eghishe Markarian. They are characterized by a large fraction of ultraviolet radiation which is seen in excess appreciably larger than that in Seyfert galaxies. In both types of galaxies, this radiation is due to a non-thermal component caused by a strong ionization source at the centre.

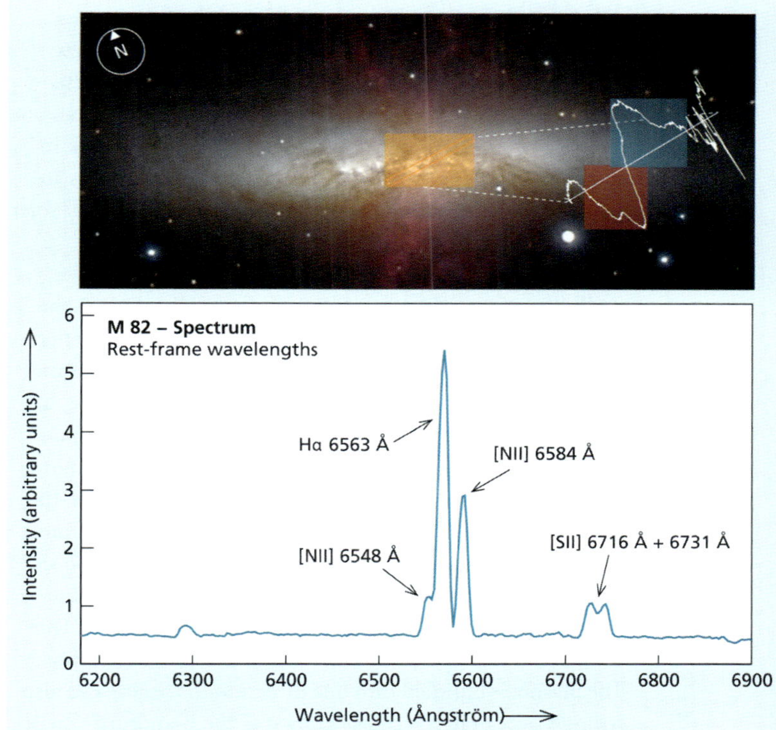

Figure 8.1: Part of the optical spectrum of the starburst galaxy M 82. Top: The slit of the spectrograph crosses the bright core of the galaxy. The red- and blue-shifted portions indicate a rotating disc of gas with a diameter of about 3000 light years. Bottom: Emission lines of the elements hydrogen, nitrogen, and sulphur (M. König, grating spectrograph Dados, 300 lines/mm, 14-inch reflector, 180 min exposure time).

By the 1970s there were discussions about a possible standard model of active galaxies, with the goal of unifying the observed anisotropic emission of various AGNs. In 1973, Nikolai Shakura and Rashid Sunyaev finally suggested a model which was later relativistically revised by Igor Novikov and Kip Stephen Thorne. This model describes a central radiation engine – a supermassive black hole – which is surrounded by a rotating disc of dust and gas. Material from outside is accreted in this disc, which loses angular momentum via friction and thus spirals slowly to the inner edge of the disc. The magnetic field of the black hole affects the edge of the disc so that material ionized by X-rays flows into the black hole in the form of plasma. The accretion disc thus fuels the central radiation engine and is so strongly heated by the gravitational energy of the infalling plasma that its inner edge becomes a strong X-ray source. Farther out, the temperature of the disc decreases, until finally it is surrounded by a torus of dense dust and gas. This torus strongly blocks the view of the centre and makes observations depend on the viewing angle. Near the centre of the active galaxy are dense clouds of ionized gas responsible for the formation of broad emission lines, the so-called broad-line region. These are visible only when the view is directly from above or below, perpendicular to the disc. If the view is from the side, the torus blocks the inner region. The view into the centre is thus no longer possible, so that only lower-density gas clouds farther out are visible. These form the narrow-line region. The dust torus functions as a reservoir of material for the accretion disc and is the link to the surrounding galaxy. Typical diameters of the toruses are 10–20 light years and their masses range from 10^5 to 10^{10} solar masses (see Figure 8.2).

The accretion raises the question of what happens to the material transported into the centre. A large part of it falls into the black hole. A smaller part, however, can escape and leave the system if the black hole is rotating. In this case, it also causes rotation of the space–time in the centre, which leads to the magnetic field of the inner disc becoming more and more twisted. The wound-up field lines form a tube-like structure along which relativistic particles can escape from the centre. Such jets thus form perpendicular to the two sides of the plane of the disc. If the magnetic-field lines become too twisted, the tube structure can become decoupled. This causes the magnetic energy to be transferred to the plasma. The plasma is bundled into one direction in the jets and also strongly accelerated – this is known as the Blandford–Znajek mechanism. This explains how the material can be transported far to the outside and why radio galaxies have lobes extending over hundreds of thousands of light years (see Figure 8.3).

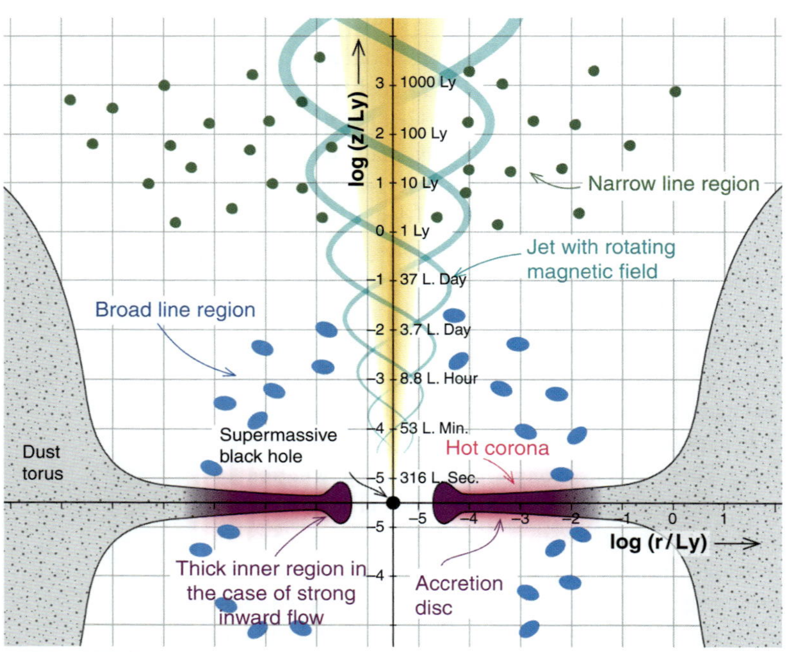

Figure 8.2: Schematic of the structure of the core of an active galaxy. The jet shown occurs in radio-loud active galaxies (above); in radio-quiet examples it is lacking (below). The logarithmically displayed quantities are valid for a typical quasar. The mass of the supermassive black hole at the centre is about 10^6 solar masses.

Active galaxies radiate from the radio through the infrared and optical to X-rays and gamma rays. They thus cover the entire electromagnetic spectrum. Their spectral-energy distribution consists of several components which contribute to the spectrum in characteristic amounts. If the centre of an active galaxy could be seen directly, the warm dust torus could be observed due to its infrared glow. The accretion disc, by contrast, is hotter than the torus and its thermal radiation is in the optical. However, the maximum of the radiation distribution in the disc is in the ultraviolet, so that it would appear blue to a visual observer. The temperatures at the centre of an active galaxy are sufficiently high to turn the central gas into plasma and thus produce a large number of high-energy electrons. If low-energy radiation, such as the thermal radiation from the dust torus or accretion disc, or even cosmic background radiation, impacts these electrons, then it is converted into energetic X-ray or gamma radiation. The high-energy radiation thus produced can cause the accretion disc to fluoresce. The X-ray spectrum displays a characteristic iron line which plays a role in X-ray astronomy similar to that of the 21-cm line of hydrogen in radio observations.

On the basis of the standard model and additional variations in scale and other geometric parameters, the described characteristics of the various types of active galaxies can be plausibly reconstructed. The three deciding components are (a) the mass of the black hole, which varies from a few million to a few billon solar masses; (b) the accretion rate, which determines the amount of material which is transformed into radiation and in the case of highly luminous AGN can be a few solar masses per year; and finally (c) the viewing angle at which the accretion disc is observed, which decides what type of AGN we see. Finer adjustments involve the rotational parameters of the black hole, the position of the accretion disc relative to the dust torus, its mass, and surrounding magnetic fields.

THE ASTROPHYSICS OF GRAVITATIONAL LENSES

The concentration of quasars at redshifts in the range $z = 2.2–2.5$, corresponding to distances of 10 to 11 billion light years, makes this time the “epoch of quasars”. The density of quasars then was 1000 times larger than today, since more frequent interaction between early galaxies in the young, more compact universe caused more active galaxies and hence more quasars to form. A peculiarity of quasars is their enormous luminosity, which makes the gravitational-lens effect possible, since only bright sources are visible at such distances. One distinguishes between two physical effects which are both due to a change in the curvature of space due to mass.

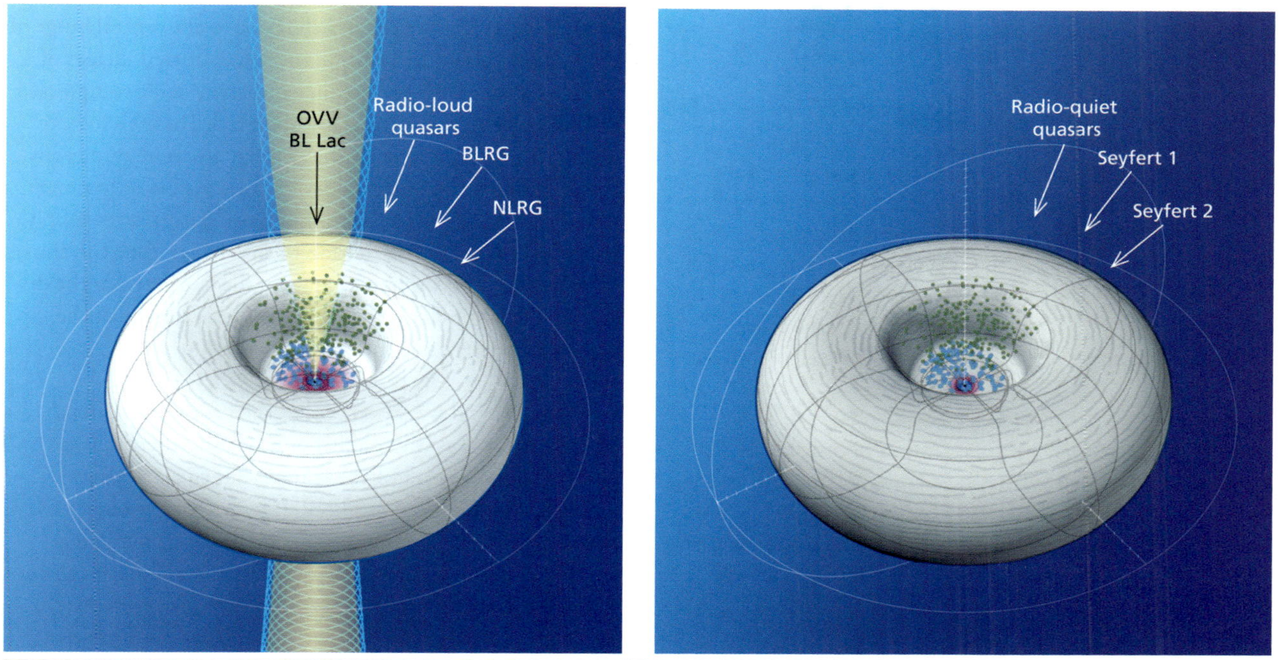

Figure 8.3: Schematics of the core region of active galaxies. The various angles under which it is seen correspond to different types of active galaxies. In radio-loud examples, the radio jets extend from a few 10 000 to 100 000 light years and exceed the dimension of the dust torus by a factor of 100 to 1000. (The abbreviation OVV stands for "optically violent variable".)

The first is the strong gravitational-lens effect which occurs when the light of a quasar on its way to us passes by a compact mass, such as a galaxy or rich galaxy cluster. The weak gravitational-lens mass, on the other hand, is the effect on the image morphology due to a less compact mass distribution. In principle, a gravitational-lens effect can occur only when the quasar, the mass acting as the lens, and the observer are aligned. Light then passes through space-time curved by the lens, where it is deflected. However, there is more than one path to the observer. This means that more than one image of the quasar is visible. The strong gravitational-lens effect causes not only multiple images but also an amplification of the quasar light. It is due to the strong focussing of the lens and can amplify the intensity by a factor of 20. Since due to their great distances quasars have a lower surface density than other types of galaxies and are also almost point-like, multiple images can be effectively studied.

It is important to distinguish multiple images from the few cases in which the source itself consists of a physical quasar pair. The members of such a pair are at the same distance, but show differences in their spectra and variability which are not compatible with a single, multiply-imaged object. The first such double quasars were detected at the end of the 1980s, though some astronomers still doubted their existence. Only with Chandra observations could it be shown that there is no galaxy cluster along the line of sight to the quasar pair which could act as a lens, since this would reveal itself via the X-ray emission from its cluster gas. Since the X-ray spectra of the two AGNs were different as well, it must be a physical pair consisting of separate sources – proof of the existence of physical double quasars.

With the help of the gravitational-lens effect it is also possible to observer quasars as if through a microscope. The special geometrical arrangement thus makes details of the quasar visible. Astronomers have also found that the Einstein rings found in the optical and infrared are not as common in the radio. This is due to the fact that the centre of the radio emission does not usually coincide with that of the quasar. Most radio emission of the quasar comes from radio lobes which are farther from the centre and thus not along the line of sight. They thus don't fulfil the requirements for the gravitational-lens effect. For this reason, optical Einstein rings usually don't have a counterpart in the radio.

As with the images formed by glass lenses in conventional optics, one can write a lens equation for gravitational lenses. Also as in conventional optics, caustics can form. These are complex diffraction figures which occur when the linear arrangement is perturbed. Astronomically, these are visible as small arcs which are oriented tangentially to the centre of mass. The extent of the arcs can be up to a few arcseconds when a large galaxy cluster is the lens responsible for the weak gravitational-lensing effect. In this case, the mass concentration is not as strong, so no multiple images are observable as with strong gravitational-lensing, but rather a distribution of arcs. These arc-shaped images often show a

REDSHIFTS, Z VALUES, AND THE RELATION TO THE COSMIC DISTANCE SCALE

In 1929, Edwin Hubble discovered that the spectral lines of distant galaxies display a redshift. This redshift is due to the expansion of the universe, in which along with the expansion of space the wavelength of electromagnetic radiation is also stretched. This cosmological redshift is measured via the corresponding shift in the wavelength and expressed as the dimensionless number z. The shift in wavelength $(\lambda - \lambda_0)$ contains the measured wavelength λ, from which the rest-frame wavelength of the emitted radiation λ_0 is subtracted. Thus z is calculated as: $z = (\lambda - \lambda_0)/\lambda_0$.

The derived recession velocity of a galaxy increases approximately linearly with distance and the constant of this relation is known as the Hubble constant H_0. Its current value is about 70 km/s/Mpc. Our current universe corresponds to the value z = 0. Objects whose light has reached us after 7.5 billion years have a redshift of z = 1. When this light was emitted, the universe was only half of its current size; at z = 2 a third, at z = 3 a quarter, etc.

The table shows light-travel time as a function of redshift. This is the time the light has needed in order to travel from the object to the Milky Way. The values on which the distances are calculated imply a current age of the universe of 13.7 billion years after the Big Bang (source: www.astro.ucla.edu/~wright/CosmoCalc.html; with H_0 = 69.9 km/s/Mpc, $\Omega_M = 0.286$, $\Omega_{vac} = 0.714$). At a redshift of z = 1000, thus about 400 000 years after the Big Bang, recombination took place, during which radiation decoupled from matter, electrons and nuclei formed atoms, and the universe became transparent to light.

Redshift	Light travel time (billion years)
z = 0.001	0.014
z = 0.005	0.070
z = 0.01	0.139
z = 0.05	0.675
z = 0.1	1.305
z = 0.5	5.071
z = 0.75	6.622
z = 1	7.784
z = 2	10.360
z = 3	11.500
z = 5	12.481
z = 9	13.113
z = 10	13.186
z = 1000	13.661

The redshift record (as of October 2010) for a galaxy, in the Hubble Ultra Deep Field, is z = 8.85 and corresponds to a light-travel time of 13.1 billion light years. The light of this galaxy, whose luminosity is about 30 mag, was emitted 600 million years after the Big Bang, shortly after the universe became transparent to light. The difficult measurement of the redshift of this galaxy (designated UDFy-38135539) needed 16 hours of exposure time on the Very Large Telescope. This cosmological expansion effect should not be confused with shifts in wavelength due to the relative motion of emitter and receiver. The latter is the Doppler effect, which can result in a redshift or blueshift depending on the direction of motion.

brightening at the centre, which is due to the central bulge of the imaged galaxy. In addition to arcs, distorted images of spiral galaxies in various colours and forms with compressed and overlapping spiral arms are observed. These arclets are not as strongly distorted as the arcs but are more numerous. As predicted by imaging theory, arcs are visible only along so-called critical curves. The amount of distortion is simply related to the mass distribution of the cluster. One can thus calculate the distribution of mass in the cluster from the distortion. This allows one to measure the total mass of the cluster and determine the amount of dark matter.

The third and weakest form of the gravitational-lens effect is known as microlensing. It occurs when a massive object passes in front of the star so that the lens effect makes the star appear brighter. The mass of the object in between, which acts as a gravitational lens, is much less than in the case of strong or weak lensing and lies in the range from planetary masses to a few solar masses. Microlensing has already been observed for stars in the core of our Milky Way, whereby the luminosity increased by about a magnitude. The process of this brightening and dimming lasts several months. Since such an event is also very rare, astronomers must monitor the brightness of millions of stars and reject the disturbing contributions from unknown variable stars.

BIBLIOGRAPHY

Courbin, F.: *ESO Messenger – Bright Quasars' Host Galaxies*, 2006, www.eso.org/sci/publications/messenger/archive/no.124-jun06/messenger-no124-32-36.pdf

Danzer, J.: *Gravitationslinsen*, 2002, http://physik.uni-graz.at/~cbl/C+P/contents/Stud-WS02/danzer/linse.htm

ESO press release, *Die entfernteste Galaxie lichtet den kosmischen Nebel*, 2010, www.eso.org/public/germany/news/eso1041/

Fendt, C.: *Aktive Galaxienkerne und Quasare*, 2003, www.mpia-hd.mpg.de/homes/fendt/Lehre/Vorlesung_AGN/index.html

Gregg, M. D. et al.: *The first bright QSO survey*, The Astronomical Journal, **112**, 1996

Gährken, B.: *Polarisation bei astronomischen Objekten*, 2007, www.astrode.de/polarise.htm

Heasarc Goddard Space Flight Center: *MILLIQUAS – Million Quasars Catalog*, 2014, http://heasarc.gsfc.nasa.gov/W3Browse/all/milliquas.html

Huchra, J. P.: *The nature of Markarian Galaxies and studies of star formation in blue galaxies*, PhD thesis, California Institue of Technology, 1997

ISDC Data Centre for Astrophysics, *3C 273 Database*, 2008, http://isdc.unige.ch/3C 273/

Jackson, N.: *A Bull's Eye for MERLIN and the Hubble*, 1998, www.merlin.ac.uk/press/PR9801/press.html

Jester, S.: *SLOAN DSS Bright Quasar Survey*, 2005, http://iopscience.iop.org/1538-3881/130/3/873/pdf/204612.web.pdf

Keel, B.: *The brightest quasar: 3C 273 and its jet*, 2002, www.astr.ua.edu/keel/agn/3C 273.html

Kudriavtcev, I.: *Specific features of the average magnitude of quasars as a function of redshift*, 2011, http://arxiv.org/ftp/arxiv/papers/1109/1109.3630.pdf

König, M.: *Zeitvariabilität in Aktiven Galaxien,* Dissertation, 1997, http://astro.uni-tuebingen.de/publications/koenig-diss.shtml

Mellier, Y. and Fort, B.: *Arc(let)s in clusters of galaxies*, Astronomy and Astrophysics Review, **5**, 1994

Müller, A.: *Astro-Wissen*, 2007, www.wissenschaft-online.de/astrowissen/lexdt_a02.html

NASA Space Telescope Science Institute: *Quasar Host Galaxies*, 2000, http://hubblesite.org/hubble_discoveries/10th/photos/slide34.shtml

Open University: *Models of active galaxies*, 2012, http://openlearn.open.ac.uk/mod/oucontent/view.php?id=398724&section=7.4

Richards, G. T. et al.: *Identification of QSOs in SDSS*, The Astronomical Journal, **123**, 2002

Rincon, P.: *Astronomers see first quasar trio*, 2007, http://news.bbc.co.uk/2/hi/science/nature/6243361.stm

Scherl, B.: *Gravitationslinsen*, 2011, http://pulsar.sternwarte.uni-erlangen.de/wilms/teach/astrosem_ss11/scherl.pdf

Schmidt, M.: *3C 273 A Star-like Object with Large Red-Shift*, Nature, **197**, 1963

Schneider, P. et al.: *Gravitational Lenses*, Springer Verlag, 1992

Schuh, P.: *Unifikationstheorien von Aktiven Galaxien*, 2011, http://app.uni-dortmund.de/~backes/AGNseminar11-Dateien/04_Schuh-Unifikation.pdf

Steinicke, W.: *Echte und unechte Doppelquasare*, 2012, www.klima-luft.de/steinicke/AGN/magellan/dq.htm

Wright, E. L.: *Redshift CosmoCalc*, 2008, www.astro.ucla.edu/~wright/CosmoCalc.html

Abell 370

The galaxy cluster Abell 370 consists of several hundred galaxies and is the most distant cluster in the Abell catalogue. Its redshift is given as $z = 0.375$, which implies a light-travel time to the Milky Way of 4.1 billion years ($H_0 = 70$ km/s /Mpc). The angular size of ABell 370 is 11′ and with a scale of 1.3 million light years per arc minute one obtains the enormous projected diameter of 14.8 million light years. The image shows the densely populated centre of Abell 370 between two foreground stars which belong to the Milky Way. One recognizes two dominant elliptical galaxies at the centre, separated by only 37″, corresponding to about 800 000 light years., Their luminosities are 18.2 and 18.9 mag. The Sb galaxy LEDA 175370 at the left edge of the image has a luminosity of 16.2 mag and is 600 million light years away.

In the image, one can study some effects of gravitational lensing. These are especially obvious in an image taken by the Hubble Space Telescope (see inset). Conspicuous are several arcs which appear as blue and red streaks. Spectacular is the arc right of the southern large elliptical galaxy, dubbed the Dragon by NASA scientists. Such an extreme distortion indicates that the arc is near the Einstein radius of the cluster. At the lower end of this arc, thus in the head of the Dragon, the lensed image of a spiral galaxy is visible; the tail is caused by the elongated image of at least one other spiral galaxy. These spiral galaxies lie at $z = 0.725$ and thus 6.5 billion years of light-travel time ($H_0 = 70$ km/s/Mpc) behind Abell 370.

The distortion of the imaged galaxy can be described mathematically by the second derivative of the gravitational potential of the cluster. The effect of the strong gravitational-lens effect consists of the concentration of light and is described by the magnification factor μ. For typical values of $\mu = 2$–5 one obtains $2.5 \times \log(\mu) = 0.8$–$1.7$ mag, which reduces the (unlensed) luminosity of just-detectable objects to fainter than 30 mag. One thus attains with the Hubble Space Telescope now a limit which is comparable to that of the planned James Webb Space Telescope.

In Abell 370, astronomers also discovered the galaxy HCM-6A with the 10-m Keck II telescope on Hawaii. Its redshift of $z = 6.56$ corresponds to a light-travel time of 12.9 billion years ($H_0 = 70$ km/s/Mpc). At the time of its discovery, HCM-6A was the most distant galaxy known.

In the last 20 years, in addition to the strong lens effect, the weak lensing effect has been studied. This is due to the light rays of distant galaxies being distorted and twisted when passing through mass inhomogeneities of the cluster. This is known as the tidal component of the gravitational field, which causes the weak gravitational-lens effect. This can't be detected by multiple images as in the strong lens effect. Rather this is done as follows: the number density of galaxies in deep exposures is about 35 per square arcminute (down to a luminosity of 25 mag), their orientation is random. The action of the tidal field changes the mean orientation of the galaxies and from the observed preferred direction one can draw conclusions about the matter distribution. One thus obtains an image of the mass of the galaxy cluster. This mass map also indicates the distribution of visible galaxies; arcs appear tangential to the cluster centre. Their axial ratio decreases with increasing distance from the centre, but their number increases. At the edge of the cluster one can hardly recognize them as distorted images. Both gravitational-lens effects indicate for Abell 370 a disc-like mass distribution and a mass of about 3.8×10^{14} solar masses for the inner region with diameter 800 000 light years.

OBJECT	Abell 370
CONSTELLATION	Cetus
RIGHT ASCENSION	$02^h\ 39^m\ 51^s$
DECLINATION	−01° 35′ 08″
TYPE	Galaxy cluster
PHOTOGRAPHERS	Philipp Keller, Konstantin Buchhold, Bernd Flach-Wilken, Johannes Schedler, Volker Wendel (Chart 32-Team)
TELESCOPE	800-mm reflector
CAMERA	FLI Proline 16803
EXPOSURE TIME	860 min
LOCATION	CTIO, Chile

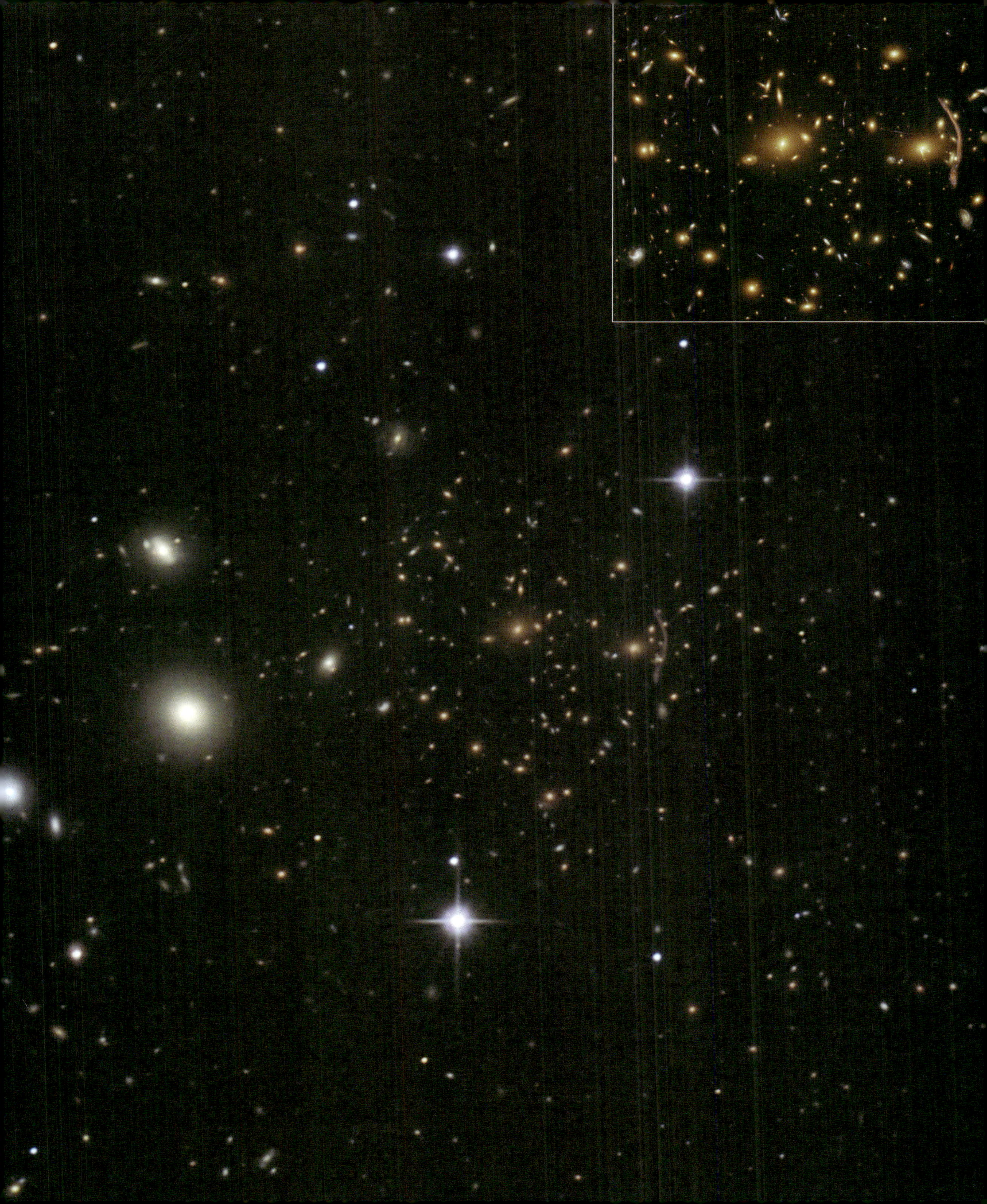

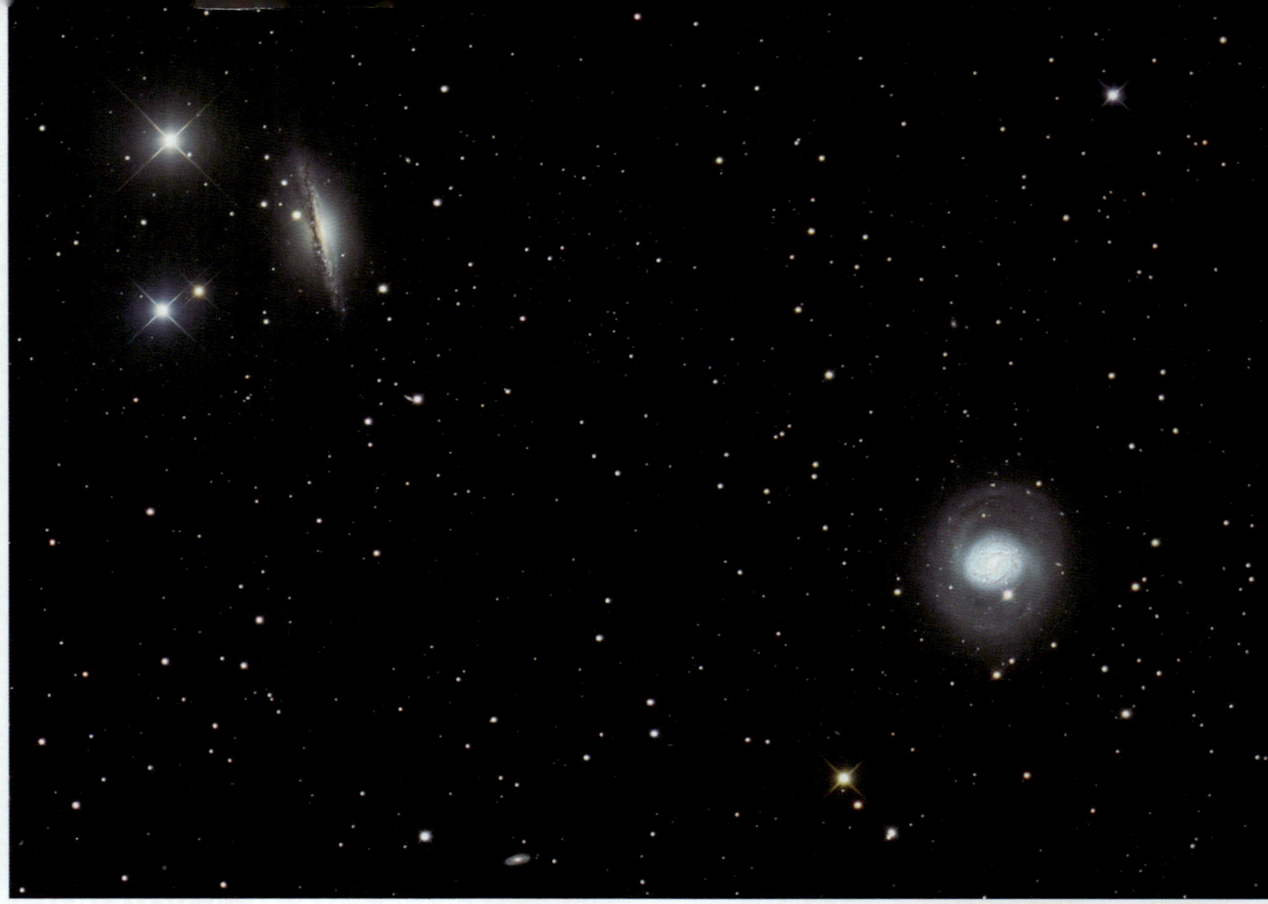

M 77

At the right in the image is Messier 77, an SA spiral galaxy, also known in the literature as NGC 1068 and Arp 37. Farther to the north near two foreground stars is the edge-on early-type barred spiral (SBb) NGC 1055. The two galaxies lie at similar distances of about 48 million light years from the Milky Way. The angular separation of 30′ corresponds to a distance of 420 000 light years, close enough for the galaxies to interact. The tidal forces involved can affect material in the galactic planes and also mutually induce an active phase due to mass accretion at the centre. In M 77, this activity is seen in its classification as a Seyfert-2 galaxy. It is the nearest and brightest Seyfert galaxy. It is also known as the radio source 3C 71 and as an ultraluminous infrared galaxy (ULIRG).

Modern methods of infrared interferometry allow high-resolution observations of molecular clouds. In M 77, these were used to detect the transport of gas into a region only 18″ from the core, which corresponds to about 45 light years and thus the size of an AGN dust torus. Due to the low dispersion of the measured gas velocities, one suspects a disc of molecular gas. This disc has a thickness of 33 light years in the inner regions and 330 light years in the outer regions, to which the spiral arms are connected. The advantageous face-on view allows for exact velocity measurements in M 77 which can be used to test complex kinematic models. These have shown that the gas disc has a central mass of about 17 million solar masses and is probably not flat but rather strongly warped.

OBJECT	M 77, NGC 1068, Arp 37
CONSTELLATION	Cetus
RIGHT ASCENSION	$02^h\ 42^m\ 41^s$
DECLINATION	−00° 00′ 48″
BRIGHTNESS	9.6 mag
TYPE	(R)SA(rs)b
PHOTOGRAPHERS	Robert Gendler
TELESCOPE	320-mm and 500-mm reflector
CAMERA	SBIG STL-11000
EXPOSURE TIME	1800 min
LOCATION	Nighthawk Observatory, New Mexico, USA

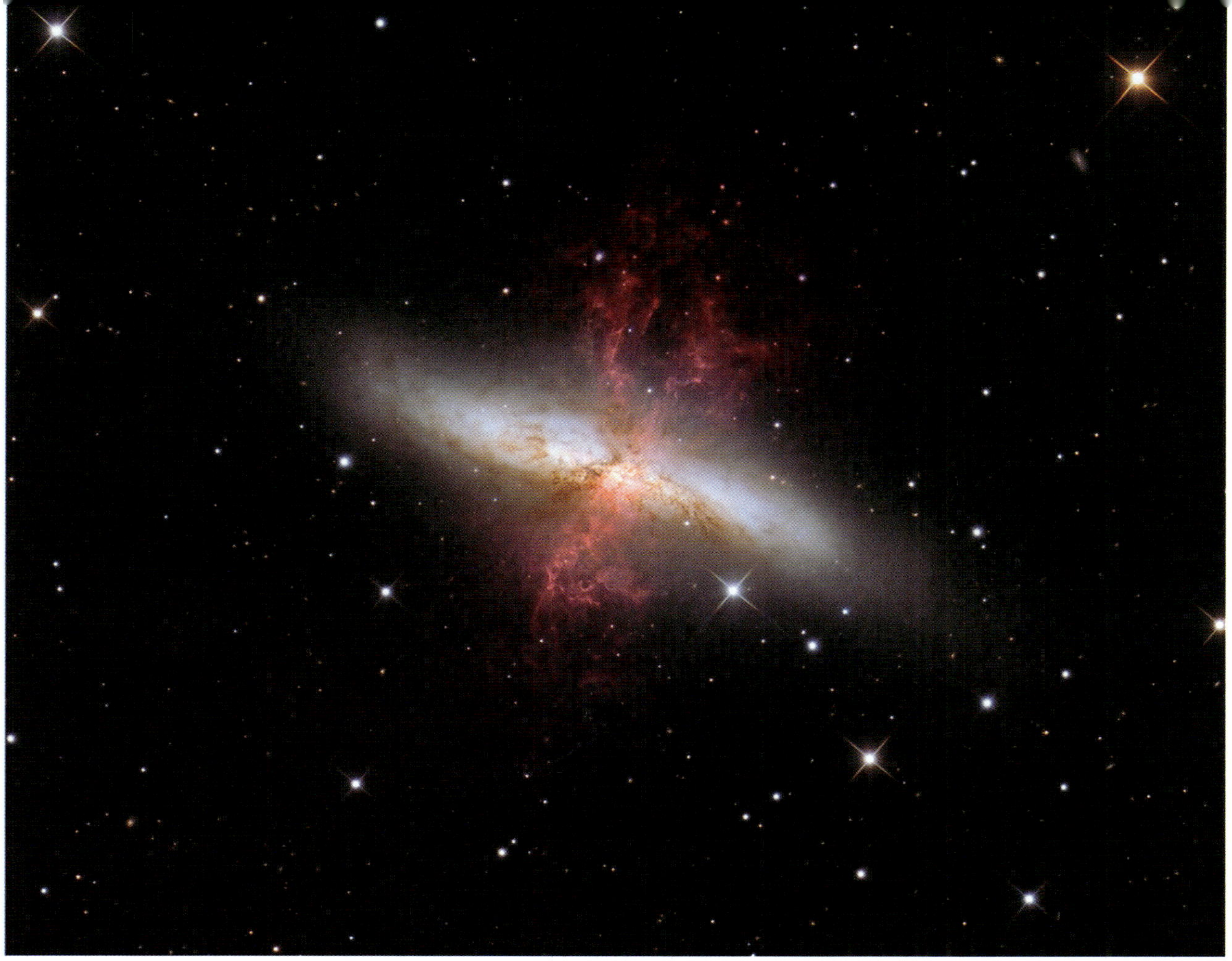

M 82

The irregular galaxy M 82 (NGC 3034) with a diameter of 11′ and a visual luminosity of 9 mag one of the largest and brightest galaxies in the *Atlas of Peculiar Galaxies* (Arp), where it is listed as number 337. Even Halton Arp couldn't unambiguously classify this galaxy, which is why Arp 337 is in the last of the 39 classes, "Miscellaneous". M 82 is the closest starburst galaxy and the activity is due to a previous interaction with the neighbouring galaxy M 81 in which it must have lost most of its spiral structure. The length scale at the distance of 15 million light years corresponds to about 4400 light years per arcminute, resulting in a diameter of 48 000 light years for M 82.

In the optical, NGC 3034 reveals prominent, filament-like structures running from north to south. The central region of this active galaxy displays a structured dust lane in which several active star-forming regions have been observed. These super star clusters contain many hot OB stars. In the central region, which covers about 1000 light years, are superwinds perpendicular to the galactic plane which fuel the outer filaments. The central starburst region lies at the colliding vertices of two cone-like structures with opening angles of 50° and sizes of a few thousand light years; these transport dust and gas to the outer regions. Thus the flow is much less focused than is the case in the jets of radio-loud active galaxies, which exceed M 82 in mass and size.

OBJECT	M 82, NGC 3034, Arp 337
CONSTELLATION	Ursa Major
RIGHT ASCENSION	$09^h\ 55^m\ 53^s$
DECLINATION	+69° 40′ 46″
BRIGHTNESS	9.3 mag
TYPE	I0 edge-on
PHOTOGRAPHERS	Adam Block
TELESCOPE	800-mm reflector
CAMERA	SBIG STX-16803
EXPOSURE TIME	2480 min
LOCATION	Mount Lemmon SkyCenter/ University of Arizona, USA

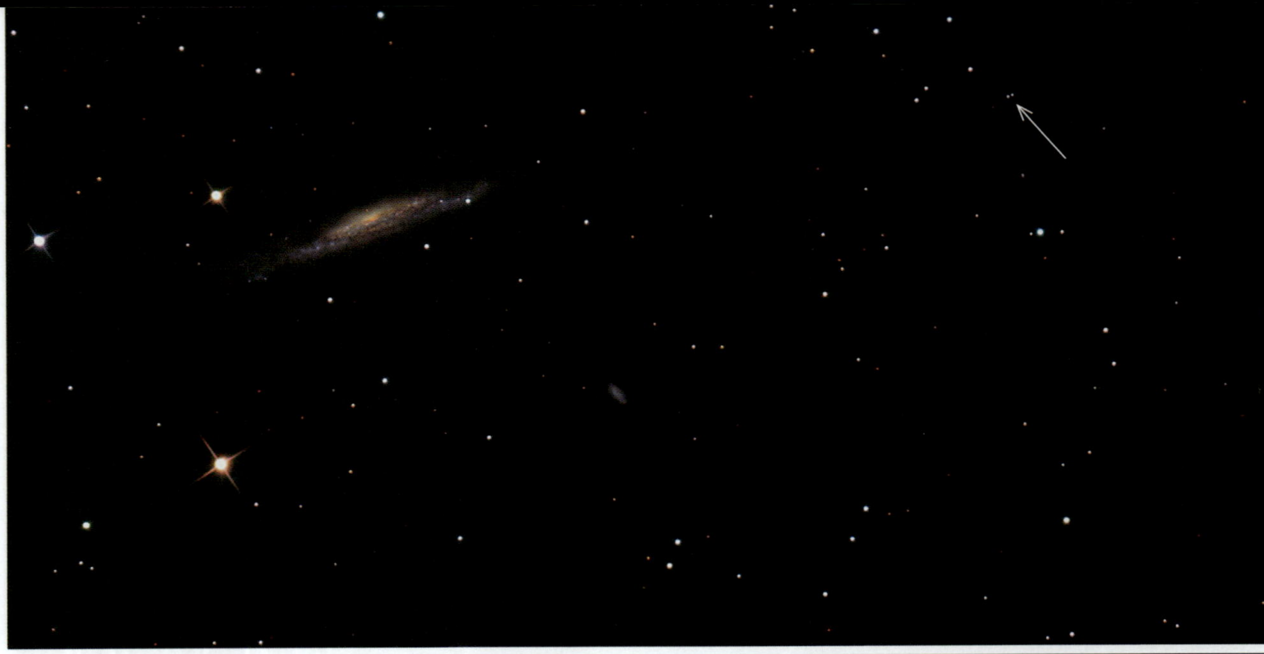

NGC 3079

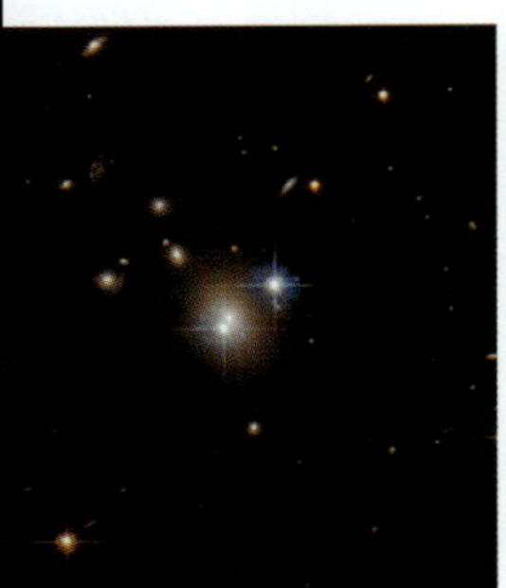

The image is in the direction of Ursa Major and shows the prominent object NGC 3079, an SB galaxy of size 8′ × 1.5′, which is listed as an active galaxy since it is a starburst galaxy with many star-forming regions in the spiral arms, but also because it has the typical emission-line spectrum of a Seyfert-2 galaxy. It is thus an active galaxy whose accretion disc is not directly observable since the surrounding dust torus blocks the view. The Hubble Space Telescope detected in NGC 3079 a large bubble structure lying in the central bulge and extending over about 3000 light years. This bubble contains hot gas from supernova explosions and the stellar winds of hot stars in the central star-forming regions. The Hubble Space Telescope also revealed Hα filaments which extend into the bubbles. The Canada–France–Hawaii Telescope was used to measure the enormous velocity of six million kilometres per second in these ascending gas filaments.

Extending the plane of the galaxy by a factor of 1.5 to the north (right) brings one to an apparent double star (see arrow). This is the gravitational-lens system GL Q0957+561, mentioned in the introduction to this chapter. The small additional image is from the Hubble Space Telescope and has the components of the quasar QSO 0957+561 at the centre. Close by to the left, southern B component, an elliptical galaxy 5–6 mag fainter is visible. The galaxy designated Q0957+561 G1 with a redshift of z = 0.36 lies in front of the quasar and acts as a gravitational lens. This is visible as a double image of the quasar with a separation of 6.2″. The distance of the galaxy from the Milky Way is 3.7 billion light years and its angular size of 0.42′ corresponds to a projected diameter of 380 000 light years. This suggests that Q0957+561 G1 is a cD galaxy within a galaxy cluster. The core of the galaxy is only an arcsecond away from the left B component of the quasar image. The redshift of the quasar itself is z = 1.41 and its distance thus 8.7 billion light years. The focussing effect of the gravitational lens makes the B component somewhat larger and brighter than the right A component, which is the unperturbed image of the quasar. Radio observations also show a one-sided jet which extends a few arcseconds out of the centre, but only in image A. This means that it lies outside of the region which is multiply imaged. However, VLBI observations with milli-arcsecond resolution are able to detect features in the smaller multiply-imaged region, and found two symmetric jets, one in each component, confirming that the double quasar Q0957+561 is the result of multiple imaging.

OBJECT	NGC 3079
CONSTELLATION	Ursa Major
RIGHT ASCENSION	$10^h\ 01^m\ 58^s$
DECLINATION	+55° 40′ 47″
BRIGHTNESS	10 mag
TYPE	SB(s)c edge-on
PHOTOGRAPHERS	Johannes Schedler
TELESCOPE	400-mm reflector
CAMERA	SBIG STL-11000
EXPOSURE TIME	150 min
LOCATION	Panther Observatory, Wildon, Austria

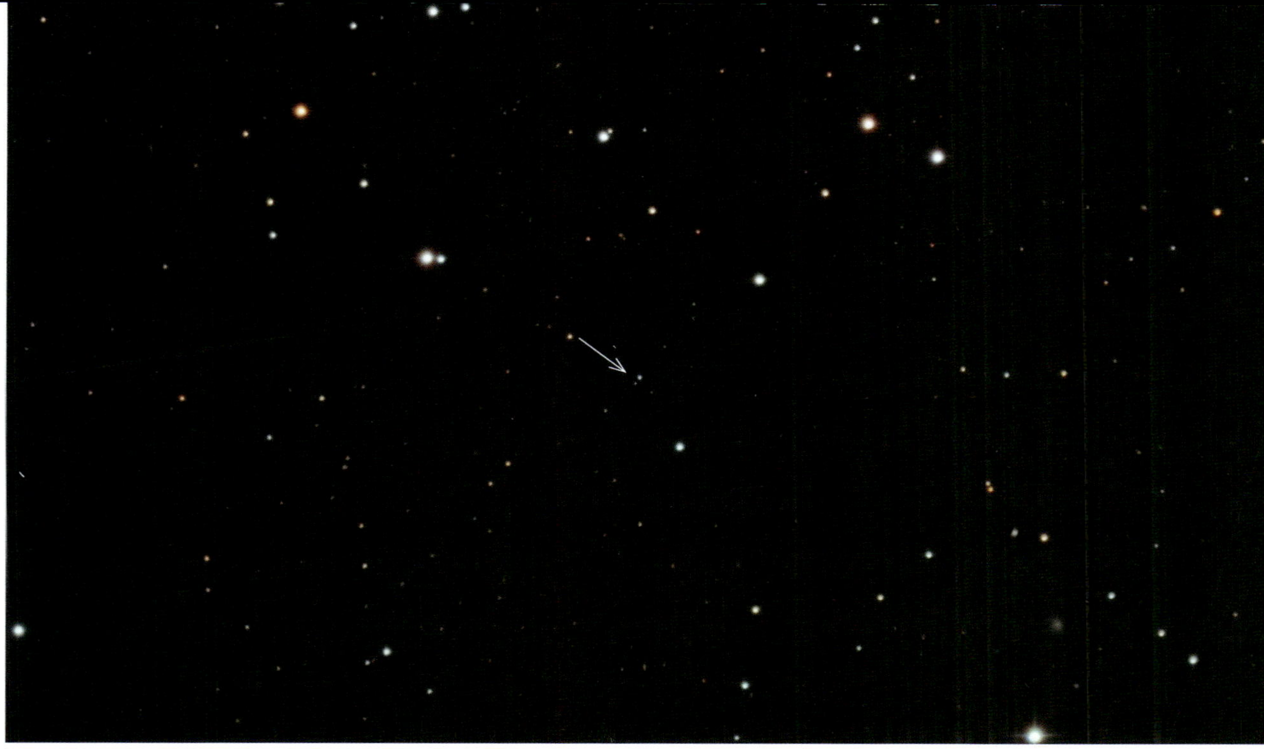

HS 1216+5032

At the centre of the image are two close, star-like objects (see arrow). These are the components A and B of the double quasar HS 1216+5032. The upper A component at 17.2 mag is brighter than the B component at 19.0 mag 9.1″ to the north. The seldom used catalogue designation HS 1216+5032 refers to the Hamburg/ESO QSO Survey which was compiled at the Hamburg Observatory in the mid-1990s with the purpose of verifying quasar candidates. During the study of the optical spectra of HS 1216+5032A/B, astronomers found almost identical spectra and a redshift of $z = 1.455$. This results in a cosmological distance of 10.7 billion light years and a projected distance of about 260 000 light years.

The images involved were not taken at the Hamburg Observatory, but with the objective-prism plate of the 0.8-m Schmidt telescope on Calar Alto in Spain. Promising candidates were followed up spectroscopically with the German–Spanish 2.2-m telescope (also on Calar Alto), resulting in some positive quasar candidates.

The possibility that HS 1216+5032 is due to a strong gravitational-lens effect can be ruled out based on comparison of the spectra of the components. Both spectra show prominent emission lines of ionized carbon and magnesium (C IV, C III, Mg II). In spite of this similarity, the spectra display differences: the spectrum of the B component is redder and shows a strong C IV absorption line in the blue wing of the C IV emission line. This absorption is due to a molecular cloud which is close to one of the quasars; the other quasar is not covered by the cloud. There are also further indications that the components are appreciably different. While the B component is radio-loud, the A component emits almost no radio emission. Also, there is no temporal correlation in the variability of the luminosity of the components. There are thus no features which are visible first in the light of one component then after a time delay in the light of the other. HS 1216+5032 is thus very probably a double quasar consisting of two especially active, interacting members of a distant galaxy cluster.

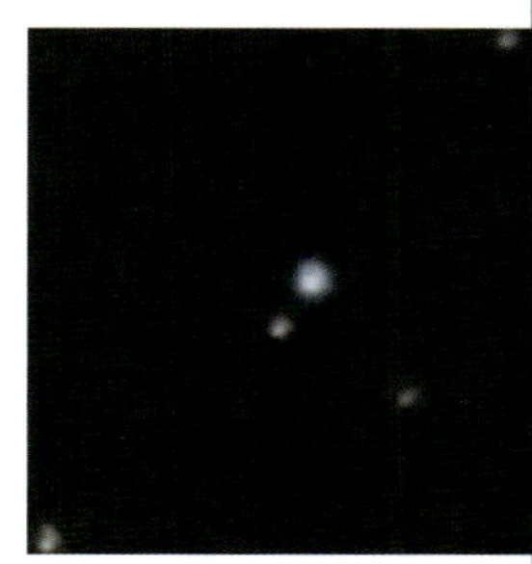

OBJECT	HS 1216+5032
CONSTELLATION	Canes Venatici
RIGHT ASCENSION	$12^h\ 18^m\ 41^s$
DECLINATION	+50° 15′ 36″
BRIGHTNESS	17.1 mag
TYPE	E/QSO
PHOTOGRAPHERS	Michael König
TELESCOPE	355-mm reflector
CAMERA	SBIG STL-11000
EXPOSURE TIME	120 min
LOCATION	Rimbach, Germany

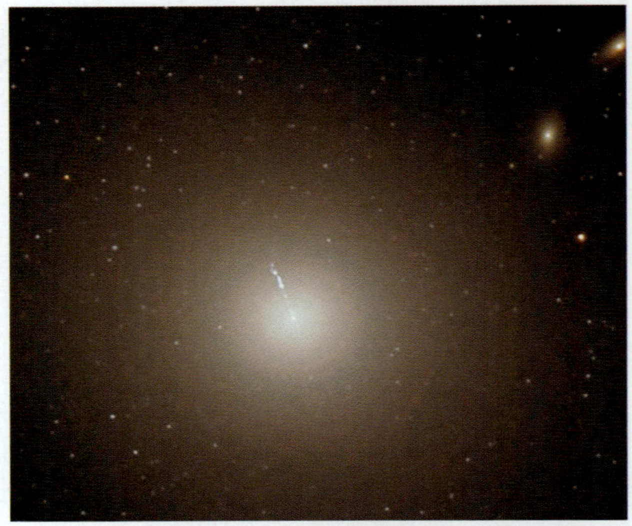

M 87

M 87 (NGC 4486) is a giant elliptical galaxy located near the centre of the Virgo Cluster, which contains about 2000 members. The fact that this is not exactly at the centre, but rather slightly offset, indicates ongoing interaction processes within the cluster. This cD galaxy is the result of many merger processes. Based on the large number of 14 000 globular clusters, one can estimate that M 87 must contain about 50 spiral galaxies. This estimate is supported by the fact that the mass of M 87 of 7×10^{12} solar masses is about 50 times that of a normal spiral galaxy $1–2 \times 10^{11}$. The large number of mergers implies a constant flow of material into the centre, supplying the accretion disc and thus the supermassive black hole with 0.1 solar masses per year. The mass of the black hole in M 87 is more than 6×10^9 solar masses, which is the record value for active galaxies, and explains why M 87 is a prominent source in the radio, X-ray, and gamma-ray portions of the spectrum.

A particular challenge for astrophotographers is the optical jet, which is readily visible in the image (see also the enlarged version in the small image). Its apparent length is about 20″, which in projection corresponds to a distance of 5000 light years. The jet has a low inclination of a few degrees; it is thus pointing almost exactly toward the observer, which allows one to see two special phenomena: on the one hand one can recognize relativistic beaming, which is an increase in brightness of the jet pointing at us. This goes hand in hand with a weakening of the opposite jet, resulting in a large contrast ratio between the two jets. This phenomenon explains why only one jet is recognizable. The other effect is known as superluminal motion, which is an apparent motion of individual structures faster than the speed of light. This phenomenon, however, does not contradict the special theory of relativity, since it is only an optical illusion. This is due to the combination of the material in the jet travelling at almost the speed of light and at a small angle towards the observer. Light emitted by points along the jet is thus approaching the observer at an only slightly higher velocity than the jet itself. Thus, light from the back of the jet arrives later than light from the front, but by an amount much less than the corresponding light-travel time between the back and front of the jet, creating the illusion that features within the jet are moving faster than light. Superluminal motion is thus a cosmic illusion and the special theory of relativity is still valid.

OBJECT	M 87, NGC 4486
CONSTELLATION	Virgo
RIGHT ASCENSION	$12^h\ 30^m\ 49^s$
DECLINATION	+12° 23′ 28″
BRIGHTNESS	9.6 mag
TYPE	cD0-1 pec
PHOTOGRAPHERS	Philipp Keller, Konstantin Buchhold, Bernd Flach-Wilken, Johannes Schedler, Volker Wendel (Chart 32-Team)
TELESCOPE	800-mm reflector
CAMERA	FLI Proline 16803
EXPOSURE TIME	220 min
LOCATION	CTIO, Chile

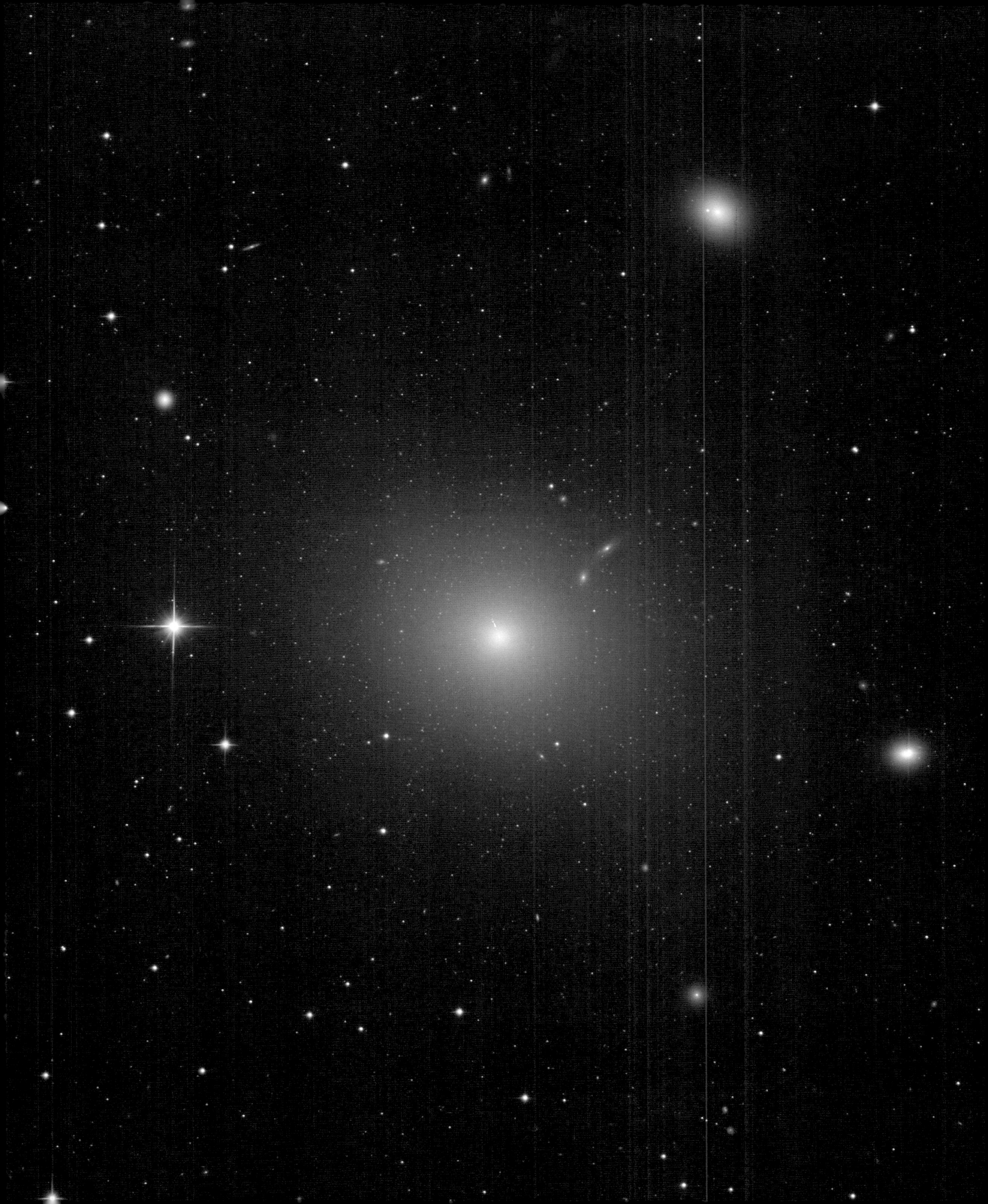

N

Abell 2218

In 1989, an expanded version of the late George O. Abell's *Catalog of Rich Clusters of Galaxies* was published by Harold G. Corwin and Ronald P. Olowin, covering portions of the southern sky below −27° in declination which had been omitted in the original catalogue (1958) and containing a total of 4073 objects. One selection criterion for a galaxy cluster was a minimum of 30 galaxies which lie within two magnitudes of the third-brightest member of the cluster. The galaxy cluster Abell 2218 is at the centre of the image. Conspicuous at a distance 4′ to the northeast of the centre is the barred spiral galaxy PGC 58586 with a diameter of just one minute of arc. Its luminosity is 15 mag, its distance is 320 million light years. Abell 2218 itself lies at a distance of 2.3 billion light years ($z = 0.175$); at this distance, an arcminute corresponds to 670 000 light years. This distance also separates the two brightest elliptical galaxies in Abell 2218, which define the two centres of mass in the cluster. The brighter of the two cD galaxies has more cluster members near it, which are also arranged appreciably more spherically, than the second, smaller cD galaxy. This second galaxy is about 70″ to the southeast, at the end of a chain of galaxies. It is thus part of a linear substructure in Abell 2218.

The detailed mass distribution in clusters can be measured by the location, form, and orientation of the delicate, faint arcs visible in the image, which are due to the weak gravitational-lens effect. The two centres of mass cause Abell 2218 to act as a double gravitational-lens telescope. In the enlargement in the inset, about two dozen arcs can be identified, as shown in this image by the Hubble Space Telescope. This shows more than 80 arcs with good signal-to-noise ratio in the range 22–25 mag. Particularly prominent is an orange arc below and to the left of the brighter cD galaxy. This is the distorted image of an elliptical galaxy at $z = 0.702$. Opposite this are two almost-touching circular arcs. The brighter arc is due to a distant galaxy at $z = 2.515$; the shorter, dimmer arc has $z = 0.532$. Also visible is an image with weak structure, to the right of the cD galaxy at the second centre of mass. One sees hints of an orange-yellow central region which indicates that it is the image of a spiral galaxy ($z = 1.034$). Without the action of the gravitational lens Abell 2218, whose mass model implies an amplification by a factor of 15 to 25, images of these distant galaxies would not be possible for amateur astronomers.

Using the Advanced Camera for Surveys (ACS) on the Hubble Space Telescope and ground-based observations at the Keck Telescope in Hawaii, a multiply-imaged source was observed which is due to a galaxy with a redshift in the range 6.6–7.1. This distance was estimated not via spectral lines but rather by the position of the spectral continuum in the infrared spectral region. This exceedingly faint galaxy has a luminosity of 28 mag. The image morphology due to the gravitational lens indicates a very compact source, only about 3000 light years in size. It is presumably the core of a star-forming galaxy such as those observed at the end of cosmic reionization. In this early phase of the universe, about 150–800 million years after the Big Bang, the first stars formed in primeval galaxies, ending the so-called dark ages.

This amateur image can be compared with one from the Wide Field and Planetary Camera (WFPC2, in use 1993–2009): WFPC2 consists of a CCD array of four chips, three of which form an L, the fourth being in the knick of the L. This fourth chip (planetary camera chip) has only a quarter of the area of the larger CCDs, but 2.2 times the resolution. The area (including the gap due to the smaller chip) measures 2.7′ × 2.7′ and completely covers the extent of Abell 2218.

OBJECT	Abell 2218
CONSTELLATION	Draco
RIGHT ASCENSION	$16^h\ 35^m\ 52^s$
DECLINATION	+66° 12′ 52″
BRIGHTNESS	17 mag
TYPE	Galaxy cluster
PHOTOGRAPHERS	Josef Pöpsel, Stefan Binnewies
TELESCOPE	600-mm reflector
CAMERA	SBIG STL-11000
EXPOSURE TIME	330 min
LOCATION	Skinakas Observatory, Crete, Greece

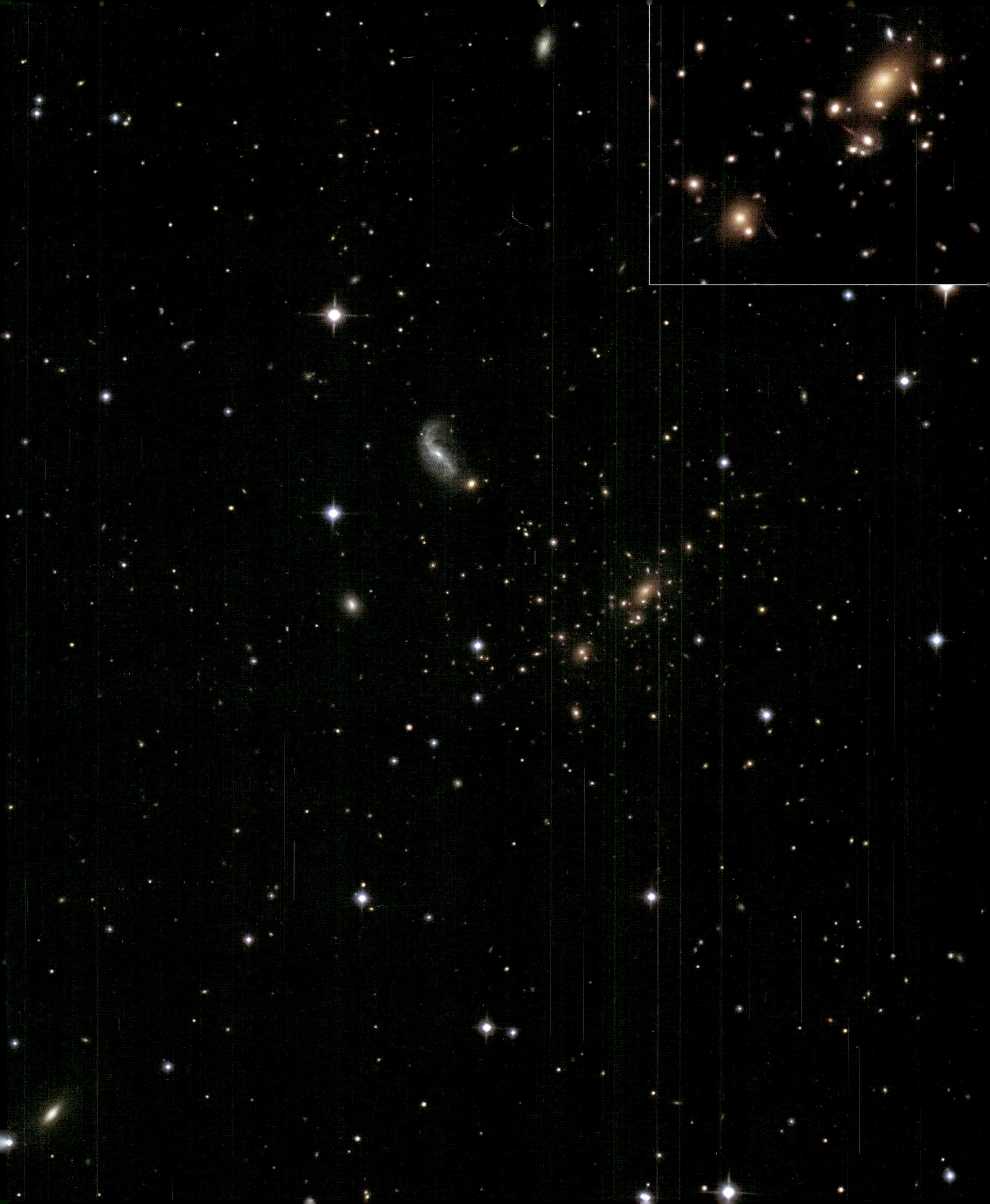

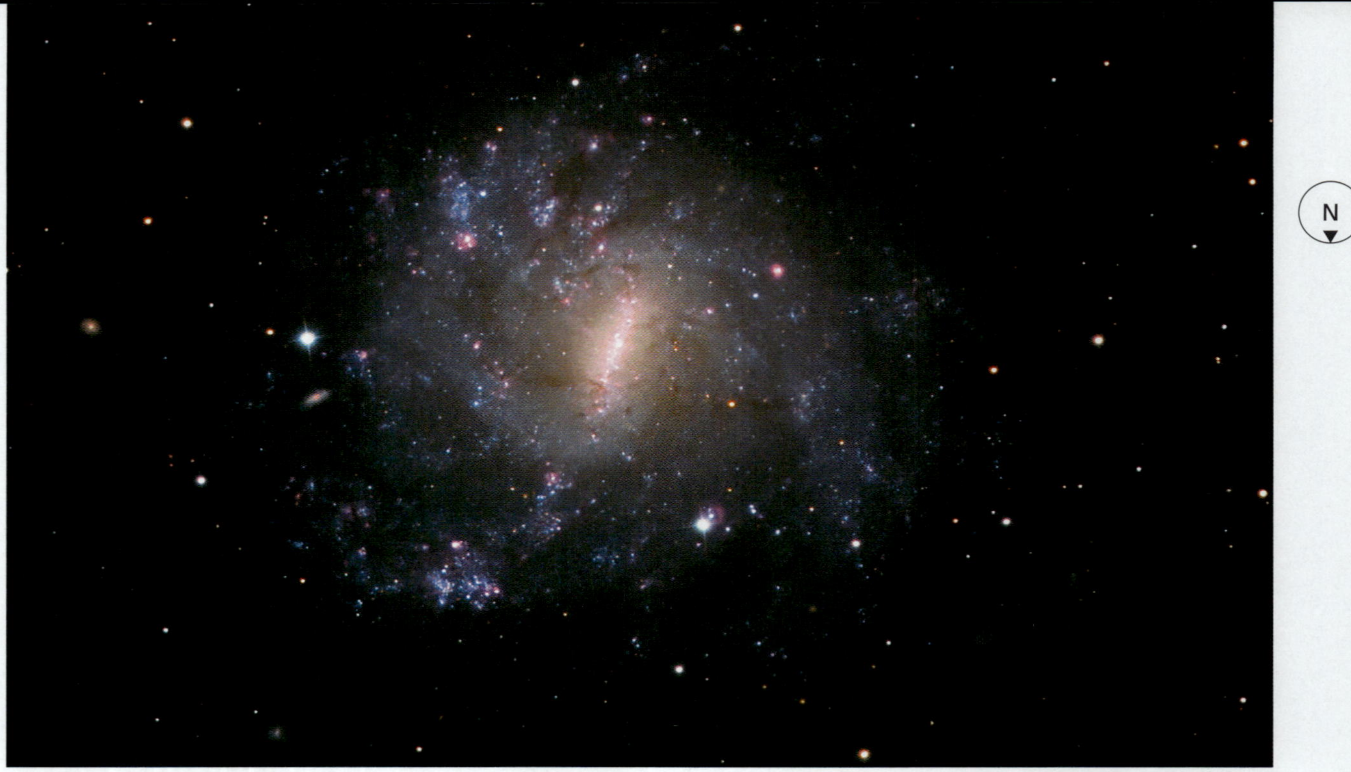

NGC 5068

NGC 5068 is an SB(s)d galaxy which is 20 million light years from the Milky Way. In the outer region of this barred spiral galaxy, the spiral structure is almost dissolved, while the central bar is easily recognizable. The centre appears very bright and compact and in the region of the bar the contrast is reduced due to the overlap of the bar and a non-bar-shaped stellar bulge. The galaxy measures 7.2′ × 6.3′ and the scale of 5800 light years per arcminute implies a bar length of 10 000 light years. NGC 5068 also displays the typical SBd characteristic of asymmetric bar ends. The northern end, lower in the image, appears broadened and has more bright H II regions than the other end of the bar. These H II regions are arranged like pearls on a string along the entire bar.

In addition to the H II regions, NGC 5068 has many bright blue OB associations in the spiral arms. The masses of most of these young, hot stars are more than eight solar masses. Their stellar winds, as well as the supernovae at the ends of their lives, determine the appearance of the entire galaxy, which is why it is termed a star-forming galaxy. NGC 5068 can thus be seen as a "mild" form of an active galaxy. In such galaxies, an important role is played by Wolf-Rayet stars. These are stars which have already left the main sequence and whose nuclear fusion has changed from central burning to shell burning. If such a Wolf-Rayet star is part of a binary system, the expansion of its outer layers causes material to flow to the partner, so that the star itself loses its outer envelope. The consequence of the radiation pressure released from the inner regions are strong stellar winds and a high mass-loss rate of about 10^{-5}–10^{-4} solar masses per year. In NGC 5068, 160 Wolf-Rayet stars were identified via spectroscopic studies. There is also a relation between the frequency of Wolf-Rayet stars and those of O stars in the associations. About 9000 O stars were found in NGC 5068, implying the presence of 270 Wolf-Rayet stars. Since the actual number is smaller, astronomers assume that their search missed fainter Wolf-Rayet stars, since these are difficult to see within the bright OB regions.

OBJECT	NGC 5068
CONSTELLATION	Virgo
RIGHT ASCENSION	$13^h\ 18^m\ 55^s$
DECLINATION	−21° 02′ 21″
BRIGHTNESS	10.5 mag
TYPE	SAB(rs)cd
PHOTOGRAPHERS	Josef Pöpsel, Beate Behle
TELESCOPE	600-mm reflector
CAMERA	SBIG ST-10XME
EXPOSURE TIME	140 min
LOCATION	Amani Lodge, Namibia

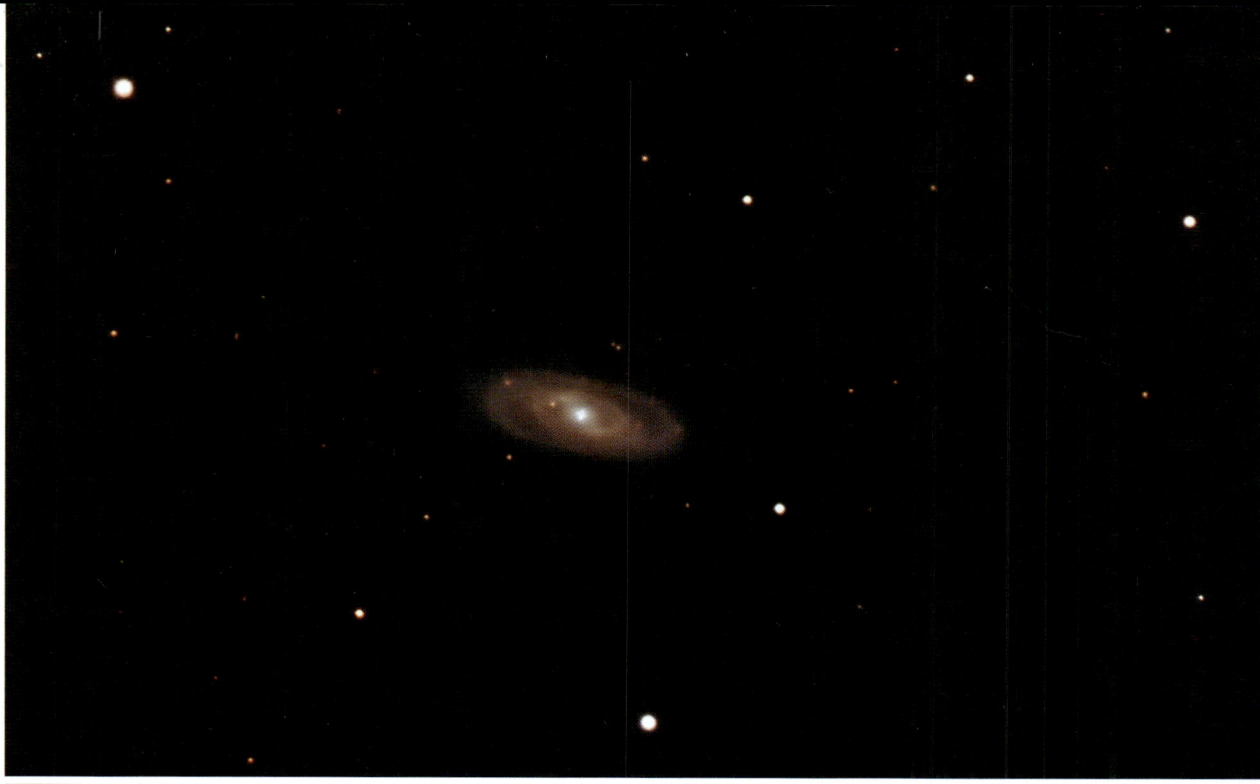

PGC 69457, The Einstein Cross

The spiral galaxy PGC 69457 (UZC J224030.2+032131, *Updated Zwicky Catalog*) is at a distance of 520 million light years from the Milky Way. The angular size of 0.87′ × 0.3′ corresponds to a projected disc diameter of 132 000 light years. PGC 69457 is listed in catalogues as an Sab type, though the bright spots at the beginning of the spiral arms suggest that the galaxy has a bar. Conspicuous at the core are four points light arranged in a cross, surrounding a somewhat fainter nucleus which appears less focussed, known in the literature as the Einstein Cross (see enlargement in the small image). This has dimensions of 1.7″ × 1.4″ and is listed as Y 88 in the gravitational-lens catalogue of H. K. C. Yee. In honour of John Huchra, who had discovered the cross in 1985, it is also known under the name Huchra's Lens. This multiply-imaged quasar is 20 times as distant as the lens galaxy PGC 69457 and is catalogued as QSO 2237+0305A.

The quadruple image of a distant quasar is due to the facts that the gravitational-lens system is not axially symmetric and the lens is not spherical. (Axially-symmetric systems produce Einstein rings; spherical but asymmetric systems produce two images.) Spiral galaxies have various mass components, and models can be fitted which reproduce the quadruple image. The best model for PGC 69457 contains an elliptic bar component which is inclined by about 20° relative to the major axis of the Einstein Cross. Calculations indicate for the most-massive components within the Einstein radius 14.3×10^9 solar masses for the bulge and 1.2×10^9 for the bar. Since the rotation curve at the centre of PGC 69457 has been measured at only a few points, detailed kinematic modelling of the bar is difficult, but the mass distribution can be determined from the lensing geometry.

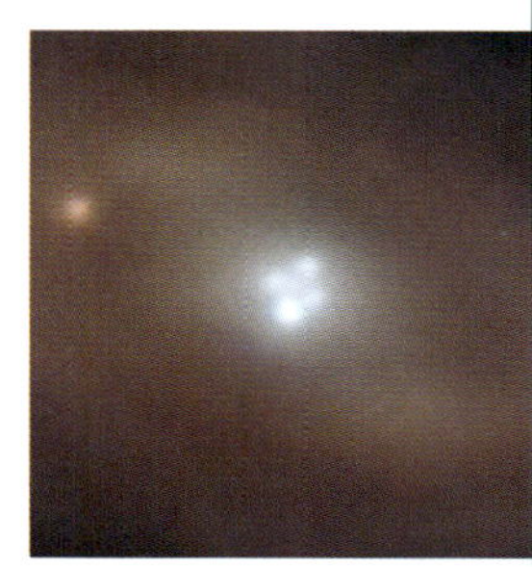

OBJECT	PGC 69457
CONSTELLATION	Pegasus
RIGHT ASCENSION	$22^h\ 40^m\ 30^s$
DECLINATION	+03° 21′ 31″
BRIGHTNESS	16.8 mag
TYPE	Sab/GL
PHOTOGRAPHERS	Makis Palaiologou
TELESCOPE	1.3-m reflector
CAMERA	SBIG STL-6303
EXPOSURE TIME	170 min
LOCATION	Skinakas Observatory, Crete, Greece

Bibliography

Abell 262

de Juan, L. et al.: *Surface photometry of low-luminosity radio galaxies,* Astrophysical Journal Supplement Series, **91**, 1994

Fath, E. A.: *Focal Pointe Observatory*, http://bf-astro.com/fath.htm

Fath, E. A.: *The Story of the Spirals*, The Century Magazine, 1912, www.unz.org/Pub/Century-1912sep-00757

Abell 370

Schneider, P: *Gravitationslinseneffekte*, 2006, https://astro.uni-bonn.de/~peter/Poster3d.html

Hu, E.M. et al.: *A Redshift z = 6.56 Galaxy behind the Cluster Abell 370*, The Astrophysical Journal, **568**, 2002

Abell 426

Abell, G. O.: *The distribution of rich clusters of galaxies. A catalogue of 2712 rich clusters found on the National Geographic Society Palomar Observatory Sky Survey*, Astrophysical Journal Supplement Series, **3**, 1958

Leslie, P. R. R. and Elsmore, B.: *Radio emission from the Perseus cluster*, The Observatory, **81**, 1961

Salomé, P. et al.: *A very extended molecular web around NGC 1275*, Astronomy and Astrophysics, **531**, 2011

Abell 1060

Fitchett, M. and Merritt, D.: *Dynamics of the Hydra I Galaxy Cluster*, The Astrophysical Journal, **335**, 1988

Abell 1367

McConnell, N. J.: *Two ten-billion-solar-mass black holes at the centres of giant elliptical galaxies*, Nature, **480**, 2011

Sanderson, A. J. R. et al.: *A statistically-selected Chandra sample of 20 galaxy clusters – I. Temperature and cooling time profiles*, Monthly Notices of the Royal Astronomical Society, **372**, 2006

Abell 1656

Fisher, D. et al.: *Kinematics of 13 brightest cluster galaxies*, The Astrophysical Journal, **438**, 1995

Nevalainen, J. et al.: *Non-thermal Hard X-Ray Emission in Galaxy Clusters Observed with the BeppoSAX PDS*, The Astrophysical Journal, **608**, 2004

Zwicky, F.: *On the Masses of Nebulae and of Clusters of Nebulae*, The Astrophysical Journal, **86**, 1937

Abell 2065

Association of Universities for Research in Astronomy: *The STScI Digitized Sky Survey*, 1994, http://archive.stsci.edu/cgi-bin/dss_form

Flin, P. and Krywult, J.: *Substructures in Abell clusters of galaxies*, Astronomy and Astrophysics, **450**, 2006

Abell 2151

Bucknell, M. J. et al.: *Studies of rich clusters of galaxies – V. Photometry and luminosity functions for eight clusters*, Monthly Notices of the Royal Astronomical Society, **188**, 1979

Cedrés, B. et al.: *Star-Forming Galaxies in the Hercules Cluster: Hα Imaging of A2151*, The Astronomical Journal, **138**, 2009

Hibbard, J. E. and van Gorkom, J. H.: *HI, HII, and R-Band Observations of a Galactic Merger Sequence*, The Astronomical Journal, **111**, 1996

Liuzzo, E. et al.: *Parsec-scale properties of brightest cluster galaxies*, Astronomy and Astrophysics, **516**, 2010

Abell 2218

Abell, G. O. et al.: *A Catalog of Rich Clusters of Galaxies,* Astrophysical Journal Supplement Series, **70**, 1989

Ellis, R.: *Testing Lens Inversion with Abell 2218*, 2015, www.astro.caltech.edu/~rse/firstlight/a2218_arcs.gif

Kneib, J. P. et al.: *A Probable z~7 Galaxy Strongly Lensed by the Rich Cluster Abell 2218: Exploring the Dark Ages*, The Astrophysical Journal, **607**, 2004

Kneib, J. P. et al.: *Hubble Space Telescope Observations of the Lensing Cluster Abell 2218*, The Astrophysical Journal, **471**, 1996

Space Telescope Science Institute: *Hubble Space Telescope WFPC2*, 2010, www.stsci.edu/hst/wfpc2

Abell 2572

Ebeling, H. et al.: *A 2572 and HCG 94 – galaxy clusters but not as we know them. An X-ray case study of optical misclassifications*, Monthly Notices of the Royal Astronomical Society, **277**, 1995

Ebeling, H. et al.: *Properties of the X-ray brightest Abell-type clusters of galaxies (XBACs) from ROSAT All-Sky Survey data – I. The sample*, Monthly Notices of the Royal Astronomical Society, **281**, 1996

Arp 65

AIS All Sky Survey: *GALEX GR6 Data Release*, 2013, http://galex.stsci.edu/GR6/

Arp 81

Schwenk, D. R. et al.: *Modeling and Analysis of the Nearby Colliding Galaxy Pair NGC 6621/22*, American Astronomical Society Meeting 207, Bulletin of the American Astronomical Society, **37**, 2005

Arp 84

Kaufman, M. et al.: *CO Observations of the Interacting Galaxy Pair NGC 5394/95*, The Astronomical Journal, **123**, 2002

Arp 104

Gallagher, J. S. and Parker, A.: *Optical Structure and Evolution of the Arp 104 Interacting Galaxy System*, The Astrophysical Journal, **722**, 2010

Arp 113

Mould, J. R. et al.: *The Hubble Space Telescope Key Project on the Extragalactic Distance Scale*, The Astrophysical Journal, **529**, 2000

Arp 143

Appleton, P. N. et al.: *A giant intergalactic HI bubble near Arp 143*, Nature, **330**, 1987

Beirao, P. et al.: *Powerful H_2 Emission and Star Formation on the Interacting Galaxy System Arp 143: Observations with Spitzer and Galex*, The Astrophysical Journal, **693**, 2009

NASA Jet Propulsion Laboratory: *Spitzer Space Telescope*, 2015, www.spitzer.caltech.edu

Arp 188

Tran, H. D. et al.: *Advanced Camera for Survey Observations of Young Star Clusters in the Interacting Galaxy UGC 10214*, The Astrophysical Journal, **585**, 2003

Arp 254

Smith, B. J. et al.: *Spirals, Bridges, and Tails: A GALEX UV Atlas of Interacting Galaxies*, The Astronomical Journal, **139**, 2010

Arp 270

Garrido, O. et al.: *GHASP: An Hα kinematic survey of spiral and irregular galaxies – I. Velocity fields and rotation curves of 23 galaxies*, Astronomy and Astrophysics, **387**, 2002

Arp 273

Zeltwanger, T. et al.: *Sloshing in High Speed Galaxy Interactions*, The Astrophysical Journal, **543**, 2000

Arp 286

Casasola, V. et al.: *The gas content of peculiar galaxies: strongly interacting systems*, Astronomy and Astrophysics, **422**, 2004

Erwin, P.: *Double-Barred Galaxies: I. A Catalog of Barred Galaxies with Stellar Secondary Bars and Inner Disks*, Astronomy and Astrophysics, **415**, 2004

Jungwiert, B. et al.: *Near-IR photometry of disk galaxies: search for nuclear isophotal twist and double bars*, Astronomy and Astrophysics Supplement series, **125**, 1997

Arp 295

Stockton, A.: *On the Tidal Origin of the Bridge of Arp 295*, The Astrophysical Journal, **190**, 1974

Copeland's Septet

Allam, S. et al.: *Far infrared properties of Hickson compact groups of galaxies. I. High resolution IRAS maps and fluxes*, Astronomy and Astrophysics Supplement, **117**, 1996

Barton, E. J. et al.: *Environments of Redshift Survey Compact Groups of Galaxies,* The Astronomical Journal, **116**, 1998

Martinez, M. A. et al.: *AGN Population in Hickson Compact Groups. I. Data and Nuclear Activity Classification*, The Astronomical Journal, **139**, 2010

Steinicke, W.: *Daten zur Entstehung des NGC/IC and seiner Revisionen*, 2012, www.klima-luft.de/steinicke/Deep-Sky/deep-sky.htm

Draco Dwarf

Biebel, O.: *Seminar zur Dunklen Materie*, 2012, http://homepages.physik.uni-muenchen.de/~Otmar.Biebel/dm.../MHerz.pdf

Odenkirchen, M. et al.: *New Insights on the Draco Dwarf Spheroidal Galaxy from the Sloan Digital Sky Survey: A Larger Radius and No Tidal Tails*, The Astronomical Journal, **122**, 2001

Einstein Cross

Trott, C. M. et al.: *Stars and dark matter in the spiral gravitational lens 2237+0305*, Monthly Notices of the Royal Astronomical Society, **401**, 2010

van de Ven, G. et al.: *The Einstein Cross: Constraint on Dark Matter from Stellar Dynamics and Gravitational Lensing*, The Astrophysical Journal, **719**, 2010

ESO 350-40

Higdon, J. L.: *Wheels of Fire. II. Neutral Hydrogen in the Cartwheel Ring Galaxy*, The Astrophysical Journal, **467**, 1996

Horellou, C. and Combes, F.: A *Model for the Cartwheel Ring Galaxy*, Astrophysics and Space Science, **276**, 2001

ESO 351-30

Carignan, C. et al.: *Detection of HI associated with the Sculptor Dwarf Spheroidal Galaxy,* The Astronomical Journal, **116**, 1998

Giuffrida, G. et al.: *BVR photometry of the Sculptor dwarf spheroidal galaxy*, Memorie della Società Astronomica Italiana, **77**, 2006

Hickson Compact Group 61

Jaffe, W. and Gavazzi, G.: *Radio Continuum Survey of the Coma/A1367 Supercluster. II. 1.5 GHz Observations of 396 CGCG Galaxies*, The Astronomical Journal, **91**, 1986

Theureau, G. et al.: *Kinematics of the Local Universe XIII. 21-cm line measurements of 452 galaxies with the Nançay radiotelescope, JHK Tully-Fisher relation, and preliminary maps of the peculiar velocity field*, Astronomy and Astrophysics, **465**, 2007

van Driel, W. et al.: *A neutral hydrogen survey of polar ring galaxies III. Nançay observations and comparison with published data*, Astronomy and Astrophysics Supplement, **141**, 2000

Hickson Compact Group 68

Elfhag, T. et al.: *A CO survey of galaxies with the SEST and the 20-m Onsala telescope*, Astronomy and Astrophysics Supplement, **115**, 1996

Sanchayeeta, B. et al.: *Detection of Diffuse Neutral Intragroup Medium in Hickson Compact Groups*, The Astrophysical Journal, **710**, 2010

Tully, R. B. and Trentham, N.: *Midlife Crises in Dwarf Galaxies in the NGC 5353/4 Group*, The Astrophysical Journal, **135**, 2008

Holmberg II

Hubble Space Telescope News: *Galaxy caught blowing bubbles*, 2011, www.spacetelescope.org/news/heic1114/

Hunter, D. A. et al.: *Mid-Infrared Images of Stars and Dust in Irregular Galaxies*, The Astronomical Journal, **132**, 2006

Croxall, K. C. et al.: *Chemical Abundances of Seven Irregular and Three Tidal Dwarf Galaxies in the M81 Group*, The Astrophysical Journal, **705**, 2009

Grise, F. et al.: *X-ray Spectral State is not correlated with Luminosity in Holmberg II X-1,* Astrophysical Journal Letters, **724**, 2010

Weisz, D. R. et al.: *The Recent Star Formation Histories of M81 Group Dwarf Irregular Galaxies*, The Astrophysical Journal, **689**, 2008

Zezas, A. L. et al.: *ROSAT observations of the dwarf star-forming galaxy Holmberg II (UGC 4305),* Monthly Notices of the Royal Astronomical Society, **308**, 1999

Holmberg 124

Kantharia, N. G. et al.: *GMRT Observations of the Group Holmberg 124: Evolution by Tidal Forces and Ram Pressure?*, Astronomy and Astrophysics, **435**, 2005

van der Hulst, J. M. and Hummel, E.: *The first detection of a radio continuum bridge between interacting galaxies*, Astronomy and Astrophysics, **150**, 1985

Holmberg 224

Schulman, E. et al.: *An HI Survey of High-Velocity Clouds in Nearby Disk Galaxies*, The Astrophysical Journal, **423**, 1994

Uppsala University: *Erik Holmberg*, 2012, www.astro.uu.se/history/holmberg.html

Vicari, A. et al.: *Large-scale star formation in galaxies. II. The spirals NGC 3377A, NGC 3507 and NGC 4394. Young star groupings in spirals*, Astronomy and Astrophysics, **384**, 2002

Holmberg 391

Barway, S. et al.: *Multicolor Surface Photometry of Lenticular Galaxies. I. The Data*, The Astronomical Journal, **129**, 2005

Erwin, P. et al.: *NGC 4340: Double Bar + Fossil Nuclear Ring*, Astrophysics and Space Science, **277**, 2001

Holmberg 800

Casasola, V. et al.: *The gas content of peculiar galaxies: strongly interacting systems*, Astronomy and Astrophysics, **422**, 2004

Dahari, O.: *The Nuclear Activity of Interacting Galaxies*, Astrophysical Journal Supplement Series, **57**, 1985

HS 1216+5032

Green, P. J. et al.: *HS1216+5032: a physical quasar pair with one radio-loud broad absorption line quasar*, Monthly Notices of the Royal Astronomical Society, **349**, 2004

Hagen, H. J. et al.: *HS1216+5032: a new double QSO separated by 9*, Astronomy and Astrophysics, **308**, 1996

IC 10

Leroy, A. et al.: *Molecular Gas in the Low Metallicity, Star Forming Dwarf IC 10*, The Astrophysical Journal, **643**, 2006

IC 239

Trentham, N. and Tully, R. B.: *Dwarf Galaxies in the NGC 1023 Group*, Monthly Notices of the Royal Astronomical Society, **398**, 2009

IC 342

Riepe, P.: *Die benachbarte Galaxie IC 342 – ihre wirkliche Ausdehnung*, VdS-Journal für Astronomie, **36**, 2011

Sheth, K. et al.: *Molecular Gas and Star Formation in Bars of Nearby Spiral Galaxies*, The Astronomical Journal, **124**, 2002

IC 1296

Italian Supernovae Search Project, 2013, http://italiansupernovae.org/en/project/description.html

IC 1613

Borissova, J. et al.: *A catalogue of OB associations in IC 1613*, Astronomy and Astrophysics, **413**, 2004

Garcia, M. et al.: *The young stellar population of IC 1613. II. Physical properties of OB associations*, Astronomy and Astrophysics, **523**, 2010

IC 2574

Cannon, J. M. et al.: *Spitzer Observations of the Supergiant Shell Region in IC 2574*, The Astrophysical Journal, **630**, 2005

Hunter, D. A.: *Star Formation in Irregular Galaxies: A Review of Several Key Questions*, 1997, http://ned.ipac.caltech.edu/level5/Hunter/Hunter1.html

Karachentsev, I. D. and Kashibadze, O. G.: *Masses of the local group and of the M81 group estimated from distortions in the local velocity field*, Astrophysics, **49**, 2006

Kewley, L. J. et al.: *Aperture Effects on Star Formation Rate, Metallicity, and Reddening*, The Publications of the Astronomical Society of the Pacific, **117**, 2005

Lick Observatory: *Crocker photographic telescope*, 2012, http://collections.ucolick.org/exhibits_on_line/E2E.1/Willard.html

Pauly, W.: *Weitere Nachrichten über den Cometen Coddington-Pauly*, Astronomische Nachrichten, **146**, 1898

Large Magellanic Cloud

Abd ar-Rahman as-Sufi: *Liber locis stellarum fixarum*, 964, www.atlascoelestis.com/alsufi%20pagina.htm

Brüns, C.: *Die dynamische Entwicklung der Großen Magellanschen Wolke*, 2010, www.astro.uni-bonn.de/~rcbruens/simulationen/lmc/

Crowther, P. A. et al.: *The R136 star cluster hosts several stars whose individual masses greatly exceed the accepted 150 M_{solar} stellar mass limit*, Monthly Notices of the Royal Astronomical Society, **408**, 2010

Harvard-Smithsonian Center for Astrophysics: *The Magellanic Clouds Are First-Time Visitors*, 2007, www.cfa.harvard.edu/news/2007/pr200722.html

Liu, L. et al.: *How common are the Magellanic Clouds?*, The Astrophysical Journal, **733**, 2011

Matsuura, M. et al.: *Spitzer observations of acetylene bands in carbon-rich AGB stars in the Large Magellanic Cloud*, Monthly Notices of the Royal Astronomical Society, **371**, 2006

Seleznev, A. F.: *The structure of the halo of the star cluster NGC 2070*, Astronomy Letters, **21**, 1995

Leo I

Blitz, L. and Robishaw, T.: *Gas-Rich Dwarf Spheroidals*, Astronomy and Astrophysics, **512**, 2010

M 31

Athanassoula, E. and Beaton, R. L.: *Unravelling the mystery of the M31 bar*, Monthly Notices of the Royal Astronomical Society, **370**, 2006

Williams, B. F. and Benjamin, F.: *Comparing young stellar populations across active regions in the M31 disc*, Monthly Notices of the Royal Astronomical Society, **340**, 2003

M 32

McConnachie, A. W. and Irwin, M. J.: *The satellite distribution of M31*, Monthly Notices of the Royal Astronomical Society, **365**, 2006

van der Marel, R. P. et al.: *Improved Evidence for a Black Hole in M32 from HST/FOS Spectra. II. Axisymmetric Dynamical Models*, The Astrophysical Journal, **493**, 1998

M 33

Chandra Publications: *Heaviest Stellar Black Hole Discovered in Nearby Galaxy*, 2007, http://chandra.harvard.edu/press/07_releases/press_101707.html

Corbelli, E.: *Dark matter and visible baryons in M33*, Monthly Notices of the Royal Astronomical Society, **342**, 2003

Verley, S. et al.: *Star formation in M33: Spitzer photometry of discrete sources*, Astronomy and Astrophysics, **476**, 2007

M 49

Battaia, F. A. et al.: *Stripped gas as fuel for newly formed HII regions in the encounter between VCC1249 and M49: a unified picture from NGVS and GUViCS*, Astronomy and Astrophysics, **543**, 2012

M 51

Dobbs, C. L. et al.: *Simulations of the grand design galaxy M51: a case study for analysing tidally induced spiral structure*, Monthly Notices of the Royal Astronomical Society, **403**, 2010

M 58

Comeron, S. et al.: *Discovery of Four New Ultra-compact Nuclear Rings in Three Spiral Galaxies, Conference Proceedings "Pathways Through an Eclectic Universe"*, ASP Conference Series, **390**, 2008

M 60

de Grijs, R. and Robertson, A. R. I.: *Arp 116: Interacting System or Chance Alignment?*, Astronomy and Astrophysics, **460**, 2008

M 61

Koda, J. and Sofue, Y.: *The Virgo High-Resolution CO Survey: VI. Gas Dynamics and Star Formation along the Bar in NGC 4303*, Publications of the Astronomical Society of Japan, **58**, 2006

M 63

Battaglia, G. et al.: *HI-study of the warped spiral galaxy NGC 5055: a disk/dark matter halo offset?*, Astronomy and Astrophysics, **447**, 2006

M 64

Braun, R. et al.: *Counterrotating gaseous disks in NGC 4826*, The Astrophysical Journal, **420**, 1994

M 65 / M 66 / NGC 3628

Chromey, F. R. et al.: *Star Formation in the Tidal Tail of the Leo Triplet Galaxy NGC 3628*, The Astronomical Journal **115**, 1998

Hogg, D. E. et al.: *Hot and Cold Gas in Early-Type Spirals: NGC 3623, NGC 2775, and NGC 1291*, The Astronomical Journal, **121**, 2001

M 66

Chemin, L. et al.: *An Hα study of the kinematics of NGC 3627*, Astronomy and Astrophysics, **405**, 2003

M 74

Sanchez, S. F. et al.: *PPAK Wide-field Integral Field Spectroscopy of NGC 628: I. The largest spectroscopic mosaic on a single galaxy*, Monthly Notices of the Royal Astronomical Society, **410**, 2011

M 77

Kamenetzky, J. et al.: *The Dense Molecular Gas in the Circumnuclear Disk of NGC 1068*, The Astrophysical Journal, **731**, 2011

Schinnerer, E. et al.: *Bars and Warps traced by the Molecular Gas in the Seyfert 2 Galaxy NGC 1068*, The Astrophysical Journal, **533**, 2000

M 81

de Blok, W. J. G. et al.: *High-Resolution Rotation Curves and Galaxy Mass Models from THINGS*, The Astronomical Journal, **136**, 2008

M 82

Yoshida, M. et al.: *Spectropolarimetry of the Superwind Filaments of the Starburst Galaxy M82: Kinematics of Dust Outflow*, Publications of the Astronomical Society of Japan, **63**, 2011

Abell 2218

In 1989, an expanded version of the late George O. Abell's *Catalog of Rich Clusters of Galaxies* was published by Harold G. Corwin and Ronald P. Olowin, covering portions of the southern sky below −27° in declination which had been omitted in the original catalogue (1958) and containing a total of 4073 objects. One selection criterion for a galaxy cluster was a minimum of 30 galaxies which lie within two magnitudes of the third-brightest member of the cluster. The galaxy cluster Abell 2218 is at the centre of the image. Conspicuous at a distance 4′ to the northeast of the centre is the barred spiral galaxy PGC 58586 with a diameter of just one minute of arc. Its luminosity is 15 mag, its distance is 320 million light years. Abell 2218 itself lies at a distance of 2.3 billion light years ($z = 0.175$); at this distance, an arcminute corresponds to 670 000 light years. This distance also separates the two brightest elliptical galaxies in Abell 2218, which define the two centres of mass in the cluster. The brighter of the two cD galaxies has more cluster members near it, which are also arranged appreciably more spherically, than the second, smaller cD galaxy. This second galaxy is about 70″ to the southeast, at the end of a chain of galaxies. It is thus part of a linear substructure in Abell 2218.

The detailed mass distribution in clusters can be measured by the location, form, and orientation of the delicate, faint arcs visible in the image, which are due to the weak gravitational-lens effect. The two centres of mass cause Abell 2218 to act as a double gravitational-lens telescope. In the enlargement in the inset, about two dozen arcs can be identified, as shown in this image by the Hubble Space Telescope. This shows more than 80 arcs with good signal-to-noise ratio in the range 22–25 mag. Particularly prominent is an orange arc below and to the left of the brighter cD galaxy. This is the distorted image of an elliptical galaxy at $z = 0.702$. Opposite this are two almost-touching circular arcs. The brighter arc is due to a distant galaxy at $z = 2.515$; the shorter, dimmer arc has $z = 0.532$. Also visible is an image with weak structure, to the right of the cD galaxy at the second centre of mass. One sees hints of an orange-yellow central region which indicates that it is the image of a spiral galaxy ($z = 1.034$). Without the action of the gravitational lens Abell 2218, whose mass model implies an amplification by a factor of 15 to 25, images of these distant galaxies would not be possible for amateur astronomers.

Using the Advanced Camera for Surveys (ACS) on the Hubble Space Telescope and ground-based observations at the Keck Telescope in Hawaii, a multiply-imaged source was observed which is due to a galaxy with a redshift in the range 6.6–7.1. This distance was estimated not via spectral lines but rather by the position of the spectral continuum in the infrared spectral region. This exceedingly faint galaxy has a luminosity of 28 mag. The image morphology due to the gravitational lens indicates a very compact source, only about 3000 light years in size. It is presumably the core of a star-forming galaxy such as those observed at the end of cosmic reionization. In this early phase of the universe, about 150–800 million years after the Big Bang, the first stars formed in primeval galaxies, ending the so-called dark ages.

This amateur image can be compared with one from the Wide Field and Planetary Camera (WFPC2, in use 1993–2009): WFPC2 consists of a CCD array of four chips, three of which form an L, the fourth being in the knick of the L. This fourth chip (planetary camera chip) has only a quarter of the area of the larger CCDs, but 2.2 times the resolution. The area (including the gap due to the smaller chip) measures 2.7′ × 2.7′ and completely covers the extent of Abell 2218.

OBJECT	Abell 2218
CONSTELLATION	Draco
RIGHT ASCENSION	$16^h\ 35^m\ 52^s$
DECLINATION	+66° 12′ 52″
BRIGHTNESS	17 mag
TYPE	Galaxy cluster
PHOTOGRAPHERS	Josef Pöpsel, Stefan Binnewies
TELESCOPE	600-mm reflector
CAMERA	SBIG STL-11000
EXPOSURE TIME	330 min
LOCATION	Skinakas Observatory, Crete, Greece

M 83
Bresolin, F. et al.: *The flat oxygen abundance gradient in the extended disk of M83,* The Astrophysical Journal, **695**, 2009

M 85
Gültekin, K. et al.: *Is There a Black Hole in NGC 4382?,* The Astrophysical Journal, **741**, 2011

M 87
Biretta, J. A. et al.: *Hubble Space Telescope Observations of Superluminal Motion in the M87 Jet,* The Astrophysical Journal, **520**, 1999

M 88
Chemin, L. et al.: *A Virgo high-resolution Hα kinematical survey: II. The Atlas,* Monthly Notices of the Royal Astronomical Society, **366**, 2006

M 89
Clark, G. et al.: *Structure and Dynamics of Elliptical Galaxies,* Proceedings IAU Symposium, **127**, Springer Verlag, 1987

M 90
Vollmer, B. et al.: *NGC 4569: Recent evidence for a past ram pressure stripping event,* Astronomy and Astrophysics, **419**, 2004

M 91
Vollmer, B. et al.: *Kinematics of the anemic cluster galaxy NGC 4548. Is stripping still active?,* Astronomy and Astrophysics, **349**, 1999

M 94
Jalocha, J. et al.: *Is Dark Matter Present in NGC 4736? An Iterative Spectral Method for Finding Mass Distribution in Spiral Galaxies,* The Astrophysical Journal, **679**, 2008

M 95
Hägele, G. F. et al.: *Subarcsecond radio continuum mapping in and around the spiral galaxy NGC 3351 using MERLIN,* Monthly Notices of the Royal Astronomical Society, **406**, 2010

M 96
Moiseev, A. V. et al.: *Structure and kinematics of candidate double-barred galaxies,* Astronomy and Astrophysics, **421**, 2004

M 98
Giovanelli, R.: ALFALFA: *HI Cosmology in the Local Universe, Dark Galaxies and Lost Baryons,* Proceedings IAU Symposium, **244**, 2007

M 99
Ganda, K. et al.: *Absorption-line strengths of 18 late-type spiral galaxies observed with SAURON,* Monthly Notices of the Royal Astronomical Society, **380**, 2007

M 100
Riepe, P. and Blauensteiner, M.: *Messier 100 im Virgo-Galaxienhaufen,* VdS-Journal für Astronomie, **56**, 2016

Sempere, M. J. et al.: *Determination of the pattern speed in the grand design spiral galaxy NGC 4321,* Astronomy and Astrophysics, **296**, 1995

M 101
Garcia-Benito, R. et al.: *The 100 Myr Star Formation History of NGC 5471 from Cluster and Resolved Stellar Photometry,* The Astronomical Journal, **141**, 2011

M 102
Capellari, M. et al.: *The SAURON project – X. The orbital anisotropy of elliptical and lenticular galaxies: revisiting the (V/σ, ε) diagram with integral-field stellar kinematics,* Monthly Notices of the Royal Astronomical Society, **379**, 2007

M 104
Bridges, T. J. et al.: *Spectroscopy of Globular Clusters out to Large Radius in the Sombrero Galaxy,* The Astrophysical Journal, **658**, 2007

M 105
Ciardullo, R. et al.: *The radial velocities of planetary nebulae in NGC 3379,* The Astrophysical Journal, **414**, 1993

Sarzi, M. et al.: *The SAURON project – V. Integral-field emission-line kinematics of 48 elliptical and lenticular galaxies,* Monthly Notices of the Royal Astronomical Society, **366**, 2006

M 106
Bregman, M. and Alexander, T.: *Accretion Disk Warping by Resonant Relaxation: The Case of Maser Disk NGC 4258,* Astrophysical Journal Letters, **700**, 2009

Wilson, A. S. and Yang, Y.: *Chandra Observations and the Nature of the Anomalous Arms of NGC 4258 (M106),* The Astrophysical Journal, **560**, 2001

M 109
Gerssen, J. et al.: *Model-independent measurements of bar pattern speeds,* Monthly Notices of the Royal Astronomical Society, **345**, 2003

Hernandez, O. et al.: *BHαBAR: big Hα kinematical sample of barred spiral galaxies – I. Fabry-Perot observations of 21 galaxies,* Monthly Notices of the Royal Astronomical Society, **360**, 2005

M 110
Cepa, J. and Beckman, J. E.: *Parameter estimates for the orbits of M 32 and NGC 205 about the centre of mass of the M 31 system,* Astronomy and Astrophysics, **200**, 1988

McConnachie, W. A. et al.: *The remnants of galaxy formation from a panoramic survey of the region around M31,* Nature, **461**, 2009

Monaco, L. et al.: *The young stellar population at the center of NGC 205,* Astronomy and Astrophysics, **502**, 2009

Richardson, J. C. et al.: *PAndAS' Progeny: Extending the M31 Dwarf Galaxy Cabal,* The Astrophysical Journal, **732**, 2011

Steinicke, W.: *Astronomie Artikel zu M 32 and M 110,* 2012, www.klima-luft.de/steinicke/Artikel/m32_m110.pdf

Maffei 1
Fingerhut, R. L. et al.: *The Extinction and Distance of Maffei 1,* The Astrophysical Journal, **587**, 2003

NGC 45
Chemin, L. et al.: *HI studies of the Sculptor group galaxies. VIII. The background galaxies: NGC 24 and NGC 45,* The Astronomical Journal, **132**, 2006

NGC 55
van de Steene, G. et al.: *Distance determination to NGC 55 from the planetary nebula luminosity function,* Astronomy and Astrophysics, **455**, 2006

NGC 134
ESO Publication: *Twisted Spiral Galaxy NGC 134,* 2007, www.eso.org/public/images/eso0749a

NGC 151
Klimanov, S. A. et al.: *A Kinematic Study of M51-Type Galaxies,* Astronomy Letters, **28**, 2002

NGC 185/ NGC 147
Deason, A. J. et al.: *Rotation of halo populations in the Milky Way and M31,* Monthly Notices of the Royal Astronomical Society, **411**, 2011

De Rijcke, S. et al.: *The internal dynamics of the Local Group dwarf elliptical galaxies NGC 147, 185 and 205,* Monthly Notices of the Royal Astronomical Society, **369**, 2006

Guha Thakurta, R.: *New tidal streams found in Andromeda reveal history of galactic mergers,* 2012, www.ucolick.org/~raja/aas_prs_rls_010610/

McConnachie, A. W. and Irwin, M. J.: *The satellite distribution of M31*, Monthly Notices of the Royal Astronomical Society, **365**, 2006

Slater, C. T. et al.: *Andromeda XXVIII: a dwarf galaxy more than 350 kpc from Andromeda*, Astrophysical Journal Letters, **742**, 2011

NGC 206

Bender, R. and Saglia, P. R.: *Blaue Sterne um das supermassereiche Schwarze Loch von M31*, 2005, www.mpg.de/324786/forschungsSchwerpunkt

Brinks, E.: *NGC 206, a Hole in M31*, Astronomy and Astrophysics, **95**, 1981

NGC 210

Sancisi, R. et al.: *Cold Accretion in Galaxies*, Astronomy and Astrophysics Review, **15**, 2008

NGC 247

Davidge, T. J.: *The Disk and Extraplanar Environment of NGC 247*, The Astrophysical Journal, **641**, 2006

NGC 253

Hlavacek-Larrondo, J. et al.: *Deep Hα Observations of NGC 253: a Very Extended and Possibly Declining Rotation Curve?*, Monthly Notices of the Royal Astronomical Society, **411**, 2011

NGC 266

Heckman, T. M. et al.: *High-resolution mapping of the giant HI envelope of the Seyfert galaxy Mkn 348*, Monthly Notices of the Royal Astronomical Society, **199**, 1982

NGC 300

Sakai, S. et al.: *The Effect of Metallicity on Cepheid-Based Distances*, The Astrophysical Journal, **608**, 2004

NGC 488

Fuchs, B.: *NGC 488: has its massive bulge been built up by minor mergers?*, Astronomy and Astrophysics, **328**, 1997

NGC 520

Neff, S. G. et al.: *UV Emission from Stellar Populations within Tidal Tails: Catching the Youngest Galaxies in Formation?*, The Astrophysical Journal, **619**, 2005

NGC 536 GROUP

Geller, M. J. et al.: *Infrared Properties of Close Pairs of Galaxies*, The Astronomical Journal, **132**, 2006

Moriondo, G. et al.: *Near-infrared observations of galaxies in Pisces-Perseus. IV. Color maps of 41 cluster spirals*, Astronomy and Astrophysics, **370**, 2001

NGC 604

Bruhweiler, F. C. et al.: *STIS Spectral Imagery of the OB Stars in NGC 604: The Most Luminous Stars*, The Astronomical Journal, **125**, 2003

de Grijs, R. and Lepine, J. R. D.: *Star clusters: basic galactic building blocks*, Proceedings IAU Symposium, **266**, 2009

Maiz-Apellaniz, J. et al.: *NGC 604, the Scaled OB Association (SOBA) Prototype. I: Spatial Distribution of the Different Gas Phases and Attenuation by Dust*, The Astronomical Journal, **128**, 2004

NGC 613

Falcon-Barroso, J. et al.: *The circumnuclear environment of NGC 613: a nuclear starburst caught in the act?*, Accepted by MNRAS, 2013, http://arxiv.org/pdf/1311.2041v1.pdf

NGC 660

Alton, P. B. et al.: *Polarimetric imaging of the polar ring galaxy NGC 660 – evidence for dust outside the stellar disk*, Astronomy and Astrophysics, **357**, 2000

Baan, W. A.: *OH and HI absorption at the nucleus of NGC 660*, The Astrophysical Journal, **401**, 1992

Karataeva, G. M. et al.: *The stellar content of the ring in NGC 660*, Astronomy and Astrophysics, **421**, 2004

NGC 678 / NGC 680 / NGC 691

van Moorsel, G. A.: *A neutral hydrogen study of interacting galaxies in the NGC 697 group*, Astronomy and Astrophysics, **202**, 1988

NGC 891

Oosterloo, T. et al.: *The cold gaseous halo of NGC 891*, The Astronomical Journal, **134**, 2007

NGC 925

Piano, D. J. et al.: *Structure and Star Formation in NGC 925*, The Astronomical Journal, **120**, 2000

NGC 1073

Kaaret, P.: *Optical Sources near the Bright X-Ray Source in NGC 1073*, The Astrophysical Journal, **629**, 2005

NGC 1097

Higdon, J. L. and Wallin, J. F.: *A Minor-Merger Interpretation for NGC 1097's "Jets"*, The Astrophysical Journal, **585**, 2003

NASA Jet Propulsion Laboratory: *The Coiled Galaxy NGC 1097*, 2009, www.spitzer.caltech.edu/images/2685-ssc2009-14a1-The-Coiled-Galaxy-NGC-1097

NGC 1187

Seigar, M. S. et al.: *Constraining Dark Matter Halo Profiles and Galaxy Formation Models Using Spiral Arm Morphology. I. Method Outline*, The Astrophysical Journal, **645**, 2006

NGC 1232

Garmire, G. P.: *X-ray discovery of a dwarf-galaxy galaxy collision*, The Astrophysical Journal, **770**, 2013

NGC 1291

Bosma, A. et al.: *Kinematics of the Barred Spiral Galaxy NGC 1291, Galaxies in Isolation: Exploring Nature Versus Nurture*, Conference Proceedings "Analysis of the interstellar Medium of Isolated Galaxies (AMIGA)", ASP Conference Series, **421**, 2010

Luo, B. et al.: *Probing the X-Ray Binary Populations of the Ring Galaxy NGC 1291*, The Astrophysical Journal, **130**, 2012

NGC 1300

Mazzuca, L. M. et al.: *A Connection between Star Formation in Nuclear Rings and their Host Galaxies*, Astrophysical Journal Supplement Series, **174**, 2008

NGC 1313

Barnes, D. G. et al.: *The neutral hydrogen environments of the nearby galaxies WLM, NGC 1313 and Sextans A*, Monthly Notices of the Royal Astronomical Society, **351**, 2004

NGC 1316

Bedregal, A. G. et al.: *S0 galaxies in Fornax: data and kinematics*, Monthly Notices of the Royal Astronomical Society, **371**, 2006

NGC 1357

Seigar, M. S.: *The connection between shear and star formation in spiral galaxies*, Monthly Notices of the Royal Astronomical Society, **361**, 2005

NGC 1365

Jorsater, S. and van Moorsel, G. A.: *High Resolution Neutral Hydrogen Observations of the Barred Spiral Galaxy NGC 1365*, The Astronomical Journal, **110**, 1995

Risaliti, G. et al.: *A rapidly spinning supermassive black hole at the centre of NGC 1365*, Nature, **494**, 2013

NGC 1399

Forbes, D. A. et al.: *HST Imaging of the Globular Clusters in the Fornax Cluster: NGC 1399 and NGC 1404*, Monthly Notices of the Royal Astronomical Society, **293**, 1998

NGC 1530
Zurita, A. and Perez, I.: *Where are the stars of the bar of NGC 1530 forming?*, Astronomy and Astrophysics, **485**, 2008

NGC 1532
Chernin, A. D. et al.: *Vorontsov-Velyaminov Rows: Straight Segments in the Spiral Arms of Galaxies*, Astronomy Letters, **26**, 2000

NGC 1961
Shostak, G. S. et al.: *NGC 1961: Stripping of a Supermassive Spiral Galaxy*, Astronomy and Astrophysics, **115**, 1982

NGC 2146
Tarchi, A. et al.: *Neutral hydrogen absorption at the centre of NGC 2146*, Monthly Notices of the Royal Astronomical Society, **351**, 2004

NGC 2207
Struck, C. et al.: *The grazing encounter between IC 2163 and NGC 2207: pushing the limits of observational modeling*, Monthly Notices of the Royal Astronomical Society, **364**, 2005

NGC 2276
Rasmussen, J. et al.: *Gas stripping in galaxy groups – the case of the starburst spiral NGC 2276*, Monthly Notices of the Royal Astronomical Society, **370**, 2006

NGC 2403
Fraternali, F. et al.: *Deep HI Survey of the Spiral Galaxy NGC 2403*, The Astronomical Journal, **123**, 2002

NGC 2442
Harnett, J. et al.: *Magnetic fields in barred galaxies III: The southern peculiar galaxy NGC 2442*, Astronomy and Astrophysics, **421**, 2004

NGC 2460
Peng, C. Y. et al.: *Detailed Structural Decomposition of Galaxy Images*, The Astronomical Journal, **124**, 2002

NGC 2537
Wilcots, E. M. et al.: *HI Observations of Barred Magellanic Spirals. II. The Frequency and Impact of Companions*, The Astronomical Journal, **127**, 2004

NGC 2595
Astronomerica: *Sloan Digital Sky Survey Images*, 2009, http://astronomerica.awardspace.com/SDSS-58/UGC4414-SDSS.jpg

NGC 2683
Kuzio de Naray, R. et al.: *Kinematic and Photometric Evidence for a Bar in NGC 2683*, The Astronomical Journal, **138**, 2009

NGC 2685
Józsa, G. I. G. et al.: *Kinematic modeling of disk galaxies III. The warped "Spindle" NGC 2685*, Astronomy and Astrophysics, **494**, 2009

NGC 2775
Hernandez-Toledo, H. M. and Ortega-Esbri, S.: *Broad-band BVRI photometry of isolated spiral galaxies*, Astronomy and Astrophysics, **487**, 2008

NGC 2841
Holwerda, B. W. et al.: *Quantified HI Morphology I: Multi-Wavelength Analysis of the THINGS Galaxies*, Monthly Notices of the Royal Astronomical Society, **416**, 2011

NGC 2859
Cameron, S. et al.: *AINUR: Atlas of Images of Nuclear Rings*, Monthly Notes of the Royal Astronomical Society, **402**, 2010

Erwin, P. and Sparke, L. S.: *Double Bars, Inner Disks, and Nuclear Rings in Early-Type Disk Galaxies*, The Astrophysical Journal, **124**, 2002

NGC 2903
Irwin, J. A. et al.: *ΛCDM Satellites and HI Companions – the Arecibo ALFA Survey of NGC 2903*, The Astrophysical Journal, **692**, 2009

NGC 2964 Group
Zinn, P. et al.: *Supernovae without host galaxies? Hypervelocity stars in foreign galaxies*, Astronomy and Astrophysics, **536**, 2011

NGC 2997
Hess, K. M. et al.: *Anomalous HI in NGC 2997*, The Astrophysical Journal, **699**, 2009

NGC 3079
Chartas, G. et al.: *Constraining H_0 from Chandra Observations of Q0957+561*, The Astrophysical Journal, **565**, 2002

Jodrell Bank Centre for Astrophysics: *New e-MERLIN radio image of the Double Quasar*, 2010, www.jb.man.ac.uk/news/2010/emerlin1/

Keel, B.: *The double QSO0957+561*, 2002, www.astr.ua.edu/keel/agn/q0957.html

Space Telescope Science Institute: *Burst of Star Formation Drives Bubble in Galaxy's Core*, 2001, http://hubblesite.org/newscenter/archive/releases/2001/28/

Stockton, A.: *The lens galaxy of the twin QSO 0957+561*, The Astrophysical Journal, **242**, 1980

NGC 3169
Möllenhoff, C. and Heidt, J.: *Surface photometry of spiral galaxies in NIR: Structural parameters of disks and bulges*, Astronomy and Astrophysics, **368**, 2001

NGC 3184
Herbert-Fort, S. et al.: *Spatially Correlated Cluster Populations in the Outer Disk of NGC 3184*, The Astrophysical Journal, **700**, 2009

NGC 3190 Group
Aguerri, J. A. L. et al.: *Diffuse Light in Hickson Compact Groups: The Dynamically Young System HCG 44*, Astronomy and Astrophysics, **457**, 2006

Rasmussen, J. et al.: *Galaxy evolution in Hickson compact groups: The role of ram-pressure stripping and strangulation*, Monthly Notices of the Royal Astronomical Society, **388**, 2008

Rubin, V. C. et al.: *Optical Properties and Dynamics of Galaxies in the Hickson Compact Groups*, Astrophysical Journal Supplement Series, **76**, 1991

NGC 3227
Mundell, C. G. et al.: *The Unusual Tidal Dwarf Candidate in the Merger System NGC 3227/3226: Star Formation in a Tidal Shock?*, The Astrophysical Journal, **614**, 2004

NGC 3239
Krienke, K. and Hodge, P.: *The Structure of the Irregular Galaxy NGC 3239*, Astronomical Society of the Pacific, **102**, 1990

NGC 3256
Förster Schreiber, N. M. et al.: *Warm dust and aromatic bands as quantitative probes of star-formation activity*, Astronomy and Astrophysics, **419**, 2004

NGC 3310
Wehner, E. H. et al.: *NGC 3310 and Its Tidal Debris: Remnants of Galaxy Evolution*, Monthly Notices of the Royal Astronomical Society, **371**, 2006

NGC 3338
Garcia, A. M.: *General study of group membership. II. Determination of nearby groups*, Astronomy and Astrophysics Supplement Series, **100**, 1993

NGC 3344
Verdes-Montenegro, L. et al.: *A detailed study of the ringed galaxy NGC 3344*, Astronomy and Astrophysics, **356**, 2000

NGC 3521

de Blok, W. J. G. et al.: *High-Resolution Rotation Curves and Galaxy Mass Models from THINGS*, The Astronomical Journal, **136**, 2008

NGC 3621

Gliozzi, M. et al.: *A Chandra view of NGC 3621: a bulgeless galaxy hosting an AGN in its early phase?*, The Astrophysical Journal, **700**, 2009

NGC 3656

Barcell, M. et al.: *HI in the shell elliptical galaxy NGC 3656*, The Astronomical Journal, **122**, 2001

NGC 3718

Sparke, L. S. et al.: *The remarkable warped and twisted gas disk in NGC 3718*, The Astronomical Journal, **137**, 2009

Pott, J.-U. et al.: *Warped molecular gas disk in NGC 3718*, Astronomy and Astrophysics, **415**, 2004

NGC 3938

Jimenez-Vicente, J. et al.: *Fabry-Perot observations of the ionized gas in NGC 3938*, Astronomy and Astrophysics, **342**, 1999

NGC 3953

Hernandez, O. et al.: *BHαBAR: big Hα kinematical sample of barred spiral galaxies – I. Fabry-Perot observations of 21 galaxies*, Monthly Notices of the Royal Astronomical Society, **360**, 2005

NGC 4038

Renaud, F. et al.: *Fully Compressive Tides in Galaxy Mergers*, The Astrophysical Journal, **706**, 2009

NGC 4088

Schmitt, H. R. et al.: *Multiwavelength star formation indicators: Observations*, Astrophysical Journal Supplement Series, **164**, 2006

NGC 4216

Martinez-Delgado, D. et al.: *Stellar Tidal Streams in Spiral Galaxies of the Local Volume: A Pilot Survey with Modest Aperture Telescopes*, The Astronomical Journal, **140**, 2010

NGC 4236

Braun, R.: *Resolved Atomic Super-Clouds in Spiral Galaxies*, Astronomy and Astrophysics Supplement, **114**, 1995

NGC 4278

Brassington, N. J. et al.: *Deep Chandra Monitoring Observations of NGC 4278: Catalog of Source Properties*, Astrophysical Journal Supplement Series, **181**, 2009

NGC 4395

Vaughan, S. et al.: *The exceptional X-ray variability of the dwarf Seyfert nucleus NGC 4395*, Monthly Notices of the Royal Astronomical Society, **356**, 2005

NGC 4438

Vollmer, B. et al.: *The influence of the cluster environment on the large-scale radio continuum emission of 8 Virgo cluster spirals*, Astronomy and Astrophysics, **512**, 2010

NGC 4449

Rich, R. M. et al.: *A tidally distorted dwarf galaxy near NGC 4449*, Nature, **482**, 2012

NGC 4490

Elmegreen, D. M. et al.: *Observations of a Tidal Tail in the Interacting Galaxies NGC 4485/4490*, The Astronomical Journal, **115**, 1998

NGC 4535

Sofue, Y. et al.: *The Virgo High-Resolution CO Survey I. CO Atlas*, Publications of the Astronomical Society of Japan, **55**, 2003

NGC 4536

Satyapal, S. et al.: *Spitzer Uncovers Active Galactic Nuclei Missed by Optical Surveys in 7 Late-Type Galaxies*, The Astrophysical Journal, **677**, 2008

NGC 4565

Dahlem, M. et al.: *Neutral hydrogen gas in 7 high-inclination spiral galaxies*, Astronomy and Astrophysics, **432**, 2005

NGC 4567

Xu, C. et al.: *Mapping Infrared Enhancements in Closely Interacting Spiral-Spiral Pairs. I. ISO CAM and ISO SWS Observations*, The Astrophysical Journal, **541**, 2000

NGC 4618

Bush, J. S. and Wilcots, E. M.: *Neutral Hydrogen in the Interacting Magellanic Spirals NGC 4618/4625*, The Astronomical Journal, **128**, 2004

NGC 4631

Rand, R. J.: *Atomic hydrogen in the NGC 4631 group of galaxies*, Astronomy and Astrophysics, **285**, 1994

NGC 4650 Group

Buta, R. J. et al.: *The Ringed Spiral Galaxy NGC 4622. I. Photometry, Kinematics, and the Case for Two Strong Leading Outer Spiral Arms*, The Astronomical Journal, **125**, 2003

Dickens, R. J. et al.: *The Centaurus cluster of galaxies – I. The data*, Monthly Notices of the Royal Astronomical Society, **220**, 1986

Powell, R.: *Atlas of the Universe*, 2006, www.atlasoftheuniverse.com/superc/shapley.html

Schechter, P. L. et al.: *NGC 4650A: The Rotation of the Diffuse Stellar Component*, The Astrophysical Journal, **277**, 1984

Sérsic, J. L. and Agüero, E. L.: *Chain of Galaxies in Centaurus*, Astrophysics and Space Science, **19**, 1972

NGC 4676

Barnes, J. E.: *Shock-induced star formation in a model of the Mice*, Monthly Notices of the Royal Astronomical Society, **350**, 2004

NGC 4698

Falcon-Barroso, J. et al.: *The SAURON project – VII. Integral-field absorption and emission-line kinematics of 24 spiral galaxy bulges*, Monthly Notices of the Royal Astronomical Society, **369**, 2006

NGC 4725

de Lorenzo-Caceres, A. et al.: *Stellar Kinematics in Double-Barred Galaxies: The σ-Hollows*, The Astrophysical Journal, **684**, 2008

NGC 4762

Wosniak, H.: *Is the edge-on galaxy NGC 4762 actually barred?*, Astronomy and Astrophysics, **286**, 1994

NGC 4945

Kotak, R. et al.: *Spitzer Measurements of Atomic and Molecular Abundances in the Type IIP SN 2005af*, The Astrophysical Journal, **651**, 2006

NGC 5033

Mediavilla, E. et al.: *Asymmetrical structure of ionization and kinematics in the Seyfert galaxy NGC 5033*, Astronomy and Astrophysics, **433**, 2005

NGC 5068

Bibby, J. L. and Crowther, P. A.: *The Wolf-Rayet population of the nearby barred spiral galaxy NGC 5068 uncovered by the Very Large Telescope and Gemini*, Monthly Notices of the Royal Astronomical Society, 420, 2012

Sellwood, J. A. and Wilkinson, A.: *Dynamics of Barred Galaxies*, Reports on Progress in Physics, **56**, 1993

NGC 5078

Condon, J. J. et al.: *A 1.425 GHz Atlas of the IRAS Bright Galaxy Sample, Part II*, Astrophysical Journal Supplement, **103**, 1996

NGC 5128
Gopal-Krishna and Wiita, P. J.: *Galaxy shells and the structure of radio galaxies: Clues from Centaurus A (NGC 5128)*, New Astronomy, **15**, 2009
Malin, D.: *Ultra Deep Image of NGC 5128*, 2005, http://ftp.aao.gov.au/images/deep_html/n5128_d.html

NGC 5153
Donzelli, C. J. and Pastoriza, M. G.: *Spectroscopic Observations of Merging Galaxies*, The Astronomical Journal, **120**, 2000

NGC 5297
Rampazzo, R. et al.: *Gaseous and stellar components in mixed pairs of galaxies. I. The data*, Astronomy and Astrophysics Supplement, **110**, 1995

NGC 5364
Haynes, M. P. and Giovanelli, R.: *Neutral hydrogen emission-absorption in the IRR II galaxy NGC 5363*, The Astrophysical Journal, **246**, 1981

NGC 5426
Fuentes-Carrera, I. et al.: *The isolated interacting galaxy pair NGC 5426/27 (Arp 271)*, Astronomy and Astrophysics, **415**, 2004

NGC 5474
Dicaire, I. et al.: *Hα kinematics of the Spitzer Infrared Nearby Galaxies Survey (SINGS) – II*, Monthly Notices of the Royal Astronomical Society, **385**, 2008

NGC 5643
Simpson, C. et al.: *A One-Sided Ionization Cone in the Seyfert 2 Galaxy NGC 5643*, The Astrophysical Journal, **474**, 1997

NGC 5701
Gadotti, D. A. and de Souza, R. E.: *NGC 4608 and NGC 5701: Barred Galaxies Without Disks?*, The Astrophysical Journal, **583**, 2003
Gil de Paz, A. et al.: *The GALEX Ultraviolet Atlas of Nearby Galaxies*, Astrophysical Journal Supplement Series, **173**, 2007

NGC 5746
Barentine, J. C. and Kormendy, J.: *Two Pseudobulges in the "Boxy Bulge" Galaxy NGC 5746*, The Astrophysical Journal, **754**, 2012

NGC 5846
Mahdavi, A. et al.: *The NGC 5846 Group: Dynamics and the Luminosity Function to* $M_R = -12$, The Astrophysical Journal, **130**, 2005

NGC 5929
Rosario, D. J. et al.: *The Radio Jet Interaction in NGC 5929: Direct Detection of Shocked Gas*, Astrophysical Journal Letters, **711**, 2010

NGC 5982
Del Burgo, C. et al.: *Spatial distribution of dust in the shell elliptical NGC 5982*, Astronomy and Astrophysics, **477**, 2008

NGC 5905
Komossa, S. and Bade, N.: *The giant X-ray outbursts in NGC 5905 and IC 3599: Follow-up observations and outburst scenarios*, Astronomy and Astrophysics, **343**, 1999

NGC 5907
Miskolczi, A. et al.: *Tidal streams around galaxies in the SDSS DR7 archive: I. First results*, Astronomy and Astrophysics, **536**, 2011

NGC 5921
Hernandez, O. et al.: *BHαBAR: big Hα kinematical sample of barred spiral galaxies – I. Fabry-Perot observations of 21 galaxies*, Monthly Notices of the Royal Astronomical Society, **360**, 2005

NGC 5945
NASA JPL-Caltech: *A roadmap to the Milky Way*, 2008, www.spitzer.caltech.edu/images/1923

NGC 5963
Hu, J. and Lou, Y.: *Collisional interaction limits between dark matter particles and baryons in "cooling flow" clusters*, Monthly Notices of the Royal Astronomical Society, **384**, 2008
Sanchez-Salcedo, F. J.: *The Dark Halo of NGC 5963 as a Constraint on Dark Matter Self-Interaction at the Low-Velocity Regime*, The Astrophysical Journal, **631**, 2005

NGC 5996
Kinney, A. L. et al.: *An Atlas of Ultraviolet Spectra of Star-Forming Galaxies*, Astrophysical Journal Supplement Series, **86**, 1993

NGC 6221
Koribalski, B. and Dickey, J. M.: *Neutral hydrogen gas in interacting galaxies: the NGC 6221/6215 galaxy group*, Monthly Notices of the Royal Astronomical Society, **348**, 2004

NGC 6300
Awaki, H. et al.: *A variability study of the Seyfert 2 galaxy NGC 6300 with XMM-Newton*, The Astrophysical Journal, **632**, 2005

NGC 6339
Pisano, D. J. and Wilcots, E. M.: *Gas-Rich Companions of Isolated Galaxies*, The Astronomical Journal, **117**, 1999

NGC 6340 Group
Chilingarian, I. V. et al.: *NGC 6340: an old S0 galaxy with a young polar disc. Clues from morphology, internal kinematics, and stellar populations*, Astronomy and Astrophysics, **504**, 2009
Hurst, G. M. et al.: *Supernova 1999bt in Anonymous Galaxy*, IAU Circular, **7142**, 1999
van der Burg, G.: *HI observations of some galaxies and their faint companions*, Astronomy and Astrophysics Supplement Series, **62**, 1985

NGC 6384
Buta, R. J. et al.: *Do Bars Drive Spiral Density Waves?*, The Astronomical Journal, **137**, 2009

NGC 6503
Greisen, E. W. et al.: *Aperture Synthesis Observations of the Nearby Spiral NGC 6503: Modeling the Thin and Thick HI Disks*, The Astronomical Journal, **137**, 2009

NGC 6580 Group
Condon, J. J. et al.: *UGC Galaxies Stronger than 25 mJy at 4.85 GHz*, The Astronomical Journal, **101**, 1991
Focardi, P. and Kelm, B.: *Compact groups in the UZC galaxy sample*, Astronomy and Astrophysics, **391**, 2002

NGC 6632
Canzian, B.: *Extensive Spiral Structure and Corotation Resonance*, The Astrophysical Journal, **502**, 1998

NGC 6744
Rydera, S. D. et al.: *HI Study of the NGC 6744 System*, Publications of the Astronomical Society of Australia, **16**, 1999

NGC 6771
Kormendy, J.: *Box Shaped Bulges*, 2013, http://ned.ipac.caltech.edu/level5/March05/Kormendy4/Kormendy4_5.html

NGC 6814
König, M. et al.: *The Seyfert Galaxy NGC 6814 – a highly variable X-ray source*, Astronomy and Astrophysics, **322**, 1997

NGC 6822

Barnard, E.: *Astronomische Nachrichten*, **110**, 1884

Cioni, M.-R. L. and Habing, H. J.: *Near-IR observations of NGC 6822: AGB stars, distance, metallicity and structure*, Astronomy and Astrophysics, **429**, 2005

Hubble, E. P.: *N.G.C. 6822, a Remote Stellar System*, The Astrophysical Journal, **62**, 1925

NGC 6845 Group

Gordon, S. et al.: *Australia Telescope Compact Array HI observations of the NGC 6845 galaxy group*, Monthly Notice of the Royal Astronomical Society, **342**, 2003

Klemola, A. R.: *Groups and Clusters of Southern Galaxies*, The Astronomical Journal, **74**, 1969

Rodrigues, I. et al.: *Study of the Interacting System NGC 6845*, The Astronomical Journal, **117**, 1999

NGC 6872 Group

Horellou, C. and Koribalski, B.: *Stars and gas in the very large interacting galaxy NGC 6872*, Astronomy and Astrophysics, **464**, 2007

NGC 6946

Larsen, S. S. et al.: *Hubble Space Telescope imaging of a peculiar stellar complex in NGC 6946*, The Astrophysical Journal, **567**, 2002

NGC 6951

van der Laan, T.: *Circumnuclear star forming rings in the barred galaxies NGC 5248 and NGC 6951*, Dissertation, Heidelberg, 2012 www.ub.uni-heidelberg.de/archiv/13872

NGC 7137

Grosbol, P. J. and Patsis, P. A.: *The Three-Armed Galaxy NGC 7137, Conference Proceedings "Galaxy Dynamics"*, ASP Conference Series, **182**,1999

NGC 7184

Marinova, I. and Jogee, S.: *Characterizing Bars at z~0 in the Optical and NIR: Implications for the Evolution of Barred Disks with Redshift*, The Astrophysical Journal, **659**, 2007

NGC 7217

Combes, F. et al.: *Molecular gas in NUclei of GAlaxies (NUGA) II. The ringed LINER NGC 7217*, Astronomy and Astrophysics, **414**, 2004

Sil'chenko, O. K. and Moiseev, A. V.: *Nature of nuclear rings in unbarred galaxies: NGC 7742 and NGC 7217*, The Astronomical Journal, **131**, 2006

NGC 7252

Chien, L. H. and Barnes, J. E.: *Dynamically-Driven Star Formation In Models Of NGC 7252*, Monthly Notices of the Royal Astronomical Society, **407**, 2010

NGC 7331

Ludwig, J. et al.: *Giant Galaxies, Dwarfs and Debris Survey. I. Dwarf Galaxies and Tidal Features around NGC 7331*, The Astronomical Journal, **144**, 2012

NGC 7424

Soria, R. et al.: *Multiband study of NGC 7424 and its two newly discovered ULXs*, Monthly Notices of the Royal Astronomical Society, **370**, 2006

NGC 7479

Laine, S. and Beck, R.: *Radio Continuum Jet in NGC 7479*, The Astrophysical Journal, **673**, 2008

NGC 7497

Welty, D. E. et al.: *On the nearest molecular clouds. III. MBM 40, 53, 54, and 55*, The Astrophysical Journal, **346**, 1989

NGC 7550

Wegner, G. and Grogin, N. A.: *Ages and Metallicities of Early-Type Void Galaxies from Line Strength Measurements*, The Astronomical Journal, **136**, 2008

NGC 7582 Group

Bianchi, S. et al.: *A multiwavelength map of the nuclear region of NGC 7582*, Monthly Notices of the Royal Astronomical Society, **374**, 2007

NGC 7741

Duval, M. F. and Monnet, G.: *Luminosity and mass models for the barred spiral galaxies NGC 7741, NGC 3359 and NGC 7479*, Astronomy and Astrophysics Supplement Series, **61**, 1985

NGC 7753

Sengupta, C. et al.: *HI content and star formation in the interacting galaxy Arp 86*, Monthly Notices of the Royal Astronomical Society, **397**, 2009

NGC 7771 Group

Alonso-Herrero, A. et al.: *The NGC7771+NGC7770 Minor Merger: Harassing the Little One?*, Monthly Notices of the Royal Astronomical Society, **425**, 2012

Mandel, S.: *The Unexplored Nebulae Project*, 2005, www.galaxyimages.com/UNP1.html

Smith, D. A. and Neff, S. G.: *The Luminous Starburst Ring in NGC 7771: Sequential Star Formation?*, The Astrophysical Journal, **510**, 1999

NGC 7793

Dicaire, I. et al.: *Deep Fabry-Perot Hα Observations of NGC 7793: A very extended Hα disk and a truly declining rotation curve*, The Astronomical Journal, **135**, 2008

NGC 7814

Fraternali, F. et al.: *A tale of two galaxies: light and mass in NGC 891 and NGC 7814*, Astronomy and Astrophysics, **531**, 2011

PGC 54559

Hoag, A.A.: *A Peculiar Object in Serpens*, The Astronomical Journal, **55**, 1950

Schweizer, F. et al.: *The structure and evolution of Hoag's object*, The Astrophysical Journal, **320**, 1987

Wakamatsu, K. I.: *On the Nature of Hoag-Type Galaxy NGC 6028 and Related Objects*, The Astrophysical Journal, **348**, 1990

Seyfert's Sextet

Da Rocha, C. and Mendes de Oliveira, C.: *Intra-group diffuse light in compact groups of galaxies. HCG 79, HCG 88 and HCG 95*, Monthly Notices of the Royal Astronomical Society, **364**, 2005

Nishiura, S. et al.: *A Multi-Band Photometric Study of Tidal Debris in a Compact Group of Galaxies: Seyfert's Sextet*, Publications of the Astronomical Society of Japan, **54**, 2002

Seyfert, C. K.: *A Dense Group of Galaxies in Serpens*, Publications of the Astronomical Society of the Pacific, **63**, 1951

Smithsonian Astrophysical Observatory: *The Harvard College Observatory Astronomical Plate Stacks*, 2012, http://tdc-www.harvard.edu/plates/

Stephan's Quintet

Fedotov, K. et al.: *Star Clusters as Tracers of Interactions in Stephan's Quintet (Hickson Compact Group 92)*, The Astronomical Journal, **142**, 2011

Gutierrez, C. M. et al.: *New Light and Shadows on Stephan's Quintet*, The Astrophysical Journal, **579**, 2002

Renaud, F. et al.: *N-Body Simulation of the Stephan's Quintet*, The Astrophysical Journal, **724**, 2010

Stephan, E.: *Nebulae (new) discovered and observed at the observatory of Marseilles*, 1876 and 1877, Monthly Notices of the Royal Astronomical Society, **37**, 1877

N

Abell 2218

In 1989, an expanded version of the late George O. Abell's *Catalog of Rich Clusters of Galaxies* was published by Harold G. Corwin and Ronald P. Olowin, covering portions of the southern sky below −27° in declination which had been omitted in the original catalogue (1958) and containing a total of 4073 objects. One selection criterion for a galaxy cluster was a minimum of 30 galaxies which lie within two magnitudes of the third-brightest member of the cluster. The galaxy cluster Abell 2218 is at the centre of the image. Conspicuous at a distance 4′ to the northeast of the centre is the barred spiral galaxy PGC 58586 with a diameter of just one minute of arc. Its luminosity is 15 mag, its distance is 320 million light years. Abell 2218 itself lies at a distance of 2.3 billion light years ($z = 0.175$); at this distance, an arcminute corresponds to 670 000 light years. This distance also separates the two brightest elliptical galaxies in Abell 2218, which define the two centres of mass in the cluster. The brighter of the two cD galaxies has more cluster members near it, which are also arranged appreciably more spherically, than the second, smaller cD galaxy. This second galaxy is about 70″ to the southeast, at the end of a chain of galaxies. It is thus part of a linear substructure in Abell 2218.

The detailed mass distribution in clusters can be measured by the location, form, and orientation of the delicate, faint arcs visible in the image, which are due to the weak gravitational-lens effect. The two centres of mass cause Abell 2218 to act as a double gravitational-lens telescope. In the enlargement in the inset, about two dozen arcs can be identified, as shown in this image by the Hubble Space Telescope. This shows more than 80 arcs with good signal-to-noise ratio in the range 22–25 mag. Particularly prominent is an orange arc below and to the left of the brighter cD galaxy. This is the distorted image of an elliptical galaxy at $z = 0.702$. Opposite this are two almost-touching circular arcs. The brighter arc is due to a distant galaxy at $z = 2.515$; the shorter, dimmer arc has $z = 0.532$. Also visible is an image with weak structure, to the right of the cD galaxy at the second centre of mass. One sees hints of an orange-yellow central region which indicates that it is the image of a spiral galaxy ($z = 1.034$). Without the action of the gravitational lens Abell 2218, whose mass model implies an amplification by a factor of 15 to 25, images of these distant galaxies would not be possible for amateur astronomers.

Using the Advanced Camera for Surveys (ACS) on the Hubble Space Telescope and ground-based observations at the Keck Telescope in Hawaii, a multiply-imaged source was observed which is due to a galaxy with a redshift in the range 6.6–7.1. This distance was estimated not via spectral lines but rather by the position of the spectral continuum in the infrared spectral region. This exceedingly faint galaxy has a luminosity of 28 mag. The image morphology due to the gravitational lens indicates a very compact source, only about 3000 light years in size. It is presumably the core of a star-forming galaxy such as those observed at the end of cosmic reionization. In this early phase of the universe, about 150–800 million years after the Big Bang, the first stars formed in primeval galaxies, ending the so-called dark ages.

This amateur image can be compared with one from the Wide Field and Planetary Camera (WFPC2, in use 1993–2009): WFPC2 consists of a CCD array of four chips, three of which form an L, the fourth being in the knick of the L. This fourth chip (planetary camera chip) has only a quarter of the area of the larger CCDs, but 2.2 times the resolution. The area (including the gap due to the smaller chip) measures 2.7′ × 2.7′ and completely covers the extent of Abell 2218.

OBJECT	Abell 2218
CONSTELLATION	Draco
RIGHT ASCENSION	$16^{h}\ 35^{m}\ 52^{s}$
DECLINATION	+66° 12′ 52″
BRIGHTNESS	17 mag
TYPE	Galaxy cluster
PHOTOGRAPHERS	Josef Pöpsel, Stefan Binnewies
TELESCOPE	600-mm reflector
CAMERA	SBIG STL-11000
EXPOSURE TIME	330 min
LOCATION	Skinakas Observatory, Crete, Greece

Small Magellanic Cloud

Goddard Space Flight Center: *Magellanic Clouds High-Mass X-Ray Binaries Catalog*, 2005, http://heasarc.gsfc.nasa.gov/W3Browse/all/maghmxbcat.html

Haschke, R. et al.: *Three-dimensional Maps of the Magellanic Clouds using RR Lyrae Stars and Cepheids. II. The Small Magellanic Cloud*, The Astronomical Journal, **144**, 2012

Hodge, P. W. and Schommer, R. A.: *The ages of the Large Magellanic Cloud blue globular clusters NGC 2133 and NGC 2134*, Astronomical Society of the Pacific, **96**, 1984

James, P. A. and Clare, F.: *On the scarcity of Magellanic Cloud like satellites*, Monthly Notices of the Royal Astronomical Society, **411**, 2011

Kozlowski, S. et al.: *The Magellanic Quasars Survey. I. Doubling the Number of Known Active Galactic Nuclei Behind the Small Magellanic Cloud*, Astrophysical Journal Supplement, **194**, 2011

Shara, M. M. et al.: *Cataclysmic and Close Binaries in Star Clusters. IV. The Unexpectedly Low Number of Erupting Dwarf Novae Detected by the Hubble Space Telescope in the Core of 47 Tucanae*, The Astrophysical Journal, **471**, 1996

UGC 6614

Mapelli, M. et al.: *Are ring galaxies the ancestors of giant low surface brightness galaxies?*, Monthly Notices of the Royal Astronomical Society, **383**, 2008

Naik, S. et al.: *An X-ray Bright Nucleus in the Low Surface Brightness Galaxy UGC 6614*, Monthly Notices of the Royal Astronomical Society, **404**, 2010

UGC 11871

Karachentsev, I. D. et al.: *Disturbed isolated galaxies: indicators of a dark galaxy population?*, Astronomy and Astrophysics, **451**, 2006

UGC 12342

Giovanelli, R. and Haynes, M. P.: *A Survey of the Pisces-Perseus Supercluster VI. The Declination Zone +15.5° to +21.5°*, The Astronomical Journal, **105**, 1993

Ursa Minor Dwarf

American Astronomical Society: *Albert G. Wilson*, 2012, https://aas.org/obituaries/albert-g-wilson-1918-2012

Karachentsev, I. D. et al.: *A Catalog of Neighboring Galaxies*, The Astronomical Journal, **127**, 2004

Strigari, L. E. et al.: *A common mass scale for satellite galaxies of the Milky Way*, Nature, **454**, 2008

Virgo Cluster

Powell, R.: *The Virgo Cluster*, 2006, www.atlasoftheuniverse.com/galgrps/vir.html

SEDS: *The Virgo Cluster of Galaxies*, http://messier.seds.org/more/virgo.html

Index